Michael Ryan's Writings on Medical Ethics

Philosophy and Medicine

VOLUME 105

CLASSICS OF MEDICAL ETHICS 3

For other titles published in this series, go to
www.springer.com/series/6414

Howard Brody · Zahra Meghani
Kimberley Greenwald
Editors

Michael Ryan's Writings on Medical Ethics

Springer

Editors
H. Brody
Institute for the Medical Humanities
University of Texas Medical Branch
Galveston
USA
habrody@utmb.edu

Z. Meghani
Philosophy Department
University of Rhode Island
Kingston
USA

K. Greenwald
Ada, USA
greekima@trinity-health.org
kgwald@aol.com

ISBN 978-94-007-3065-6 e-ISBN 978-90-481-3049-8
DOI 10.1007/978-90-481-3049-8
Springer Dordrecht Heidelberg London New York

Softcover reprint of the hardcover 1st edition 2010

Printed on acid-free paper

Springer is part of Springer Science+Business Media (www.springer.com)

Acknowledgements

We began this project as a team of four. Sara Schuman assisted us in our early stages until other responsibilities forced her to withdraw.

During most of the process of preparing this volume we benefited from the facilities provided by the Center for Ethics and Humanities in the Life Sciences, Michigan State University. We are especially grateful to Jan Holmes and the late Beth McPhail for their assistance in scanning the Ryan texts.

We could not have completed this volume without assistance from a great many people, most of whom are listed in the Endnotes to the Introductory Essay in connection with their specific contributions. Our primary research source was the Wellcome Library in London with its magnificent history of medicine collection, and we are grateful to the various staff of the Wellcome who assisted us in numerous ways.

As none of us are historians of medicine by training, we have relied all along on advice from our better-educated colleagues, including Peter Vinten-Johansen, Declan O'Reilly, Harold Vanderpool, Jason Glenn, and Dayle DeLancey. Our editors, Bob Baker and Larry McCullough, provided considerable advice and guidance on matters both historical and philosophical.

We wish especially to acknowledge the steadfast assistance of Mary Hope Griffin, without whom the many perplexing Latin passages in Ryan's text would have refused to divulge their secrets. She soon discovered that her extensive knowledge of classical Latin was of limited help in deciphering what apparently passed for that language among English medical practitioners in the early nineteenth century. On one occasion she explained to us that the only way she could make sense of some of it was to imagine that it was very bad Italian.

Acknowledgements

Contents

Part Two. Selected Writings by Ryan

Part I
Introductory Essay: Michael Ryan and Medical Ethics

Howard Brody, Zahra Meghani, and Kimberley Greenwald

I Medical Ethics in the English and Scottish Enlightenment

His contemporaries remembered Michael Ryan (d. 1840) as an Irishman, a learned and kind-hearted colleague, a prolific writer, editor of the *London Medical and Surgical Journal*, and a colorful disputationist who regularly went toe-to-toe with Thomas Wakley (1795–1862), the acerbic and outspoken founding editor of *The Lancet*. Later historians identified him as one of the founders of modern medical ethics, and the first to link the medical ethics of his own day with the historical tradition of the Hippocratic Oath. Ryan was a seminal figure in the development of medical jurisprudence and an original thinker on research ethics, apparently the first to publicly defend the necessity of obtaining the informed consent of the research subject.

Ryan appears to have been the only person in Great Britain, during the middle years of the nineteenth century, to attempt to produce a systematic account of medical ethics, for use especially by medical students. While relatively little of his thinking was original, Ryan is noteworthy for addressing medical ethics under the broader rubric of medical jurisprudence, identifying some important ethical issues in his other special-interest area, obstetrics, and breaking some new ground with his ethical reflections. He also advocated addressing medical ethics as a critical portion of the curriculum for medical students.

When, around 1830, Michael Ryan began to compile a work on medical ethics, the subject was dominated by two figures from the end of the eighteenth century, John Gregory (1724–1773) and Thomas Percival (1740–1804). These writers, in turn, were heavily influenced by the philosophy of the Enlightenment period, especially as it was formulated in Scotland.[1] The ethics of the Scottish Enlightenment generally, and the work of Gregory and Percival in particular, therefore provide a context for Ryan's role in the history of medical ethics.

H. Brody et al. (eds.), *Michael Ryan's Writings on Medical Ethics*,
Philosophy and Medicine, Vol 105,

Understanding the ethical writing of the Scottish Enlightenment is very challenging for us at the start of the twenty-first century. It is difficult not to see the Enlightenment through the prism of whatever issues most concern us today. For example, we imagine that writings on ethics can be divided generally into theoretical and practical works, and that practical works can be subdivided into those that are clearly based on or derived from theory, and those that are not. We also imagine that a fundamental feature of medical ethics is the physician-patient (or patient–physician) relationship, and that the extent to which the physician does or does not respect the autonomous wishes and choices of the patient is a major line of demarcation between an "old" ethic, such as is suggested in the Hippocratic Oath, and a "modern" ethic suitable for today's practice.

Adopting this presentist perspective, we might conclude several things about Gregory and Percival. Since Gregory seems to attend more to the physician–patient relationship, while Percival focuses on relationships among physicians, we might conclude that Gregory is a pioneer of "modern" medical ethics while Percival is primarily a conservative figure from the Hippocratic past. We might also see in Gregory more explicit mention of various theoretical strands of ethical thought and conclude that his ethical writings are theoretically grounded in a robust way, while Percival represents a "merely" practical or applied medical ethics.[2]

While it is ultimately impossible to stand free from our present vantage point, we may at least approximate an interpretation that better situates medical ethics within the intellectual life of the late eighteenth century. This latter interpretation might see Gregory and Percival, despite their differences, as allied within a single tradition of thought. Both are transition figures at the intersection between an old, guild-based medicine and modern institutional and professional medicine.[3]

The educational background of Gregory and Percival shaped their thinking on medical ethics. They both did some of their medical training at the University of Edinburgh, which Ryan attended several decades later. Edinburgh was the home of the Scottish Enlightenment culture. The medical school had inherited the mantle previously held by Herman Boerhaave (1668–1738) and the University of Leiden, as the dominant school for teaching modern medicine.[4] The emphasis in both medicine and religion was on tolerance and free rational inquiry, associated with a loathing of sectarian dogmatism. The medical professors called this "Hippocratic teaching," but this was a modernized Hippocratic approach that stressed critical thought, a Hippocrates of the Enlightenment. "Eventually a new Hippocrates emerged who spoke in Baconian cadences and who, not surprisingly, espoused the popular mechanist philosophy of Newton and Boyle adapted by Boerhaave and his Edinburgh disciples."[5] Since English and Scottish thinkers in the eighteenth century were distrustful of grand systems, Hippocrates was held up as a model of the physician who allowed himself to be taught by nature and not by theory.

Edinburgh influenced its graduates in several more concrete ways. The Edinburgh system allowed access to medical school for religious dissenters, while Oxford or Cambridge accepted only those who would subscribe to the Church of England faith. Lectures were given in English rather than in Latin, which remained the language of instruction in the English as well as many continental universities. Professors were not paid on a salary basis, but supplemented their clinical practice incomes with fees collected from students for admission to their courses of lectures. As a result the professors tended to give more practical and appealing lectures, and were actively involved in the life of the university and community.[6]

Gregory and Percival shared an immersion in the ethical thinking and writing of the Scottish Enlightenment, led by figures including Francis Hutcheson (1694–1746) and Gregory's cousin, Thomas Reid (1710–1796). Their ethical thinking flowed seamlessly into a natural-law view of both science and religion. Neither Gregory nor Percival thought it possible to have ethics without a religious basis. Both were apolitical and represented the conservative Enlightenment. They had no wish to overturn the social order that seemed to provide a stable home both for the medical profession and the advancement of medical science, and for civilized life generally.

Today's label, "practical ethics" as opposed to "theoretical ethics," does not track very well the Enlightenment goal of producing a smooth integration of practice and speculation. "To their contemporaries, … Gregory [and] Percival … were qualified to write medical ethics only because they were practical men of speculative bent. Not only did they have theoretical expertise in their own fields but they were also practical moralists."[7] "[Both] were recognized by their contemporaries as moral philosophers, a reputation based on their previously published and internationally well-received works on literary, moral and medical topics."[8] Gregory and Percival believed that medical science must follow the recipe for the inductive method laid down by Francis Bacon. In their ethics, they followed another of Bacon's teachings – that there are two types of duties, general duties owed by all persons to all others, and specific duties owed by virtue of one's office or station in life. Bacon argued further that ethics of the special duties are best written about by those with the most intimate practical understanding of that office. Hence, physicians are best situated to write about medical ethics. Bacon, in *The Advancement of Learning*, realized there would be tendency to magnify the importance of one's own office, leading to a danger of distorting one's ethical pronouncements. His proposed corrective was to require that anyone writing on special duties must address the vices of that office as well as its virtues.[9]

Duties peculiar to special offices or roles could be viewed as social contracts, since the natural order included the organization of society and the interrelationships among its parts. To Thomas Reid, the physician–patient contract was blatantly (but not solely) commercial. The patient voluntarily seeks the physician's advice and pays a fee, and the physician agrees to prescribe according to the best of his skill. "This idea of the implied contract which binds physician and patient together in a moral association that is a small constituent of the common good, is the central point in Enlightenment medical ethics."[10]

The importance of duties incumbent upon social roles help to explain why politeness and decorum are so frequently mentioned in the ethical writings of the period. Some have complained that these "ethical" works are primarily about "medical etiquette" and not about ethics at all.[11] Politeness may easily be misunderstood as "merely" etiquette when in fact it was seen as a core virtue of the scientist. The gentleman-scientist who debates rationally, with liberality and candor, was contrasted with the disputatious, pedantic sectarian who did nothing to advance true scientific knowledge. Some of Gregory's and Percival's comments on the decorum of the physician harked back to Cicero, who agreed that mere fashion (in clothing, for instance) is not of ethical interest. Nevertheless, one's decorum in part establishes what role one asserts for oneself and whether one is going about it honestly and devotedly. Hence some discussion of decorum is appropriate as part of the moral analysis of special duties owed by virtue of offices.

IA John Gregory

Gregory was born to an aristocratic Scottish family, 9 years after the next-to-last Scottish uprising was put down by England. Gregory's philosophical style was engendered by the Scottish national identity crisis. From his involvement in the Scottish Enlightenment philosophical tradition, he developed the ethical underpinnings that would later form his concept of medical ethics.[12] Women in the Gregory family were articulate and intelligent, influencing his future view of women and how they should be treated in medicine and in society at large.[13]

Gregory's father and half-brother were doctors who studied and taught at Edinburgh. Gregory himself attended Edinburgh from 1742 to 1745. He traveled to Leiden for further study at the school that was viewed as Edinburgh's ancestral home. He was then called back to King's College, Aberdeen, where his grandfather had been principal and where he had received his pre-medical education, and for 3 years taught math, natural history, and moral philosophy. He left teaching to start a private medical practice in Aberdeen in 1749. In 1754 Gregory married and moved to London to practice medicine.[14] He spent only 2 years in London before he returned to King's College to take up a post vacated by the death of his physician-brother, James. While in London he met Elizabeth Montagu, "Queen of the Bluestockings" and her circle of women intellectuals.[15] Mrs. Montagu and Gregory became good friends, and she was a confidant to his four daughters after his wife died.[16]

Teaching at King's College for the decade, 1756–1766, Gregory tried and failed to create a modern medical school there, but succeeded in forming the Aberdeen Philosophical Society with the help of his cousin, Thomas Reid. Gregory's most important popular work, *A Comparative View of the State and Faculties of Man with Those of the Animal World* (1765), grew out of discourses delivered to the Society.[17] By this time Gregory had a clear philosophical position, sharing Reid's view on importance of religion in morals and on the methods of the Common Sense philosophy of the Scottish Enlightenment. He relied on Bacon's rules of induction for the empirical foundations of any science – under which heading he included morals – and was skeptical of any hypothetical system not grounded in induction. He also shared Reid's interest in the mind–body connection.[18]

In 1766, Gregory went back to Edinburgh, where he occupied the chair in medicine until his death 7 years later. As a medical lecturer he was overshadowed by his more famous and influential colleague, William Cullen (1710–1790). But he established an excellent local reputation both as a physician to the elite and as a writer-intellectual, characterized by Boswell, "as amiable as he was benevolent."[19] His reputation and his writings therefore made it easy after his death to place Gregory on a pedestal as a paragon of virtue.[20]

The medical reformers, first at Leiden and then at Edinburgh, advocated learning at the bedside of the patient from concrete experience, rather than blindly following the dictates of doctrine.[21] In Gregory's day, the most obvious voices of outmoded doctrine came from the medical corporations, chiefly the Royal College of Physicians of London and its sister organization, The Royal College of Physicians of Edinburgh. The corporations' overriding concern was self-interest and the preservation of their ancient privileges. They issued strong rules against criticizing a colleague or consulting with any physician who practiced without their license, however impressive that physician's training and credentials. In the period we are considering, only a graduate of Oxford or Cambridge could become a Fellow of the Royal College of Physician of London, even though all dispassionate observers regarded the curriculum at Edinburgh as far superior to that of the English universities. Gregory's goal was to improve medicine from the ground up "by relieving its estate from the reactionary, anti-scientific and therefore harmful forces of the medical corporations."[22]

Gregory understood Bacon's method of investigation to consist of observation, or experience. Experience should be obtained in a disciplined manner to act as the appropriate progenitor of new thoughts, thus making it experimental in nature. Gregory agreed with the direction both Bacon and Boerhaave thought medicine should take, beginning with bedside observations of the sick and then proceeding to analytical thought into the cause of disease. He contrasted this approach with the synthetic way medicine was being taught at Universities, relying on rote acceptance of the humoral teachings of Galen and Aristotle. Gregory believed that only the analytic, experimental method would advance scientific concepts within medical practice.

The Scottish Enlightenment philosophers sought to study scientifically what we would call the contents of the mind and what contemporaries called 'ideas' or beliefs. They tried to make intelligible the external world, God, and the self, as a way of understanding the principles of morality. Gregory's morals laid special emphasis on the concept of sympathy, which he (following Hume) viewed in Baconian, experimental terms. Sympathy functions analogously to the long nerves and muscles that, according to the physiologists, connected the mind to the body. On this view, the mind is active, the body (initially) inert. The nerves and muscles become an apparatus whereby the active mind can impart activity to the body. The Scottish Common Sense philosophers applied a similar active/inactive distinction to reason and passion (emotion), with reason seen as inactive and passion as active. Passion was thus a more powerful moral motivator than reason alone. The "experiment" occurs when the physician (e.g.) tries to imagine himself in the place of the patient and feels or seems to feel the patient's suffering. The activity of sympathy can therefore be seen as a moral and social nervous system, naturally binding everyone together. The physician is more powerfully moved to respond in a useful way to the patient's distress through the exercise of sympathy than he ever could via reason alone divorced from passion. For Gregory, sympathy is central to the moral character of the physician.[23]

Gregory himself did not practice a specific religion, but had a strong belief in God and agreed with most of his fellow Scottish Enlightenment philosophers that religion played a key role in the underpinnings of morality. Religion is important as it admits of the Supreme Being who supports the structure of the world by laws that are "steady and uniform" but that are often unknowable to mere humans. A man with sound understanding realizes that extensive knowledge is the truest teacher of humility. If a man has a weak understanding, he is bound to get too caught up with secondary causes and overlook the "first and great cause" (God).[24]

Gregory first lectured to the Edinburgh students on philosophy of medicine and medical ethics. In a later part of the curriculum, students went to the Royal Infirmary of Edinburgh to see their patients.[25] The Royal Infirmaries of Edinburgh and Britain had a structured environment regarded as perfect for study of the "natural history of disease and the effects of medical and surgical interventions."[26] Many of the variables that influenced the disease process could be controlled within the infirmary setting. Infirmary patients ate the same food, kept the same hours of sleeping and waking, breathed the same air and drank the same water. This allowed physicians to observe the effects of the disease process upon the patient in a setting that controlled for many ancillary factors.

If the Infirmary setting had major advantages from the standpoint of Baconian science, it also created a new set of ethical concerns. McCullough argues, "The Royal Infirmaries made a crucial contribution to the social transformation of the *patient-physician* relationship into the *physician-patient* relationship."[27] When the physician visited the patient at home and the patient paid for the physician's services, the power relation was dominated by the patient. Patients in the Royal Infirmaries were charity rather than paying patients, and had left their homes to enter the life of a regimented institution. The new power relation between the physician and patient was dominated by the physician. This shift in power dynamics challenged Gregory to write medical ethics in a scholarly manner not just for paying outpatients, but for non-paying Infirmary patients as well. Gregory's main moral concept was sympathy; and sympathy could serve as an antidote to class and religious differences. Thomas Percival, coming a generation later, more fully developed a medical ethics to grapple with the new institutional settings in which medicine was increasingly practiced.[28]

Gregory approached the duties and office of the physician first by addressing temperament and understanding.[29] He was interested in two applications of temperament – what best suited an individual to medical practice, and what best prepared a person to advance medical science through new discoveries. Each physician must rest on his own judgment and experience alone. Gregory was fully aware that the physicians of his day had a very uneven foundation of scientific knowledge, and that market conditions encouraged them to denigrate their rivals' methods of practice.[30] A virtuous physician would not participate in such collegially derogatory practices, but would instead dispassionately try to establish which method of treatment worked best. One implication of this view was that Gregory tended to see the gentleman of independent means as a more scientifically inclined individual than one who needed to make his living through medical practice.[31]

Gregory listed the virtues appertaining to the role of physician under the heading of "moral qualities." The most important quality is that of humanity, or sympathy. "Sympathy produces an anxious attention to a thousand little circumstances that may tend to relieve the patient; an attention which money can never purchase: hence the inexpressible advantages of having a friend for a physician."[32] In keeping with an Aristotelian model of virtue, Gregory worried about excesses as well as deficiencies of virtuous qualities. Physicians often had to act under emergency conditions, in which too much sympathy for the patient's suffering could impede the decisive action that might save the patient's life. Gregory argued, "Doctors can feel what is amiable in pity without suffering it to enervate or unman them."[33] Similar balancing was required in giving instructions to patients – avoiding both undue flexibility and undue rigidity – and in responding to apparently trifling symptoms, avoiding both negligence and sarcastic ridicule on the one hand, and excessive attention and anxiety on the other. Gregory warned the students of the common falling off of virtue as the physician matures in practice. In his early years, a physician might be solicitous and attendant, but once established, often becomes "haughty, rapacious, careless, and sometimes perfectly brutal in his manners."[34] Physicians should maintain temperance and sobriety at all times. Because a physician knows the private issues of families, it is important that he remain circumspect in his public utterances, especially where women are concerned. A physician should never take advantage of his office to seduce women in his care.

Gregory next addressed physician decorum. A physician should make considerable effort to fit in with the community where he is practicing. Adopting the manners and dress of the area will allow a physician to further his practice and create a comfortable situation for the patient who is receiving medical care. Once again, Gregory was thinking about both the physician's duties, as a practitioner and as a scientist. The physician who dressed in an excessively gaudy or an excessively slovenly fashion would not have a comforting and reassuring effect on the vulnerable patient, and the physician who imagined that ostentation in dress would impress the public more than real scientific attainments failed in his duty to advance the profession.[35] A physician should listen to his patient's suggestions and consider them rather than imagining that his dignity has been assaulted. Gregory agreed that it was sometimes best to withhold the truth of a dangerous prognosis from patients at a time of crisis, when emotional shock could by itself bring about the feared outcome.[36] He opposed concealing the real situation from the relations of the patient, or abandoning the patient at the hour of their death. Gregory agreed with Bacon that bringing comfort to the dying was as much a defining duty of the physician as curing disease, and was also in keeping with his moral ideal of physician-as-sympathetic-friend.[37]

Gregory approached the contentious issue of consultations among practitioners from his usual dual perspective – what best serves the needs of the patient? What best advances medical science? The medical corporations attacked any public dissension among physicians as bringing dishonor upon the professional body. Gregory also opposed open dissension, but for other reasons. It was highly unlikely that either the patient would get better, or that medical science would be advanced, if the physicians gathered in consultation began bickering. To assure that consultations truly served the needs of the patient rather than the false pride of some physicians, Gregory urged that any medical practitioner who possessed a skill relevant to the treatment of the disease be allowed a role: "If a surgeon or apothecary has had the education, and acquired the knowledge of a physician, he is a physician to all intents and purposes, whether he has the degree or not, and ought to be respected and treated accordingly."[38] Gregory here made perhaps his most radical proposal with regard to the typical practices of his own day. He took to heart Bacon's admonition that anyone who would write of the ethics of a profession would have to address the profession's defects as well as its virtues.[39]

IB Thomas Percival

Ryan thought the work of Percival sufficiently important to reprint most of the core of Percival's *Medical Ethics* nearly verbatim. Percival supplemented that core with copious Appendices that make up about half of the final volume and expand our understanding of Percival's approach to ethics.

Percival was born in Warrington, near Manchester.[40] Orphaned in infancy and raised by his eldest sister, Percival chose a medical career after his uncle Thomas, a physician, died when the young man was 10 and left him an inheritance sufficient for his education. At age 17, Percival enrolled in the new Warrington Academy set up by his former tutor, Rev. John Seddon. Seddon had left the Anglican Church to found a Unitarian congregation. Although non-Anglicans (Dissenters) were still denied many legal rights, the Unitarians were a powerful and wealthy group in the Warrington area, and had considerable influence on the development of Manchester as the first modern industrial city in Europe.

The Warrington Academy welcomed students from any faith, and prepared them for careers in commerce as well as the professions. All its tutors were graduates of Scottish universities and were heavily influenced by (and indeed actively participated in) the Scottish Enlightenment. Haakonssen thinks it important that Warrington was the first academy founded on institutional principles. Previously, a school was more or less a large house, with a chief tutor as head of the family. The

Warrington experience may have later helped Percival realize the need to develop a medical ethics suitable for institutional practice.[41]

In 1761, Percival traveled to Edinburgh to enroll in the medical school. There he met and admired David Hume, and while Hume's atheism challenged some of his beliefs, Percival did not abandon his Christian faith. Rather than graduating at Edinburgh, he went on to Leiden, where he took his M.D. in 1765. He visited Paris and then walked the wards of the great London hospitals before he returned to Warrington, married, and started practice. He moved to Manchester in 1767, set up a new practice and began his writing career. He later served 21 years as president of the Manchester Literary and Philosophical Society, which he had helped to found. Percival also assisted in establishing two local colleges.

Like Gregory, Percival declared his allegiance to the Baconian ideal for medical and scientific advances. He wrote numerous essays, mostly oriented toward public health, and experimented on the uses of cod liver oil; but his biographer-son had to admit that he had little real talent for experimental work. In 1773, Percival and some colleagues completed a study of bills of mortality in Manchester and Salford, which they presented to the Royal Society. The study formed the basis for several reform proposals, the early creation of a board of health, and the opening of a fever hospital. Percival was also involved in a failed project to promote mass inoculation against smallpox.

The Manchester Infirmary and institutions like it were funded by private charity and staffed on a voluntary basis by medical men and clergy. They provided limited, short-term care for the laboring poor who were not covered by the existing Poor Law relief system. They excluded the majority of the sick poor who had chronic diseases and could not readily return to employment. Percival viewed these institutions' relationship with the poor as a social contract. Justice required that since the poor were willing to work to make articles of value to the rich, the rich had a reciprocal duty to assure the poor a decent wage, assist them in sickness, and provide education for their youth so that a general upward movement in society could continue. Unlike some other charities of this era, the hospitals were successful; they never had a serious financial failure or a crisis of public confidence.

With new medical knowledge came expanded spheres of duty for physicians. By 1803 the Manchester Infirmary included a lunatic asylum, an out-patient dispensary, and the fever hospital. By this time the hospital was emerging as an essential laboratory for clinical medicine, both for observation and experimentation. Percival as a public health expert had to admit the downside – admission to the hospital carried with it an increased risk of dying from contagious fevers. Percival favored home births and out-patient management of many disorders to avoid the risks of hospitalization.[42]

According to his biographer-son, Percival epitomized the virtue of candor, defined as a habitual disposition to form fair and impartial judgments and to avoid bigotry and prejudice. He was an expert at conversing rationally and bringing disagreeing parties into concurrence. When the Manchester Infirmary suffered a crisis among its medical staff in 1790, "[I]t is hardly surprising that [his associates] should have asked him to be Manchester Infirmary's Justinian.... He had a scholarly reputation as one of Britain's best-known medical moralists; he had previously written extensively on hospital policy; and his views had been in circulation for many years."[43] The 1790 crisis was precipitated by an old guard of Tory physicians and surgeons whose families had long enjoyed a relative monopoly in Manchester and who opposed the Trustees' plans to expand the medical staff so as to increase services to the poor. In 1792 the institution asked Percival to prepare a set of rules that would prevent future disputes of this nature, and the associated unseemly publicity.[44]

When Percival prepared the Preface to the second (1803) edition of his code of professional conduct for physicians and surgeons, he noted that the code written for his Infirmary colleagues in 1792 formed the first chapter of the new work. He now wished to expand that work into a general system of medical ethics "that the official conduct, and mutual intercourse of the faculty, might be regulated by precise and acknowledged principles of urbanity and rectitude."[45] He proceeded to list a diverse and distinguished set of readers who had reviewed his manuscript, including Thomas Gisborne (1758–1846) and William Paley (1743–1805). Percival saw his work as a supplement to Gregory, Gisborne's chapter "On the Duties of a Physician," and the ethical proclamations of the Royal College of Physicians – works he saw as helpful but "not sufficiently comprehensive for the existing sphere of medical and chirurgical duty."[46]

Percival explained in his Preface that the earlier draft of his code that he had circulated in 1794 was "originally entitled 'Medical Jurisprudence'; but some friends having objected to the term Jurisprudence, it has been changed to Ethics."[47] In this manner, less than 30 years before Ryan wrote on the subject, Percival became the first writer to employ the phrase "medical ethics."[48]

Percival, like Gregory, devoted a good deal of his attention to the virtues and character of the physician. Unlike Gregory, Percival devoted attention to the institutional character of the hospital. Both issues were informed by the Baconian ideal of scientific progress. When Percival spoke of the reputation and dignity of the medicine, he was not defending a professional monopoly. Rather he was tacitly admitting that in previous years medicine's reputation had been based on an inadequate foundation, and greater scientific accomplishment was required to shore up that foundation. The hospital was the laboratory where that scientific advance could best occur. Gregory had assumed that only the elite gentleman-physician could properly move medical science forward. Percival, living among the commercial class in a modern industrial city, looked instead to cooperation among all classes of practitioners to provide the basis for medical progress. According to Haakonssen, Percival "realized that creation of medical knowledge would no longer be the domain of independently wealthy gentlemen but 'interested' ones who, like himself, lived by their profession and nevertheless remained committed to the Baconian ethos of the Royal Society."[49] Percival wished to show that an "interested" professional could still be a Christian gentleman and scholar.[50]

While the ideal physician was a gentleman, Percival saw the need to demarcate medical ethics from the gentleman's code of honor. The uproar at the

Manchester Infirmary in 1792, like most unseemly clashes within the medical fraternity, had occurred precisely because the "gentlemen" of the faculty felt their honor had been threatened and insulted, and because their code of behavior gave them no other framework within which to settle disputes. Percival's ethics can be seen as an effort to reconceptualize the physician from a private gentleman to a public professional, and to separate questions of personal honor from those of professional ethics.[51]

For Percival, as for Gregory, "[T]he ideal physician was, as ever, an attentive, humane, indulgent, confidential, tactful, faithful, punctual, honourable, candid, and, not least in an age when sobriety was not always considered a virtue, a strictly temperate man."[52] Gregory argued that the physician should be the friend of the patient, not as a social equal (especially in the case of infirmary patients), but as the patient's protector. In Percival's day, the increasing numbers of poor patients, and the rise of institutional settings, further complicated the notion of the physician as "friend." Percival wrote an essay, "False Opinions Concerning Friendship," to show that friendship could never be a duty because it was always particular and interested, whereas true duties were disinterested. Justice and probity were universal and disinterested social virtues, and so they always took priority over friendship. As historian Lisbeth Haakonssen characterizes Percival's ethics, the physician owed every person an equal duty of care and consideration; but as people had very different stations in society, this "equal" duty was compatible with unequal treatment, modified according to custom and decorum in how social unequals were to treat each other:[53]

According to Percival's scheme of ethics, the patients' status was inseparable from their position and role in society. To respect a patient, or to treat a patient as a person, was, essentially, to respect that patient's capacity for fulfilling the duties of his or her particular station. In this view, a physician might best demonstrate respect for his patients by assisting them to return to the scene of their active duties, a goal equally compatible with benevolence and the general welfare of society.[54]

If the physician was to be a friend to the patient while acknowledging social inequality, what does this imply for disclosing the truth about a grim prognosis? Percival addressed this problem in one of his Appendices, which is perhaps today the best-remembered part of his *Medical Ethics*. (Since the detailed discussion appears in an appendix and not in the main work, Ryan did not see fit to reprint it, though he did address the issue.) Percival began by surveying the views of the most important moralists of the time on whether one may lie to produce a good end. Gisborne, for instance, argued not, while Hutcheson said yes. Percival then offered his own analysis, based on the idea that the duty to be truthful was a personal duty for the physician, but only a relative duty within the physician-patient dyad. By appeal to the Golden Rule, duties to patients took priority over duties to oneself. In the patient setting, the most important duty was not to cause harm. Percival interpreted the overriding nature of the duty not to cause harm as effectively eliminating any duty of truthfulness to the patient in circumstances of extreme danger.

A duty to tell the truth is then reduced solely to a duty the physician owes to himself, which he would then unselfishly renounce in favor of the greater obligation to serve his patient. The modern reader tends to focus on Percival's eventual rationalization of physician paternalism.[55] This reader may then miss Percival's method of approaching the question, which epitomized the sort of "candid" free and rational inquiry that he argued was the sure way toward medical advancement. In that regard, the essay on deception might have been intended as much as a model of ideal medical reasoning and inquiry as an action guide to behavior.[56]

Having described the duties of the virtuous physician, Percival proceeded to the nature of the hospital as an institution. The notion of ethical duties within and about institutions was relatively novel at that time. Percival included as an appendix to his volume the "Discourse on Hospital Duties" delivered by his since-deceased son, Rev. Thomas B. Percival, to the Liverpool Infirmary in 1791. While the younger Percival saw the hospital as public and social in a way that made it quite different from the traditional family, he was unsure how to conceptualize it except as a large home with its managers acting *in loco parentis*. He argued that it was critical that patients' family members be allowed into the hospital and encouraged to do as much as they could for the sick: "Domestic virtue was the basis for social virtue, and the hospital should be wary of breaking this tie [between patient and family]."[57] The younger Percival claimed that modern physicians were doing Christ's work within an institution that was an extension of the "sacramentally founded family."[58]

The elder Percival realized, however, that the hospital was clearly something other than a large household. He saw that new institutional relationships required a new ethical analysis. The domestic ethic of the family, and the traditional code of conduct of the English gentleman, would no longer serve medicine's needs. One pressing issue was the power balance between the physicians and the hospital's lay managers. On the one hand, physicians needed and generally received considerable discretion to prescribe as they wished for each patient. On the other hand, physicians were of necessity subservient to budgetary decisions as to how much and what kind of medicine and food the hospital would purchase.

Medicine practiced in the hospital, unlike traditional practice in the patient's home, was fully visible to one's peers. This visibility created an opportunity for disputes among practitioners that could undermine public faith in the hospital and dry up the wellspring of philanthropy. Percival was in favor of retaining the three traditional medical orders (physicians, surgeons, and apothecaries), and much of his ethical code was devoted to mechanisms to produce a harmonious collaboration among them while maintaining a useful division of labor.[59] While Percival was not concerned about the physician–patient relationship in the modern sense, he was extremely concerned about patient safety, in a time of high hospital mortality. Hygiene and fever control, argued Percival the public health reformer, were critical *ethical* issues in hospital medicine and required a collaborative orientation among the three orders.

Percival's proposed rules to govern consultations among practitioners maintained both the hierarchy of the three medical orders, and also the preferential treatment of practitioners long settled in town. True to his Edinburgh education, however, he also sought to assure that any practitioner with intelligence and experience would receive a polite hearing from others, whether or not that individual had a degree. He emphasized that the goal was the well-being of the patient, not the honor or pride of any physician. Percival the ideal Baconian scientist might have hoped for a system in which errors were candidly admitted so as to lead the way to new discoveries and improved future practice. Percival the diplomat and peacemaker designed a practical system for people to cooperate and resolve the inevitable disputes. The compromise required that most disputes be kept in-house among the medical staff.

Percival recognized the tension between individual patient benefit and the advance of medical science. He recognized that the public at large had a horror of being subjected to experimentation. He thought that the right sort of education and rhetoric, continued for a long enough period, might convince them that medical experimentation was in the general public interest. He believed that the poorer classes treated in hospitals, who were most likely to be subjected to experiments, were also the most likely to benefit from the experimental results. He favored consultation among the medical staff prior to any experiment, in effect anticipating today's system of committee review of research. As one would expect in a day when there was no concept of patient autonomy in any ethical writing, Percival saw no need for explicit patient consent for participation in research. He did, however, urge that hospitals try whenever possible to accede to the preferences of patients, such as to be admitted under the care of one physician rather than another. He argued that it was important to attend to feelings and emotions of patients as well as to their symptoms.

Percival, in one of his most often quoted passages, wrote: "[Physicians] should study, also, in their deportment, so to unite *tenderness* with *steadiness*, and *condescension* with *authority*, as to inspire the minds of their patients with gratitude, respect, and confidence."[60] Haakonssen urges us to read this passage with an understanding of the intellectual environment within which Percival practiced. True condescension was guided by education and training. The right sort of condescension allowed the physician calmly and without inner turmoil to attend carefully to the patient's situation and needs, and wisely to employ the best means to relieve them. Percival contrasted the "turbulent emotion" of pity with the "calm principle" of condescension.[61] For Percival, "sympathy," "condescension," "fellow feeling," and "benevolence" were all allied concepts that undergirded a physician's legitimate authority over his patients.[62]

Nevertheless, focusing too much on the physician's *authority* risks missing the truly radical aspects of Percival's thought. The authority of the medical practitioners over the sick poor in hospitals and infirmaries was guaranteed by custom and social status. Percival, Gregory, and Gisborne were nearly alone in their day in stressing the physician's *obligations* to their patients. Ryan would later extend this reform movement in ethical thinking.

Gregory and Percival succeeded both in addressing some of the most pressing ethical issues specific to the practice of medicine in their day, and also in situating that ethical discussion within the larger framework that characterized ethical inquiry in eighteenth century England and Scotland. We can now turn our attention to Ryan's life and writings.

II Michael Ryan: A Biographical Summary

IIA Origins

Ryan's birth and early life remain mysterious. Many authors continue to give his dates as 1800–1841, following the *Dictionary of National Biography*. The 1841 date for his death is demonstrably incorrect, as his death certificate shows that Ryan died on 11 December 1840.[63] His death certificate also indicates that he died in his 40th year, but an anonymous obituary, which appears otherwise to be reasonably accurate, states clearly that he died in his 46th year.[64] We believe Ryan to have been admitted as a Member of the Royal College of Surgeons in 1819.[65] The official rules of the college in those days required a minimum age of 22, so that would seem to argue against a birth date of 1800.[66] Adrian Desmond, in his study of radical reform in British medicine, accepts the obituary account and gives Ryan's dates as 1794–1840.[67]

The records of the Royal College of Surgeons and of the University of Edinburgh Medical School agree that Michael Ryan's address in 1818–1819 was Barris O'Leigh (today Burrisoleigh) in County Tipperary, Ireland. Ryan dedicated his M.D. thesis at Edinburgh to a clergyman at Glankeen, which is near Burrisoleigh, and he returned to that locale for at least a year following graduation from Edinburgh. We conclude that Burrisoleigh, Tipperary was most likely the location of Ryan's birth and family home. All contemporary birth records for that parish have been destroyed, so it may be impossible today to trace any further information about Ryan's origins.[68] When Ryan was brought to insolvent debtor's court in London in 1836, he was reported to have previously had income from land in Ireland to the sum of about £150 annually, raising the possibility that his family was of the landowning class.[69] Ryan once referred to the "views of his relative, Dr. O'Ryan, on 'Consumption,' " suggesting that other members of his family may have practiced medicine.[70] Ryan also at some point became sufficiently well versed in both French and Latin so as later to be able to translate volumes written in those languages.[71] These characteristics would be consistent with higher rather than lower class origins.

Other recent authors have assumed from available evidence that Ryan was Roman Catholic.[72] We believe it more likely that Ryan was affiliated with the Church of Ireland (Anglican). Most importantly, he dedicated his 1821 M.D. thesis at Edinburgh to Rev. Richard Boyle Bernard, rector of Glankeen, a Church of Ireland clergyman.[73] Also, if his family were indeed Irish landowners, this status at that time would be more likely to be enjoyed by a family of Protestant affiliations. Finally, if Ryan had indeed been Catholic, he might have had even more trouble establishing himself in the world of London medicine than he later did.[74]

IIB Medical Studies

Ryan wrote much later that he began his career as a medical student in 1815.[75] He also mentioned in passing that he had been clinical assistant in the practice of

Professor Dease at the Charitable Infirmary, Jervis Street, Dublin.[76] The Charitable Infirmary was the first voluntary (non-religious) hospital in the British Isles, opening in 1718, 1 year before the first such hospital in London, the Westminster Hospital. A small medical school was attached to the Infirmary between 1808 and 1833. Richard Dease (1774–1819) held the chairs of Surgery and Anatomy in the medical school of the Royal College of Surgeons in Ireland, besides attending and lecturing at the Infirmary.[77]

Apparently on the strength of his Dublin medical training, Ryan submitted his credentials to the Royal College of Surgeons of England and was admitted Member (MRCS) on 2 April 1819. The College archives list his address at the time as "Burris O'Leigh, Tipperary."[78]

Ryan next sought to enhance his credentials by matriculating at the medical school of Edinburgh University, where John Gregory had taught some 40 years before. The Edinburgh M.D. was not recognized by the Royal College of Physicians of London and gave one no legal entitlement to practice in England. Nevertheless the Scottish university retained the same high prestige that it had enjoyed in Gregory's time, though it would soon be supplanted by Paris as the preferred academic destination of American physicians seeking wider educational and research opportunities in Europe.[79]

Ryan apparently began his studies at Edinburgh in 1818 or 1819.[80] He graduated in 1821, the title of his M.D. thesis being, "De Genere Humano ejusque Varietatibus," which he later translated as, "on Man, and the apparent varieties of the Human Species."[81] The anonymous obituarist in the *Provincial Medical and Surgical Journal* stated, "During his collegiate career he attracted the attention of the then professor of midwifery, Dr. Hamilton, from whom he received many marks of kindness, and it was this intimacy which laid the foundation of that taste for the theory and practice of the obstetric art which Dr. Ryan so much excelled in."[82] In his M.D. dissertation, Ryan included sections on human reproduction and birth, as well as on the division of humanity into discrete racial groups.[83]

IIC Irish Medical Career

Following completion of his Edinburgh studies, Ryan returned to his native Ireland to practice. Apparently he spent at least 1 year back in Tipperary before relocating to Kilkenny.[84] The 1824 Kilkenny City directory listed Michael Ryan as practicing in Patrick Street.[85] Ryan also began his voluminous literary production. He published papers on topics including the various causes of aphonia, the use of prussic acid in heart and lung conditions, and blindness caused by cataract.[86] He also expanded one article into a 46-page book entitled, *A Treatise on the Most Celebrated Mineral Waters of Ireland containing an Account of the Waters of Ballyspellan, Castleconnel, Ballynahinch, Mallow, Lucan, Swadlinbar, Goldenbridgre, Kilmainham, &c., &c.: and of the spa lately discovered at Brownstown, near Kilkenny, with Plain Directions during the Use of Mineral Waters and an Account of some of those Diseases in which they are Most Useful.*[87]

Mineral Waters bore several features suggesting an ambitious young physician eager to make his name known to a wider audience. He stressed that the book was intended for both professional and lay readers, and made note of the fact that he had been selected by the local authorities to do an analysis of the Brownstown mineral waters.[88] He commented that the water "has been used by an immense number of persons last season, as soon as my opinions were known on its nature…"[89] He also mentioned the attractions of Kilkenny town in an apparent ploy to draw potential patients from a distance: "From its central situation, it is well adapted for the reception of invalids…"[90]

In two ways this relatively slim pamphlet provided a model that Ryan would utilize repeatedly in his subsequent books. Most importantly, Ryan began his discussion of mineral waters with a historical review, citing instances of water used for healing purposes in "Heathen Mythology," the Bible, and the works of Hippocrates and Aristotle.[91] Ryan also defended himself from possible accusations that he had slighted the better-known mineral spa at Ballyspellan in order to sing the praises of his local Kilkenny waters by adopting for himself the motto, "amicus Socrates, amicus Plato, sed magis amica veritas" ("Socrates and Plato are my friends, but truth is a greater friend still").[92] This phrase (minus the word "sed") appeared in the *Medicus-Politicus* of Roderic a Castro, a Renaissance writer on medical ethics from whom Ryan would later borrow extensively.[93] Ryan would repeat this phrase in the preface or introduction to many of his subsequent books, claiming that he personally always wished to avoid controversy, but that the facts of the matter forced him to disagree with the statements of others.

We have no record of the precise "circumstances [that] induced him to seek a larger field for his exertions," as his obituary stated.[94] By July 1827, Ryan was residing in Hatton Garden, London, and serving as a physician to the Central Infirmary in adjacent Greville Street.[95] He announced in 1828 that he was offering lectures on midwifery and on medical jurisprudence at the Infirmary.[96] Since his 1828 *Manual on Midwifery* was characterized as "an analysis of my lectures," it is likely that he had been lecturing for at least a few months prior to that date.[97]

In 1828, Ryan announced his presence within the London medical community by the publication of two books. One, *Remarks on the Supply of Water to the Metropolis…*, was an opportunistic cut-and-paste. His "remarks" on London water – a comment on the polluted state of the Thames, with a recommendation to pipe purer drinking water from a distance, as had been done successfully in Edinburgh – occupied all of three pages. The next portion of the volume reprinted a journal article that he had published, discussing the chemistry of water and listing all known mineral waters in Europe with notes as to their composition.[98] He then inserted the comment, "As the mineral waters of Ireland have been noticed by few writers, and are consequently little known, the author deems it necessary to add a brief account of them, in order to complete this work."[99] This was Ryan's cue to append, verbatim, his entire 1824 pamphlet on the Irish mineral waters.[100] In this way he managed to produce what purported to be a new book while preparing exactly five pages of new material.

IID New Directions – London, 1828

Ryan's second work of 1828, based on his midwifery lectures, was a more substantial production. Its title page hinted that Ryan was playing with the idea of a unique career direction. The courses of lectures at the London medical schools, like the lectures he had attended at Edinburgh earlier, followed a rigidly prescribed formula: anatomy, physiology, chemistry, botany, surgery, principles and practice of medicine, materia medica (therapeutics), and, bringing up the rear in prestige, Medical Jurisprudence or Forensic Medicine.[101] The mere fact that a professor happened to be interested in a given subject did not justify offering a formal course of lectures in a topic not part of the standard list. But Ryan boldly broke the mold. On the title page of his *Manual on Midwifery*, he characterized himself as "Lecturer on Midwifery, Medical Ethics, and Medical Jurisprudence" (Fig. 1). As Robert Baker has noted, this made Ryan the first person anywhere in the world to claim the status of lecturer on medical ethics.[102] Ryan, it appears, was exploring in his own mind the relationships among medical jurisprudence, medical ethics, and midwifery (or obstetricy, as he preferred to call it).[103] He laid stress on the fact that his manual of midwifery would thoroughly discuss "obstetrico-legal medicine [which] has had no place previously in systematic works on midwifery."[104] He perhaps already envisioned the "obstetrico-legal" portion of the present work as later to form part of a manual specifically devoted to medical jurisprudence. He also, it seems, envisioned that later manual to include a detailed discussion of medical ethics. He promised the reader of the midwifery manual, "The moral conduct, and professional duties of the medical practitioner, are clearly defined, and will be reduced to a few simple precepts."[105]

As he would later make clear in his writings on medical jurisprudence (reprinted in this volume), Ryan regarded it as scandalous that so little attention was paid to medical ethics in an educational program designed to train young men to become physicians. Moreover, he proposed also to discuss as part of his midwifery manual "the moral and religious question – whether one life should be sacrificed to save the other. The opinions of the profession in this country, on the continent of Europe, and in America, are fully detailed."[106] Finally, he advised prospective students: "Dr. Ryan's lectures on Medical Ethics and Forensic Medicine ... are intended to Explain the Moral Duties and Obligations of Medical Men in public and private Practice, in Hospitals, Infirmaries, Dispensaries, in the Military and Naval Service, in private life, and also in Courts of Justice, on whose Evidence the lives, liberty, honour, reputation, property, happiness or misery of all classes, so often depend..."[107] The physician's role in offering evidence to the courts, the traditional focus of medical jurisprudence, was therefore but one of the public roles of the professional life that might entail moral duties.

One other aspect of the midwifery manual suggested that Ryan had finally determined his special niche within the medical literature:

> Such are the advantages attempted to be affected in this work, which is intended as a Pocket Companion for students, and the junior members of the profession, who are actively engaged in practice, and whose time will not permit them to consult the larger works upon

A

MANUAL ON MIDWIFERY;

OR, A

Summary of the Science and Art of

OBSTETRIC MEDICINE;

INCLUDING THE

ANATOMY, PHYSIOLOGY, PATHOLOGY,

AND

THERAPEUTICS, PECULIAR TO FEMALES;

TREATMENT OF

Parturition, Puerperal, and Infantile Diseases;

AND AN EXPOSITION OF

OBSTETRICO-LEGAL MEDICINE.

BY

MICHAEL RYAN, M.D.

Member of the Royal Colleges of Surgeons, in London and Edinburgh, Member of the Association of Fellows and Licentiates of the Royal College of Physicians in Dublin, one of the Physicians to the Central Infirmary, and Dispensary, Greville Street, Lecturer on Midwifery, Medical Ethics, and Medical Jurisprudence.

LONDON:

PRINTED FOR MESSRS. LONGMAN AND CO. PATERNOSTER ROW; AND J. ANDERSON, MEDICAL BOOKSELLER, SMITHFIELD.

1828.

Fig. 1 Title page of Ryan's *A Manual on Midwifery*, 1828 edition, on which Ryan lists as one of his credentials "Lecturer on Midwifery, Medical Ethics, and Medical Jurisprudence" (reproduced with the kind permission of the Wellcome Library, London)

the subject. To the student it affords great advantages, for he may fairly consider it a review of the ancient and most modern works on obstetric medicine.[108]

Ryan would henceforth focus on writing the pocket handbook or medical manual, not the learned, systematic treatise or the original monograph. And to those who would criticize such works for what they necessarily omitted, or for lack of originality, he had a ready excuse:

In justice to myself, I must allude to the style of this work, and briefly explain how it was arranged and sent to press. It has been composed within a few weeks, since the cessation of my summer courses of lectures; amidst the daily discharge of official duties, at one of the most crowded Dispensaries of this metropolis, and amidst the innumerable interruptions attendant on private practice. Allowance ought, therefore, to be made for … want of polished style…[109]

Ryan's determination of his niche within the medical literature led directly to another recurring feature of his work – lack of appreciation from reviewers and critics. The reviewer for the *London Medical Gazette* took off sarcastically after Ryan's tone of self-promotion. It saves the reviewer time and trouble, he noted wryly, when the author himself so loudly announces to the world the exalted quality of his own work. The book itself "does contain a great deal of information, ill arranged, and mixed up with much extraneous matter."[110] (As if to prove the reviewer's point, when Ryan quoted from this review in advertisements to later editions of his works, he repeated only the phrase, "contain[s] a great deal of information.")

The reviewer concluded by giving Ryan the author some detailed advice:

Should Dr. Ryan's work come to a second edition, we would advise him to devote three times as long a period to its correction as he represents himself to have done in its original composition. Let him not be persuaded, either by 'talented friends,' or impatient pupils, to give it prematurely to the world, for the world will wait without complaining till he is ready. Let him expunge all passages laudatory of his own productions. Then – and we fear not till then – with the industry he has displayed, may we hope for a work entitled to receive at the hands of impartial judges a degree of praise equivalent to that which the present edition has met with from a less disinterested source [i.e., Ryan himself].[111]

Ryan, unfortunately, gave little indication of ever heeding this advice.[112]

After the publication of his midwifery manual, Ryan moved a few doors down from the Central Infirmary to 61 Hatton Garden; and by October, 1829, the *Lancet's* listing of his lectures suggested that he was giving them at his private residence.[113] The proposed *Manual of Medical Jurisprudence* duly appeared in 1831. By then, Ryan had apparently rethought his titles, if not his career direction. The title page to this new *Manual* listed him, among other distinctions, as "Lecturer on Practice of Medicine, Obstetrics, and Medical Jurisprudence," with no mention now of Medical Ethics.

IIE The Medico-Botanical Society

The medical societies formed an important part of the social and cultural network that made up the world of medicine in London. A young physician, arriving in the metropolis without formal letters of introduction to people of influence, might hope to establish himself by becoming active in one or more of these societies.[114] Ryan became affiliated particularly with the Medico-Botanical Society, a group organized to study the medical uses and the toxic properties of substances derived from plants, and to support the discoveries of new plants of potential medical use. He would appear to have joined this society fairly soon after his arrival in London, since by 1831 he had already served as a member of the Society's Council.[115]

The President of the Society was Philip Henry, 4th Earl Stanhope (1781–1855), who is best remembered today for his involvement with the case of the Bavarian 'wild boy,' Kaspar Hauser.[116] Ryan dedicated the first, 1831 edition of his *Manual of Medical Jurisprudence* to Earl Stanhope, one of only three formal dedications we have been able to discover in his books.[117] Later Stanhope was a relatively generous contributor to funds aiding Ryan during periods of financial embarrassment (such as in the wake of the Ramadge lawsuit, discussed below). It is possible that Stanhope in some way became a patron of the young Ryan and aided his entry into the world of London medicine. It would otherwise be difficult to explain, for example, how an unknown from Kilkenny could have received an appointment as physician to a London dispensary so soon after arriving in the metropolis.

While apparently an active member of the Society, Ryan seems not to have contributed much to its meetings. A review of 38 reports of Society meetings in *The Lancet* during Ryan's London years shows him mentioned as a speaker only twice.[118]

Ryan appears to have attended meetings of at least one other medical society, the Medical Society of London. He was reported to have made comments following the main presentation at a meeting of this society held 13 April 1829.[119] He wrote a letter to the editor of the *London Medical Gazette* complaining that his reasoning had been inadequately captured by the journal's reporter.[120]

IIF The London Medical and Surgical Journal

When Ryan took over editorship of the monthly *London Medical and Surgical Journal (LMSJ)* in 1829, radicalism and reform were the big issues of the day. Very generally, being "pro-reform" meant several things. One opposed the traditional power of the medical graduates of Cambridge and Oxford, schools which had deteriorated into medieval backwaters of medicine compared to the modern schools at Leiden and Edinburgh, and compared to the schools of Paris where statistical analyses of autopsy series were leading to rapid advancement of knowledge. One opposed the nepotism of the Royal Colleges and the London teaching hospitals, where positions were usually awarded based on patronage and not on merit. And one tended to favor the interests of the new, rapidly expanding group of general practitioners (officially, dually trained surgeon-apothecaries) against the restrictive and hierarchical practices of the College of Physicians.[121]

The *LMSJ* had been founded the previous year by John Davies, John Epps, and Joseph Houlton. In two ways the editors of the *LMSJ* set out to compete with Thomas Wakley's weekly journal, *The Lancet*, which had begun publication in

1823. First, the *LMSJ* attempted to position itself as a pro-reform journal, but without the vituperative personal attacks that characterized the *Lancet's* editorial style. Second, the *LMSJ* sought to undercut the *Lancet* in price and so appeal to a general practitioner readership.[122]

Ryan's work as a medical editor ultimately had a strong influence on all of his writing. For one thing, he must have been extremely busy to have kept up a full schedule of lectures, a reasonable medical practice, and his editorial duties all at once. More specifically, he initially wrote many of his longer works, including his *Manual of Medical Jurisprudence*, as series of brief articles for his journal.[123] The articles were crafted to meet a deadline and to fit into available space, rather than being designed as fully developed, thoughtful scholarship. The journalistic style and its limitations explains many of the less desirable qualities of Ryan's book-length works.

James Fernandez Clarke, MRCS, who published a memoir many decades later, assisted both Ryan and Wakley at different times during his early professional career. Clarke reported:

> Ryan, however, at this time, did valuable service to the cause of the Profession. He was editor of the *London Medical and Surgical Journal*, then published by Renshaw, in the Strand. This journal might be regarded as an *equipoise* between the *Lancet* and the old *Medical Gazette*; it avoided the personalities and virulence of the *Lancet*, and the tameness and milk-and-water contents of the *Gazette*.... I have reason to believe that if the *Medical and Surgical Journal* had been carried on in the manner in which it was originally framed, it would have had a most beneficial influence on the Profession. While it excluded from its pages the disgraceful attacks on personal character which then were a prominent feature of the *Lancet*, it afforded to a large class of contributors to periodical literature a medium of expressing or publishing to the world their experiences in the science and practice of Medicine.[124]

Wakley in the *Lancet* included Ryan, as a rival editor, as fair game for his character attacks, notwithstanding their shared views on medical reform. In April, 1832, when the *LMSJ* had published the amounts paid to members of the Central Board of Health and intimated that their pecuniary interests influenced their views on the contagiousness of the then-prevalent epidemic cholera, Wakley replied with a strong editorial condemnation and referred to the *LMSJ* as a "filthy repository."[125] Ryan in turn gave back as good as he got, arguing when he introduced the transition of *LMSJ* from a monthly to a weekly journal in 1832, "We deliberately believe that the coarse and violent importunity with which medical reform in this country has been insisted on for the last few years, has materially retarded the application of a wise and efficient measure for the removal of all our grievances."[126]

In March, 1832, Ryan had indulged in a bit of self-congratulation, noting that the *LMSJ* was increasing its circulation and that many of the top authorities were writing for it. He promised to continue to try to rise above petty personal attacks, and suggested that he could not completely avoid the appearance of *ad hominem* arguments when momentous issues were discussed.[127] The overall pattern of Ryan's writing shows that his claim to avoid personal attacks unless he was driven to them was disingenuous at best. He was driven to such attacks amazingly often for a man who claimed that he never sought them out. For example, when the *Lancet* indulged in a lengthy and totally negative review of the first edition of the *Manual of Medical Jurisprudence*, Ryan replied with a letter that occupied more than two pages, which

Wakley happily printed, taking the opportunity to add two further pages of anti-Ryan invective.[128] Ryan reprinted in the *LSMJ* the whole of this letter to *The Lancet*, and proceeded to grace the pages of his own journal with a lengthy follow-up letter in his own defense which Wakley did not see fit to print, as well as letters from several sympathizers.[129]

One may excuse Ryan for feeling obliged to leap to the defense of one of his major works when it was attacked by an unfriendly critic. It is somewhat harder to excuse an earlier exchange of letters in the *Lancet* in which Ryan traded accusations with Dr. John Gordon Smith, whom he recognized in his *Manual of Medical Jurisprudence* as one of the few major British authorities on that subject. Smith accused Ryan of publishing material from their private correspondence and said that as a result the "acquaintance between them necessarily terminates."[130] Ryan replied with a long diatribe in which among other things he insisted that Dr. Smith "ought to have discovered that our acquaintance had ceased" at Ryan's own insistence some days before Smith's letter appeared in *The Lancet*.[131]

Perhaps the most striking example of Ryan's penchant to trade invective with any real or supposed critic was provided by his concluding comments in his volume on *Prostitution in London* (1839). Ryan asserted, "My strenuous endeavour, and most anxious wish have been to diminish immorality, and crime, as well as disease, in this and every other country; and I am perfectly regardless of the dishonest, partial, unjust, and venal criticisms of many of the medical reviewers, or ignorant critics of this kingdom…"[132] He went on to claim that as a former editor of *LMSJ*, he wrote his share of negative reviews, though he was always ready to praise works of real merit. Those whom he criticized before, he now suggested, took full advantage of the fact that Ryan was no longer editor of his own journal: "They are now armed, and cowardly attack one unarmed."[133] He added gleefully, "Their unjust censures have not, however, prevented repeated editions of the several standard works of mine, which they so loudly, unsparingly, and unjustly abused and condemned."[134] After arguing that this problem was peculiarly English, and that other European nations were not harassed by a similarly unprincipled pack of reviewers, Ryan turned now to the work that he was in the process of completing: "I enter into these particulars to prepare the reader for the attacks of the critics on this work. He must know, that every volume I produce, however useful or successful, is unsparingly assailed; although the profession and the public induce me to continue my feeble and humble exertions for the promotion of the interests of science and humanity."[135] Ryan thus launched a pre-emptive strike against his future critics before they even had an opportunity to put pen to paper.[136]

IIG The Ramadge Case

Whatever the benefits of editing a journal, Ryan soon got a full taste of the disadvantages. A Dr. Francis H. Ramadge was known to have frequent social contacts with St. John Long, regarded by regular physicians as a quack.[137] Long was accused in the deaths of two patients and Ramadge wrote a long letter to *The Lancet* defending Long.[138] This letter apparently prompted the London Medical Society to expel Ramadge. Ramadge responded to the motion of expulsion with threats and vituperations against the Society members.[139]

In September, 1831, Ryan's *LMSJ* printed the following article:

> Tweedie v. Ramadge. – Dr. Ramadge was in attendance on a case of typhus; the patient (a young lady) was bled from the arm on a Friday, and eight dozen leeches (96) applied to the head and neck. On Saturday both temporal arteries were opened–the patient fainted, and the apothecary (who was likewise in attendance) left her; the nurse brought her round with wine and water. On the Sunday, another dozen leeches were applied, and immediately she became delirious; when Dr. Tweedie's advice was requested by the relatives.
>
> Dr. Tweedie having spoken apart with Dr. Ramadge, addressed Mrs. Reynolds, the sister of the patient, and said, 'That having attended before, he should be happy now to give his assistance to the young lady, but that Dr. Ramadge's conduct in a late correspondence with John Long had been such, that no medical man of respectability could call him in or consult with him without injuring himself in the eyes of his brethren. That he (Dr. Tweedie) bore no private pique against Dr. Ramadge, – he believed him indeed to be clever; but his character, as respected the above transaction, rendered it imperative for all medical men to decline acting with him; and Mrs. Reynolds must choose which she would entrust. Dr. Ramadge replied in great anger, that he was a gentleman by birth, education, and profession, but that Dr. Tweedie was neither. *** Dr. Tweedie answered him by turning cooly on his heel and walking out of the room. Dr. Tweedie was retained, and cured the patient by exactly opposite treatment. Dr. Ramadge (it is said) frequently is at supper with John Long. – *Lancet.*
>
> Dr. Tweedie has honourably and faithfully discharged his duty to his medical brethren; and we hope every one else will do the same. We are well aware who it is–and a medical man to boot – that makes the trio in these family suppers. Let him be warned in time. He takes upon him to defend this nefarious quack and manslaughterer in the face of the whole profession: let him take warning, or we will not spare him. – *Ed. Med. Surg. J.*

Thus Ryan primarily lifted his account of the Tweedie–Ramadge affair directly from *The Lancet*, but added his own further editorial comment attacking Ramadge, Long, and an unnamed third associate.[140]

Ramadge brought suit for libel individually against both Wakley and Ryan. The trials were held respectively on 25 and 26 June 1832. In the first case the jury found against Wakley and assessed damages of one farthing (1/4 cent), plus costs. In the second case the jury found against Ryan and assessed damages of £400 plus costs. It appears that the court was heavily influenced by the additional editorial comment that Ryan had appended; his defense that he merely passed on to his readers an article already printed in another journal was not accepted. There was also some confusion engendered by the fact that Ryan had at one point apparently offered Ramadge an apology for what he had published, raising questions as to whether he had already tacitly admitted that it was libellous.[141]

Wakley commented at some length editorially on the outcome of the trials, and indicated that despite his editorial disagreements with Ryan, he viewed the jury verdict against his rival as both unduly harsh and illogical in light of the very different verdict in the *Lancet* case. A number of physicians, including some luminaries among the London medical establishment, contributed to a fund to help defray Ryan's judgment.[142]

IIH Later History of the LMSJ

Ryan's association with the *Journal* – and, perhaps, his tendency to get himself into conflicts with others – continued to bring him financial woe even after the Ramadge case was over. James F. Clarke recalled the later fate of the publication:

> … [U]nfortunately, a dispute between the editor [Ryan] and the publisher [Renshaw] led to a disruption between them.[143] The consequence was, Renshaw carried on the *Journal*, which was edited by John Foote, then a general practitioner in Tavistock Street, Covent Garden. Foote was a man of undoubted ability, but conceited and pragmatical. Ryan published an opposition Journal, to which he appended the title of "original." Foote was not a match for his learned and cumbrous antagonist.
>
> … Unfortunately, the quarrel between Renshaw and Ryan terminated in a collapse of both the journals, each having deteriorated in every way, and Ryan's "original" was badly printed on bad paper…. Ryan had become involved in pecuniary difficulties, and the journal was neglected by him.[144]

According to Desmond, Ryan was actually imprisoned for nonpayment of debts.[145] Ryan was taken to Insolvent Debtors Court in 1836 by his new publisher and printer, Mr. Henderson, for the sum of £1,000 (though Henderson "claimed to be a creditor for £1,431"). Ryan had given up his furniture and medical library as payment to his creditors. However, Henderson took Ryan to court because "the whole of the books were not given up." The court found that no part "of the furniture or books had been unfairly kept back." The court also found that "the work (Ryan's journal) had not been a losing concern, and that the insolvent had laid out a large sum upon it." However, the court came to the conclusion that "There were a few books that must be given up."[146]

III Ryan the Medical Lecturer

Ryan's career as a lecturer is depicted in Table 1.

Ryan 's affiliations were solely with the smaller, private medical schools rather than with any of the large teaching hospitals. The private schools experienced sinking fortunes and a precarious existence during this decade. The medical schools attached to the large hospitals were in the ascendency, and within another 10 years most of the private schools had closed.[147]

Ryan apparently left the Central Infirmary and began offering his own private lectures down the street.[148] In 1830 he joined the staff of a new school set up in Brewer Street, near the much-better-known Windmill Street school that had been founded by the great eighteenth-century anatomist, John Hunter. He may have gone there in company with a young Edinburgh graduate, John Epps (1805–1869), one of his immediate predecessors as editor of *LMSJ*. A problem with one of the other lecturers led to this school dissolving, and Ryan and Epps then threw in their lot with a lecturer on anatomy and surgery, George Dermott.[149] According to Desmond, the trio became the nucleus of another medical school in Gerrard Street, Soho, which gained a reputation

Table 1 Ryan's lectures and medical school affiliations

Dates[261]	School	Subjects
1828	Central Infirmary, Greville Street, Hatton Garden	Midwifery, medical jurisprudence
1829	Private residence, 61, Hatton Garden	Medicine, midwifery
1830	New Medical School, 34, Brewer Street, Windmill Street, Golden Square; and ? Private residence[262]	Medicine, midwifery, medical jurisprudence
1831[263]	Westminster School and Dispensary, 9 Gerrard Street, Soho; and Private residence, 61, Hatton Garden	Medicine, midwifery, medical jurisprudence
1831–1835	Westminster School and Dispensary, 9 Gerrard Street, Soho	Medicine, midwifery, medical jurisprudence
1836	Hunterian School, 16 Great Windmill Street	Medicine, midwifery
1837–1839	North London School, 20 Charlotte Street, Bloomsbury	Medicine, midwifery
1840	Charlotte Street School of Medicine, 15 Charlotte Street, Bloomsbury	Midwifery, medical jurisprudence

as a center of radical medical reform.[150] George Darby Dermott (1802–1847) was impulsive and unpredictable, but was widely reputed to be one of the best teachers of anatomy in the kingdom. He also lectured on surgery and physiology. Epps was well regarded as a lecturer in botany, chemistry, and materia medica.[151]

Dermott had first set up his own school in 1825 in Little Windmill Street, also trying to borrow some of the reflected glory of the longer-established Hunterian school in Windmill Street. When Ryan joined, the school was situated in Gerrard Street and shared some quarters with the Westminster Dispensary. Dermott continually had problems paying the rent and the school next had to move to Charlotte Street in Bloomsbury. For a brief time, Ryan and Epps broke off and allied themselves with the Old Hunterian School in Great Windmill Street, but Ryan then rejoined Dermott in Charlotte Street.[152] Just before Ryan's death, a rival "Hunterian" school took over the Charlotte Street address, and Dermott and Ryan moved to a new building a few doors down.

James Fernandez Clarke enrolled as a student in Dermott's school in 1833. Since we have located no graphic portrait of Ryan, Clarke's verbal portrait of him gives us the most vivid picture of the man available:

> Dr. Michael Ryan … was a genuine Irishman.[153] His lectures were of a somewhat "rigmarole" kind, or, more properly speaking, compilations of almost every author from the time of Hippocrates downwards. Ryan was totally deficient in originality, and was a very inferior Practitioner; but he was learned, had most agreeable manners, and was strongly impressed with the importance of appearance in a Physician.[154] He dressed in black, wore a white "choker," carried a large gold watch, as big as a turnip, to which was appended a massive chain, with seals and keys. It is true, he did not carry a gold-headed cane, but he had a huge gold snuffbox, the contents of which were frequently applied to his olfactory organ.[155]
>
> Ryan was one of the Physicians of the Western Dispensary, in Charles Street, Westminster, and most of his students at the Gerrard Street School were also his pupils at

that institution… . Ryan had a curious way of eking out a lecture. He would select for his theme some case which he had treated in the Dispensary, and if any remarkable case occurred there, he had a fine opportunity of availing himself of this privilege. I recollect, on one occasion, he had treated a case of hypertrophy of the heart "successfully" with the iodide of potassium. He did not think it sufficient to state the fact; he would take from his pocket a list of the names of students who attended his practice at the Dispensary. "Gentlemen," he would say, "I have the honour of giving you the names of those who witnessed this remarkable case;" and he read from the list the names of 150 of those who were, or ought to have been, present. The reading of this list necessarily occupied a considerable portion of his lecture, and he would conclude with the remark – "With such a host of witnesses, gentlemen, I think that I am entitled to assert that my diagnosis was correct, and my treatment most successful." His audience, of course, cheered and laughed.[156]

… Ryan was just above the middle height, and inclined to corpulency; he was of fair complexion and red haired. His face was truly Milesian.[157] He had a large head. He was a good-natured, kind-hearted man; but too facile and with little consistency and firmness. He was a favourite with his pupils.[158]

IIJ Ryan's Medical Writings

Table 2 summarizes Ryan's book-length publications.

Several of these works are purely of medical interest. *Lectures on Population* is more of a pamphlet (72 pages) and consists of lectures from his midwifery course. The two "vademecum" volumes are handbooks of aphorisms on medicine and obstetrics, respectively, initially compiled by other authors, then edited and expanded by Ryan.[159] The *Practical Formulary* fully achieves Ryan's goal of writing a "pocket guide," since the book measures only 5 by 3.5 in. (and is 1.5 in. thick). *The Medico-chirurgical Pharmacopœia* was, most likely, a reissue of the *Practical Formulary* with only minor modifications. The remaining works have some ethical interest, and we will discuss them individually below.

According to booksellers' advertisements, Ryan was at the time of his death preparing two more volumes for the press – an atlas of midwifery, and a book on auscultation and on internal medicine more generally.[160] Earlier he had given notice of a volume titled, "A Treatise on the Physical Education and Diseases of Infants, from Birth to Puberty, compiled from the best National and Foreign Works, and arranged as a Manual of the Practice of Medicine." Such a work would have been quite innovative for its time, when pediatrics was unknown as a specialty, and when lectures on diseases of children were tacked onto the midwifery course almost as an afterthought. However, even though Ryan stated that this book was "In the Press" in 1836, nothing more was apparently heard of it.[161]

IIK The Metropolitan Free Hospital

The formation of the Metropolitan Free Hospital was announced at the end of March, 1836. The new institution differed from existing hospitals in opening its doors to people "whose only recommendations are poverty, destitution and disease."

Table 2 Ryan's published books[264]

First British edition	Title	First American edition	Later British editions	Later American editions
1821	Tentamen medica physica inauguralis, de genere humano, ejusque varietatibus (M.D. thesis, Edinburgh)			
1824	A treatise on the most celebrated mineral waters of Ireland containing an account of the waters of Ballyspellan, Castleconnel, Ballynahinch, Mallow, Lucan, Swadlinbar, Goldenbridge, Kilmainham, &c. &c.: and of the spa lately discovered at Brownstown, near Kilkenny, with plain directions during the use of mineral waters and an account of some of those diseases in which they are most useful			
1828	Remarks on the supply of water to the metropolis:with an account of the natural history of water in its simple and combined states, and of the chemical composition and medical uses of all the known mineral waters being a guide to foreign and British watering places			
1828	A manual on midwifery; or a summary of the science and art of obstetric medicine, including the anatomy, physiology, pathology, and therapeutics, peculiar to females; treatment of parturition, puerperal, and infantile diseases; and an exposition of obstetrico-legal medicine	1835(?)	1830, 1831, 1841 (title varies in later ed.)	
1831	A manual of medical jurisprudence, compiled from the best medical and legal works comprising an account of (I) the ethics of the medical profession; (II) the charters and statutes relating to the faculty; and, (III) all medico-legal questions, with the latest decisions: being an analysis of a course of lectures on forensic medicine, annually delivered in London. And intended as a compendium for the use of barristers, solicitors, magistrates, coroners, and medical practitioners	1832	1836	

1831	Lectures on population, marriage, and divorce, as questions of state medicine, comprising an account of the causes and treatment of impotence and sterility, and of the morbid and curative effects of marriage…	1835	
1833	The physician's vade-mecum, by Robert Hooper/improved by M. Ryan	1837, 1838	
1835	A new practical formulary of hospitals of England, Scotland, Ireland, France, Germany, Italy, Spain, Portugal, Sweden, Russia, and America; of MM. Magendie, Lugol, etc. Or a conspectus of prescriptions in medicine, surgery, and obstetrics. With the doses of all new and ordinary medicines/translated from the new French ed. of Milne Edwards and P. Vavasseur, and considerably augmented		1836, 1839[265]
?	The medico-chirurgical pharmacopoeia; or, a conspectus of the best prescriptions in medicine, surgery, obstetricy, and infantile medicine; with a table of the doses of all medicines in use; the additions in the London Pharmacopoeia, 1836; M. Magendie's formulary; aphorisms on the treatment of poisoning; reduction of dislocations and fractures; and on natural and difficult parturitions, with puerperal diseases[266]		1838
1836	The obstetrician's vademecum; or aphorisms on natural and difficult parturition; the application and use of instruments in preternatural labours; on labours complicated with hemorrhage, convulsions, etc. [by Thomas Denman], Considerably augmented and arranged according to the present state of obstetricy, by M. Ryan[267]	1848 (title varies in American ed.)	1837(?)

(continued)

Table 2 (continued)

First British edition	Title	First American edition	Later British editions	Later American editions
1838(?)	The philosophy of marriage: in its social, moral, and physical relations; with an account of the diseases of the genito-urinary organs … with the physiology of generation in the vegetable and animal kingdoms; being part of a course of obstetric lectures delivered at the North London School of Medicine		1839, 1843	1844, 1870, 1886, 1974 (title varies in later ed.)
1839	Prostitution in London, with a comparative view of that of Paris and New York … with an account of the nature and treatment of the various diseases, caused by the abuses of the reproductive function			

It did not, as was the general practice at the time, require a letter of recommendation from one of the governors of the hospital, thus making its services much more accessible to the larger mass of the poor.[162] Prince Albert was one of its early supporters.[163] The hospital was located at 29 Carey Street, adjacent to Lincoln's Inn and a short distance from Lincoln's Inn Fields where the headquarters of the College of Surgeons was located; it was a relatively brief walk from Hatton Garden and Greville Street, where Ryan had lived and worked when he first arrived in the metropolis. Despite Royal support, the hospital had serious financial difficulties in its early years, having trouble paying its tradesmen's bills, and because of limited funds it was unable to add beds for inpatients until near the end of its first decade of existence.[164]

Ryan attended meetings of the hospital committee during March and April, 1836, when the plans for the institution were being laid down. He ceased to serve on the governing committee after that but along with a Dr. Uwins was appointed as a Physician to the Hospital on 7 April 1836.[165] Ryan listed this affiliation under his name on the title pages of some of his later books.[166] In June, 1839, when Ryan was in the process of revising the final edition of his *Manual of Midwifery*, he mentioned a case of a pregnant woman with a severe prolapse of the uterus "under my care at the Metropolitan Free Hospital."[167]

Ryan no doubt hoped that this small hospital would grow and that eventually his affiliation would lead to more lucrative connections in the London medical world. The hospital, however, remained a small, struggling operation during the few remaining years of Ryan's life.

IIL Ryan's Death

Ryan died on 11 December 1840 following what his anonymous obituarist called a "lingering illness."[168] Nine days earlier, on 2 December 1840, the General Committee of the Metropolitan Free Hospital had discussed Ryan's lack of attendance at meetings, and the fact that he had sent in as a substitute a Mr. Earles, who was not yet fully qualified. It was decided to send a respectful letter to inquire about the situation.[169] The Committee's reaction suggests that however "lingering" the illness may have been, at least some of Ryan's associates had no inkling of its severity even a short time before his death.

If James Clarke's recollections are to be believed, we know rather more about Ryan's death than we do about many aspects of his life: "He was attended at his house in Charlotte Street, Bedford Square, by Drs. Bright, Roots, and myself. The post-mortem examination was made by me at 7 o'clock on a January morning, the two Physicians being present at that early hour. I never saw so much general visceral disease in any body before or since."[170] Unfortunately there is evidence that Clarke's memory, recalling these events decades later unaided by notes, was faulty.[171] Since Ryan died in early December there is no reason to expect that the autopsy would have been postponed until January. The death certificate lists "John

Daniel Earles, surgeon" as having reported the death and as having been present at the time of death, with no mention of any other physicians. The cause of death according to the certificate was "chronic bronchitis with dilatation of the heart."[172]

The physicians and surgeons of the Metropolitan Free Hospital officially reported to the Committee on 16 December 1840:

> The Medical Officers beg to announce to the Committee and Governors of the Hospital, the melancholy and premature decease of their colleague, Dr. Ryan. The circumstances attending this sad event are calculated to excite in their minds universal feelings of regret, for they have not only been deprived of an active, intelligent, and kind-hearted colleague, and one which they feel will be difficult to replace, but they have also to deplore the hopeless and destitute condition of his widow and children, who were entirely dependent upon Dr. Ryan's professional and literary exertions, and who are now thrown upon the wide world without a refuge unless speedy relief be afforded them by the benevolence of the Profession and the public. The long and faithful services rendered by Dr. Ryan to this Institution constitute, in the opinion of the Medical Officers, a very strong reason for pressing upon the Committee and the Governors at large the expediency of taking such measures as may testify their regret at the loss of their invaluable Physician, and their sympathy in the deplorable state of his widow and family.[173]

Wakley in *The Lancet* saw no reason to mention Ryan's death. A sympathetic obituary appeared in the *Provincial Medical and Surgical Journal*, and a short notice of death in the *Gentleman's Magazine*.[174] The *Lancet* did, however, later open its pages to an announcement of the subscription taken up "for the widow and children of the late Dr. Ryan." As of 10 March 1841, approximately £150 had been collected.[175]

III Ryan's Writings on Medical Ethics

IIIA Manual of Medical Jurisprudence

Ryan's major writing on medical ethics went through three versions. First, it appeared in the form of filler interspersed among other articles in his *LMSJ* in 1829–1830.[176] He then reprinted these articles as the first four chapters of his *Manual of Medical Jurisprudence*, in its first London edition (1831). This volume, as we have seen, was the object of a long, highly critical review in *The Lancet*.[177] The reviewer took Ryan to task for, among other things, plagiarizing and inadequately indicating the differences between his own writing and direct quotations from other works; for his apparently gratuitous diatribe against phrenology; and for his antipathy to medical experiments on patients. The reviewer also accused Ryan of padding the book by reprinting large sections of his earlier-published *Manual on Midwifery*.

While Ryan discounted both the content and the motives of the *Lancet* review, he gave evidence in his second (1836) edition that he had taken at least some of the criticisms to heart. He included in his Introduction an extremely long list of the works on medical jurisprudence that he had consulted in compiling the present book. He also appeared to devote more concern to distinguishing quotations and giving credit to other authors.

The second edition of *Manual of Medical Jurisprudence* differed from the first in a number of other ways. Ryan expanded his discussion at several points. In the first edition he had been content to offer a summary and paraphrase of Percival; in the second edition he reprinted the major portions of Percival's ethics verbatim. He also added two chapters – one on "American Medical Ethics," the second a miscellany directed at the student or new graduate – that had not appeared in the first edition. Therefore the second edition seems to represent Ryan's more carefully constructed and considered work, and we have chosen to use that edition for reprinting in this volume. In the preface to the second edition, Ryan stated that he had "considerably enlarged the first part, on Medical Ethics, by introducing the codes of continental Europe and America, and also those of Dr. John Gregory and Dr. Percival."

Ryan intended, by producing a manual of medical jurisprudence in which the essentials of the topic were compressed into a relatively small space, to meet the needs of a number of audiences – "legislators, barristers, magistrates, coroners, private gentlemen, jurors, and medical practitioners," as he stated somewhat ambitiously on his title page. Nonetheless, in the Preface to his first edition as reprinted in the 1836 edition, Ryan suggested that he had one particular audience in mind when he included the material on medical ethics: "[The author] hopes that the promulgation of Medical Ethics, or the institutes of professional conduct, will contribute in no small degree to maintain and support the honour, dignity, character, and utility of the profession, by impressing the minds of medical students with a sense of the noble and virtuous principles which have always characterized their predecessors, and which ought ever to distinguish the scientific cultivator of medicine."[178] Since his writings on medical jurisprudence grew out of the scattered articles in his *LMSJ*, which in turn were most likely derived from Ryan's notes from his own lectures to medical students, it would be natural for him focus on a student audience.

The American historian, Chester Burns, argued that Ryan's particular contribution to the history of medical ethics lay in his having incorporated ethics into medical jurisprudence:

> Ryan is singularly significant because he attempted to correlate medical ethics, health legislation, and forensic medicine. He realized that any society could incorporate its values about professional behavior into civil statutes and, consequently, impose both moral and legal obligations on professional persons. Moreover, practitioners could not satisfactorily discharge professional obligations without understanding community expectations embodied in laws, and a satisfactory fulfillment of certain community obligations involved a special knowledge of law as well as medicine. Thus, Ryan sustained Percival's emphasis on the moral import of laws as well as his understanding of the forensic responsibilities of practitioners.[179]

Following an Introduction which covered the work as a whole, Ryan then devoted the first six chapters of his *Manual* to medical ethics. We will now follow him through the organization of his material.

The Introduction pays little attention to medical ethics and focuses rather on the larger subject of medical jurisprudence. Ryan discusses various definitions and synonyms of this term and also indicates how its contents may be subdivided. He ends with a long laundry list of the important works written on medical jurisprudence through the ages, tending to confirm Clarke's assessment that whatever he may have lacked in originality, he made up for with comprehensive reading and learning. Scattered through this list of works are a few that we would now classify as pertaining to medical ethics.

Chapter 1 is intended to drive home two important points. The first Ryan selects to begin this section of his work: "All medicine is derived from God, and without his will it cannot exist or be practised. Hence the healing art, if disunited from religion, would be impious or nothing."[180] Ryan's view of the relationship between religion and medical ethics seems to have two components. First, he views all ethics as beginning with humankind's duty to God, and commonly refers to the Bible (often elliptically as the "sacred book") as the repository of God's word. Second, he takes for granted a natural-law view of both ethics and science. To do what is in accord with God's law is to do both what is ethically right, and also what promotes the highest level of human health. So physicians have a dual reason to dedicate their activities to furthering the laws and dictates of God.

Ryan's second major point in Chapter 1 is to claim (based on a selective winnowing of the historical evidence) that physicians were uniquely esteemed in virtually every civilization, and at every historical period, in which medicine was practiced. "The concurrent testimony of historians of all ages proves [medicine] to be the noblest and must useful of human pursuits; and hence the esteem and veneration universally entertained for its cultivators by mankind. The dignity of medicine arises from the nobleness of its subjects and its end; its subject is the human body, which excels all other material bodies; its end is health, which is the greatest temporal blessing of man."[181] There seems to be implicit in this discussion a version of a social contract view – since society expresses such high esteem for physicians, physicians ought to feel obligated to adhere to the highest ethical standards so as to be worthy of this esteem:

> That this confidence and esteem should be fairly merited, the father of medicine, and all his eminent successors, to the present period, required an oath of their disciples, the principal obligations of which were, the cultivation of every virtue that adorns the human character. A code of professional duties, or ethics, was arranged, which all were obliged to obey, and which still governs those who have been properly educated.[182]

Given the importance of this implicit social contract, Ryan thought it scandalous that medical schools of his day spent so little time teaching medical ethics:

> But there never was a period in medical history, in which ethics was so neglected and violated as in this "age of intellect," nor the dignity of the science so degraded and disregarded. It is therefore necessary to inform the rising members of the profession, of those virtuous and noble principles which regulated the professional conduct of their predecessors, and procured that unbounded confidence and universal esteem, which was bestowed on them by society in every age and country. I shall, therefore, describe the ethics of the founder and father of medicine, and those of his successors to the present time, with the hope of exciting my readers to imitate their example, and practise those precepts which have always characterized the erudite and scientific portion of the profession.[183]

Baker and McCullough argue that by electing to organize his "medical ethics" within a historical framework, Ryan was making one of his few original contributions to this field. The very term "medical ethics" dated back only to Percival's use of that title in 1803. Ryan was thus the first writer to offer a history of medical ethics. Ryan, moreover, claimed for the first time that what he (and Gregory and Percival before him) were addressing was not a set of issues or concerns that had recently arisen due to the particular circumstances of medicine in England and Scotland in the Enlightenment era. Rather, Ryan was making the much grander claim that he and his immediate predecessors were participants in a very long historical tradition that extended back to the earliest days of medicine on Earth, and certainly extended as far back as Hippocrates. (By contrast, Gregory and Percival had not referred explicitly to the Hippocratic Oath in their writings, nor did they regard that oath as in any way constituting an authority.)

Ryan thus became the first in a long line of physician writers on ethics to wrap himself in the mantle of venerated tradition, and to imply that his own work gained authority and credence based on that tradition.[184] As we have seen, while this may have been a new departure for the nascent "field" of medical ethics, it was, for Ryan, merely standard operating procedure. He did here for ethics little more than he had done previously when writing about mineral healing waters, obstetrics, and caesarian section. Ryan tended to begin a discussion of any medical topic with the implicit assumption of continuous historical progression.[185]

Ryan ostensibly devotes Chapter 2 to the ethics of Hippocrates, but finds ways to supplement this discussion with further observations of a religious nature. He summarizes the Hippocratic Oath, and notes with approval that his own medical school, Edinburgh, was the only school in Britain to require that its graduates take the Oath. Ryan supplemented the Oath with a list of physicianly virtues derived from other Hippocratic writings, addressing how the good physician treats both patients and colleagues. He included here the injunction, "Some, on account of friendship or acquaintance, expect attendance gratuitously, but these deserve neglect."[186] This passage, in the first edition, had brought down on his head the wrath of the *Lancet* reviewer, so Ryan tries to make more clear here that he is merely citing Hippocrates.

Ryan has no problem creating a segue from Hippocrates to the Bible, insisting as he will again later that in the pages of "the sacred volume … is laid the foundation of all ethics, 'do unto others as you would they should do unto you.' "[187] Ryan adds, "Hoffman was right when he said, a physician ought to be a Christian."[188] He then lists distinguished physicians who have written learnedly on philosophy and morals, including the philosopher, John Locke.

Despite having been chastised by *The Lancet*, Ryan cannot resist repeating here his diatribe against phrenology: "Vain attempts have been lately made by materialists, among whom are certain phrenologists, to prove that the mind is a mere function of the brain, and consequently dies with it; that the embryo has no mind or soul either in the uterus or at birth, but that its mind is built up by the five external senses, and is annihilated by death."[189] Ryan here stuck to his guns and dared the anonymous reviewer to do his worst: "I shall repeat my denunciation of this false philosophy as published in the former edition of this work, and advise certain commentators not again to misquote and misrepresent my language, as on a former occasion."[190] Ryan's ire was apparently aroused by what he took to be the claim of phrenologists that the mind was but a secretion of the brain, or in some way was totally dependent upon the physical function of the brain, and so ceased to exist when the body died. Ryan interpreted this as a denial of the immortality of the soul. By denying the immortality of soul, materialists undermine all religion and threaten the foundations of morality, which alone can restrain the passions. Therefore, Ryan argued, if a person who cared about his fellow beings should by misfortune become a materialist, he would be obliged never to divulge that he held that opinion: "Let the patrons of the revived and long refuted philosophy persuade their wives that their souls die with their bodies – let them instil the same doctrine into the minds of their children – let the doctrine become generally received – unfaithful wives, unchaste daughters, rebellious sons, and general confusion and anarchy will be the blessed fruits of their philosophy."[191]

Chapter III is called "Medical Ethics of the Middle Ages," and consists of a discursive listing of aphorisms loosely organized around a list of the virtues displayed by the good physician. The reader would naturally wonder what the Middle Ages have to do with all this, as one would imagine that here we are reading Ryan's own views on the subject. It is only at the start of the next chapter that Ryan states that he has adapted virtually the whole of Chapter 3 from the *Medicus-Politicus* of Roderic a Castro (Rodrigo de Castro, 1546–1627), published at Hamburg in 1614.[192] Winfried Schleiner has argued that Castro's book is "a milestone in the history of medical ethics" and that Castro and some of his contemporaries deserve credit for "creating something that can be called medical ethics, rather than a general (and usually Christian) ethics applied to medical practice."[193] Unfortunately, Ryan's third chapter as a whole is rather confused, since he periodically inserts comments that pertain to nineteenth century practice and that could not possibly had appeared in Castro's original work.

In keeping with most other ethical writings of both the seventeenth and the nineteenth centuries, little distinction is drawn here among duties owed to the patient, duties owed to other physicians, and prudential advice that will make the physician more successful in attracting fees. The behavior of the ideal physician is discussed in the form of both virtues to cultivate and vices to avoid. Ryan addresses one perennial ethical question – whether the physician is justified in lying to a dangerously ill patient about his condition – by appealing to his favorite authority. After acknowledging that truthfulness is a general moral duty, Ryan gives a number of examples from the Bible in which famous persons lied to get themselves out of difficulties. Ryan agrees with the assessment, common in the heyday of humoral medicine, that the mind powerfully influences the body: "a tone, a word, a look, will destroy life in delicate and dangerous cases."[194] He then concludes, "Official mendacity is less pernicious than that which is malicious."[195] The physician who lies to spare the patient a potentially fatal shock is lying in his "official" role and intends no harm; therefore his behavior does not fall under the usual injunction against telling falsehoods.

Ryan devotes the last several pages of this chapter to the subject of fees. He quotes a number of classical sources to illustrate the perennial problem of the patient who would give his fortune to be cured while in the throes of illness, but once recovered begrudges the physician the most reasonable fees. (Ryan seems here to have forgotten his arguments in Chapter I, about how all societies hold physicians in the highest possible esteem.) He argues against the British practice of the standard "guinea fee" in favor of the Continental habit of varying fees according to the experience of the practitioner. This is fair both to the patient and also to the less established medical practitioners; Ryan argues that so long as the most experienced physicians are willing to accept low fees, the less experienced ones are likely never to earn a living. Finally, Ryan briefly addresses the profession's role in policing malpractice among practitioners, and argues that the tarnishing of the individual's own reputation through his misdeeds is sufficient punishment.

Chapter IV, "Ethics of the Present Age," is Ryan's longest. The bulk of the chapter is taken up with a precis of Gregory's writings on the duties of the physician, and a mostly verbatim reprinting of Percival's *Medical Ethics* (minus the extensive Appendices). Between them Ryan adds verbatim transcripts of the Edinburgh University oath; the "promise" and Statutes of Morality of the Royal College of Physicians; and the oath of the Royal College of Surgeons.[196]

Ryan gives as his excuse for merely reprinting extracts from his predecessors: "Want of leisure precludes me from making sufficient research for the compilation of a complete system of ethics."[197] Additional commentary is nevertheless needed for several reasons – neither Gregory's nor Percival's work was regarded as authoritative;

medical students seldom studied them; and Gregory's work was now half a century out of date and so did not reflect the modern conditions of practice.[198]

Ryan has few critical comments on any of Gregory's injunctions. On the physicians' duty of secrecy and discretion, Ryan takes for granted that civil law overrides medical ethics, and that the physician is helpless when the revelation of patients' private confidences is demanded by a court. This point is very important as it appears to be the first acknowledgment in modern medical ethics that the duty of confidentiality is constrained by law.[199] But Ryan does note that some French authorities are of the opinion that physicians ought to risk civil punishment in order to remain true to the traditional Hippocratic promise to maintain confidentiality.[200]

On one important issue, Ryan disagrees with both Gregory and Percival. Both of the older authors, full of admiration for Baconian scientific advance, spoke indulgently of experimentation on patients. Percival's relevant passage is as follows:

> XII. Whenever cases occur, attended with circumstances not heretofore observed, or in which the ordinary modes of practice have been attempted without success, it is for the public good, and in an especial degree advantageous to the poor (who, being the most numerous class of society, are the greatest beneficiaries of the healing art) that *new remedies* and *new methods of chirurgical treatment* should be devised. But in the accomplishment of this salutary purpose, the gentlemen of the faculty should be scrupulously and conscientiously governed by sound reason, just analogy, or well authenticated facts. And no such trials should be instituted, without a previous consultation of the physicians or surgeons, according to the nature of the case.[201]

In one way, Percival's admonition parallels the modern view of the ethics of research on human subjects, by calling for a special sort of peer review. Percival clearly realized that the same moral foundations that justified therapeutic interventions by physicians failed to extend so far as to justify research aimed at scientific advancement. This led Henry Beecher, the American anesthesiologist and advocate for increased scrutiny of the ethics of research during the 1960s, to credit Percival as a pioneer.[202]

Ryan's own argument goes like this:

> The duty of *caution* in practice, means 'care not to expose the sick to any unnecessary danger.' The best rule of conduct on this important point, is the simple and comprehensive, religious and moral precept, 'Do unto others as you would they should do unto you.' Whatever the practitioner does or advises to be done for the good of his patient, and what he would do in his own case, or in the case of those who are dearest to him – if he or they were in the same situation – is not only justifiable on his part, but it is his indispensable duty to do. The patient should have the chance, whether it be a hundred to one, or only one in a hundred in his favour. Whatever may be the result, the practitioner has the greatest of all consolation–the consciousness of rectitude – "mens conscia recti;" this will be his solace, should the case terminate unfavourably, when the vulgar, the ignorant, the envious, the malicious, and the interested, will not fail to blame him for the death of his patient. But if he administers a dangerous medicine, merely to gratify his own curiosity, or zeal for science – to ascertain the comparative advantage or disadvantage of some new remedy, either proposed by himself or suggested by others – he is held guilty of a breach of ethics, and of a high misdemeanour, and a great breach of trust towards his patient; and if the patient dies, he might be severely punished.
>
> Medical men have tried the most dangerous experiments upon themselves, from their zeal for science, and even sacrificed their lives; but patients, in general, have no such zeal for science – no ambitions for such a crown of martyrdom – and generally employ and pay

> their medical attendants for the very opposite purpose. It must be admitted, that men who would try experiments on themselves, would be very apt to try experiments on their patients. It is a melancholy truth, but cannot be denied. The profession, however, has always reprobated such conduct; and the medical phrase of reproach and contempt for it, "corio humano ludere," to play with the human hide, abundantly testifies in what abomination it has been held by the faculty. It is unnecessary to dwell on this point in this age, because all experiments are made upon the inferior animals ... But every man of common understanding well knows, that neither physic nor surgery can be practised without some danger to the sick. It is universally known, that many surgical operations are dangerous to life, and that our most powerful remedies are highly dangerous; and more especially when improperly employed, or when they cannot be borne. A safe medicine is often extremely dangerous, from the peculiarity of constitution: and the great and urgent danger, in many diseases, requires the immediate use of dangerous remedies. It is admitted, by the best practitioners, that many remedies are still wanted for the cure of disease, and this want leads us most justifiably, and almost inevitably, to try new remedies on many occasions; and such experiments are not blamable, for they are necessary: – *sic enim medicina arta; subinde aliorum salute, aliorum interritu perniciosa discernans a salutaribus.* From these causes there results much inevitable danger in the practice of physic. From this acknowledged danger, results the important duty of caution in a physician, or care to make the danger as little as possible.[203]

Aside from his historical frame of reference, this passage is perhaps Ryan's most important *original* contribution to medical ethics. He acknowledges that all medical therapy has an inherently experimental nature, because patients have idiosyncrasies and one can never predict with full confidence how any individual will react to any treatment. But he distinguishes this dangerousness due to medicine's inherent uncertainties, from added risk brought about from the physician's curiosity or scientific zeal. This foreshadows the important ethical distinction between *research* and *experimentation* made by the national commission in the U.S. that authored the Belmont Report on the ethics of research in the 1970s.[204] Ryan even implicitly admits that his favorite fount of ethical wisdom, the Golden Rule, might fail here. Physicians in their zeal for knowledge often try dangerous experiments on themselves; so the injunction, "do unto others" might erroneously suggest that this is equally acceptable for their patients. Ryan appeals to an implicit form of contract theory once again: patients do not pay physicians to put their lives at risk by experiments that are designed rather to advance scientific knowledge than to bring about any real hope for cure. Ryan discerns, in a way that goes beyond even the patient-centered ethics of Gregory and Percival, that a fundamental duty to the patient prohibits at least some forms of medical research.[205] Why he imagined that experiments on animals had made this question obsolete in his day is never adequately explained.[206]

Ryan dissents from Percival on one other, minor point. Gregory, Percival, and Ryan all concurred that consultations among practitioners were a prime flash point for disputes that generally served none of the patients' interests and that therefore needed strong regulation to keep them amicable. Percival had addressed this need by stipulating in great detail the order in which all involved parties would give their opinions. In the process he suggested that seniority should be determined by how long one had practiced in the local community, not how many years of practice experience one had. He apparently had somewhat exaggerated fears of physicians who had no intention to settle permanently in the area coming into town, disrupting local relationships, and then moving on. Ryan very reasonably offers the counter-argument: If Sir Astley Cooper (one of the most prominent London lecturers on surgery) were to move to Manchester, and were called into a consultation, it made no sense to give him the same ranking as a surgeon just 6 months out of training.[207]

Ryan devotes Chapter 5 to "American medical ethics." One would have imagined that he would here spend time discussing the writings of Benjamin Rush (1745–1813), who had attended Gregory's lectures in Edinburgh, corresponded with Percival, and written *Observations on the Duties of a Physician* (1789), which was widely reprinted in England.[208] Instead the entire chapter is a verbatim reprint from a lecture by Dr. John D. Godman of Rutgers Medical College, New Jersey. Ryan notes that the volume of Godman's collected lectures had been "politely sent to me."[209] Apparently Ryan chose this work to "place American Medical Ethics before British readers" simply as a matter of convenience.[210]

The lecture Ryan chooses to reprint is intended for medical students and is on the subject of "professional reputation." Godman is mainly concerned here to distinguish between a reputation built up slowly over many years from virtuous behavior and solid accomplishments, and a fame based on a few lucky chances that is destined to fade as quickly as it came. He urges students to bear poverty and privation in their early years of practice with fortitude, arguing that almost all successful practitioners had to pass the test of going through such a period in their own lives.

Godman also argues forcefully that a classical education, including languages, is an essential preparation for any physician who would wish for true professional success. Among the virtues of the ideal practitioner, Godman lays special emphasis on truthfulness; neither he nor Ryan comment on the apparent contradiction between the moral duty to be truthful and the oft-cited injunction against truthful disclosure of prognosis to the seriously ill patient.

Chapter 6, dealing further with medical education, is also a new feature of the second edition, but appears to be a hodgepodge of miscellaneous topics, including long sections on physical diagnosis and the "art of prescribing" that seem especially out of place to today's reader. Ryan takes up the topic that Godman had addressed, the professional reputation of the young physician, and inserts a further (rather repetitious) quotation from the American author. Indulging in a bit of self-interest, perhaps, as a journal editor, Ryan advises the student that buying medical periodicals is cheaper than buying books, and by that means one can assemble a very complete medical library.

Ryan next addresses the medical reform proposals then before Parliament, and notes the general testimony to the effect that the required period of medical education in Britain is too short, so that too many illiterate and unqualified persons have managed to earn diplomas. "It is now generally known, that almost all medical appointments in England, Scotland, and Ireland, are filled by jobbing and bribery. The majority of the physicians and surgeons selected, do not possess any, or sufficient, or superior merit; and obtain their situations by intrigue, family influence, or purchase. Such is the case as regards our court, hospital, army, navy, and all other public medical appointments."[211] Ryan much prefers the French system by which professional juries decide who gets medical appointments. After some additional swipes at the privileges and self-interested behavior of the medical corporations in England, Ryan notes with regret that almost all new scientific advances were then coming from the Continent; and he hopes that with the anticipated Parliamentary reforms, Britain would once again take its place in the vanguard of medicine.[212]

The general public is the next target of Ryan's criticism: "If the public do not derive the usual advantages from medical practitioners, they should blame themselves for encouraging illiterate pretenders to medical knowledge, nominal doctors, and empirics. Such imposters are very numerous in all countries; and if the public do not take the trouble of distinguishing the learned from the ignorant, the fault is their own."[213] He fails to ask whether the profession's own traditional secrecy (e.g., concealing the contents of prescriptions by the use of Latin) might itself further public confusion about what counts as good medicine. Indirectly, also, the public is to blame for some of the non-virtuous behavior among physicians: "When a thirst for gold is the only object of professional reputation, it leads to meanness and disreputable behaviour. We see empirics, illiterate and professional, amassing great wealth at the expense of every virtue which adorns the true medical character. The public is the cause of this, as it awards reputation by caprice, and this is the reason that charlatans share, in common with educated physicians, reputation and renown."[214]

Ryan continues here to address some of the dilemmas of character that Godman had discussed – how properly to balance desire for virtue, fame, and money. Desire for fame, if it could be directed into the right channels, should be encouraged, since the behaviors it promotes among physicians contributes to the betterment of society at large. Ryan admits that some distinguished physicians had also amassed great wealth, and again cites Astley Cooper as an example, but he is generally disapproving of wealth-seeking and thinks that the physician of moderate means is more likely to be a person of virtue and professional attainment.

The young physician is instructed on various ways to achieve public acknowledgment without stepping over the line into charlatanism – how, for instance, to spend enough time in society to develop pleasing and easy manners, without wasting time in aimless amusements and detracting too much from scientific study. Ryan's strong interest in obstetric practice comes through in his admonition to the young physician to marry, as a married man is much more likely to be trusted when care is needed for a female patient.

A set of guidelines for the medical writer comes next:

- Don't write at all unless you have something really new to say.
- Don't use hurry as an excuse for error or poor style; the public would have happily waited for a better quality work.
- Praise others and be very sparing with vituperation, but also be modest and ready to reveal errors and untruths. "When criticism is unjustly severe, it often becomes pointless, and only serves the circulation of a work."[215]
- Don't write or try to get mentioned in the press merely as a means of self-advertising.

Perhaps Ryan, in a moment of self-awareness, was offering advice that he wished he could have followed. Sadly, his impetuous and disputatious nature prevented him from adhering to these guidelines very often.

Ryan admonishes the physician to speak well always of his colleagues, something he perhaps neglected to do in the Gordon Smith and Ramadge case. He also praises the use of mental influences as a mode of healing, thereby reminding his readers that the attitude and manner of the physician are of therapeutic importance. While the latter portion of this chapter, on the methods of prescribing, seem quite out of place to us in a volume supposed to be about medical ethics, a few passages appear in which Ryan addresses the extent to which different types of information ought to be shared with patients, and to which patients ought to be allowed to make their own decisions. Incongruously, Ryan appends at the very end a couple of paragraphs about the care of the dying patient.

IIB Manual on Midwifery

Ryan included in his *Manual on Midwifery* (1828 and later editions) passages that by today's standards would be viewed as having to do with ethical issues that arise in obstetric practice. He offered no explanation of why he omitted these same passages when he set out to write about medical ethics in his *Manual of Medical Jurisprudence*.[216] He noted in the latter work that he had failed to find sufficient time to write up a complete "system" of medical ethics; and perhaps he imagined that in a more comprehensive work, these ethical issues related to midwifery would be included. More likely, he was simply dependent on his distinguished predecessors, Gregory and Percival, for his sense of what belonged or did not belong in a work on medical ethics. Since midwifery-related issues had not been discussed (or much discussed) in their earlier works, Ryan defined the contents of these passages as "midwifery" and not "ethics." We have elected to reprint two threads from the *Manual of Midwifery* in this volume because regardless of how Ryan categorized them, the passages do in fact show something of how he reasoned about ethical matters.

The first set of passages comes both from the *Manual on Midwifery* of 1828, and the later (1841) edition. The "Professional Approach of the Obstetrician" describes the physician's manner of dealing with the patient in active labor. The long passage from 1828 is heavily dependent on Gregory, in some instances following the Scottish physician almost word for word. Ryan mixes a discussion of the attitude and demeanor with which the obstetrician should approach the patient, with technical details as to how to conduct the examination and prepare the chamber. "The weakness and peculiar delicacy of the female constitution, call forth our greatest tenderness and compassion," Ryan advises, "and on no occasion so powerfully as in the agonies of child-bed, when she is stretched upon the rack on which she is laid by nature."[217] He would here seem to have accepted fully the paternalistic attitudes common in his day, regarding labor as one of the medical crises described by Percival, during which one should never bluntly tell the patient her true situation if it is in any way frightening, but do all to spare her any emotional shock.

The continuation of the equivalent passage from the 1841 edition raises at least the possibility that Ryan was gradually modifying these attitudes. Consistent with the generally accepted formula, Ryan argues that *the family and friends* of the laboring woman have a right to know if any danger threatens, so that they may seek a second opinion or other aid. But he now adds that no instrumental operation should be performed without advising the family of its necessity, or discussing the matter with "[the woman] herself, when she possesses a strong mind."[218] He then adds that it is often desirable to ease the woman's fears of instruments such as the forceps, by showing them openly to her and demonstrating how they are used, rather than trying to conceal them from her or keep her in the dark about their application (a tip that he credits to his predecessor, Dr. Denman, whose book of aphorisms he had edited).

These passages regarding more openness and truth in the obstetrician's relationship with the laboring woman can certainly be interpreted in a solely instrumental fashion. Ryan may simply have been worried about the woman moving at the wrong time and injuring herself when forceps were used, and have determined that these techniques assured a patient who would reliably lie still when needed. He may have had no idea whatever that the woman as patient had any *right* to this information. But it is harder similarly to dismiss his statement about telling a woman of the impending need for an operation. By distinguishing between a woman of "strong mind" and other women, Ryan seems to imply that at least some women, even in the throes of labor, lack the characteristic that seemed to justify withholding an unwelcome truth from the sick person. Absent the "weak mind," with its attendant risk of emotional over-reaction to grim news, on what grounds could a physician who *generally* was devoted to truthfulness recommend concealing vitally important information from the patient? It was not, then, too much of a stretch from the patient's representatives' undoubted *right* to know the truth, so that they could seek other aid if necessary, to the patient's own "right" to know. We might wonder what further thoughts Ryan would have had on this subject had he lived to write yet later editions of this work.

The next long passage we have reprinted, this time only from the later edition, addresses the relative values of preserving the life of the newborn or the mother when there is a risk that both cannot be saved. Ryan contemplated this question in relation to four obstetrical procedures, three of which he considered appropriate options:

1. Gastro-hysterotomy, or caesarian section: An incision is made in the abdominal wall and carried through the wall of the uterus, to deliver the infant via the incision, and then to suture the edges. We must recall that in Ryan's day, such a procedure would necessarily be done without either anesthesia or antiseptic precautions.
2. Embryotomy, craniotomy, or crochet: The fetus's head is perforated, skull contents removed, and the head (and sometimes the body) crushed, so as to allow passage through a pelvic opening too narrow to admit the intact fetus.
3. Induction of premature labor: Uterine contractions and cervical dilatation are stimulated before their natural onset, to allow the passage through the pelvis of a smaller fetus, when the full-term fetus would presumably be too big to pass.
4. Symphyseotomy: The cartilage between the two halves of the pubic bone is incised surgically to allow greater separation of the pubic bones and hence a wider pelvic outlet for passage of the fetus. This procedure was favored by some

French obstetricians but was not recommended as an option for English practice. Ryan alludes to it only in passing.

Ryan's sense of Divine law leads him then to a number of conclusions about the use of these procedures. Induction of premature labor, if used as a method of abortion before the fetus is viable, would be absolutely prohibited, as is abortion in any form. (In a passage we do not reprint, Ryan describes the technical aspects of this procedure in Latin, as an extra precaution against uneducated persons adopting the method as an abortifacient.) Used in hopes of delivering a living infant, when there is good reason to believe that the maternal pelvis would never allow a full-term infant to pass, it should be considered an acceptable option.

English obstetricians routinely recommended embryotomy on a still-living fetus in cases where delivery appeared impossible, arguing that it was better to lose the infant's life than the mother's – by an adage Ryan wryly quoted, "the tree should be preferred to the fruit."[219] Ryan forthrightly rejected this position. As a matter of principle, no one has the authority to claim that the life of one human being is worth less than the life of any other human being.

Ryan adopts an "out" in cases where the fetus cannot descend through the pelvis in the natural way – he notes that the forces of attempted expulsion will probably lead to fetal demise long before the mother's life is truly in danger. There is no moral objection to embryotomy performed to remove a dead fetus. Ryan also believes that with the examinations available, the life or death of the fetus can be accurately determined.

In cases where the mother's life is endangered while the fetus is still alive, or when the pelvis is so contracted that one can predict with certainty that even embryotomy would fail, Ryan advocates caesarian section, even while admitting its high mortality rate. On the one hand, it is such a dangerous operation that Ryan strongly criticizes the physician who would employ it when any less radical means is available. On the other hand, Ryan suggests that if physicians were willing to perform it when truly indicated, without waiting until the mother had been weakened by a prolonged labor, its mortality rate would probably be a good deal less.

Obstetricians had encountered cases where a woman with a severely contracted pelvis, who previously had been delivered only by means of caesarian section, repeatedly became pregnant. A few of Ryan's obstetric colleagues advocated in such cases either dividing the fallopian tubes at the time of caesarian section, to prevent future pregnancy, or operating on the woman in early pregnancy effectively to abort the fetus. We will see below that Ryan strongly opposed contraception as contrary to the Divine plan. Ryan is therefore equally condemnatory of any attempt to interfere with natural fertility even in this extreme instance. He proposes sardonically that if the goal is to prevent fertility, why do not obstetricians recommend sterilizing the woman's husband, since he is at least equally responsible if not more so?

Ryan supplements appeals to Divine law with consequentialist reasoning. For example, he mentions that a consequence of the English practice of performing embryotomy on the still-living infant, to save the life of the mother, is that occasionally infants are born partly dismembered yet still alive. He suggests that the suffering thereby inflicted on such an infant becomes yet another reason not to perform the procedure while life remains. A further argument against embryotomy is that the woman, as well as the fetus, may be injured by the sharp instruments. He also strongly condemns the use of invasive techniques to deliver an infant simply so that it can be baptized. He argues that on the one hand, scriptural requirements can be met by other modes of baptism, such a using a syringe to inject water against the amniotic membranes; and on the other hand, the risks and damage caused to the mother can never justify an operation such as caesarian section, when the infant very likely would die soon no matter what was done. Ryan also occasionally appeals to authority, such as a decision of the theological faculty of the Sorbonne in 1648, that it would be better for both mother and baby to die without assistance, than that the physician should be the direct cause of death of the infant.[220]

IIIC Prostitution in London and Philosophy of Marriage

Prostitution in London and *Philosophy of Marriage* (in passages not reprinted in this volume) provide further insight into Ryan's moral thought. *Prostitution in London* focuses on what he considered to be the two primary social causes of venereal diseases – poorly conceived legislation contributing to increase in prostitution, and immoderate overuse of the reproductive organs (masturbation and sexual excess). *Philosophy of Marriage* provides in addition a fairly comprehensive account of cross-cultural reproductive norms. Since he assembled the volume from his medical student lectures, *Philosophy of Marriage* also repeats many of Ryan's observations on the physician's approach to the patient in labor.

The main lessons we learn from these volumes about Ryan's ethics are, first, that he attributed to physicians a moral responsibility to reveal the Divine laws of nature to both their own students and educated lay persons, and to act as social critics where current practices deviated from those laws; and second, that he advocated compassionate treatment of some groups of patients whose behavior deviated from those Divine laws and who, as a consequence, were commonly mistreated by physicians as well as by society.

The Physician as Social Critic. The diversity of topics covered in *Philosophy of Marriage* is indicative of what Ryan considered to be the scope of his moral responsibility:

> [It is] the [ethical] duty of every lecturer of on physiology, on obstetric and on legal medicine, to describe … [the function of reproduction] according to ancient and modern conclusions; because it materially influences population, morals, public health, disease, mortality; as well as personal reputation, property, legitimacy, and even life, together with a vast number of other questions …"[221]

Physicians should reveal the laws of nature to the masses and to exhort them to act in accordance with them: "the sole object of the medical inquirer is to display nature in her true character, to defend her laws, and to expose the errors and follies of mankind in her violation."[222]

Ryan particularly believed that sexual overindulgence was inconsistent with the natural purpose of the reproductive organs, compromising the ability to produce healthy offspring. That in turn would increase crime and disease, undermining the well-being of any empire:[223]

> Nothing is more certain that this, that animals and plants shorten their existence by multiplied sexual enjoyments. It was to secure vigour of the mind and body that the founders of certain religious sects prescribed chastity and celibacy to their ministers. This rule is in some degree accordant with physiology; for it is well known that our moral and physical powers are diminished by coition, because we impart a portion of our endowments to our offspring, and diminish them in ourselves.
>
> A most important train of consequences to society and government follow from this prodigious influence on the population of empires.[224]

Ryan urged his students to encourage their patients to practice sexual moderation in conformity with the Divine edict. He also vehemently opposed the use of birth-control measures advocated by so-called Malthusians:

> The most zealous disciples of Malthus were said to be the Westminister political economists, including Bentham, Ricardo, Place, (James) Mill, Tooke, Brougham, Miss Martineau, and others of minor note. A number of grossly immoral men followed their example, and in 1822 distributed the most infamous handbills throughout the large manufacturing districts in England, which purported to contain 'the important information for the working classes, how to regulate the number of a family.'[225] Various abominable means were proposed, which few if any one would follow, for all were contrary to the dictates of nature, to the precepts of revealed religion, to morals, to the divine and primitive command – 'go forth and multiply.'
>
> It must be scarcely necessary to observe, that the doctrine of limiting population is based upon a most irreligious doubt in the conservative power of the Divine Creator; which regulates, preserves, and reproduces the illimitable number of organized beings in the animal and vegetable kingdoms.
>
> It was, however, most erroneously contended by the advocates of this cold-hearted and immoral doctrine, the consequences of controlling the faculty of reproduction would be moral, civilizing, would prevent much crime and unhappiness, that they would improve the manners and moral feelings, alleviate the burdens of the poor, diminish the cares of the rich, and lastly, they would enable parents more comfortably to provide for, and educate, their off-spring. But to these conclusions it may be unanswerably replied, that the limitation of offspring is based upon principles severely condemned and reprobated in the sacred volume, which are subversive of every virtue, and holding out inducements and facilities for the degradation of our daughters, sisters, and wives.[226]

From Ryan's vantage point, the neo-Malthusian proposals to limit fertility could not count as true social reform, since they deviated from Divine law as related in the "sacred volume" (the Bible). Because all biological laws had to be in conformity with God's design, it therefore followed logically that the beneficial results promised by the Malthusians *could not* be produced by the means they suggested. Virtually by definition, only human misery could result from disobeying the natural law.[227]

Ryan was not, however, unresponsive to Malthus's contention that population growth surpassing food supply promised misery. Speaking of the multitudes of the poor in Ireland, Ryan argued that late marriage was the solution to the dilemma of over-population. Since sex within marriage, the institution of marriage itself, and the practice of the sexual moderation and self-restraint necessary to postpone marriage were all seen as in accord with Divine law, Ryan saw no conflict between the advice to marry late and the Biblical injunction to "be fruitful and multiply." Noting that those with secure earnings tended to marry late and have fewer children than those with little or no income, he contended that alleviating poverty would result in late marriages, leading to a reduction in births:

> The evidence proves that the sons of farmers, and even the few of the labourers who are in better circumstances than the generality of their class, are much more cautious in contracting marriage than their poor brethren, and that they wait until they meet with a woman who has some little property or other means of assisting to maintain a family. In short, it appears quite certain that in every part of Ireland the more destitute the labouring population are, the more recklessly, nay, the more eagerly, and at a much earlier age, do they marry… . In this county of Clare, M'Mahon, a labourer, stated that, "It is always the poorer man that marries first, because he knows he cannot be worse off by it; it is better for him to marry early than to seduce the girls, who are so poor and wretched that this would often happen. Besides, we poor people have a strange idea that it is a good thing to have children as soon as possible, to help support us when we both begin to grow old."[228]

Ryan espoused late marriage for reasons beyond the problem of over-population. He held that men ought to make a careful, considered decision about whom to marry because it determined, at least in part, the quality of their progeny.[229] Citing John Gregory, he lamented the fact that while men invested considerable effort to improve the quality of their livestock by carefully breeding them, they were all too thoughtless with respect to their own offspring.

> Dr. [James] Gregory maintains, in his Conspectus Medicinae Theoreticae, that children often resemble their parents, and are certainly like them, not only in features, but in forms, mind, virtues, and vices.[230] Dr. John Gregory, the father of the last-named distinguished author, makes the following remarks in his Comparative View of the State and Faculties of Man with those of the Animal World.
>
> Thus by a proper attention we can preserve and improve the breed of horses, dogs, cattle, and indeed of all other animals. Yet it is amazing that this observation was never transferred to the human species, which were it would be equally applicable. It is certain, that notwithstanding our promiscuous marriages, many families are distinguished by peculiar circumstances in their character. This family character, like a family face, will often be lost in one generation and appear again in the succeeding. Without doubt, education, habit, and emulation, may contribute greatly in many cases to preserve it; but it will be generally found that, independent of these, nature has stamped an original impression on certain minds, which education may greatly alter or efface, but seldom so entirely as to prevent its traces from being seen by an accurate observer. How a certain character or constitution of mind can be transmitted from a parent to a child, is a question of more difficulty than importance. It is, indeed, equally difficult to account for the external resemblance of features, or for bodily diseases, being transmitted from a parent to a child. But we never dream of a difficulty in explaining any appearance of nature, which is exhibited to us every day. A proper attention to this subject would enable us to improve not only our constitutions, but the character of our posterity. Yet, we every day see very sensible people, who are anxiously attentive to preserve or improve the breed of their horses, tainting the blood of their children, and entailing on them, not only the most loathsome diseases of the body, but madness, folly, and the most unworthy dispositions; and this too when they cannot plead being stimulated by necessity, or impelled by passion …[231]

The inclusion of these sentiments in both *Philosophy of Marriage* and *Manual of Midwifery* suggests that Ryan considered it a part of his social reform agenda to argue that only the physically and mentally healthy should reproduce.[232] His commitment to the idea that physical, mental, and moral traits were hereditary may also be found in his account of Wrolick's and Velpeau's thesis. Foreshadowing Darwin's theory of evolution, Wrolick and Velpeau contended that the various human races were in a hierarchical relationship with each other and non-human primates. Ryan summarized their views:

> There is a vast variety in the pelvic structure of animals, which as we gradually ascend the zoological scale, becomes insensibly more perfect. Thus, the pelvis of the monkey and ourang outang have a strong resemblance to that of the human species; and we can trace the shades between these and the Boshiemans, who are considered by some to be the connecting link between monkeys and man, the Ethiopians, Negroes, Malays, Japanese, and Caucassians, the last of whom are most different from the other mammiferae; and thus, we perceive, that the parturition is more painful as the species is more perfect, and *vice versa*; a most admirable and singular provision of nature, that the dangers are increased and most accumulated, according to perfection and degree of intelligence of the animal.[233]

Given Ryan's commitment to the ideas of racial superiority and inferiority, it is surprising that in *Philosophy of Marriage* he extols the benefits of interracial alliances:[234]

> Nature also condemns [marriage between close relatives]; she wishes marriages between families and nations, because these intermarriages, or crossing of the races, are the true means of improving and invigorating the species. Some of the ancient legislators framed their laws on this ground (Plutarch, Quaest. Roman. St. Augustin, City of God, &c. Vandermonde, Essai sur le Perfect. De l'Esp. Hum., Buffon, &c.). Daily examples confirm the validity of the opinion. Pallas adduces the fact that the intermarriages of the Mogul Tartars with the Russians produce very fine individuals; while Humboldt has observed, that the offspring of a negro and European is more robust and active than that is a white with an American; because the best mode of effacing hereditary diseases, gout, scrofula, phthisis, mania, epilepsy, &c., is by the commixture of the species in intermarriages, as this corrects the defects of one individual, by the soundness of the another. The Jews, who refuse to intermarry with other nations, transmit many hereditary diseases to each other; and they preserve, by this custom, their Hebrew cast of countenance in all countries (For other proofs of this position, the reader may consult *Mr. Combe's Essay on the Constitution of Man, considered in Relation to External Objects*, 1835).[235]

Ryan addressed topics such as the abuse of the sexual organs and prostitution in his works although he was well aware that he would be condemned by those who believed that such matters were inappropriate for discussion. He described such critics as "… possessed of a weak, erotic, or prurient imagination … or those who pretend to excessive modesty and rigid chastity," and made it clear that their censure would not sway him from attending to his responsibilities as a medical teacher.[236] He argued, for example, that strategies used by the corrupt to coerce female children and adolescents into prostitution should be made public so that parents would be forewarned:

> [H]ow … could parents, ignorant of the artifices employed and the extent of the evil referred to, give the needed warning [to their children]? … I most fully agree with Mr. Talbot [of the London Society for the Protection of Young Females and Prevention of Prostitution], and therefore place the information with which he has kindly supplied me

> before the reader, convinced as I am, that every moral and religious individual will approve my motive, which I am sure cannot fail to be productive of a vast deal of good, by assisting in urging the legislature of this empire to correct or diminish, the horrible state of immorality amongst us.[237]

Ryan's conviction that it was his responsibility as a physician to bring social ills to light and to argue for legislative reform is particularly evident in *Prostitution in London*, modeled on Dr. Parent-Duchatelet's *On Prostitution in Paris* (1836).[238] Parent-Duchatalet reported on the prevalence and causes of prostitution in Paris, which he attributed in large part to poorly-conceived legislation. Agreeing with his French colleague, Ryan intended that his work be read by legislators, magistrates, and philanthropists as well as by physicians and surgeons.[239]

Prostitution in London is a scathing critique of laws that allowed men to seduce and impregnate lower-class women and then abandon them.[240] The 1834 Poor Law Bill ended the old system by which poor unwed mothers could receive financial support from the parish while remaining in their own homes. Hoping to decrease moral delinquency, the framers of the new law demanded that such women now be admitted to houses of correction. But rather than be incarcerated in an institution, a number of these unwed mothers resorted to prostitution.

> Here is legislation with a vengeance. All the expenses, the odium, and the ruin, are placed on the mother, who is, in ninety-nine cases in a hundred, the victim of seduction, by some heartless, vicious, and unprincipled scoundrel. It is proved by the host of evidence on the state of prostitution in this and other countries, in the work already quoted, that for the most part girls and women are decoyed and ruined by men, and not by reason of inherent licentiousness … It is, therefore, in my opinion, a most glaring defect in our legislation, to exonerate the male sex from all responsibility and punishment for seduction and bastardy, because this sex is by far the most guilty and vicious; for all physiologists admit, that the amorous impulse is stronger in the males than in the females of all grades of the mammiferae, from the lowest to the human species, because the male imparts vitality and perpetuates the species.[241]

Ryan had no doubt that the male's stronger sexual impulses were biologically determined, but did not see this fact as a moral excuse for men who victimized and abandoned women. Properly designed legislation (as well as religion), he thought, could serve as a check against men's biological impulses.

Ryan considered it his moral duty as a physician who was familiar with the plight of unwed mothers to argue for legislative reform that would give these women better options than the house of correction or prostitution.[242] He noted that the trades open to females did not pay a living wage. For instance, he argued that seamstresses ought to be paid enough so that they could support themselves and their children, without having to resort to prostitution.[243] Ryan argued that legislation on the subject should be formulated in consultation with medical practitioners acquainted with the social causes and the pathology of venereal diseases.[244] Finally, Ryan wrote, "… I must take leave to observe, that there can never be sound legislation unless in strict accordance with physiology or nature, and the sacred scriptures," which of course for him amounted to the same thing.[245]

Ryan admitted one wrinkle in his generally rosy view, that violations of Divine law always went hand in hand with unnatural practices and deleterious consequences, while adherence to God's will always ensured good health. He admitted that in one instance, society was forced to choose the lesser of two evils. The issue was what he delicately called 'profligately licentious' behavior in sex-segregated penal colonies for males. Ryan considered the presence of female prostitutes to be a potential solution to the problem of sex between males, which was a clear violation of the true purpose of generative organs, resulting in disease. Faced with a forced choice between homosexual acts and prostitution, Ryan favored the latter:

> If notwithstanding laws, punishments, public contempt, even gross brutality, and frightful disease, the inevitable consequences of prostitution, prostitutes still exist, is not this an evident proof that they cannot be put down, and that they are unavoidable in society. Remove them from among mankind if you can, and all will become profligately licentious … The truth and force of these conclusions are, alas! too apparent in all communities where women are not allowed, or only a few admitted; as in penal colonies, barracks, ships, prisons, modern workhouses, schools, colleges, &c. &c. It must be acknowledged by everyone conversant with physiology, or the history of nature, that all legislation which enforces such conditions, is contrary to nature, reason, and morality; and is productive of the most horrible crimes … It is horrible to contemplate the consequences of compelling the most depraved men, in the prime of life, in perfect health, and too often of most vicious and wicked minds, to be congregated in prisons, barracks, ships, workhouses, and even chained together in penal settlements- wholly deprived of the other sex.[246]

Still, as a more satisfactory longer-term solution, Ryan hoped that convicts would not be banished to sex-segregated penal colonies and that existing settlements would be well provided with clergy.[247]

Compassionate Treatment of Stigmatized Groups. Ryan, as we have noted, had sympathy for prostitutes, the majority of whom he thought had been driven into their plight by "seduction, by some heartless, vicious, and unprincipled scoundrel."[248] His own work in charity hospitals had enlarged his experience among classes of society shunned by the average citizen. *Prostitution in London* and *Philosophy of Marriage* are notable for Ryan's clear statements that certain stigmatized groups of patients deserve more compassionate treatment than they often received.

Ryan, for example, generally followed Parent-Duchatelet's ideas in *Prostitution in London*, but categorically rejected the Frenchman's argument that prostitutes had no claim to civil rights because they had violated the law of the land by virtue of their profession. Besides the sheer impossibility of designing laws that would protect the rights of all women but that would specifically exclude prostitutes, Ryan argued that it would be immoral to deny prostitutes civil and legal protection:

> [H]umanity dictates, that such persons must be protected by the laws, as well as others. It would be very difficult to enact laws to deprive this class of persons of civic or social rights, to allow them, as unfortunate human beings, to be maltreated or murdered …. Be it remembered, that our laws most wisely protect the most abandoned prostitute as well as the most virtuous woman. The violation of either would be equally punished.[249]

Ryan also advocated the creation of asylums for female prostitutes as an exercise of Christian charity:

> Such institutions exist in every civilized country. It is to be recollected, that their unfortunate inmates are members of the human family, and that they have erred like all their species; but, 'to err is human, to forgive, divine.' If these unfortunates have contributed to demoralize society, they have been most severely punished by an unfeeling, wicked, and depraved world; and if some few good Samaritans pour oil and wine into their bleeding wounds, they but imitate a Christian and humane example.[250]

Another group of patients that Ryan viewed as deserving more compassionate treatment than they often received were the sufferers from venereal diseases. At one level his concern was the respectful treatment of these patients as fellow human beings: "For my own part, I cannot but consider that such patients require great commiseration as well as relief, and that their distressing disorders deserve as much attention from medical practitioners as any other class of human infirmities."[251] In a similar vein he expressed respect for the privacy of these patients: "I might easily introduce several cases which fell under my own care, in exemplification of the statements made in the preceding volumes; but I could not do so, without, in some instances, giving pain to many of my patients now living."[252] But a more practical level of concern was that venereal sufferers, fearing disrespectful treatment, would be driven away from the physicians who could truly help them and into the hands of quacks:

> Every individual who entertains a doubt as to his capabilities for generation, is anxious to obtain medical advice on his condition; and it is much to be regretted that it is too often the practice of the profession to treat the matter with levity or derision. Hence few of the faculty are consulted, an unreserved disclosure of the symptoms is seldom given, and the inquirer is often fearful that his condition may be made known to his acquaintances. Every duly educated physician is bound to secrecy, in all delicate matters, by an oath, and so far from treating his patients with levity or carelessness, should consider his case as attentively as any other that may come before him. Were this line of conduct generally adopted by the medical profession, an immense number of the public would not be driven to seek advice from the low, ignorant, and unprincipled empirics, who not only defraud them of immense sums of money, but also destroy, what is far more important, their health.[253]

IIID Ryan's Later Influence

Ryan spent all of his short life following one bit of his ethical advice, strenuously seeking fame and public acclaim through his accomplishments as a writer, lecturer, and practitioner. Leisure was something with which he probably had little familiarity. It is a shame that he lacked the "leisure," as he put it, to conduct "sufficient research for the compilation of a complete system of ethics." He apparently recognized that what he had written fell short of a "system." One would like to have known what form he imagined a system of medical ethics would take if fully developed, since he implied that neither Gregory's nor Percival's work rose to that level of organization.

The next "system" of ethics to appear on the scene in Anglo-American medicine after Ryan was the Code of Medical Ethics of the American Medical Association (1847). Based primarily on Percival, the Code organized its content under the headings: duties of physicians to patients; obligations of patients to physicians; duties of physicians to each other and to the profession; duties of the profession to the public.[254] The next British author to attempt to write a code of medical ethics, Jukes Styrap (1815–1899), also relied heavily on Percival and adopted a set of internal categories much like that of the AMA code.[255]

Burns notes that Ryan's *Manual of Medical Jurisprudence* was widely reprinted in the U.S., and therefore, along with Gregory's and Percival's work, was in a position to influence the development of medical ethics in America.[256] Nonetheless, Ryan's writings seem to have had little influence in his own adopted country, Great Britain. Ryan's six chapters on medical ethics may have been the most extensive writing on this subject in Britain during the long drought between Percival's 1803 edition and Styrap's code of 1878. But even the few writers who attended to the topic during the middle years of the century appeared to pay little attention to Ryan's work.[257]

IIIE Ryan's Place in the Development of Modern Medical Ethics

A final way to assess Ryan's legacy is to evaluate Robert M. Veatch's important recent work on the history of Anglo-American medical ethics between 1770 and 1980.[258] Veatch argues, following previous writers such as McCullough, that medical ethics as practiced by Gregory and Percival was characterized by open dialogue between physicians and the moral philosophers and other humanities scholars of the day. Soon after Percival died, this dialogue was broken off, and to the extent that physicians addressed ethical issues in the intervening years, they did so in near-total isolation from the ideas of contemporary humanists. Due to a variety of factors that emerged during the 1960s, this dialogue was rather abruptly re-established around 1970 and has characterized the new field of bioethics since then.

Veatch proposes two grand traditions in medical ethics, that he calls the Gregory and Monro views. The Gregory view entails the close connection between work in medical ethics and work in moral philosophy, whether religious or secular or both. The Monro view, by contrast (named for the three generations of Alexander Monros who taught anatomy at Edinburgh before and after Gregory's day), calls for physicians to spend all their time and energy mastering the increasing intricacies of medical science. As physicians disengaged from the philosophical and humanist dialogues of their own time, they needed something to retain a sense that there was a moral core or commitment to medical practice, and the Monros and their followers found that a simplistic, mostly ceremonial attachment to the Hippocratic Oath would serve. In this they contrasted with Gregory and Percival, who were fully immersed in the intellectual discourse of the Scottish Enlightenment and accordingly saw no need to draw any ethical inspiration from the teachings of a two-millenia-old Greek mystery cult.

Veatch assesses briefly the work of Ryan and treats him as a typical nineteenth-century devotee of the Monro view.[259] It is easy to see how Veatch could come to view Ryan in this way. Clearly Ryan was a derivative, not an original thinker; while he quoted copiously from Gregory and Percival, he added little of substance. He spoke fulsomely of Hippocratic ethics and implied, at least, that the words of Hippocrates could be pertinent in his own time. And Ryan was ignorant of most of the work being done in his own day in moral philosophy and theology. But a strong case can be made that this dismissal of Ryan as a *mere* Monro disciple misses the mark – despite all the weaknesses in Ryan that we have catalogued above. In four ways, Ryan could lay more claim than Veatch acknowledges to being a disciple of Gregory and Percival.

First, none of the Monros or their progeny seemed concerned about the neglect of medical ethics in the curriculum of the nineteenth- or early twentieth-century medical school. So long as one could consult the less-than-one-page Hippocratic Oath, apparently, no sustained study was required. Ryan, by contrast, strongly condemned the neglect of this topic. While he agreed that the works of Gregory and Percival had become somewhat out of date by the 1830s, Ryan nevertheless thought it scandalous that medical students of his day knew so little of these leading thinkers.

Second, it is shortsighted to see in moral philosophy or theology all that is worth studying in the "humanities." True, Ryan did not examine the moral-philosophy literature of his day to determine how best to do ethics. But, as we have seen, he is given credit for the first thorough-going application of a different humanities discipline – history – to the study of medical ethics. His claim that medical ethics forms a historical continuum from pre-Hippocratic times to his own day may be, from our point of view, a poorly-supported and amateurish hypothesis. Ryan was, after all, a physician and medical journalist, not a trained historian – though it is important to recall that in his day, no one became a historian by getting a doctorate of philosophy in history, and that in some sense, all historians of that time were amateurs. But nevertheless it is a highly original way of looking at medical ethics for his time, and it is, arguably, a humanities-oriented approach.

Third, Veatch assumes that physicians would naturally look to work on moral philosophy for guidance in ethics, because the material provided by medicine itself is insufficient to develop a satisfactory ethical framework. This assumption does not apply to Ryan in any meaningful way because of his deep religiosity. For Ryan, it was quite clear that physicians needed to look outside medicine for moral guidance, and it was just as clear where that guidance was to be found – in the Bible, and more specifically, in his favorite Golden Rule. It would have been foolish and indeed incoherent for someone of Ryan's religious views to consult the works of philosophers such as Hume, Reid, Adam Smith, or Bentham, as if he was going to find something in those works that provided *superior* ethical wisdom to his preferred Biblical sources. To accuse Ryan of attending inadequately to moral theological discussions of his time would be more to the point, but that would in turn assume that the moral theological scholars of his day were doing work of a sort that was directly helpful to medical–ethical issues. Like Gregory and Percival, Ryan *did* look outside medicine to find ethical guidance. He found it in a different place, which, according to his lights, was much sounder. His disagreement with Gregory and Percival over where to look for moral wisdom did not blind Ryan to the value of their contributions to medical ethics, which is all the more to Ryan's credit.[260]

Fourth, the Monro disciple retreated from any engagement or dialogue with the humanities mostly because the task of keeping up with a burgeoning medical-scientific literature was becoming all-consuming. For the same reason, that physician would usually retreat from any broader social agenda save for the care of individual patients, with the possible exception of anything that could clearly be labeled "public health." Ryan, as we have seen, was of the opposite opinion. He repeatedly called upon his colleagues to remain engaged in wider social movements, to influence jurisprudence and legislation, and to share their understanding of Nature's laws with the populace toward the development of enlightened policies.

The most accurate way to characterize Ryan's role in the unfolding of Anglo-American medical ethics from the later eighteenth century to the present day, therefore, is as a minor disciple of Gregory and Percival. To see him as falling under the Monro tradition highlights his limitations while ignoring his relatively few real strengths.

Notes

1. Our treatment here follows very closely Lisbeth Haakonssen, *Medicine and Morals in the Enlightenment: John Gregory, Thomas Percival, and Benjamin Rush*. Atlanta: Rodopi, 1997:1–35. Haakonssen surveys a number of historical and philosophical interpretations of the works of John Gregory and Thomas Percival, and proposes a largely persuasive account that situates these writings within the intellectual world of the Scottish Enlightenment. We are grateful to Robert Baker and Laurence McCullough for a better understanding of the limits of Haakonssen's analysis.
2. Examples of such interpretations can be found in Chauncey D. Leake, Preface to Percival's *Medical Ethics*, Baltimore, MD: Williams & Wilkins, 1927; Jeffrey Berlant, *Profession and Monopoly: A Study of Medicine in the United States and Great Britain*, Berkeley, CA: University of California Press, 1975; and Laurence B. McCullough, "Virtues, Etiquette, and Anglo-American Medical Ethics in the Eighteenth and Nineteenth Centuries," in *Virtues and Medicine: Explorations in the Character of Medicine*, ed. Earl E. Shelp, Boston, MA: D. Reidel, 1985. The purported contrast between Gregory and Percival is inaccurate, as both addressed physician-patient relationships as well as interactions among practitioners.
3. Gregory and Percival rejected the "traditional guild ethos of the Royal College of Physicians [of London] in favour of one which, they believed, better reflected the social and moral goals of their society as well as those of the new inductive science." (Haakonssen, *Medicine and Morals*: 17) In their day, Haakonssen asserts, "professional" was a tainted term, seen as allied with "interested." This compared unfavorably to the "disinterested" amateur gentleman of science who pursued truth for its own sake and not because he was tied to the interests of a sect or guild. Gregory, Percival, and Ryan all tended to show their disdain for the physician who merely wished to make money and had no devotion to the cause of advancing medical science.
4. Early in the eighteenth century Leiden (in the Netherlands) assumed prominence, thanks to Boerhaave, who became known as the medical instructor of Europe. Boerhaave's goal was to cultivate good clinicians and his first lecture was a call to return to the study of Hippocrates. Leiden had a 12-bed charity ward that was set aside for clinical teaching, and this ward in effect became the world's first teaching hospital. (Generally, hospitals were prohibited from placing patients in a teaching environment because of their religious ethos.) Boerhaave's principal innovation in medical was to stress the importance of the students seeing real patients, both as a source of experience and as illustrations for material covered in lectures. Other teaching centers began to mimic Leiden's manner of teaching medicine. Edinburgh became the first such school in Britain; its medical school dated from 1726 when Leiden-trained Alexander Monro (the first) was appointed as professor of anatomy. Edinburgh was not known for its bedside teaching approach even though it did pioneer infirmary-based teachings. Edinburgh's strength in medical education lay in its more modern, practical orientation, compared to the classical, highly theoretical curriculum then taught at Oxford and Cambridge. After 3 years at Edinburgh, a medical student who had been trained in medicine and surgery was ready to practice medicine in the capacity of general practitioner. Roy Porter, *The Greatest Benefit to Mankind*, New York: W.W. Norton, 1997: 290–291.
5. Haakonssen, *Medicine and Morals*: 14.
6. Gregory and Percival defended the Edinburgh system against critics who viewed it as "a free-for-all and the school a degree-mill" (Haakonssen, *Medicine and Morals*: 16). To some extent their lectures on ethics were a defense of the Edinburgh approach of free inquiry and egalitarianism, as contrasted with monopolies enjoyed by Royal College of Physicians of London, which they viewed as opposed to best interests of patients and the profession. Ryan, as an advocate of radical reform in the 1830s, strongly concurred with these sentiments.
7. Haakonssen, *Medicine and Morals*: 21
8. Haakonssen, *Medicine and Morals*: 22. Moral exhortation was a well known and popular genre of writing; and Gregory and Percival both viewed their writings as appealing to a wider

audience beyond physicians and medical students. Both had written in this genre elsewhere, notably Gregory on the duties of women (*A Father's Legacy to his Daughters,* published posthumously, 1774, 1779).

9. There was indeed an important distinction between moral theory and moral practice in Scottish Enlightenment thought; but it did not track the modern distinction. Moral theory was seen as branch of "theory of mind, or pneumatology: what, in modern terms, is generally referred to as moral psychology or moral epistemology" (Haakonssen, *Medicine and Morals*: 25). Moral theory was contrasted with practical principles of action, often organized as virtues viewed as interchangeable with duties (again defying twentieth century categories). Since little was known empirically at the time about workings of the mind, the moral theory of the day was highly speculative and necessarily, by today's standards, extremely muddled. Thus it made sense to say that moral theory appeared to be quite unnecessary for the practical moralist. A precise theory of moral faculties and moral psychology "did not matter, as long as it was understood that the active power was cognitive, educable, and potentially authoritative over the passions" (Haakonssen, *Medicine and Morals*: 26).
10. Haakonssen, *Medicine and Morals*: 31.
11. Leake, Preface to Percival's *Medical Ethics* (1927).
12. Laurence B. McCullough, *John Gregory and the Invention of Professional Medical Ethics and the Profession of Medicine*, Hingham, MA: Kluwer, 1998: Introduction. An early mentor of Gregory was his cousin, Thomas Reid, a prominent philosopher in the Scottish Enlightenment; Haakonssen, *Medicine and Morals*: 46–48.
13. McCullough, *John Gregory.*
14. Haakonssen, *Medicine and Morals:* 48–54.
15. McCullough, *John Gregory*: 57.
16. "She became his exemplar of a woman of learning and virtue, a concept that he championed and made central to his medical ethics." McCullough, *John Gregory*: 69. Haakonssen places somewhat less stress on Gregory's association with Montagu; *Medicine and Morals*: 48–54. Gregory held relatively advanced views for his day on the status and education of women, but nothing in his medical ethics suggests views that would today be termed feminist. In his posthumously published *A Father' Legacy to his Daughters*, he suggested that women be passive and amiable and hide from men any scholarly inclination they might have; Haakonssen, *Medicine and Morals*: 50–54.
17. Haakonssen characterizes Gregory's *Comparative View* as very popular in both Britain and America, considered "ingenious and elegant" (*Medicine and Morals*: 51). Gregory recommended demystifying medicine and disseminating medical knowledge to laymen. (Gregory would later assist William Buchan in preparing one of the first works that tried to make medicine accessible to the general public, his *Domestic Medicine*, 1769 (65–70). *Comparative View* "shares with the writings of Gregory's fellow literati, particularly the moderate clergy, the ethical preoccupations, literary style, and scientific methodology of the Scottish Enlightenment. Like many physicians and educators, Gregory adopted the polemical strategies of the more reform-minded clergy and adapted them to his own campaign to reform medicine. He attacked the character of the professional physicians as 'tyrannical expert men,' just as the clerics assailed 'zealous orthodox divines.' In *Comparative View* he laid the groundwork for his creation of the ideal physician – 'the man of science and art … [with] a liberal, knowing or disinterested spirit' – who was the medical equivalent of the reforming moderate clergy." (52)
18. Haakonssen, *Medicine and Morals:* 48–54.
19. Quoted in Haakonssen, *Medicine and Morals:* 54.
20. Gregory's writings comprise *A Comparative View of the State and Faculties of Man with those of the Animal World* (1765, 1772); *Observations on the Duties and Offices of a Physician* (1770); *On the Method of Prosecuting Enquiries in Philosophy* (1770); *Elements of the Practice of Physic: For the Use of Students* (1772); and *Lectures on the Duties and Qualifications of a Physician* (1772). *A Father's Legacy to his Daughters* (1774, 1779) was published posthumously and also became immensely popular. His son, James Gregory, edited

a revised edition of *Lectures on the Duties and Qualifications of a Physician* in 1805, and subsequent editions of this work appeared in 1817 and 1820. McCullough, *John Gregory*: 44.

21. Haakonssen, *Medicine and Morals:* 14–16.
22. McCullough, *John Gregory*: 187.
23. Laurence B. McCullough, ed., *John Gregory's Writings on Medical Ethics and Philosophy of Medicine*, Boston, MD: Kluwer, 1998:112. There is some dispute between McCullough and Haakonssen as to whether Gregory's concept of sympathy followed David Hume or Adam Smith. In Hume's conception of sympathy, A sympathizes with B and that leads A to have same sentiments as B. In Smith's conception of sympathy, A sympathizes with B and imaginatively puts self in same situation as B, but this may or may not lead A to have same sentiments; see also Haakonssen, *Medicine and Morals*: 70–74. Fortunately we need not resolve this dispute in order to understand Gregory's medical ethics as one source used by Ryan. Haakonssen suggests further that Gregory seems to view sympathy for the suffering patient as "feeling for" that individual's distress, not feeling the distress itself. This makes his sense of sympathy equivalent to benevolence, a standard view found among all Scottish moralists of his time. Haakonssen further suggests that Hume was "moral anatomist," concerned with theoretical underpinnings of morality in a theory of mind. Gregory by contrast was a "moral painter" (73) more interested in practical morality, trying to bring out good behavior and concrete actions by exhortation and example. A final reason for Gregory to have resisted too close an adherence to Hume's teaching was Hume's openly atheistic posture; Gregory "loved Mr. Hume personally as a worthy agreeable man" (50) but "detested Hume's philosophy" for its atheism (50).
24. McCullough, *Gregory's Writings:* 127.
25. McCullough, *Gregory's Writings:* 4.
26. McCullough, *John Gregory:* 133.
27. McCullough, *John Gregory:* 134.
28. See for example Haakonssen, *Medicine and Morals*: 101–102.
29. Gregory called his Edinburgh lectures "The Duties and Offices of a Physician" in reference to the idea from Cicero and the Stoics, of the implied contract embodied in a social role or office. This idea was taken up by Scottish natural law philosophy, so that each social role had associated virtues and duties seen as intrinsic to it. Haakonssen, *Medicine and Morals*: 56–59.
30. Haakonssen, *Medicine and Morals:* 59–62.
31. Haakonssen, *Medicine and Morals:* 54–56.
32. McCullough, *Gregory's Writings:* 102.
33. McCullough, *Gregory's Writings:* 103.
34. McCullough, *Gregory's Writings:* 104.
35. McCullough, *Gregory's Writings:* 107; Haakonssen, *Medicine and Morals:* 76–79.
36. We are grateful to Robert Baker and Laurence McCullough for suggesting that Gregory's position here could not properly be characterized as paternalistic. They note that Gregory treated telling the patient unpleasant prognostic news in these circumstances as an unpleasant duty. According to his sense of role responsibilities, it was a duty that family members, not the physician, ought to discharge. Shielding the patient for his own good (paternalism) was not properly part of the analysis.
37. Haakonssen, *Medicine and Morals:* 76–79.
38. Haakonssen, *Medicine and Morals:* 81; quoting Gregory, *Lectures*, 49–50; McCullough, *Gregory's Writings*: 69.
39. Haakonssen, *Medicine and Morals:* 81–82.
40. Again we rely especially on Haakonssen, *Medicine and Morals*: 94–173. Her primary source in turn was Percival's biography written by his son Edward.
41. Haakonssen, *Medicine and Morals*: 100–102. Percival may also have owed to the Academy the original choice of "medical jurisprudence" as the title for his work. One of his tutors, John Aikin Sr., was a student of "natural jurisprudence" as part of larger system that included "natural religion" and "private morality." These more or less matched the tripartite division of duties (to God, self, and others) as later written about by Thomas Gisborne, whom Percival viewed as an

important authority on practical ethics. Professional ethics logically fell under the duty to others, and hence under the division of "jurisprudence." Haakonssen, *Medicine and Morals*: 102–107. Ryan would later argue that medical ethics should be subsumed under the larger category of medical jurisprudence.

42. Haakonssen, *Medicine and Morals*: 141–145.
43. Haakonssen, *Medicine and Morals*: 129. Modern critics explain the incompleteness and lack of philosophical foundations in Percival's *Medical Ethics* as due to the fact that the origins of the work lay in the response to a specific local problem. Haakonssen claims that the deficiencies are less in the work itself and more in the eye of today's reader: "Percival claimed in his preface that his work was up-to-date, comprehensive, original, and that it had the stamp of approval of the most eminent physicians and philosophers. The relevant public seems to have accepted this, and it undoubtedly did so because the work was in fact a judicious mixture of the traditional and the innovative" (129).
44. J.V. Pickstone and S.V.F. Butler, "The Politics of Medicine in Manchester, 1788–1792: Hospital Reform and Public Health Services in the Early Industrial City," *Medical History* 28: 227–249, 1984.
45. Thomas Percival, *Medical Ethics; or, a Code of Institutes and Precepts adapted to the Professional Conduct of Physicians and Surgeons…* (Manchester: Russell, 1803; modern reprint by the Classics of Medicine Library, Birmingham, AL: 1985): 1. "Urbanity and rectitude" might sound off-putting to the modern reader; Haakonssen argues that in Percival's day this simply meant the "whole duty of the upright Christian character" (Haakonssen, *Medicine and Morals*: 127).
46. Percival, *Medical Ethics*: 6.
47. Percival, *Medical Ethics:* 7.
48. Robert B. Baker and Laurence B. McCullough, "What is the History of Medical Ethics?" In: *Cambridge World History of Medical Ethics*, ed. Robert B. Baker and Laurence B. McCullough, Cambridge, UK: Cambridge University Press, 2009:3–15. Baker and McCullough emphasize the historical problem this creates for arguing (as Ryan was in fact the first to do) that "medical ethics" represents a long historical tradition embracing both the Hippocratic Oath and today's bioethics. Percival himself wrote as if previous works on "medical ethics" were in no way authoritative for his project.
49. Haakonssen, *Medicine and Morals*: 136.
50. At the end of the eighteenth century, the existing "gentleman's" code of honor, especially in matters such as dueling, was coming under criticism from practical moralists like Gisborne. There was a call for a new code of gentlemanly conduct more suited to the role of reason and religion. Percival felt this debate important enough to insert a discussion of dueling into his Appendix; Percival, *Medical Ethics*: 214–28. "[Gregory] had been attracted to the ideal combination of liberal education and independent fortune in the make-up of the professional gentleman. Percival's vision not only was much more middle-class, it also had a different institutional focus in the hospital. The pursuit of medical science and its application to practical medical care were gradually united within the one institutional framework, and the profession's mastery of this new medicine could indeed make it into a gentlemanly elite, unsurpassed by any previous generation of scholarly or philosophical physician" (Haakonssen, *Medicine and Morals*: 139).
51. Robert B. Baker, "The Discourses of Practitioners in Nineteenth- and Twentieth-Century Britain and America," in *Cambridge World History of Medical Ethics*, ed. Robert B. Baker and Laurence B. McCullough, Cambridge, UK: Cambridge University Press, 2009:446–464.
52. Haakonssen, *Medicine and Morals*: 154.
53. Haakonssen, *Medicine and Morals*: 154–158. Haakonssen quotes Percival: "Every man who enters into a fraternity engages, by a tacit compact, not only to submit to the laws, but to promote the honour and interest of the association, so far as they are consistent with morality, and the general good of mankind" (Haakonssen, *Medicine and Morals*: 123, quoting Percival, *Medical Ethics*: 45–46). Therefore the implied social contract, according to Haakonssen, was

aimed in two directions – physicians owed duties both to the profession and to other physicians, and to society generally for the honest and faithful carrying out of their duties. Later critics who claim that Percival's ethics has no sense of duties owed to patients and is simply an etiquette of how doctors should treat each other omit the latter part of this contract, in her view. We are grateful to Robert Baker and Laurence McCullough for offering a somewhat different interpretation of Percival's ethics, noting that as a Unitarian, Percival would have held more socially egalitarian views than Haakonssen attributes to him.

54. Haakonssen, *Medicine and Morals*: 149.
55. We are grateful to Robert Baker and Laurence McCullough for a novel reading of this passage in Percival's ethics, that does not seem to have been considered by other recent writers on the history of the ethical debate on truthful disclosure. They suggest that Percival intended all along that the patient should be told of the dire prognosis. He merely intended that the family members, rather than the physician, be the ones to make this communication, arguing that such words coming from the physician would have much more dire emotional consequences. We suggest in turn that this alternative reading presumes that as a matter of fact, it was customary for families to tell seriously ill patients the truth about their prognoses at the end of the eighteenth century, if the physician left this office to their discretion. We know of no empirical data to indicate one way or the other.
56. Haakonssen, *Medicine and Morals*: 167–173. It is important to recall that Percival was not defending deception as general medical practice, but the much more specific issue of whether one should tell the patient a possibly shocking truth when the patient is in the midst of a medical crisis and is presumed to be in an especially vulnerable state. Percival, *Medical Ethics*: 166–167.
57. Haakonssen, *Medicine and Morals*: 132.
58. Haakonssen, *Medicine and Morals*: 134. We are grateful to Robert Baker and Laurence McCullough for the reminder that Percival, as a Unitarian, was very careful not to confuse his own ethical code with any sort of sectarian Christian document. Rev. Thomas B. Percival might well have been a rebellious son who broke with his father over religion, and it would be a mistake to attribute the younger Percival's religious views to his father.
59. Specifically, he claimed that three measures could manage the potential rivalry among the orders: (1) a strong faculty committee; (2) improved dispute arbitration; and (3) new, detailed consultation guidelines. Ryan reprinted these relevant passages. Haakonssen, *Medicine and Morals*: 141–145.
60. Percival, *Medical Ethics*: 9; quoted in Haakonssen, *Medicine and Morals*: 152.
61. Percival, *Medical Ethics*: 125–126.
62. Haakonssen, *Medicine and Morals*: 148–154.
63. England General Register Office, Certified Copy of an Entry of Death, District of St. Giles in the Fields of St. George Bloomsbury, copy certified 6 January 2004.
64. "Death of Dr. Ryan," *Provincial Medical and Surgical Journal*, 1 (1840–1841): 207. A handwritten "0" or "6" could easily be confused one for the other, perhaps accounting for this discrepancy.
65. Confirmed by electronic mail communication, Tina Craig, Royal College of Surgeons, to HB, 1/13/04. We will see later, however, that there seem to have been two Michael Ryans, each MRCS, practicing in Kilkenny in 1824, and the University of Edinburgh lists a number of Michael Ryans as having matriculated during the period 1816–1822; so some confusion cannot be entirely excluded.
66. The Wellcome Library catalogue entry for Ryan's medical school thesis agrees with the MRCS date of 1819 and adds the useful information about the minimum age requirement.
67. Adrian Desmond, *The Politics of Evolution: Morphology, Medicine, and Reform in Radical London*, Chicago, IL: University of Chicago Press, 1989: 427.
68. Electronic mail communications, Paul Gorry to HB, 8/6/04.
69. "Insolvent Debtor's Court, November 2." *The Times of London*, 4 November 1836: 6.
70. "Essays, Papers, &c. Published by Dr. Ryan," endpaper advertisement included in Ryan, *Prostitution in London with a Comparative View of That of Paris and New York, as Illustrative of the Capitals and Large Towns of All Countries; and Proving Moral Depravation to Be the*

Most Fertile Source of Crime, and of Personal and Social Misery; with an Account of the Nature and Treatment of Various Diseases, Caused by the Abuses of the Reproductive Function. London: H. Bailliere, 1839.

71. See below for Ryan's writings, including *A new practical formulary of hospitals … Translated from the new French ed. of Milne Edwards and P. Vavasseur,* London: John Churchill, 1835. Besides the translations of Latin passages within *Manual of Medical Jurisprudence* (below), Ryan also translated William Harvey's celebrated work on the circulation of the blood, in a series of articles printed in the *LMSJ*, 1832–1833.
72. Electronic mail communication, Robert Baker to HB, 8/2/04. Baker and McCullough have written that Ryan had an "Irish-Catholic education"; Baker RB, McCullough LB. "What Is the History of Medical Ethics?" In: *Cambridge World History of Medical Ethics*, ed. Robert B. Baker and Laurence B. McCullough, Cambridge, UK: Cambridge University Press, 2009:3–15. Admittedly, Ryan might have had a Catholic-influenced education even though he and his family were not of that faith; in fact we know no details of his education prior to medical school. We are grateful to Harold Vanderpool for pointing out several phrases in Ryan's ethical writings that suggest a close knowledge of Catholic teachings.
73. Rev. Richard Boyle Bernard, born Sept. 4, 1787, was second son of the first Earl of Bandon. He served as rector of Glankeen, 1817–1822; rector of Shankill, Leighlin, 1826–1833. He was Dean at Leighlin from 1822 until he died on March 1, 1850, leaving a number of munificent charitable bequests to church institutions. Personal communication, Dr. R. Refaussé, Librarian, Church of Ireland, to HB, 2/15/04.
74. Against these arguments, Robert Baker has urged that Ryan quoted with apparent approval the Roman Catholic teaching on abortion and on the destruction of fetal life to save the live of the mother; electronic mail communication, Baker to HB, 8/2/04. Ryan was, in that context, surveying the extant opinions of various authorities in Europe; *A Manual on Midwifery* (1828): ii, 269–270. We think it much more likely that a non-Catholic physician should have quoted from Catholic opinions on these subjects, than that a Catholic person would have dedicated his M.D. thesis to a non-Catholic clergyman.
75. Michael Ryan, *A Manual of Midwifery and Diseases of Women and Children…*, 4th ed. London: (self-published), 1841: 7.
76. Ryan, *Prostitution in London* (1839): 322.
77. Eoin O'Brien, ed., *The Charitable Infirmary, Jervis Street, 1718–1987: A Farewell Tribute,* Dublin: Anniversary Press, 1987; J.D.H. Widdess, *The Royal College of Surgeons in Ireland and Its Medical School, 1784–1984*, Dublin: Royal College of Surgeons in Ireland, 1984.
78. In a number of his books, beginning in 1824, Ryan gave as one of his qualifications on the title page, "Member of the Association of Fellows and Licentiates of the Royal College of Physicians in Dublin" (or in some cases, the "King's and Queen's College of Physicians in Ireland"). The current keeper of records, however, states that no evidence of Ryan's membership now exists; electronic mail communication, Robert Mills to HB, 10/9/03.
79. Porter, *Greatest Benefit*: 306–320.
80. The archives at Edinburgh list two Michael Ryans, one from "Barrisoleigh, Tippy." matriculating in 1818–1819, one from "Co. Tipperary" matriculating in 1819–1820. Electronic mail communication, Irene Ferguson to HB, 11/27/03.
81. Ryan discussed his Dissertation in his *A Manual on Midwifery…* (London: Longman and Co., 1828):127. At this point there is some confusion in the Edinburgh archives. The Ryan who graduated in 1821 is described further: "Asst Surgeon 51 Foot 1825; Surgeon 1841. Died at Ireland Island, Bermuda, 1853." This would not fit our Michael Ryan who, so far as we know, never had any military connections and died in 1840. It appears that the Archives mixed up the correct graduation date and thesis title of our Michael Ryan with the later biography of a different Michael Ryan. The second Ryan is described in the catalog of the Wellcome Library as "Michael Ryan of Cork, d. 1853." The Library holds a copy if his M.D. thesis from Edinburgh, titled "De Variis Ictericis Morbis," 1819. The Edinburgh archives includes a Michael Ryan of Cork who matriculated in 1817–1818. It seems quite possible

that such a transposition of records occurred: electronic mail communication, Irene Ferguson to HB, 1/13/04.

Further confirmation as to the correct identity of our Ryan is provided by the minutes of the Royal College of Physicians of London, to whom Ryan applied in 1830 for the status of Licentiate (the highest recognition possible to someone who had not graduated from Oxford or Cambridge): "Dr. Michael Ryan produced a Diploma from the University of Edinburgh by which he was created Dr. of Medicine on 1st of Augt. 1821." Comitiis Minoribus Ordinariis, 7 May 1830. This Michael Ryan is almost certainly our Ryan, as he applied for this status around the date when our Ryan was known to have taken up residence in London and was seeking appointments as a lecturer (for which the RCP Licentiate status would be very useful). By contrast, the Michael Ryan of Cork, who apparently served as a military surgeon, would have no reason to seek an RCP qualification.

82. "Death of Dr. Ryan," *Provincial Medical and Surgical Journal*, 1 (1840–1841): 207. The professor of midwifery at Edinburgh between 1800 and 1839 was James Hamilton (1767–1839), son of the preceding professor, Alexander Hamilton. The younger Hamilton was best known for his *Practical Observations* (1839). He was said to have been very kindly toward his patients and notably egalitarian in how he treated patients of different social classes, but rude and argumentative toward his colleagues, against two of whom he filed lawsuits at various times. Alexander Grant, *The Story of the University of Edinburgh during its First Three Hundred Years* (London: Longmans, Green & Co., 1884): ii:417–419. Ryan possibly emulated his favorite professor in more than his choice of specialty.

83. Ryan, *Tentamentum Medico-Physicum Inaugurale, de Genere Humano ejusque Varietatibus...*, Edinburgh: P. Neill, 1821. The titles of the sections within the dissertation are: Generis humani historia; de generatione; de partu humano; de infantia; de forma generis humani externa; de calore; de frigore; de situ; de colore; de statura.

84. The *List of the Members of the Royal College of Surgeons in London* is available in the Wellcome Library, London for the years 1822, 1825, and 1833. The 1822 edition lists one Michael Ryan, with the address, "Burris O'Leigh, co. Tipperary." The 1825 edition lists *two* persons, each named Michael Ryan and each practicing in Kilkenny (see note 85). The 1833 edition lists our Michael Ryan in Hatton Garden, London, and the second Michael Ryan is no longer listed.

85. Electronic mail communication, Declan Mcauley to HB, 1/21/04. This Ryan listed his credentials in the directory as both an M.D. from Edinburgh and MRCS status, consistent with our Michael Ryan. As previously noted, there would appear to have been two Michael Ryans practicing in Kilkenny as of 1825 if the *List* of the Royal College of Surgeons is to be believed. Ryan later wrote in a case report of a 21-year-old woman whom he treated on 8 July 1823 for angina pectoris (with the highly toxic-sounding remedy of prussic acid): "Leeches and blisters were applied to the left side, and many internal remedies administered, by my late much respected namesake, but with only temporary advantage." "Remarks on the Use of Hydrocyanic Acid in Angina Pectoris and other Diseases of the Heart, and in Bronchitis, Phthisis, and Dyspepsia," *London Medical and Physical Journal* 51 (1824): 369. This suggests the presence of another physician named Michael Ryan in the same city. We are grateful to Paul Gorry for unearthing the obituary of this second Michael Ryan, M.D., who appears to have died November 26, 1822; death notice in the *Leinster Journal*, Kilkenny, 30 Nov 1822. (We assume that the Royal College of Surgeons had not yet received notice of this Dr. Ryan's death in time to omit him from the 1825 edition of their List.) The other Michael Ryan also had a son named Michael Ryan who in 1837 was living in 47 Gower Place, Euston Square, London, according to a deed dated 30 March 1837, in which the son agreed to sell his share of the property he had inherited on the death of his father; Registry of Deeds, Dublin, 1837 6 295. This other Michael Ryan living in London was apparently not a physician and seems to have had no connection with our Michael Ryan.

86. From a list of "essays, papers, &c. published by Dr. Ryan," included in the printer's endpapers to Michael Ryan, *A Manual of Midwifery and Diseases of Women and Children...*, 4th ed. London: (self-published), 1841. These reportedly appeared in journals including the

London Medical and Physical Journal and the *Transactions of Dublin College of Physicians*.

87. The title page bore the imprint, "Kilkenny: Printed and published by J. Reynolds, for Hodges, M'Arthur, and J. Cummings, Dublin…, 1824."
88. Ryan, *Treatise on the Most Celebrated Mineral Waters* (1824): 18–19. These chemical analyses were totally qualitative; Ryan apparently lacked either the know-how or the equipment to conduct quantitative chemical analyses.
89. Ryan, *Treatise on the Most Celebrated Mineral Waters* (1824): 20.
90. Ryan, *Treatise on the Most Celebrated Mineral Waters* (1824): 21.
91. Ryan, *Treatise on the Most Celebrated Mineral Waters* (1824): ix–xii.
92. Ryan, *Treatise on the Most Celebrated Mineral Waters* (1824): viii.
93. Roderic a Castro (Rodrigo de Castro, 1546–1627), *Medicus-politicus: sive de officiis medico-politicis tractatus, quatuor distinctus libris: in quibus non solum bonorum medicorum mores ac virtutes exprimuntur, malorum vero fraudes et imposturae deteguntur* … (The Politic Physician: or a Treatise on Medico-Political Duties …), Hamburg, 1614; quoted in Winfried Schleiner, *Medical Ethics in the Renaissance* (Washington, DC: Georgetown University Press, 1995): 71. The quotation, however, was in common use, and Ryan may have taken it from another source. We are indebted to Debra Nails for tracing various forms of the quotation back as far as Plato (*Republic* 595b–c) in Greek, and in Latin to Roger Bacon, *Opus Magnum* I..vii ("Amicus est Socrates magister meus sed magis est amica veritas"). Cervantes included in *Don Quixote* II. Chap. 51, "Amicus Plato, sed magis amica veritas." Isaac Newton was said to have headed remarks in his commonplace book, "Amicus Plato amicus Aristoteles magis amica veritas" (reportedly in turn quoting Walter Charleton, b. 1619).
94. "Death of Dr. Ryan," *Provincial Medical and Surgical Journal*, 1 (1840–1841): 207.
95. Ryan published "Delirium Tremens, Treated as a Nervous Disease," in *The Lancet* 2 (1827–1828): 791–793. The article describes a case that he treated, beginning 23 July 1827, and contains information on the status of the patient 1 year later. Ryan is described in the masthead as "one of the Physicians to the Central Infirmary, Greville Street, Hatton Garden, Lecturer on Midwifery." While the article does not specifically say so, every indication suggests that he had treated this patient in London as part of his infirmary practice. Had he begun treating the patient before leaving Ireland, it is not clear how he would know about the patient's later clinical course. This case report would therefore suggest July, 1827 as the latest possible date for his arrival in London. However, when Ryan was later hauled into Debtor's Court (1836), it was reported that he had left Ireland in 1828, and that at the time he had an income of £156 annually from property there; "Insolvent Debtor's Court, November 2." *The Times of London*, 4 November 1836: 6.
96. The advertisement of these lectures appears in the endpapers of Ryan, *A Manual on Midwifery; or a Summary of the Science and Art of Obstetric Medicine, including the Anatomy, Physiology, Pathology, and Therapeutics, Peculiar to Females; Treatment of Parturition, Puerperal, and Infantile Diseases; and an Exposition of Obstetrical-Legal Medicine*. London: Longmans, 1828 (following page 353). In those years, a number of lecturers unaffiliated with medical schools offered lectures on various topics in their private rooms. All lecturers, whether or not affiliated with medical schools, received reimbursement by selling individual tickets to their lecture courses.
97. Ryan, *Manual on Midwifery* (1828): i.
98. Ryan, "An Essay on the Natural, Chemical, and Medical History of Water, in its simple and Combined States; including an Account of the Chemical composition, and Medical Effects, of the principal Mineral Waters in the United Kingdom of Great Britain and Ireland, and also the Continent of Europe," *London Medical and Physical Journal* 54 (1825): 442–461. Ryan had authored this paper while still in practice in Kilkenny.
99. Ryan, *Remarks on the Supply of Water to the Metropolis* (1828): xii.
100. Ryan, in fairness, did add two additional pages at the end of his earlier work, "Observations on Bathing" (45–46).
101. This formula was set for the most part by the requirements for certification by the Society of Apothecaries and the Royal College of Surgeons. Since students bought tickets individually

for each course of lectures or term of clinical experience in the wards or dispensaries, there was no incentive for the student to attend any lectures that were not part of the prescribed curriculum leading to the LSA and MRCP certificates.

102. "Introduction," Robert B. Baker, Arthur L. Caplan, Linda L. Emanuel, and Stephen R. Latham, eds., *The American Medical Ethics Revolution: How the AMA's Code of Ethics Has Transformed Physicians' Relationships to Patients, Professionals, and Society*, Baltimore, MD: Johns Hopkins University Press, 1999: xix. We have, however, been unable to find any documentation to confirm Baker and colleagues' statement that Ryan claimed to be a *professor* of medical ethics, or that he was a professor affiliated with University College, London.
103. Ryan claimed to have originated at least the term "obstetrician" to replace the awkward usages, "male midwife" or the French "accoucheur"; *Manual on Midwifery*: v–vi.
104. Ryan, *Manual on Midwifery* (1828): ii.
105. Ryan, *Manual on Midwifery* (1828): ii.
106. Ryan, *Manual on Midwifery* (1828): ii.
107. Ryan, *Manual on Midwifery* (1828): endpaper advertisement (following 353).
108. Ryan, *Manual on Midwifery* (1828): iii–iv.
109. Ryan, *Manual on Midwifery* (1828): iv–v.
110. [Anonymous review of Ryan, *A Manual on Midwifery*,] *London Medical Gazette* 2 (1828): 714–715, quote p. 715.
111. [Anonymous review of Ryan, *A Manual on Midwifery*,] *London Medical Gazette* 2 (1828): 714–715, quote p. 715.
112. The same reviewer noted as much when he came to review the 3rd edition of this volume that appeared in 1831. He did, however, repeat and even expand his compliment, that in the book could be found "more information on the subject of midwifery ... than in any other work of three times its size to which we can refer." [Anonymous review of Ryan, *A Manual of Midwifery*, 3rd ed.] *London Medical Gazette* 9 (1831–1832): 52.
113. *Robson's London Commercial Directory* ... London: Robson, Blades & Co., 1828; *Robson's Classification of Trades and Street Guide* ... London: Robson, Blades & Co., 1829; *Robson's London Commercial Directory* ... London: Robson, Blades & Co., 1830. His lecture notices appeared in "Practice of Physic and Midwifery," *Lancet* 1 (1829–1830): 18. Admission to one course of lectures in either subject cost the student 3 guineas (3 pounds, 3 shillings).
114. For example, John Snow, coming to London as an apprentice-trained unknown from Yorkshire in 1836, was able to use the Westminster Medical Society as an effective springboard to a career in medical research; Peter Vinten-Johansen, Howard Brody, Nigel Paneth, Stephen Rachman, and Michael Rip, *Cholera, Chloroform, and the Science of Medicine: A Life of John Snow*, New York: Oxford University Press, 2003: 69–72, 84–86.
115. Ryan, *Manual of Medical Jurisprudence*, 1st ed. London: Renshaw and Rush, 1831 [title page].
116. Kaspar Hauser (d. 1831) was found in 1828, apparently aged about 16 or 18 but talking like a 3–4 year old child, and gave evidence of having been imprisoned for most of his youth. His case attracted widespread attention in Europe and there were rumors that he was actually the son and rightful heir of the Prince of Baden. Stanhope, a friend of Baden's family, initially became very interested in Hauser, befriended him, and helped to provide for his support. After a while Stanhope apparently became bored and convinced himself that Hauser was a fake. Stanhope's interventions had the result of cutting Hauser off from those who were best able to care for him and placing him among others who treated the youth badly. A good overview is to be found at http://www.mysteriouspeople.com/Hauser1.htm (accessed August 25, 2004).
117. The other dedications are to Rev. Boyle Bernard in his M.D. dissertation, already noted; and to Dr. John Elliotson, professor of medicine at University College, London, in the *Physician's Vademecum*, 1836.
118. "Dr. Ryan also made some observations on the catechu and cardamoms," "Medico-Botanical Society," *Lancet* 1 (1835–36): 341 (society meeting of Nov. 24, 1835). The *Lancet* for December 10, 1836 (*Lancet* 1 (1836–37): 414) describes a meeting (no date given) at which Ryan made some extended comments on croton oil and buchu leaves.

119. "Medical Society of London," *London Medical Gazette* 3 (1828–1829): 646–647. "Dr. Ryan did not think the case very peculiar. It was not, he thought, an hour-glass contraction. He concurred with Mrs. Ashwell and Waller. Dr. Milligan some time ago gained great reputation because he turned when no one else could. This he affected by opium." The case being discussed was one of difficult labor; "turning" referred to version of the fetus, to change the fetal position to affect an easier delivery.
120. Ryan, "To the Editor of the *London Medical Gazette*," *London Medical Gazette* 3 (1828–1829): 684. We have not searched through all the reports of medical society meetings in the *Lancet* and *London Medical Gazette* to see whether Ryan is listed as a speaker on any other occasion. The journals of that time frequently gave very full reports of these meetings, hiring medical students to serve as short-hand reporters.
121. On radicalism and reform generally during this period see Adrian Desmond, *The Politics of Evolution: Morphology, Medicine, and Reform in Radical London* (Chicago, IL: University of Chicago Press, 1989).
122. Desmond, *Politics of Evolution:* 15, 170–171.
123. Ryan explained in the Preface to the first edition, *Manual of Medical Jurisprudence* that the text was originally published in the "lacunae" of *London Medical and Surgical Journal* as an aid for students (x).
124. James F. Clarke, *Autobiographical Recollections of the Medical Profession*, London: J. and A. Churchill, 1874: 134–135.
125. *Lancet* 2 (1831–1832): 92, 21 April 1832. At this time Wakley favored a contagious explanation of the transmission of cholera, as did the Board of Health, while Ryan favored a miasmatic, non-contagionist hypothesis. As the epidemic progressed the majority of medical opinion swung into the non-contagion camp, giving Ryan numerous opportunities to crow over Wakley's downfall; see for instance *LMSJ* 1 (NS) (1832): 411–413.
126. *LMSJ* 1 (NS) (1832): 17–18. Ryan here referred to the Anatomy Act, and blamed Wakley and the *Lancet* for retarding its passage in a desirable form.
127. *LMSJ* 1 (NS) (1832): 289–291.
128. "Letter from Dr. Ryan," *Lancet* 1 (1831–1832): 222–226.
129. *LMSJ* 7 (1831): 433, 512, 522–526.
130. "Drs. Gordon Smith, Ryan, and A. Thomson," *Lancet* 1 (1830–1831): 72–73, quote p. 72. Smith died in 1833, while incarcerated in the Fleet Prison. The author of his obituary in *Gentleman's Magazine* described Smith as having become "involved in pecuniary difficulties, combined somewhat with irregular habits," leading to his imprisonment, "where he gradually sunk and expired"; H.W.D., "Obituary: John Gordon Smith, M.D., F.R.S.L.," *Gentleman's Magazine*, September 1833, 278–279. In fairness to Ryan, perhaps Smith's "irregular habits" manifested themselves several years before his death and caused him to act unpredictably. The obituary author credited Ryan and Smith jointly with having prevailed upon Apothecaries Hall to include the subject of medical jurisprudence in their required examination, thus compelling all medical schools to provide courses of lectures in that "neglected but truly important subject."
131. "Drs. Ryan and Gordon Smith," *Lancet* 1 (1830–1831): 106–108, quote p. 108. It is quite surprising to the modern reader that a supposedly legitimate medical journal of that day would devote this many columns to a spat over who stopped speaking to whom first.
132. Ryan, *Prostitution in London* (1839): 433.
133. Ryan, *Prostitution in London* (1839): 433.
134. Ryan, *Prostitution in London* (1839): 434.
135. Ryan, *Prostitution in London* (1839): 434. Indications of the sort of plot to malign his work that Ryan imagined–and perhaps evidence that he was not far wrong in so thinking–are provided by an exchange in *The Lancet* in 1839. An anonymous (as always) *Lancet* reviewer had criticized a volume, *A Manual of Diseases of the Eye; or, Treatise on Ophthalmology*, by S. Little, revised and enlarged by Hugh Houston, as "got up for the express purpose of bespattering with praise Michael Ryan, Esq., in the execution of which laudable task the original work has been most disgracefully disfigured"; *Lancet* 1 (1838–1839): 494. Ryan

could never resist such an opening, and wrote back a lengthy letter defending Houston's book. He claimed, "The article of which I complain was surreptitiously concocted, by persons who are most ungrateful to me for many acts of kindness, and now, without just cause, my personal enemies; and whose conduct, if exposed, would do them little credit in the estimation of the profession"; *Lancet* 1 (1838–1839): 706. Wakley took delight in appending an editorial comment: "Dr. Ryan, we are well aware, may have a sufficient ground for believing that persons who were at one time connected with him in the publication of a medical journal are now his bitter and malignant enemies; but he may be assured, with reference to the notice of Mr. Houston's work, ... that it was not the production of his secret foes, but of the sub-editor of this Journal"; *Lancet* 1 (1838–1839): 706.

136. In anticipating negative reviews of *Prostitution in London*, Ryan proved prescient. The anonymous reviewer for the *British and Foreign Medical Review* insisted that he was no prude, and that Parent-Duchatelet's *On Prostitution in Paris* had indeed been very favorably reviewed by that journal. Ryan's volume, however, was of a very different nature: "The greater portion of the volume is filled with details of the most disgusting kind ... without any necessary connexion to the professed subject of the work, unredeemed by the slightest tincture of utility, and only calculated ... to pander to the vilest and most depraved tastes"; [Anonymous review of Ryan, *Prostitution in London*], *British and Foreign Medical Review* 7 (1839): 540.
137. None of the discussion of the case in *The Lancet* or the *LMSJ* mentions Dr. Ramadge's full name; but it is given as "Francis H. Ramadge" in an editorial, "Dr. Ramadge and St. John Long," *London Medical Gazette* 8 (1831): 117–120. In *A Manual on Midwifery* (1828) Ryan in numerous places speaks glowingly of a colleague at the Central Infirmary in Greville Street, Dr. Ramadge, who maintained an excellent collection of pathological specimens. This Dr. Ramadge appeared in several of the annual lists of London medical lecturers in *The Lancet*. Presumably the Dr. Ramadge of the present libel case was a different individual.
138. Letter from Dr. Ramadge addressed to "[John] Long the Quack" as printed in *Lancet* 2 (1830–1831); No. 398 (16 April 1831): 90–93; No. 400 (30 April 1831): 154–156.
139. "London Medical Society: Expulsion of Dr. Ramadge," *Lancet* 2 (1830–31): 251–252.
140. The original version, "Tweedie v. Ramadge," appeared in *Lancet* 2 (1830–31), no. 419 (10 Sept. 1831). The version with Ryan's editorial comment appended would have appeared in *LMSJ* 7 (1831), which we were unable to examine in the original. That version in turn was reprinted later in *Lancet* 2 (1831–32): 408–9, which we use as our source.
141. A summary account of both trials (Ramadge v. Wakley; Ramadge v. Ryan et al.) may be found in *Lancet* 2 (1831–32): 408–409. Ryan published a brief account of the trials, with little commentary; *LMSJ* 1 (NS) (1832): 705–6. See also *The Times of London*, 27 June 1832, 7 November 1832, 16 November 1832, and 21 November 1832.
142. "Subscription for Dr. Ryan," *Lancet* 1 (1832–33): 351–352. The list of donors included staff physicians from many of the major London hospitals, fellow medical editors and authors, and a number of professors from Dublin; Earl Stanhope, president of the Medico-Botanical Society, contributed £5, one of the larger donations. (Two other donors were names frequently seen in meeting reports of the Medico-Botanical Society: Drs. Sigmond and Hancock.) The total raised at that time was approximately £160; it is not known how much of the judgment Ryan himself eventually had to pay, and how much this contributed to his later problems with debt.
143. Ryan's account of the quarrel, typically, depicted him as having wished to remain silent, and attacking Renshaw only after the latter had circulated false advertisements. Ryan accused Renshaw of interfering in editorial decisions, and trying to convince Ryan to publish more articles on various nostrums so as to boost sales of advertisements. He claimed that as he was the sole proprietor of the *LMSJ* when he first entered into partnership with Renshaw, the dissolution of that partnership left him again as proprietor and retaining all rights to the *Journal*. *LMSJ* 6 (NS) (1834): 153–155.
144. Clarke, *Autobiographical Recollections* (1874): 135–136.
145. Desmond, *Politics of Evolution:* 170.

146. "Insolvent Debtor's Court, November 2." *The Times of London*, 4 November 1836: 6. The account mentioned in passing that a "Mr. Wakley" was also a creditor of Ryan, but did not further identify this person. It seems unlikely that Ryan would have become financially indebted to his bitter editorial rival. Since the dispute at the time of this hearing was over a few books, it seems somewhat questionable that this should have been sufficient grounds to send Ryan to prison, though there may have been an earlier legal proceeding of which we could find no record.
147. Zachary Cope, "The Private Medical Schools of London, 1746–1914." In *The Evolution of Medical Education in Britain*, ed. F.N.L. Poynter, Baltimore, MD: Williams and Wilkins, 1966: 89–109.
148. For at least 1 year after Ryan's departure, some lectures were still offered at the Central Infirmary and Dispensary; *Lancet* 1 (1829–1830): 12. So the dissolution of the school was not, apparently, the immediate reason for Ryan's setting himself up to lecture on his own.
149. Elizabeth Epps, ed. *Diary of the Late John Epps*. London: Kent & Co., 1875: 181.
150. Desmond, *Politics of Evolution:* 166–170.
151. Ryan viciously attacked phrenology as a materialist movement that saw the human mind as nothing but a secretion of the brain, that died with the body, and thus implicitly denied the immortality of the soul, see especially *Manual of Medical Jurisprudence*, 2nd ed. London: Sherwood, Gilbert, and Piper, 1836: 14–20, reprinted in this volume. Epps wrote a book on phrenology and was sympathetic to the movement. His writing on phrenology was "chiefly intended to prove that phrenology was intimately connected with religion"; Elizabeth Epps, ed., *Diary of the Late John Epps*, London: Kent & Co., 1875: 168, quoting the editor of the *Monthly Gazette of Practical Medicine*. (Epps was the author of *Internal Evidences of Christianity Induced from Phrenology*, Edinburgh: J. Anderson, 1827.) Dermott wrote a book in which he argued that the mind and the soul were two distinct concepts so that the former could be material while the latter was immortal and immaterial; *A Discussion on the Organic Materiality of the Mind, the Immateriality of the Soul, and the Non-Identity of the Two* (London: Callow & Wilson, 1830). The three must have had some interesting conversations.
152. Cope, "The Private Medical Schools of London, 1746–1914," 105–106.
153. This comment highlights the stereotypes with which the Irish were viewed in that day, and the difficulties Ryan very likely experienced in being accepted in London medical circles. Any behavior of Ryan's that seems excessively self-promoting must be interpreted within that context as he might well have felt that he had to work especially hard to overcome these prejudices. We have discovered no evidence that Ryan was of the Roman Catholic faith, and have seen some evidence that he was Anglican (Church of Ireland); if so, he would at least not have had to deal with religious prejudice on top of ethnic prejudice. It may be of interest that despite the various *ad hominem* attacks launched against Ryan by his rivals of *The Lancet*, we have discovered no blatant ethnic slurs.
154. Indirect support for the assertion that medical practice was not Ryan's strong suit comes from the anonymous obituary in the *Provincial Medical and Surgical Journal*: "With all these avocations, it was not to be expected that Dr. Ryan could have had leisure to toil much in the practice of his profession; the zeal and delight which he took in intellectual pursuits must have left him little time and have given him but little relish for what he appropriately designated 'the trade of his profession;' and often may be seen among his writings allusions to the degraded position of the majority of its members, who, in place of prosecuting medicine as a science, dealt in it as a trade." "Death of Dr. Ryan," *Provincial Medical and Surgical Journal* 1 (1840–1841): 207.
155. The gold-headed cane was, at the time, a distinctive accessory for physicians.
156. Clarke, *Autobiographical Recollections* (1874): 133–134.
157. We are grateful to Jason Glenn for pointing out that "Milesian" refers to the mythical account of the origins of the Irish people from the sons of Mil Espáine.
158. Clarke, *Autobiographical Recollections* (1874): 136.
159. The word "vade-mecum" (literally, "come with me," Latin) would probably be rendered today as "companion."
160. Endpaper advertisements, *Prostitution in London*, 1839.

161. The notice announcing this volume appears in endpaper advertisements for Ryan's books and lectures, appended at the back of the second (1836) British edition of *Manual of Medical Jurisprudence*; see these on-line at http://books.google.com/books?id=vToEAAAAQAAJ&printsec=frontcover&dq=michael+ryan+manual+of+medical+jurisprudence#PPA556,M1. The copy of this volume that we possess lacks these endpaper advertisements.
162. "Metropolitan Hospital." Archives in London and the M25 Area. http://www.aim25.ac.uk/cgi-bin/search2?coll_id=5160&inst_id=51 (accessed January 23, 2005).
163. "Voluntary Hospitals," http://www.nhshistory.net/voluntary_hospitals.htm (accessed January 23, 2005).
164. "Metropolitan Hospital." The hospital was forced to move several times between 1836 and 1885 as land was taken up by expanding railways. In the 1880s it changed its name to "Metropolitan Hospital," and finally closed in 1977.
165. Minute books of the General Committee, Metropolitan Free Hospital, 10 March through 7 April 1836 and 11 May 1836; electronic mail communication, Samantha Farhall (St. Bartholomew's Hospital archivist) to HB, 1 February 2005. The archives of the Metropolitan Free Hospital are now lodged with those of St. Bartholomew's Hospital.
166. For example, "Senior Physician to the Metropolitan Free Hospital," title page of *Prostitution in London* (1839).
167. Ryan, *Manual of Midwifery* (4th ed. 1841): 35.
168. "Death of Dr. Ryan," *Provincial Medical and Surgical Journal* 1 (1840–1841): 206–207.
169. Minutes, General Committee, Metropolitan Free Hospital, 2 December 1840; electronic mail communication, Katie Ormerod (St. Bartholomew's Hospital archivist) to HB, 4 February 2005.
170. Clarke, *Autobiographical Recollections* (1874): 136.
171. Clarke admitted, "The entire work is written from memory, for I have had no documents or notes to refer to." Clarke, *Autobiographical Recollections* (1874): vi.
172. England General Register Office, Certified Copy of an Entry of Death, District of St. Giles in the Fields of St. George Bloomsbury, copy certified 6 January 2004. We cannot tell whether the Mr. Earles, surgeon, who signed the death certificate was the same as the supposedly unqualified "Mr. Earles" whom Ryan apparently sent to substitute for himself at the Metropolitan Free Hospital. Presumably someone listed on a death certificate as a "surgeon" had at least an MRCS qualification.
173. Minutes of the Metropolitan Free Hospital MHA 1/1 1836–1850, entry for 16 December 1840, Archives, St. Bartholomew's Hospital. We are grateful to Samantha Farhall and Katie Ormerod of the St. Bartholomew's Archive for their assistance in locating and copying this material. Despite the rather obvious hint contained in the report from the medical staff, the Committee of the Hospital simply had no spare cash to address the needs of Ryan's widow and children. They proceeded at that meeting to officially express their regret and condolences, and to place an advertisement for a replacement physician.
174. "Death of Dr. Ryan," *Provincial Medical and Surgical Journal* 1 (1840–1841):206–207. "Dec. 11. Aged 41, Dr. Michael Ryan, editor of the Medical and Surgical Journal. He was an amiable and clever man, and has left a young family wholly unprovided for." *Gentleman's Magazine* 15 (1841): 105. Another, very brief obituary notice stated that Ryan left four infant children; *London Medical Gazette*, January 1, 1841: 560.
175. "Subscription for the Widow and Children of the late Dr. Ryan," *Lancet* 1 (1840–1841): 903. The list of contributors is very limited; previous to this notice £137 had already been collected. In this short list were included Sir J. Clarke (certainly not James F. Clarke), Mr. Copeland, Dr. Crucifix (probably a pseudonym), Plymouth Medical Society, Dr. Ramsbotham, and Mr. Churchill (probably the medical publisher, John Churchill). The treasurer of the fund was Dr. Klein Grant, who lived at 18 Charlotte Street and was thus a near neighbor of the Ryans's; and lectured on Therapeutics at the nearby North London School (20 Charlotte Street); "North London School" (catalogue of courses), *Lancet* 1 (1838–39): 13.
176. See the Bibliography for the list of articles published in *LMSJ* which were later collected into the *Manual of Medical Jurisprudence*, 1st ed. (1831).

177. The negative review appeared as "*A Manual of Medical Jurisprudence* ... by Michael Ryan..." *Lancet* 1 (1831–1832): 137–143. Wakley printed one letter he received from Ryan in response to this review, and added at the end of his own reply to that letter that he had received a follow-up letter from Ryan, "occupying no less than seven closely written pages of letter paper," and accompanied by numerous positive notices of his book clipped from other periodicals. Wakley stated that, with apparently newly-found concern for the patience of the reader, he would not publish this second letter. [Wakley, Thomas.] "Letter from Dr. Ryan," *Lancet* 1 (1831–1832): 222–226.[0]
178. Michael Ryan, *A Manual of Medical Jurisprudence and State Medicine, Compiled from the Latest Legal and Medical Works, of Beck, Paris, Christison, Fodere, Ofila, etc.* ... 2nd ed. London: Sherwood, Gilbert, and Piper, 1836: v.
179. Chester R. Burns, "Reciprocity in the Development of Anglo-American Medical Ethics," in *The Codification of Medical Morality*, ed. Robert Baker, Boston, MD: Kluwer, 1995: 135–143; quote p. 137. To our reading, Burns may have over-emphasized the degree to which Ryan could be said to have an overarching theory of the relationship between medical ethics and law.
180. Ryan, *Manual of Medical Jurisprudence* (2nd ed): 3. Ryan's reliance on the Bible as his main source of ethical wisdom, coupled with the political prominence of Evangelical Anglicanism during the era when he was writing, make it reasonable to wonder whether Ryan was himself an Evangelical or influenced by those beliefs. We are grateful to Harold Vanderpool for identifying several passages in Ryan's works that would argue against an Evangelical background. These passages include Ryan's frequent reference to the Bible as the "sacred book," and his insistence that denying the immortality of the soul inevitably leads to immorality.
181. Ryan, *Manual of Medical Jurisprudence* (2nd ed): 6. Ryan elected not to address problems with reconciling his assertion of the great esteem in which the populace holds medicine and physicians, with his later complaints of poor behavior among physicians, or of the tendency of the public to ignore the advice of wise physicians and instead to seek out quacks.
182. Ryan, *Manual of Medical Jurisprudence* (2nd ed): 8.
183. Ryan, *Manual of Medical Jurisprudence* (2nd ed): 8.
184. Robert B. Baker and Laurence B. McCullough, "Introduction." In: *Cambridge World History of Medical Ethics*, eds. Robert B. Baker and Laurence B. McCullough, Cambridge, UK: Cambridge University Press, 2009:3–15. Burns also alluded to Ryan's priority in organizing his discussion of ethics within a historical framework; Chester R. Burns, "Reciprocity in the Development of Anglo-American Medical Ethics," in *The Codification of Medical Morality*, ed. Robert Baker, Boston, MD: Kluwer, 1995: 135–143. The way Ryan organized the Introduction to this volume–listing the works on the topic published in each historical period–suggests that a historical approach to any given topic came naturally to him and was not unique to his approach to medical ethics, per se.
185. An additional bit of evidence in support of this assertion is found in Ryan's letter to the editor of *The Lancet*, dated May 21, 1828, introducing his "Introductory Lecture to the Theory and Practice of Midwifery": "Another reason that induces me to send this communication is, that it contains a concise, though perfect, history of the rise, progress, improvement, importance, and utility of this branch of medicine [midwifery or obstetrics], and also the universal sanction it has received from mankind in the past ages, and present most enlightened times."*Lancet* 2 (1827–28): 394. With only a slight change in wording, this could have served as a preface to the ethics portion of the *Manual of Medical Jurisprudence*.
186. Ryan, *Manual of Medical Jurisprudence* (2nd ed): 10.
187. Ryan, *Manual of Medical Jurisprudence* (2nd ed): 11–12.
188. Ryan, *Manual of Medical Jurisprudence* (2nd ed): 13. Ryan may be referring to Friedrich Hoffmann (1660–1742), who among other topics wrote works on medical ethics (*Medicus Politicus*, 1738) and hygiene.
189. Ryan, *Manual of Medical Jurisprudence* (2nd ed): 14.
190. Ryan, *Manual of Medical Jurisprudence* (2nd ed): 15–16.
191. Ryan, *Manual of Medical Jurisprudence* (2nd ed): 19.
192. The full title is: *Medicus-politicus: sive de officiis medico-politicis tractatus, quatuor distinctus libris: in quibus non solum bonorum medicorum mores ac virtutes exprimuntur, malorum vero*

fraudes et imposturae deteguntur … (The Politic Physician: or a Treatise on Medico-Political Duties…) Winfried Schleiner has noted that Roderici á Castro Lusitani (1564–1627), to give one variant spelling of his name, was one of a group of "Lusitani," Jewish physicians who had nominally converted to Christianity to escape persecution in the anti-Semitic Europe of their day. This group wrote on morality from a secular-humanistic perspective. Winfried Schleiner, *Medical Ethics in the Renaissance* (Washington, DC: Georgetown University Press, 1995): 49–86. It is ironic that Ryan, given his emphasis on how all morality arises from Christianity, would have adopted Castro as a source for such a large portion of his text. According to Schleiner, later scholarship has tended to downplay or deny the fact that this group of physicians were Jews; hence Ryan might simply have been unaware of the fact.

193. Winfried Schleiner, *Medical Ethics in the Renaissance* (Washington, DC: Georgetown University Press, 1995): 50, 86.
194. Ryan, *Manual of Medical Jurisprudence* (2nd ed): 35.
195. Ryan, *Manual of Medical Jurisprudence* (2nd ed): 36.
196. While Ryan, as an avid medical reformer, argued strongly against the dogmatism and unmerited privileges of the medical corporations, he has nothing here but praise for the "admirable" Statutes of Morality of the Royal College of Physicians; *Manual of Medical Jurisprudence* (2nd ed): 67.
197. Ryan, *Manual of Medical Jurisprudence* (2nd ed): 47. Apparently mindful of the charges of plagiarism launched by the *Lancet* reviewer of his first edition, as well as of his own legal difficulties in the Ramadge case, Ryan took pains to claim that the works of Gregory and Percival were in the public domain and that he had every legal right to reprint them (46–47).
198. Ryan also had severe criticism for the anonymous editor of the most recent edition of Percival's *Medical Ethics,* whom he accused of adding "inappropriate comments"; *Manual of Medical Jurisprudence* (2nd ed): 47. The edition in question was *Medical Ethics; or, a Code of Institutes and Precepts, Adapted to the Professional Conduct of Physicians and Surgeons…by the late Thomas Percival …,* London: W. Jackson, 1827. A review in *The Lancet* noted that the editor had omitted Percival's supplementary notes and illustrations, replacing them with extensive, sarcastic comments on contemporary medical quackery; [review of *Medical Ethics…*, by Thomas Percival], *Lancet* 12 (series 2) (1826–1827): 696.
199. We are grateful to Robert Baker and Laurence McCullough for stressing this point. This also seems further to reinforce Chester Burns's observation that Ryan made important contributions to medical ethics by situating it within the field of medical jurisprudence (note 186).
200. Ryan, *Manual of Medical Jurisprudence* (2nd ed): 72.
201. Thomas Percival, *Medical Ethics,* Manchester, S. Russell, 1803: 14–15.
202. Robert B. Baker and Laurence B. McCullough, "Introduction." In: *Cambridge World History of Medical Ethics*, eds. Robert B. Baker and Laurence B. McCullough (Cambridge, UK: Cambridge University Press, 2009:3–15, quoting Henry K. Beecher, *Research and the Individual: Human Studies*, Boston, Little, Brown, 1970:218.
203. Ryan, *Manual of Medical Jurisprudence* (2nd ed): 70–71.
204. U.S. National Commission for the Protection of Human Subjects of Biomedical and Behavioral Research. *The Belmont Report: Ethical Principles and Guidelines for the Protections of Human Subjects*. Washington, DC: U.S. National Commission…, 1978. The distinction is concisely explained by Robert Levine, "Clarifying the concepts of research ethics," *Hastings Center Report* 9 (3) (1979): 21–26. We must recall that Gregory, Percival, and Ryan all wrote in an era when the concept of informed consent was virtually unknown within medicine; so all imagined that experimenting on patients would be done without their explicit consent.
205. In an era in which we have focused so much on informed consent as the central component of research ethics, we might ask whether Ryan suggests anything akin to patient consent as a requirement for the ethical conduct of research. One might argue that any use of a form of contract theory implicitly invokes the concept of consent, since parties to a contract can modify the contract by mutual, explicit agreement. Martin Pernick, reviewing the history of informed consent in nineteenth and twentieth centuries medicine, argued that physicians in Ryan's day advocated educating and informing the public about medical matters, and in specific contexts

indicated considerable respect for the patient's decision on accepting or rejecting specific treatments. But the underlying logic focused much less on any concept of respect for patients' rights or autonomy, and more on the prevalent concept that respecting patients' choices could be therapeutic; Pernick, "The Patient's Role if Medical Decisionmaking: A Social History of Informed Cosent in Medical Therapy," in: President's Commission for the Study of Ethical Problems in Medicine and Biomedical and Behavioral Research, *Making Health Care Decisions: The Ethical and Legal Implications of Informed Consent in the Patient-Practitioner Relationship; Volume Three: Appendices : Studies on the Foundations of Informed Consent*, Washington, DC: U.S. Government Printing Office, 1982: 1–35. Consider, for example: "The *feelings* and *emotions* of the patients, under critical circumstances, require to be known and to be attended to, no less than the symptoms of their diseases. Thus, extreme *timidity*, with respect to venæsection, contraindicates its use, in certain cases and constitutions. Even the *prejudices* of the sick are not to be contemned, or opposed with harshness. For though silenced by authority, they will operate secretly and forcibly on the mind, creating fear, anxiety, and watchfulness." Thomas Percival, *Medical Ethics,* Manchester, S. Russell, 1803: 14–15; quoted in Ryan, *Manual of Medical Jurisprudence* (2nd ed.): 75.

206. Ryan's unexplained assertion that animal experimentation had replaced human experimentation became a bone of contention between him and *The Lancet*. The negative reviewer for that journal expressed frank incredulity that anyone could suggest that animal experiments would suffice to establish the curative powers of any treatment intended for human beings. Ryan on his side invoked the Golden Rule once again and asked whether the reviewer would volunteer himself to be the subject of the experiments. The exchange is nicely summarized by Robert B. Baker, "The Discourses of Practitioners in Nineteenth- and Twentieth-Century Britain and United States" (Chapter 36, Section VIII), in *Cambridge World History of Medical Ethics*, eds. Robert B. Baker and Laurence B. McCullough, Cambridge, UK: Cambridge University Press, 2009:446–464. Baker also reminds us that Gregory had proposed a Golden Rule test for medical experimentation: "I would ask that a man if he have done [dangerous Experiments] to his Child, if he would not do it to his own Child, why should he indanger the lives of other people"; *John Gregory's Writings on Medical Ethics and Philosophy of Medicine*, ed. Laurence B. McCullough, Dordrecht: Kluwer, 1998: 249. Ryan must certainly have appreciated this appeal to his favorite moral rule, and one wonders why he did not highlight it in his own discussion.
207. Ryan, *Manual of Medical Jurisprudence* (2nd ed): 80.
208. On Rush's medical ethics see Haakonssen. *Medicine and Morals*: 187–226. See also Chester R. Burns, "Reciprocity in the Development of Anglo–American Medical Ethics," in *The Codification of Medical Morality*, ed. Robert Baker, Boston, MD: Kluwer, 1995: 135–143.
209. Ryan, *Manual of Medical Jurisprudence* (2nd ed): 105. The volume sent to Ryan appears to have been John D. Godman, *Addresses Delivered on Various Public Occasions*, Philadelphia, PA: Carey, Lea, and Carey 1829. Godman (1794–1830) lectured at a number of U.S. medical schools and was considered a first-rate anatomist and naturalist.
210. Ryan, *Manual of Medical Jurisprudence* (2nd ed): 105.
211. Ryan, *Manual of Medical Jurisprudence* (2nd ed): 125.
212. Truly comprehensive reform would not occur till 1858, when the traditional distinctions among physicians, surgeons, and apothecaries were abolished in favor of uniform educational standards.
213. Ryan, *Manual of Medical Jurisprudence* (2nd ed): 130–131.
214. Ryan, *Manual of Medical Jurisprudence* (2nd ed): 131.
215. Ryan, *Manual of Medical Jurisprudence* (2nd ed): 133.
216. As we have already seen, Ryan was probably not motivated by any compunction about duplicate publication. He was never loath to borrow liberally from already published works when he came to write a new book. Even had he been finicky about not duplicating his earlier writing, he could at least have called readers' attention to these earlier passages, if he truly believed that "medical ethics" described their contents.
217. Ryan, *Manual on Midwifery* (1828): 238.
218. Ryan, *Manual of Midwifery* (1841): 226.

219. Ryan, *Manual of Midwifery* (1841): 255.
220. Ryan mentions that the view of the Sorbonne theologians was still the official stance of the Roman Catholic Church; *Manual of Midwifery* (1841): 255. This passage apparently led Robert Baker and Chester Burns to conclude that Ryan himself was Catholic. We offer reasons above (note 74) to discount this view.
221. Ryan, *The Philosophy of Marriage in its Social, Moral, and Physical, Relations; with an Account of the Diseases of the Genito-urinary Organs, Which Impair or Destroy the Reproductive Function, and Induce a Variety of Complaints; with the Physiology of Generation in the Vegetable and Animal Kingdoms; Being Part of a Course of Obstetric Lectures Delivered at the North London School of Medicine, Charlotte Street, Bloomsbury, Bedford Square*. [unnumbered edition "from the last London edition."]. Philadelphia, PA: Barrington & Haswell, 1848:3.
222. Ryan, *Philosophy of Marriage*, 31. Ryan seems to have used the term "nature" as a shorthand for God. In a later chapter discussing reproduction within the animal kingdom, he wrote, "I have described, in my Lectures on Midwifery and the Diseases of Women and Children, published in the London Medical and Surgical Journal, 1836, the function of reproduction in all the classes in zoology, with a view of showing the analogy of all, and the consummate wisdom of the Creator in regenerating organized beings"; *Philosophy of Marriage*: 172.
223. Ryan, *Philosophy of Marriage*: 53–57, 116–17, 126–28, 246.
224. Ryan, *Philosophy of Marriage*: 128.
225. The posters Ryan alludes to appear to have advocated a device similar to a contraceptive sponge. They were distributed in 1823, rather than 1822, and appear to have been the work of Francis Place, whose volume, *Illustrations and Proofs of the Principle of Population: Being the First Work on Population in the English Language Recommending Birth Control*, had appeared in 1822 (a reprint appeared in 1994, published in London by Routledge/ Thoemmes). We note that Place and his compatriots were "so-called" Malthusians because even though they invoked Malthus's theory of population, Malthus himself, an Anglican clergyman, disapproved of birth control for the same reasons argued by Ryan; John C. Caldwell, "Malthus and the Less Developed World: The Pivotal Role of India," *Population and Development Review* 1998; 24: 675–96.
226. Ryan, *Philosophy of Marriage*: 19.
227. Malthus argued that the Poor Laws ought to be abolished, leaving virtually all of the poor to swim or sink on their own; he believed that guaranteeing relief for the poor actually worked against their interests by perpetuating and expanding poverty; T. Robert Malthus, *An Essay on the Principle of Population as It Affects the Future Improvement of Society with Remarks on the Speculations of Mr. Godwin, M. Condorcet, and Other Writers*. 1798. (http://www.faculty.rsu.edu/~felwell/Theorists/Malthus/Essay.htm#35) Ryan by contrast sought to reform the Poor Laws but not to abolish them completely.
228. Ryan, *Philosophy of Marriage*: 63, 64, 65.
229. It must be noted that Ryan occasionally speaks approvingly of early marriages. Condemning the dissoluteness of city life, whilst praising the morality of those living in the countryside, he writes, "It is proved beyond doubt that the population increases much more in the country, and in villages, than in large cities. The citizens pass their youth in dissipation, and marry late for the sake of interest. But in the country, illegitimate unions cannot occur without exposure, as everyone knows the conduct of his neighbour; a man marries early in life, he is stranger to luxury and effeminacy, and his offspring is generally healthful, vigourous, and numerous." *Philosophy of Marriage*: 128.
230. James Gregory (1753–1821), son of John, wrote *Conspectus Medicinae Theoreticae ad Usum Academicum (Views on the Theory of Medicine…)*, which appeared in numerous editions from 1780 through the 1830s.
231. Ryan, *Philosophy of Marriage*, 153. The same passage may be found in *Manual of Midwifery* (1841): 87.
232. However, he did acknowledge that even healthy parents could produce defective infants. He believed that moral and physical traits had hereditary and environmental components.

For example, he quoted extensively from Moore's *Marriage Customs of All Nations* in *Philosophy of Marriage*; Moore attributed the different characteristics of nations to climatic variances.

233. Ryan, *Manual of Midwifery*: 34–35; he cites "Wrolick's Essay on the Diversity of the Pelvis, Bull. De Sciences Med. Fev., 1827. Velpeau, 1835." Alfred Armand L.M. Velpeau (1795–1867) was the author of *Traité complet de l'art des accouchemens*, published in Paris in 1835. It is unclear whether Ryan is quoting or just referencing Wrolick and/or Velpeau. For more information on Velpeau, see http://www.general-anaesthesia.com/images/velpeau.html.
234. Ryan appears to use the terms "race" and "nation" interchangeably.
235. Ryan, *Philosophy of Marriage*: 83.
236. Ryan, *Philosophy of Marriage*: 10. Ryan responded to the anticipated censure from these prudes with typical self-promotion, boasting that earlier editions of his works had garnered approval from "enlightened individuals, including statesmen, judges, divines, lawyers, and all classes of society"; *Philosophy of Marriage*: 16.
237. Ryan, *Prostitution in London* (1839): 168.
238. A.-J.-B. Parent-Duchatelet, *De la Prostitution dans la Ville de Paris : Considérée Sous le Rapport de l'Hygiène Publique, de la Morale et de l'Administration; Ouvrage Appuyé de Documents Statistiques Puisés dans les Archives de la Préfecture de Police; et Précédé d'une Notice sur la Vie et les Ouvrages de l'Auteur par Fr. Leuret*, Paris: J.-B. Baillière, 1837. Ryan may initially have intended merely to translate Parent-Duchatelet's work, but then decided rather to expand it considerably by adding observations on prostitution in London and New York. For more on Parent-Duchatelet, see Ann Elizabeth Fowler La Berg, *Mission and Method: The Early Nineteenth-Century French Public Health Movement* (Cambridge History of Medicine Series), Cambridge: Cambridge University Press, 2002.
239. Ryan, *Prostitution in London:* 42.
240. Ryan's critique of the gendered, unjust effect of the poor laws co-existed comfortably with his conviction that that men and women should behave in a manner dictated by their reproductive physiology; *Philosophy of Marriage*: 108, 236. He echoed the predominant view of his era, that "…the present mode of female education is highly injurious to health, predisposes to spinal curvature, and consequently, to deformity of the hip and other bones, thereby often rendering parturition highly dangerous and fatal"; *Philosophy of Marriage*: 236. Yet he expressed no doubts that women could act as obstetricians and midwives, provided they received adequate training; *Philosophy of Marriage*: 209, 230.
241. Ryan, *Philosophy of Marriage*: 27.
242. Ryan might appear at first glance to be inconsistent in arguing for specific legislative reforms, since he advised the physician, as part of his survey of "ethics of the present period," to avoid all involvement in politics; Ryan, *Manual of Medical Jurisprudence*, 2nd ed.. (1836): 69–70. It is likely that Ryan distinguished between the physician's personally becoming involved in politics (or worse yet, engaging in political harangues with his patients), as opposed to offering his expert advice to legislators via the medical press or other means.
243. Ryan, *Philosophy of Marriage*: 28–29. Ryan quoted an unnamed "public journal," describing the case "of a mother and her three daughters who are toiling for only 8 s. a-week, is, we are sure, not a solitary one, and that many unfortunate girls are literally compelled, at the close of their ill-requited honest daily work to walk the streets for money to procure the necessities of life, is a notorious and a very deplorable fact" (28). Ryan further noted that unwed mothers would not be able to secure employment as domestics – another line of work available to women – because potential employers considered them a likely immoral influence upon their families and home.
244. Ryan, *Prostitution in London*: vi.
245. Ryan, *Philosophy of Marriage*: 27. Ryan also wanted to marshal religion and medicine in the efforts to curb prostitution because he believed they could be relied on to inspire fear in men, restraining them from engaging in venereal excesses and abuses that resulted in disease; *Prostitution in London*, 392–393.
246. Ryan, *Prostitution in London*: 85–86.
247. Ryan, *Prostitution in London*, 39.

248. Ryan, *Philosophy of Marriage*: 27.
249. Ryan, *Prostitution in London*: 82.
250. Ryan, *Prostitution in London*: 87. In Belgium, it appears, proposals had recently been made to encourage the use of contraceptives among prostitutes, in hopes of reducing the incidence of unwed births even if the evil of prostitution itself appeared to be immune to elimination. As one would expect from his comments on the Malthusian contraceptive proposals, Ryan utterly opposed any such measures on the grounds that contraception was contrary to Divine edict.
251. Ryan, *Prostitution in London*: 9.
252. Ryan, *Prostitution in London*: 433.
253. Ryan, *Philosophy of Marriage*: 46–47. His attacks on "empirics" or quacks were generated both by compassion for the patient, and also by his desire to stand up for what he considered proper medical science and appropriate training and qualifications; *Prostitution in London*, 3.
254. See generally Robert B. Baker, Arthur L. Caplan, Linda L. Emanuel, and Stephen R. Latham (eds.) *The American Medical Ethics Revolution*. Baltimore, MD: Johns Hopkins University Press, 1999.
255. Jukes Styrap, *A Code of Medical Ethics*, reprinted in *The Codification of Medical Morality*, ed. Robert Baker, Boston, MA: Kluwer, 1995: 149–171. See also Peter Bartrip, "An Introduction to Jukes Styrap's *A Code of Medical Ethics* (1878)," in Baker, *Codification*: 145–148. Bartrip notes that despite Styrap's efforts, the British Medical Association refused to adopt his Code as an official statement, but by 1896 the *British Medical Journal* had pronounced it "the usually accepted authority on ethics in the BMA" (145).
256. Chester R. Burns, "Reciprocity in the Development of Anglo-American Medical Ethics," in *The Codification of Medical Morality*, ed. Robert Baker, Boston, MA: Kluwer, 1995: 135–143.
257. We base this conclusion on two brief works that appeared at mid-century. W. Fraser published a 14–page booklet, *Queries in Medical Ethics, Read before the Medico-Chirurgical Society of Aberdeen, 5th April, 1849* (apparently a reprint from the *London Medical Gazette*, 2 (1849): 181–187, 227–232). This took the form of a list of 27 ethical queries with answers ranging from a single paragraph to a page or so. Fraser generally followed Percival's aphoristic style, but referred only once to Percival, and never to Gregory or to Ryan. Perhaps more pertinent was an anonymous *Review of Works on Medical Ethics*, attributed to John Brown (1810–1882), Edinburgh: Murray and Gibb, 1850, a 12-page pamphlet reprinted from the *Monthly Journal of Medical Sciences*. Brown offered a reading list of about a dozen items for the student of medical ethics, and included Gisborne, Gregory, Percival, and the new American Medical Association code, but omitted any mention of Ryan's work.
258. Robert M. Veatch, *Disrupted Dialogue: Medical Ethics and the Collapse of Physician-Humanist Communication (1770–1980)*. New York: Oxford University Press, 2005.
259. Veatch, *Disrupted Dialogue*: 72–74.
260. Veatch admits that all through the years of "disrupted dialogue," religious ethicists, most notably Roman Catholics, were engaged in writing moral guidance for the physicians who happened to be of their religious sect. So Veatch could, alternatively, have classified Ryan in this manner. Such would still be inaccurate because clearly Ryan saw himself as writing an ethics for all physicians, not just for those of one denomination. We have already seen reasons to doubt that Ryan himself was Catholic.
261. The medical school calendar began at the beginning of October; there was traditionally a long winter session (October–April) followed by a short summer session (May–July). Medicine (physic) and midwifery as major topics would typically be offered during winter, while medical jurisprudence was viewed as a subsidiary topic to be covered during summer. The date shown is that of the beginning of the academic year. The 1828 information is taken from endpaper advertisement, Ryan, *A Manual on Midwifery* (1828) (following page 353). In 1829, *The Lancet* began to print a complete listing of all London medical school courses and independent lectures in its initial number for each volume (issued around October 1, as *The Lancet* tied its publication schedule to the schools' calendar), and the remaining information is from those listings.
262. The *Lancet* medical school listings (25 September 1830) show Ryan both as affiliated with the New School in Brewer Street and as a private lecturer, but no specific address is given for

the latter. He is shown as lecturing on Medicine and Midwifery at the school, and Medicine, Midwifery, and Medical Jurisprudence privately. *Lancet* 1 (1830–1831): 16, 18.

263. The *Lancet* listing for October, 1831 lists Ryan as lecturing at two different sites – 9 Gerrard Street (the Westminster Dispensary school) and 61 Hatton Garden (his private residence); *Lancet* 1 (1831–13): 14.

264. From catalogues of the Wellcome Institute Library, London; National Library of Medicine, Bethesda; title pages of available copies; and booksellers' lists. The information on later editions appears incomplete, as editions are alluded to in some texts that are not accounted for in the available catalogues. Besides the titles listed here, the Yale University medical library contains a volume, "The anatomical exercitations of … on the motion of the heart and circulation of the blood [by William Harvey], translated from the original Latin, by M. Ryan, M.D. To which is prefixed, a biographical sketch of the illustrious author," 1832–1833. However, this appears to be a series of articles by Ryan published in the *LMSJ* and then privately bound, rather than a true book publication.

265. The 1839 or 3rd London edition of this work ("Considerably Enlarged and Improved") has the variant title, *The Universal Pharmacopœia, or, a Practical Formulary of Hospitals both British and Foreign, Including All Medicines in Use, Translated from the Last Edition of Mm. Mile Edwards, and P. Vavasseur.*

266. We find this volume mentioned both in the catalogue of the Wellcome Library, London, and in an end-paper advertisement for books published by John Churchill in Delabere Blaine, *Outlines of the Veterinary Art…*, London: Longman, Orme & Co., and J. Churchill, 1841. The 1841 advertisement does not mention the number of the current edition. The Wellcome Library copy is a second edition, dated 1838. We have been unable to find notice of the first edition.

267. *The London Catalogue of Books Published in Great Britain, with their Sizes, Prices, and Publishers' Names, 1816 to 1851* (London: Thomas Hodgson, 1851) (http://books.google.com/books?id = 6a8YAAAAIAAJ&pg=PT938&dq = % 22Obstetric + aphorisms% 22 + ryan, accessed December 14, 2007) lists as a work of Michael Ryan "Obstetric Aphorisms," a 32mo (very small dimension) volume published by Brown. In the endpapers of one copy of *Prostitution in London*, in a list of Ryan's works, mention is made (with no further details) of a volume, "Obstetric Aphorisms on the management of natural and difficult Parturitions, Puerperal Diseases, and the Physical management of Infants" (http://books.google.com/books?id = zHsEAAAAQAAJ&pg = PP 9 &dq = % 22obstetric + aphorisms% 22 + ryan#PPP9,M1, accessed December 14, 2007). As the Wellcome Library and other libraries with extensive historical collections appear to lack a volume with this title, we suggest that this is a variant title of *The Obstetrician's Vademecum, or Aphorisms on Natural and Difficult Parturition…* by Denman as edited and updated by Ryan.

Part II
Selected Writings by Ryan

Explanatory Note on Selections from Ryan's Works

We have attempted to reprint the following selections from Ryan's writings in a form as closely as possible identical to the original publications. The pagination in the original is noted by pairs of numerals within square brackets, so that, for example, "[76/77]" indicates the point within the text of the page break between pages 76 and 77. The original footnotes are retained and are identified as "*Author's note*." In the original, these are not numbered but are indicated by typographic symbols such as asterisks. By contrast, notes that we have added for this edition are designated by "*Editors' note*." In a few instances, a footnote is neither Ryan's nor ours; rather, Ryan copied them verbatim from the original source that he is quoting. These notes occur in Chapters IV (where Ryan extensively quotes Percival's *Medical Ethics*) and V (where Ryan quotes extensively from a lecture by Dr. John Godman). Such notes are labeled with the original source from which Ryan copied them.

Many of the Editors' notes offer translations of Latin passages in the original text. Ryan assumed that these passages would be understood by the medical students and physicians who made up his intended audience. We have tried to follow the policy of providing a translation in those instances where Ryan's text does not, by itself, proceed to render the meaning of the Latin passage in English. We are immensely grateful to Mary Hope Griffin for her assistance in wrestling with the debased form of language that nineteenth century British medical practitioners took to be Latin.

When Ryan is merely copying material from other sources, we have taken the liberty of omitting some of the Latin passages for which translations are immediately provided in the text, and quotations from other authors that do not add significantly to the meaning of the passages.

Section I
I. A Manual of Medical Jurisprudence

A

MANUAL

OF

MEDICAL JURISPRUDENCE,

AND

STATE MEDICINE,

COMPILED FROM THE LATEST LEGAL AND MEDICAL WORKS,

OF BECK, PARIS, CHRISTISON, FODERE, ORFILA, ETC.

CONTAINING

PART I

MEDICAL ETHICS OF ANCIENT AND MODERN TIMES.

PART II

LAWS RELATING TO THE MEDICAL PROFESSION, FROM PARIS AND FONELANQUE, SCULLY, WILCOCK, AND CHITTY.

PART III

MEDICAL JURISPRUDENCE AND STATE MEDICINE FROM THE MOST CELEBRATED ANCIENT AND MODERN MEDICO-LEGAL WRITERS.

PART IV

LAWS RELATING TO THE PRESERVATION OF PUBLIC HEALTH–MEDICAL POLICE–STATE MEDICINE–PUBLIC HYGIENE.

INTENDED FOR THE USE OF

LEGISLATORS, BARRISTERS, MAGISTRATES, CORONERS, PRIVATE GENTLEMEN, JURORS, AND MEDICAL PRACTITIONERS.

By MICHAEL RYAN, M.D.,

Member of the Royal College of Physicians in London; Member of the Association of Fellows and Licentiates of the King's and Queen's College of Physicians in

Dublin; Member of the Royal Colleges of Surgeons in London and Edinburgh; Consulting Physician to the East London Midwifery Institution; Lecturer on Medical Jurisprudence in the Medical School, Gerrard Street, Soho Square, &c. &c.

Second Edition considerably Enlarged and Improved.

LONDON:
SHERWOOD, GILBERT, AND PIPER,

PATERNOSTER ROW.

1836.

[i/ii]

LONDON:

JOHN HADDON AND CO. DOCTORS' COMMONS.

[ii/iii]

CONTENTS

PART IV – LAWS FOR THE PRESERVATION OF PUBLIC HEALTH

[iv/v]

PREFACE

TO THE

FIRST EDITION

The object of the author of these pages, is to give a concise and comprehensive view of the received Principles of Medical Jurisprudence, and to collect the scattered and isolated facts from the standard works of legal and medical writers. He is inclined to hope that this volume contains all that is valuable in the systematic works upon the subject; and numerous other topics of vast importance to medical practitioners, which have no place in similar productions. In proof of this assertion, he refers the reader to the ethical and legal parts of this work, and to the articles on medical evidence, adulterations of alimentary matters, and public medicine. He hopes that the promulgation of Medical Ethics, or the institutes of professional conduct, will contribute in no small degree to maintain and support the honour, dignity, character, and utility of the profession, by impressing the minds of medical students with a sense of the noble and virtuous principles which have always characterised their predecessors, and which ought ever to distinguish the scientific cultivators of medicine.

Another novel and important feature in this production is, the exposition of the laws relating to the different orders of the faculty in these kingdoms, which cannot fail to prove instructive and useful to medical practitioners, by informing them of the immense power, vast influence, and high privileges, conferred upon them by the legislature, in exempting them from the performance of many civil duties, and in deeming their evidence conclusive in an immense number of civil and criminal proceedings, which affect the lives, the liberty, the honour, reputation, and property, of every class of society. For many facts mentioned in this part of the subject, the author is indebted to the works of Dr. Paris, and Mr. Fon- [v–vi] blanque, of Mr. Willcocks, of Mr. Scully, the celebrated Irish barrister; of Mr. Phelan and Mr. Chitty.

The medical part has been compiled from the standard systems on Forensic Medicine, both domestic and foreign, and illustrated by the opinions and experience of the author. The works of Drs. Male, Gordon Smith, Duncan, Paris, Beck, Christison, Foderè, Mahon, Chaussier, Orfila, Briand, Sedillot, &c. have been laid under large contribution, and the extracts duly acknowledged.

The chapter on medical evidence has been condensed from the valuable lectures delivered in the London University by Professor Amos, as published in the *London Medical Gazette*. Under this head are considered the powers of coroners, and *the propriety and expediency of appointing medical men to the office of coroner*. The addition of the forms of medical certificates, and the method of examining recruits, were considered useful for the guidance of young practitioners.

It has been the author's anxious wish to compress the fullest information in the smallest space, and in the most familiar language, in order to simplify the subject, and render it intelligible to every class of medical practitioners, to barristers, solicitors, coroners, magistrates, private gentlemen, and general readers. Every medico-legal fact which can become the subject of judicial inquiry, is accurately detailed, so that this Manual may be fairly looked upon as a text-book for the practitioners of the law and medicine. It is intended to save both classes much trouble and research. Whether the author has executed his task in a satisfactory manner, must be decided by his contemporaries; but he consoles himself with the reflection, that his design and intention were good, had he sufficient ability to execute it as he wished.

In exposing the absurd distinctions, the defective state of the laws relating to the profession, and the gross abuses of its constituted authorities, a love of freedom and of equality, with an ardent desire to promote the interests of his favourite science, and of humanity, have impelled him to declare the truth, however unpalatable it may be in certain quarters, or to the different orders of the faculty. His motto has been, *"amicus Socrates, amicus Plato, sed magis amica veritas."* [vi/vii] He has not been the advocate of any party, of any order, of any corporation, but the advocate of the whole profession. He has examined the whole of the charters and statutes relating to the practice of medicine in these countries, and has found them defective, contradictory, oppressive; and, since the legislative Unions between England, Scotland, and Ireland, highly unjust and impolitic. In proof of this statement, the reader has only to recollect the fact, that a Scotch physician, surgeon, or apothecary, cannot practise legally in England; an English member of the faculty cannot practise in Ireland; and an Irish member can neither practise in England nor in Scotland. Such is the state of the law, which was just and right when the three countries were distinct nations, but is unjust and preposterous since they have formed one united empire. The author has, therefore, considered that there never was a more auspicious period than the present, for exposing the absurd state of the laws relating to the faculty in the United Kingdom, when the propriety of constitutional, ecclesiastical, and legal reforms, is the subject of national discussion; and when medical reform is loudly demanded by the whole profession, with the exception of the few interested monopolists. It need scarcely be stated, that the principles and practice of medicine and therapeutical agents, are identically the same in every part of these countries.

Medical jurisprudence is at length included in the course of study required by the Apothecaries' Company of London; but it is strangely excluded in the course required by the Royal College of Surgeons. Had not the University of London appointed a professor of Forensic Medicine, chiefly at the recommendation of the philanthropic Dr. Birkbeck, this branch of medical science would as yet have no place in the course of medical education required in England.

Nevertheless, the most casual observer must be convinced, by daily examples, that the ignorance of ethics and medical jurisprudence, impedes or arrests the career of the medical practitioner, and frequently destroys professional reputation altogether. If proof were demanded of the validity of this position, I need only refer to the newspaper reports of legal [vii/viii] proceedings, in which we daily observe ample attestations of the fact. We peruse the most absurd and unscientific medical evidence, more especially in the reports of coroners' inquests, which could never have appeared, had the witness possessed a proper knowledge of forensic medicine, or had the coroner been a medical practitioner. Such displays of ignorance excite the pity and contempt of the scientific portion of the faculty, and the ridicule of the legal profession, and of the public. Mr. Amos has well illustrated this defect, in his admirable cautions to medical witnesses, in which he has cited many cases to prove, that medical evidence has been censured by the bench, ridiculed by the bar, derided by the auditory, and has entailed obloquy and disgrace upon the unfortunate witness.

Medical practitioners should be aware, that all the rising barristers of our criminal courts attend lectures on legal medicine; and often does forensic fame arise from the ability with which an advocate examines a medical witness. A knowledge of medico-legal science is almost as indispensable to the one as to the other; and the coroner who is ignorant of it, is evidently incompetent to discharge his duty to the public, or to secure impartial justice to the accuser or accused. This has been incontrovertibly proved in the article on medical evidence.

It has been erroneously stated by some writers, that the science of medical jurisprudence, is nothing more than the application of the elementary branches of medicine, to the elucidation of judicial investigations; and, consequently, that a scientific medical practitioner must necessarily be a good jurist. This is not correct, inasmuch as the most scientific physicians and surgeons have too often proved to be the worst jurists; because they could not derive the requisite information on medico-legal science from the common systems of medicine or surgery, as it is only to be obtained from works exclusively devoted to the subject. In no lectures or works, except those upon this science, is a student informed of the laws relating to his profession; of his rights, privileges, and immunities; of the cases, civil and criminal, on which be is liable to be called to give evidence: of the received opinions [viii/ix] upon these cases; of the danger of wounds, and contusions, and of injuries prejudicial to health or destructive of life; of the analysis or mode of detecting the numerous poisons; of the manner of giving evidence, or of his ethical duties in public and in private practice. It is therefore manifest, that medical jurisprudence is a distinct science, and one of the greatest importance and utility to the members of the medical and legal professions. If it could be taught by the professors of the elementary branches of medical education, there would be no need of a separate professorship; which, therefore, exists in all the medical schools of these and foreign countries.

In conclusion, the author has to remind the critical reader, that in attempting to compress the extensive information comprehended in the narrow limits of these pages, brevity of expression, and too much conciseness, may have rendered the style occasionally obscure or inelegant. This perhaps may be pardoned in a work chiefly intended for medical students, and which was originally published in one of the periodicals (*The London Medical and Surgical Journal*), to fill up lacunae, and arranged as the suddenness of the occasion demanded; which in general afforded little time for attending to the beauties of style, to euphonious sentences, or to the other qualities of literary composition. If the work supply a want, or contribute to the maintenance of the character and utility of medicine, or in any way benefit the interests of mankind, or the administration of justice, the object of the author will be fully attained. [ix/x]

[illegible]

PREFACE

TO THE

SECOND EDITION

The favourable reception which this work has received in this and other countries has compelled the author to endeavour to improve a new edition. He has accordingly enlarged many chapters introduced much new matter, and carefully revised every part of his work. He has referred to the last editions of all standard works on the subject of which he treats, and appended all that is valuable since his former edition. He has considerably enlarged the first part, on Medical Ethics, by introducing the codes of continental Europe and America, and also those of Dr. John Gregory and Dr. Percival. Under this head, he has commented on Medical Education, Degrees, Diplomas, Medical Appointments – Success – Reputation – Eminence – Moral and Physical Medicine – Clinical Medicine – Rules for Prescribing Medicines – Action of Medicines on the Economy – Posology, or Fixation of Doses of Medicines – Pharmacology.

Part II. Laws relating to the Medical Profession is brought down to 1836. Few additions are made in this section, as the author is of opinion that the whole of the laws relating to Medical Polity in this country, will be remodelled during the present or next Session of Parliament.

Part III. Medical Jurisprudence is considerably enlarged, and in the opinion of the author, improved. Several additional Medico-legal questions have been made in this section.

Part IV. Laws for the Preservation of Health-Medical Police-State Medicine-Public Hygiene. This section [x/xi] occupies 60 pages of new matter. It comprises a vast number of subjects of great interest to public health and happiness.

Chapter I. Laws for the preservation of Public Health – Quarantine, Boards of Health—Contagious Diseases – Disinfection – Purity of Air, Water, and Situation – Guardians of the Public Health – College of Physicians, as inspectors of Apothecaries' Shops, officers of Public Charities – Commissioners of Paving, Sewers, Cleansing the Streets, &c – Framers of the Bills of Mortality.

Chapter II. Inhumation – Burial of the dead in Cities – Searchers and Inspectors—Danger of Exhumation and Opening of Graves – Premature Interment – Uncertainty of the Signs of Death – Account of Individuals buried alive—Vivisections – Proper period of Inhumation – Custom in France and England.

Chapter III. Signs of Real Death; aspect of the Face; Absence of Heat, and Lividity of the Skin; Absence of Circulation and Respiration; Cadaverous Rigidity of Stiffening; Physiological Causes of it; Physical Proofs of Death; Surgical Proofs; Incisions, Decapitation, &c.

Chapter IV. Putrefaction of Animal Matters; Putrefaction in the open Air, Water, and different Earths; modified by Age, Constitution, Sex, State of Thinness or

Obesity; Mutilation or Integrity of the Body; Genus and Duration of Disease; Phenomena before Death; Period Of Inhumation; Appearance and Cause Of Insects; Humidity and Dryness of Earth; Chemical Composition of Earth; Depth of the Grave; Naked or Clothed state of the Body; Atmospheric Influences; Conversion of the Body into Adipocere; Dr. Fletcher's Physiological Views of Putrefaction, 1836.

Chapter V. Nuisances Legally and Medically considered; Trades and Manufactures; Filth in the Streets; Noises by Day or Night; Transitory Nuisances; Physical Effects of; Arrangement of by Paris and Fonblanque, including the Principal Trades and Manufactures; Purity of Air, Water, and Situation.

In addition to the new matter, there is prefixed a history of the rise, progress, and present state of Public Medicine, and also a copious Index, which may, perhaps, be considered a [xi/xii] Medico-Legal Dictionary. It is the fullest hitherto published in our language. Such are the features of this work. The author extracted from all available sources, both domestic and foreign; added the result of his own experience; and has presented to his readers a variety of information, not so far, as his researches enable him to state, to be found in any other elementary work in one volume, of equal size, on Medical Jurisprudence and State Medicine. He has not prefixed or added a bibliography, but leaves his readers to form their own conclusions on the number of works to which he has referred, and which he has duly acknowledged.

Great Queen Street
St. James's Park, Westminster
February, 1836

[xii/xiii]

Introduction

Medical Police, Political Medicine, State Medicine, Public Hygiene, Police of Health, and Medical Jurisprudence, comprise the acts of a legislature or Government, and magistracy, for the conservation of public health, and also the enactment of laws for the regulation of the practice of the medical profession, and the duty of medical practitioners in aiding the legislature in forming just laws, and public tribunals in the administration of justice.

Medical Jurisprudence, or Legal Medicine, is a science by which medicine and its branches, are rendered subservient to the construction, elucidation, and administration of the laws for the preservation of public health. This term is considered by some writers as best calculated to express, in the most comprehensive manner, the application of the medical sciences to the purposes of law.

It has been divided into Forensic, Legal, Judiciary, and Judicial medicine, comprising the opinions and evidence and into Medical Police or State Medicine, comprising all medical opinions and precepts which inform the legislature and magistracy in constructing the laws, and in enforcing them for the preservation of the public health. Both these divisions are included by the Germans in the term *State Medicine*. I have employed both in the construction of this work, as I have enumerated the laws relating to the profession, and to the preservation of public health, together with all medicao-legal inquiries. I have preferred this arrangement to any other proposed by medico-legal writers in this country, as it is the most comprehensive. M. Foderè, of Strasburg, who has executed the most elaborate modern work on State Medicine, has divided the science [xiii/xiv] into Medical Jurisprudence and Public Hygiene,[1] and published three octavo volumes on each subject. The term Medical Jurisprudence is incorrect, as it may imply a knowledge of the laws relating to medical subjects, or a knowledge of the medical sciences as applied to legislation, forensic inquiries, and the Police of Health. It may also unite the legal and medical sciences, and in this sense it is employed by some modern writers (Paris and Fonblanque). Dr. Beck used the term in his excellent work in the restricted sense, and defined Medical Jurisprudence, Legal Medicine or Forensic Medicine, a science which applies the principles and practice of the different branches of medicine to the elucidation of doubtful questions in courts of justice

(Elements of Medical Jurisprudence, 1823). He designed to offer another treatise on Medical Police, which has not as yet been published.

Dr. Gordon Smith, our first systematic author on this science, followed the example of Dr. Beck, and published his work entitled, "The Principles of Forensic Medicine, Systematically arranged and applied to British Practice;" it was also his intention to write a separate production on Medical Police, which he did not accomplish. I have included Medical Ethics and Jurisprudence, and State or Public Medicine, in one work, and thus differ from M. Foderè, Dr. Beck, Dr. Gordon Smith, and Dr. Paris. This was the arrangement of some ancient medico-legal authors: Medicus-Politicus, sive de officiis Medico-Politicis tractatus, quatuor distinctus libris: in quibus non solum bonorum medicorum mores ac virtutes exprimuntur, malorum vero fraudes et Imposturæ deteguntur: verum etiam pleraque alia circa novum hoc argumentum utilia atque jucunda exactissime proponuntur: opus admodum utile medicis, ægrotis, ægrotarum assistentibus, et cunctis aliis literarum, atque adeo politicæ disciplinæ cultoribus. Roderici à Castro, Lusitani Philos. ac Med. Doct. per Europam Notissimi. Hamburgi, 1614.[2]

I have also included the codes of Ethics of Dr. John Gregory and Dr. Percival, of the Royal College of Physicians in London, and of the Americans.

The history of Hygiene is coeval with the origin of society. The first wants of man were aliment, and he was instructed in what was salutiferous and noxious to his condition. He soon discovered the influences of exertion, repose, sleep, waking, &c., upon his constitution. The knowledge of the laws, [xiv/xv] morals, and police of the people relating to the preservation of health, constitutes public hygiene or medicine. It necessarily existed in the remotest ages. The ancients were of opinion, that there was an intimate and mutual dependence between the moral and physical states, and of the necessity of enacting laws for the promotion of temperance and wisdom, as well as for the punishment of excesses and of crimes.

All eastern nations had their systems of medical police; we trace it in their legislation, their manners and customs, and in the rules of their public police.

The hygiene of Moses embraced three principal objects: the prohibition of certain aliments, ablutions for certain impurities, and the sequestration of certain diseases reputed contagious, as lepra, &c. The Levitical law, ch. xiii. xiv., commanded the priests to visit the houses infected with the plague of leprosy, or with other contagious disease; to examine the inhabitants, to establish quarantine, to purify the houses, to shut them up, and pull them down in certain cases. We find in Deuteronomy, ch. xxii., that in questions of doubtful virginity, the elders were to be consulted, and enjoined to deliver judgment according to the physical evidence of the case.

It is easy to conceive, the object of legal purifications in warm climates, where the tendency of animal foods to putrefaction, and the copious perspiration of the people, were the causes of insalubrity, which frequent ablutions destroyed. The chief hygienic means of the ancients, in addition to those already mentioned, were gymnastics, baths, and repose after the latter.

The principal rules relative to public police among the ancients, referred to the salubrity of habitations, the locality of towns, cities, aind streets.

The Father of Medicine wrote a Treatise on Air, Water, and Situation, and this work has influenced all civilized nations for nearly 3,000 years.[3] It led to the draining of marshes, and improvement of situations. Most persons know that the Roman emperors Julius and Augustus Cæsar, ordered the Pontine marshes to be dried; to which Horace alluded:

> **** Sterilis diu palus, aptaque remis
> Vicinas urbes alit, et grave sentit aratrum.[4]

Hygiene was, as already observed, cultivated long before the time of Hippocrates; and the rules regarding it, were deduced from the universal laws of the animal economy, and referred [xv/xvi] to age, constitution, intelligence, to wants, or pleasures.

Moses enforced laws as to the use of all aliments, fruits, grains, herbage, bread, milk, fish, flesh, wine, &c. According to Herodotus, the Egyptians invented beer, which was subsequently mentioned by Moses,[5] and they were the most healthful individuals. The Egyptian law ascribed to Menes, ordained, according to Plutarch, that no pregnant woman should suffer afflictive punishment. The Jews made a distinction between mortal and dangerous wounds; and had laws relating to virginity, marriage, adultery, embalming and interring the dead, and other subjects of public medicine. Joseph found the Egyptian priests physicians, and he ordered his father's body to be embalmed (Gen. 1. 6. A.C. 1747).

The ancient Greeks highly estimated the power of morals in preserving health, and perfecting the species; and we find this opinion prevalent in all ancient nations. Their primary object was to give to the country a robust people and vigorous defenders. It was this principle that gave origin to customs, which, in our time, appear to be inhuman and barbarous. Thus the destruction of delicate infants by the Spartans – a horrible custom of the Chinese at present with respect to their female offspring.

The laws of Lycurgus on the physical education of infants and of girls to the time of marriage, with a view of transmitting a good constitution to their offspring, and many more hygienic ordinances, afford ample proof of the state of public medicine among the ancient Greeks. The physical legislation of Pythagoras and Plato was also intended for the conservation of the public health.

It is remarkable, however, that there are scarcely any traces of the union between law and medicine in the works the ancient Greek physicians. There is some portion of hygiene and state medicine scattered through the voluminous writings attributed to Hippocrates; and some crude speculations on the nature and growth of the infant, the period of pregnancy; and the abolisliment of some rude obstetricy. The opinions of the father of physic on the perpetuation of the species, the animation of the fœtus, and on many other physiological questions had great influence on all ancient legislators. Those of Aristotle have had a like tendency; and even to this time. Both these authors held that the fœtis inanimate in the womb for a certain time, during which it was no crime to cause abortion. The canon law was founded [xvi/xvii] on this opinion, and has influenced all legislators to the present period.[6] The above authorities also maintained, that pregnancy might be protracted, beyond the ordinary term of 9 months; which is now the received opinion of physiologists.

Medical men, both among the Greeks and Romans, were consulted by the magistrates most frequently on questions of *medical police*, as on the salubrity of cities, and on the means of assuaging the virulence of epidemic diseases: and to subjects of that nature the public functions of the *Archiater,* or state physician, were confined in the lower ages of the empire. That they were summoned to give evidence in a court of justice no where appears; and had they been called, their testimony, in ages when human anatomy was proscribed, must, in a majority of difficult cases, have been of comparatively little value.

The laws of ancient Rome were chiefly enacted on the same grounds as those of Greece; but they embraced more important medico-legal questions. In the reign of Numa Pompilius, before the Christian era, 600 years, and 140 before the time of Hippocrates, it was enacted, "de inferendo mortuo," (lex regia)[7] that the bodies of all women who were in the last months of pregnancy, should be opened immediately after death, so that infants might be saved if possible. History also informs us, that the elder Scipio Africanus, Marcus Manlius, and the first Cæsar,[8] were brought into the world by incision of the abdomen, or gastro-hysterotomy; and hence the origin of the term, Cæsarian operation. It was also enacted in the Twelve Tables, A. C. 452, that the infant in the mother's womb was to be considered as living, and all civil rights secured to it. The legitimacy of an infant was limited to 300 days after the death or absence of his putative father, or within ten months of the supposed time of impregnation; but the period was afterwards extended to eleven months. Most of the Roman laws were framed in accordance with the opinions of the ancient philosophers and physicians, and many legal decisions were given; *propter auctoritatem doctissimi Hippocratis.*[9]

Several laws were constructed during the reign of Severus, Antonine, Adrian, and Aurelius, on the authority of Hippocrates and Aristotle; the crime of procuring abortion was limited to those cases in which the fœtus had exceeded 40 days, as before that time it was not considered a living [xvii/xviii] being;[10] and the Emperor Adrian extended the term of pregnancy from ten months – the period of legitimacy according to the Decemvirs – to eleven, in accordance with the physiological opinions of the times.[11] It is, also recorded by Suetonius, that the bloody remains of Julius Cæsar, when exposed to public view, were examined by a physician named Antistius, who declared that out of twenty-three wounds which had been inflicted, one only was mortal, which had penetrated the thorax between the. first and second ribs. Gerike has collected some curious instances of the inspection of murdered bodies by medical witnesses, from the writings of Suetonius, Tacitus, and Plutarch. The body of Germanicus was also medically inspected; and, by indications conformable to the superstitions of the age, it was decided that it had been poisoned.[12] The bodies of Agricola and others were also examined by physicians, according to Tacitus.

It does not appear that there was any positive law which required the inspection of wounded bodies by medical practitioners, or that legislators required medical opinions before making laws. It is, however recorded, that before the time of Philumenos, A.D. 80 – who first studied obstetricy at Rome – the midwives were ordered by the prætors to examine pregnant women in judicial inquiries. Augustus had previously favoured the profession of medicine, A.D. 10, and exempted its members from public burthens and taxes. About 100, valetudinaria and veterinatia were established in the Roman camps, for wounded soldiers and horses.[13]

Galen was born in 131, and was physician to the gladiators at Pergamos, A.D. 159. He alluded to some questions. of legal medicine; he remarked the difference which subsists between the lungs of an adult and the fœtus; admitted the legitimacy of seven months children, and offered remarks on the manner of detecting simulated diseases. He related the case of a Greek lady who escaped punishment for adultery on the opinion of a physician. She had borne an infant very unlike the supposed father; and the physician accounted for this, by referring its resemblance to that of a picture which hung in the chamber of the mother.

The promulgation of the Justinian Pandects, A.D. 529, [xviii/xix] which laid the foundation of the legal codes of most of the modern states of Europe, may be considered the origin of the connexion between the sciences of law and medicine, or of forensic medicine. Under the titles, *De Statu Hominum*; *De Pœnis et Manumissis*; *De Sicariis et Veneficis*, of the *Lex Cornelia*; *De Inspiciendo Ventre custodiendoque Partu*, of the *Lex Aquilia*; *De Hermaphroditis*; *De Impotentia*; *De Muliere quæ peperit undecimo mense* – some questions in medical jurisprudence are discussed; but the decisions of the judges were directed to be guided on such matters, not on the oral testimony of living Physicians, examined on the particular point then under consideration, but *Propter auctoritatem doctissimi Hippocratis*.

If the Justinian code was an improvement in many respects on more ancient usages, in *one* point of medical jurisprudence, it must be allowed to be more erroneous than that which preceded it, namely, on the time of utero-gestation.

Aulus Gallus informs us, that a decree of Adrian acknowledged the legitimacy of a child born eleven months after the death of its reputed father; and this period has been adopted by Justinian (Novella 39 c. 2, Nov. 89).

From this period to the year 555, the Roman empire was overrun by the Goths and Vandals. The most savage customs now prevailed in the west of Europe, the principles of legal medicine were totally neglected, and the trial by the ordeals of fire, water, and duel, were introduced as appeals to the Deity, in cases which, to rude minds, appeared above human investigation.

> The trials by ordeal in the dark ages of modern Europe, when the decision of the most important questions was abandoned to chance or to fraud, when carrying in the hand a piece of red hot iron, or plunging:the arm in boiling water,[14] was deemed a test of innocence, and a painful or fraudulent experiment, supplanting a righteous award, might consign to punishment the most innocent, or save from it the most criminal of men, have ever been deemed a shocking singularity in the institutions of our barbarous ancestors. We are ready to admit the justice of this charge generally; and yet we fancy that, upon some occasions, we are enabled to discern through the dim mist of credulity and ignorance, a ray of policy that may have been derived from the dawning of a [xix/xx] rude philosophy. Trials by ordeal, as we are informed by Mr. Mill, hold a high rank in the institutes of the Hindus. It appears that there are no less than nine different modes of trial, but that by water in which an idol has been washed, and the one by rice, are those which we shall select, as well calculated to illustrate the observations which we shall venture to offer.
>
> The first of these trials consists in obliging the accused person to drink three draughts of the water in which the images of the sun and other deities have been washed; and, if within fourteen days he has any indisposition, his crime is considered as proved.

In the other species of ordeal alluded to, the persons suspected of theft are each made to chew a quantity of dried rice, and to throw it upon some leaves or bark of a tree; they from whose mouth it comes dry, or stained with blood, are deemed guilty, while, those who are capable of returning it in a pulpy form, are at once pronounced innocent.

When we reflect upon the superstitious state of these people, and at the same time, consider the influence which the mind, under such circumstances, is capable of producing upon the functions of the body, it is impossible not to admit, that the ordeals above described are capable of assisting the ends of justice, and of leading to the detection of guilt. The accused, conscious of his own innocence, will fear no ill effects from the magical potation, but will cheerfully acquiesce in the ordeal; whereas the guilty person, from the mere uneasiness and dread of his own mind, will, if narrowly watched, most probably discover some symptoms of bodily indisposition, before the expiration of the period of his probation. In the case of the ordeal by rice, a result, in correspondence with the justice of the case, may be fairly anticipated on the soundest principle of physiology. There is, perhaps, no secretion that is more immediately influenced by the passions than that of saliva. The sight of a delicious repast to a hungry man is not more effectual in exciting the salivary secretion, than is the operation of fear and anxiety in repressing and suspending it. If the reader be a medical practitioner, we refer him for an illustration to the feelings which he experienced during his examination before the medical colleges; and if he be a barrister, he may remember with what a parched lip he gave utterance to his first address to the jury. Is it then unreasonable to believe, that a person under the influence of conscious guilt, will be unable, from the dryness of his mouth, to surrender the rice in that soft state, which an innocent individual, with an undiminished supply of saliva, will so easily accomplish? – *Paris and Fonblanque.* [xx/xxi]

It was supposed, even in this country, that the agitation of a murderer, on seeing his victim, was sufficient to prove his guilt; and, in some cases, he was obliged to place his hand on the dead body. It is also a popular opinion, that the countenance of a murdered corpse will change when gazed on by the murderer, or any of his near relatives. The following extraordinary cases are noticed by Dr. Beck, from Hargrave's State Trials, vol. x. Appen. p. 29 – reign of Charles II:

An ancient and grave person, *minister to the parish were [sic] the fact was committed*, sworn to give evidence, according to custom, deposed, "that the body being taken up out of the grave, thirty days after the party's death, and lying on the grass, and the four defendants being present, were required each of them to touch the dead body. Okeman's wife fell upon her knees, and prayed God to show tokens of her innocency. The appellant did touch the dead body, whereupon the brow of the dead, which before was of a livid and carrion colour (in terminis, *the verbal expression of the witness*), began to have a dew, or gentle sweat arise on it, which increased bv degrees, till the sweat ran down in drops on the face; the brow turned to a lively and fresh colour; and the deceased opened one of her eyes, and shut it again; and this opening of the eye was done three several times. She likewise thrust out the ring or marriage finger three times, and pulled it in again, and the finger dropped blood from it on the grass." Sir Nicholas Hyde, chief-justice, seeming to doubt the evidence, asked the witness, Who saw this besides you? – Witness. I cannot swear what others saw. But, my Lord (said he), I do believe the whole company saw it, and if it had been thought a doubt, proof would have been made of it, and many would have attested with me. Then the witness, observing some admiration in the auditors, spake further, "My Lord, I am minister of the parish, and have long known all the parties, but never had occasion of displeasure against any of them, nor had to do with them, or they with me; but as I was minister, the thing was wonderful to me. But I have no interest in the matter, but as called upon to testify the truth; and this I have done." [This witness was a very reverend person, as I guessed, of about seventy years of age. His testimony was delivered gravely and temperately, to the great admiration of the auditory.] Whereupon, applying himself to the chief-justice, he said, "My Lord, my brother here present, is minister of the next parish adjacent, and I am sure saw all done that I have affirmed." Therefore, that person was also sworn to give evidence, and did depose in every point – 'the sweating of the brow, the change of the colour, thrice opening the [xxi/xxii] eye, the thrice motion of the finger, and drawing it in again:' only the first witness added, that he himself dipped his finger in the blood which came from the dead body, to examine it, and beswore he believed it was blood. I conferred afterwards with Sir Edward Powell, barrister-at-law, and others, who all concurred in the observation. And for myself, if I were upon oath, can depose that these depositions (especially the first witness) are truly reported in substance" (Ibid. p. 29).

"In the trial of Standsfield, for the murder of his father, a similar charge was brought. It is stated, that when the son was assisting in lifting the body of his father into the coffin, it bled afresh, and defiled all his hand. The opposite lawyers observe, that "this is but a superstitious observation, without any ground either in law or reason. Carpsovius says, he has seen a body bleed in the presence of one not guilty, and not bleed when the guilty were present." They assign, with great probability, as the cause of this bleeding, the fact, that the surgeons had made an incision about the neck, and that the motion of the body, in removing it, caused the fresh hemorrhage from that part. Hargrave, vol. 4, p. 283 –

I will refer those who are curious on this subject, to Metzger (p. 328); and Valentini Novellæ, App. 3. *De stillicidio sanguinis in hominis violenter occisi, cadavere conspicui, an sit sufficiens præsentis homicidæ indicium?*"

The many causes of sudden or suspicious death, could be very imperfectly discovered, even by medical practitioners, in the dark ages, when prejudices and superstition prevented the study of anatomy. It was no sooner than the year 1315, that Mondini gave the first public demonstrations on the human subject in the University of Bologna; and it was not until 1334, that human dissection was permitted in France, in the celebrated medical school of Montpellier. About this period, Frederick II. of Naples, authorized one annual dissection in his capital. This legalization of anatomy, was the revival of medical science in Europe.

The first cases in which medical witnesses were summoned to assist the judge, were after the establishment of the inquisition by Pope Innocent III. A.D. 1204, when torture was inflicted to extort confession. This mode of arriving at truth, was too favourable to screen the indolence or incapacity of the judges, not to be adopted by other tribunals. Accordingly the rack, and other modes of inflicting the extreme of suffering compatible with life, passed into the codes of almost all European nations, from which they are not yet altogether expunged; and medical men were sum- [xxii/xxiii] moned by the tribunals to superintend these horrible scenes, to mark how far fiend-like ingenuity could protract human sufferings, without allowing the victim the last refuge of human misery. Accordingly, we read in the works of the early medical jurists, observations on the mode of applying the torture in criminal trials.

It is humiliating to acknowledge, that even now a species of torture is enforced by British laws in the army and navy, and in civil cases, by means of the tread-mill, too often it is to be feared without proper discrimination of age, sex, or state of health.

But medical jurisprudence, as a science, cannot date farther back than the middle of the sixteenth century. The Emperors Rudolf, Sigismund, Albert, and Maximilian I., had successively attempted reforms of the criminal codes of Germany, by the introduction of a uniform system of legislation; but this idea was first reduced into practice by George, Bishop of Bamberg, who published, in 1507, a German penal code, drawn up and systematized by John, Baron of Schwartzemburg, which was speedily adopted in the petty States of Bayreuth and Anspach, and afterwards in Brandenburg. The other States of Germany, from narrow views of self-interest, and the jealousy of neighbouring innovations, for some time resisted the introduction of this code; and even Charles V., in the plenitude of his power, made two unsuccessful attempts to persuade the diet of Nuremberg to adopt the project of a similar code; but his perseverance finally prevailed, and the diet of Ratisbon, in 1532, proclaimed, as the law of the empire, the criminal code since well knowm by the title of *Constitutio Criminalis Carolina*. This celebrated body of jurisprudence, first published in 1553, was founded on that of Bamberg, and both were remarkable as requiring the evidence of medical men in all cases where their testimony could enlighten the judge, or assist the magistrate, as in cases of personal injuries, murders, pretended pregnancy, abortion, infanticide, hanging, drowning, poisoning, &c. This is the true era of the dawn of legal medicine; and we must regard Baron Schwartzemberg as the father of medical jurisprudence, and Germany as the country which gave it birth.

To the same nation we must attribute the glory of having produced the man who first dared to throw the shield of medical science over the victims of a barbarous and dark fanaticism.

The belief in the powers and influence of witches and sorcerers was in full force in the sixteenth century.

It is computed, that in Lorraine, nine hundred persons, of [xxiii/xxiv] both sexes, were burnt alive in fifteen years for the imputed crime of sorcery; and that, in the Electorate of Treves alone, within a few years, six thousand five hundred individuals had perished in the flames for the same imaginary crime. In various parts of Germany and France, instances of supposed demoniacal possession were perpetually occurring; and at Freidberg public prayers were ordered to assuage this dire calamity. Weiher, physician to William, Duke of Cleves, had the boldness openly to impugn these superstitious notions, in a curious work entitled, *De Præstigiis Dæmonum et Incantationibus*, printed at Basil, in 1568. He undertook to prove, that magicians and demoniacs ought to be considered as unfortunate persons, subject to hysteria and hypochondriasis, and that they should rather excite pity for their infirmities, than be obnoxious to punishment; and he ridiculed the ordinary modes of persecution to which these unhappy beings were subjected.

This attack on a popular superstition, and on a powerful and lucrative engine of clerical influence, aroused alike indignation and vengeance against the daring innovator; and Weiher was himself arrested as a magician, dragged before a tribunal, and owed his liberty and life to the fluence and earnest solicitations of his ducal patron. Weiher was too enlightened for the age in which he flourished; for we find his opinions attacked in no measured terms by two of his contemporaries, Erastus and Scribonius.[15]

The belief in diabolical illusions and witchcraft was entertained in this country so late as the seventeenth century. Sir Thomas Browne, the author of the Religio Medici, who was called on by Lord Chief Baron, Sir Matthew Hale, to give evidence in the cases of two unfortunate persons indicted for having bewitched two children, and caused them to have fits, who were tried and executed at Bury St. Edmonds; deposed, "He was clearly of opinion that the fits were natural, but heightened by the devil, co-operating with the malice of the witches, at whose instance he did the villanies," and he added, "that in Denmark there had been lately a great discovery of witches who used the very same way of inflicting persons by conveying pins into them." This relation of Sir Thomas Browne, says the historian of the case, made that good and great man, Sir Matthew Hale, doubtful; but he would not so much as sum up the evidence, but left it to the jury with prayers that the great God of heaven would direct their [xxiv/xxv] hearts in that weighty matter. The jury accordingly returned a verdict of guilty; and the executions were amongst the latest instances of the kind that disgrace the English annals.

Sir Thomas Browne was born in 1605, was a celebrated physician in London in 1646, and was knighted by Charles II.

The publication of the Caroline Code very naturally attracted the attention of the medical profession, and many of its members commenced the study of forensic medicine. The kings of France soon felt the necessity of a similar code. In 1556, Henry II. promulgated a law, by the virtue of which death was inflicted on any woman who concealed her pregnancy, and destroyed her offspring. In 1606, Henry IV. presented letters patent to his first physician, authorizing him to appoint two surgeons in every city and important town, whose exclusive duty it should be to examine all wounded or murdered persons, and make reports thereon. This law was amended by Louis XIV in 1667, who declared that no report should be received, unless sanctioned by one of these surgeons. These were named medical councillors to the king, they were often appointed by court intrigue, and became so corrupt that they were suppressed in 1790. There are still, however, district medical reporters in France, as will appear hereafter (Foderè op. cit.). In 1692, physicians were associated with surgeons.

Various detached treatises on different branches of legal medicine appear to have been produced towards the close of the sixteenth century. Ambrose Paré wrote on monstrous births, simulated diseases, and on the art of drawing up medico-legal reports, 1575. In 1598 Severin Pineace (or Pinæus) published at Paris his treatise *De notis Integritatis et Corruptionis Virginum*; a book still quoted on that subject.

The earliest systematic work on legal medicine is unquestionably the book *De Relationibus Medicorum* of Fortunato Fidele, published in 1598 in Sicily. As might be expected in his age and country, the opinions of this physician are greatly warped by his servile deference to the dicta of the canon law, and his submission to its clerical expounders. It consists of four books, of which the following may be received as an outline of the contents, viz. I. On Public Food; the Salubrity of the Air; Pestilence. II. Wounds; Pretended Diseases; Torture; Injuries of the Muscles; Medical Errors. III. Virginity; Impotence; Hereditary Diseases; Pregnancy; Moles; the Vitality of the Fœtus; On Birth; Monsters. IV. Life and Death; Mortality of Wounds; Suffocation; Death by Lightning and Poisoning. [xxv/xxvi]

The rapid progress of anatomy in the commencement of the seventeenth century, by the labours of Sylvius, Vesalius, Fallopius, and Eustachius, had a surprising influence on the progress of medico-legal investigations, which became speedily apparent in the publication of the great and invaluable *Quæstiones Medico Legales* of Paulo Zacchia; which appeared in successive volumes from 1621 to 1635.

Zacchia was one of the most eminent physicians of Italy. His celebrity brought a vast body of facts on medical jurisprudence under his cognizance; and his work, which has gone through many editions, is still regarded as the great magazine of medico-legal lore, illustrated by immense erudition, and by a subtile and refined discrimination. Though Zacchia be tinctured with the superstition of his age, in what relates to demoniacal possession, and some other subjects, yet his judicious remarks on the gross injustice and impropriety of classing as demoniacs those females in whom suppressed catamenia had produced mania, or men in whom frenzy had originated in a melancholic temperament, place his physiological reasoning and his medical indications in a favourable point of view; and, perhaps, we ought to place his recommendation in such cases, not to neglect the prayers of the church, while suitable remedies are prescribed, as much to his anxiety to avoid offence, as to his faith in the efficacy of the former.

With the imperfections that were unavoidable in an age in which physiology was in its infancy, chemistry imperfect, and ere our illustrious countryman Harvey had demonstrated the circulation of the blood, the work of Zacchia will ever be considered as one of the landmarks of medical jurisprudence, and remain a stupendous monument of his learning and sagacity. It was written at different times; abounds in repetitions, and occasionally in contradictions; circumstances which render it better suited for the consultation of the learned than the information of the student.

The following is an outline of the contents of the first volume. *First Book*: Age; Legitimacy; Pregnancy; Superfœtation and Moles; Death during Delivery; Resemblance of Children to their Parents. *Second Book*: Dementia; Poisoning. *Third Book*: Impotence; Feigned Diseases; The Plague and Contagion. *Fourth Book*: Miracles; Rape. *Fifth Book*: Fasting; Wounds; Mutilation; Salubrity of Air, &c. The second volume is filled with casuistical questions. This work is highly esteemed as one of great authority. Zacutus Lusitanus, in alluding to its value, exclaims: emi, vidi, legi, obstupui.[16] It is justly considered the most celebrated of all the Italian works on the subject. [xxvi/xxvii]

The next important step in the history of our science is due to Harvey, who investigated the difference between the lungs of the fœtus and of an infant that had breathed, and pointed out how this fact might be available in alleged cases of infanticide. It should, however, be admitted, that long before, the difference in colour, consistence, and weight between the fœtal and adult lungs of animals, had been remarked by Galen, but who had not applied it to any practical use.

Almost at the same time appeared two valuable treatises by Melchior Sebiz, at Strasburg, viz. *De Notis Virginitatis* (1630) and *Examen Vulnerum* (1638). In the first he maintained that the existence of the hymen was the indisputable mark of virginity; an inference which was warmly denied by Orazio Augenio, and defended by the celebrated Pietro Gassendi (*Opera* VI.), &c. In the second he drew an important distinction between wounds necessarily fatal, and those which become so incidentally.

We must give the next important step to the Danish physician, Thomas Bartholin, who carefully investigated the period of human utero-gestation; and, having confirmed the opinions of Galen and Harvey respecting the difference between the lungs before and after the first respiration, proposed the hydrostatic test for solving the question, and pointed out the best method of rendering it available in medico-legal investigations (*De Pulmonum substantia et Motu*. Hafniæ 1663). The rationale of this process was more fully explained by Swammerdam in a tract, published at Leyden, in 1677 – and its first practical application was made in 1682, by Jan Schreyer, after it had been investigated by Thurston and Carl Rayger.

This process is the celebrated *Docimasia Pulmonum*; which, till within a comparatively recent period, was considered as an irrefragable test of the question whether or not the child had breathed; but which has given rise to a keen controversy which I shall have afterwards an opportunity of examining. See *Infanticide*. About this period the work of Ambrose Paré, published 1575, which was regarded for nearly a century the only standard authority in France, was superseded by the more comprehensive treatises of Gendri, of Augers, in 1650, of Blegni, of Lyons, in 1684, and of Devereux, of Paris, in 1693. Louis is, however, with great justice, considered as the first who promulgated a just idea of the science to his countrymen. He flourished about the middle of the last century, and will be noticed in the history of that period.

While these important steps were in progress, Germany, the country where medical jurisprudence took its rise, had [xxvii/xxviii] instituted public prelections on this important branch of medical and forsenic education. About the middle of the seventeenth century, Michaëlis gave the first lectures on this subject at Leipsic, which were followed by those of Bohn, Professor of Anatomy and Surgery in that city, who, before the end of that century, had published his valuable works *De Vulnerum Renunciatione*, and *Dissertationes Medicinæ Forensis* (1689), and these were speedily followed by a tract *De officio Medici duplici, Clinico et Forensi* (1704). In the same period appeared the interesting investigations of Gottfried Welsch, and of P. Amann, on the fatality of wounds.[17] The celebrated work of Licetus, *De Monstris,* appeared in 1669 at Amsterdam; and it would be unpardonable to omit the vast accessions to medico-legal knowledge unfolded in that noble pathological collection, the *Sepulchretum* of Bonnet. (Lugd. 1700, 3 tom. folio). The mode of conducting medico-legal investigations obtained some attention in France during the seventeenth century, but its institutions were extremely imperfect compared to those of Germany even to a much later period.

The following were among the German authors of the eighteenth century: John Bohn, De Renunciatione Vulnerum, 1689, 4to. Amsterdam. Valentini, Pandectæ Medico-Legales, 4to. Francof. 1702. Fred. Boerner, Prof. Med. Wirtemburg, 1723.

Several Dissertations. Kannegeiser, Inst. Med. Leg. Michael Alberti, Prof. Med. Hall. – Systema Jurisprudentiæ Medicæ Schneeberg, 4to. 1725. tom. vi. Zittman, Medicina Forensis, 4to. Francofurti. Richter, Decisiones Medico-Forenses. Teichmeyer, Institutiones Med. Leg. 4to. Jenæ 1740. Stark, De Medicinæ Utilitate in Jurisprudentia, 4to. Helmont, 1730. Hebenstreit, Anthropologia Forensis, 8vo. Lipsiæ, 1753. Ludwig. Institutiones Medicinæ Forensis. Fazellius, Elementa Medicinæ Forensis.

The first German work of authority was Bohn's (1689), in which he attempted to discriminate what wounds were necessarily fatal. The next was Valentini's *Pandects*, 1702. Michele Bernardo Valentini published his *Pandectæ Medico-legales* in 1702 (Francofurte), and his *Novellæ* in 1711, which were incorporated, in 1722, into his *Corpus Juris Medico-legale*. This great work contains an excellent review of all that had been done before his time, and is considered by some as equal in value to the work of Zacchia. Valentini was fully aware of the importance of his subject, and strenuously insists, [xxviii/xxix] in the preface, on the necessity of cultivating this branch of medical knowledge. Several Professorships for teaching juridical medicine were about this period founded in the German Universities; and very useful though sometimes too diffuse treatises on the subject issued from the German press. The number of successive authors becomes now so great, that I cannot, in this sketch, attempt to mention all, far less to characterise them.

Zittman, Boerner, Kannegeiser, and Teichmeyer, each published systems of various, yet conspicuous merit. The *Institutiones Medicincæ Legalis* of the latter, long formed the manual of the student and the text-book of the professor. The clear and forcible reasoning of Stoerck (*De Medicinæ Utilitate in Jurisprudentia*, Helmont 1730) vindicated the high importance of this branch of knowledge. This author very ably advocated the importance of medical knowledge in legal investigations. But the *Systema* of Alberti, the professor of medical jurisprudence in the University of Halle, in six quarto volumes, was the most laborious and complete work of that century. The writings of this learned person are obscured by his devoted attachment to the mysticism of the Stahlian school; yet the industry with which he has collected an immense body of facts, renders his work a precious mine of medico-legal information. His example appears to have stimulated other physicians to publish collections of cases; and curious additions were made to our knowledge by Loesve, Reichter, Budæus, Fritch and Wolff, Hermann, Clauder, Herzog, and Parmeon, which are particularly valuable to those who may be called to exercise the profession in Germany.

The next author of celebrity was Plenck, who published his "Elementa Medicinæ et Chirurgiæ Forensis, ["] 1781. Then followed the lectures of the illustrious Haller, on Juridical Medicine, which were based on the Institutes of Teichmeyer, but greatly amplified and improved. These celebrated lectures were afterwards published in three volumes, octavo, entitled, "Vorlesungen über die gerichtliche Arzneiwissenchaft, ["] 1782. Daniel published his Bibliotheca of State Medicine at Halle, 1784, entitled "Bibliothek der Staatsarzneykunde." Several other important German works appeared towards the close of the eighteenth century, some illustrative of particular questions of legal medicine, and some elementary productions which deserve notice. Among these are Sikora's "Conspectus Medicinæ Legalis,"

Pragæ, 1792; Loder's first lines, "Anfangsgründe der Medicinischen Anthropologie und der Staatsarzneykunde," 1793; Metzer's [xxix/xxx] system, "System der Gerichtlichen Arzneiwissenschaft," Koningsb., 1793; Latin, by Keup, 1794; Muller's Delineations, "Entwurf der Gerichtlichen Arzneiwissenchaft," Francf.

Those who may be desirous of obtaining the titles of the German works or essays, omitted in this sketch, most of which refer to wounds, poisoning, and infanticide, will find the necessary information in the Bibliotheque Medicale of Plouquet, in the works of Struvius and Goelicke, and in the "Collectio Opusculorum Selectorum ad Medicinam forensam Spectantium, Curante. Dort," F. C. T. Schlegel, Leipsiæ, 1785–1800. Dr. Beck gives a list of the titles inserted by Schlegel; but these, and a vast number of others, will be found in the last works to which I have referred, and these are in the possession of most medical practitioners throughout the world – The Cyclopædia of Practical Medicine, London, 1834; and Forbes' Medical Bibliography, 1835.

The eighteenth century commenced with happier prospects for this branch of science, and the press teemed with important works on legal medicine. As early as 1,700 we find an admirable treatise on the diseases of artificers by Bernardino Ramazzini, of Padua. His attention had been called to this subject by observing how frequently a fatal asphyxia occurred to persons employed in clearing sewers and privies; and this accomplished man extended his inquiries to the more remarkable diseases of artizans of various denominations. Ramazzini's work, *De Morbis Artificum Diatriba* is still a standard work, and has been commented on by several writers even so late as 1822. It has also been translated into many languages, and is a standard work in all civilized countries.[18]

During the last century little had been done, in Italy, towards the advancement of medical jurisprudence or police. Beccaria's work appeared in 1749, entitled "Scriptura Medico-legalis," and soon after Bonni's "Instituzioni Teorico Pratiche di Chirurgia. ["]

The science had made little progress in France until after the middle of the eighteenth century. In 1788 Professor Louis commenced the publication of his Memoirs on the Certainty of the Signs of Death, on drowning, on the means of distinguishing suicide from homicide in cases of hanging. [xxx/xxxi] His consultations on the cases of Calas, Montbaillet, Syrven, Cassagneux, and Baronet, which are preserved in the Causes Cèlébres, are models of medico-legal reasoning, and exalt him as a jurist. He also examined, most minutely, the signs of real and pretended pregnancy, and was the first who publicly taught jurisprudence in France.

The Courts of France had decided for the legitimacy of several persons born twelve, and even thirteen months, after the death of their putative father. The famous case of Villeblanche, born 320 days after such an event, called forth the Mémoire of the celebrated Louis, President of the Royal College of Surgeons in Paris. In this dissertation, Louis attacked the legitimacy of those pretended cases of retarded birth with powerful arguments, in which he was seconded by Astruc and Bouvart; but was vehemently opposed by Le Bas and Antoine Petit. This gave rise to very able publications on both sides, in which Bertin Pouteau and Vogel bore a part; but the victory remained with Louis and his adherents.

Winslow discussed the moral, political, and religious relations of the Cæsarean Operation; Lorry investigated the subject of survivorship with success. Salin examined the question of poisoning, and undertook to prove that one Lamotte, who had been buried for 67 days, had died from the effects of corrosive sublimate.[19] This inquiry led to important results. About this period Lafosse endeavoured to distinguish the phenomena caused by death, from the effects of injuries inflicted on the living body. He also described the positive signs of pregnancy and parturition.

In the year 1789 Professor Chaussier read his excellent Memoir before the Academy of Dijon, on the great irnportance of juridical medicine, and in it he treated of several important forensic questions. In it he enforced the necessity of the careful personal examination, by the medical man, in all cases of violence, as blows and wounds; and pointed out the precautions he should adopt in such visits. He gave admirable models of reports, and shewed the attention which was necessary to arrive at truth: and to enforce his precepts still more, next year he delivered a full course of lectures on legal medicine to numerous pupils. Towards the close of the last century the preceding authors, in conjunction with Professor Mahon, compiled the elaborate articles on Medical Jurisprudence in the Encyclopoediæ Methodique. Such were the materials, says Foderè, which enabled me to publish my first systematic work in 1796. The subject now claimed the [xxxi/xxxii] attention of the existing government, and became generally patronised.

In 1796, Foderè published the first edition of his celebrated work, in three octavo volumes, entitled, "Les lois eclairèes par les Sciences Physiques, ou Traite de Medicine Legale, et d'Hygiene publique." This learned professor was the author of many other works of high reputation.

In the present century, France has taken the lead in this science. The great work of Foderè was soon followed by the establishment of a professorship of forensic medicine in Paris, another at Strasburg, and a third at Montpellier. The first was conferred on Mahon, who obtained great celebrity as a teacher; but the posthumous publication of his lectures has not confirmed his reputation. The arrangement of the whole is defective, and the matter is very slovenly put together, being in a great measure made up of excerpts from French authors, and translations from German writers. It is no doubt, the duty of a public teacher to collect his facts from all sources, and it is his duty rather to aim at giving a fair view of the present state of his subject than to affect originality; but those who take the trouble of examining the small volume of Mahon, will be surprised how little pains he has taken to clothe the facts in his own words.

The chair of Montpellier was conferred on René, whose advanced age rendered him little able to perform its duties.

The first Professor at Strasburg was Noel, who is said to have left interesting manuscript lectures, which, I believe, have not been published.

Several valuable works appeared in the early part of the present century. Vigné, of Rouen, published his valuable production in 1805. In 1807, the posthumous work of Professor Mahon was edited by Fautrel. *Medicine Legale et Police Medicale de P. A. O. Mahon,* Prof. de Med. Leg. &c. avec quelques notes de Fautrel. About this date, Belloc, a surgeon at Agen, published a small but sensible treatise, *Cours de Medicine Legale, Theorique et Pratique,* de J. J. Belloc, Chirurgien à Agen.

In 1808, Marc translated from the German, the Manuel of Rose on Medico-legal Dissection, and enriched it with many original observations; to which he also subjoined two instructive dissertations – one on *the docimasia pulmonum*, and the other on *death by drowning. Manuel d'Autopsie cadaverique Medico-Legale, &c.* In 1812, Ballard translated from the German, of Metzger's *Principles of Legal Medicine,* already noticed, and his notes are replete with information.

In 1813, Professor Foderè enlarged his work, which may [xxxii/xxxiii] be considered a new one, in six octavo volumes; four on legal medicine, and two on medical police. The renowned author divides his work into three parts; viz. the *First*, comprehending subjects of a mixed nature, or those, which admit of application to civil as well as criminal cases, "*Medecine Legale mixte*." The *second* exclusively relating to criminal jurisprudence, "*Medecine Legale Criminelle*;" and the *third,* to medical police, "*Medecine Legale Sanitaire*."

The work opens with a learned introduction, in which the importance of the science is fairly examined, and its history pursued with much detail, from its origin, to the period at which the author wrote. The qualifications of the forensic physician are also considered, and the different circumstances opposed to the success of his labours, enumerated and appreciated. Then follow in succession the subjects of the first division, viz. the different ages of human life, puberty, minority, majority, with the anomalies to which the natural growth and development of the body are liable. Personal identity and resemblance. The relative and absolute duration of life. The grounds of prohibition in testatorship, such as habitual, periodical, and temporary insanity; suicide; deaf and dumb state; somnambulism; intoxication. The qualifications of testators and witnesses. Marriage and divorce. Pregnancy, true and false. Parturition, and the signs denoting the death of the fœtus in utero. Paternity and filiation. Premature and retarded births. Monsters. Hermaphrodites. Survivorship. Signs of real and apparent death. Treatment of the different varieties of Asphyxia. Certificates of exemption, and diseases which exempt. Feigned, dissimulated, and imputed maladies.

The *second* division commences with the third volume, and includes, in their respective order, chapters on the examination of bodies found dead. The distinction of assassination from suicide. Wounds. Poisoning. Rape. Abortion. Concealment and substitution of the offspring; and Infanticide.

The *third* division, with which the fifth volume commences, successively treats of the preservation of the human species, and of the means of remedying its physical degeneracy. Contagions, hereditary, and epidemic diseases, and the precautions to be adopted against them. The medical police of cities, with regard to aliment, arts, manufactures, and attention to the sick. Military and naval hygiene; and, lastly, the medical police of hospitals and prisons.

This is universally allowed to be the most valuable systematic treatise in the French language; it evinces great research, learning, judgment, and ability; but it is not calcu- [xxxiii/xxxiv] lated to instruct the British jurist. "It is unnecessarily prolix and minute, and is adapted only to the judicial courts of the continent." In 1817, Bertrand published his valuable manual. It is not calculated for British practice in courts of justice, though it is replete with scientific information.

The next work of great value which has appeared in France since 1800, is the system of Toxicology of M. Orfila. *Toxicologie Générale considérée, sous les Rapports de la Physiologie, de la Pathologie, et de la Medicine Legale*; and another—*Leçons faisant Partie du Cours de Medicine Legale, de M. Orfila*, 1821. This celebrated Professor has published many other valuable works, which are laid under contribution in the following pages. His Traité des Exhumations, &c. 1833, is largely quoted in the article Putrefaction, in this work. The author is now the most celebrated medical jurist in existence. In 1821, Professor Capuron published his admirable work relating to all obstetric inquiries. – *La Medecine Legale, relative à l'art des Accouchemens, par J. Capuron,* Doct. en Med., Professor des Accouchemens, &c. This production contains every thing valuable in the branch of medicine, of which it treats.

The various essays in the four voluminous French dictionaries of Medicine, and in the many periodicals of France, abound with medico-legal information. The treatises on infanticide, by Lecieux, on the proper manner of examining bodies for legal inquiries, by Renard; on the perforation of the stomach, by Laisné; on ecchymosis and sugillation, by Rieux, published in one volume in Paris, in 1819, ought to be in the possession of every one engaged in the practice of medicine. The works of Briand, 1828 – *Manuel complet de Medecine Legale, &c.,* and of Sedillot, 1830, contain all the valuable conclusions of French jurists, on legal or forensic medicine. The treatise of M. Devergie, just published, January, 1836, ought to be in every medical library.

In Germany, this century has produced the excellent compends of Schmid and Müller (Landshut 1804), Metzger (Kœnigsberg, 1805), Masiers (at Rostock, 1810), and of Wilberg (Berlin, 1812.) Rose published a very valuable tract on Medico-legal dissection. The useful Toxicologia of Plenck appeared in one volume, at Vienna, in 1801. An useful compend on Pharmaco-chemical Medical Police, which forms the best treatise we have on the preparations and adulterations of food – on the proper kinds of culinary vessels – on the venom of snakes, insects, &c. and the police of apotheca- [xxxiv/xxxv] ries' shops, was published by Professor Remer of Kœnigsberg in 1811, and was translated into French by Lagrange and Vogel, in 1816.

During this century, Italy has contributed some admirable treatises. That of Tortosa is the most scientific and elaborate of his country in the present age. It was written under the sanction of Caldani Franck, of Pavia, and Plouquet of Turin. It only embraces forensic medicine, and is entitled – *Istituzioni di Medicina Forense, di Guiseppe Tortosa,* Professore Medico della Commissione Dipartimentale di Sanita del Bacchiglione, 1809. It is divided into three parts, viz. (1) Comprehending all the principal objects of *Ecclesiastical* jurisdiction. (2) Subjects relating to the *Civil* courts. (3) Those which relate to the *Criminal* courts. The subdivisions of each part are arranged in the following order. Part I. Conjugal Impotence – Conjugal Rites – Monstrous Births – Hermaphrodites – Magic – of Persons possessed of Spirits – Miracles – Ecclesiastical Fasting. Part II. Age – Pregnancy – Birth – Superfœtation – Cæsarean Operation – Simulated and Dissimulated Diseases. Part III. Of Deflowering – Sodomy – Torture – Legal Examination of Wounds, and Dead Bodies – Poisoning – Infanticide – Homicide by wounding – Fœticide – Accidental Death.

Another valuable work is by Barzelotti, entitled *Medicina Legale*, in the form of questions, which embraces the most important points of medical jurisprudence. The best edition of it is by Rossi of Bologna, which was published in 1823.

During the present century, America has contributed to advance the Science of Medical Jurisprudence. In 1823, Dr. Theodoric Romeyn Beck published his *Elements of Medical Jurisprudence*, of which two editions have since appeared in this country. This work is a monument of research, industry, and talent. The author confines himself to Medical Jurisprudence or Forensic Medicine, and excludes State Medicine, or Medical Police. The work consists of 19 chapters: I. Feigned diseases; II. Disqualifying diseases; III. Impotence and Sterility; IV. Doubtful sex; V. Rape; VI. Pregnancy; VII. Delivery; VIII. Infanticide; IX. Legitimacy; X. Presumption of Survivorship; XI. Age and Identity; XII. Mental Alienation; XIII. Persons found dead; XIV. Wounds on the Living Body; XV. Poisons; XVI. and XVII. Mineral Poisons; XVIII. Vegetable Poisons; XIX. Animal Poisons. There is also an Appendix. All the subjects treated by Dr. Beck, are considered most minutely and generally illustrated by numerous cases which [xxxv/xxxvi] occurred in courts of justice. Dr. Duncan, jun. and Dr. Male, two of our best medical jurists, have declared the work as one of the best hitherto published in this or in any other country. I fully agree with them in this opinion, and have accordingly laid it under great contribution in arranging the following pages. So far as it extends, there is no work in the English language equal to it for laborious research and scientific information. It is, however, confined to forensic medicine, and necessarily omits a vast number of questions on State Medicine, a knowledge of which is indispensably necessary to the medical practitioner; and which the justly celebrated author intended to elucidate in another work on Medical Police. The present production is an admirable system of forensic medicine, which, as Dr. Duncan observed, "embraces almost every valuable fact relating to it. Each of its diversified departments has been investigated so minutely, that few cases can occur in practice, on which it will be necessary to seek elsewhere for farther information." This criticism is fair, candid, and just; but only refers to one department of State Medicine, and not to the whole. As a book of reference and authority, and as a report of medico-legal decisions, the work is unequalled in our language. It ought to have a place in every medical library. It is not, however, a perfect production. It contains a vast number of judicial opinions and law cases, of little or no interest to the great mass of medical practitioners in this country. It omits many subjects of interest to the profession, such as Medical Ethics, the laws relating to the education and Practice of the Medical Profession, and the laws relating to Public Health. These, it is true, are subjects of Medical Police or Public Medicine, but they are as deeply important to all engaged in the practice of Medicine, as pure forensic questions. This being my firm conviction, I have accordingly introduced them in this work.

Such is the history of medical jurisprudence and state medicine, from the earliest age to the present period, except as regards this kingdom, and this I shall now introduce.

It is a matter of no small surprise, that this science has scarcely been patronised in this country. Britain, which has so distinguished itself in almost every science; which has made legislation its study and its boast, has, hitherto, neglected a branch of medicine, so highly cultivated, in all other civilized nations. In proof of this statement I must observe, that the science of medical jurisprudence formed no part of medical education in this country until the present century; nor was there a single work published on the subject before the last quarter of the past century. Many medico-legal [xxxvi/xxxvii] questions were discussed in our periodicals, but no work appeared in the United Kingdom, expressly treating of the subject, previous to the small and imperfect production of the first Professor Dease, of Dublin, the author of the work on Injuries of the Head, which was publisbed in 1783, entitled, "Remarks on Medical Jurisprudence," dedicated to Lord Clonmel, Chief Justice of Ireland. "This was the first attempt to write on medical jurisprudence in our native language."[20]

The next in succession was also an imperfect production, by Dr. Samuel Farr, in 1788, entitled, "Elements of *Medical Jurisprudence*," and was an abstract of the work of Faselius, already noticed. In or about this period the first Professor Duncan, of Edinburgh, delivered a private course of lectures on the subject, and was chiefly instrumental in directing the attention of the profession to its importance. In 1808 Dr. Robertson published a *Treatise on Medical Police,* in two volumes. In 1815, Dr. Bartley, of Bristol, published a most imperfect work, entitled, a *Treatise on Forensic Medicine*. The first respectable and original production that this country contributed to medical jurisprudence, was "*An Epitome of Juridical or Forensic Medicine, for the Use of Medical Men, Coroners, and Barristers,*" *by Dr. Male, of Birmingham,* who is justly considered the father of the science in Great Britain and Ireland. This work was published in 1816. It embraces the following subjects: Poisons, wounds and contusions, infanticide, pregnancy, abortion and concealed birth, pretended delivery, rape, hanging and strangulation, drowning, dangerous inebriety, insanity, pretended diseases, imputed diseases, apparent death, impotence, hermaphrodites. These subjects are described in 1,999 octavo pages.

It is true, that there are many valuable discussions on various medico-legal questions in the writings of Mead, Monro, the Hunters, Denman, Percival, and others; but these eminent individuals have no claim to rank among our authors on medical jurisprudence. Dr. John Gregory's Duties and Qualifications of a Physician, and Dr. Percival's Medical Ethics, contain much valuable information on the moral duties of the profession, and which naturally comprise the rules that ought to guide them in public and private practice, and consequently in state and forensic inquiries. There are also admirable essays on certain medico-legal questions, such as Dr. Hunter's "On the Uncertainty of the Signs of [xxxvii/xxxviii] Murder in the Case of Bastard Children," 1783. Dr. John Johnstone's essay on *Medical Jurisprudence in Cases of Madness*, 1800; Dr. Haslam's admirable essay on Medical Jurisprudence, as relates to Insanity according to the Law of England, 1817; re-published with other Tracts by Dr. Cooper, Philadelphia, 1819; and Dr. Hutchinson's elaborate "*Dissertation on Infanticide*," 1819.

The next, and by far the most comprehensive work that had appeared in this country, to the date last-mentioned, was Dr. Gordon Smith's, entitled, Principles of Forensic Medicine, systematically arranged, and applied to British Practice, 1821. This work rapidly passed through three editions. Its contents were arranged as follows: (1) Questions relating to the extinction of life. (2) Questions arising from personal injuries not involving a fatal issue. (3) Disqualifications for the discharge of offices, or the exercise of social functions. (4) Miscellaneous questions. This production was most favourably received by the profession. The author commenced a course of lectures, and a few followed his example. In 1823, the conjoint treatise of Dr. Paris and J. S. M. Fonblanque, Esq., Barrister at Law, was published in three volumes, octavo, entitled, "*Medical Jurisprudence*."

In 1825, Dr. Gordon Smith published his "Analysis of Medical Evidence," in 1827 Dr. Lyall his little volume "on the Duration of Pregnancy;" in 1829 Dr. Gordon Smith his "Hints for the Examination of Medical Witnesses," and the same year, Mr. Forsyth his "Synopsis of Modern Medical Jurisprudence, Anatomically, Physiologically, and Forensically illustrated, for the Faculty of Medicine, Magistrates, Lawyers, Coroners, and Jurymen."

Towards the end of 1831, the first edition of this work appeared, entitled, "A Manual of Medical Jurisprudence, compiled from the best Medical and Legal works, comprising an account of: I. The Ethics of the Medical Profession; II The Charters and Statutes relating to the Faculty; and, III, All Medico-legal Questions, with the latest decisions: being an analysis of a course of lectures annually delivered in London, and intended as a compendium for the use of Barristers, Solicitors, Magistrates, Coroners, and Medical Practitioners."[21] This work contained an account of medical ethics, the laws relating to the medical profession, and all forensic subjects noticed by preceding writers. It was very favourably received by the public as well as the medical press, and republished by Professor Griffith in America, as will appear by the notices appended to the present edition.

The last work, before the present edition of this, published [xxxviii/xxxix] in England, is by Mr. Chitty, the celebrated and voluminous legal author, entitled, "A Practical Treatise on Medical Jurisprudence, with so much of the Anatomy, Physiology, Pathology, and the Practice of Medicine and Surgery, as are essential to be known by Members of Parliament, Lawyers, Coroners, Magistrates, Officers in the Army and Navy; and Private Gentlemen, and all the Laws relating to Medical Practitioners, with explanatory Plates, Part 1. ["]

I feel it due to the Profession to which I have the honour to belong, to state, that the present part of his work is almost entirely medical, anatomical, physiological and pathological, with some legal references; all admirably arranged and executed, and most instructive to non-medical readers. It is a valuable treatise, and one well calculated to interest the legislature, the legal profession, the magistracy, and private individuals, and cannot fail to advance the interests of medical jurisprudence and state medicine.

In the preceding sketch of the history of MEDICAL JURISPRUDENCE AND STATE MEDICINE, I have necessarily confined myself to our monographic and systematic writers on the subject; and I deemed it a digression to allude to all essayists, and to those who have elucidated certain parts of it. Among these are the late Professor Duncan, of Edinburgh, the first public lecturer on medical jurisprudence in this country, whose valuable essays and reviews in *the Edinburgh Medical and Surgical Journal*, have contributed largely to advance the science in this country. His successor, Professor Christison, has given us a work on Toxicology, that will transmit his name to the latest posterity, as one of the most laborious, erudite, and experienced medical jurists that this or any country has produced. This work appeared in 1829, entitled, "*A Treatise on Poisons, in relation to Medical Jurisprudence, Physiology, and practice of Physic,*" the third edition of which is lately published.

The medico-legal writers in the *Cyclopædia of Practical Medicine*, and in our numerous medical periodicals, have also elucidated many questions of forensic medicine, and are entitled to much praise for their contributions. I shall briefly allude to the writers in the *Cyclopædia of Medicine* in alphabetical order: Dr. Apjohn, of Dublin, on Spontaneous Combustion and Toxicology; Dr. Arrowsmith, of Coventry, On Infanticide; Dr. Beatty, of Dublin, on Impotence, Persons found Dead, Rape, Doubtful Sex, Survivorship, and Death after Wounds; and Dr. Montgomery, of Dublin, on Personal Identity, on the Signs of Pregnancy and Delivery, and on Legitimacy. [xxxix/xl]

There are numerous other recent writers in the French Medical Dictionaries, Dictionnaire de Medecine et de Chirurgie Pratiques, and the Dictionnaire de Medecine, in the American Cyclopædia of the Medical Sciences, and in the celebrated German Dictionary of Medicine, Encyclopädisches Wörterbuch der Medicinischen Wissenchaft, now in course of publication. The Traité des Exhumations Juridiques, and the Leçons de Chimie appliquée à la Medicine Pratique et à la Medicine Legale, by M. Orfila, 1836, and Nouvelles Rescherches, sur les Secours à donner aux Noyés et Asphyxiés, by M. Marc, 1836, are valuable additions to medico-legal literature.

Notes

1. *Author's note*: Traité de Medecine Legale, et d'hygeine Publique ou de Police de Santé adapté aux Codes de l'Empire Francais, &c. Par F. E. Foderè Docteur in Med. Professeur de Med. Legal. et de Police Medicale al a Faculté de Medicine de Strasbourg. T. vi. 1813.
2. *Editors' note*: A full translation of the title is: "The State Physician: Treatise concerning the office of State Physician in four separate books in which not only are good, medical mores and virtues described, but even deceptions of evil and deceit are revealed: truly also a new argument is most precisely proposed by another method both useful and agreeable. The full work being fit for a physician, student of illness, attendant upon the ill, and by all others of letters, and also by supporters of state education."
3. *Author's note*: Hippocrates flourished and wrote before Christ, 460 years.
4. *Editors' note*: "[O]r the lake, for a long time infertile and fit for oars, now maintains its neighboring cities and feels the heavy plow." Horace, "The Art of Poetry: To the Pisos," in *The Works of Horace*, trans. C. Smart. New York: Harper & Brothers, 1863.
5. *Author's note*: Leviticus ch. x. Numbers ch. vi.
6. *Author's note*: The present law of this country awards a different punishment against the perpetrators of abortion before and after quickening. See INFANTICIDE.
7. *Editors' note*: "concerning reducing death (royal law)."
8. *Author's note*: Pliny, Hist. Nat. L. vii.
9. *Editors' note*: "according to the authority of the most learned Hippocrates."

10. *Author's note*: See INFANTICIDE.
11. *Author's note*: Foderè. Introduction, p. xiv.
12. *Author's note*: Foderè, op.cit.
13. *Author's note*: Hospitals were first established in Jerusalem by Eudocia, wife of Theodosius, A.D. 440; but medical practitioners did not attend their inmates.
14. *Author's note*: Priests were among the earlier chemists, and it is asserted, that they frequently instructed the accused, either from a conviction of his innocence, or from less disinterested motives, in some of those means of resisting the action of fire, by which modern jugglers are still enabled to amuse and astonish the vulgar.
15. *Author's note*: Erasti Disputatio de Lamiis. *Basil*, 1572. 4to. Scribonius de Sagarum Natura et Potestate, *Helmstadt*, 1584. 4to.
16. *Editors' note*: "Buy! Look! Read! Be amazed!"
17. *Author's note*: Welschii Rationale Vulnerum Letbalium judicium, 1660. Ammanni Irenicum Numæ Pompilii c. Hippocrate, 1698. – Praxes Vulerum Lethalium, 1701.
18. *Author's note*: This production had led the late Mr. Thackrah to publish his instructive and valuable work on the same subject, which passed through two impressions in a year. (The Effects of Arts, Trades, and Professions, and of Civic States and Habits of Living, on Health and Longevity: with suggestions for the removal of many of the Agents which produce Disease and shorten the duration of Life. Second Edition, 1832.)
19. *Author's note*: Recueil Periodiqtie de la Sociéte de Med. T. vii.
20. *Author's note*: Dr. Gordon Smith's Analysis of Medical Evidence, 1825. Appendix, p. 181.
21. *Editors' note*: The inconsistent punctuation in the title is taken directly from the original.

Chapter I
Origin of Medicine – Veneration for Its Cultivators

All medicine is derived from God, and without his will it cannot exist or be practised. Hence the healing art, if disunited from religion, would be impious or nothing. Illness requires us to implore the Deity for assistance and relief, and humbles human pride. The seeds of the art, the wonderful cures, and the powers of remedies, are in the hand of God. He has beneficently supplied various remedies, and pronounces with our tongues, the fate, life and death, of man. Whence, we see the dignity of medicine, and what reverence is due to the Divine Author of it. Sacred history confirms this sentiment, "Every cure is from God." "The Most High created medicines out of the earth." Every thing we enjoy are the gifts of God: none but the impious ever doubted this truth; none but fools dared to deny it.

It is recorded that Jesus, the son of Sirach, was one of the first who attributed the origin of medicine to the Deity. And we also read in Scripture, "Honour the physician for the need thou hast of him, for the Lord hath created him."

"All medicine is from God, and without him it cannot exist or prosper; our art, disunited from religion, is either impious or nothing."[1] Such is the first precept in the moral statutes of the Italian Universities, and it is that of Roderic a Castro, in his Medico-Politica, and of the profession in all countries.

The fate of the sick and the success of medicines are in [3/4] the hand of the Deity: "in him we live, move, and have our being;" and our curative means, and our knowledge of the nature and treatment of diseases, are subservient to his divine will and pleasure. He works with his own, and not with our hands; his power begins where ours ceases.[2] Without him man can do nothing good for himself or his species. His first duty is, therefore, due to the Creator and Conservator of the universe. In the first and unerring history of the world, we find that human aid was given during parturition by the midwives of Egypt, which proves the early practice of obstetric medicine. Some even contend that Adam was the first obstetrician. We likewise discover, that Solomon wrote very largely on the nature of animals and plants, from "the hyssop that grows out of the wall to the cedar of Lebanon;" and we are informed by profane authors of his time, that his writings were copied by the Greeks, Arabians, and finally by the Romans. Again we find, that Joseph commanded the physicians to embalm the body of his father (Gen. I. A.C. 1769). Some have pretended that Moses was initiated into the medicine of the Egyptians, and that to them he was indebted for a knowledge of the precepts he laid down in his writings.

H. Brody et al. (eds.), *Michael Ryan's Writings on Medical Ethics*,
Philosophy and Medicine, Vol 105,

The antiquity of medicine is proved by reference to the sacred volume, in which we find many of its precepts inculcated, the natural history of animals and plants recorded, diseases and their remedies described, and hygienic rules laid down for the preservation of health and prevention of human infirmities.[3]

The instinct of man and animals directs them to select what is salutary both in health and disease, and to avoid that which is pernicious. It is recorded, that the first of our species was, soon after his creation, rendered liable to diseases; and as he was also gifted with the knowledge of the nature of all the things that surrounded him, he was necessarily apprized of their noxious and sanative properties. Medicine was, therefore, almost coeval with man; and must have occupied his [4/5] attention in the first days of his existence. The vicissitudes of season, the varieties of climate, the influence of the circumambient atmosphere, the action of surrounding bodies, and the construction of the human fabric, must have rendered diseases nearly coetaneous with mankind. The presence of bodily infirmity produces pain, and impels man to seek immediate alleviation, and to employ means for that purpose, either by instinct, experiment, or spontaneous exertion. The many accidents and injuries to which he was exposed in the early and rude ages, must have frequently obliged him, to suppress hæmorrhage, to remove the deformity of dislocation, and to adjust the painful fracture. Thus necessity conceived the art of medicine, reason nourished it, long use promoted it, and experience at length completed it, and made it absolute. The foundations of this art among mankind, were first laid by chance, instinct, and unforeseen events; these were improved by the success and recollection of former experiments; the results of observations, experiments, and remedies were carefully recorded, and a comparison was instituted between events already observed, with those of daily occurrence. The sick were exposed in the streets and highways; and inquiries made of passengers if they knew any remedy. The names of diseases, their remedies and events, were handed down from father to son among the ancestral traditions; and were at length inscribed on the walls and paintings of the public temples: a system of medical instruction was formed from these sacred records, which all were obliged to follow in the cure of diseases; a law alone that was the cause of the extermination of thousands, until the healing art was practised by certain individuals, as a peculiar calling. Such was the manner of cultivating medicine in the early ages, even after the deluge, by the Egyptians, Babylonians, Grecians, Germans, Indians, Assyrians, Chaldeans, and Portuguese; and afterwards in the islands of Cnidos, Cos, and Epidaurus.[4]

Ancient medicine, like all the early transactions of mankind, is a mixture of monsters, giants, demi-gods, and fables; and, therefore, I shall not notice the legends and dark archives of antiquity relative to the first epoch of the healing art, except by adverting to Ovid's allusion to Apollo, as the [5/6] inventor of physic, in the following lines, which now grace the armorial bearings of our society of apothecaries:

> Inventum medicina meum est, opiferque per orbem Dicor; et herbarum subjecta potentia nobis.[5]

All the ancient heathens agree in one point, that some Deity was its inventor, which is also confirmed by Hippocrates (*Liber de Vet. Med.*), and by Cicero, in *Tusc.* Lib. iii. "Deorum immortalium inventioni consecrata est ars medica."

The concurrent testimony of historians of all ages proves it to be the noblest and most useful of human pursuits; and hence the esteem and veneration universally entertained for its cultivators by mankind. The dignity of medicine arises from the nobleness of its subject and its end; its subject is the human body, which excels all other material bodies; its end is health, which is the greatest temporal blessing of man. The importance and dignity of medicine were felt and acknowledged in all ages and countries, both civilized and savage, because the severity of pain and dread of death, were almost coeval with the existence of the human race.

The heathens also highly esteemed the healing art. Democritus said it was the sister of philosophy; the latter removed affections of the mind, the former diseases of the body. Hippocrates said the ancient cultivators of medicine ascribed its origin to a divine source, in which he concurred, as also did Galen. This was likewise the belief of the primitive Christians: St. Austin observes, "Medicina non invenitur unde ad homines manare potuerit, nisi a Deo" (*De. Civ. Dei*).[6] But the healing art was rendered pre-eminent by the divine Redeemer having practised it, while he avoided all other human pursuits.

The excellence and pre-eminence of the healing art were admitted by the Roman orators and moralists. Cicero said, "In nulla re homines propius ad Deos appropinquant, quam salutem hominibus dando."[7] Seneca observes, "quædam pluris esse quam emuntur; emis a medico rem inestimabilem vitam ac valetudinem bonam. ["][8]

"A physician of genius," says Monfalcon, "is the most magnificent present that nature can make the world." Me- [6/7] dicine was encouraged, cultivated, and practised by kings, princes, and pontiffs; the highest, wisest, and best of men. In proof of this position it may be stated, that Solomon, Saphoris, and Gyges, kings of Persia; Habidus, and Mithridates, kings of Pontus; Mesue, king of Damascus; Avicenna, prince of Cordova; Isaac, the adopted son of the king of Persia; Nicholas V., and John XXII., Roman pontiffs, illustrated medicine by their writings. Homer records the great esteem entertained for Machaon and Podalirius, the sons of Esculapius, in the Grecian army; Virgil, that for Japis, physician to Æneas; and Silius Italicus, that for Synalaus, the physician to Hannibal. All civilized nations conferred the highest privileges and honours on the practitioners of medicine. They were exempt from the performance of all civic duties; they were supported by the state in many countries; and ample, nay, prodigal rewards were bestowed upon them throughout the civilized world. The history of medicine affords abundant proof of this assertion.

All wise and prudent emperors and kings, duly estimated the utility and excellence of the healing art, and were extremely desirous of having learned and experienced physicians, to preserve their own and their subjects' health; and hence the honours, immunities, and privileges bestowed in every civilized nation upon the faculty. It is unnecessary to enumerate the vast number of temples dedicated to the early founders of physic, or refer to the rank so signally conferred on Hippocrates by the Athenians, or to the honours bestowed by all nations in succeeding ages on the profession. I may briefly mention, that the court physicians and surgeons of our own country have had titles and emoluments amply conferred upon them; and such is the custom of all nations in Europe.

Medical practitioners are those men to whom we confide our health, which is above all earthly concerns. To them are intrusted the existence of those that are most dear to us, and in their hands are placed the lives of our nearest connexions, and of the friends to whom we are most attached. All gradations of society are alike dependent on them, and must, sooner or later, require their assistance; for, from the earliest period of life to the last moment of existence, their [7/8] skill may be exerted to preserve health, to arrest the progress of disease, or to smooth the approach of death. It is to physicians that all classes commit their health, which is above all treasure, the father confides his child, the husband his wife, the monarch as well as the peasant; they preserve the feeble infant from all the dangers to which it is exposed on coming into the world; their cares protect infancy, childhood, adolescence, manhood, and old age. At every period of life, man calls for the assistance of a physician, and he rarely implores in vain. The confidence placed in him is unbounded: health and life are committed to his care, and on him depend the comfort or misery of families. Hence it is, that no class of men enjoys the respect of every rank of society to such an unlimited degree, as the practitioners of the healing art; and hence it is that they have ever acquired the esteem and veneration of mankind.

That this confidence and esteem should be fairly merited, the father of medicine, and all his eminent successors, to the present period, required an oath of their disciples, the principal obligations of which were, the cultivation of every virtue that adorns the human character. A code of professional duties, or ethics, was arranged, which all were obliged to obey, and which still governs those who have been properly educated. But there never was a period in medical history, in which ethics was so neglected and violated as in this "age of intellect," nor the dignity of the science so degraded and disregarded. It is therefore necessary to inform the rising members of the profession, of those virtuous and noble principles which regulated the professional conduct of their predecessors, and procured that unbounded confidence and universal esteem, which was bestowed on them by society in every age and country. I shall, therefore, describe the ethics of the founder and father of medicine, and those of his successors to the present time, with the hope of exciting my readers to imitate their example, and practise those precepts which have always characterized the erudite and scientific portion of the profession.

Notes

1. *Author's note*: "Omnis medicina a Deo est. Cœlitus delapsa non sine Dei consilio vivit agitque. Hine ars nostra sine religione, vel impia vel nihil." – La Politica del Medico nell' esercizio dell' arte sua. Dal celebre Alessandro Knipps Macoppe, Professore di Medicina nell' J. R. Università di Padova. Milano, 1826.

2. *Author's note*: "Deus propriis non nostris agit manibus. Incipit artis Dei potentia, ubi tua desinit." *Op. cit.*
3. *Author's note*: Gen. xl. Exod. xxi. Lev. xii. Ecelias. xlvii.
4. *Editors' note*: This paragraph very closely follows Ryan's "Introductory Lecture to the Theory and Practice of Midwifery," *Lancet* 2 (1827–28): 394–400, quoting p. 395.
5. *Editors' note*: "The art of medicine is my invention, and the healing power of herbs, which are declared useful throughout the world" (loosely quoting Ovid, *Metamorphosis*, 1:523).
6. *Editors' note*: "Medicine would not have been invented unless humans were able to derive it from God." The reference is to St. Augustine, not Austin.
7. *Editors' note*: "Humans approach the gods in no things more than when giving health to people."
8. *Editors' note*: "Some things are worth more than they seem; you buy from a doctor inestimable things—life and good health."

Chapter II
Medical Ethics of Hippocrates

The duties and qualifications of medical practitioners were never more fully exemplified than by the conduct of Hippocrates, or more eloquently described than by his pen.[1] He admitted no one to his instructions without the solemnity of an oath, the chief obligations of which were, "the most religious attention to the advantages and cure of the sick, the strictest chastity, and most inviolable secrecy about private or domestic matters, which might be seen or heard during attendance, and which ought not to be divulged."[2]

The father of physic strongly inculcated the necessity of the cultivation of piety and virtue; and held that his disciples should excel in religion and morals. He also maintained that they should acquire the most perfect knowledge of every form of disease, and of the best mode of treatment. He considered calumny and illiberality disgraceful, and the disclosure of the errors of a contemporary highly culpable. He was of opinion, that the morals of a medical practitioner should be excellent and unexceptionable, conjoined with gravity and humanity. He ought to be correct in every custom of life; and demean himself honourably and politely towards every rank in society, and thus will he promote the glory of his profession. To these precepts nature is the best guide. He is to retain in his recollection all remedies, their mode of preparation and application, and the use of all mechanical means which are employed for the cure of diseases. This is the beginning, middle, and end of medicine. Let him be cautious in his prognosis, and predict only those events sanctioned by observation and experience. In his approach to the sick, let his [9/10] countenance be mild and humane, not rough, proud, or inhuman; and let him evince a sincere desire to afford relief; and employ all remedies with diligence and caution. He should be ready to answer all questions, establish constancy in perturbations of mind, allay tumult by reason, and be ever ready to afford relief in all emergencies. He is never to exhibit an improper or dangerous remedy, even to a common malefactor; but try those medicines approved by the majority of the profession. When a patient desires popular remedies, he is to be cautioned against them, but left to his own discretion. But if attendance is commenced without remuneration, the sick must not be abandoned. Any discussion relative to pecuniary matters is injurious, especially in acute diseases. When disease is rapid, there is no time to arrange concerning reward; it has no influence on a good practitioner, who is only

H. Brody et al. (eds.), *Michael Ryan's Writings on Medical Ethics*,
Philosophy and Medicine, Vol 105,

anxious to preserve life, and enjoy the more noble gratification, the universal esteem of mankind. It is much better to accuse those cured of disease of ingratitude, than deny them aid when in danger.

Some, on account of friendship or acquaintance, expect attendance gratuitously, but these deserve neglect. In all cases the best remedies ought to be employed; and the reward should be in accordance with the custom of the faculty, and with the wealth or means of the sick. In some cases aid is to be afforded gratuitously, which will ensure more renown than if remuneration had been awarded. When an opportunity offers in the case of a stranger or necessitous individual, afford succour immediately. Some men are so avaricious as to extract riches from the most indigent; but these men seldom prosper, while the humane and good practitioner is considered an honour to his art. Consultations cannot be refused, even to the most necessitous. "This I affirm by an oath, that one medical man should never invidiously calumniate another, or rob him of his merit, or diminish the confidence of his patient." Such were the leading features of the code of ethics, inculcated by the immortal founder of physic;[3] the chief obligations of which have been [10/11] recommended by all his successors down to our own time.

It must be conceded that the first duty of a medical practitioner in common with his species, is to God, for his sovereign majesty, supreme excellence, and infinite goodness. The history of medicine, of all other sciences, comprises the most intimate acquaintance with the works of nature, and elevates the mind to the most sublime conceptions of the supreme Being, and expands the heart with the most pleasing ideas of Providence. It nearly extends through the whole range of the creation; and no other profession requires so extensive a knowledge of the works of Providence. The objects which engage the attention of the medical practitioner are the influence of the sun, moon, and heavenly bodies, the laws of their uninterrupted revolutions and various movements; the various productions of the earth, including the vegetable and mineral kingdoms, the innumerable variety of living creatures that fill the air, the earth and the waters; and the microcosm of the human body, with its wonderful organs, functions, and immortal principle. Now when we consider the creation, and conservation of all these objects, and their subserviency to human happiness, we can never reflect without the profoundest veneration upon the attributes of that Being from whom all these things have proceeded. We cannot help acknowledging what an exertion of benevolence creation was, of a benevolence how minute in its care, how vast in its comprehension!

The most eminent of the faculty have been distinguished for their piety; among whom we find the names of Harvey, Sydenham, Locke, Boerhaave, Arbuthnot, Winslow, Haller, Hoffman, Stahl, Baglivi, Steno, Helmont, Riverius, De Hean, Vesalius, Ruysch, Lancisi, Gaubius; and in our own day, Gregory, Baillie, Bateman, Davy; and many other illustrious men now in existence. They studied the sacred volume, which to use the language of Cicero when speaking of the twelve tables, is, "A little book that alone exceeds the libraries of all the philosophers in the weight of its authority, and in the extent of its utility." In it is to be learned the only complete history of the universe, the divine precepts of religion, which refer all honour and glory to the Author of all [11/12] things, and good will to mankind.

There is laid the foundation of all ethics, "do unto others as you would they should do unto you." This is the great and first principle of human conduct, it commands man to observe justice and benevolence towards his fellow man; duties which by their reciprocity lead to his own and the general good. Of all the calumnies launched against medical practitioners, there is none so diffused and so odious as that of irreligion. It is broadly asserted that medical men are Deists or Atheists. This is a popular error. A medical practitioner cannot be an anatomist and an Atheist, as nothing presents such strong proof of the existence of God, as the wonders of our organization and its functions. The wonderful relation between the structure and functions of all parts of the human body; the marvellous disposition of the bones and muscles; the distribution of the arteries and veins, and their anastomoses; the functions of the brain, heart, lungs, &c., every thing in the study of anthropology attests a superior intelligence. The origin and cessation of life manifest the peculiar power of the Deity. The reproduction of the species, the endowment of life and of mind, can only be ascribed to the same Being. The anatomy of a hair is sufficient to overthrow all the reasonings of materialists.

All the universities, schools, and colleges, inculcate and enforce the cultivation of religion. "Such is the connection between the Deity, religion, and a physician, that without God and religion, no physician can be successful."[4] This reflection of Broesiche's is a great moral truth. A religious physician will not arrogate to himself an absolute empire over the lives and health of men; he will not pretend to govern the progress of diseases; he will not consider himself the God of nature; but he will refer all things to the Supreme Being. "It is from Him he derives his light – it is on Him he calls for succour."

The practice of medicine requires the exercise of all the virtues. Whoever fulfils all the duties it imposes on him, [12/13] obeys the most rigid moral laws. To cultivate religion, to love his country, and to be a truly good man, form the true character of the medical practitioner. It may be urged that there are exceptions; but they are few in number.

In observing the different phenomena of life, physicians are continually obliged to recognize the omnipotence, goodness, and other attributes of the Deity; they cannot consider their art a sovereign power; because they too often find it to fail. It is to the Divinity they must attribute their success or failure. Such is true medical wisdom. Who knows so well as the medical practitioner the miseries of man, his infirmities, and the dangers which every moment threaten his life? Who so well knows the suddenness of death in the most robust constitutions? The whole history of man, and of the healing art, recalls to his mind the idea of a Supreme Being. He discovers in religion the powers opposed to all the annoyances inseparable from his ministry, and the real consolations against the ingratitude of mankind. It enables him to bear all the weaknesses and follies of the sick, to perform the beneficent powers of his art in the most disgusting instances, and to be the instrument in the hands of the Author of all, of affording relief, and diminishing the sufferings of humanity. Hoffman was right when he said, a physician ought to be a Christian – "medicus sit Christianus."

There are intimate relations between religion, morals, philosophy, and medicine. It is well observed by the ancients, that there is a relation between certain physical states, certain characters of the intellectual faculties, and certain passions; that is to say, that according to the habits of the body, the proportion of the members, the colour of the skin, the disposition of blood vessels, lymphatics, nerves, &c., will thought and the train of ideas correspond. Many of the ancient sages found in the organization of man, compared with the functions of life, the solution of the most important moral phenomena.

Many physicians have written esteemed works on philosophy, and on the human mind. It is scarcely necessary to mention the names of Locke, Dugald Stewart, Thomas Brown, &c., nor can I omit the splendid work of Cabanis, on the relations of the Physical and Moral States of Man. He has [13/14] given a most luminous view of the relations between the physical and moral conditions, and followed Locke and Condillac.

Moral philosophy, medicine and philosophy, have numerous relations. Medical philosophy informs us of the formation of ideas, the rules which ought to regulate life, the course that leads to happiness, the influence of the different climates on the physical state of man, and on the institutions of society, that of regimen upon the intellectual faculties and passions, and that of diseases on the operations of the judgment. It embraces the operations which constitute the functions of intelligence, and determine the power of volition; it comprises the different characters of the passions, and supplies the basis of the moral condition. The faculties of the mind are intimately connected with the organization of man, and nothing can separate the study of his physical and moral states. I may briefly illustrate this position.

Vain attempts have been lately made by materialists, among whom are certain phrenologists, to prove that the mind is a mere function of the brain, and consequently dies with it; that the embryo has no mind or soul either in the uterus or at birth, but that its mind is built up by the five external senses, and is annihilated by death. Here is mere assertion without the slightest proof. According to this doctrine, life and mind are distinct, as the living infant, before or at birth, has no mind; but the materialists have not proved that the living brain of the infant in the womb is unconnected with soul: they merely assert it. If their conclusion were admitted, it would be no crime to destroy the fœtus in the womb, or even at birth, no more than the young of any other animal; and fœticide or infanticide might be expunged from the catalogue of crime. These philosophers forget that the first of the human species was perfectly formed and organized before the breath of life and soul were breathed into his nostrils; and it is manifest that life and soul were added to his organization. It, therefore, follows, that life and soul, and matter or brain, were different in the first man, and were not made up by the five external senses. I cannot agree with a recent phrenological physiologist, who maintains "to call the human mind positively a ray of the divinity – *Divinæ particula auræ ex* [14/15] *ipso Deo decerptus,* ex universa mente *delibatus*,[5] appears to me to be absolute nonsense." Neither can I assent to another absurd position, which denies the interposition of an omnipresent Deity at the time of generation, as if the divine power was not in constant operation in every atom of the universe. Who knows how life is communicated, or what becomes of it, when the body is deprived of it? "That life, or the assemblage of the functions, is immediately dependant on organization, appears to me, physiologically speaking, as clear as that the presence of the sun above the horizon causes the light of the day. Mind is the functional power of the living brain. As I cannot conceive *life* any more than the power of attraction, unless possessed by matter, so I cannot conceive *mind* unless possessed by a brain, or by some nervous organ, whatever name we may choose to give it, endowed with life. I speak of terrestrial or animal mind; with angelic and divine nature we have nothing to do, and of them we know, in the same respects, nothing." I ask, was life and mind distinct from organization in the first of the human species? Is the life and soul of the Deity connected with organization? Nevertheless the terrestrial mind, which we presume differs from the angelic or celestial, is material; and thus the human mind is a compound of terrestrial and celestial, it is material and immaterial according to the doctrine of Dr. Elliotson. Such is the illogical and unphilosophical conclusion of some materialists and phrenologists.[6] The pseudo-science of phrenology or materialism, has led to the too prevalent opinion of the materiality of the mind in this age, though some of the disciples of the newly-revived and fantastic hypothesis of moral philosophy stoutly deny it. There are some phrenologists who disavow materialism: there are more who avow it. I agree with those who consider phrenology and materialism synonymous; and as I hold both one and the other to be totally unworthy of belief, which is the opinion of the greatest physiologists in existence, in this and every other country, I shall repeat my denunciation of this false philosophy as published in the former edition of this work, [15/16] and advise certain commentators not again to misquote and misrepresent my language, as on a former occasion.

The cultivators of medicine have been generally, and most unjustly accused of favouring infidelity and a contempt of religion; though of late it must be acknowledged, with pain, that an abortive attempt had been made which has given too much proof of the justness of the accusation; – this was the revival of materialism. The attempt, however, was as impotent as it was wicked – it was attacking youth on the weak side; and it was extinguished by the universal voice of the most eminent members of the faculty; for it was observed to have unhinged all the bonds of society in a neighbouring nation, and to have produced a degree of anarchy, confusion, and atrocity unequalled in the annals of mankind. It contained not a single argument that had not been urged and refuted a thousand times before; and after all the schemes of a reluctant and erring philosophy, the indispensable resort must be to a Deity and his ordinances. How appropriate the following distich to such philosophers:

> Know thou thyself, presume not God to scan,
> The proper study of mankind is *man*.

The doctrine is repudiated by the medical profession throughout the civilized world; and out of the whole, amounting to many thousands, there are not ten men of any eminence, either phrenologists or materialists. That philosophy which assails the attributes of the Supreme Being is fallacious: that wild hypothesis which is in direct opposition to the principles of revealed religion can have few, if any disciples, in an age so enlightened as the present. The really learned in the medical profession, have never been infected by the poison of infidelity.

The doctrine of the materiality and mortality of the soul, which is that of materialism and phrenology, "should for ever be exploded as totally false, and unworthy of all regard, as subversive of the fundamental principles of all religions, as introducing civil anarchy into the political economy of legislation, as substituting disorder for harmony, despair for hope, and eternal darkness for everlasting light."

If materialism tended to promote the happiness of society, [16/17] to assist our hopes, to subdue our passions, to instruct man in the happy science of purifying the polluted recesses of a vitiated heart, to confirm him in his exalted notion of the dignity of his nature, and thereby to inspire him with sentiments averse to whatever may debase the excellence of his origin, the public and the medical profession would be deeply indebted to the phrenologists. But the tendency of phrenology, however disguised, is to make mind a secretion or function of the brain, and thus to deny the immortality of the soul. Unbelievers, in general, wish to conceal their sentiments; they have a decent respect for public opinion, are cautious of affronting the religion of their country, and fearful of undermining the foundations of civil society. Some few have been more daring, but less judicious; and have, without disguise, professed their unbelief and again retracted their opinions. In denying the immortality of the soul, they deny the authenticity of the Bible; they sap the foundations of all religions; they cut off at one blow the merit of our faith, the comfort of our hope, and the motives of our charity. In denying the immortality of the soul, they degrade human nature, and confound man with the vile and perishable insect, and overturn the whole systems of religion, whether natural or revealed. In denying religion, they deprive the poor of the only comfort which supports them under their distresses and afflictions, and wrest from the hands of the powerful and rich, the only bridle to their injustice and passions; and pluck from the hearts of the guilty the greatest check to their crimes – that remorse of conscience which can never be the result of a handful of organized matter – that interior monitor which makes us blush in the morning at the disorders of the foregoing night, and which erects in the breast of the tyrant a tribunal superior to his power. Such are the consequences naturally resulting from the principles laid down in some of the late phrenological writings. It is no intention of mine to fasten the odium of infidelity on any members of my profession; but it surprises me that men, whose understandings have been enlightened by the Christian revelation, and enlarged by the study of medicine (the most extensive and varied of all human sciences), should broach tenets which [17/18] equally militate against the first principles of reason and the oracles of the Divinity; and which, if true, would be of no service to mankind. Of what benefit to humanity would be the establishment of phrenology or materialism? I answer none; but, on the contrary, the greatest injury. If any man be so unhappy as to work himself into the conviction that his soul is a function or a secretion of the brain, and of course must perish with it, he would still do well to conceal his horrid belief with more secrecy than the Druids concealed their mysteries. In doing otherwise, he only brings disgrace upon himself, for the notion of religion is so deeply impressed on our minds, that the bold champions, who would fain destroy it, are considered by the generality of mankind, and our profession, as public pests, spreading disorder and mortality wherever they appear; and in our feelings we discover the delusions of a cheating and unmedical philosophy, which can never introduce a religion more pure than that of the Christians', nor confer a more glorious privilege on man than that of an immortal soul. In a word, if it be a crime to entertain such a doctrine, it is consummate folly to boast of it. Whence this eagerness to propagate systems, the tendency whereof is to slacken the reins that curb the irregularity of our desires, and restrain the impetuosity of our passions? It must proceed from a corruption of the human heart, averse to restraint, or from the vanity of the mind, which glories in striking from the common path, and not thinking with the multitude. In vain are the phrenological materialists informed by the anatomist, that he can find bile in the liver, urine in the kidneys, but none of the faculties of the mind in the brain. In vain are they told that after death, when volition has ceased, the motions of the muscles can be excited by galvanism, and that though muscular motion be restored, we cannot recal volition, or the other mental faculties; rather a strong proof that motion and volition are not exactly the same thing. In vain have materialists examined dead bodies to explain the most important phenomena of life – their imagination only has answered; in vain have they mutilated the brain, in a hundred different ways, to discover the seat of the intellectual faculties – the vainest hypotheses are the result of [18/19] their researches. What do they know about life, about the astonishing phenomena ascribed to sympathy, or the impenetrable mystery of generation? God reserves these secrets to himself. Facts are observed every instant in practice, which science does not explain; and it was this that made the father of medicine declare, there was something divine in diseases – that is to say, incomprehensible to man. These and ten thousand other proofs are lost on the phrenologists. They set up the proud idols of their own fancies in opposition to the received opinions of their profession, and in opposition to the oracles of the Divinity; and in endeavouring to display absurdities in the Christian religion, fall into much greater. To them we can, with due deference, and without disclaiming our title to good manners, apply the words of St. Paul to the philosophers of his time – "They became vain in their imaginations; professing themselves wise they became fools."

Let the patrons of the revived and long refuted philosophy persuade their wives that their souls die with their bodies – let them instil the same doctrine into the minds of their children – let the doctrine become generally received – unfaithful wives, unchaste daughters, rebellious sons, and general confusion and anarchy will be the blessed fruits of their philosophy. To those philosophers, the words of an able and learned prelate very forcibly apply: "The Bible has withstood the learning of Porphyry and the power of Julian, to say nothing of the manichean Faustus; it has resisted the genius of Bolingbroke and the wit of Voltaire, to say nothing of a numerous herd of inferior assailants; and it will not fall by your force. You have barbed anew the blunted arrows of former adversaries; you have feathered them with blasphemy and ridicule; dipped them into your deadliest poison, aimed them with your utmost skill; shot them against the shield of faith with your utmost vigour; but like the feeble javelin of the aged Priam, they will scarcely reach the mark, will fall to the ground without a stroke [.]"[7]

The doctrine of materialism is not more discordant with [19/20] the principles of revealed religion, than with the opinions of the greatest men who have ever adorned the science of medicine; men who were, and still are, as great ornaments to the literary world in general, and medical literature in particular, as they are useful to mankind. On the other hand, the soul shrinks within itself at the horrors committed by materialists, as they pretended to disbelieve a future state of rewards and punishments, and assented to every evil suggestion of their unhappy minds.

"The first duty," says Napoleon, "of a medical man is to his God, the next to his king and country, and the next to his patients."[8] This was the rule of duty of the greatest ornament of the medical profession, and comprises the whole institutes of professional conduct. The man who pays due homage to his Creator, and obeys the laws laid down for his guidance, must act honourably and justly towards his fellow-creatures. A religious man cultivates all the virtues that adorn the human character, and justly place man the lord of the creation. He will be distinguished for humanity, sympathy, politeness, a large share of good sense, and knowledge of the world. Dr. Gregory considered the chief of the moral qualities peculiarly required in the character of a physician was humanity, "that sensibility of the- heart which makes us feel for the distresses of our fellow-creatures, and which of consequence, excites us in the most powerful manner to relieve them; sympathy produces an anxious attention to a thousand little circumstances that may tend to relieve the patient – an attention which money can never purchase; hence the inexpressible comfort of having a friend a physician. Sympathy naturally engages the affection and confidence of a patient, which, in many cases, is of the utmost consequence to his recovery. The patient feels the approach of a man who possesses it, like that of a guardian angel ministering to his relief, while the approach of a rough unfeeling man is like that of an executioner. A certain command of the temper and passions must be natural or acquired to medical men, as sudden emergencies frequently occur in practice, which may flutter [20/21] the spirits and judgment of the best practitioner; and the weakness and bad behaviour of patients and their attendants are well calculated to ruffle the temper of the mildest individual, and cloud his judgment, and make him forget propriety and decency of behaviour; hence the necessity of presence of mind, composure, and steadiness." Let him be remarkable for the humane and liberal exercise of compassionate philosophy, Dr. Gregory has lucidly described the genius, understanding, temper, and qualifications, which are required for the duties and office of a physician, which will be inserted hereafter. He observes, "to excel in this profession requires a greater compass of learning than is necessary in any other." This assertion is borne out by the most exalted testimony. Judge Blackstone gave physicians pre-eminence for "general and extensive knowledge." Dr. Johnson was scarcely less favourable in his estimation. "Whether," he observes, "what Temple says be true, that physicians have had more learning than any other faculties, I will not stay to inquire; but I believe every man has found in physicians, great liberality and dignity of sentiment, very prompt effusions of beneficence, and willingness to exert a lucrative art where there was no hope of lucre."[9] The late Dr. Parr, the justly celebrated philologist, remarked, "while I allow that peculiar and important advantages arise from the appropriate studies of the three liberal professions, I must confess, that in erudition and science, and in habits of deep and comprehensive thinking, the pre-eminence, in some degree, must be assigned to physicians."[10] Rousseau spoke as follows of the faculty: "Il n' y a pas d'etat qui exige plus d'etudes que leur: par tous les pays ces sort les gens les plus veritablement utiles et savans."[11] "In addition to all other means of augmenting your true fame," said Professor Godman,

> "the observance of one circumstance will be of great importance; this is the unremitted exercise of humanity towards those who seek your professional [21/22] aid, whatever be their conditions. The character of a truly good physician, is one of surpassing excellence, and his reputation is the most exalted we can hope for. He is the friend of the wretched and woe-worn; the cheerer of the despondent; the solacer of the broken-hearted. His soul is the empire of benevolence; his actions the result of a principled charity, and unaffected good-will. He is the blessing conferred on the society in which he lives, and an honour to the human race. Wherever the afflicted dwell – wherever the voice of suffering is heard, he is to be found. The diseased find cheering and consolation from his presence, and the sounds of sorrow are stilled. Even when hopes of life can no longer be given, he calms the tumultuous grief of relatives, by recalling their thought to that better world, where – sickness and sorrow are to be no more – 'where the wicked cease from troubling, and the weary are at rest.'
>
> "Such are the common offices, and frequent exercise, in which he is engaged. His character, even under ordinary circumstances, may be contemplated with gratifying emotions. But there are conditions in which he is presented in a more sublime aspect. It is when the lurid breath of pestilence scatters destruction, desolation, and dismay, throughout the land, and death tramples with indiscriminate fury over the people – when the ties of relationship and affection are sundered by the violence of fear, and utter selfishness seizes on the hearts of men; then the good physician, unmoved by such examples, untouched by terror, regardless of himself, is seen actively discharging every duty. Then he becomes the father, the brother, the friend of the destitute; his steadfast attention smooths the pillow of the dying; he inspires the desolate with hope; and, like a beneficent angel, wherever he goes, is a dispenser of good. Who can estimate the feelings, or measure the fame of such a man? Who would not imitate his example for such a reward? What is there in death's most frightful forms that could withhold us from attempting to deserve it? It is a glorious privilege which our profession confers, of inscribing our memories, not on perishable marble, but in the living affections of our fellow-men, to be cherished as long as [22/23] our race shall endure."[12]

It is unnecessary to illustrate this truism by other citations, as it is universally admitted by those capable of judging. As to the observance of ethics, it may be fairly stated, that there is much room for improvement in this section of the empire.

The following graphic and elegant description of the moral conduct of medical practitioners, is so accurate and so just, that I quote it with much pleasure. The author defended the study of anatomy before the passing of Mr. Warburton's Act, which legalized it.

> "Nevertheless, anatomical pursuits are neither criminal in themselves, nor yet fraught with dishonour or disrespect to the dead; they outrage no feelings but such as are of a superstitious nature; neither do they in any way deteriorate or brutalize the character of those who pursue them. Insinuations of the latter kind are, indeed, frequently thrown out, but we repel them with meet and honest indignation. We appeal to observation for the truth of our statement, when we assert, that society does not present another class of individuals more numerous and respectable than that of the medical profession; and one, at the same time, against the general moral conduct of which so little reproach can be made. There is none, too, possessed of more varied and valuable information. In these respects physic has no occasion to shrink even from a comparison with divinity. With temptations

infinitely stronger and more diversified, practitioners in medicine do in no wise cede to the clergy, taken generally, in the morality of their conduct. It ought moreover to be borne in mind, that the very vocation of the latter, by abstracting them from temptation, diminishes the merit attached to the rectitude of their walk. At the same time we need not hesitate to affirm, that more benevolent, more truly kind and charitable individuals, than those which have adorned the profession of physic can no where be found. No description of men, whatever their calling or station in life, render such valuable services to the poor and [23/24] needy sick; none expose themselves to dangers equally numerous and great, without the remotest prospects of pecuniary remuneration. They work silently, yet not the less effectually. They make use of no ostentatious preconization of their good deeds, which are of unsolicited and spontaneous origin; and whilst others are idly preaching the duty of charity, they exemplify it in their daily converse with man. No one is better acquainted with the distresses of poverty and sickness than the physician; and no one, therefore, can more fully and deeply sympathize with the afflicted. What a bright galaxy of medical philanthropists does history exhibit to us! Of men who have conferred lasting and invaluable blessings on society; who have laboured through evil and through good report, for the benefit of their fellow-creatures! And do they not still labour in the same cause? Do they not pursue the same undeviating path of benevolence; gratuitously devoting their time and talents to the indigent sick? We will say nothing of what is privately wrought in this respect; let our public hospitals, our dispensaries, and asylums be consulted: let them speak. There are very few such institutions, in which those who have the care of the soul are not adequately remunerated for their trouble; whilst, universally, those who cure the body bestow their time and ability gratuitously. And yet, time is infinitely more precious to the latter than to the former.

"Again, no body of men in the community can boast of brighter ornaments to science than are to be found amongst the members of the profession of physic. Where shall we find more truly liberal and enlightened philosophers? Individuals, that have more effectually contributed to dissipate error and superstition, or more zealously promoted the general good of mankind? Where, in fine, shall we meet with men who have united higher cultivation of mind with a more truly virtuous nobility of character?

"We repeat, then, neither in mental nor moral attributes does the profession of physic yield to that of theology. Let no one imagine we are instituting an invidious comparison, with the intention of exalting the merits of one body in the community, by depreciating those of another. We have no [24/25] aim but that of evincing the general worth, industry, and acquirements of medical practitioners. We wish to show, that the study of anatomy does not exert any baneful influence on their characters; that it does not deprive them of the distinguishing sensibilities of humanity, and thus render them callous to the sufferings of their fellow-men. No! they pursue an honourable and dignified vocation, and are urged on in their career by the noble ambition of achieving the utmost possible good. It would be difficult, indeed, to point out in society, individuals of a more laborious, persevering, and indefatigable character. At all hours, at the table of repast, on the couch of repose, amid the inclemency of weather, the harass of an anxious mind, and the oppression of bodily fatigue, they must be ready to obey each capricious call! And yet how ill-treated and ill-requited! Patients rigorously exact an assiduous attention; whilst with all latitude which may suit their fancies, they will follow the advice of a medical attendant, yet immediately suspect the extent of his skill, should the amelioration demanded not ensue. But this is not all; they even seek at the hands of the law to obtain compensation for any supposed deficiency of skill, to the attainment of which, nevertheless, both themselves and the law are equally opposed!"[13]

They cease to suit their own convenience and to attend to their private affairs, as they are always ready to wait on suffering humanity; they allow nothing whatever to prevent them from attending to the sick, they are always ready; they bear with the injustice, the caprices, and ingratitude of men; they expose their lives in the most dangerous circumstances – when pestilence devastates the earth, in all times, in all places; they possess courage, exemplary patience, and an entire abnegation of self. Such are the virtues of medical practitioners.

They are, however, *amply rewarded* for all annoyances and ill-treatment. They have numberless opportunities of giving that relief to distress which is not to be purchased with the wealth of India. This, to a benevolent mind, is one of the [25/26] greatest pleasures. "Is there any thing in the world more estimable," asks Voltaire, "than a physician; who, having studied nature in his youth, knows the relations of the human body; the diseases that torment it, the remedies that may assuage it, exercises his art in defiance of it, takes equal care of the rich and the poor, who does not receive remuneration but with regret, and employs it to the succour of the indigent? Men," continues he, "who are occupied in affording health to other men, exercise the only principles of beneficence, are far above all the great ones of the earth, they partake of the divinity."

They are now duly appreciated by kings, nobles, poets, and literary and philosophical characters. All the virtues are displayed in the exercise of the functions of a physician – his ministry commands the respect of men, and the admiration of sages.

As there is no perfect code of ethics to guide the profession in this country, a succinct detail of the rules of conduct prescribed in the different works of authority may be given, and then the leading points can be fully discussed under separate heads. The following is an imperfect epitome of the ethics of the last century, collected from various sources. To this will be appended the ethical code of the present period.

Notes

1. *Author's note*: He flourished about 460 years before the present era.
2. *Author's note*: The Edinburgh University requires this oath on conferring the degree of M.D. and is the only university or college in this empire that binds its members by so serious and necessary an obligation.
3. *Author's note*: De Medico, de decenti Habitu aut decoro, Prenotiones.
4. *Author's note*: "Tanta est inter Deum, religionem, et medicum connexio, ut sine Deo et religione nullus exactus medicus esse queat" (*Macoppe op. cit. Editors' note:* See Footnote 1, Chapter 1).
5. *Editors' note*: "a divine ray of light derived from God alone extracted from a universal mind."
6. *Author's note*: See Mr. Roberton's Strictures on Dr. Elliotson's *Physiology. – Medical Gazette*, November, 1835.
7. *Author's note*: Bishop Watson's Apology for the Bible.
8. *Author's note*: O'Meara's Voice from St. Helena.
9. *Author's note*: Lives of the Poets, Garth.
10. *Author's note*: Remarks on the Statement of Dr. Charles Coombe, pp. 82, 83.

11. *Author's note:* Letters.

12. *Author's note:* Address delivered on Professional Reputation, before the Philadelphia Medical Society, 1826, by John D. Godman, M.D., Professor of Anatomy and Physiology in Rutger's Medical College.

13. *Author's note*: Dr. Corden Thompson's Letter on the Necessity of Anatomical Pursuits, 1830.

Chapter III
Medical Ethics of the Middle Ages[1]

First of all things, a medical man ought to exercise piety, and give due honour to the Supreme Being. Next, he ought to render to every one his due; obedience to his superiors, concord to his equals, and equity to his inferiors. He ought to preserve a clean heart, and silent tongue, and cultivate every virtue. The whole praise of virtue consists in action. He is to avoid anger, and suppress all its perturbations, intemperance and insolence, having always before his eyes the great deformity of mind produced in those who give way to them, and the amiableness and gracefulness of those who avoid them. Sensuality, intemperance, and dissipation, produce concupiscence and carnal gratification, which increase rapidly, and would eventually ruin a medical practitioner. These are to be strenuously avoided, as well as every luxury. Continence consists in moderating pleasure; gluttony, debauchery, and ignominy, in abusing it. An incontinent, or an intemperate man never rose to eminence, and is completely unfit for medical practice. Men of loose and dissolute habits, and of excursive amours, debase themselves to the rank of the brute creation, and render the mind stupid and inert, and totally unfit for the pursuits of science. Such profligate and abandoned characters cannot be found in the history of the medical profession – in truth, men so vitiated could not long pursue the practice of medicine. What man would commit the care of his wife, daughters, or female relatives to a medical practitioner, if such could be found, of so debased and brutal character – to a man burning with desire of violating the conjugal and vestal honours of his neighbour's family. Hence the necessity of practising chastely and honourably; and hence the preference which is given to those members of the profession who have entered into the sacred bonds of matrimony, especially in obstetric practice. Every one is [27/28] bound to support his own and the professional dignity, with noble sentiments, probity, and humanity. Sadness and fear depress the mind and body, and unman the practitioner. Fortitude is opposed to sadness and fear, and is often necessary to enable us to bear patiently the calumny and contumely, to which no class of men are more exposed than the professors of medicine. All these passions should be expunged from the medical character, and an ardent desire of fame and glory be substituted in their place. Avarice, pride, and envy, are evils which must be carefully avoided. Avarice was considered the chief of all improbity by the ancients; and it is highly cruel in medical men, when it

H. Brody et al. (eds.), *Michael Ryan's Writings on Medical Ethics*,
Philosophy and Medicine, Vol 105,

precludes aid from the sick. But those do not consider its cruelty whose sole object is the accumulation of riches. Well might the poet exclaim

Quid non mortalia pectora cogis
Auri sacra fames?[2]

The sick should never want aid on account of pecuniary consideration; and the practitioner ought to be satisfied according to the affluence of the patient. Above all things, pride is to be avoided. It is odious in the sight of God and man, as it excites an inordinate desire of excellence, and induces one to think, that he enjoys from himself all the gifts of nature, talent, intellect, memory, power, and science, which are bestowed on him by the Deity: he despises others, and thinks they are to submit to him, although his superiors; and hence follows his insatiable desire of praise, fame, honour, glory, and reverence, which is but vain glory. Physicians ought never to be guilty of such an error, nor of presumption, ambition, nor curiosity. On the contrary, let them display humility without sordidness. Envy at the prosperity or success of another, is an evil which ought not to be named among the profession. An envious man is pusillanimous, of a narrow mind and abject talent, for he shows by envying others, that he is inferior to them; he envies what he does not possess, but vehemently desires. Envy is a compound of hatred, dissimulation, avarice, pusillanimity, mendacity, and ambition; and is opposed to friendship, liberality, truth, mag- [28/29] nanimity, and prudence. Medical men must avoid this most pernicious evil. All ought to enjoy fortune happily, and no man should be sad at the prosperity of another. Having thus pointed out some of the vices which are to be avoided by medical men, it is right to enumerate the virtues which they ought to cultivate. These are prudence, circumspection, foresight, caution, perspicacity, continence, sobriety, mildness, modesty, taciturnity, veracity, gravity, magnanimity, liberality, and honesty – friendship towards acquaintances, affability and civility to strangers, and decorum according to age, sex, and condition. The contrary vices are imprudence, stupidity, precipitation, enmity, cunning, curiosity, and all the excesses of the will and desire, as irascibility, concupiscence, which are to be avoided by medical practitioners as the most mortal pestilence. About which let them consult the writers on morality, especially Plato, Aristotle, Plutarch, and Seneca; and, above all, let them peruse the Sacred Volume, which will lead to virtue and prudence, from which they will learn how to be good men and prudent practitioners, and to think piously. The conversation is to be moderate and veracious; the morals must be grave, benign, and cheerful; the diet temperate and frugal; the apparel respectable and professional; witticisms to be free but few, and to these may be added the ancient precept –

Mens humilis, studium quærendi, vita quieta,
Scrutinium tacitum, paupertas, terra aliena.[3]

A medical man should be affable and hospitable; friendly to his relations and neighbours: polite to all without any moroseness; he is to relieve the sadness of the suffering patient with placid and mild discourse. Let him not be peevish, but ingenuous, affable, familiar, and enforce his authority without any disdainful gesture. Let nothing fictitious, nothing simulated be in him, nothing low or base, but let his mind soar with sublimity above all the vicissitudes of fortune. Let the studies of his life be the meditation on the delight and riches of science, virtue, and honour. Let him shun litigation and vain popular applause: he ought candidly to praise the good, always avoid detraction, tolerate the bad, [29/30] indulge the inferior, agree with his equals, and obey his superiors; he is to injure no one; he is to live in the greatest harmony with his family, for this tends to support dignity, and a good reputation; he is to live on the best terms with all, and endeavour to obtain universal esteem. Let him maintain his opinions with modesty and eloquence, and always with veracity, but without obstinacy. He is not to be proud or haughty, but cure rich and poor, slaves and free, of whatever nation, for medicine is the same to all. He must be careful to observe, that his remedies and directions shall be faithfully exhibited and attended to. *He is to esteem as hidden mysteries whatever is said or done in the house of his patient*, and thus he and his art will acquire more praise. By the observance of these institutes a medical practitioner will obtain a distinguished place among the wise and good.

The display of a diploma is to be left to jugglers.[4] The greatest cleanliness is necessary; and particular attention to the hand and fingers is requisite. In visiting the sick, let the countenance be meditative, not melancholy or peevish, which is odious to all. Risibility and hilarity are deemed intolerable, from whence the axiom, "medicus garrulus ægrotanti alter morbus."[5] His whole gesture is not to be so humble as to excite contempt, nor so proud and arrogant as to excite hatred for him. No perfumery is to be used, but if there be any unpleasant odour in the breath, it is to be corrected, lest the patient have reason to exclaim, "cure thyself." The visits to the sick should not be too frequent or too rare, twice a day, except in extreme cases, is quite sufficient. Dr. Gregory maintains an opposite opinion, but reason and common sense are against him. Frequent visits are deemed troublesome by some, while they are esteemed angelic visitations by others. They are to be regulated according to the wish of the patient, or danger of his disease. By gravity and affability we can best learn the nature of disease.

In visiting the sick we should recollect a proper approach, [30/31] authority, silence, and answers to all inquiries, promptitude in prescribing, that nothing be done precipitately. It is proper to enter the chamber calmly, and not boisterously, not with petulance, noise, garrulity, nor an elevated voice, so that in approach, aspect, or any other manner, there can be nothing indecorous. In conversation with the sick, our questions must be grave, plain, and intelligible, without solecisms, pedantry, harshness, not foolishly advising them to be of good hope; neither are we to condemn the former errors of life too severely, which might lead the patient to suppose his case hopeless. The sick are generally pusillanimous and suspicious; and many medical men are so imprudent and loquacious, that they express their opinion openly, rendering the patients melancholy and timorous, who fear that which is related is their own condition. We should be extremely cautious in our expressions, and duly consider what is proper to be said to the sick and the attendants; and enforce with mildness and confidence the necessity of obeying all directions and injunctions, and of faithfully administering the remedies prescribed. In this way the sick can never be alarmed, and we have the best chance of success from the due exhibition of medicines. Some men are harsh and overbearing to the sick, and others too accommodating and flattering; both err exceedingly. Humanity, moderation, and suavity are indispensable, because, unless the patient looks on his physician almost in the light of a deity, he will never obey his precepts. Let him therefore be careful of his appearance, voice, manners, and actions, if he wish compliance with his precepts. These requisites captivate society in every rank and condition.

On entering the patient's chamber, a medical man is recommended to fix his eyes on the floor, or prudently on the countenance of the sufferer, to salute the attendants and patient politely; and then he is patiently to inquire of all, the age, temperament, habit, constitution, symptoms, and causes of the present illness; and he is also to recollect the season and state of the weather. All these things were comprehended in a single distich by the ancients—

Ars, ætas, forma, complexio, virtus,
Mos et symptoma, repletio, tempus et usus.

[31/32] After he shall have patiently heard the complaints and narration of the patient, and however tedious, of the attendants, he should consider the state of the apartment, the ventilation, warmth, moisture, and so on. He is next to learn the order of symptoms, and examine the state of the cerebral, respiratory, circulatory, and assimilatory functions. The countenance, tongue, respiration, and pulse, are to be examined; the condition of the bowels and urinary secretion ascertained; and the state of the uterine evacuation in women. He is to inspect all the egesta, expectoration, alvine, uterine, and urinary discharges. If these things be well observed, you will explain the feelings and state of the patient better than he can do himself; and, therefore, he will admire and proclaim your skill and excellence, and will repose more confidence in your opinion. The confidence of the patient is often a more certain cure, than can be accomplished by the practitioner and his medicines. In prognosticating the event to attendants and friends of the sick, we must duly consider his former and present condition; and only predict those things which we know must eventually happen. Thus, they will perceive, that you are not the cause of death, should it occur, but of health, should it be restored. It is from prognosis the world extols a physician, and celebrates his name with utility and honour.

In prescribing remedies, it is necessary to inquire the form which the patient prefers. In this, as in all other desires which are not injurious, he is to be gratified; and all medicines are to be made as palatable as possible. In some cases, the parents must persuade the child, and the child the parents, or one friend another, to take certain remedies. There is no objection to this, on the contrary, we must avail ourselves of it as an auxiliary, or life may be sacrificed. In many diseases the sick must be kept tranquil, and all visitors excluded. In this injunction we must be peremptory, and no one can be so unreasonable as to require admittance, at the risk of destroying the life of the patient. Some patients are exceedingly sad and melancholy, and these must be preserved from the visits of all who would further afflict them. The intemperate zeal of the clergy, and even when there is no necessity for their interference, is often fatal to the patient. Every remedy sanc- [32/33] tioned by the major part of the faculty ought to be faithfully employed. Too much physic is highly injurious to the constitution; and disease can often be cured as well without, as with medicines.

Those whom nature cures, are injured by medicines. Administer nothing timorously or rashly, and never change medicine without having given it a fair trial. A man who is constantly changing his remedies will not possess the confidence of his patient, because he affords strong evidence that he is ignorant of the means of cure.

Cleanliness, ventilation, and regulation of temperature in the sick chamber, are as valuable as medicines. Should the disease be incurable, and prove fatal, the medical attendant may pay a visit of condolence to the relatives which always looked on as a proof of his sincerity and friendship. In some cases he may opportunely mention patients whom he had cured of a disease similar to that under treatment; if these were the friends or acquaintances of the sufferer, so much the better; he serves the sick by inspiring confidence, and also increases his own reputation. If spoken to about compensation for his attendance, he should declare that the recovery of the patient was his chief concern; but in many cases he must stipulate for his fees, or he will be cheated. This may be done through the friends and attendants, but not with the patient when dangerously ill, unless required by himself. A medical man should be affable to all, but not too familiar with any, especially with the illiterate; and his conversation should be always dignified and reserved.

A physician should be always ambiguous in his prognosis, unless the most certain and infallible indications of health, or death be manifest. "Ambigue futura pande," says Macoppe, "predict the future ambiguously." The sick, their relations, domestics, and neighbours, are too desirous of knowing the result of disease. They wish us to be oracles and demi-gods instead of men. He who expects inheritance or title; he who loves a father, mother, wife, child, friend, or acquaintance; and he who hates an enemy, are all anxious to learn our opinion on the issue of disease. It is extremely difficult to form an accurate prognosis; and in a vast majority of cases it must be doubtful. Professional fame is often [33/34] injured by a prognosis. When it is doubtful, if the patient perishes, it is the fault of the disease and not of the practitioner; if he recovers, a new Esculapius is lauded to the skies. If, on the contrary, the patient dies after his recovery had been predicted, the medical attendant will be censured, and it will be said, he mistook the nature of the disease and caused death. He is to be moderate in his promises, and always rather pronounce a hope of recovery, than certain death. If a patient is deserted, and afterwards recovers by nature or chance, which often happens, the practitioner will incur much blame. But if he had afforded a hope of health, and death occurs, the ignominy is not so great, because many errors and excesses may take place, or even a new disease; and the commutation is much easier from health to death, than from death to health, which, in the course of nature, is almost impossible. A medical man should not interfere in recommending the patient to make a will, it is the strongest proof of despair, and a matter with which he has no concern. When he sees it necessary, he ought to apprize the friends or attendants, and let them advise the patient to arrange his affairs, in order to prevent disputes and litigation among his heirs. Even this requires great caution, for most probably other aid will be had, and should the patient ultimately recover, it will be to the injury of the first attendant. The friends, relatives, or clergy, are the most proper persons to advise the patient to arrange his affairs; but if the medical attendant be present, he, with a composed countenance, mixed with sincere hope, is to console the patient, and tell him, that the making of a will cannot affect his recovery, for many have done so, who have recovered and lived for many years, or are now in existence. When the patient is labouring under a disease that may speedily suspend the mental faculties, or endanger life, it is right to apprize his relatives or attendants, and advise them to propose sending for a clergyman, to afford the consolations of religion. In Catholic countries the law is, that the medical attendants should recommend a clergyman at the fourth visit, when the disease is acute. I feel convinced, that I have known some individuals whose lives were destroyed by the premature recommendation to see a clergyman; [34/35] and others who shared a similar fate when advised to settle their affairs by making their last will and testament. Others have fallen victims in consequence of the indiscreet zeal of the clergyman, more especially when he professes, as is sometimes the case, without any real pretension, a knowledge of medicine. Again, a decided opinion delivered by a medical practitioner, in whom great confidence is placed, may turn the balance in favour of life or death. The greatest caution is therefore necessary in forming a prognosis. Health or recovery ought never to be promised, lest it offend the Deity, who alone can decide the fate of the patient: if the healthy and robust are uncertain of tomorrow, how much more are the diseased and infirm. In making promises, many natural and auxiliary causes are to be considered, the compliance and obedience of the sick, the diligent application of remedies, and the sudden and unaccountable changes of diseases. We may state, there is or is not danger at present, that there are many changes in diseases, and that the case is in the hands of Divine Providence. Lastly; that every thing which art affords shall be carefully tried. How powerful the first aphorism of Hippocrates, "judicium difficile;" and also that of Macoppe, "ambigue futura pande."

Plato, Hippocrates, and Galen, maintained, that "medical men were justified in deceiving the sick for the cause of health." They held statesmen and physicians excusable for this breach of ethics, but no other class of society. The Christian religion, however, is opposed to mendacity in any circumstance. To the physician deception is however almost necessary, because his patients are often timid, suspicious, and observant of his words and aspect, and drawing unfavourable conclusions from his very appearance. David, in order to escape from king Lachish, changed his dress, and simulated insanity. It is a fact, that a tone, a word, a look, will destroy life in delicate and dangerous cases. How often must we disguise medicine, in order to conquer the aversion and idiosyncrasy of the sick. But mendacity is an evil by which we misrepresent the order of the Creator and of nature. We express not that which we think. We are not to do evil, that good may follow. Nevertheless, Abraham said he was his [35/36] wife's brother; Jacob said to his father, he was Esau, his eldest son; and the midwives of Egypt said they could not kill the Hebrew infants, a murderous injunction, for the breach of which they were justly commended. Official mendacity is less pernicious than that which is malicious. We must never promise recovery, without premising that it depends on the Divine will, on the due performance of all medical precepts, both by the patient and the ordinary attendants; and even then we must pronounce our prognosis with the greatest caution. The learned physician will bear in mind the great influence of change of weather on the human body; and the many changes a disease of the most favourable aspect may undergo. Presumption must be avoided. The presumptuous practitioner knows every thing, has read all works, and seen all diseases; the most difficult cases do not alarm him, the most delicate operations are undertaken by him with alacrity; in fact, nothing embarrasses him. He speaks of himself in flattering terms, he knows more of the profession than the whole body, he can cure consumption, cancer, hydrophobia; the first aphorism of Hippocrates does not apply to him; in fine, he believes he possesses the genius and power of Esculapius. Such is the presumptuous practitioner, though he is, in truth, very superficially informed, and generally an ignorant individual. He is an imitator of the uneducated and daring empiric, who professes to cure every one, and proclaims "all diseases incurable to the faculty cured here." No well-informed practitioner can be accused of such ridiculous vanity; he duly appreciates the difficulties and dangers in practice, the mutations and incurability of many diseases, and he is cautious in his prognosis.

Timidity, on the contrary, is a bad feature in the character of a medical practitioner. Some are frightened at every case that comes under their care, they have not courage to employ active or powerful remedies, they do not kill their patients, but allow them to die. They accuse medicine of the unfavourable issue, and not their own ignorance and culpable neglect, in not applying proper means of treatment.

There are others who may be considered fanatics in medicine, the partizans of this or that doctrine. In vain have [36/37] new discoveries changed the face of science, they are and remain incredulous. Such are the disciples of Abernethy, of Broussais, and of Hanhemann. They forget that diseases are as numerous as the sands of the ocean; that they depend upon a variety of morbid changes, and that they are successfully treated by the most opposite remedies. Thus the inflammations and fevers are cured by antiphlogistic measures, and sometimes by powerful stimulants; while the painful, spasmodic and nervous affections are best relieved by sedatives, tonics, and improvement of the general health.

A few words may be said on the duties of the sick, and their ordinary attendants during the cure of disease.

Hippocrates laid down, in his first aphorism, that the physician should not only excel in the due exercise of the best of his skill, but also direct his attention to the conduct of the patient and his attendants, as well as to all extraneous affairs. He said, the medical art consisted in three things, in disease, in the sick, and in the physician, the minister of nature and art. We cannot cure disease, unless the patient assists us. If he refuses remedies, the disease must in general remain, or prove fatal. Hence, the mortality of violent maniacal patients. The assistants are the minor order of medical men and domestics; and the physician must be acquainted with all their duties. Thus, Galen was of opinion, that he must be conversant with pharmacy, venesection, obstetricy, culinary affairs, and even menial servitude. Thus qualified, he can discover artifice, amend, and avoid it. The ordinary attendants should be chosen from those who are acquainted with the patient, and are most likely to pay implicit obedience to the medical directions, and who are accustomed to wait on the sick. They are to report all the symptoms at each visit, and preserve the apartment in the best order. They are to amuse the sick with cheerful conversation, and never indulge in frightful or ominous narrations; they are to preserve all the egesta, sputa, alvine, and urinary discharges, and whatever is rejected from the stomach, or is passed by hæmorrhage. Of all things they must avoid quackery, and never administer either diet or medicine, unless that which has been prescribed. They must not indulge in the use of ardent or inebriating liquors; they [37/38] must avoid informing the patient of any thing that may distress him; they must not exhibit any of their own food or drink to the sick; and, in order to compel them to comply with their duties, a faithful servant ought to be placed as a watch over them. They should not divulge any thing they may see or hear, but adopt the old adage – "Hear and see, and say nothing." Let the young physician duly consider, how much the event of the disease depends on the proper performance of the duties of nurses; and he will be very cautious in his prognosis.

This precept is to be chiefly observed by a medical man; never to visit a patient unless requested to do so. Should he volunteer his services, he renders himself liable to be suspected of ambition or of avarice. When one goes unsolicited, he may be certain he will not have the confidence of the sick, and will be deemed indigent or necessitous; and he cannot expect any reward for his spontaneous visitation. If known to the sick, he will be despised for his intrusion; if unknown, his conduct will be suspected. A visit unsolicited is not only useless and indecent, but suspicious. Another medical practitioner may be already in attendance, and will view any intrusion as highly unprofessional. The only exception to this conduct, is when we visit a friend; but even then we are to come as friends, and not in a professional capacity. Again, a patient might be alarmed at seeing a strange practitioner, having one in whom he reposes all confidence already in attendance. It is also derogatory to the dignity of the profession to be called in, through the recommendation or persuasion of nurses and others. Attendance even on friends must not be given, unless regularly requested. It is also improper to visit a patient at the request of a relative, unless the person affected, if rational, concurs in the request.

There are some avaricious and eccentric men who refuse medical aid, because it is expensive, but their friends solicit attendance in the usual way. It may be given, provided due compensation be made to the practitioner, but not otherwise. It is monstrously improper for eminent professional men to give advice gratuitously to the affluent at hospitals; it is a direct injury to the rising members of the profession. [38/39].

If a relation or contemporary who has quarrelled with a medical man be ill, ought we to visit him? Most certainly, even if he has called in another. A good and generous man will conquer evil with good, not considering the deserts of the patient, but what is right and meet for himself, and to promote the honour and glory of the Deity.

A patient is to be visited once a day in chronic cases, twice in acute, and oftener in the most acute. After a lapse of 40 days, two visits in the week may be sufficient; but in all cases they must be regulated by the patient or his friends. In acute and dangerous cases the friends are to be informed of the perilous condition of the sufferer, and should they not speak of the next visit, they are to be apprized of its necessity, however painful, to the practitioner. After convalescence attendance may be occasionally given, and a gratuity is not to be refused; but must not be accepted, if the patient declare his restoration to health. In such cases some complimentary visits ought to be made. Medical aid is to be afforded to every human being, friend or foe, native or stranger. A physician should not undertake the cure of himself, his wife, or his children in dangerous diseases, as his mind and reason will be perturbed, and unfit him for the arduous and difficult treatment of violent cases. A lawyer who pleads his own cause is said "to have a fool for his client."

Some men suppose if they once pay a physician he is to attend them through a succession of diseases; but who will be imposed on in this manner? Others who pretend friendship, will not call a practitioner, but should he visit them, will enjoy his labour without recompence. This is very often the conduct of relations, and of many towards young practitioners; in such cases attendance ought to be withheld, unless in the regular manner. What other class of society will act without reward? Will the clergy, or lawyers, or military, or kings, or any other class of society act so generously? Surely the labourer is worthy of his wages.

A patient is not to be deserted in the most hopeless condition, for recovery may still happen. Let the practitioner ever recollect the adage; "dum anima est, spes est," while there is life, there is hope. With respect to consultations, [39/40] those who form them should be regularly qualified, and only look forward to the welfare of the patient, and in preference to fame and lucre. The sick have a right to select those they please. No medical practitioner is justified in refusing to meet another in consultation, if both be duly educated and legally qualified, whether in the same university or another. The patient is not to suffer by their disputes, with which he can have little or no concern; the great object of a consultation is the cure or relief of the patient, and not disputation. Consultations are highly necessary in dangerous cases; and no man should ever object to them. A remedy may occur to one, which has escaped the recollection of another, "Et quod tu nescis fortassis novit Ofellus." The physician called in should be older and more eminent in the art than the former attendant. It is the duty of the attendant to communicate the history and treatment of the case, that both may consult what may be added to the cure. If a new remedy be proposed, it is to be tried if either of the attendants would use it himself, if afflicted in a similar manner to the patient. A few words may be said on the manner in which a medical man ought to behave on receiving compensation.

It is unnecessary to enumerate the arguments in proof of the justice of rewarding medical practitioners for their labours. An art that is not purchased, is disesteemed, therefore the correctness of the axiom,

> Exige dum dolor est, nam postquam pœna recessit,
> Audebit sanus dicere, multa dedi.[6]

And of another,

> Accipe, dum dolet.[7]

What will not a man give, to one, who liberates him from the greatest danger? Or is there any one so insane, as to spare his riches, when life is in danger? All that a man possesses, says Job, he will give for his life. Thus Philip, the father of

Alexander, said to his physician, after he had reduced his dislocated clavicle, "Accipe omnia quae voles, [40/41] quando quidem clavem habes."[8] Many similar examples might be quoted; for,

Medicis in morbis totus promittitur orbis,
Mox fugit a mente medicus, morbo recedente: [9]

A physician was said to possess three casts of countenance. When he converses with the healthy, he displays his ordinary one; when he approaches the sick, labouring under acute and painful disease, he is said to display his angelic one; and when he visits his patient after cure, and seeks compensation, then his aspect is satanic,

Dum præmia poscit medicus, Satan est. [10]

As health is above all temporal blessings, no one can give an adequate reward for its recovery. Nor has a medical man ever received sufficient compensation, for the labours and troubles which he experiences for the calamities and miseries of others, which he makes his own. What can compensate him for the continued anxiety, inconvenience, loss of rest, deprivation of ordinary pleasures and comforts, which he always experiences? The members of the church and law have time for recreation and amusement; they are not always employed, but the medical man must be ever at his post, his motto is "semper paratus;"[11] and it was for this reason, Dr. Johnson defined the duties of a medical man thus, "a truly melancholy attendance on misery, a mean submission to peevishness, and a continual interruption to rest and pleasure."[12] Soranus said, "if rewards be given, let them be accepted and not refused if they be not given, let them not be required; because however much any one can give, is inadequate value for the benefits conferred by medicine." The remuneration of medical practitioners ought to be regulated according to custom and to the rules of the profession. It ought, however, to be proportionate to the circumstances or pecuniary means of the sick, and to the standing and eminence of the practitioner. The fee of a guinea a visit is much too high for a [41/42] preponderating majority of society in most countries, and it has led even the most eminent physicians and surgeons in London, to pay two visits to those in the middle ranks in life for one *honorarium* or fee. It is unreasonable to expect, that a junior will be remunerated on the same scale as a senior; nor is this done in the church, at the bar, nor in the army or navy. The custom in France, Germany, Italy, and most foreign countries, of regulating fees according to the standing of the practitioner and means of the sick, is, in my opinion, much wiser and more beneficial to the junior members of the profession, than our system of guinea fees. In the countries now mentioned, the fee of a junior practitioner, a doctor in medicine and surgery, is, at first, about a shilling a visit; but, as practice increases, the fee is gradually raised to a sum equal to a guinea. The system of fees in this country has originated general practitioners, or surgeon-apothecaries, who generally receive no fee, but are remunerated by their medicines. This system was declared to be objectionable, by the heads of that body before Mr. Warburton's Parliamentary Committee, 1834, and all were in favour of small fees, which will most probably be enforced by law. Many eminent physicians and surgeons in this metropolis are in the habit of visiting a patient in the middle ranks of life, as often as three times for one fee; and they also cheerfully receive half a sovereign for a consultation at their own habitations. This being a fact, what chance has a junior physician or surgeon, of competing with them in practice, if even he was satisfied with the same remuneration? For these and many other reasons, it is expedient and necessary to regulate and proportion the fees of medical practitioners, so as to protect the junior members and surgeon-apothecaries. This subject will come under full consideration in the section following the chapter on education, and in that relating to the laws of the medical profession.

It is contrary to medical morals to bargain for a certain fee before a cure is effected. This is the system of empirics who make a contract, "no cure no pay." But physicians and surgeons in extensive practice, will not go several miles distant, or perform capital operations, without specifying the exact remuneration they require. They do not act in the [42/43] above manner, nor violate the ethical rule – "de mercede ne pasciscaris. Hoc impostoribus et circulatoribus relinque."[13]

Many presents are given to medical practitioners in addition to fees, and sometimes a large amount of property both in money and estates. Others bestow books, carriages, &c. In former times, it was considered derogatory to the profession to accept gifts of this kind in place of fees. Many instances of ample and magnificent fees and gifts to medical practitioners in past and modern times might be quoted.

A question has been discussed in all ages, namely, whether a medical practitioner ought to be punished for bad practice? The most eminent philosophers and legislators of old, were unanimously of opinion, that the errors of a medical man, if involuntary, should be forgiven. The father of medicine maintained, the only punishment should be his own ignominy. Plato de Republica says, "quivis medicus, si is, qui ab eo curatur, moritur, invito ipso purus sit secundum legem."[14] This opinion obtains in some civilized countries.[15] Pliny, lib. 29. is of the same opinion – "Nulla lex est, quæ puniat inscitiam capitalem medicorum, nullum exemplum vindictæ, discunt periculis nostris, et experimenta per mortes agunt, medicoque tantum hominem occidisse summa impunitas est, quando in hac artium sola eveniat, ut unicuique medicum se profitenti statim credatur."[16] Barbarians held an opposite opinion, and punished medical men with death, if they had failed in curing the sick. Manes promised to cure the son of the king of Persia, and having failed, he was ordered to be flayed, which sentence was executed. Zerbus having failed to cure Bassa, the emperor of the Turks, was instantly sabred by the soldiers. The experienced and talented Avenzoar, was thrown into chains for a similar failure.[17] But I must proceed to consider the other duties of medical practitioners. [43/44]

Notes

1. Editors' note: Only in the next chapter, p. 46 in the original, does Ryan acknowledge that all the ethical reflections in this chapter "were compiled from a work by Roderic a Castro, entitled 'Medicus-Politicus sive de officiis Medico-Politicis tractatus, quatuor distinctus libris: in quibus non solum bonorum medicorum mores ac virtutes exprimuntur, malorum vero fraudes et impostures deteguntur: verum etiam pleraque alia circa novum hoc argumentum utilia atque

jucunda exactissime proponuntur. (Hamburg. anno 1614.)'" As noted in the Introduction, a full translation of this title might be rendered as: "The State Physician: Treatise concerning the office of State Physician in four separate books in which not only are good, medical mores and virtues described, but even deceptions of evil and deceit are revealed: truly also a new argument is most precisely proposed by another method both useful and agreeable. The full work being fit for a physician, student of illness, attendant upon the ill, and by all others of letters, and also by supporters of state education."

2. Editors' note: "Why, sacred longing for gold, do you not constrain the hearts of mortals?" (Virgil, Aeneid, 3.57–8).
3. Editors' note: "A humble mind, zeal for learning, a quiet life, silent investigation, poverty, a foreign land" (From Bernard of Chartres, Policraticus).
4. Author's note: It is a common custom in England to frame and glaze diplomas.
5. Editors' note: "a garrulous doctor causes another sickness."
6. Editors' note: "Collect while the pain still persists, for after it recedes, he will dare to say, once healthy, 'I suffered much.' "
7. Editors' note: "Accept [the offered fee] while [the patient] still suffers [from the throes of pain]."
8. Editors' note: "Take anything you wish, once you have indeed reduced the clavicle."
9. Editors' note: "In sickness the patient will promise the whole universe to the physician; but the physician soon slips from the mind when the illness subsides."
10. Editors' note: "When the physician demands recompense, he is Satan."
11. Editors' note: "Always ready."
12. Author's note: Thus the axiom, Medicè vivere est miserè vivere.
13. Editors' note: "Concerning price, do not grow fat on high fees. Leave such practices to imposters and peddlers."
14. Editors' note: This is a paraphrase rather than an actual quotation from Plato's Republic: "I argue that any physician is blameless according to the law if he attempts to cure someone who later dies."
15. Author's note: Not in Great Britain and Ireland, a civil, and even a criminal action, may be instituted for mala praxis, if death ensues.
16. Editors' note: "There is no law in existence whereby to punish the ignorance of physicians, no instance before us of capital punishment inflicted. It is at the expense of our perils that they learn, and they experimentalize by putting us to death, a physician being the only person that can kill another with sovereign impunity. To all this, however, we tend to pay no attention, [due to the seductive hope of recovery each person feels when ill]." This is a rather fractured, out-of-order quotation from Pliny the Elder, Natural History, Chap. 8, trans. John Bostock and H.T. Riley.
17. Author's note: This horrible and barbarous law is still in force in Turkey.

Chapter IV
Ethics of the Present Age

I will never set politics against ethics, for true ethics are but as a handmaid to divinity and religion.

– Bacon

It is not easy to conceive the reason why the cultivation of ethics, a matter of primary importance to the success of medical practitioners, in the commencement of their career, should be almost totally neglected in the medical schools of an age so enlightened as the present. The fact is so, however incomprehensible it may appear. It is now the custom to initiate men into the mysteries of medicine, without the slightest allusion to the duties they owe each other or the public; or to the difficulties to be encountered on the commencement of their practice. Hence arise the frequent misunderstandings, disputes, and improper behaviour between medical practitioners, which are so disreputable and injurious to the dignity and interests of science. From some cause which remains to be explained, the majority of medical professors have excluded the discussion of ethics from their instructions; the faculties of physic and surgery have acted in like manner, so that there is no code of ethical institutes to be referred to, in the daily violations of those high moral principles, which have always characterized the true cultivators of medicine. The moral statutes and obligations which are required by some of our colleges, are so few, and so little known, that they are nearly useless; they are seldom observed, obeyed, or enforced. Indeed, the only works we have on the subject are those of Dr. John Gregory and Dr. Percival; but these are not deemed authority, nor are they perused by medical students. The former was published above half a century since; the latter previous to the changes made in the constitution of the profession by recent legislation – and both unsuited to the state of the profession at the present time. [44/45] There is, therefore, a fair field for further observations upon the subject.

In the subsequent remarks it will be necessary to describe the present state of the profession, and necessarily comprehend the condition of every class of practitioners. To execute this task in a satisfactory manner to all parties, is not to be contemplated, as the conduct of all richly deserves animadversion and censure. Truth, justice, impartiality, and an ardent desire to promote the dignity of my profession, shall guide me in the execution of my subject; and I am confident that I cannot be accused of partiality towards any denomination of the profession. My motto is,

H. Brody et al. (eds.), *Michael Ryan's Writings on Medical Ethics*, Philosophy and Medicine, Vol 105,

"amicus Socrates, amicus Plato, sed magis amica veritas." It is an old adage, that "it is impossible to please all parties;" neither shall I attempt it, nor endeavour to please one party more than another. Here I follow the steps of Dr. Gregory, who fearlessly avowed his determination to expose abuses, whatever might be the animadversions of his contemporaries. "Whatever opposition," says the editor of the Duties of a Physician, in 1770,

> this part of the work may meet with, from those who find their own foibles, or rather vices, censured with a just severity, the ingenuous part of mankind, however, will not fail in bestowing that degree of applause so justly due to its merit. At present, there seems to be a general disposition in mankind to expose to their deserved contempt *those quackish, low, and illiberal artifices,* which have too long disgraced the profession of medicine. It is therefore hoped, that the general spirit will have a remarkable tendency to promote this laudable end; and that it will excite men of influence and of abilities to exert themselves in crushing that arrogance which has frequently served to cover the ignorance of many practitioners of medicine, by means of which alone they acquire such a share of practice as they are by no means entitled to.

The same defects and abuses on which Dr. Gregory animadverted still remain uncorrected, and consequently are deserving of further exposure; in fact, they are too glaring to be defended, except by those whose personal interests render them insensible to the important advantages of reformation and improvement. A faithful picture of modern medical [45/46] ethics is certainly much wanted; a comparison of what the profession is and was, may be entertaining and instructive to a large majority of my readers.

The only works on medical ethics which can be cited, are those of Drs. Gregory and Percival; and to these will be added the oaths required by one or two of our colleges, and the moral statutes sanctioned by them. These must be laid under free contribution. The first entitled, "Lectures on the Duties and Qualifications of a Physician," has justly received the universal approbation of the profession for more than half a century; and is an excellent abridgement of the maxims of preceding writers, and on many occasions their language is quoted *ipsissimis verbis*. In proof of this assertion, it is only necessary to compare the author's observations with the ethics in the last article, which were compiled from a work by Roderic a Castro, entitled "Medicus-Politicus sive de officiis Medico-Politicis tractatus, quatuor distinctus libris: in quibus non solum bonorum medicorum mores ac virtutes exprimuntur, malorum vero fraudes et imposturæ deteguntur: verum etiam pleraque alia circa novum hoc argumentum utilia atque jucunda exactissime proponuntur. Hamburg. anno [...]1614."[1] The comments and original views of Dr. G. are, however, numerous and valuable, and will be esteemed so long as medicine will be cultivated. As his work is almost obsolete, I trust that an analysis of its contents may be presented in a form which cannot fail to meet the eye of the industrious student. The author's remarks on the dignity and importance of medicine, and on the genius and education required for its proper cultivation, need not be inserted, as these points have been already discussed in a former section. The extracts I shall give do not obviate the necessity of referring to the original work – a production that ought to be in the hands of every practitioner.

Dr. Percival's valuable and truly classic work was re-published in 1803, and completed what Dr. Gregory had omitted. It consisted of the following chapters:

(I) On Professional Conduct in Hospital Practice; (II) In Private Practice; (III) In relation to Apothecaries; and (IV) In Cases Which Require a Knowledge of the Laws; to which were added, a Discourse on Hospital Duties, being the substance of a Sermon preached [46/47] by the Rev. T. B. Percival, and some valuable notes and illustrations by the author himself. It was his intention to have treated of the powers, privileges, honours, and emoluments of the faculty, an object he did not accomplish. The work was arranged in 1792, as a code of institutes and precepts for the professional conduct of the physicians and surgeons of the Royal Manchester Infirmary, and afterwards extended into a system of Medical Ethics, "in which," says his biographer, "he has drawn a portrait of himself, by tracing, with his own hand, what sort of character a physician ought to be." The work consists of several aphorisms, on which very inappropriate comments have been made by the anonymous editor of the last edition. Of the original opinions of Dr. Percival I shall avail myself; and therefore cannot be considered to interfere with the copyright of the last and worst edition of his work.

In citing the opinions of Dr. Percival and others, I claim the usual privilege of all writers, to collect and arrange materials from all available sources, which are public property. This privilege is sanctioned by the laws of the country. In a word, I shall endeavour to prove that the most eminent members of the profession are the strongest advocates of a branch of education which has been most preposterously overlooked and neglected. Want of leisure precludes me from making sufficient research for the compilation of a complete system of ethics; but I trust I shall be able to accumulate a mass of facts, which will perfect the subject they so ably and laudably commenced.

The following is a condensed detail of Dr. Gregory's opinions on the duties of medical practitioners:

Physicians, considered as a body of men who live by medicine as a profession, have an interest separate and distinct from the honour of the science. In pursuit of this interest, some have acted with candour, with honour, with the ingenuous and liberal manners of gentlemen. Conscious of their own worth, they disdained every artifice, and depended for success on their real merit. But such men are not the most numerous in any profession. Some impelled by necessity, some stimulated by vanity, and others anxious to conceal [47/48] ignorance, have had recourse to various mean and unworthy arts to raise their importance among the ignorant, who are always the most numerous part of mankind. Some of these arts have been an affectation of mystery in all their writings and conversations relating to their profession; an affectation of knowledge, inscrutable to all, except the adepts of the science; an air of perfect confidence in their own skill and abilities; and a demeanour solemn, contemptuous, and highly expressive of self-sufficiency. These arts, however well they might succeed with the rest of mankind, could not escape the censure of the more judicious, nor elude the ridicule of men of wit and humour. The stage, in particular, has used freedom with the professors of the salutary art; but it is evident, that most of the satire is levelled against the particular notions or manners of individuals, and not against the science itself.

The practice of the healing art affords a vast field for the exercise of humanity. A physician has numberless opportunities of giving that relief to distress, which is not to be purchased with the wealth of India. This, to a benevolent mind, must be one of the greatest pleasures. But, besides the good which a physician has it often in his power to do, in consequence of skill in his profession, there are many occasions that call for his assistance as a man; – as one who feels for the misfortunes of his fellow-creatures. In this respect he has many opportunities of displaying patience, good nature, generosity, compassion, and all the gentler virtues that do honour to human nature. A physician endowed with this virtue diffuses consolation and comfort; he employs his talents, his time, and his fortune, in removing misery. He is the friend of the poor and unfortunate. The victims of misery, those of disease and of death, offer a distressing, a shocking picture. Here is an opportunity of doing good; in such cases man can assist man without any witness; it is here that generosity, benevolence, and tender pity, may be exerted to dry up tears and wailings. Is there any other class of citizens that perform these duties with so much zeal and courage as medical practitioners? These labours, these pleasures, are those of almost all the practitioners of medicine; they have their first lessons from the poor and miserable, who feel their [48/49] benevolence and virtue. The disinterested cares given to the sick are always recompensed, the practitioner invariably finds, that his benevolence and charity are the foundation of his celebrity. When he acquires a high reputation, he must not forget the means by which he has acquired it; the poor and miserable are the base of his fame and prosperity. When he acquires fame and wealth, he still gives his aid to the poor at hospitals, dispensaries, or at his own residence. The faculty has often been reproached with hardness of heart, occasioned, as is supposed, by their being so much conversant with human misery. I hope and believe the charge is unjust; for habit may beget a command of temper, and a seeming composure which is often mistaken for absolute insensibility. But, by the way, I must observe, that, when this insensibility is real, it is a misfortune for a physician, as it deprives him of one of the most natural and powerful incitements to exert himself for the relief of his patient. On the other hand, a physician of too much sensibility may be rendered incapable of doing his duty, from anxiety, and excess of sympathy; which cloud his understanding, depress his mind, and prevent him from acting with that steadiness and vigour, upon which perhaps the life of his patient in a great measure depends.

Though a physician possess that enlarged medical genius already described, yet talents of another kind are necessary. He has not only for an object the improvement of his own mind, but he must study the temper, and struggle with the prejudices of his patient, of his relations, and of the world in general; nay, he must guard himself against the ill offices of those, whose interest interferes with him; and it unfortunately happens, that the only judges of his medical merit are those who have sinister views in concealing or depreciating it. Hence appears the necessity of a physician's having a large share of good sense, and knowledge of the world, as well as medical genius and learning.

Such are the genius and talent required in a physician; but a certain command of the temper and passions, either natural or acquired, must be added, in order to give them their full advantage. Sudden emergencies often occur in practice, and diseases often take unexpected turns, which are [49/50] apt to flutter the spirits of a man of lively parts and of a warm temper.

Accidents of this kind may affect his judgment in such a manner as to unfit him for discerning what is proper to be done; or, if he does perceive it, may nevertheless render him irresolute. Yet such occasions call for the quickest discernment, and the steadiest and most resolute conduct; and the more, as the sick so readily take the alarm, when they discover any diffidence in their physician. The weaknesses too and bad behaviour of patients, and a number of little difficulties and contradictions which every physician must encounter in his practice, are apt to ruffle his temper, and consequently to cloud his judgment, and make him forget propriety and decency of behaviour. Hence appears the advantage of a physician's possessing presence of mind, composure, steadiness, and an appearance of resolution, even in cases where, in his own judgment, he is fully sensible of the difficulty.

I come now to mention the moral qualities peculiarly required in the character of a physician. The chief of these is humanity; that sensibility of heart which makes us feel for the distresses of our fellow-creatures, and which of consequence incites us in the most powerful manner to relieve them. Sympathy produces an anxious attention to a thousand little circumstances that may tend to relieve the patient; an attention which money can never purchase: hence the inexpressible comfort of having a friend for a physician. Sympathy naturally engages the affection and confidence of a patient, which in many cases is of the utmost consequence to his recovery. If the physician possesses gentleness of manners, and a compassionate heart, and what Shakspeare so emphatically calls "the milk of human kindness," the patient feels his approach like that of a guardian angel ministering to his relief; while every visit of a physician who is unfeeling, and rough in his manners, makes his heart sick within him, as at the presence of one who comes to pronounce his doom. Men of the most compassionate tempers, by being daily conversant with scenes of distress, acquire, in process of time, that composure and firmness of mind so necessary in the practice of medicine. They can feel whatever is amiable in pity, without [50/51] suffering it to enervate or unman them. Such physicians as are callous to sentiments of humanity, treat this sympathy with ridicule, and represent it either as hypocrisy, or as the indication of a feeble mind. That sympathy is often affected, I am afraid, is too true; but this affectation is easily seen through. Real sympathy is never ostentatious, on the contrary, it rather strives to conceal itself. But what most effectually detects this hypocrisy is a physician's different manner of behaving to people in high and people in low life; to those who reward him handsomely, and those who have not the means to do it. A generous and elevated mind is even more shy in expressing sympathy with those of high rank, than with those in humbler life; being jealous of the unworthy construction so usually annexed to it. The insinuation that a compassionate and feeling heart is commonly accompanied with a weak understanding and a feeble mind, is malignant and false. Experience demonstrates, that a gentle and humane temper, far from being inconsistent with vigour of mind, is its usual attendant; and that rough and blustering manners generally accompany a weak understanding and a mean soul, and are indeed frequently affected by men void of magnanimity and personal courage, in order to conceal their natural defects.

There is a species of good humour different from the sympathy I have been speaking of, which is likewise amiable in a physician. It consists in a certain gentleness and flexibility, which makes him suffer with patience, and even apparent cheerfulness, the many contradictions and disappointments he is subjected to in his practice. If he is rigid and too minute in his directions about regimen, he may be assured they will not be strictly followed; and if he is severe in his manners, the deviations from his rules will as certainly be concealed from him. The consequence is, that he is kept in ignorance of the true state of his patient; he ascribes to the consequence of the disease what is merely owing to irregularities in diet, and attributes effects to medicines which were perhaps never taken. The errors which in this way he may be led into are sufficiently obvious, and might easily be prevented by a prudent relaxation of rules that cannot well be obeyed. The [51/52] government of a physician over his patient should undoubtedly be absolute; but an absolute government very few patients will submit to. A prudent physician should therefore prescribe such laws, as though not the best, are yet the best that will be observed; of different evils, he should choose the least; and, at no rate, lose the confidence of his patient, so as to be deceived by him as to his true situation. This indulgence, however, which I am pleading for, must be managed with judgment and discretion; as it is very necessary that a physician should support a proper dignity and authority with his patients, for their sakes as well as his own.

There is a numerous class of patients who put a physician's good nature and patience to a severe trial; those I mean who suffer under nervous ailments. Although the fears of these patients are generally groundless, yet their sufferings are real; and the disease is as much seated in the constitution as a rheumatism or a dropsy. To treat their complaints with ridicule or neglect, from supposing them the effect of a crazy imagination, is equally cruel and absurd.[2] They generally arise from, or are attended with, bodily disorders obvious enough; but supposing them otherwise, still it is the physician's duty to do every thing in his power for the relief of the distressed. Disorders of the imagination may be as properly the object of a physician's attention as those of the body; and surely they are, frequently, of all distresses the greatest, and demand the most tender sympathy; but it requires address and good sense in a physician to manage them properly. If he seems to treat them slightly, or with unseasonable mirth, the patient is hurt beyond measure; if he is too anxiously attentive to every little circumstance, he feeds the disease. For the patient's sake therefore, as well as his own, he must endeavour to strike the medium between negligence and ridicule on, the one hand, and too much solicitude about every trifling symptom on the other. He may sometimes divert the mind, without seeming to intend it, from its present sufferings, and from its melancholy prospects of the future, by insensibly [52/53] introducing subjects that are amusing or interesting; and sometimes he may successfully employ a delicate and good-natured pleasantry.

This class of patients are in general extremely unreasonable. They are constantly complaining, nothing does them good, and every thing injures them. They suppose that they labour under one or many incurable diseases. They complain of every disease, and more than every disease; or, as it is well expressed, "de omnibus rebus, et quibusdam aliis."[3] They do not know, nor can they understand, that disease in any part of the body may derange the whole organs, in consequence of the universal nervous connection; or that all their unpleasant, and to them unaccountable sensations, are as obvious as the noon-day sun to every well-educated medical practitioner. They are excessively, or morbidly sensitive; they watch every cast of countenance; they put the most unfavourable construction on every sentence uttered by their medical attendant; they fear the worst; and, should he deliver an ambiguous or doubtful prognosis, they fear him much more than they are pleased with him. Their esteem and respect are very variable and inconstant, and very readily converted into hatred and contempt. No matter how much you relieve them, even after the most eminent practitioners have failed, you are suddenly dismissed without the slightest reason, more especially when women are your patients. I have been repeatedly treated in this way, and after many other practitioners have been consulted, again recalled. But we must not forget the adage, "*varia et mutabilis fœmina,*" in plain language, women are variable and changeable; and I must add, when excessively nervous or hypochondriacal, very often most unreasonable. Disease, however, is their excuse; and every humane physician will forget and forgive their inconsistencies, however remarkable. Such patients may imagine the most improbable conditions of constitution, as being made of glass, butter, &c., being a grain of wheat, pregnant of a hen and her chickens, & c.; examples of which have fallen under my own observations. The fact is, that many nervous persons and hypochondriacs do not possess perfect reason; and, therefore, great allowances must be made for them. [53/54]

We sometimes see a remarkable difference between the behaviour of a physician at his first setting out, and afterwards when fully established in reputation and practice. In the beginning, he is affable, polite, humane, and assiduously attentive to his patients; but afterwards; when he has reaped the fruits of such a behaviour, and finds himself independent, he assumes a very different tone; he becomes haughty, rapacious, and careless, and often somewhat brutal in his manners. Conscious of the ascendancy he has acquired, he acts a despotic part, and takes a most ungenerous advantage of the confidence which people have in his abilities.[4] A physician, by the nature of his profession, has many opportunities of knowing the private characters and concerns of the families in which he is employed. Besides, what he may learn from his own observation, he is often admitted to the confidence of those who perhaps think they owe their life to his care. He sees people in the most disadvantageous circumstances, very different from those in which the world views them – oppressed with pain, sickness, and low spirits. In these humiliating situations, instead of wonted cheerfulness, evenness of temper, and vigour of mind, he meets with peevishness, impatience, and timidity. Hence it appears, how much the characters of individuals, and the credit of families, may sometimes depend on the discretion, secrecy, and honour of a physician. Secrecy is particularly requisite where women are concerned. Independently of the peculiar tenderness with which a woman's character should be treated, there are certain circumstances of health which, though in no respect connected with her reputation, every woman, from the natural delicacy of her sex, is anxious to conceal; and, in some cases, the concealment of these circumstances may be of consequence to her health, to her interest, and to her happiness.

Temperance and sobriety are virtues peculiarly required in a physician. In the course of an extensive practice, difficult cases frequently occur, which demand the most vigorous exertion of memory and judgment. I have heard it said of some eminent physicians, that they prescribed as well when [54/55] drunk as when sober. If there was any truth in this report, it contained a severe reflection against their abilities in their profession. It showed that they practised by rote, or prescribed for some of the more obvious symptoms, without attending to those nice peculiar circumstances, a knowledge of which constitutes the great difference between a physician who has genius and one who has none. Drunkenness implies a defect in the memory and judgment; it implies confusion of thought, perplexity, and unsteadiness; and must therefore unfit a man for every business that requires the lively and vigorous use of his understanding.

An obstinate adherence to an unsuccessful method of treating a disease, must be owing to a high degree of self-conceit, and a belief in the infallibility of a system. It has been the cause of the death of thousands. Patients ought to be indulged in every thing consistent with their safety; and if they are determined to try an improper or dangerous medicine, a physician should refuse his sanction, but he has no right to complain of his advice not being followed. A physician is often at a loss in speaking to his patients of their real situation, when it is dangerous. A deviation from truth is in this case both justifiable and necessary. It often happens that a person is extremely ill, but he may recover if he is not informed of his danger. Again, a man may not have settled his affairs, though the future happiness of his family depends on his making a settlement. In such cases the physician may apprize the friends, and occasionally the patient, of the necessity of the arrangement and disposal of his property. In all dangerous cases, the real situation of the patient should be communicated to his nearest relatives, as it gives them an opportunity of calling other assistance, if they think it necessary.

The patient is not to be deserted when his case is despaired of; it is as much the duty of a physician to alleviate pain, and to smooth the avenues of death, when inevitable, as to cure diseases; his presence and assistance as a friend may be both agreeable and useful, where his skill is of no further avail. In some cases we should caution the indiscreet enthusiasts among the clergy against too much zeal, as they often [55/56] terrify the patient, and contribute to shorten a life which might otherwise be saved.

Medical men should never involve their patients in private and professional quarrels, in which the sick can have little or no concern. All personal feelings should be forgotten in consultations, the good of the patient ought to be the chief and only consideration. The quarrels of the faculty, when they end in appeals to the public, generally hurt the contending parties, discredit the profession, and expose it to ridicule and contempt. Nothing can justify the refusal to consult but want of temper, nor can such circumstances as the university where a person has taken a degree, "or whether he had any degree at all, justify the refusal." This assertion, I may observe, is at variance with the usages of the profession, though society has sanctioned it. Fellows of the College of Physicians refuse to meet graduates of all the British and Foreign universities in consultation, until admitted into the College. But of this hereafter. It becomes young practitioners to be particularly attentive to the propriety of their behaviour when consulting with their seniors. Besides the respect due to age, these are entitled to a particular deference from their longer and more extensive experience, provided it be scientific.

The revolutions indeed of medical hypotheses and systems are so quick, that an old and a young physician seldom reason in the same way on subjects of their profession; although the difference is sometimes rather apparent than real, when they use only a different language to express sentiments essentially the same. But it generally happens, that the speculations which principally engage the attention of young physicians seldom in any degree affect their practice; and therefore, as they are in a great measure foreign to the business, they should never introduce them in medical consultations. They show equal want of sense and good manners, when they wantonly take opportunities of expressing contempt for opinions as antiquated and exploded, in which their seniors have been educated, and which they hold as firmly established. A little reflection might teach them, that it is not impossible but in the course of a few years, their own most favourite theories [56/57] may be discovered to be as weak and delusive as those which have gone before them: and this should lead them to consider how sensibly they may be hurt themselves, when they find those idols of their youth attacked by the petulant ridicule of the next generation; when, perhaps, they are arrived at a time of life when they have neither abilities nor temper to defend them.

There are, however, many old practitioners who have not kept pace with science, and who despise all new remedies. There are many who oppose and ridicule the stethoscope, the use of the alkaloid medicines, such as iodine, strychnine, &c.; as if these discoveries were not made known by Divine Providence, and as if medicine was a perfect and positive science. If a practitioner is ignorant of the value of new discoveries, he is not justified in abusing them; and, if they have been amply proved to be valuable, they ought to be adopted for the relief of the sick. Macoppe has ably commented on this point: "Si nova inventa, si elegantes recentiorum ignoras eas spurco lividoque non vituperes ore. Tuas liceat extollere, pedibus non alias calcare."[5] It is a just opinion, that a man who is ignorant of new remedies or discoveries ought not to despise them; he may extol his own, but must not condemn those of others. Our predecessors have done much; we have done much; but a vast deal remains to be done by our successors. It is to be recollected, that diseases are cured with remedies, and not by disputation. "When you are an old practitioner, you must not be ashamed to praise the new remedy of a junior, when more efficacious than your own. You must not look on the young practitioner with a jaundiced eye; he has left the schools, and is generally more conversant with the actual state of science than you are. Praise what is just, be silent of what is doubtful or inefficacious. Do not praise first, and vituperate afterwards."

A routine practitioner is ignorant of the progress of science, and is consequently self-opinionated. He is like a machine that always performs the same evolutions. He dispenses with all study, and he is content with what the vulgar consider experience.

This experience is necessarily false; for how can any man [57/58] exercise an art of whose principles he is ignorant ? Dewees has well observed, "that an obstetrician may have thirty years experience, but, without a knowledge of first principles, he is as ignorant and dangerous as he who has had the slightest experience." A question has been asked, whether a senior or junior physician ought to be preferred? It has been answered, that, in common disorders, a senior is preferable; but a junior in difficult and complicated diseases. "I hold," says Kyper, "that a physician cannot be experienced, unless he is learned; because an ignorant man does not know how to make judicious observation, or to deduce correct experience. No one can be a good practitioner unless he has been properly educated in public schools and hospitals, and has kept pace with the progress of science." Students are now better informed than their teachers were 50 years ago; they are taught all the new doctrines, and must excel those who are content with old ones. The ablest physician is he who is of middle age, and possesses real scientific knowledge, and whose judgment is formed on science, after sober and extensive observation on the poor, and at hospitals and dispensaries. He is not less informed than a young practitioner, and has the superior advantage of great experience. He is judicious in consultations, intrepid in dangers, competent to anticipate results, fertile in his resources, and endowed with great sagacity. Knowledge makes a young man old; and ignorance makes an old man a student. Talent compensates for the want of age,

> Quid numeras annos, vixi maturior annis;
> Acta senem faciunt, hæc numerandi tibi.[6]

It is not a head adorned with grey hairs that establishes merit; it is superiority in consultation, and at the bed-side of the sick. A young man may be a great physician, but an old man can scarcely be a great surgeon, and decidedly not a great physician. Talent, and not years, makes a good physician or surgeon.

A young man, gifted with genius for medicine, will in a few years be a great practitioner; but a man of 60 years of age, though he has seen 100,000 patients, will never be a good physician or surgeon, if deprived of this valuable gift of nature. [58/59] It is therefore an error to imagine, that the best practitioner is he who has seen the most cases. This is a popular prejudice; but, as Zimmerman observes, they do not inquire whether such a practitioner has received a proper medical education, whether he is a man of penetration and genius, though he is grey-headed. It is for this reason that many prefer an old to a young practitioner; they consider senescence and experience inseparable. They cannot distinguish scientific experience from common routine. It is also a fact, that old practitioners entertain the vulgar notion; in their opinion, a young man, of the greatest talent, is only a young man, and they can never believe that he can be equal to themselves. In their consultations and writings they maintain their superiority, even when the public confidence is being withdrawn from them, and when the world thinks they ought to retire. They forget, or perhaps have never seen the saying of Galen, that men of mere experience, without previous education, may be considered idiots; but the axiom of this venerated author is so apposite, that I must quote it: "Medicos qui solam experientiam sequuntur, non admittimus; quoniam ipsi idiotæ faciunt, quæ vident inspicientes, et rerum quidem eventum continentes, causam autem ignorantes."[7] This remark applies to uneducated practitioners only, and not to the scientific and experienced.

The latter, when advanced in life, are esteemed and cherished by the public and the profession, they are the Mentors of their juniors; their sentiments are received with profound respect; and they are heard with a religious emotion in the lecture-room and in consultations.

But as age advances the intellectual faculties become enfeebled. Horace said,

> Multa senem circumveniunt incommoda.[8]

And Virgil expressed the same opinion:

> Tarda senectus Debilitat vires animi mutatque vigorem.[9]

On the arrival of senescence the most renowned physicians have lost all their practice. They have often proclaimed their prerogative of experience, and fruitlessly endeavoured to deprecate their juniors. They have said that these had little patience, [59/60] no assiduity, circumspection; their impetuosity confuses them, they cannot coolly observe nature, judge accurately, preserve constancy, or form an accurate opinion of the nature or treatment of diseases.

It must, however, be obvious, that these assertions are groundless, because there is no reason why young practitioners, who are properly educated, who are conversant with the old and new opinions, and who are in the prime of life, when the intellectual faculties are fully developed, should not observe, think, reason, and practise, as well as their seniors. Many proofs might be adduced to support this statement, but a few shall suffice. Harvey discovered the circulation of the blood at an early age; Baglivi, who was the restorer of medicine, died at the age of 39; Bichat died at the age of 31 years.[10]

Indeed, a man who is not a good physician or surgeon at the age of 30, provided he has enjoyed ample opportunities, will never be one; for it is not years, but knowledge, that constitutes such a character.

A young physician of talent, who has received a good and extensive education, who has studied with assiduity all the elementary sciences, as well as the practice of medicine, surgery, obstetricy, and pharmacy; who has read all the best works in the vernacular and other languages, who has rigidly observed the effects of medicines for years, will far surpass his contemporaries in the aggregate, and leave them, whether old or young, immeasurably behind him. It is true, he may learn wisdom and caution from his seniors, when scientific and acquainted with the actual state of the science and practice of the healing art; but this very seldom happens. Those advanced in age, and actively engaged in practice, are generally unacquainted with the rapid progress of medicine; and their opinions are too often erroneous. Nevertheless, the junior is bound to respect his senior, because to honour a master is to honour one's self. Age is honourable, and ought to be always respected.

Dr. John Gregory defended the necessity of medical men being versed in all the branches of the healing art, and concludes by observing

> Every department of the profession is respectable, when exercised with capacity and integrity. I only contend for an evident truth, either that the different [60/61] branches should be separately professed, or, if one person will profess all, that he should be regularly educated to, and thoroughly master of all. I am not here adjusting points of precedence, or insinuating the deference due to degrees in medicine. As a doctor's degree can never confer sense, the title alone can never command regard; neither should the want of it deprive any man of the esteem and deference due to real merit. If a surgeon or apothecary has had the education, and acquired the knowledge of a physician, he is a physician to all intents and purposes, whether he has a degree or not, and ought to be respected and treated accordingly. In Great Britain, surgery is a liberal profession. In many parts of it, surgeons or apothecaries are the physicians in ordinary to most families, for which trust they are often well qualified by their education and knowledge; and a physician is only called where a case is difficult, or attended with danger. There are certain limits, however, between the two professions, which ought to be attended to: as they are established by the customs of the country, and by the rules of their several societies. But a physician of a candid and liberal spirit, will never take advantage of what a nominal distinction, and certain privileges, give him over other men who, in point of real merit, are his equals; and will feel no superiority, but what arises from superior learning, superior abilities, and more liberal manners. He will despise those distinctions founded in vanity, self-interest, or caprice; and will be careful that the interests of science and of mankind shall never be hurt, on his part, by a punctilious adherence to formalities.

Much stress has been laid on the formality of a physician's dress, but there is no reason in preferring one garb to another. In some cases there is great impropriety in having any distinguishing formality in dress and manners. Negligence and luxuriousness of dress are extremes, and ought to be avoided. A respectable appearance, propriety, convenience, and elegance without pretension, are the characters which ought to mark the costume of a medical practitioner. Carelessness or foppishness in dress are unworthy of the followers of the healing art. The vulgar attach great importance to [61/62] personal appearance. The professions of theology and law take advantage of this weakness; and still continue to wear peculiar habiliments, while the scarlet robes, velvet caps, and gold-headed canes of the professors of physic are abandoned. The divines and juris-consults are, perhaps, as wise as the Esculapians, and though the latter may dress as they please, I cannot help agreeing in opinion with those who maintain that external appearance is every thing with the world. For this reason the dress of medical men should be grave and elegant, not showy or *à la mode*. A cravat tied in the newest fashion, a coat of the colour and cut of the day, and all the other frivolities of the passing hour, are unworthy of the professors of the noble science of healing. A philosophic physician allows his tailor to equip him. Macoppe makes some happy remarks on moustaches, and contends that knowledge does not depend on hair, which characterizes the head or hope of the flock – vir gregis, ipse caper, but is no proof of philosophical or medical knowledge. Perfumery and odours were also condemned by Hippocrates, as they are often disagreeable and hurtful to delicate patients, especially to nervous and delicate women in whom they may excite fainting, vomiting, spasms, &c. The father of medicine taught, that a medical practitioner ought to avoid all odoriferous apparel, and that he was most perfumed who was not at all perfumed: optime olet medicus quum nihil olet. Septal, Roderic a Castro, Triller, and other medical moralists, inculcate this doctrine. Musk, amber, and other perfumes, are disagreeable to many persons, cause head-ache, vertigo, convulsions, hysteria, &c.; and he who uses them is the author and not the remover of diseases.

The attendance should be in proportion to the urgency and danger of the disease. A patient or his friends have a curiosity to know the nature of the medicine prescribed, which it is often very improper to gratify; but other cases occur in which it may be proper to acquaint the patient with the nature of remedies, as the peculiarities of constitution require great attention, both as to the quantity and quality of certain medicines. Such are the chief of the duties of medical men, according to the amiable and revered Dr. John Gregory; [62/63] the observance of which cannot fail to promote the honour and dignity of the profession. He included many minor topics, which need not be recorded at the present period.

There are certain duties belonging to the learned professions which are supreme, and which no individual and no set of men can, either for themselves or their successors violate, renounce, or neglect, without substantial injustice. These duties, so far as they relate to physicians, are comprised in the oaths required by the Universities, Colleges of Physicians, and in one of the Colleges of Surgeons, in this empire. The substance of these oaths is that proposed by Hippocrates nearly 2000 years ago, and the oath was formerly administered in all Universities in which medicine was taught, to those who were created doctors, and to those who were licensed to practise by the Colleges of Physicians. The oath required by the Edinburgh University is in the following words. After an invocation of the Deity, the graduate pronounces these words:[11] "To practise physic *cautiously, chastely, and honourably*; and faithfully to procure all things conducive to the health of the bodies of the sick; and lastly, never, without great cause, to divulge any thing that ought to be concealed, which may be heard or seen during professional attendance. To this oath let the Deity be witness."[12] I believe no similar oath is required by the Universities of Oxford, Cambridge, Dublin, Glasgow, Aberdeen, or Saint Andrews, or by any of the Colleges of Physicians or Surgeons, except those of London. The Royal College of Physicians requires the following promise (not an oath) from its members, fellows, and licentiates, and prescribes a code of moral statutes: – [13][63/64] "You faithfully promise that you will observe the statutes of the College, and that you will promptly discharge all fines imposed on you for the breach thereof, and that you will do everything in the practice of medicine for the conservation of health, to the honour of the College, and the good of the public." The following are the *Statutes of Morality* of the Royal College of Physicians of London in 1835:[14]

1. No fellow, candidate, or licentiate, shall accuse a fellow, candidate, or licentiate, of ignorance or mala praxis of his art, unless before legitimate judges, or before those concerned. If it be known to the president and censors, or the majority of them, that any person shall so act, he shall pay £4. for the first offence, and the fine will be doubled for the second; but if he transgress a third time, and be convicted in the manner mentioned, if he is a fellow or candidate he [65/66]

shall be expelled from our society, or from the order of candidates; and if he is a licentiate he shall pay £10, and we ordain, that licentiates shall be fined a like sum for every similar transgression.

2. No fellow, candidate, or licentiate, shall afford medical aid or prescribe for a patient whom he knows is under the care of another physician, whether fellow, candidate, or licentiate, and to whom he has not been regularly called.
3. If any one be convicted of this vice, besides the known ignominy which we wish him to suffer, he shall be fined £2. by the president and censors.
4. If any one shall bargain with apothecaries for any percentage on prescriptions, if a fellow or candidate, and if convicted in the manner before mentioned, he shall be expelled from the fellowship, or from the order of candidates.
5. If a licentiate, he shall be fined £10. for each offence.
6. Every physician, whether fellow, candidate, or licentiate, shall inscribe his initials, the date of the prescription, and name of the patient, on every prescription, unless some cause intervenes which shall be approved by the president and censors.
7. If many physicians be called to a patient, they are to consult with great modesty, and in the absence of witnesses or unprofessional persons. Nor shall any one prescribe or insinuate what is to be done to the sick or attendants, before he has stated his method in consultation. But, as medical men have different opinions, so that they may not agree in the plan of treatment, they are to conduct themselves with the greatest prudence and moderation; the ordinary attendant shall signify to the sick and attendants their dissension, so that it may appear as trifling, and as slightly disagreeable to the patient or his friends as possible.
8 Whoever will not obey these rules of consultation, and is convicted by the president and censors, shall be fined £5.
9. Finally, no physician, fellow, candidate, or licentiate, shall consult in the city of London, or within 7 miles thereof, unless with a fellow, candidate, or licentiate, under a [66/67] penalty of £5 as often as convicted by the president and censors, or majority of them.
10. All fines imposed by these statutes must be paid.

It is much to be regretted, that the great bulk of the profession – university graduates in medicine, surgeons, and apothecaries – have no opportunity of being acquainted with these admirable statutes, or have nothing similar to inform them of the etiquette they owe to each other. In printing these statutes, and placing them before the medical public, I hope and trust I may add to the honour and dignity of the profession. The majority of the tenets maintained in them are highly conducive to the fame of every class of medical men; and, if duly observed, would extinguish that base and unprofessional and ungentlemanly behaviour, which of late has characterized too many medical practitioners, and has debased and degraded the profession. The disputes and calumnies of medical men have been so frequent, so violent, so notorious of late, that the character of the profession is lowered in the estimation of the public to a degree unequalled in the history of medicine. Actions against medical men by their contemporaries, or their patients, are now frequent in our courts of justice. This degeneracy of the profession is not confined to this country, it extends throughout Europe, and has even crossed the Atlantic Ocean; and it arises from the exclusion of medical ethics from the prescribed courses of professional education. This malignant spirit pervades every branch of the healing art; the physicians, the surgeons, and the apothecaries, are the most prominent of litigants in our courts of justice. What a falling off is here! If we turn to private practice, we find those uninfluenced by the statutes under consideration, vituperating each other, "by look, gesture, and suspicious silence," and often without any disguise; and the injured individual has no remedy afforded him by the body to which he belongs, and which gravely promises him rights, privileges, immunities, and protection in the discharge of his vocation; his only remedy is an appeal to the laws of his country. But the fact is, our Colleges of Surgeons, and Companies of Apothecaries, have no power to protect their members; nor is there any country in [67/68] the world, in which the laws relative to the practice of the medical profession are so imperfect and defective as in the British empire.

But to return to the subject immediately under consideration: I have to insert the oath required by the Royal College of Surgeons in this city, which is as follows:

> You swear that while you shall be a member of the Royal College of Surgeons in London, you will observe the statutes, bye-laws, ordinances, rules, and constitutions thereof; that you will obey every lawful summons issued by order of the court of assistants and examiners of the said college, or of either of them, having no reasonable excuse to the contrary that you will pay such contributions as shall be legally assessed upon and demanded of you; that you will demean yourself *honourably* in the practice of your profession; and to the utmost of your power maintain the dignity and welfare of the college – So help you God.

It is to be feared that some surgeons forget to demean themselves honourably in the practice of their profession, more especially as their rivals, the general practitioners, are under no such obligation. From the open violation of our laws relative to the practice of medicine, the surgeons act as physicians, and must become apothecaries in self-defence; the apothecaries act as physicians and surgeons, while the chemists and druggists, without any medical education whatever, act as physicians, surgeons, and apothecaries; and as to quacks, they are allowed to flourish to an illimitable extent, and to destroy more than the sword, famine, and pestilence united. Such is a true picture of the medical profession in the greatest nation upon earth – in a country pre-eminent for literature, the sciences and the arts: – such is the state of medical practice in England.

But I leave this part of the subject to another opportunity, and return to the topic more immediately under consideration[.] That medical men should practise *cautiously, chastely*, and *honourably*, and observe strict secrecy in all delicate cases, and in all domestic affairs, which may fall under their notice during professional attendance, is not only consonant to the usage of the profession, but to common sense and justice. It would be highly improper to divulge the nature of certain [68/69] diseases, or expose the affairs of families, to gratify idle curiosity, impertinence, or serve the purposes of an interested knave. The law, however, compels us to violate these principles; and hence the exception in the Edinburgh oath, "not to divulge without weighty reasons." In such cases the violation or renunciation of our moral and professional duties is compulsory.

Chastity and honour are general moral duties, and not peculiarly belonging to any one profession. No profession commands a greater purity of morals than the medical. The intimate confident of the other sex – the adviser in her mental and corporal diseases; but never the abuser of these advantages. Never has a properly educated physician employed his ascendancy to seduce innocence, which places herself in his hands; and scarcely ever has his voice been employed in corrupt discourses towards women who have selected him as their consoler and friend. A medical practitioner is often placed between his duties and vice; his station almost daily exposes him to sacrifice honour to interest; he frequently feels, without danger, the strongest passions; – but he thinks it more glorious and virtuous to conquer them. It is for the good of society that he employs his powerful influence in the most virtuous and honourable manner. Men who confide to him all that is dearest to them – their wives and children – have a just right to require of him a pure heart and untarnished morals. Such is the moral character of the entire profession in all countries, and it is but rarely, scarcely ever tarnished, notwithstanding the immense number of medical men, and the certainty that there must of necessity be a few exceptionable, alias unprincipled individuals, amongst them. Were medical practitioners to act immorally, they would cease to be employed; and society would place the signet of reprobation upon them. I am certain that there have not been half-a-dozen actions against medical practitioners in this country, for immoral conduct, since the first century to the present date. My position is therefore proved by historical records.

The multiplied studies of a truly informed medical practitioner; his various duties; the practice of his profession; the care of his reputation – all prevent him from participating [69/70] in the commotions which occupy societies and empires. He avoids politics and political assemblies, while engaged in the peaceful pursuits of his profession. He takes no interest in the quarrels of sovereigns. There have been a few exceptions, but wise practitioners avoid all political and religious discussions. Their patients may entertain the most opposite opinions on these subjects; and no good ever results from medical practitioners involving themselves in disputes concerning them. In time of war, medical men feel equally bound to aid friends and enemies, according to the law of nations.

The duty of *caution* in practice, means "care not to expose the sick to any unnecessary danger." The best rule of conduct on this important point, is the simple and comprehensive, religious and moral precept, "Do unto others as you would they should do unto you." Whatever the practitioner does or advises to be done for the good of his patient, and what he would do in his own case, or in the case of those who are dearest to him – if he or they were in the same situation – is not only justifiable on his part, but it is his indispensable duty to do. The patient should have the chance, whether it be a hundred to one, or only one in a hundred in his favour. Whatever may be the result, the practitioner has the greatest of all consolation – the consciousness of rectitude – "mens conscia recti;" this will be his solace, should the case terminate unfavourably, when the vulgar, the ignorant, the envious, the malicious, and the interested, will not fail to blame him for the death of his patient. But if he administered a dangerous medicine, merely to gratify his own curiosity, or zeal for science – to ascertain the comparative advantage or disadvantage of some new remedy, either proposed by himself or suggested by others – he is held guilty of a breach of ethics, and of a high misdemeanour, and a great breach of trust towards his patient; and if the patient die, I apprehend, he might be severely punished.

Medical men have tried the most dangerous experiments upon themselves, from their zeal for science, and even sacrificed their lives; but patients, in general, have no such zeal for science – no ambition for such a crown of martyrdom – [70/71] and generally employ and pay their medical attendants for the very opposite purpose. It must be admitted, that men who would try experiments upon themselves, would be very apt to try experiments on their patients. It is a melancholy truth, but cannot be denied. The profession, however, has always reprobated such conduct; and the medical phrase of reproach and contempt for it, "*corio humano ludere*," to play with the human hide, abundantly testifies in what abomination it has been held by the faculty. It is unnecessary to dwell upon this point in this age, because all experiments are made upon the inferior animals; and the just reproach entertained by the faculty, in former times, is now inapplicable. But every man of common understanding well knows, that neither physic nor surgery can be practised without some danger to the sick. It is universally known, that many surgical operations are dangerous to life, and that all our most powerful remedies are highly dangerous; and more especially when improperly employed, or when they cannot be borne. A safe medicine is often extremely dangerous, from the peculiarity of constitution: and the great and urgent danger, in many diseases, requires the immediate use of dangerous remedies. It is admitted, by the best practitioners, that many remedies are still wanted for the cure of disease, and this want leads us most justifiably, and almost inevitably, to try new remedies on many occasions; and such experiments are not blamable, for they are necessary: *sic enim medicina arta; subinde aliorum salute, aliorum interritu perniciosa discernans a salutaribus*.[15] From these causes there results much inevitable danger in the practice of physic. From this acknowledged danger, results the important duty of caution in a physician, or care to make the danger as little as possible. Whatever is best for the sick, it is the indispensable duty of a medical man to do for them. It is his duty and obligation, "faithfully to do all things conducive to the health of his patients;" and this is so complete and indefeasible, that it cannot be set aside by any cause whatever.

The last obligation imposed on a medical practitioner, is discretion, or secrecy. The depositary of family secrets – often the possessor of the reputation of those who have [71/72] reposed in him their confidence – to what disgrace would he not expose them, were he to divulge the mysteries he knows, which ought to be hidden from the public! Here, the unfortunate victim of seduction implores his succour and his silence; there, a father or a mother avows to him the unfortunate consequences of youth abandoned to passion. The confidence reposed in him, and revelations made to him, during his professional attendance, are such that honour commands him not to abuse the one, or publish the other, unless in our courts of justice, which have the power to compel him. Thence the phrase, *non sine gravi causa, in* the doctors' oath; that is, such secrets are not to be divulged without the greatest necessity; but the French medical moralists contend, that they ought not to be divulged even at the risk of liberty and life. Hippocrates inculcated this conduct in the oath he required of his disciples: "Quæ vero inter curandum aut etiam medicinam minime faciens, in communi hominum vitâ, vel videro, vel audiero, quæ minime in vulgus efferi oporteat ea arcana esse ratus, silebo."[16] The Edinburgh University exacts a similar obligation.

Prudence is indispensably necessary in the practice of medicine – not only in the selection of medicines, but in moral conduct. The greatest care, and constant attention are required, to preserve the integrity of reputation. We are consulted in cases of the greatest delicacy and difficulty. We are the advisers of parents, husbands, wives, youths, and children. The most scrupulous morality is required in cases of girls and women. The presence of the mother, some near relation, or intimate female friend, is required in those delicate and frequent cases, in which the most secret charms of nature are subjected to indispensable examination, when the timid and blushing virgin is compelled to place at her feet the last veil of modesty. In these obstetric, and all female cases, let the following rules be rigidly observed. Modestissimus in curandis mulieribus existas; et si pectus, venter imus, aut alinæ arcanæ partes tangendæ, aut pertractandæ sunt, te fidelem, aut ita dicam, marmoreum et gelidum, animo constanti, vel effinge, vel efforma. Pessima, jure merito, tuo nomini fama inuretur, si lubrica manus, impurus animus, corrupti sermonis [72/73] castitatem per solas etiam aures violabunt. Quorundam morborum, præcipue paellas, matronas, principes divexantium, aut eos, quibus ex his nomen aliquo modo periclitatur, labes ne detegas, secreto ac fido pectore, naturæ mortalium errores conde. Caute tamen in viduarum domos, ubi sunt virgines, petilcæ uxores, pulchræ ancillulæ juventutem intrude. Vidi mulierculas puellasque insano practicantium amore captas, ac dulcæ rabidæ tentigini remedium quærentes; vidique ex virginibus factas matres cum summo incauti præceptoris dedecore.[17]

Such is the line of moral conduct which guides all religious and respectable medical practitioners. It is highly creditable to so large a portion of society in all civilized countries, that their moral conduct has been almost invariably so excellent. If medical practitioners were so immoral as to insul[t] feminine dignity by soliciting the chastity of their patients, or of seducing them, they could not escape detection; and their punishment would be public execration, the severe inflictions of the offended law, and the utter ruin of their professional character. They would be instantly compelled to relinquish the profession. Women, when labouring under disease, or in the agonies of child-birth, are objects of sympathy and compassion, just as all mankind are when suffering from painful infirmities. Sensual impulse and severe pain are contrary influences, setting aside all moral feeling; and gallantry and brutality are equally incompatible. Amorous desire is extinguished by pain, and the brute deprived of reason is restrained by this law of nature. It is this law, as well as a moral obligation, that enables medical practitioners to act in the most virtuous manner towards their female patients, and to relieve their weaknesses and infirmities in the same manner as all other human diseases.

Such is the code of ethics which ought to influence medical men, both in public and in private practice; "but it is matter of question," says Dr. Gordon Smith, "whether it has in reality an existence."[18] This is a truism that cannot be doubted; and yet the rising members of the profession are [73/74] expected to support the honor and dignity of the faculty, without any rules to guide them, without having heard a single word upon the subject, during their education. Hence the cause of that improper conduct which has degraded the profession to a degree unparalleled in the annals of British medicine. I shall not prosecute this subject at present, as it will be more properly considered in my account of the laws relative to the practice of every branch of medicine in this country, and of the constitution of the faculty.

I shall now conclude the subject with the Code of Ethics of Dr. Percival, and append some notes.

Dr. Percival's Medical Ethics[19]

Section 1. Of Professional Conduct Relative to Hospitals, or Other Medical Charities

1. Hospital physicians and surgeons should minister to the sick, with due impressions of the importance of their office; reflecting that the ease, the health, and the lives of those committed to their charge depend on their skill, attention, and fidelity. They should study, also, in their deportment, so to unite tenderness with steadiness, and condescension with authority, as to inspire the minds of their patients with gratitude, respect, and confidence.
2. The choice of a physician or surgeon cannot be allowed to hospital patients, consistently with the regular and established succession of medical attendance. Yet personal confidence is not less important to the comfort and relief of the sick-poor, than of the rich under similar circumstances; and it would be equally just and humane, to inquire into and to indulge their partialities, by occasionally calling into consultation the favourite practitioner. The rectitude and wisdom of this conduct will be still more apparent, when it is recollected that patients in hospitals not unfrequently request their discharge, on a deceitful plea of having received relief; and afterwards procure another recommendation, that they may be [74/75] admitted, under the physician or surgeon of their choice. Such practices involve in them a degree of falsehood, produce unnecessary trouble, and may be the occasion of irreparable loss of time in the treatment of diseases.
3. The feelings and emotions of the patients, under critical circumstances, require to be known and attended to, no less than the symptoms of their diseases. Thus, extreme timidity with respect to venesection, contra-indicates its use in certain cases and constitutions. Even the prejudices of the sick are not to be contemned, or opposed with harshness. For though silenced by authority, they will operate secretly and forcibly on the mind, creating fear, anxiety, and watchfulness.
4. As misapprehension may magnify real evils, or create imaginary ones, no discussion concerning the nature of the case should be entered into before the

patients, either with the house-surgeon, the pupils of the hospitals, or any medical visiter [*sic*].

5. In the large wards of an Infirmary, the patients should be interrogated concerning their complaints, in a tone of voice which cannot be overheard. Secrecy, also, when required by peculiar circumstances, should be strictly observed. And females should always be treated with the most scrupulous delicacy. To neglect or sport with their feelings is cruelty; and every wound thus inflicted tends to produce a callousness of mind, a contempt of decorum, and an insensibility to modesty and virtue. Let these considerations be forcibly and repeatedly urged on the hospital pupils.
6. The moral and religious influence of sickness is so favourable to the best interests of men and of society, that it is justly regarded as an important object in the establishment of every hospital. The institutions for promoting it should, therefore, be encouraged by the physicians and surgeons whenever seasonable opportunities occur. And by pointing out these to the officiating clergyman, the sacred offices will be performed with propriety, discrimination, and greater certainty of success. The character of a physician is usually remote either from superstition or enthusiasm; and the aid which he is now exhorted to give, will tend to their exclusion from the [75/76] hospital, where their effects have often been known to be not only baneful, but even fatal.
7. It is one of the circumstances which softens the lot of the poor, that they are exempt from the solicitudes attendant on the disposal of property. Yet there are exceptions to this observation: and it may be necessary that an hospital patient, on the bed of sickness and death, should be reminded, by some friendly monitor, of the importance of a *last will* and *testament* to his wife, children, or relatives, who, otherwise, might be deprived of his effects, of his expected prize-money, or of some future residuary legacy. This kind office will be best performed by the house surgeon, whose frequent attendance on the sick, diminishes their reserve, and entitles him to their familiar confidence. And he will doubtless regard the performance of it as a duty. For whatever is right to be done, and cannot by another be so well done, has the full force of moral and personal obligation.
8. The physicians and surgeons should not suffer themselves to be restrained, by parsimonious considerations, from prescribing *wine,* and *drugs,* even *of high price,* when required in diseases of extraordinary malignity and danger. The efficacy of every medicine is proportioned to its purity and goodness; and on the degree of these properties, *caeteris paribus,* both the cure of the sick, and the speediness of its accomplishment must depend. But when drugs of inferior quality are employed, it is requisite to administer them in larger doses, and to continue the use of them a longer period of time; circumstances which, probably, more than counterbalance any savings in their original price. If the case, however, were far otherwise, no *economy,* of a *fatal* tendency, *ought to be admitted into institutions*, which, founded on principles of the purest beneficence, in this country, when well conducted, can never want contributions adequate to their liberal support.

9. Hospital affairs ought not to be incautiously revealed.[20]
10. Professional charges are to be made only before a meeting of the faculty.
11. Medical and surgical cases to be distinguished.
12. Principles which authorize the use of new medicines and operations. (See page 71, ante.)[76/77]
13. To advance professional improvement, a friendly and unreserved intercourse should subsist between the gentlemen of the faculty, with a free communication of whatever is extraordinary or interesting in the course of their hospital practice. And an account of every case or operation, which is rare, curious, or instructive, should be drawn up by the physician or surgeon to whose charge it devolves, and entered in a register kept for the purpose, but open only to the physicians and surgeons of the charity.
14. Hospital registers usually contain only a simple report of the number of patients admitted and discharged. By adopting a more comprehensive plan, they might be rendered subservient to medical science, and beneficial to mankind. The following sketch is offered, with deference to the gentlemen of the faculty. Let the register consist of three tables; the first specifying the number of patients admitted, cured, relieved, discharged, or dead; the second the several diseases of the patients, with their events; the third the sexes, ages, and occupations of the patients. The ages should be reduced into classes; and the tables adapted to the four divisions of the year. By such an institution, the increase or decrease of sickness; the attack, progress, and cessation of epidemics; the comparative healthiness of different situations, climates, and seasons; the influence of particular trades and manufactures on health and life; with many other curious circumstances, not more interesting to physicians than to the community, would be ascertained with sufficient precision.[21]
15. By the adoption of the register, recommended in the foregoing article, physicians and surgeons would obtain a clearer insight into the comparative success of their hospital and private practice; and would be incited to a diligent investigation of the causes of such difference. In particular diseases it will be found to subsist in a very remarkable degree: and the discretionary power of the physician or surgeon, in the admission of patients, could not be exerted with more justice or humanity, than in refusing to consign to lingering [77/78] suffering and almost certain death, a numerous class of patients, inadvertently recommended as objects of these charitable institutions. "In judging of diseases with regard to the propriety of their reception into hospitals," says an excellent writer, "the following general circumstances are to be considered."

    "Whether they be capable of speedy relief; because, as it is the intention of charity to relieve as great a number as possible, a quick change of objects is to be wished; and also because the inbred disease of hospitals will almost inevitably creep, in some degree, upon one who continues a long time in them, but will rarely attack one whose stay is short.

    "Whether they require, in a particular manner, the superintendance of skilful persons, either on account of their acute and dangerous nature, or any singularity or intricacy attending them, or erroneous opinions prevailing among the common people concerning their treatment.

"Whether they be contagious, or subject in a peculiar degree to taint the air, and generate pestilential diseases.

"Whether a fresh and pure air be peculiarly requisite for their cure, and they be remarkably injured by any vitiation of it."[22]

16. But no precautions relative to the reception of patients, who labour under maladies incapable of relief, contagious in their nature, or liable to be aggravated by confinement in an impure atmosphere, can obviate the evils arising from close wards, and the false economy of crowding a number of persons into the least possible space. There are inbred diseases which it is the duty of the physician or surgeon to prevent, as far as lies in his power, by a strict and persevering attention to the whole medical polity of the hospital. This comprehends the discrimination of cases admissible, air, diet, cleanliness, and drugs; each of which articles should be subjected to a rigid scrutiny, at stated periods of time.
17. The establishment of a committee of the gentlemen of the faculty, to be held monthly, would tend to facilitate [78/79] this interesting investigation, and to accomplish the most important objects of it.[23] By the free communication of remarks, various improvements would be suggested; by the regular discussing of them, they would be reduced to a definite and consistent form: and by the authority of united suffrages, they would have full influence over the governors of the charity. The exertions of individuals, however benevolent or judicious, often give rise to jealousy; are opposed by those who have not been consulted; and prove inefficient by wanting the collective energy of numbers.
18. The harmonious intercourse, which has been recommended to the gentlemen of the faculty, will naturally produce frequent consultations, viz., of the physicians on medical cases, of the surgeons on chirurgical cases, and of both united in cases of a compound nature, which falling under the department of each, may admit of elucidation by the reciprocal aid of the two professions.
19. In consultations on medical cases, the junior physician present should deliver his opinion first, and the others in the progressive order of their seniority. The same order should be observed in chirurgical cases; and a majority should be decisive in both: but if the numbers be equal, the decision should rest with the physician or surgeon, under whose care the patient is placed. No decision, however, should restrain the acting practitioner from making such variations in the mode of treatment, as future contingences may require, or a farther insight into the nature of the disorder may show to be expedient.
20. In consultations on mixed cases, the junior surgeon should deliver his opinion first, and his brethren afterwards in succession, according to progressive seniority. The junior physician present should deliver his opinion after the senior surgeon; and the other physicians in the order above prescribed.
21. In every consultation, the case to be considered should be concisely stated by the physician or surgeon, who [79/80] requests the aid of his brethren. The opinions rela-

tive to it should be delivered with brevity, agreeably to the preceding arrangement, and the decisions collected in the same order. The order of seniority, among the physicians and surgeons, may be regulated by the dates of their respective appointments in the hospital.[24]

22. Due notice should be given of a consultation, and no person admitted to it, except physicians and surgeons of the hospital, and the house-surgeon, without the unanimous consent of the gentlemen present. If an examination of the patient be previously necessary, the particular circumstances of danger or difficulty should be carefully concealed from him, and every just precaution used to guard him from anxiety or alarm.
23. No important operation should be determined upon, without a consultation of the physicians and surgeons, and the acquiescence of a majority of them. Twenty-four hours' notice of the proposed operation should be given, except in dangerous accidents, or when peculiar occurrences may render delay hazardous. The presence of a spectator should not be allowed during an operation, without the express permission of the operator.[25] All extra-official interference in the management of it should be forbidden. A decorous silence ought to be observed. It may be humane and salutary, however, for one of the attending physicians or surgeons to speak occasionally to the patient; to comfort him under his sufferings; and to give him assurance, if consistent with truth, that the operation goes on well, and promises a speedy and successful termination.[26]

    As an hospital is the best school for practical surgery, it would be liberal and beneficial to invite, in rotation, two [80/81] surgeons of the town, who do not belong to the institution, to be present at each operation.
24. Hospital consultations ought not to be held on Sundays, except in cases of urgent necessity; and on such occasions an hour should be appointed, which does not interfere with attendance on public worship.[27]
25. It is an established usage, in some hospitals, to have a stated day in the week, for the performance of operations. But this may occasion improper delay, or equally unjustifiable anticipation. When several operations are to take place in succession, one patient should not have his mind agitated by the knowledge of the sufferings of another. The surgeon should change his apron, when besmeared; and the table or instruments should be freed from all marks of blood, and every thing that may excite terror.
26. Dispensaries afford the widest sphere for the treatment of diseases, comprehending, not only such as ordinarily occur, but those which are so infectious, malignant, and fatal, as to be excluded from admission into infirmaries. Happily, also, they neither tend to counteract that spirit of independence, which should be sedulously fostered in the poor, nor to preclude the practical exercise of those relative duties, "the charities of father, son, and brother," which constitute the strongest moral bonds of society. Being institutions less splendid and extensive than hospitals, they are well adapted to towns of moderate size; and might even be established, without difficulty, in populous country districts. Physicians and surgeons, in such situations, have generally great influence: and

it would be truly honourable to exert it in a cause subservient to the interests of medical science, of commerce, and of philanthropy.

The duties which devolve on gentlemen of the faculty, engaged in the conduct of Dispensaries, are so nearly similar to those of hospital physicians and surgeons, as to be comprehended under the same professional and moral rules. But greater authority and greater condescension will be found [81/82] requisite in domestic attendance on the poor; and human nature must be intimately studied, to acquire that full ascendancy over the prejudices, the caprices, and the passions of the sick, and of their relatives, which is essential to medical success.

27. Hospitals, appropriated to particular maladies, are established in different places, and claim both the patronage and the aid of the gentlemen of the faculty. To an asylum for female patients, labouring under syphilis, it is to be lamented that discouragements have been too often and successfully opposed. Yet whoever reflects on the variety of diseases to which the human body is incident, will find that a considerable part of them are derived from immoderate passions, and vicious indulgencies. Sloth, intemperance, and irregular desires are the great sources of those evils, which contract the duration, and embitter the enjoyment of life. But humanity, whilst she bewails the vices of mankind, incites us to alleviate the miseries which flow from them. And it may be proved, that a *lock hospital* is an institution founded on the most benevolent principles, consonant to sound policy, and favourable to reformation and to virtue. It provides relief for a painful and loathsome distemper, which contaminates, in its progress, the innocent as well as the guilty, and extends its baneful influence to future generations. It restores to virtue and to religion, those votaries whom pleasure has seduced, or villainy betrayed; and who now feel, by sad experience, that ruin, misery, and disgrace, are the wages of sin. Over such objects pity sheds the generous tear; austerity softens into forgiveness; and benevolence expands at the united pleas of frailty, penitence, and wretchedness.[28]

A few remarks on the preceding axioms are necessary.

The relief afforded by hospitals, though they are institutions of the most benevolent kind, is procured with difficulty; patients are admitted only 1 day in the week, no matter how dangerous their cases (unless they be accidents), fees [82/83] are often required, the sick are removed from their families, the nurses are strangers. These defects are so manifest that the public have wisely established dispensaries to obviate them. In these establishments medical assistance is obtained with the greatest facility every day; it is afforded to one parent, without removing him from the means of earning support for himself and family, and to the other without withdrawing her from the superintendence of her domestic concerns. Besides, the natural affections, which every philanthropic mind must wish to see cherished, are reciprocally called into exercise, and strengthened, where the parent is the patient, where the wife becomes the nurse, and the children assistants; and medical aid is rendered more efficacious when the mind is relieved from the anxieties necessarily attendant upon a separation from family, and a removal from home. The early application on the first feelings of indisposition prevents the diffusion of contagious diseases; and pestilence, which once stalked forth, spreading terror and desolation around, is now arrested in its progress, or strangled in its birth; and, it is not too much to assert, that the general healthiness of the metropolis, and the less frequent recurrence of contagious disorders, are to be in a great measure attributed to their early suppression in the abodes of poverty, by the activity and vigilance of the medical officers of dispensaries. It is also apparent, that, without the medical assistance thus afforded the poor, the demand on parochial rates would be increased in a very considerable degree, and the medical establishments of every parish would be increased to double their ordinary expenditure.

But it is quite contrary to the objects for which hospitals and dispensaries are founded, to render them subservient to those in affluent circumstances; an abuse which exists in every one of them. This is an imposition on charity, and a direct injury to the profession; yet the medical officers cannot prevent it. It is a fact which cannot be controverted, that a large proportion of the patients admitted into the hospitals (especially of this metropolis), and relieved at dispensaries, are not real objects of charity, and are often the relatives, or personal friends, or servants, of the governors or subscribers; [83/84] and thus the junior members of the profession are seriously injured. This abuse exists in every part of the empire, but to a vast extent in this metropolis. I have often remonstrated with my colleagues, and with governors, on this impropriety; but their reply was ready – "These things are tolerated in every public institution."

It has been a maxim with the faculty, that a practitioner of standing, a senior, should be called over the ordinary attendant. This rule is often violated, and indeed it is not an easy matter to observe it on many occasions. The late eminent Dr. James Gregory, of Edinburgh, has commented with his usual force on this point. He says,

> but mere *standing,* or seniority, superadded to the most complete and regular education in the profession, will neither procure confidence from the public, nor success and employment to any person. We are well accustomed to see many juniors surpass, and most deservedly surpass their seniors, perhaps even their own instructors; and leave them so far behind, that, before half their race is run, they can have no farther hopes of success.
>
> Some individuals soon shew by their talents, and the use which they make of them, that they can profit more by seven years of observation and experience, than others could do in the longest life. And very many soon show that they are incapable of ever improving; from a real natural want of those faculties which would enable them to observe accurately, to compare different observations together, to reason acutely and fairly, and ultimately to draw just and useful practical inferences from what they had observed. Many, not naturally deficient in their intellectual powers, become so from defects or improprieties in their education; especially the want of that general preliminary education, which improves the faculties, while it extends the sphere of knowledge, and directs the attention to proper objects. And many more, who have no such excuse either from natural or accidental defects, never improve, and soon show that they never will, purely by their own fault. They think the knowledge or improvement they had acquired, when they first entered on the exercise of their profession, sufficient for all purposes, or, at least, for their purpose; they find the effort of attention in [84/85] observing, comparing, reading, and thinking, too laborious; and, as they flatter themselves it is unnecessary for them, they soon cease to make it.
>
> Of course, all chance of improvement in them is at an end; they grow older, and yet grow never the better or wiser. On the contrary, as they often become more negligent, they

grow worse in every respect, and really become more ignorant, forasmuch as they acquire no new knowledge, and forget much of what they had formerly learned.

They become a kind of drones, content to do their business in a humdrum workman-like sort of a way; by which they have the best chance of escaping reflections or censure. Their faults are much more frequently sins of omission than of commission. For once that they do any thing positively and immediately pernicious, they miss, from negligence or ignorance, or both, an hundred opportunities of doing good. None but those in the secret have any notion how faithfully many physicians and surgeons go on for thirty or forty years, or longer, if they live longer, employing, even in the commonest diseases, the remedies which they were taught when young, though useless at best, if not pernicious; how faithfully many great and grave writers have transcribed from their predecessors; from generation to generation, the same frivolous, absurd, or dangerous precepts, the same useless or pernicious prescriptions, and the same silly remarks; how tenaciously many practitioners adhere to old recipes, so extravagantly absurd as to contain perhaps fifty or a hundred ingredients, of which probably not more than three or four are of any use; and how manfully they fight against the introduction of other remedies, the most simple, powerful, and safe; which they reprobate, and will not employ, for no other reason but because they are new.

Men of such talents, characters, and habits, whether physicians or surgeons, can neither improve by experience themselves, nor contribute to the instruction of others, and the improvement of their art. They are peculiarly unfit to practise in an hospital, where, on account of the great number and urgency of the cases to be treated, the greatest extent and accuracy of knowledge, the greatest quickness, precision, [85/86] and discrimination in applying it; and, in one word, the greater effort of attention and thought is required. Any deficiencies in them, which in private practice might well have escaped observation and censure, must soon become conspicuous on so public a stage; just like those of a lawyer at the bar; and will not only bring on themselves reproach and contempt, but will in some measure affect the character of the hospital itself. Whatever lessens the confidence of the public in the administration of it, and of the patients who resort to it in the skill of those to whose care their health and lives are entrusted, tends strongly to frustrate the benevolent purpose of the institution, and is, in truth, a very great injury to the public. Some men, naturally of good sense and quick discernment, and active, vigorous minds, who attend accurately to what passes around them, are distinguished even at an early period of life for sagacity, prudence, decision, and quickness in conduct, and a thorough knowledge of the characters of men, and the management of business. They are accordingly respected in the world, and often consulted on nice and difficult occasions by those who are acquainted with them, and who very wisely rely more on the judgment of such men than they would do on their own.

But such men are not the majority of mankind. An infinitely greater number are either so deficient in natural talents, or so culpably negligent in the use they make of them, that they appear to acquire no improvement at all by their experience of men and things. At the age of fifty or sixty, they are a good deal more dull, but not a jot wiser, than they were at twenty-five or thirty. They become as arrant drones in common life as any are in law, or physic, or surgery. No man of sense, who knows them, would ever think of consulting them, or relying on their judgment, in any business whatever, any more than he would think of consulting a lawyer when he was sick, or a physician when he was engaged in a law-suit.

A man of such a character can never deserve respect, or confidence, or employment, even in his own profession: and there are many such in law, in physic, in surgery, and in all the employments of life.[29] [86/87]

It must be unnecessary to enter into serious proofs of the importance of consultations. The mere want of medical assistance, says the distinguished physician whom I have just quoted, is in many cases so bad, as to imply almost certainly very pernicious, if not fatal consequences. In such cases, to withhold it voluntarily would be almost as criminal as to suffer a wretch to perish by withholding food from him. This point being proved, a few words may be said on the utility of numerous consultations. The opinion of Dr. James G. is so excellent upon this topic, that it must be quoted.

> With respect to physicians and surgeons both, and their patients, it is plain that all the good that can be expected from a consultation may be obtained from one of two, or three, or four, at the utmost, at least as well as from one ten times as numerous; and I should think it almost as plain, that much of that good may be prevented, and much positive evil done, by a very numerous consultation.
>
> On this point I presume, without vanity, to know as much as most men. For full half of my life, I have been a professor of physic in the University of Edinburgh, during which time consultations have been a great part of my business, to the number certainly of some thousands. Nineteen times out of twenty, at least, I have been the youngest physician of the consultation; and, of course, when any written directions were to be given to the patient, have had the honour to put them in writing, to the number, I presume, of two or three hundred at least. I can say with confidence, in point of fact, that I never knew any good come of a very numerous consultation; and I doubt much, whether any physician or surgeon of competent experience will give a different account of the result of what he has observed. The conduct of physicians and surgeons, when themselves or any of their families are sick, affords a still better proof and illustration of the same truth, and is indeed supreme and decisive authority with respect to what is useful, or what is useless, or worse than useless, in medical consultations. With us all considerations of economy are out of the question. Bad as we may be thought, we are not such cannibals as to prey on one another. We may all have, for nothing, to ourselves and our families, [87/88] as much assistance in point of physic and surgery as we choose. We feel strongly, that we have not sufficient calmness and firmness to judge and act properly, when the lives of those are at stake in whom we are most tenderly interested: and, as to ourselves when sick, we all know, for it is a long settled point in physic, that every man who doctors himself has a fool for his patient.
>
> For these reasons, we are all accustomed, when ourselves or our families are sick, to ask the assistance, not of all, but of some of our professional brethren. A numerous consultation is a kind of debating society, in which the patient's welfare, which ought to be the only object in view, is nearly forgotten. The illustrations of such consultations by Moliere, Le Sage, Fielding, and many others, were just, though inapplicable at present. In former times the *odium medicum* was as violent as the *odium theologicum,* even matters went so far that the disputants resorted to arms; but there is little danger of modern theorists taking the field in support of their opinions, though they war with words fully as bitterly as their predecessors. – *Op. Cit.*[30]

Section II. Of Professional Conduct in Private or General Practice

1. The moral rules of conduct, prescribed towards hospital patients, should be fully adopted in private or general practice. Every case, committed to the charge of a physician or surgeon, should be treated with attention, steadiness, and humanity: reasonable indulgence should be granted to the mental imbecility and caprices of the sick: secrecy and delicacy, when required by peculiar

circumstances, should be strictly observed. And the familiar and confidential intercourse, to which the faculty are admitted in their professional visits, should be used with discretion, and with the most scrupulous regard to fidelity and honour.

2. The strictest temperance should be deemed incumbent on the faculty; as the practice both of physic and surgery at all times requires the exercise of a clear and vigorous understanding: and, on emergencies, for which no professional man [88/89] should be unprepared, a steady hand, an acute eye, and an unclouded head, may be essential to the well-being, and even to the life, of a fellow-creature.[31]
3. A physician should not be forward to make gloomy prognostications; because they savour of empiricism, by magnifying the importance of his services in the treatment or cure of the disease. But he should not fail, on proper occasions, to give to the friends of the patient, timely notice of danger, when it really occurs, and even to the patient himself, if absolutely necessary. This office, however, is so peculiarly alarming, when executed by him, that it ought to be declined whenever it can be assigned to any other person of sufficient judgment and delicacy. For the physician should be the minister of hope and comfort to the sick, that by such cordials to the drooping spirit, he may smooth the bed of death; revive expiring life; and counteract the depressing influence of those maladies, which rob the philosopher of fortitude, and the Christian of consolation.
4. Officious interference, in a case under the charge of another, should be carefully avoided. No meddling inquiries should be made concerning the patient; no unnecessary hints given, relative to the nature or treatment of his disorder; nor any selfish conduct pursued, that may directly or indirectly tend to diminish the trust reposed in the physician or surgeon employed. Yet though the character of a professional busy-body, whether from thoughtlessness or craft, is highly reprehensible, there are occasions which not only justify but require a spirited interposition. When artful ignorance grossly imposes on credulity; when neglect puts to hazard an important life; or rashness threatens it with still more imminent danger; a medical neighbour, friend, or relative, apprized of such facts, will justly regard his interference as a duty. But he ought to be careful, that the information on which he acts, is well founded; that his motives are pure and honourable; and that his judgment of the measures pursued is built on experience and practical knowledge, not on speculative or theoretical differences of opinion. The particular circumstances of the case will suggest the most proper mode of conduct. In general, however, a personal and confidential [89/90] application to the gentlemen of the faculty concerned, should be the first step taken, and afterwards, if necessary, the transaction may be communicated to the patient or to his family.
5. When a physician or surgeon is called to a patient, who has been before under the care of another gentleman of the faculty, a consultation with him should be even proposed, though he may have discontinued his visits: his practice, also, should be treated with candour, and justified, so far as probity and truth will permit. For the want of success in the primary treatment of a case is no impeachment of professional skill or knowledge; and it often serves to throw light on the

nature of a disease, and to suggest to the subsequent practitioner more appropriate means of relief.

6. In large and opulent towns, the distinction between the provinces of physic and surgery should be steadily maintained. This distinction is sanctioned both by reason and experience. It is founded on the nature and objects of the two professions; on the education and acquirements requisite for their most beneficial and honourable exercise; and tends to promote the complete cultivation and advancement of each. For the division of skill and labour is no less advantageous in the liberal than in the mechanic arts: and both physic and surgery are so comprehensive, and yet so far from perfection, as separately to give full scope to the industry and genius of their respective professors. Experience has fully evinced the benefits of the discrimination recommended, which is established in every well regulated hospital, and is thus expressly authorized by the faculty themselves, and by those who have the best opportunities of judging of the proper application of the healing art. No physician or surgeon, therefore, should adopt more than one denomination, or assume any rank or privileges different from those of his order.[32]
7. Consultations should be promoted, in difficult or protracted cases, as they give rise to confidence, energy, and more enlarged views in practice. On such occasions no rival- [90/91] ship or jealousy should be indulged: candour, probity, and all due respect should be exercised towards the physician or surgeon first engaged; and as he may be presumed to be best acquainted with the patient and with his family, he should deliver all the medical directions agreed upon, though he may not have precedency in seniority or rank. It should be the province, however, of the senior physician, first to propose the necessary questions to the sick, but without excluding his associate from the privilege of making farther inquiries, to satisfy himself, or to elucidate the case.
8. As circumstances sometimes occur to render a special consultation desirable, when the continued attendance of another physician or surgeon might be objectionable to the patient, the gentleman of the faculty whose assistance is required, in such cases, should pay only two or three visits; and sedulously guard against all future unsolicited interference. For this consultation a double gratuity may reasonably be expected from the patient, as it will be found to require an extraordinary portion both of time and attention.

   In medical practice, it is not an unfrequent occurrence, that a physician is hastily summoned, through the anxiety of the family, or the solicitation of friends, to visit a patient, who is under the regular direction of another physician, to whom notice of this call has not been given. Under such circumstances, no change in the treatment of the sick person should be made, till a previous consultation with the stated physician has taken place, unless the lateness of the hour precludes meeting, or the symptoms of the case are too pressing to admit of delay.
9. Theoretical discussions should be avoided in consultations, as occasioning perplexity and loss of time. For there may be much diversity of opinion, concerning speculative points, with perfect agreement in those modes of practice, which are founded not on hypothesis, but on experience and observation.

10. The rules prescribed for hospital consultations, may be adopted in private or general practice.[33] And the seniority of [91/92] a physician may be determined by the period of his public and acknowledged practice as a physician, and that of a surgeon by the period of his practice as a surgeon, in the place where each resides. This arrangement, being clear and obvious, is adapted to remove all grounds of dispute amongst medical gentlemen: and it secures the regular continuance of the order of precedency, established in every town, which might otherwise be liable to troublesome interruptions by new settlers, perhaps not long stationary.[34]
11. A regular academical education furnishes the only presumptive evidence of professional ability, and is so honourable and beneficial, that it gives a just claim to pre-eminence among physicians, in proportion to the degree in which it has been enjoyed and improved: yet, as it is not indispensably necessary to the attainment of knowledge, skill, and experience, they who have really acquired, in a competent measure, such qualifications, without its advantages, should not be fastidiously excluded from the privileges of fellowship. In consultations, especially, as the good of the patient is the sole object in view, and is often dependent on personal confidence, the aid of an intelligent practitioner ought to be received with candour and politeness, and his advice adapted, if agreeable to sound judgment and truth.
12. Punctuality should be observed in the visits of the faculty, when they are to hold consultation together. But as this may not always be practicable, the physician or surgeon, who first arrives at the place of appointment, should wait 5 min for his associate, before his introduction to the patient, that the unnecessary repetition of questions may be avoided: no visits should be made but in concert, or by mutual agreement: no statement or discussion of the case should take place before the patient or his friends, except in the presence of each of the attending gentlemen of the faculty, and by common consent; and no prognostications should be deli- [92/93] vered, which are not the result of previous deliberation and concurrence.
13. Visits to the sick should not be unseasonably repeated; because, when too frequent, they tend to diminish the authority of the physician, to produce instability in his practice, and to give rise to such occasional indulgences, as are subversive of all medical regimen.

    Sir William Temple has asserted, that "an honest physician is excused for leaving his patient, when he finds the disease growing desperate, and can, by his attendance, expect only to receive his fees, without any hopes or appearance of deserving them." But this allegation is not well founded: for the offices of a physician may continue to be highly useful to the patient, and comforting to the relatives around him, even in the last period of a fatal malady, by obviating despair, by alleviating pain, and by soothing mental anguish. To decline attendance, under such circumstances, would be sacrificing, to fanciful delicacy and mistaken liberality, that moral duty which is independent of, and far superior to, all pecuniary appreciation.
14. Whenever a physician or surgeon officiates for another who is sick or absent, during any considerable length of time, he should receive the fees accruing from such additional practice; but if this fraternal act be of short duration, it should

be gratuitously performed, with an observance always of the utmost delicacy towards the interest and character of the professional gentleman, previously connected with the family.

15. Some general rule should be adopted by the faculty, in every town, relative to the pecuniary acknowledgments of their patients; and it should be deemed a point of honour to adhere to this rule, with as much steadiness, as varying circumstances will admit. For it is obvious, that an average fee, as suited to the general rank of patients, must be an inadequate gratuity from the rich, who often require attendance not absolutely necessary; and yet too large to be expected from that class of citizens, who would feel a reluctance in calling for assistance, without making some decent and satisfactory retribution.

    But in the consideration of fees, let it ever be remembered, [93/94] that though mean ones from the affluent are both unjust and degrading, yet the characteristic beneficence of the profession is inconsistent with sordid views, and avaricious rapacity. To a young physician, it is of great importance to have clear and definite ideas of the ends of his profession; of the means for their attainment; and of the comparative value and dignity of each. Wealth, rank, and independence, with all the benefits resulting from them, are the primary ends which he holds in view; and they are interesting, wise, and laudable. But knowledge, benevolence, and active virtue, the means to be adopted in their acquisition, are of still higher estimation. And he has the privilege and felicity of practising an art, even more intrinsically excellent in its mediate than its ultimate objects. The former, therefore, have a claim to uniform pre-eminence.

16. All members of the profession, including apothecaries as well as physicians and surgeons, together with their wives and children, should be attended gratuitously by any one or more of the faculty, residing near them, whose assistance may be required. For as solicitude obscures the judgment, and is accompanied with timidity and irresolution, medical men, under the pressure of sickness, either as affecting themselves or their families, are peculiarly dependent upon each other. But visits should not be obtruded officiously, as such unasked civility may give rise to embarrassment, or interfere with that choice, on which confidence depends. Distant members of the faculty, when they request attendance, should be expected to defray the charges of travelling. And if their circumstances be affluent, a pecuniary acknowledgment should not be declined; for no obligation ought to be imposed, which the party would rather compensate than contract.

17. When a physician attends the wife or child of a member of the faculty, or any person very nearly connected with him, he should manifest peculiar attention to his opinions, and tenderness even to his prejudices. For the dear and important interests which the one has at stake supersede every consideration of rank or seniority in the other; since the mind of a husband, a father, or a friend, may receive a deep and lasting wound, if the disease terminate fatally, from [94/95] the adoption of means he could not approve, or the rejection of those he wished to be tried. Under such delicate circumstances, however, a conscientious physician will not lightly sacrifice his judgment; but will urge, with proper confidence,

the measures he deems to be expedient, before he leaves the final decision concerning them to his more responsible coadjutor.

18. Clergymen, who experience the *res angustæ domi,*[35] should be visited gratuitously by the faculty. And this exemption should be an acknowledged general rule, that the feeling of individual obligation may be rendered less oppressive. But such of the clergy as are qualified, either from their stipends or fortunes, to make a reasonable remuneration for medical attendance, are not more privileged than any other order of patients. Military or naval subaltern officers, in narrow circumstances, are also proper subjects of professional liberality.
19. As the first consultation by letter imposes much more trouble and attention than a personal visit, it is reasonable, on such an occasion, to expect a gratuity of double the usual amount. And this has long been the established practice of many respectable physicians. But a subsequent epistolary correspondence, on the further treatment of the same disorder, may justly be regarded in the light of ordinary attendance, and may be compensated, as such, according to the circumstances of the case, or of the patient.
20. Physicians and surgeons are occasionally requested to furnish certificates, justifying the absence of persons who hold situations of honour and trust in the army, the navy, or the civil departments of government. These testimonials, unless under particular circumstances, should be considered as acts due to the public, and therefore not to be compensated by any gratuity. But they should never be given without an accurate and faithful scrutiny into the case; that truth and probity may not be violated, nor the good of the community injured, by the unjust pretences of its servants. The same conduct is to be observed by medical practitioners when they are solicited to furnish apologies for non-attendance on juries; or to state the valetudinary incapacity of persons appointed to execute the business of constables, churchwardens, or overseers of the [95/96] poor. No fear of giving umbrage, no view to present or future emolument, nor any motives of friendship, should incite to a false, or even dubious declaration. For the general weal requires that every individual, who is properly qualified, should deem himself obliged to execute, when legally called upon, the juridical and municipal employments of the body politic. And to be accessory, by untruth or prevarication, to the evasion of this duty, is at once a high misdemeanour against social order, and a breach of moral and professional honour.
21. The use of quack medicines should be discouraged by the faculty, as disgraceful to the profession, injurious to health, and often destructive even of life. Patients, however, under lingering disorders, are sometimes obstinately bent on having recourse to such as they see advertized, or hear recommended, with a boldness and confidence which no intelligent physician dares to adopt with respect to the means that he prescribes. In these cases some indulgence seems to be required to a credulity that is insurmountable. And the patient should neither incur the displeasure of the physician, nor be entirely deserted by him. He may be apprized of the fallacy of his expectations, whilst assured, at the same time, that diligent attention should be paid to the process of the experiment he is so unadvisedly making on himself, and the consequent mischiefs,

if any, obviated as timely as possible. Certain active preparations, the nature, composition, and effects of which are known, ought not to be proscribed as quack medicines.[36]

22. No physician or surgeon should dispense a secret nostrum, whether it be his invention, or exclusive property. For if it be of real efficacy, the concealment of it is inconsistent with beneficence and professional liberality. And, if mystery alone give it value and importance, such craft implies either disgraceful ignorance, or fraudulent avarice.
23. The *Esprit du Corps* is a principle of action founded in human nature, and when duly regulated, is both rational and laudable. Every man who enters into a fraternity engages, by a tacit compact, not only to submit to the laws, but to promote the honour and interest of the association, so far as they are consistent with morality and the general good of [96/97] mankind. A physician therefore, should cautiously guard against whatever may injure the general respectability of his profession; and should avoid all contumelious representations of the faculty at large; all general charges against their selfishness or improbity; and the indulgence of an affected jocularity or scepticism, concerning the efficacy and utility of the healing art.
24. As diversity of opinion and opposition of interest may in the medical, as, in other professions, sometimes occasion controversy, and even contention; whenever such cases unfortunately occur, and cannot be immediately terminated, they should be referred to the arbitration of a sufficient number of physicians or of surgeons, according to the nature of the dispute; or to the two orders collectively, if belonging both to medicine and surgery. But neither the subject-matter of such references, nor the adjudication, should be communicated to the public; as they may be personally injurious to the individuals concerned, and can hardly fail to hurt the general credit of the faculty.
25. A wealthy physician or surgeon should not give advice gratis to the affluent; because it is an injury to his professional brethren. The office of a physician can never be supported but as a lucrative one; and it is defrauding, in some degree, the common funds for its support, when fees are dispensed with which might justly be claimed.
26. It frequently happens, that a physician, in his incidental communications with the patients of other physicians, or with their friends, may have their cases stated to him in so direct a manner as not to admit of his declining to pay attention to them. Under such circumstances, his observations should be delivered with the most delicate propriety and reserve. He should not interfere in the curative plans pursued; and should even recommend a steady adherence to them, if they appear to merit approbation.
27. A physician, when visiting a sick person in the country, may be desired to see a neighbouring patient, who is under the regular direction of another physician, in consequence of some sudden change or aggravation of symptoms. The conduct to be pursued, on such an occasion, is to give advice [97/98] adapted to present circumstances; to interfere no farther than is absolutely necessary with the general plan of treatment; to assume no future direction, unless it be expressly

desired; and, in this case, to request an immediate consultation with the practitioner antecedently employed.

28. At the close of every interesting and important case, especially when it hath terminated fatally, a physician should trace back, in calm reflection, all the steps which he had taken in the treatment of it. This review of the origin, progress, and conclusion of the malady; of the whole curative plan pursued; and of the particular operation of the several remedies employed, as well as of the doses and periods of time in which they were administered, will furnish the most authentic documents on which individual experience can be formed. But it is in a moral view that the practice is here recommended, and it should be performed with the most scrupulous impartiality. Let no self-deception be permitted in the retrospect; and if errors, either of omission or commission, are discovered, it behoves that they should be brought fairly and fully to the mental view. Regrets may follow, but criminality will thus be obviated. For good intentions, and the imperfection of human skill, which cannot anticipate the knowledge that events alone disclose, will sufficiently justify what is past, provided the failure be made conscientiously subservient to future wisdom and rectitude in professional conduct.
29. The opportunities which a physician not unfrequently enjoys, of promoting and strengthening the good resolutions of his patients, suffering under the consequences of vicious conduct, ought never to be neglected. And his counsels, or even remonstrances, will give satisfaction, not disgust, if they be conducted with politeness; and evince a genuine love of virtue, accompanied by a sincere interest in the welfare of the person to whom they are addressed.
30. The observance of the Sabbath is a duty to which medical men are bound, so far as is compatible with the urgency of the cases under their charge. Visits may often be made with sufficient convenience and benefit, either before the hours of going to church, or during the intervals of public worship. And in many chronic ailments, the sick, together with their [98/99] attendants, are qualified to participate in the social offices of religion; and should not be induced to forego this important privilege, by the expectation of a call from their physician or surgeon.
31. A physician who is advancing in years, yet unconscious of any decay in his faculties, may occasionally experience some change in the wonted confidence of his friends. Patients who before trusted solely to his care and skill, may now request that he will join in consultation, perhaps with a younger coadjutor. It behoves him to admit this change without dissatisfaction or fastidiousness, regarding it as no mark of disrespect but as the exercise of a just and reasonable privilege in those by whom he is employed. The junior practitioner may well be supposed to have more ardour than he possesses in the treatment of diseases; to be bolder in the exhibition of new medicines, and disposed to administer old ones in doses of greater efficacy. And this union of enterprize with caution, and of fervour with coolness, may promote the successful management of a difficult and protracted case. Let the medical parties, therefore, be studious to conduct themselves towards each other with candour and impartiality; co-operating by mutual concessions in the benevolent discharge of professional duty.

The commencement of that period of senescence when it becomes incumbent on a physician to decline the offices of his profession it is not easy to ascertain; and the decision on so nice a point must be left to the moral discretion of the individual. For one grown old in the useful and honourable exercise of the healing art may continue to enjoy, and justly to enjoy, the unabated confidence of the public. And whilst exempt, in a considerable degree, from the privations and infirmities of age, he is under indispensable obligations to apply his knowledge and experience in the most efficient way to the benefit of mankind. For the possession of powers is a clear indication of the will of our Creator concerning their practical direction. But, in the ordinary course of nature, the bodily and mental vigour must be expected to decay progressively, though perhaps slowly, after the meridian of life is past. As age advances, therefore, a physician should, from time to [99/100] time, scrutinize impartially the state of his faculties, that he may determine, *bona fide,* the precise degree in which he is qualified to execute the active and multifarious offices of his profession. And whenever he becomes conscious that his memory presents to him with faintness those analogies on which medical reasoning and the treatment of diseases are founded; that diffidence of the measures to be pursued perplexes his judgment; that, from a deficiency in the acuteness of his senses, he finds himself less able to distinguish signs, or to prognosticate events; he should at once resolve, though others perceive not the changes which have taken place, to sacrifice every consideration of fame or fortune, and to retire from the engagements of business. To the surgeon under similar circumstances this rule of conduct is still more necessary. For the energy of the understanding often subsists much longer than the quickness of eye-sight, delicacy of touch, and steadiness of hand, which are essential to the skilful performance of operations. Let both the physician and surgeon never forget, that their professions are public trusts, properly rendered lucrative whilst they fulfil them; but which they are bound, by honour and probity, to relinquish as soon as they find themselves unequal to their adequate and faithful execution.

Section III. Of the Conduct of Physicians Towards Apothecaries

1. In the present state of physic, in this country, where the profession is properly divided into three distinct branches, a connection peculiarly intimate subsists between the physician and the apothecary; and various obligations necessarily result from it. On the knowledge, skill, and fidelity of the apothecary depend, in a very considerable degree, the reputation, the success, and usefulness of the physician. As these qualities, therefore, justly claim his attention and encouragement, the possessor of them merits his respect and patronage.
2. The apothecary is, in almost every instance, the precursor of the physician; and being acquainted with the rise and [100/101] progress of the disease, with the hereditary constitution, habits, and disposition of the patient, he may furnish very

important information. It is in general, therefore, expedient, and when health or life are at stake expediency becomes a moral duty, to confer with the apothecary before any decisive plan of treatment is adopted; to hear his account of the malady, of the remedies which have been administered, of the effects produced by them, and of his whole experience concerning the juvantia and lædentia[37] in the case. Nor should the future attendance of the apothecary be superseded by the physician: for if he be a man of honour, judgment, and propriety of behaviour, he will be a most valuable auxiliary through the whole course of the disorder, by his attention to varying symptoms; by the enforcement of medical directions; by obviating misapprehensions in the patient, or his family; by strengthening the authority of the physician; and by being at all times an easy and friendly medium of communication. To subserve these important purposes, the physician should occasionally make his visits in conjunction with the apothecary, and regulate by circumstances the frequency of such interviews: for if they be often repeated, little substantial aid can be expected from the apothecary, because he will have no intelligence to offer which does not fall under the observation of the physician himself; nor any opportunity of executing his peculiar trust, without becoming burthensome to the patient by multiplied calls, and unseasonable assiduity.

3. This amicable intercourse and co-operation of the physician and apothecary, if conducted with the decorum and attention to etiquette, which should always be steadily observed by professional men, will add to the authority of the one, to the respectability of the other, and to the usefulness of both. The patient will find himself the object of watchful and unremitting care, and will experience that he is connected with his physician, not only personally, but by a sedulous representative and coadjutor. The apothecary will regard the free communication of the physician as a privilege and mean of improvement; he will have a deeper interest in the success of the curative plans pursued and his honour and reputation will be directly involved in the purity and excellence of the [101/102] medicines dispensed, and in the skill and care with which they are compounded.
4. The duty and responsibility of the physician, however, are so intimately connected with these points, that no dependence on the probity of the apothecary should prevent the occasional inspection of the drugs, which he prescribes. In London, the law not only authorizes, but enjoins a stated examination of the simple and compound medicines kept in the shops. And the policy that is just and reasonable in the metropolis, must be proportionally so in every provincial town throughout the kingdom. Nor will any respectable apothecary object to this necessary office, when performed with delicacy, and at seasonable times; since his reputation and emolument will be increased by it, probable in the exact ratio, thus ascertained, of professional merit and integrity.
5. A physician called to visit a patient in the country, should not only be minute in his directions, but should communicate to the apothecary the particular view, which he takes of the case; that the indications of cure may be afterwards pursued with precision and steadiness; and that the apothecary may use the discretionary power committed to him, with as little deviation as possible from the general plan prescribed. To so valuable a class of men as the country apothecaries, great attention and respect is due. And as they are the guardians

of health through large districts, no opportunities should be neglected of promoting their improvement, or contributing to their stock of knowledge, either by the loan of books, the direction of their studies, or by unreserved information on medical subjects. When such occasions present themselves, the maxim of our judicious poet, is strictly true, "the worst avarice is that of sense." For practical improvements usually originate in towns, and often remain unknown or disregarded in situations, where gentlemen of the faculty have little intercourse, and where sufficient authority is wanting to sanction innovation.

6. It has been observed, by a political and moral writer of great authority, that "apothecaries' profit["] is become a bye-word, denoting something uncommonly extravagant. This great apparent profit, however, is frequently no more than the [102/103] reasonable wages of labour. The skill of an apothecary, is a much nicer and more delicate matter than that of any artificer whatever; and the trust which is reposed in him is of much greater importance. He is the physician of the poor in all cases, and of the rich when the distress or danger is not very great. His reward, therefore, ought to be suitable to his skill and his trust, and it arises generally from the price at which he sells his drugs. But the whole drugs which the best employed apothecary, in a large market town, will sell in a year, may not, perhaps, cost him above 30 or 40 lb. Though he should sell them, therefore, for 300 or 400, or a 1,000% profit, this may frequently be no more than the reasonable wages of his labour charged, in the only way in which he can charge them, upon the price of his drugs.
7. Physicians are sometimes requested to visit the patients of the apothecary, in his absence. Compliance, in such cases, should always be refused, when likely to interfere with the consultation of the medical man usually employed by the sick person, or his family. It would be for the interest and honour of the faculty to have this practice altogether interdicted. Physicians are the only proper substitutes for physicians; surgeons for surgeons; and apothecaries for apothecaries.
8. When the aid of a physician is required, the apothecary to the family is frequently called upon to recommend one. It will then behove him to learn fully whether the patient or his friends have any preference or partiality; and this he ought to consult, if it lead not to an improper choice. For the maxim of Celsus is strictly applicable on such an occasion; Ubi par scientia, melior est amicus medicus quam extraneus.[38] But if the parties concerned be entirely indifferent, the apothecary is bound to decide according to his best judgment, with a conscientious and exclusive regard to the good of the person, for whom he is commissioned to act. It is not even sufficient that he selects the person on whom, in sickness, he reposes his own trust; for in this case, friendship justly gives preponderancy; because it may be supposed to excite a degree of zeal and attention, which might overbalance superior science or abilities. Without any regard to any personal, family, or [103/104] professional connections, he should recommend the physician whom he conscientiously believes, all circumstances considered, to be best qualified to accomplish the recovery of the patient.
9. In the county of Norfolk, and in the city of London, benevolent institutions have been lately formed, for providing funds to relieve the widows and children of apothecaries, and occasionally also members of the profession, who become

> indigent. Such schemes merit the sanction and encouragement of every liberal physician and surgeon. And were they thus extended, their usefulness would be greatly increased, and their permanency almost with certainty secured. Medical subscribers, from every part of Great Britain, should be admitted, if they offer satisfactory testimonials of their qualifications. One comprehensive establishment seems to be more eligible than many on a smaller scale. For it would be conducted with superior dignity, regularity, and efficiency; with fewer obstacles from interest, prejudice, or rivalship; with consider-able saving in the aggregate of time, trouble, and expence; with more accuracy in the calculations, relative to its funds; and, consequently, with the utmost practicable extension of its dividends.[39]

Dr. Percival recommends the formation of district dispensaries, and, in an excellent pamphlet on the Farming of Parishes by surgeons, a practicable plan, now acted on in Warwickshire, for superseding that wretched mockery of medical aid, called "Parish Doctoring," by District Infirmaries, is suggested by Mr. H. L. Smith.

Mr. Phelan, of Clonmell, proposes to establish district infirmaries and dispensaries in Ireland, in the proportion of one to every 40,000 inhabitants, and that each ought to be attended by a physician, surgeon, and resident apothecary, as in the English hospitals and dispensaries. (On the Medical Charities of Ireland, 1835.)

The Commissioners of the New Poor Law Act of 1834, have divided several parishes into unions, and appoint the lowest medical bidders, however young and inexperienced, as medical officers. The average pay allowed is half-a-crown a head; but in cases of cholera, or other epidemics, the pay is only allowed for a certain number of sick, and the rest are to be attended for nothing! This Act must be amended speedily, as it is most injurious to the sick poor. [104/105]

Notes

1. *Editors' note:* See Footnote 1, Chapter 3.
2. *Author's note:* It will generally be found that there is disease in the brain or spinal marrow (cerebro-spinal system), or in some of the organs in the abdomen; as the stomach, liver, bowels, spleen, kidneys, womb, &c.
3. *Editors' note:* "About everything in general and another thing in particular."
4. *Author's note:* This is rarely the case at present; it, however, sometimes happens.
5. *Editors' note:* "If you wish to ignore new inventions, if you wish to ignore their elegance of novelty, then do not criticize with utterances of defilement and envy. It is permissible for you to extol, but not to [literal translation] put shoes on the feet of others." (The sense here appears to be: send others packing or into exile.)
6. *Editors' note:* "What number of years I have lived are the more mature years? Actions create (respected) old age; the actions and not the years are what you should enumerate."
7. *Editors' note:* The literal translation of this phrase is somewhat misleading. The general sense is as Ryan relates, that those who rely solely on experience, without appropriate scientific education, cannot really make sense of what they observe and hence remain ignorant.
8. *Editors' note:* "Many inconveniences encompass a man of advanced years" (Horace, *De Arte Poetica* 3.60.153).

9. *Editors' note:* "Old age, arriving late, debilitates the virility of the soul and alters vigor" (Virgil, *Aeneid,* 9.610–11).
10. *Editors' note:* Giorgio Baglivi (1668–1706) experimented in musculo-skeletal physiology; Marie-François-Xavier Bichat (1771–1802) was a prominent anatomist and surgeon. By including an example whose career came after Gregory's death, Ryan shows here that he has freely interpolated his own comments among those taken directly from Gregory.
11. *Editors' note:* Latin version omitted; Ryan's own translation follows.
12. *Author's note:* The obligations of this oath will be fully explained after the insertion of the whole of the British Moral Statutes.
13. *Editors' note:* Latin version omitted; Ryan's own translation follows.
14. *Editors' note:* Latin version omitted; Ryan's own translation follows.
15. *Editors' note:* "Thus the medical arts, undaunted, discern the workings of remedies, that sometimes bring forth health and other times are pernicious."
16. *Editors' note:* This portion of the Hippocratic Oath is commonly rendered as, "And whatsoever I shall see or hear in the course of my profession, as well as outside my profession in my intercourse with men, if it be what should not be published abroad, I will never divulge, holding such things to be holy secrets."
17. *Editors' note*: "You should proceed with the utmost modesty in treating women; and if you must touch or examine the breast or genitals, ensure trust by speaking smoothly and professionally, with a calm spirit, either by custom or formula. The worst rumors that could be attributed to you by corrupt gossip, understandably, are that you have a slippery hand, an impure spirit, and violate chastity. The very names of certain diseases that principally afflict girls and women, might be dangerous to them if you reveal their disgrace. Secretly, as if confided into your heart, hide any knowledge of sexual transgressions (*literally:* the wanderings of human genitalia). Be cautious in entering the homes of widows, of youthful virgins, of wanton wives, and of attractive servant women. I have witnessed foolish women and girls practically captive to insane lust, searching for a cure to their passions; and I have seen mothers disgraced by the deeds of virgins incautiously entrusted to the supervision of male tutors."
18. *Author's note:* Analysis of Medical Evidence.
19. *Editors' note:* Ryan's "sections" correspond to chapters in the original work by Percival (1803).
20. *Editors' note:* Paragraphs 9–12 of this section are shortened from Percival's original and the general topic only is given.
21. *Author's note:* There are no such case-books or registries in the London hospitals, even in 1835, and such as are kept are locked up, and students are seldom allowed to see them.
22. *Footnote in Percival (1803), copied here by Ryan:* See Dr. Aikin's Thoughts on Hospitals, p. 21. *Added by Ryan:* This suggestion is now universally adopted throughout the United Kingdom.
23. *Author's note:* This is now done by the medical committees of hospitals and dispensaries.
24. *Author's note:* Seniority, in my opinion, ought to be determined by the date of admission into the Colleges or Apothecaries' societies. Suppose Sir Astley Cooper [a prominent London surgeon of Ryan's day] were to go and reside in some provincial city, he could not be considered junior to some young provincial surgeon.
25. *Author's note:* It is usual in London, that all members of the profession, who choose, and all medical students, may be present.
26. *Author's note:* The barbarity of some old hospital surgeons during operations, even on children, is disgraceful to such individuals, and to a civilized country. *Editors' note:* Both Percival and Ryan practiced in the days before anesthesia was introduced.
27. *Author's note:* It is most improper to interfere with the religion of the sick, by circulating tracts, bribing patients, giving them flannel, &c. The sick ought to have free will in the selection of clergymen of their own.

28. *Footnote in Percival (1803), copied here by Ryan:* See two Reports, intended to promote the establishment of a Lock Hospital in Manchester, in the year 1774, inserted in the author's Essays Medical, Philosophical, and Experimental. Vol. II. p. 263. 4th edition. *Editors' note:* "Locked Hospitals" referred to asylums for the insane. Ryan here omits a further paragraph of 27 and 28–31, all of which relate to insane asylums.
29. *Editors' note:* Ryan here fails to cite the source for this long quotation from James Gregory, even though he later references another quotation "op. cit." as if he had already given the source.
30. *Editors' note:* Ryan here resumes quoting nearly verbatim from Percival.
31. *Editors' note:* Ryan here omits a brief anecdote about Philip of Macedon that is included in Percival's original text.
32. *Author's note:* This rule is right, though seldom observed. In London all the surgeons act as physicians, and also throughout the British dominions.
33. *Footnote in Percival (1803), copied here by Ryan:* See rules xix. xx. xxi. pp. 80, 81.
34. *Author's note:* Seniority is, in general, determined by the date of admission into the respective colleges. An individual, who is not a physician or surgeon according to law, may be the oldest resident, but does not belong to the profession.
35. *Editors' note*: narrow circumstances at home; poverty.
36. *Editors' note:* From this point on, Ryan, without explanation, omits the Roman numerals used by Percival to order his ethical pronouncements. We have reinserted the numerals according to Percival's 1803 edition.
37. *Editors' note:* benefits and harms.
38. *Editors' note:* "Where equal in knowledge, a friendly doctor is better than a stranger." We could not locate this passage in the works of Celsus.
39. *Editors' note:* This concludes the verbatim quotation from Percival.

Chapter V
American Medical Ethics

Professional Reputation

The following description of professional reputation was published by Dr. Godman, Professor of Anatomy and Physiology in Rutger's Medical College in 1829, and is so graphic and admirable, that I am induced to copy it. This essay was read, by appointment, before the Philadelphia Medical Society, Feb. 8, 1826, and is extracted from a volume, entitled, "Addresses delivered on Various Occasions. By John D. Godman, M. D., &c.," and politely sent to me. It affords me much pleasure to place American Medical Ethics before British readers.

Our profession is coeval with the distresses and sufferings of the human race, and its respectability is as universal as the benefits it is capable of conferring, when rightly administered – those engaged in the discharge of its duties having always been tacitly considered by their fellow-men, as beings peculiarly set apart from the rest of mankind, and worthy of an estimation, not conceded to persons employed in merely secular affairs.[1] The real excellence and usefulness of our art – when worthily practised – has always tended to increase the confidence and admiration of the public; and, if medicine have not attained a degree of perfection and immunity from censure, equal to its venerable age and importance to society, this results from circumstances, which, however they may have injured, are entirely extrinsic to the profession.

[105/106] Yet our useful and excellent science presents a great number of obstacles and difficulties to her votaries, which are only to be surmounted by well-directed and most persevering efforts. A mistake made in the outset, may exert its prejudicial influence on the whole of your subsequent course; therefore it is desirable, that your principles and views should be both early and correctly formed.

The members of our profession are subjected to many temptations from ambition, which are scarcely to be resisted. Few, perhaps none of us, are willing to look upon our art as a mere mode of obtaining subsistence, whatever be our situations. We hope to gain a reputation, or fame, by the exercise or improvement of it; and this is the unseen, but ever operative cause, which urges us forward in our variously deviating careers. This desire of fame – this hungering after the approbation of the wise and good of our species – this wish to be singled out and placed above the great mass of our fellow-creatures – is a perfectly natural feeling, and of kin to immortality. To this cause we are indebted for the noblest exertions of human genius; it was this feeling which incited all the great of former days to the actions which still live on the page of history; – and the same breath will continue to enkindle from their ashes, fires which shall warm, cherish, and enlighten, universal society.

There are two kinds of fame, between which it is necessary for you to know the distinctions. The first, and only excellent, is that which tempts the wise and good man to become *great;* whose influence is not only felt during the existence of the possessor, but leaves behind it a holy light, undimmed and undiminished by the lapse of ages. This fame is built upon the solid basis of usefulness, genuine worth, and high desert. Its growth is not rapid, but its maturity is perfect; at first, it is the applause of those who are emphatically called "the few;" it is not gained until many privations and toils have been endured; yet, like the ascending sun, it surely attains a meridian altitude, and disperses by the potency of its irradiations, all clouds which would obscure or intercept its brightness.

The other kind of fame, is "base, common, and popular." [106/107] It is never the result of great intellectual exertion – often it is produced by accident, and it frequently is awarded to great vice. At first, it may *appear* bright and dazzling; but this light is the phosphorescent gleam hovering over putrefying substances, compared with the intense, steady, and sun-like ray of that first mentioned. This second fame, is the clamorous plaudit of the deceived or ignorant crowd; it is sustained solely by the breath of the vulgar herd, and would sink for ever in a purer atmosphere.

The fame that you should desire to win, is that which rewards the exertions of generous and virtuous minds. But, you should not only feel the proper emulation – you must be aware of the best mode of attaining your object. Let the intellectual capacity be what it may, or the impulse of ambition never so strong, much time may be wasted in ill-directed and desultory efforts, without the proper training and preparation; even giant strength may be rendered worse than useless, for want of skill to direct its exertions.

A first requisite to your success, is a good education, concerning the best mode of gaining which, wise men have differed in opinion. As the great object is to enlarge the mind, stock it with images, and train it to habits of investigation and sound reasoning, "a classical education" may be stated, as of the various modes, one of the best adapted for the discipline and development of the intellectual powers[.] Of this education we consider the study of those languages, whence not only our technical phrases, but our mother tongue itself are derived, as a most essential and vitally important part.

In speaking thus, we are conscious of advancing an opinion directly opposed to notions which, of late, are becoming very general and fashionable. It is easier, however, to declaim against the ancient languages, than to learn and employ them; as to the indolent, it is far more agreeable to demolish a noble edifice, than to erect even a comfortable shed.

The correctness of the opinion we have advanced, is not supported by assertion merely: it will bear close examination – and equally resist the subtilties of sophistry, or the ruder shocks of ignorance. Other preparatory branches of educa- [107/108] tion have their specific value by the aid of mathematics, the mind is sobered, sharpened, subtilized – taught to abstract itself and become concentrated on a point: to reach out and grasp the almost inconceivable combinations of numbers, or the ineffable extensions of space. But it is with man that physicians have to do – in all his varieties – his excellence, his errors, and his sufferings; it is with the hidden springs of the passions and emotions of our race that we wish to become acquainted; it is with the defaced, not destroyed, image of the Creator, that we are to be continually engaged. We cannot comprehend man better, than by understanding the manner in which he communicates his sensations and wishes to those around him; learning from the context of his thoughts and modes of expression, the nature of the mind whence they spring; and having gained thus much, become better able to make ourselves and our profession more useful and acceptable.

We can neither acquire nor impart knowledge, without the use of words. These, however imperfect, are the signs of our ideas; hence, he who is acquiring a language – if taught aright – is, at the same time, accumulating a vast store of objects for the future exercise of his intellect, and is also forming habits of reflection and discrimination rarely to be attained in any other way.

Independent of other advantages, the language of Judea, Greece, and Rome, are particularly worthy of regard, as containing the most sublime exertions of genius – the most valuable body of truth – and, moreover, as being the fountains, whence the now widely flowing streams of knowledge gushed forth to animate and adorn the world, after the prolonged and dreary periods of its cheerless gloom. In the tongue first mentioned, we see language in its ancient and original form, venerable alike for its simplicity and force. By it are we instructed of the origin of our race, and the commencement of human society. In the Greek, we see language refined to the highest degree, and are furnished, through it, with models in almost every exercise of human intellect. It is not only the tongue by which the invaluable observations of the primitive father of our science are preserved; but we have also [108/109] delivered to us, in the same language, the words of Him "who spoke as never man spake." In the Latin tongue, we have an inexhaustible storehouse of intellectual gratification; it is, moreover, the true language of science – the ideas attached to the words being fixed, and freed from the mutations to which a living language must always be subject – it is the key to a great number of living dialects, forms a large part of the body and substance of our own tongue, and constitutes, along with the Greek, almost the whole mass of the language consecrated to the use of our profession. Hence, those who enter upon the study of medicine without having learned either of these languages, necessarily meet with numerous difficulties, which the instructed have not to encounter; and, even with the most assiduous attention, a large amount of their professional reading must remain unintelligible.

Opposers of classical education object, that the time necessary for the acquisition of learning, might be more profitably employed, as if the student were not learning to think and judge correctly; at the same time, filling his mind with ideas, and becoming well versed in the history and characters of men. Let it be remembered, that any or all of these languages may be studied while the memory is vigorously retentive and the judgment unformed. The exercise afforded by such studies, develops every faculty of the mind; the memory is replete with words, and if the studies be correctly pursued, the mind becomes acquainted with the things to which they relate; the habit of patiently investigating, and understanding the philosophy of modes of expression, teaches proper care, and gives us greater skill in our own language; and the attention is awakened to its true value and meaning, which otherwise might be neglected from habit.[2] [109/110]

You may inquire how the acquisition of such knowledge is to assist *you* in becoming distinguished in your profession. The answer is easy; nothing is more essential to the success of a physician, than a facility of communicating his own sentiments, as well as of understanding the sentiments of these who consult him. He must approach persons of every rank in society, and commune with every variety of intellect. Possessing a well-grounded knowledge of language – fully acquainted with the true value and nature of his own, which is derived from various other tongues – he is always prepared; whether by speech or writing, he addresses those with whom he is concerned successfully, because he is sure of making his wishes or opinions plain and intelligible to all. Classic learning has another influence, not less powerful or beneficial, on the human mind. The books, which should be studied, continually present the most excellent sentiments and morals, conveyed in the most refined style; and the superiority of such refinement over coarseness and vulgarity, will imperceptibly lead the student to an habitual imitation of them. The virtues of the good and the wise, and the examples of the truly great, will invite to a similarity of behaviour; while the conduct of an elegant scholar, will be a perpetual recommendation to the intelligence and acquirements of the physician.

Every commender of the learning we have spoken of, exposes himself to the charge of being prejudiced, or having too much veneration for mere antiquity. Instead of attempting to disprove such an unfounded accusation, let us employ the words of a celebrated author on the same subject. "I have," says he, "a great reverence for posterity; nor do I think lightly of the learned men of the present day. There are many, I know, who adorn our age, who would have ornamented any period; yet among the whole number, I have not known one who did not cultivate and honour ancient learning, whose wisdom was not similar to that of the ancients, or who did not admire and observe their precepts; from which, in proportion as you depart, you wander from nature and truth."[3] [110/111]

Many of the younger members of the society now present, are, ere long, to receive the honours of the profession, as a testimonial of their diligence and faithfulness as students, and will then be preparing to take their stand among the guardians of the public health. The boisterous sea of the world is attended with comparatively few dangers, to those who have not trifled away their time, and who set forth under the guidance of correct principles. Though you know not in what haven you may ultimately cast anchor, the possession of a sound moral and professional education, will insure your safety through all the turbulence you may be exposed to, during your voyage.

The greatest evil to be guarded against, when you commence your efforts for professional distinction, is impatience and instability of purpose. It will be wrong in you to anticipate that business can, in any situation, immediately follow your annunciation of being ready to receive it, more especially, where you are to meet with competition from a member [111/112] of the profession already established. The first half-year of a young physician's residence in a strange place, is the most trying part of his probation; for, should he mistake fewness of calls for neglect of his merits, or suppose that he will never be employed, because he is not immediately preferred to another, he is in danger of becoming unsettled, restless, neglectful of his books, society, and acquaintances, thus sacrificing the very means of his eventual success. If, however, we recollect how much people are prejudiced by education and habit, we shall find no fault with them for not employing a stranger on his first arrival, neither should we suppose their prejudices to be immovably fixed. A proper degree of patience, and an improvement of those opportunities that are every where presented of winning confidence, will, in no long time, yield us that opening, which is the great requisite to future profit and eminence. Wherever we attempt to establish ourselves – except under very extraordinary circumstances – some time must be passed in acquiring the confidence of those around us, by the recommendations of our friends, by our own deportment, and the event of such cases as may be incidentally thrown into our hands. However unimportant any such particulars of conduct may individually appear, they are of great moment collectively viewed; as every circumstance in the appearance, conversation, and character, of a newly-arrived physician, is of deep interest to those among whom he wishes to remain.

It will be unjustifiable to trust your success, in the slightest degree, to accident, because accident has occasionally given currency to men neither remarkable for education, talent, nor judgment. Accident *has*, at times, given a man of the highest merit an introduction to extensive business; but we must never forget, that accident cannot sustain our reputation, nor minister to our continued success. Our greatest care must be to acquire reputation by a diligent cultivation of our talents; though we should never neglect to improve any accidental success, in all honourable ways, to forward our professional views. If a character for skill and discernment be acquired suddenly, we must not attempt to increase it, by endeavouring to extend this reputation to the utmost stretch of possibility, [112/113] but by displaying new instances of talent and intellectual strength, thus substantially augment our capital of fame. The fortuitous elevation of men destitute of true greatness of character, is almost universally followed by reverses, as sudden and severe as this elevation was great. This is frequently exemplified in the fate of those who have a great air-built reputation, and much verbal fame; who, instead of modestly refusing a part of the honours proffered them, and exerting themselves *to prove* that they are worthily praised, receive the whole as no more than their right, and leave their admirers in a short time to discover, that their extraordinary pretensions are unfounded, and that their reputation is nothing better than the gratuitous offering of ill-judging and partial friends.

Next to an acquaintance with the principles[4] of the profession, and correct moral feeling, the young physician should most rely on the exercise of what may be summed up by the word *manners*. This embraces his general intercourse with society; in which he should display that habitual ease and cheerfulness, which results from correct habits of thought and action, and that kind attention to the wishes, prejudices, and necessities of those he associates with, which shows that he possesses the most generous and elevated feelings. To be accessible and attentive, without familiarity or cringing – to be mild, gentle, and forbearing, without sinking into tame submissiveness – to be ever ready to act when called on, without being officious or intrusive – and to do full justice to all those with whom he is professionally concerned, will insure a physician a degree of public respect, that may at length [113/114] amount almost to idolatry; filling every bosom with kindness towards him, and every mouth with his praise.

The sagacious Lord Bacon has given a rule for increasing our knowledge, and insuring conversation with all sorts of persons, which is one of the best that could be devised, and one more positively conducive to popularity can scarcely be imagined. This is, to learn something from all persons, whatever is their occupation, when we chance to be with them; this is always to be accomplished by inciting them to speak of what they know. As every man is better acquainted with his own business than we can possibly be, by inducing him to converse on the subject, we not only gain some valuable ideas, but we win his regard by manifesting an interest in what so peculiarly interests him. By adopting this method of Lord Bacon's, you need never suffer from tedium in the company of unlettered men, nor need you in the slightest degree to descend from your place, while judiciously exciting their remarks. This is not merely applicable to your intercourse with persons of inferior standing in society; the most learned, refined, and accomplished men, are equally pleased to find, that their pursuits, avocations, and interests, are interesting to you. You may frequently induce such persons to display before you, a stock of knowledge which otherwise would have been withheld, and you will part mutually satisfied, instead of being in ignorance of, or prejudiced against each other.

This rule may be observed with the utmost sincerity, and without the slightest approach to the meanness of flattery. The information thus to be acquired, will, in general, be far more easy of attainment, as well as more valuable, than can be gathered from many books; and you will, at the same time, be forming a more profound acquaintance with human nature, and also gain friends. Kindness, uniformly produces kindness; confidence inspires confidence. If we examine ourselves, we shall find that we are as excitable in this way as others, for we never deliver our thoughts with more force and feeling, than when we reply to interrogations on subjects in which our minds are most deeply engaged.

In all your intercourse with society, as well as in all your [114/115] thoughts and actions, cultivate an habitual tenderness of regard for TRUTH. By this, I would not pretend to warn you against the disgrace of falsehood; but, that you should guard against a habit, which is almost as common as the human family is numerous, of suffering apparently harmless exaggerations to escape. Truth and falsehood, like light and darkness, are opposite extremes – the one is as excellent as the other is base. But there are a great many aberrations from truth, which the world does not consider to be absolutely false; as there are many deviations from honesty, which, by a similar laxity, are not considered as positively dishonourable? If my wishes could influence, you should begin your career with resolving to adhere to the full purity of truth, and the perfect honourablenesss of honesty; so that when the day of your success arrives, you may look back on the means by which it was attained, without breathing a sigh of regret, or suffering one blush of shame.[5]

Our profession has long been subjected to the charge of "envy, hatred, malice, and all uncharitableness," among its members; and unfortunately, too much of the charge is well-founded. We cannot at present enter into an investigation of the causes by which this state of things has been produced, although it does not affect the profession to the degree which persons commonly suppose.[6] To lessen this evil, and avoid meriting such an accusation, make it a rule never to [115/116] speak of a professional rival, unless you can speak to his advantage; if he have merit, allow him the whole of it, and give your sentiments of his talents with the unaffected earnestness of truth. Do not imagine that your acknowledgment of his merits, will hide his defects, or obscure your own good qualities. Grant that he adopts a contrary course, speaks ill of you, or throws out insinuations intended to be prejudicial to your interests; – then is your triumph complete. Think you that men will not contrast his mean and soulless conduct, with your manly and honest candour? Think you that he will not more deeply condemn himself, by attempting to misrepresent you – that society will not visit his ungenerous conduct on his own head, while the profession silently spurn him from their confidence?

Should you be eminently successful after others have failed, avoid pushing your triumph so far as to wound the feelings and outrage the pride of your less fortunate competitors. Your success is sufficient for you, and by judicious deportment, you may compel a man to respect, if he does not esteem, who might otherwise cherish against you, a spirit as stern as hate – as inexorable as the grave. If, after such success as we have alluded to, you hear of disparaging suggestions made against you, by one you have set aside or over-shadowed, you are neither obliged to know, nor resent it[7]; you would owe it to the dignity of your own character, to recollect that some allowances are to be made for mortified feeling, as well as that no malicious insinuations can stand against the daily repetition of actions, which prove you exempt from a grovelling and miserably irritable disposition.

That you will not attain the professional elevation you desire, without struggling against hosts of difficulties, and encountering every degree of opposition, is most certain. "It may be, that the iron grasp of poverty, for a considerable time, will impede your progress and enfeeble your efforts. Against rivalry and opposition, your armour of principles and determined perseverance will afford every security, and [116/117] poverty itself may be made to minister to your success, by urging you to the display of your noblest powers.[8] Look at the men of talents who now lead the van of our profession, and are considered as its ornaments. Who are they? Men born to fortune and reared in the lap of luxury? No. Men who have been elevated by protection and patronage? – who have been favoured by circumstances, or raised by accident? No. They are, most frequently, those who have emerged from poverty, if not obscurity. Many of them have been nursed in sorrow, and baptized with tears; – they have protected and patronized themselves, until the great and powerful have become proud to rank as their friends; they have *made* the very circumstances, which superficial observers suppose to have been the *causes* of their elevation. It is the triumph of talent, of genius, to rise in proportion to the magnitude of difficulties; to trample the opposition of malignant mediocrity into the dust; and gaining its merited elevation, to raise the profession it has chosen to a corresponding degree of eminence.

Since the commencement of the present session of the society, some of our young friends, who entered on the career with hopes as warm, and eyes as bright as ours, have been called away to the "narrow house," and their spirit-stirring bustle of youthful expectation, has been exchanged for the solemn quiet of the tomb.[9] While we sympathize with such as mourn over youth snatched away in its blooming – and warm hopes chilled by the icy hand of death; while we sorrow over the mental anguish of those, whose far distant parents were not permitted to minister to their last earthly wants, or receive their dying sighs – let us not forget to be thankful, that we are still spared to usefulness and virtue. Yet a little [117/118] while, and the mighty ocean of oblivion will whelm us in its fathomless depths, sweeping away every trace of our existence. This is not matter of regret. All nature tends to one common point – disintegration and change of form; – the cloud-capped and tempest-braving mountains, towering in seemingly indestructible grandeur, are hourly yielding their atoms to the earth and the air. Virtue alone survives all change; the immortal mind bids defiance to the destructibility of matter.

Build then your monuments, imperishably, on the love of mankind, by sincerely devoting yourselves to the cause of humanity – to the honour of your profession and country – to the faithful service of your friends – to the humble worship of God: thus, the necessary evils of life will pass over you unheeded; and the inevitable shaft of death, while it stills for ever all mortal disquiets, shall be unable to disturb the serene and ecstatic composure of your intellectual being. [118/119]

Notes

1. *Editors' note:* Omitted here is a footnote copied from Godwin's original text, appending a quotation in Spanish from Don Juan de Zabaleta.
2. *Footnote copied from Godman's original:* In urging the importance and necessity of classical learning to those destined for the profession of medicine, it is no part of my intention to state, that the manner in which the languages are most generally taught is either the best or even the true one. It is but too common to ascribe the faults of teachers to the thing taught; to prejudge and condemn an unexplored region, because the ways leading thereto chance to be foul.
3. *Editors' note:* Omitted here is a footnote copied from Godman's original, giving this quotation in Latin, and crediting it to "And. Dacier." This may be André Dacier, who published a French translation of Hippocrates' works in 1697. Following this quotation Godman appended a further quotation in English, credited to "Harris's *Hermes*," extolling the study of ancient languages. This would appear to be a reference to James Harris, *Hermes, or a Philosophical Inquiry Concerning Universal Grammar*, London, 1771.
4. *Footnote copied from Godwin's original:* "I would not be understood, in what I have here said, or may have said elsewhere, to undervalue *experiment,* whose importance and utility I freely acknowledge in the many curious nostrums and choice receipts with which it has enriched the arts of life. Nay, I go further – I hold all justifiable practice in every kind of subject, to be founded in *experience*, which is no more than the results of many repeated experiments. But I

must add, withal, that the man who acts from experience alone, though he acts ever so well, is but an empiric or quack; and that, not only in medicine; but in every other subject." – Harris's *Hermes*, 352.

5. *Editors' note:* Here is omitted a footnote copied from Godman's original containing a quotation in Latin from Marcus Antoninus.
6. *Editor's note:* Here is omitted a footnote copied from Godman's original containing a quotation in Spanish from la Zabaleta.
7. *Editors' note:* Here is omitted a footnote copied from Godwin's original, containing a quotation in Latin from Marcus Antoninus.
8. *Editors' note:* Here is omitted a footnote copied from Godwin's original, containing a Latin quotation attributed to "Reinwardt," whom we have been unable to identify further.
9. *Footnote copied from Godwin's original:* Several had then recently died of small-pox.

Chapter VI
Medical Education

Degrees – Diplomas – Medical Appointments – Success – Reputation – Eminence – Moral and Physical Medicine – Art of Prescribing

It is now universally admitted by all our medical institutions connected with education, that those intended for the study of the healing art should receive a good general and classical education. The professors of all our universities and medical schools, the examiners of all our colleges and apothecaries' societies, delivered their evidence in proof of this point, before Mr. Warburton's Parliamentary Committee on Medical Education and Practice, in the summer of 1834. All the witnesses proved that a knowledge of the Greek, Latin, English, French, German, and Italian languages; of mathematics, logic, moral philosophy, natural history and philosophy; in fine, that the course of general education required by the universities for degrees in arts, is indispensably necessary to those intended for the medical profession. This extensive course of preliminary instruction has long been required by the universities of all candidates for admission into the learned professions; and is now exacted by most of the colleges of surgeons, and societies of apothecaries, of those intended for the practice of medicine.

No class of men stand in so much need of extensive erudition and knowledge. They may attend all ranks of society from the lowest to the highest, and ought to be exceedingly well-informed in general as well as in medical literature.

A greater character for learning and science cannot exist than that which constitutes an accomplished physician; the most extensive study, and the most comprehensive mind, are therefore requisite. The youth intended for medicine should have all his senses acute, and in the utmost degree of perfec- [119/120] tion; because it is from the perceptibility of the senses alone, that the human mind is stored with all those sublime ideas which shine so conspicuously in the future life of a medical practitioner. When the senses are obtuse, dulness is the consequence; and a dull student never made a brilliant physician. The medical student should, therefore, have the clearest powers of perception, so as to receive accurate impressions, and possess tenacity in their retention.

He ought to have a good memory and great reflection. The intellectual faculties should be copiously enriched by indefatigable industry, and unceasing study; with the

H. Brody et al. (eds.), *Michael Ryan's Writings on Medical Ethics*,
Philosophy and Medicine, Vol 105,

most extensive ideas of sensation and reflection; with memory, invention, and genius – so as to be capable of arranging a variety of ideas in strict logical order. When all demonstrative facts relating to our art are acquired, reflection, acute reasoning, and profound judgment, will decide the manner in which they are to be applied in the treatment of diseases. What a vast field of knowledge is comprehended in the medical sciences! How many days and years of labour and industry are required! What sedulous diligence is absolutely necessary! We must, therefore, pursue with an ambitious zeal the various branches of the medical and collateral sciences; we should study all, as all united form, with experience, the greatest medical characters.

We set out in pursuit of professional distinction when the buoyancy of youth, and the vigour of imagination lift us over every impediment, and break down every barrier. Hope tints the distance with the most glowing and flattering colours, and the mind revels in delightful anticipations of pleasure, fortune, and renown. A moderate experience in the cold realities of life proves that we have been dreaming, and teaches that if these good things are ever to be realized it is only when years of patient endurance have elapsed, and after the fires of youth have been well nigh expended in the service of our fellow-creatures. Accident may sometimes realize the expectations of youth, but the most universal rule is, that wealth and fame from professional exertion is the slow, though sure reward of long labour and persevering industry. This circumstance is of the greatest advantage to society, and to [120/121] our profession; but those who have yielded too much to the dominion of hope and fancy, are frequently so much affected by discovering the truth, as to suffer an entire revulsion of feeling, and sink from the most brilliant flights of imagination to the lowest depths of despair. This despondence is permitted sometimes to prey on the mind until it produces neglect of business or harsh misanthropy; and the unfortunate sufferer is continually tortured with notions of the ingratitude of mankind, the neglect of merit, the low state of professional character; while he is letting slip the best opportunities to convince himself of the contrary by efficiently performing those duties his profession enjoins, and society requires. Be then prepared to discover, that the world yields neither wealth nor distinction except as the price of industry and great deservings. Stop not to consider whether men are ungrateful or merit is neglected; but perform the actions that create a claim to their gratitude; declare your merits by the faithful discharge of your duties: and then you will find such complaint impossible.

If such were not to be the result, policy would dictate the propriety of concealing our mortification. The voice of repining and discontent is ever painful and offensive to others; and the same persons who warmly sympathize with a noble spirit struggling against misfortune, and, though broken-hearted, looking calmly on the approach of inevitable fate, despise the creature who is continually vexing their ears with fruitless and peevish complaints, or venting selfish ejaculations against the characters of those who have lived beneath a brighter sky, or been wafted along by more propitious gales.

Of this you may feel perfectly assured, that really meritorious conduct cannot go altogether unrewarded; neither can the fire of true genius be entirely smothered. The time must come when perseverance in the conscientious discharge of high duties will secure the remuneration and respect it is entitled to; the mind that has been wrought up by the study of proper objects, and is sustained by a determined enthusiasm, to effect great purposes, may for a time be weighed down by poverty or misfortune; but, like the giant of ancient fable, its struggles will convulse the superincumbent mass, [121/122] and must eventually shake off every hindrance to perfect success.

If, in offering these considerations to you on the present occasion, we appear diverging too far from the beaten track, we trust you will pardon the zeal that urges the laying before you, what reason and. experience induce us to hope may be to your advantage. Being exclusively devoted to the service of those who are engaged in the study of medicine, we

> may be allowed in some degree to identify our feelings with theirs, and be anxious to spare them suffering, not less than to aid in insuring their success. Whatever defect there may be in manner, there is none in feeling; nor is there the slightest departure from fact in stating
> For you, ye studious, I strive,
> For you, I tame my youth to philosophic cares, And grow
> still paler o'er the midnight lamp.[1]

We may, however, practise any one department suited to our taste or inclination; but we should study and know the whole. One practitioner prefers surgery, another obstetricy, a third aural, a fourth ophthalmic, a fifth dental surgery; others devote themselves to diseases of the brain and mind; some to maladies of the chest; more to those of the heart; many to disorders of the digestive system – stomach, intestines, liver, &c.; more to genito-urinary diseases; others to local affections – lithotrity, gout, &c. But every physician and surgeon should be conversant with the nature and treatment of every species of human disease. It is perfectly impossible to separate diseases into medical, surgical, and obstetrical; as any one disease, no matter what may be its situation, can and does derange and disorder all the organs and functions in the human body. Away, then, with the absurd and untenable distinctions of medicine, surgery, and obstetricy; the profession as a science, is one, and indivisible.

It is highly advantageous to the cultivator of the healing art, to visit different universities and schools, and to learn the principles and practice of medicine under the most celebrated [122/123] professors. It is for this reason, that students who commence in London, repair to Edinburgh, Dublin, Paris, Vienna, or Pavia. When a man of superior merit is to be found in any part of the world, his fame has great influence on strangers, and remote countries send him disciples and admirers. Few are ignorant of the great afflux of students from many nations to the lectures of Boerhaave, Morgagni, Monro, Hunter, Frank, Cullen, Scarpa, Abernethy, Cooper, &c., &c. It has often been a great advantage to have studied under such masters; and, even in our own time, under an Abernethy, a Cooper, a Dupuytren, a Bell, a Colles, &c. Lucrative and important appointments have been given to the pupils of these eminent individuals. There is, however, a more independent spirit abroad in our day; and few are ready to swear by the word of a master. Moreover, the opinions of all eminent practitioners are generally published, and may be procured with the utmost facility. This incalculable improvement was effected by the medical periodical press, and is now general throughout the world. The periodical press publishes the lectures and opinions of the most celebrated and eminent practitioners of different countries, and presents them on such low terms that every student and practitioner may possess them. It saves a vast expense which would otherwise be incurred, in the purchase of numerous original works; and affords the greatest facility of learning the opinions of celebrated physicians and surgeons of all countries.

The medical periodicals also form a complete encyclopaedia of the healing art, and prevent the necessity of procuring many expensive works. It is for this reason that every scientific practitioner, and every industrious medical student, possess them. A good library is, however, indispensable to medical practitioners. It ought to contain the best systematic, monographic, and elementary works, of this and foreign countries.

A well-selected museum is also a source of much instruction and information. It is likewise essential to associate with literary and scientific individuals, and to join the medical societies and academies. Such are a few of the requisites for [123/124] obtaining a proper knowledge of the principles and practice of medicine.

Medical Degrees, Diplomas, and Medical Appointments

It is now universally admitted, that the present curricula of medical education and practice in the British dominions, require the interference of the legislature for their improvement. This position is amply attested, by the appointment of a parliamentary committee in 1834, to examine into the actual state of medical education and practice, before which all the representatives of the universities, colleges of physicians and surgeons, and societies of apothecaries, in this kingdom, were examined; and all admitted that there was the greatest need of improvement. It has been long complained of, and with reason, that there is too much facility in obtaining degrees and diplomas, and that the period of study was much too short. A great number of professors attested this fact before the Parliamentary Committee. The period of 2.5 years, required for acquiring medical knowledge by the Royal College of Surgeons, in London, and also by the Apothecaries' Company, is much too short; and the examinations are by no means probationary, nor the diplomas a sufficient guarantee to the public, of the possessors' competency as medical or surgical practitioners. Some of the universities are equally liable to the same charge of neglect and dereliction of duty. The title of doctor is too often obtained by incompetent individuals, who necessarily dishonour it. Some of the foreign universities absolutely sell their degrees, and it is now of daily occurrence to pay a few pounds into a certain bank, and receive a doctor's degree by post, from some of the German universities. This was the custom at some of the Scottish universities, until a recent period, (1826). It cannot be denied, that too many illiterate persons are still admitted into the profession in the different schools of this country, and such indiscriminate admission has considerably lowered the character of the faculty. This observation particularly applies to the College of Surgeons and Apothecaries' Company of London. These bodies require the most superficial preliminary and medical education, [124/125] and often allow the most illiterate persons to obtain their testimonials. No experienced physician or surgeon will declare; that 1 year's attendance on the medical or surgical practice of an hospital or dispensary is sufficient for acquiring a practical knowledge of medicine and surgery? Who is it that will admit, that the science of medicine and surgery can be properly studied in two or three winter sessions of six months each? The time has nearly arrived when the good of the public and profession will enforce a uniform system of education, and one sufficiently comprehensive. Such will most probably be the result of the intended Medical Reform Act, which will be proposed by Mr. Warburton in the next session of parliament. It is now generally known, that almost all medical appointments in England, Scotland, and Ireland, are filled by jobbing and bribery. The majority of the physicians and surgeons selected, do not possess any, or sufficient, or superior merit; and obtain their situations by intrigue, family influence, or purchase.

Such is the case as regards our court, hospital, army, navy; and all other public medical appointments. Men of talent and genius can never succeed in obtaining places in this kingdom; while political and religious feeling, with intrigue and money, throw them into the shade. In France, a concours or medical jury, decide who is best qualified, among several candidates, for professorships and other medical situations. This excellent custom excites emulation, and rewards superior merit with a proper recompense. There is no means so favourable to the progress of science, and nothing so likely to foster that ambition which is so laudable in medical practitioners. What efforts are impossible to a young physician or surgeon, which inspire the hope of an honourable and profitable appointment, and the desire of conquering able rivals by extent and variety of knowledge? We find, accordingly, that some of the French professors had successively contested for appointments, so often as six different times, before they succeeded in attaining one of the objects of their ambition. It has been urged against concours, or public examination, that a timid individual might become so confused as to acquit himself so badly in his answers, that one of inferior talent or acquirement might be [125/126] elected. It has certainly happened, that the public voice has not always approved of the decision of the jury. Nevertheless the decision of a medical jury after the public disputation, examination, and competition of rival candidates, would be infinitely superior to the election of professors and medical officers by private interest, and non-professional governors, which is the usual mode in this country. It is also notorious, that many individuals have often purchased their situations, though totally incompetent to discharge the duties belonging to them. It has also been long remarked, that the medical situations, filled by the government and public societies, are disposed of through favour and interest, and seldom on the grounds of merit or talent. Thus it was the custom, until within a few years, to place the fellows of the London College of Physicians in hospitals, dispensaries, and other public appointments, though, in general, far inferior to rival candidates.

Again, the surgeons appointed to all lucrative situations, as hospitals, &c., are mostly the relatives or apprentices of former hospital surgeons, who transmit appointments as ancestral property. Moreover, we find the colleges, excluding all who practise obstetricy or pharmacy from their council and examining boards. We observe, too, that the apprentices of the Dublin College of Surgeons, preclude surgeons of all other colleges, from infirmaries, and all situations of fame and emolument in Ireland. The result of all this monopoly is a carelessness and indifference on the part of the bulk of the faculty to the proper cultivation of science; they reasonably declare we have not the least chance of appointment under the present system, and there is no stimulus to exertion and study. The consequence is, that fewer improvements have been recently made in medical science in this kingdom, than in any other in the world. What have we done in introducing new and powerful medicines? Our neighbours have introduced all of them. What discoveries have we made, except those of Sir Charles Bell? The stethoscope and lithotrity do not belong to us. It is true, that we can boast of Sir Charles Bell's splendid discovery on the functions of the nerves, and of the application of transfusion, by Dr. Blundell, to the preservation of life after hemorrhage; but with these exceptions our [126/127] improvements in physiology, pathology, and therapeutics, bear no comparison to those of vicinal nations. It is also true, that, Sir Astley Cooper tied the aorta, but *cui Bono ?*

Such are a few of the evil results of monopoly and abuse in the medical profession in Great Britain and Ireland. Talent and merit are unrewarded. But better times approach, a complete reformation in education, and the practice of medicine is at hand.

Entry of a Junior Medical Practitioner into the World

A young physician or surgeon who has passed some of the best years of his life in the lecture rooms, hospitals, and libraries, expects, on receiving his degree or diploma, to acquire fame and fortune on commencing his professional career. He is well acquainted with the principles of his art, and closely follows them. He employs all that he has heard and read in the investigation of the nature and treatment of disease. He is a scrupulous observer of the rules of art, and fears to break them in their application; he examines with great care, and gives his opinion with diffidence or fear. He is cautious and timid in the use of remedies. The better informed he is, the less presumptuous or rash he will be in practice. He is in general diffident in himself, hesitates or is too anxious about his patient, and is for some years before he acquires that confident assurance, which is the result of experience. He generally supposes that he has mistaken the disease, and has not done enough for its removal. This is the ordinary feeling of the best informed medical practitioners in the commencement of their career; but experience will, in time, remove it. The complication or intensity of disease may be such that no remedial means can preserve life. The contrast between well-informed and ignorant practitioners is very remarkable.

It has been long observed that the most uninformed in every science and art are, in general, the most confident and self-sufficient. This is also the case with pretenders to medical knowledge. No case is too dangerous for them; they can cure all diseases, though they kill a vast number of patients. They attribute all their failures to the intensity of disease, and not to their own ignorance. As they have had no pre-[127/128] liminary or medical education, which is so necessary to prepare the mind for the investigation and estimation of morbid phenomena, they treat diseases by names, and not according to their real nature. Men of unlettered ignorance usually pride themselves, on imagined practical experience, or on what Cullen aptly termed "false facts," and blindly pursue the same erroneous system without a particle of sound judgment. In fact, nothing appears beyond their ability, or too difficult for them to attempt. They can cure all diseases – a power only possessed by empirics; they ascribe their success, in cases in which nature effected a cure, to their active interference, but they never suppose that their failures are in any degree the result of their incapacity. They do not know that experience, without a proper scientific education, is useless.

It cannot be denied, that unless a medical practitioner knows the anatomy and physiology of the human body, or, in other words, the phenomena of the animal economy, his observation and experience are based on ignorance, and must be erroneous. Such is the advantage of experience, unaccompanied by proper medical information. On the other hand, the well educated junior or senior practitioner, who has had the most extensive experience and information, is always diffident in his own powers, and doubtful of the result of the disease. He never professes to cure every patient.

But I must return to the career of the young physician or surgeon. On receiving his diploma, he is most anxious for fame, and thinks the world ought to be instantaneously aware of his qualification to practise. He, however, speedily discovers that the world esteems him as a student, and that he is not a more important individual than he was, the day or month before he received his qualification. He speedily discovers that patients will not consult him. Some suppose him inexperienced, and too youthful; others have their own medical advisers; and all will prefer practitioners of standing to our young Aesculapius. He now thinks that his youth is a great or invincible obstacle to his success, and he sighs for the time in which the public will reward him for all the labour and expense he has incurred in the study of his art. Nevertheless, every other member of the profession had the same difficulties [128/129] to contend with, and only surmounted them by time, attention, skill, and unexceptionable morality. He does not imagine that a physician, extensively engaged in practice, has any cause to complain. He does not know that such an individual is deprived of all pleasure, recreation, or amusement, and regrets the period of his studies, when he enjoyed liberty. Thus, man is never content with his condition.

The first success or failure of a physician depends on the opinion entertained by the public of his talent. How embarrassing and difficult is the debut of a physician in the world! How careful he must be in gaining the confidence of his first patients, in investigating the symptoms of their diseases, and in employing remedies! He must recollect, that others of his faculty have been consulted on former occasions, and that their manners and conduct will be compared with his own. When the disease is dangerous, and he effects a cure, he will greatly benefit his reputation. His religion and zeal compel him to attend the poor and miserable, as well as the comfortable and affluent; and he may gain reputation by both. He will often succeed after his seniors have failed; because they in general belong to the old school, and he to the new. He should not expose their errors, unless glaring and dangerous to the life of the patient. He is always ready to afford relief to suffering humanity, with or without remuneration. In the same manner, a surgeon should never refuse to perform a doubtful operation when it is required. The patient ought to have every chance. Either physician or surgeon can explain to the relations the danger of any case.

The essential requisites of a physician are, in addition to the possession of real virtue, knowledge of the world, and polished manners, an earnest desire for fame, a great love of study, a good knowledge of charlatanism, and what is vulgarly termed humbug, a good share of small talk, and an audacity which nothing can disconcert. The critical reader will inquire, why a medical practitioner should be acquainted with charlatanism and humbug? The answer is, because it is the nature of mankind to esteem quackery, as it professes to effect cures of incurable diseases. All classes of society encourage and patronize empiricism; but an immense crowd [129/130] of fools constitute the public. In all countries, the most absurd quackeries have been patronized; but this is easily explained, when we dispassionately observe the varied degrees of intelligence, and the love of life and of the marvellous which influences all ranks of society.

It must be confessed, though a stain on the dignity of the medical profession, but equally remarkable in every class of society, that some Esculapians have acquired success, fame, and opulence, by fostering the follies and prejudices of the world; and by acting as arrant humbugs. Such men are, however, despised by their brethren, who, in general, are actuated by the love of humanity and reputation. The former are despised by the honourable cultivators of medicine, notwithstanding their titles, station, and influence.

The medical profession has great reason to complain of the follies and injustice of the public. The mass of fools forming the public, with few exceptions, will ascribe professional success to chance or to nature; and, if the patient dies, some will not hesitate to declare that the doctor killed him. They will unhesitatingly proclaim the incompetency of the medical practitioner, unmindful of the well-known unfortunate truism, that many diseases and many constitutions are utterly incurable.

M. Lorry, a celebrated member of the profession has happily exposed the folly of the public in this respect. He says—"A physician fond of study has spent many years in the schools, hospitals, and anatomical rooms; he has passed the best years of his life in the infected air of hospitals: the pallor of his countenance, and the meagreness of his figure, attest the multiplicity of his vigils; and what remuneration compensates for these labours?"

Here the ignorant man of the world declaims against the stability of one of the noblest of human sciences, and unblushingly confounds medicine and empiricism. Thus there are many, whose lives have been saved by medical practitioners, who forget it, and even denounce the medical faculty.

If the public do not derive the usual advantages from medical practitioners, they should blame themselves for encouraging illiterate pretenders to medical knowledge, nom- [130/131] inal doctors, and empirics. Such impostors are very numerous in all countries; and if the public do not take the trouble of distinguishing the learned from the ignorant, the fault is their own. If the nobility and gentry prefer the St. John Longs and such basket makers, the Hahnemanns, and other empirics; the woodmen of Molière, to well-educated and eminent medical practitioners, they must not decry the benefits of the healing art, and depreciate the Bells, Coopers, Dupuytrens, &c., &c.

In general, young medical practitioners are most attentive to their patients, and render them the greatest services. So do all duly-educated members of the medical profession, who are influenced by the code of ethics laid down for their guidance. There are a few, however, among the eminent in all countries who are a disgrace to the profession. They take advantage of public credulity, which in all ages has rendered the worthy part of mankind dupes to the artifices of the knavish; who, unrestrained by principle, are ever eager to profit by the unsuspicious disposition of generous minds. Among the various kinds of imposture practised on society, quackery has always been the most successful, in consequence of the extreme respect paid to the professors of the healing art. These have now and then availed themselves in acquiring reputation by the foibles of the public.

Reputation of a physician. When the desire of reputation is inspired by the love of fame, it is allowable, because it tends to the good of society. It has been often observed, that those physicians who have not a laudable ambition to rise to eminence by their talents and honourable conduct, seldom or ever become renowned. They remain every-day characters, and never obtain the highest places in the profession. When a thirst for gold is the only object of professional reputation, it leads to meanness and disreputable behaviour. We see empirics, illiterate and professional, amassing great wealth at the expense of every virtue which adorns the true medical character.

The public is the cause of this, as it awards reputation by caprice, and this is the reason that charlatans share, in common with educated physicians, reputation and renown. They [131/132] announce, with unblushing effrontery their marvellous powers, and all ranks of society, both high and low, are ready to believe them. Kings, nobles, and commons, patronize the empiric, no matter how preposterous his pretensions. Let him be an animal magnetizer, a homeopathist, an extractor of all diseases with a stimulating liniment, or metallic tractors, it is all the same. Witness the patrons of a Long and a Hahnemann, of a Graham and his celestial beds, from which a noble race of offspring was to follow; witness a scion of nobility attesting, three years since, that he saw a quack extract quicksilver from a lady's brain! Look to the noble patrons of Hahnemann, who believe the most salutiferous doctrine, that the millionth part of a grain of magnesia or of sugar could not only cure the inhabitants of a city, but an empire, however extensive. What difficulties, on the contrary, has the modest man to surmount before he can obtain the smiles of public favour.

There, is, however, in medicine, different kinds of reputation which lead to fame and fortune. One gives himself up to scientific pursuits, another distinguishes himself by writing or teaching, a third is renowned as an excellent physician, a fourth as an operative surgeon or obstetrician, &c.

It rarely happens that one individual possesses these different claims to celebrity. The healing art is too abstruse and extensive for the greatest genius to comprehend it in all its parts. One is superior in medicine, another in surgery, and another in obstetricy, chemistry, botany, &c.

When a physician acquires reputation by merit, he should never attempt to increase it by resorting to the artifices of charlatanism, which invariably lower him in the estimation of his brethren, and, through them, of the public. He should avoid all species of puffing, whether having nurses or relations to trumpet his abilities, or superficial works to circulate gratuitously among his connexions, and their friends. If his fame be founded on talent and merit, it rapidly increases without the artificial helps – it enlarges as it progresses – *vires acquirit eundo*.

If the cacoethes scribendi should seize one, let him write something rare, new, or – nothing: *scribe rara aut nova, vel nihil*. The style and composition ought to be correct and [132/133] elegant. Citations may be numerous and acknowledged, though some writers too often give the whole of other men's works and opinions as their own. These are generally detected and exposed by reviewers and critics. It is no excuse to the reader that a work is written or printed in a hurry, and full of blemishes; the public would have waited for it with patience, and seldom care if it had never been published. An author, should deliver his opinions with modesty, and expose what is true, dubious, or erroneous. He must rarely condemn others, or censure the living or dead with undue severity. He must praise sparingly, and vituperate more sparingly. There are some writers and critics who praise no one, and vituperate every one. They find fault with every thing – style, composition, opinions, arguments, observations, and conclusions. They hold, that every thing is to be found in the works of the ancients, and that there is nothing new under the sun. These critics raise a host of enemies against them, and seldom have any friends. They should adopt the ethical axiom: – "it is not right always to praise or dispraise"—"Nec laudare semper, nec semper vituperare decet. Lauda parce, vitupera parcius. Lauda recentes, lauda veteres, vel utrosque carpe, si licet. Stet sua cuique reverentia, suus honos. Non viventes auctores enormiter laudes, ut vicissim lauderis. Stet verbis, stet titulis, stet sua encomiis mensura. Unico libello scriptores omnes, omnes amicos non alliga. Nec æmulorum, nec mortuorum laudes dissimules, nec excedas."[2]

When criticism is unjustly severe, it often becomes pointless, and only serves the circulation of a work.

There is no work free from some objection, and there is a vast number of modern publications deserving of severe criticism. Thus one man writes a pamphlet or book, for the express purpose of making himself known by advertisements in the public journals; another does not state a fact that was not well known for centuries before his time; another gets himself puffed in a sly paragraph of some newspaper; and in these, and a thousand other ways, do medical practitioners attempt to captivate the public. These schemes, *ad captandum vulgus*,[3] are derogatory to the dignity of medicine, though [133/134] they are now unblushingly laid by some of our highest physicians and surgeons. We daily observe their advertisements by the side of those of empirics, and their paragraphic puffs are most abundant.

Few of the many medical works now published in this country are successful, and few pass through future editions. Many of them are still-born from the press; others speedily die. This failure of success has influenced publishers so much, that they rarely purchase a copyright of a medical work, and will take no part in the risk incurred in bringing out books of this description. It was lately stated by one of the most extensive medical publishers in London, that few monographs had a circulation of five hundred copies. Does this want of circulation arise from the mediocrity of books, or from the distaste for medical literature? I think from the former. But a second edition is called for; a new title-page, preface, or chapter or two are added; the public is gratefully thanked for its patronage; and this species of charlatanism is invariably discovered and censured.

Besides these modes of gaining fame, we find physicians and surgeons recommending apothecaries, and vice versa. The female portion of their own and other families become puffers of their skill, and often succeed in increasing their fame. A lady of title has been known to make the reputation and fortune of her medical attendant.

People of fashion are led by notoriety and imitation of each other, and thus the physicians to royalty are indispensable to the nobility. No other will do.

In other instances, a reputation and a name are made by many petty artifices – by gossiping, a talent for flattery, the influence of medical and other patrons, and, lastly, by religious partisanship. These means will often raise a medical man in public estimation, though he does not possess a ray of talent; while the man of genius, of erudition, and of practical knowledge may, without them, remain all his life in "illustrious obscurity." He sees numbers of his inferiors who, by their address, policy, and worldly wisdom, rapidly pass him by, and leave him in the shade.

Nevertheless, what a galaxy of men of genius and knowledge [134/135] have risen to eminence and fame. It is unnecessary to multiply examples; but be it observed, that no physician arose to greater eminence since Hippocrates than Boerhaave. A letter was addressed to him from China with this superscription – "To the Great Boerhaave, in Europe." It reached him.

To make oneself known. It has been long observed, that great talents are not the surest or speediest way to acquire much medical reputation. It is, in general, a matter of the greatest difficulty, for a young physician, or surgeon, to make himself known in a large city, where there is a prodigious number of doctors of one sort or another. What difficulty, what address, and what labour are necessary, before a junior can obtain a prominent place among the multitude; or raise the edifice of his reputation! Years must elapse, in general, before he has accomplished his object. In fact, as Dr. Baillie truly observed: "he cannot earn bread until he has not teeth to eat it."

Let us examine the means which are to be employed to succeed in practice, and those that ought to be avoided. The credulity and folly of the public are the causes which impede medical practitioners in their career. The public never distinguish between the educated and uneducated professors of the healing art. Those who make the greatest promises – and they are the most ignorant – are surest of favour and success; while those who ought to be patronized, are entirely neglected. Such is the general result, but there are some few exceptions.

It sometimes happens, that distinguished or celebrated individuals have patronized a young medical practitioner, more on account of vanity, than a love of science. They are not always particular in their selection, and often prefer the ignorant to the learned; indeed, it generally happens, that genius is neglected, while ignorance finds the most powerful protectors.[4] The natural protectors of a young physician, or surgeon, [135/136] are his masters, or those practitioners who have acquired eminence and celebrity. The estimation in which they are held, easily allows the junior to clear the foundation of his reputation. Every medical practitioner who wishes to succeed, must be determined to accomplish his object, and to surmount a host of difficulties. He ought to have a good knowledge of the world, and accommodate himself to the follies and foibles of those with whom he comes in contact. He must possess policy, politeness, tact, and, as I have already observed, some knowledge of charlatanism, and perhaps, "humbug." Such are the means of acquiring great celebrity and fortune. Indeed, it rarely happens that talent – the enemy of artifice – leads to celebrity.

I cannot agree, however, with M. Monfalcon (Dict. Des. Sciences Med.) in his advice, which is similar to that of Lord Chesterfield to his son, as regards society in general: "that all young medical practitioners who wish to gain a reputation rapidly, should bespeak the good opinion of the fair sex – be assiduous, complaisant, gallant to them, but nothing more. They should study every thing to please, and flatter feminine vanity. The first thing necessary, is a perpetual adulation; the second, an absolute devotedness. There are some whose love of study and self-denial, unfit them for paying proper attention to the fair sex, and they can never become doctors a la mode. When their vanity is flattered by a fashionable doctor, they proclaim him in all places as a charming man and a learned physician. He takes care to sympathize with them on account of their nervousness from imaginary disorders; he advises amusements, change of residence, travelling, &c. and often assists in rendering the husband the very humble slave of the will of the wife. Some medical practitioners have enjoyed the faculty of addressing small-talk most agreeably to the other sex. It would be invidious to mention names, though it might be easily done. But there is no reason why grave, sad, and discreet doctors should not possess amiability as well as all other individuals."

Medicine is a science by no means incompatible with the usages of the world; it does not exempt those who practise it from being distinguished for that politeness, amenity, and those graces which charm society. [136/137] It has long been admitted, that the art of pleasing and the art of healing have intimate relations to each other.

If a medical practitioner does not possess politeness, and those graces which distinguish a gentleman, he will be very much impeded in his career. Manners and personal appearance are every thing with the majority of the world. No man ought to be more polite or possess a greater knowledge of the world than a medical practitioner. He should avoid eccentricity, or peculiarity of manners.

There are some who are so passionately fond of study, that they cannot live without their books, and they withdraw themselves from society to pursue their learned researches. This constant study gives them an embarrassed air and a timid demeanour in company, and is very injurious to their success in the practice of their profession. No professional man ought to neglect every legitimate means which is calculated to increase his reputation, or to obtain public approbation. He must renounce all amusements and pleasures, as society calls for him every instant; he cannot enjoy repose without interruption, and his sleep only continues so long as others do not disturb it; but the incessant toil of his profession requires relaxation to preserve his health. He must occasionally frequent places of public amusement, though he is advised by his seniors, to leave balls, fetes, routes, &c., to the unthinking portion of society. A medical practitioner, who frequents such amusements too much, is usually considered by the public, as having little to do in his vocation. It is true, that more serious matters generally require his attention; *urgentiora ilium vocent negotia*. The world is so severe, that it will sometimes censure a physician or surgeon who cultivates those arts and sciences which have no immediate connexion with his profession. It also requires him to avoid all other pursuits except his vocation. This is often a great hardship, as when a medical practitioner can scarcely support his family by his calling, and might acquire wealth by following other business. It is for this reason that many members of the profession are now obliged to engage in various commercial and other speculations.

"Scientific physicians," says M. Monfalcon, "sacrifice [137/138] your health and fortune to become learned and clever; sit over your books, become pale in the hospitals, meditate night and day on the most difficult points of your art, study like Boerhaave, fourteen hours a day for sixty years, renounce all the pleasures of life, and the charms of society, observe a complete self-denial, and if you disdain to make yourself known, you will be forgotten, neglected, and unappreciated; and you will never approach the fame of an empiric."

Charlatanism will always be successful, as it governs man by the first of all his interests, the love of life and the fear of death. Its origin commenced with man, and every one is in turn a doctor.

Fingunt se medicos quivis idiota sacerdos,
Judæus, monachus, histrio, rasor, anus,
Miles, mercator, cerdo, nutrix et orator,
Vult medicus hodie quivis habere manus.[5]

But the surest means of success are profound knowledge of science and of the world. Some obtain fame and reputation by the first, others by the second, and some by possessing both. Another means of obtaining success is forming a matrimonial engagement.

Marriage of a medical practitioner. There are many reasons to prove the necessity of marriage for a medical practitioner; it gives him more importance and maturity; it leads the world to overlook his youth and age, and it gains the confidence of families and mothers, who, unless he had entered into this state of life, would often decline his services. Hoffman was of opinion that a medical practitioner ought not to marry too early, unless he did so very advantageously, because a wife and establishment would occupy a large portion of the time, which ought to be devoted to study. This last objection is futile, as we find Racine was married, and was alternately pleased with his children and his books; Haller was assisted in his literary pursuits by his wife, and was one of the most voluminous authors of his time.

It is, however, the general opinion at present, that a medical practitioner ought to marry as soon as he can do so prudently and advantageously. This was the subject of a [138/139] thesis, sustained by Dr. Treyling at the University of Ingolstad, in which he discussed it in the form of two questions: (1) Ought a physician marry? (2) What sort of a wife is suited to him? This writer declaimed against marriage, cited all the arguments against it, and contended that the public prefer unmarried physicians! He passed in review all the disagreements of married persons, and dwelt more especially on the danger of female infidelity.

Accidit et hoc viro præsertim medico, quod si juvenculam sibi junxerit, hancque formosam, habeat, quod metuat, illud Epicteti dicentis: qui formosam duxerit, habebit communem. Cum enim medicus densa praxi obrutus, nec domûs nec uxoris custos esse valeat, quid? Si hæc interim hospitalis sit, et Dianam æmulata cornifica metamorphosi maritum cervina superbum corona in Actæonem transformat, hæredesque ipsi aperat, non nisi adamitico cum ipso sanguine conjunctos? Ita ut non semel saltem tacitè secum murmurare querelus debeat: haud ego mi uxorem duxi, tulit alter amorem. Sic vos non vobis.[6]

Such was the opinion of Epictetus. It is not, however, true, that he who marries a beautiful woman, enjoys her in common with others, nor does it happen that the husband always wears horns, or has a spurious heir. Handsome women are objects of general admiration, but when their minds are properly cultivated, they spurn the slightest equivocal compliment, or impertinent remark. It must be admitted, however, that if, on the other hand, there is a want of high moral feeling, the wife of a medical man, actively engaged in practice, has ample opportunity and time to degrade herself and dishonour her husband. Nevertheless, there are no more indiscretions in the families of medical practitioners than in all others; and, therefore, marriage is as suited to them as others; and, indeed, much more so, for the reasons already assigned. It is a state prescribed for all men, for "it is not good that man should be alone."

In fine, a medical practitioner seldom succeeds as an obstetrician, or is consulted in diseases of women until he is married, or considerably advanced in life. The virtuous of the other sex have a great objection to very young practitioners during parturition; and even the poor, who are compelled to [139/140] obtain medical attendance from public charities, very frequently decline it, when young medical students are ordered to attend them. "Marriage is honourable in all," and particularly so to the members of the medical profession.

Means of obtaining the confidence of the sick. It is useless for a medical practitioner to assume a grave exterior, and to possess the most profound scientific knowledge, unless he obtain confidence of the sick. Without this the greatest talent loses much of its power; and with it, every thing is possible to mediocrity. The principal means of obtaining this confidence, is the art of persuasion; which is a gift not always possessed by genius. The medical practitioner ought to be a man of the world, as well as minister of health. Some will prefer a grave, others a cheerful physician; but whatever may be the talents and acquirements of a medical practitioner, he cannot preserve the confidence of his patients, except by success; and a few unfavourable events may injure or ruin a reputatation, however well established. The public is, in general, disposed to attribute the impotence of medicine to its practitioners.

Another cause of loss of confidence, is when a practitioner is unable to explain the nature of disease, or fails to remove or alleviate it. The sick, or their friends, will not be satisfied without further advice; and if recovery happens, the whole merit is awarded to the new attendant. Thus it is manifest, that "knowledge is power," and often triumphs over all other means of securing the confidence of the sick.

Rewards of medical practitioners. – If the office of physician, or surgeon, exposes every day to the disdain of ignorance, to the forgetfulness of ingratitude, and to the outrages of calumny; if reputation entirely depends on the caprices of the multitude; if to practise, requires a medical man to renounce all pleasure, and even his liberty – he finds, however, in the exercise of his profession, many rewards for his sacrifices. The esteem of a small number of learned and sensible men, and the firm conviction that he has done his duty towards the sick, console him against the jealousies of his rivals, and the unreasonableness of the world. A quiet conscience is the [140/141] recompense of a medical practitioner, who honourably discharges his noble functions. He is delighted with the good he has done, and with the conviction that he cannot do too much. The poor implore his aid; he is the harbinger of hope and consolation in the abodes of misery and distress; and is blessed by the afflicted whom he relieves.

When, under Divine Providence, he rescues a patient from the jaws of death, and conducts a convalescence after a dangerous disease or operation, his success is an ample reward for his cares and anxieties. He that is saved becomes his friend for life, and lauds his skill and judgment.

It has long been remarked, that physicians and surgeons, in general, with few exceptions, seldom acquire a large fortune; nevertheless, the fruit of their labours is not subjected to such sudden vicissitudes as affect commercial men, and precipitate them from extreme opulence to extreme poverty. They are placed in tranquility and a happy mediocrity, which, of all states of life, is that which is most accordant with happiness. Respected by society – esteemed by men of learning – desired by the rich and poor – the cultivators of medicine to the remainder of life are venerated, loved, and honoured. Many of the profession have, however, acquired immense wealth. Sir Astley Cooper was in the receipt of £20,000 a year; and it is said, that one year his income, by his profession alone, amounted to £25,000. Few deserved more success, but fewer, as the world now exists, if any, will ever acquire so much.

Relations of medical practitioners towards each other. The etiquette of the profession requires that its members should live on friendly terms with each other, a reciprocal indulgence ought to make them excuse the errors they commit, and especially those of junior practitioners. They ought to be always on good terms. It would be shameful and scandalous in one member to expose the errors, or depreciate the merits of another; for no one would like this to be done to himself. Calumny and detraction are base vices, which ought not to characterize any medical practitioner. No one ought to injure another "by look, word, or suspicious silence," or under any pretext whatever interfere with the patients of [141/142] another. It is an axiom, that we should refuse attendance on patients under the care of another practitioner, unless in co-operation with him, or until his visits are discontinued, or when there is danger of death. Nevertheless, this vice is too common with necessitous practitioners. They are often jealous at the success of others, and depreciate, or calumniate their rivals. There is no envy so great as that of medical practitioners: "non est invidia supra medicorum invidiam." We rarely observe true friendship existing among them: envy and interest oppose it. But the real philosophical physicians despise such views, and entertain the strongest friendship for each other. A little reflection will convince any rational practitioner, that no man can be treated badly without resenting such conduct; and that never was there a wiser maxim than, "do unto others as ye would they should do unto you."

Some cannot speak without wounding the reputation of others. Many are bitter in their remarks, prodigal in their criticisms and sarcasms, imperious and cutting in their decisions; and these make a host of enemies for themselves. The wounds inflicted on self-love never heal; and it shows little common sense or discretion, to lose a friend for a remark. This however is often done; and some few of our most scientific practitioners have suffered severely by it. They have lost much consultation practice, and very justly, on account of such unprofessional behaviour. Thus, some engaged in consultations act uncourteously towards the first attendant; wound his vanity and self-love; lower, or endeavour to lower him in the opinion of his patients, his friends, or attendants. Others arrive before the hour appointed for the consultation, or several hours afterwards; and then act in the same disgraceful manner. A few encourage their trumpeters to sound their praises to the relations, acquaintances, or menials of the sick; or bribe nurses, with the view of depreciating the practitioner in attendance. These, and numerous other acts of immorality, are disgraceful to all who are guilty of them. They are, however, of too frequent occurrence in all large cities, and among unprincipled practitioners in all [142/143] situations. But no respectable or honourable physician or surgeon would be guilty of resorting to them.

Medicine of the mind. One of our celebrated poets has asked,

> Canst thou minister to a mind diseased,
> Or pluck from memory its rooted sorrow?

The answer may be given in the affirmative in many cases. There are cases in which we minister to diseased minds with perfect success. Indeed, the art of reading the human heart is indispensable to the medical practitioner; and it is often all that he can do. Medicines do not always cure disease; sage advice or discourse, which convinces the reason; and the proof of friendship, which touches the heart, are often the most powerful means of preserving health and life. He who knows the characters of the passions, can best moderate them, diminish their influence, and often prevent their fatal results. Those who have studied the human mind, can mostly discover the cause of the emotions of patients, the value of their answers, and the confidence that ought to be placed in them. Without this quality, we cannot satisfy or govern hypochondriacs, or remove their fears and apprehensions. Every one knows the influence of the mind over the body in health and disease. It was a profound knowledge of the human heart which led Hippocrates to discover the love of Perdicas for Phylla; and Galen, that of the Roman lady for Pylades. The importance of the moral influences in therapeutics is so great, that the ancients considered the mind, with philosophy and rhetoric, as medicinal remedies; and that their influence on the body was most advantageous. It cannot, indeed, be denied, that moral medicine is in many cases more powerful than physical. Who does not know, that the declaration of an eminent physician or surgeon, "that recovery is certain," will do more good than the whole materia medica ? We often see this in the practice of obstetricy – the assurance of safety and speedy relief has the most astonishing effect in expediting parturition.

It was on this account that philosophy and medicine were cultivated by the ancient physicians, and that modern universities require an excellent preliminary education of those [143/144] intended to study medicine. It is really a melancholy reflection to any one of education and acquirements, to think, that nine-tenths of the profession in this section of the British empire are admitted as medical practitioners if they can translate the first chapters of Gregory's "*Conspectus Medicinæ Theoreticæ,*" and those of "*Celsus de Medicina.*" And such is the classical ignorance of those entering the profession, that divers translations of the above are published, and even some that are absolutely interlineal! No knowledge of English composition, of logic, of rhetoric, of natural or moral philosophy, of mathematics, and the other studies included in humanity are required; and every one who can translate the above is admitted by the great manufactory for doctor-making, the Apothecaries' Society in London. What notion of moral medicine have individuals thus educated? Can persons thus educated be influenced by that consummate knowledge of the human mind so essential in the practice of medicine? What will foreigners say to the preliminary education which extends to a power of translating the first and third books of "Celsus de Medicina," and the first nine chapters of "Gregory's Conspectus de Medicina?" Surely, no one can consider this a sufficient proof of classical or general knowledge. I grieve to indite these remarks, but truth obliges me.

It is different, however, in the universities. In these institutions the preliminary education is as extensive, and indeed more so, than that required in other countries. They adopt the adage

> Adde quod ingenuas didicisse fideliter artes,
> Emollit mores, nec sinit esse feros.[7]

Physical medicine. We now approach the usual occupation of medical practitioners. We bring them to the bed-side, to investigate the nature and treatment of disorders and diseases. All general and medical studies – all our former remarks, are subservient to this most important ministry.

The physician or surgeon bases his diagnosis, on the history of his complaint given by the patient. He considers the age, sex, temperament, habit, occupation, residence, and constitution of the patient, and the season; and he next inquires, [144/145] what is the most troublesome or painful sensation? The patient replies, that he has pain in some particular part, which at once directs the attention of the inquirer to the seat of his malady. It is in the head, chest, abdomen, or extremities. A momentary glance at the countenance, chest, or abdomen, will enable a scientific practitioner to determine the seat of the disease. He can also conclude in a few minutes, whether the malady is a disorder of function, an unpleasant sensation, pain, spasm, &c.; or a disease, a change of structure, congestion, inflammation, or the usual consequences of the last.

In either case, disorder or disease of one organ will derange the whole, and may cause painful sensations in all, and a multitude of symptoms. In every possible case, there is disorder of function, change of structure, or a combination of both.

In disorders of function, the nervous power, or innervation, is anormal, [*sic*] or deranged; and the internal and external use of sedatives is indicated.

In disease, the circulatory system is affected; there is congestion, inflammation, suppuration, ulceration, cicatrization gangrene, &c. Disorder or disease may affect all parts of the body, in consequence of the universal distribution of nerves and blood-vessels. It is most important to determine the existence of the one or the other, as the treatment must differ. In disorder of function, without change of structure, there will be some unnatural sensation, which may cause pains in all parts of the body, and be removed by the internal and external employment of sedatives. In disease, general and local bleeding, low diet, counter-irritation, purgatives, and diaphoretics, will be necessary. In complicated cases, both classes of remedies must be employed.

It is not understood by society, that a learned and scientific practitioner examines the head, chest, and abdomen, in a minute or two; and patients suppose that they must detail every sensation they have felt for months or years. All complaints are either disorders or diseases, or both combined; and the latter are acute or chronic. The simple question, what is the matter? – where have you pain? – what distresses you most? – will enable the inquirer to discover the seat of the malady. He then determines whether it is disorder or dis- [145/146] ease, and prescribes the appropriate remedies. The fundamental principles of medicine are now happily fixed, or it would be impossible to arrive at a correct notion of disease.

Many patients cannot express their ideas correctly, and jumble the most opposite symptoms together; they enter into long digressions, and confound the most different objects. They complain of all diseases. The question, where have you most pain? – or, where did it commence? will lead the practitioner to the seat of the complaint. In my hospital and dispensary practice, I inquire, before a large number of students, which of all your complaints is worst? and what was the first symptom? I also adopt this plan of interrogation in private practice. It affords a light which guides me through great darkness; and by it I distinguish the circumstances that preceded the attack, as well as the nature of the complaint itself. I have fully described the method of examining patients, and distinguishing diseases, in my edition of "*Hooper's Physician's Vade-mecum*," and "*Lectures on the Physical Education of Children*, 1835," already quoted.

There are some patients who exaggerate their sufferings, and the expression of pain is not always real. Others put insidious questions to their medical adviser, in order to learn his opinion of their condition, as hypochondriacs, nervous, and parturient women. The former suppose that they labour under one or many fatal diseases; the last are anxious to learn, are they in a safe condition, and how soon they will be well, or delivered.

We also observe many who simulate a variety of diseases with great success. But the pathological symptoms which belong to the functions, however modified by the influence of the brain, are the signs on which we can place absolute confidence. Physical signs will never deceive us, and we should always judge by them. The statements of patients, or of medical practitioners, cannot always be relied on, while physical evidence is invariably correct. I might mention numerous instances in which patients and practitioners would have deceived me, had I not solely depended on the physical phenomena.

In examining patients, we should inflect the voice, and [146/147] avoid putting our questions too abruptly, lest we confuse and embarrass them. There is an excellent chapter in the work of M. Double on Semeiotics, as to the method of examining patients, and further information may be derived from the production of M. M. Merat, Creuveilher, and Martinet (see Professor Quain's translation). M. Double divides the subject into two distinct parts – the knowledge of the history preceding the disease, and the circumstances which appertain to the disease itself. The physician ought to investigate all circumstances relative to surrounding objects, season, temperature, and medical topography where he practises. He then determines the age, sex, profession, occupation, or mode of life, of the patient; the passions, habits, general health, or ordinary state of functions, state of health before the invasion of the disease, the hereditary diseases of the father, mother, and family; the antecedent diseases from the period of infancy: and, lastly, the general effects of medicines upon the constitution.

The practitioner should sit by the bed-side, so as to see the patient's countenance, and then examine the external appearance of the body, attitude, movements, coloration of the skin, &c. He then compares the existing and natural states of the functions, and proceeds to examine the external senses, respiration, circulation – including the state of the heart and pulse, digestion, secretions, excretions, generation, sensibility, irritability, voice, voluntary motions, sleep, intellectual faculties, and temperature of the body. Every part of the human body is thus examined and passed in review. We also know all that belongs to the symptoms, mutations, and sympathies – all the circumstances that can modify the signs and prognosis of diseases. The nature and treatment of the complaint are then easily determined.

It is right to visit the patient at the paroxysmal periods, or when he is worse; and also at other hours.

It is worthy of remark, that individuals who suffer from the effects of libertinage, are often ashamed, and disposed to mislead their medical advisers. A large experience warrants me in stating, that the abuse of the generative function, is a most common cause of indigestion, hypochondriasis, neuralgic, or [147/148] painful, and nervous complaints. It requires great delicacy and caution in alluding to this function at all times, but especially when youths are our patients. The excitation of the generative organs at puberty, and often during the whole time of celibacy, leads to the adoption of natural or other means for its subdual. There are, of course, exceptions to this general proposition. Those who abuse or over-exert this function, are affected with severe indigestion, flatulence, lowness of spirits, fear of death, dread of impotence, and innumerable unpleasant sensations. Such cases are of frequent occurrence from the age of puberty to the decline of life, and sometimes even in old age. The celebrated Tissot has given numerous illustrations in his work on Onanism, and a vast number has fallen under my own observations. I have slightly alluded to such cases in my work on population, marriage, impotence, and sterility.[8]

I am convinced, by multiplied experience, that the genital function is, in general, too much excited, and consequently debilitated. The profession, in general, overlook this fact, and very improperly fall into the vulgar error, of ridiculing those who have abused this function, and rendered themselves impotent or sterile. Such persons hesitate to consult many, and those who have not offspring, are too often derided by their acquaintances. They avoid consulting respectable medical practitioners – for the reason above stated – and apply to unprincipled empirics. They are, as a matter of course, duped and deceived by these wretches, and often seriously injured, or their health destroyed. Several have applied to me under such circumstances. I feel convinced that the generative function is one of the most powerful in the body, and influences mankind in general, more than any other. Every male, high or low, makes it the theme of conversation, and the most modest and virtuous women are greatly annoyed at infecundity.

In delivering an opinion to a patient, the physician, or medical practitioner, ought, in general, to predict a favourable [148/149] result when circumstances permit; avoid alarming him, and hope for the best. He may, however, express his real opinion to the relatives, so as to give them an opportunity of consulting others, if they think proper. In most cases, the prognosis must be guarded, as so much depends upon the mutations of disease, punctuality of attendants in following directions, on the seasons, situation, and compliance of the sick. Medical practitioners have often been most unjustly ridiculed by the world, on account of the ambiguity of their opinions as to the issue of acute diseases; but no man can be positive as to the prognosis. He may state, he is most happy to say, that at this moment there is no danger: but diseases undergo numerous changes, and that it is utterly impossible to predict, with certainty, the result.

This is the fact and truth, in the majority of human infirmities. The diagnosis or distinction of diseases, is also a matter of grave importance. If the disease is mistaken, censure will assuredly follow. Thus, suppose it is stated, that a young female of irreproachable character is affected with gonorrhœa, or is pregnant, when there is really no such disease or condition; the practitioner would be most severely and justly censured. It is a common error to assert, that female children from the age of six to fourteen years have been violated, and infected with gonorrhœa, when their condition depends on physical diseases. I have fully described such cases in another part of this work; and they are noticed by Underwood, Hamilton, Dewees, Sir Astley Cooper, Jewel, and numerous French authors.

In fine, I have to observe, that I have already described the moral qualities which are necessary to a medical practitioner. I may, however, remark here; that he ought to possess sensibility, humanity, amenity, amiability, and compassion. He must make the greatest allowances for the conduct of those in pain and suffering. Thus, a medical practitioner must display all the finer feelings of human nature, and all the moral excellencies which distinguish the true cultivators of his art, while he is affording aid at the painful process employed by nature at the nativity of his species. Women are naturally timid at this period, and require all the sympathy [149/150] and kindness that can be shown towards them. I have fully dilated on the duties of the profession under such circumstances, in my Manual of Gynæcology, or Obstetricy.

I may state, in this place, that medical practitioners ought to be most patient, humane, benevolent; and never resent any offensive remarks that may be applied to them by the sick. Such remarks are invariably apologised for by patients.

Lastly, we ought to be cool and undisturbed in the worst cases. Perturbation of mind betrays weakness and ignorance. If agitation and indecision characterize us, patients will lose all confidence in us.

One who knows his profession well, will be never agitated. He will employ the resources of his art; and if these fail, he will remember, that all medicine is from God, and that it may not be accordant with his divine will, that recovery should happen in every case of disease.

Rules for prescribing. The art of prescribing medicines demands especial attention, and requires to be well understood. Diseases are cured with remedies, and not with words. There are certain rules for prescribing, which ought to be followed as closely as possible, and ought not to be deviated from, except in certain circumstances. The precepts of Gaubius are simple, and I have added some remarks in my translation of the New Practical Formulary of Hospitals, 1835.[9]

1. The practitioner, whom prudence ought always to guide, should never prescribe a medicine without being able to give satisfactory reasons for so doing. The first question he ought to put to himself should be, is it or is it not necessary to administer medicines in the present case?

When it is thought that the powers of nature are sufficient to effect a cure; that the disease is absolutely incurable, or that the cure of the disease would produce a greater disease, it is generally thought proper to abstain from prescribing medicines, as well to prevent injuring the patient as uselessly tormenting him. Medici plus interdum quiete, quam movendo et agendo, proficient.[10] Nevertheless, as in some cases it would be inhuman to abandon the patient, and in others impolitic to shew the imperfection of our art, under such circumstances those substances which, if they are not really useful, are not [150/151] injurious, ought to be administered. This precaution should principally be attended to in the treatment of women and young girls.

2. When on the contrary it is necessary for the practitioner to prescribe, he ought, in the first place, to determine what he should do, and the medicines he ought to employ, &c.; questions, the solution of which he will find in his therapeutic knowledge. He should always recollect, that his end is the cure of his patient, as promptly and in as agreeable a manner as possible.
3. He ought always to make choice of the most efficacious remedies, and those best calculated to the attainment of his end; and this not only with respect to their nature, but likewise to the forms under which he prescribes them. Those medicines ought always to be employed whose actions are the most certain, and which are not likely to cause any bad effects. In certain hopeless cases, a medical man may try extreme remedies, but always with reserve, and be careful to announce to the relatives, the uncertainty of their results.
4. Medicines which are but imperfectly known ought never to be used, if the same effects can be obtained by the employment of those substances which have been

sanctioned by usage; and if it is necessary to prescribe new remedies, it should be done with the greatest prudence.

5. The use of substances which, by being kept, easily become changed, and which age has rendered inert or prejudicial, ought to be carefully avoided. In acting in this manner, a practitioner runs no risk of not obtaining the desired effects, of uselessly fatiguing the patient, and even producing serious accidents. It is for this reason that a practitioner, if he does not compound his own medicines, should have his prescriptions compounded in those houses where the sale is rapid, by which means he would stand every chance of having them well prepared, and their ingredients fresh.[11]
6. All things being equal in other respects, he should prefer indigenous medicines to those of foreign countries; they [151/152] are more easily known, and less likely to be adulterated.
7. The use of medicines of a low price should be preferred, provided they are as efficacious as those of a higher. Nevertheless, when a practitioner has to treat a rich patient, who thinks, as it sometimes happens, that medicines are only efficacious inasmuch as they are costly, he ought, in a certain degree, to comply with this ridiculous prejudice; because, as we have before said, the influence of the imagination is not to be despised.[12]
8. There are likewise cases in which, on account of prejudices, or individual repugnances, the practitioner is obliged to disguise, in different manners, the substances he prescribes. At one time it is their name he is obliged to change; at another, their taste and odour he will find necessary to mask by proper mixtures. But he must always be careful to be perfectly intelligible to the pharmacopolist, and not alter the therapeutical properties of the medicines he orders.
9. It is advisable, as much as possible, not to make use of those medicines whose odour, taste, &c. are very disagreeable; or, at least, to use them in small quantities, and disguised as we have already mentioned. And here it may be proper to remark, that this plan should be strictly followed in treating women and children, and in every case.
10. Before prescribing any medicine it is indispensable to find out, by every possible means, the idiosyncrasy of the patient. For it sometimes happens that a medicine, in other respects judiciously chosen, may, on account of certain individual dispositions, impossible to be foreseen, become useless or even prejudicial. For example, castor oil, one of the mildest purgatives, and one most commonly used, even for children, acted as a poison on all the individuals of a whole family which one of us was called in to attend. Gaubius relates an example of a man upon whom a small dose of the powder of crabs' eyes produced all the symptoms of poisoning by arsenic. Instances of this kind are too common to require enumeration. [152/153].
11. It sometimes happens that the patient is strongly prejudiced in favour of certain medicines, either because he has seen them administered with success in cases which he considers similar to his own, or for some other cause. If the practitioner thinks that the use of the medicine desired will not be injurious, he ought to comply with the wish of the patient; in other cases, without positively refusing, he should endeavour to make his patient comprehend that there would be danger

in complying with his request, and he should endeavour to gain time until his patient has changed his mind, or that his state will allow of the administration of the medicine desired.[13]

12. Temperaments, which modify in so powerful a degree the progress and character of diseases, likewise merit a particular attention with regard to the therapeutical means employed. In strong and robust individuals, endowed with a sanguine temperament, the sanguineous evacuations, diluents, in a word, the antiphlogistic treatment is much oftener employed than in persons of a weak, and irritable constitution, and of a lymphatic and nervous temperament; in the last-mentioned cases, tonics and antispasmodics are more frequently administered. It will nevertheless be conceived, that this is far from being an invariable rule.[14]
13. Attention ought also to be paid to the effects of habit; and it should be remembered, that organs frequently submitted for any length of time to the influence of a medicinal substance, become so accustomed to it as to be insensible to its effects. This is forcibly exemplified, amongst other examples, in the enormous quantities of opium which certain individuals can take, without experiencing any immediate accidents, as is seen in numerous instances in the east. Therefore, when it is necessary to apply the same substance for a long time, the dose should be gradually increased for it to make any impression on the organs. It is especially in medicines which [153/154] act upon the nervous system that this phenomenon is remarkable. There are, on the contrary, some medicines, whose action is slow and gradual, which require some time for them to develop themselves, and their effects are not manifest until after they have been administered for a long time. Their effects are much less weakened by use than those whose actions are more prompt. Nevertheless, after some time, quantities may, without any danger, be administered, which, in the first instance, would have been followed by alarming symptoms.
14. In prescribing a medicine, the consideration of the circumstances which may tend to favour or modify its action should not be neglected. Thus, in administering a sudorific, the patient should be placed in a warm situation; because, if he is exposed to cold, diaphoresis will not be produced.
15. Before introducing a medicinal substance into the intestinal canal, the practitioner ought to examine attentively the pathological and physiological state of this organ, the nature and extent of the diseases of which it is the seat, &c.; for a medicine which would be inoffensive, and even salutary, if the stomach were in health, may become fatal if this organ is diseased; it is necessary in this case to associate the medicine with others which weaken its local action, or even abstain entirely from its use.[15]

I shall now proceed to examine the modes by which medicinal substances are made to act on the economy.

1. When medicines have only a decided action upon those organs with which they are put in contact, they ought, as far as is practicable, to be applied to the diseased part, at least, where it is not desirable to obtain general effects by revulsion; and in that case it is always a healthy part that ought to be acted on.

2. When the influence of a medicine can be propagated through continuity of organs, the nearest parts to those affected are to be acted on, in order that their effects may be as marked as possible; because the influence of these [154/155] substances is as much less strong, as the parts to which they are applied are distant from those of which a change of their actual state is required.
3. Those medicines which act by sympathy are generally introduced into the stomach; because this organ has the most direct sympathetic connections with all other important organs.
4. When medicinal substances act through the absorption of their molecules, they are generally administered through the medium of the stomach. But they may sometimes be introduced into the economy, by putting them in contact with some other part of the mucous surface of the alimentary canal; it is on this account that enemata are sometimes administered, and frictions made on the gums, &c.
5. Formerly, advantage was taken of the absorbent power of the mucous membrane which covers the aerial passages, to cause the same result; and the patient was made to respire the vapour of those substances, under the influence of which it was desirable to bring him.
6. In fine, there are cases in which medicines are caused to penetrate into the economy by applying them to the skin. But as the presence of the epidermis is a powerful obstacle to the absorption of medicinal molecules, their action would be very slow, and even almost useless, if a simple application alone, was pursued. To obviate this inconvenience, it is necessary to make them penetrate the pores of the epidermis, by means of frictions more or less violent, or by raising this membraneous layer to a certain extent, and thus putting them in immediate contact with the surface of the dermis or skin. The first of these methods, that of friction, has been for a long time known, and is called iatraleptic. M. Christien, of Montpellier, has much extolled it, and put it in practice with success in a great many cases. The second is named, by M. Lembert, the methode endermique, who, conjointly with M. Bailly, has made numerous experiments, at the hospitals of La Pitié and Cochin, in applying upon a blistered surface, different medicinal substances, capable of acting by absorption. The result of. these experiments – repeated since by a great number of French and foreign practitioners – leaves no doubt of the efficacy of this mode of applying medicines, which [155/156] appears to be very advantageous when their irritating action is dreaded upon the mucous membrane of the gastro-intestinal canal, or when it is wished to prevent the alteration which the digestive faculties may produce in them. It would be, nevertheless, advisable to employ, in this manner, those medicines only which are susceptible of acting effectively in very small doses, such as morphia, strychnine, &c.

The knowledge of the doses in which medicines are administered, is called Posology; but it is impossible to determine them.

The doses in which medicines are administered, differ according to their nature and their degree of activity. It would be difficult to establish fixed rules in this respect, as experience alone must be our guide.[16] We shall only observe, that the doses of the same medicine ought to vary according to the effects that are wished to be produced, and according to the age, sex, and temperament, &c. of the patient.

The effects of a medicine frequently differ, according to the quantity in which it is administered. It is thus that the greater part of astringent, tonic, and exciting substances have only a local action, when given in small doses; whilst, on the contrary, in large doses, they extend their influence over the whole of the economy. Opium, taken in small quantities, is a very energetic sedative; in larger doses, it becomes excitant, and produces cerebral congestion when the dose administered is too large. Digitalis, in large doses, acts directly on the intestinal canal, as is proved by the vomiting and alvine evacuations which follow its administration. In smaller doses, on the contrary, its local effects are no longer observable, but are replaced by general phenomena, such as quickening the action of the heart, and augmenting the secretions, especially the urine. Some antimonial preparations are, according to the doses in which they are administered, alternately emetic, purgative, diaphoretic.

The doses of medicines ought always to be proportioned to [156/157] the age and strength of the patient. It is worthy of observation, that, in general, the weaker a patient is, and the more under the adult age, the more characteristic are the effects of a determined quantity of a medicine. It therefore follows, that to obtain similar effects on an adult and an infant, very different doses must be employed.

The following table, drawn up by Gaubius, may serve as a guide to young practitioners in the administration of active substances at the different epochs of life;

For an adult where the dose is 1 dr	
Under 1 year	1–15th to 1–12th
2	1–8th
3	1–6th
4	1–4th
7	1–3rd
14	1–half
20	2–3rds
From 20 to 60	1

although it should always be remembered, that this rule allows of deviation according to circumstances.[17]
Above this age, the inverse gradation must be followed.

The constitution of women is, in general, less strong than that of men; it will therefore be seen, from what has been already said, that the doses administered to them must be less; but it would be impossible to state in what exact proportion.

The doses of medicines ought likewise to be modified according to the temperament and idiosyncrasy of the patient; for it will readily be conceived that a very irritable person, endowed with what is called a nervous temperament, could not, without inconvenience, bear the dose of certain medicines, of excitants, for example, which could be given with advantage, to one of a lymphatic constitution.

It is, therefore, highly important to adapt the doses to different constitutions. There are certain individual dispositions, at first unknown in their nature, the whole of which form idiosyncrasy, and which prevent the [157/158] same substances – given in the same doses, and under the same circumstances, from acting in the same manner, and with the same energy in all individuals. It is in this way, that a small quantity of opium will produce in some persons all the symptoms of narcotism, whilst in others it would act insensibly. Half an ounce of a neutral salt of any kind, in some cases, produces abundant evacuations, and even superpurgation; whilst, in other cases, two ounces of the same substance would have scarcely any effect.

In fine, the effects of medicines being modified by habit, as I have before observed, it is of importance to have regard to this consideration, as often as it is necessary to continue, for any length of time, the use of a medicinal substance, or when we wish to administer, in large doses, certain very energetic preparations. Some of our popular pharmacological writers, have given the preceding as their own, thinking that the old standard medical works are seldom seen: These authors are to be pitied, and ought to remember, "palmam qui meruit, ferat."[18]

Following the precepts I have laid down, I think that every one will be enabled easily to modify the doses, according to the exigence of the case, and the observations he must have made individually in this respect.

Medicines are either simple or compound. Those are called simple which can be administered in such manner as nature offers them, or which are formed of one substance, of which the intimate nature may, in other respects, be more or less complex – such as ether, the acetate of morphia, &c. The second, on the contrary, are the result of a combination of several simple medicines.

Simple medicines in general ought to be preferred to compound and, when recourse is had to the latter, simplicity should be sought after as much as possible. The following maxim ought always to be present in a practitioner's mind: *Superflua nunquam non nocent*,[19] and those substances only should be united, whose reciprocal action and influence on the animal economy are well understood.

Medicinal substances are mixed together, or *compounded*, for the attainment of divers ends: [158/159]

First. To augment the action of the principal medicine which is intended to be exhibited. This may be attained:

(A) in mixing different preparations of the same substance. When all the active principles of a medicine are not soluble in the same liquid, and when it cannot be administered in substance, recourse should be had to this kind of combination. It is in this manner that infusions and decoctions are made more active by the addition of a small portion of the tincture or extract of the same plant.

(B) In combining medicines of the same species, that is to say, those which taken separately produce the same effects, but with less energy than when combined. This augmentation of activity is only evident in a certain number of medicines. According to the observations of Valisniéri, twelve drachms of cassia produce a purgative effect almost equivalent to that of four ounces of manna. But if eight drachms of cassia and four of manna are united, the effects obtained are much more marked, and even may be said to be double. The mixture of diffusible aromatic substances is equally susceptible of modifying the action of each individually.

(C) In uniting a medicine with a substance of a different nature, which exercises no action on it, but which renders the economy in general, or the stomach, or any other organ, more sensible to its influence. It is much easier to prove this than to explain it; therefore I shall content myself with giving a few examples. The mixture of ipecacuanha and jalap render the purgative effects of the latter much more energetic. The action of certain purgatives is increased by the addition of a bitter. Cullen remarks, that, in mixing a bitter substance with an infusion of senna, the same effects would be obtained in administering a small dose of this purgative as in employing a large dose of it alone. The influence that opium has over mercury is likewise very remarkable. It appears in some cases, that after the general effects of mercury have completely ceased, they reappear under the influence of opium.

Secondly. To diminish, or to correct in some degree, the too irritating effects of a medicine. This indication is fulfilled:

(A) by mixing a medicine with another which augments or diminishes its solubility. It is by this means that [159/160] the addition of a small quantity of an alkali diminishes the tendency of certain drastics to produce colic; and in mixing camboge with an insoluble substance, nausea is prevented, by rendering the solution more difficult.

(B) By its mixture with a substance susceptible of preserving the stomach, or the economy in general, from deleterious effects. There is a great number of substances which, when they irritate the intestinal canal too violently, cannot be absorbed, and are expelled without producing the desired effects. Squills and antimonial preparations, for example, do not act as diuretics or diaphoretics when they cause purging and alvine dejections. In such cases, it is necessary to know how to combine substances capable of remedying this local action, and of correcting such effects. Opium frequently fulfils this indication; at other times aromatic stimulants are used, or mucilaginous and emollient substances, which envelope, in some degree, the active ones, and thus diminish the local action which is dreaded.

Thirdly. To obtain, at the same time, the effects of two or more medicines:

(A) In employing substances which, though they act differently, produce frequently the same result when combined. To augment the secretion of urine, for example, medicines, whose modes of action on the economy are entirely different, are combined, such as calomel and squills. The former acts, as most mercurial preparations do, as an active absorbent; whilst the latter acts principally on the urinary organs.

(B) In combining substances of which the action is entirely different, and which are designed to fulfil several indications at the same time. It is with this view that purgatives are frequently united with antispasmodics, narcotics, tonics, mercurials, &c. The use of tonics often occasions constipation, and it is necessary to combine a purgative medicine to counterbalance this effect. In the treatment of ascites, and chronic dropsies in general, there are cases in which the practitioner finds it necessary to support the strength of his patient, while at the same time he causes abundant evacuations. This is effected by uniting tonics and excitants with drastic purgatives. [160/161]

Fourthly. To obtain effects which, if taken separately, would not result:

(A) In uniting, medicines whose actions are essentially different, and which, by their combination, produce other effects than those they would have produced singly, without acting chemically on each other. This effect appears to me inexplicable; but examples of the kind are too numerous for a doubt to be entertained on the subject. We see that opium and ipecacuanha, administered together, produce neither the narcotic effects of the one, nor the emetic effects of the other, but act as a powerful diaphoretic.

(B) In combining substances which act chemically on each other, and which give rise to new compounds, or which render the active principles of one of them null. In making, for example, acetic acid act upon ammonia, a new product is formed, the action of which is very different from that of the two other bodies taken separately. In the anti-emetic potion of Riviere, citric acid is mixed with the carbonate of soda. This latter is decomposed by the citric acid, and disengages the carbonic acid which it contains.

(C) In mixing substances which augment or diminish the solubility of the principles which contain the medicinal properties. This indication may be fulfilled by the aid of substances which act either chemically or mechanically. Thus the tartaric acid, or cream of tartar, becomes more soluble, and, consequently, more active, by the addition of the acid of borax.

Fifthly. In fine, to give them a form more agreeable or efficacious. Substances mixed with medicines, with a view either to render their taste or odour less disagreeable to the patient, or to prevent a too prompt decomposition, or in order to facilitate their action, vary according to the nature of the medicines employed, their degree of solubility, the end proposed, and, to a certain point, the caprice of the patient. Nevertheless, a choice ought to be made of such as would not annul the efficacy of the principal medicines. We shall have occasion to revert to this subject hereafter.

Such are the different objects that are had in view, when a mixture of several simple medicines is made to form a compound. According to the effects that these different sub- [161/162] stances are wished to produce, they are called by the following names the *base*, the *adjuvant* or *auxiliary*, the *corrective*, and the *excipient or intermediate*. The base is the principal medicine; the adjuvants are those added to facilitate and accelerate its action; the correctives are destined to reduce the too energetic action of the base; the excipients serve as a vehicle in which it may be taken; and the intermediate, a kind of excipient, is intended to render it miscible in water.

It is often useless to employ at the same time the whole of these elements in the formation of a compound medicine. Many substances want no adjuvant to facilitate their action, and others are administered very well without any corrective, or even without any vehicle. It also frequently happens, that the same substance fulfils at the same time several of these indications. For example, the adjuvant may serve both as a corrective and a vehicle. These last considerations are so much the more important, as simplicity is one of the most essential conditions in the composition of medicines.

Pharmaceutical preparations are divided into two great classes.

First, *Officinal preparations*, that is to say, those whose composition is laid down in the pharmacopoeias, and which are generally those kept in shops.

Second, *Magistral preparations* are those whose composition is indicated by the practitioner, and which the apothecary prepares from the formula given.

A formula, or pharmaceutical prescription, is the indication of the names and doses of substances which enter into the composition of a magistral preparation, to which is generally added instructions for its administration.

First. Clearness and conciseness are two essential conditions in writing prescriptions.

Second. They ought to be written in a legible hand, and in Latin, or perhaps in the vulgar language.[20]

Third. At the commencement of the first line, the sign Rx, or R, which is an abbreviation of the Latin word *recipe*, should be placed. [162/163].

Fourth. Each substance should be indicated by its scientific or pharmaceutical name, according as the one or the other is more generally known, and less liable to be mistaken for any other. The names of medicines ought always to be placed under each other, taking care to put but one in the same line.

Fifth. The order of arranging them is of little importance; nevertheless it would be well to place the most active ingredients first.

Sixth. The quantity of the dose ought always to follow the name of the medicine, and be placed in the same line, leaving a small interval between them. The following signs are those which have been established by use:[21]

The quantity of each of these weights is generally indicated by Roman cyphers. When the same dose of several different substances is used, they are united by a brace, and the letters ana, or āā are placed before the designation of the common quantity of all.

Seventh. The prescription should he ended by indicating the mode of the preparation of the medicine, and the manner of its administration. When the preparation presents nothing particular, it is merely necessary to write the letters F.S.A. *(fiat secundum artem)*. In other cases, the mode of preparation, should be indicated as briefly as possible; then it is to be dated and signed with the initials of a physician, and the name of a surgeon.

Before examining the different pharmaceutical preparations, and the forms under which they are administered, I think it necessary to call the reader's attention to the errors that may be committed in compounding magistral preparations; errors which arise from three principal sources, namely:

First. The association of substances which do not combine, or do not form compositions of a proper consistence. Many [163/164] substances, insoluble in water, cannot be administered in a liquid form, without the aid of an *intermediate*, such as a mucilaginous or albuminous substance, which keeps their molecules in suspension. If the *intermediate* is neglected, the formula will not accomplish the desired effects. This would be the case in ordering camphor and the balsam of capaiba in pills, without adding a proper *intermediate;* because these two substances, mixed together, would form a syrupy consistence, and it would be impossible to make them into pills, unless a small quantity of the coagulated yolk of an egg was added.

Second. The association of substances which mutually decompose each other, by which means their action is changed, or entirely destroyed. Every time that two salts are mixed in solution, which, by an exchange of their bases or acids, may be formed into one soluble and one insoluble salt, or into two insoluble salts, a decomposition necessarily takes place.

Third. The method indicated for the preparation of medicines is insufficient to attain the end proposed, or is of a nature to change or destroy the action of some of the substances employed. Certain medicines are only soluble in alcohol, ether, or oil; and others are only soluble with the aid of caloric; and some lose their active principles by ebullition. It is therefore of the highest importance not to order in an infusion of cold water a substance which is only soluble in warm, and not to order a decoction of medicines which ebullition alters, and which lose their virtue by this process, &c.

The forms under which medicines are administered vary according to the nature of the substances, and the use that is wished to be made of them. These forms are solid, soft, liquid, and gaseous, and most of the pharmaceutical preparations have a special destination; some are always employed externally, and others internally, whilst there is a certain number which serves at the same time both for external and internal uses.

In the preceding remarks, I have offered but a very brief outline of those principles which guide scientific practitioners in prescribing medicines; as I consider this information of great importance to students and junior practitioners. I could not treat of them more fully in a work of this kind; but refer the reader to the excellent Pharmacologia of Dr. Paris, in [164/165] which he will find the most ample account of this branch of medical science in our language.

It has been truly observed, that disease terminates in health, death, or another disease. I may observe, that, when the period of convalescence arrives, it must be managed with judgment and skill. Relapses are more dangerous than the original disease, as the constitution is generally too much enfeebled when they occur. The greatest attention must therefore be paid to diet, regimen, and other hygienic precepts, and curative means. When death approaches, the medical practitioner has certain duties to perform which are indispensable, and have been already described.

Duty of Physicians to the Dying. If a physician have judgment, and all the other requisites essential to his character, if he take a true interest in the health of the sick, he will afford consolation to the dying. The respect due to humanity, and to the dying, imposes the obligation of consoling those departing from life, while their senses remain; and in reminding them of a better world, in which suffering and sorrow will be no more. We ought not to quit a patient in danger, unless with a calm and serene air. When the fatal hour is about to strike, the physician is bound to recommend the consolations of religion. The sublime doctrines of Christianity calm agitation, and smooth the avenue of death. Imprudent zeal, on the contrary, has often hastened the fatal event.

It is also right to inform the patient, his near relatives, or the clergyman who attends him, of the necessity of arranging his affairs, so as to secure his family their rights, and prevent litigation. This is a delicate affair; but it must not be forgotten by the medical attendant. Let him also take care that his patient, "is of sound mind, memory, and recollection, when disposing of his property." Instances have fallen under my own observation, in which testators did not really know what they were signing; and I am convinced that such wills would be annulled, if contested.

Notes

1. *Author's note:* Godman's Addresses, 1824. *Editors' note:* Ryan here takes a long quotation from another of Godwin's addresses, "Monitions to Students of Medicine," dated November 1824; John D. Godman, *Addresses Delivered on Various Public Occasions* (Philadelphia: Carey, Lea & Carey, 1829):16–18.
2. *Editors' note:* "It is not right to praise or criticize. Praise sparingly; criticize more sparingly. If permissible, praise new things and old. Carp on both sides. Your respect remains, as does your honor. Do not praise a living author with enormity in order that you in turn will be praised. Your measure depends on words, titles, and praise. Do not praise or implicate all writers with a single booklet, neither because of jealousy, nor lack of praise or excesses."
3. *Editors' note:* for luring the public.
4. *Author's note:* Gresset has well remarked on this custom: "Des protégés si bas, des protecteurs si bêtes."
5. *Editors' note:* "Any amateur imagined themselves doctors. A priest, a Jew, a king, an actor, a barber, a hag, soldier, merchant, laborer, wet nurse, and an orator. Anyone at all today wished to be a surgeon."
6. *Editors' note:* "And it came to pass that especially for a medical man, being married to a young and beautiful girl was a cause for anxiety. Epictetus described this: he whose wife is beautiful, has her in common with others. When a physician is deeply occupied by his practice, how will he be a successful guardian of either his home or his wife? If meanwhile she is hospitable, and by emulating Diana transforms her arrogant husband like Actaenon into a stag wearing a crown of horns, and by herself produces heirs, not of the husband's blood? Just as at least not once silently with to be responsible for murmuring complaint. By no means I commanded my wife, another bears love. Thus you all not by you all." The latter part of the passage is difficult to translate due apparently to the author's garbled version of nineteenth century medical Latin. The allusion is to the myth of the hunter Actaenon coming upon Diana naked, and the enraged goddess turning him into a stag who is then consumed by his own hounds. Here, though, the "crown of horns" alludes metaphorically to the cuckolded husband. Epictetus wrote in Greek rather than Latin and a corresponding passage could not be located among his works.
7. *Editors' note*: "Moreover, because the noble and more trusted arts are the most extensive, they should not be undertaken except by those with more refined manners" (somewhat freely translated).
8. *Author's note*: See also "De l'Onanisme et des autres Abus Veneriens, consideres clans leur rapports avec la Sante. Par M. Deslandes." 1835.
9. *Editors' note*: Hieronymus David Gaubius, 1705(?)–1780.

10. *Editors' note*: "Sometimes quiet and rest are as effective as the movements and actions of the physician."
11. *Author's note*: Some writers think it high treason to act on this rule. They shall be nameless, though they deserve exposure and censure.
12. *Author's note*: Nothing will do with some of the asinine part of society, but expensive medicines.
13. *Author's note*: Every patient, according to our English notions of liberty, is allowed to play the fool as much as he pleases. He may swallow two or two hundred of Morrison's pills should he think proper, and poison himself.
14. *Author's note*: Some physicians bleed every patient excessively. Every disorder or disease with them is inflammation. They have gone deservedly to the tomb of all the capulets [*sic*].
15. *Author's note*: The routine practice is to prescribe strong aperients or purgatives in all cases. This is a grievous error.
16. *Author's note*: All tables of doses are absurd; they merely mean, that the smallest quantity is prescribed. Ten times the amount is often given, according to the violence of the disease, or power of the constitution.
17. *Author's note*: The safest plan is to prescribe moderate and repeated doses, until the desired effect is produced.
18. *Editors' note*: "Who deserves honors, has earned them."
19. *Editors' note*: "Superfluous ingredients sometimes harm"
20. *Author's note:* I do not assent to this doctrine, because patients would be horrified on being ordered the poisons, mercury, &c.
21. *Editors' note:* We omit a table in which Ryan indicates the approved symbols for writing pounds, ounces, drachms, scruples, grains, and pints, as many of the typographic symbols are no longer in use. Ryan also indicates the conversion factors among these units of measure.

Section II
Selections from Ryan's Obstetrical Writings

A Professional Duties of the Obstetrician, or Accoucheur

1 From Manual on Midwifery, *1828 Edition*

[237] The first rule to be observed by the practitioner of obstetric medicine is, to obey the summons to give his personal attendance to the parturient female, as soon as possible. This duty admits of no compromise, for it is of the first importance to ascertain the presenting part, before the disruption of the membranes; because, if preternatural, it can be more easily rectified before the completion of the first stage; for in such a case, the practice is to rupture the membranes, when the os uteri is sufficiently dilated, to seize the feet, and deliver the infant by the operation of turning. If the membranes have burst, the uterus is in close contact with the body of the infant; and if the pains be violent, as it mostly happens, the operation of turning cannot be performed; as the womb would be lacerated, which is a most fatal occurrence. Again, an early examination is most necessary, as flooding may destroy the patient, or the feet of the infant may be expelled, the body retained in the vagina, and such pressure made on the navel string as to impede the circulation of the blood between the mother and child, and kill the latter.

The obstetrician, on his arrival, should ascertain all delicate inquiries from the nurse or other female attendant; and not in the sick-chamber. He should learn the duration of labor, the state of the bowels, and the history of the case, and impress the necessity of admittance to the patient as soon as possible. After admission into the sick chamber, he should approach the patient with his countenance contemplative, cheerful, nor grave and melancholy; his look pleasing, mixed with mildness and humanity, and expressive of a sincere desire of affording her alleviation. He ought to be polite and attentive, never proud, insolent, haughty or grave, which would alarm the patient, [237/238] who is generally timid and dejected. There should be nothing indecorous in his aspect or conduct, nothing rough or uncouth. He should display those good manners and politeness that characterize the well educated medical practitioner and the polished gentleman. Sympathy engages the confidence and affection of the patient; she feels the approach of the practitioner, who displays it, like that of an angel; while that of an unfeeling man, as of an executioner.

Of all diseases incident to humanity, those of the female are most deserving of sympathy and attention. The weakness and peculiar delicacy of the female constitution, call forth our greatest tenderness and compassion, and on no occasion so powerfully as in the agonies of child-bed, when she is stretched upon the rack on which she is laid by nature. On entering the parturient chamber, the practitioner should approach the patient with all the mildness and serenity of manners he may possess, and assure her, that from the history he has heard from the nurse, he is perfectly confident of her safety. After a few pains have occurred, he should pass his hand over the bed or body clothes on the abdomen, in order to ascertain whether there be pregnancy; for it has often happened that females and their medical attendants have expected delivery for days, when there was no pregnancy. Drs. Hamilton and Blundell record such cases, and I can bear testimony to the same fact. I have known a medical man remain four successive days and nights in attendance on a patient, who was not pregnant, so confident was he of delivery; yet the pains gradually ceased, and 8 years afterwards she expelled an immense quantity of hydatids from the uterus. After having ascertained the fact of pregnancy, we should next proceed to discover all the essential characters of labor, by an examination through the vagina, called by the French "the touch." Before we proceed to examine, the proposal ought to be made through the nurse, its impor- [238/239] tance strongly dwelt on, the length of time the labor has continued, the urgency of the pains, and above all, the necessity of ascertaining whether the labor be natural. Some women of high delicate feelings will not permit an examination, until the labor becomes severe; but then we need little persuasion or reasoning, to induce them to comply. A medical friend informed me, that a patient of his on entering her chamber, covered her face with her hands and exclaimed, "You shall not touch me; I'll lose my life first." He withdrew, but in a short time, when the pains became severe, he was anxiously recalled. In such cases, an intelligent nurse will be able to report whether the labor were natural or not. Other women are greatly terrified at the approach of the medical attendant, if never attended previously by an accoucheur; they are alarmed, lest they should experience violence and harsh treatment, which are now unknown, and those foolish fears are carefully fomented, rather than appeased, by the nurse, whose peculiar province she considers invaded. This is mostly the case in first labors, for women once attended by a medical man, who can never be guilty of rudeness, violence, or harsh treatment, without a gross dereliction of duty and breach of established obstetric rules, will never submit to the ignorance and temerity of female attendance afterwards. Surely every educated female must be convinced, that an illiterate and ignorant nurse knows nothing of the mode in which nature accomplishes the wonderful mechanism of labor. How then can she render assistance, in cases of difficulty and danger? What idea can such a one form of the almost innumerable difficulties that may impede labor? Hence the extensive employment of medical practitioners in every civilized nation in modern times. All doubts and fears, which a parturient female entertains must be obviated by reason and good sense, and the examination by the vagina instituted as soon as may be convenient. The patient is to lay on her left side, a [239/240] coverlet being thrown over her, and her hips as near the edge of the bed as possible, her knees drawn up towards the abdomen, and the bosom bent downwards and forwards, towards her knees. The nurse and another female attendant should be in the apartment, the light is to be excluded in some measure, and the curtains drawn close. The index and middle finger of the right hand, the nails being pared closely, are to be lubricated with some oleaginous substance, which is generally prepared by the attendants, as lard, butter, pomatum, olive oil, and passed under the right thigh, during the next pain from the perineum into the vagina, when they are to be directed downwards and backwards, as the orifice of the womb in the first stage of labor is low down, towards the sacrum. The orifice of the womb will be found dilated or not; if it admit the point of the finger, it is to be considered dilating, and will be pushed down during the pains. If it be dilated as large as the surface of a shilling, the head or other presenting part of the infant may be guessed at; and if the pains be those of labor, the vagina will be more or less lubricated with the increased mucous of parturition. Then the hand should be withdrawn, when the pain ceases, and wiped under the bed-clothes with a napkin, previously in readiness by the nurse.

This introduction of the fingers is to be accomplished as speedily and gently as possible, and the greatest delicacy is to be observed. The examination gives no pain, and hence removes the dread which many women, either from some misconception or former harsh treatment, entertain of this operation. A vast deal of useless suffering and fatigue will be saved the patient by an early examination; for if the labor be preternatural, it can be rectified readily before the rupture of the membranes. If labor be about to commence, the practitioner is generally consulted, and his superintendance requested, concerning the arrangements of the sick chamber, and of the bed especially. The nurse usually attends to the adjustment of the bed, but often [240/241] enquires, especially of young men, how the bed is to be arranged, in order to discover whether they have had much practical experience. In the better ranks, the mattress is covered with a skin of red or brown leather, and over this a folded sheet or blanket is applied, to absorb the moisture. A sheet is pinned over all, to keep them in their places. Others recommend a folded blanket alone, having turned up the pallet and blankets towards the head of the bed, and this is most prevalent among the middle and lower classes. A coverlet is to be thrown over the patient, or more covering, if the season require it. The woman is to put on a night wrapper, or other loose dress; but when the head of the child has descended low down into the cavity of the pelvis, she must undress and remain in bed. The chemise may be folded up on the hips, and its place supplied by a loose petticoat. The bedclothes ought to be light and comfortable, the chamber to be kept cool; no fire in summer, although a small fire tends to ventilate the chamber; and no more persons should be admitted into the apartment than are absolutely necessary. One female friend, to whose kind and sympathizing ear the poor sufferer may communicate her sorrows and anxieties, and the nurse, are quite sufficient for every useful purpose. The relation of all bad cases and frightful stories should be avoided, especially during a first confinement; as they do infinite mischief, by depressing the mind of the suffering patient, who naturally becomes alarmed, lest her case might be equally unfavorable; she loses confidence in her own powers, and retards, or perhaps entirely impedes, the process of labor. All nurses, and some medical men, are guilty of this imprudence, in narrating such stories. Dr. Dewees observes, "the poor suffering woman is entitled to all the consolation, a well grounded assurance of a happy termination of her case can afford; yet she must not be betrayed by false promises, as to a speedy issue; for it requires great experience to be able to state with certainty, [241/242] when any case will terminate. Her mind should be kept as free from anxiety as the nature of her situation will permit; therefore no conversation should be indulged in, which might for an instant excite her apprehensions. Conversation should be cheerful, and free from idle discussions of danger from similar situations, and should be as void of levity or want of feeling and sympathy, as of gloominess. Levity ill suits the situation of a woman in labor, and moroseness or ill humour is quite brutal, when the poor sufferer has a reasonable claim to pity and compassion."[1] The patient's mind should be diverted by a well chosen and general conversation, always combined with a confident assurance of her safety. The practitioner can afford no assistance during the dilatation of the womb in natural labor, and therefore he ought to withdraw from the parturient chamber, as his presence is a restraint; and his absence will allow the evacuation of the bowels and bladder, and also abridge the period of his watching. He may even visit other patients, as the first stage of labor may occupy several hours. The woman may walk about, sit, lay in bed, or sleep, during the dilatation of the womb; and she may have any light food, such as tea, toast, coffee, water gruel, sago, arrow root, tapioca, broths, &c., but no cordials, unless she be really debilitated, which is not the case once in a thousand instances. The exhibition of ardent liquors, although a popular custom, is highly improper; for no woman ever dies of weakness, during natural labor. The middle and lower classes are greatly prejudiced in favour of their use, or rather the abuse, of ardent and fermented liquors during labor; and hence the frequency of inflammations, fevers, and deaths amongst them. There is no medicine necessary during the first stage of labor, except the bowels be confined, when they are to be regulated by some castor oil, or a clyster. If the pains be trifling and very inert, for twelve to twenty hours, a dose of tincture of opium will be of advantage. Every accoucheur should [242/243] carry about him some tincture of opium, some ergot of rye, a catheter, a tracheal pipe, and a lancet. The opium is invaluable in allaying irritability, produced by false or spasmodic pains, and the other articles are equally valuable under certain circumstances. During the dilatation of the uterus, or first stage of labor, it is an established rule, that the practitioner need not examine by the vagina, more than two or three times. Frequent examinations, or attempts to dilate the vagina or uterus prematurely, a common practice with nurses, are highly injurious, by inducing irritation, inflammation, and swelling of these parts, which not only oppose invincible obstacles to delivery, but also lay the foundation of many of the fatal fevers and inflammations so common after parturition. The late Dr. Osborne, of this metropolis, used to remark, "that the practitioner should sit quietly and observe nature." The less manual interference in natural labor the better. Nature is the best obstetrician – "a meddlesome midwifery is bad." How many thousand women are delivered annually in the face of the globe, without any assistance![2]

2 From Manual of Midwifery, *1841 Edition*

[226] No obstetric instrument or manual operation ought ever to be performed until the necessity is explained to the relatives of the parturient woman or to herself, when she possesses a strong mind; so that an opportunity may be afforded of calling in another medical practitioner, if convenient, should it be desired by the woman, her relatives, friends, or acquaintances. Every medical practitioner, whether old or young, ought to observe this precept.

The woman or her friends have an undoubted right to have the opinions or assistance of as many medical practitioners as they please.

It is scarcely necessary to state, that the medical attendant may be considered too old or too young – or that there may not be sufficient confidence reposed in him. [226/227]

It is always to be remembered, that women in labour have, in general, the greatest dread of manual or instrumental operations; and that religion and humanity command every thing to be done to appease their fears and alarms, as these are always more or less prejudicial to their condition, as well as to restoration to health.

It may be necessary, as was well observed by Denman, to explain the object of using the forceps, or other blunt instrument, and to prove that no injury can be inflicted upon the woman or infant, when the instrument is employed with caution and judgment.

> In some cases of great apprehension, I have also showed them, upon one of my knees, all I intended to do with the forceps. – *Denman's Aphorisms*[3]

I have long followed this excellent advice, and observed to women, that the instrument was blunt, and not cutting, that it could pass where the hand could not, and that its branches might be considered as artificial hands, and would not injure the woman or infant, and that the latter might or would be born alive.

This explanation almost invariably removes all fear and apprehension.

There are, however, some nervous, hysterical, irritable, and impatient women, who, impelled by their fears, impatience, or sufferings, implore us to deliver them with instruments, long before there is a real necessity.

Such individuals may be encouraged, by fixing some remote period at which it is most likely they may be delivered, as six to twelve hours, and unless they are, and when twenty-four hours of actual and regular labour have elapsed, that then instruments will be employed.

They should also be told, that most women are delivered within twenty-four hours, without instruments; that others remain in labour a day or two, and are safely delivered without artificial interference; and that the good old remedies, time and patience, ought to have a fair trial – and that nature is the oldest and safest obstetric practitioner.

The forceps, or any other obstetric instrument, ought never to be used clandestinely, or without the knowledge of the woman, or of her friends.

There is always great caution required in the use of obstetric instruments, as great fears are entertained by most women, and great blame attached to the obstetrician, unless the woman and infant do well ultimately, and unless he possess a high reputation.

It is always necessary to impress upon the woman the imperious necessity of remaining quiet during the use of obstetric instruments, for when she is restless, there is more or less danger of injuring her, however experienced the obstetrician may be, unless extremely cautious.

When she moves suddenly, or changes her position, though the operator removes his hand off the instrument, it may contuse her slightly, and give her pain.

It is therefore manifest, that instruments ought not, as a general rule, to be applied without her knowledge, and not until having impressed her with the indispensable necessity of her remaining quiet and motionless.

Every woman endowed with common sense, will obey the advice of her medical attendant, when she places proper confidence in him.

B Conflicts Between Preserving the Lives of the Infant and Mother: from *Manual of Midwifery*, 1841 Edition

[241] This operation [induction of premature labour] has been performed, when the length of the superior aperture of the pelvis, from before backwards, the antero-posterior, or sacro-pubic diameter, measures only from two to three inches instead of five, as in the ordinary sized woman. The object in such cases is laudable, and not criminal; namely, to save the life of the infant, and the operation totally differs from the induction of criminal abortion, which is never attempted by the medical faculty, who justly consider it foeticide or homicide, but cases occasionally occur, in which women are so deformed, that although their infants may be extracted alive at seven months and a half, they must be destroyed either by the efforts of nature in strong labour at the full time, which cannot expel them, or by art; and it has frequently happened, that as many as seven and more infants have been thus successively destroyed. A most important question here arises, as to the morality of such women having children at all, and if they have, whether the former ought not be saved by the induction of premature labour. When they cannot be saved by this means, it was the opinion of the late revered and virtuous Dr. Denman, and most others of his contemporaries and successors, that however dangerous the caesarean operation, which was almost mortal in his time, it should be performed upon a woman who had so many infants successively destroyed, when in a future labour. The induction of premature labour, is, however, preferred at present.

I may here observe that there are not theologians, juris-consults, or physicians, even of these enlightened times, who will venture to determine the propriety of separating husbands and wives, or of their using any means to prevent the great object of matrimony. But is infant after infant to be destroyed, either by omission or commission? The result is the same – the death of human beings. In a moral point of view, it matters not whether an infant is starved, exposed to cold, strangled, or so treated or neglected, that its life is destroyed. A woman is assured by competent medical authority that she can never have a living infant at the full time, though she may have seven or eight infants in succession, every one of whom will be destroyed by the efforts of nature or by art, unless saved, when possible, by the induction of premature labour. [241/242]

I have given this subject very mature consideration, and my deliberate conviction is, that husbands and wives are, in general, equally culpable: sometimes the one, and sometimes the other, but mostly the former; in causing such repeated destruction of human life. I have repeatedly known instances of both, in which either party was perfectly indifferent on the subject; but there are many exceptions in which most persons deeply deplore the want of offspring. But however extreme the cases under consideration may be, the medical faculty will not interfere with the dictates of nature, nor violate the laws of religion. What a contrast exists between them and the modern anti-populationists, who insanely propose to limit families, or, in plain English, prevent procreation among healthful individuals, on the fallacious grounds of want of sufficient nutriment, although there is not a single civilized country on the face of the globe at present, to which this immoral, erroneous, and inhuman doctrine can apply. I have fully exposed and denounced this infamous doctrine in another work, "*The Philosophy of Marriage*," 1839, 3rd edition, and its baneful consequences in another volume, "*Prostitution in London, Paris, New York, &c.*", already quoted. I need scarcely observe, that if the medical faculty were to recommend such a doctrine, even in the extreme instances which gave rise to these remarks, they would, in all probability, do much more harm than good to the interests of society at large. They, therefore, always follow nature's laws and dictates, and assist her as much as possible.

[245] Embryotomy is required when the pelvis is so small by conformation or diseased contraction, that a full-grown infant cannot be extracted through the natural passage, either with the hand alone, or with any blunt instrument. The opinion in Continental Europe is, that the infant ought to be dead or destroyed by the pressure of the womb during its contractions or labour pains, before perforation of the skull is attempted, and in this opinion I most fully concur; for I boldly deny that there is any text in either the Old or New Testament which justifies the destruction of the infant in the womb, under any circumstances whatever. But, nevertheless, the obstetric rule in this country is, that the head of a living infant should be perforated with a sharp instrument, its brain pierced, lacerated, and evacuated, the skull reduced in size (see *Craniotomy*), to save the life of the mother; and lamentable cases have happened in which, after the partial destruction of the upper part of the skull and brain, the mutilated infant was brought into the world alive, to perish in a few minutes. Most persons ignorant of medical knowledge may doubt this statement, but it is still positively correct, as every well-informed medical practitioner must admit.

I must here observe, that in all cases requiring craniotomy, the pressure of the womb in forcing the infant's head against the contracted passage of the mother, will very speedily and inevitably destroy life, by causing congestion of the brain, or apoplexy, and in a preponderating majority of instances long before the life of the mother is, or can be in danger of destruction; and it is therefore manifest that a living infant should not be invariably destroyed in general, under the false and unjustifiable impression of saving the life of the mother, which I repeat, is seldom endangered for hours or days; and when it really is, in extreme cases, the destruction of the infant will seldom, if ever, preserve it.

It is well known to obstetricians, that in all difficult labours in which the infant cannot easily pass into the world, it is sooner or later, almost if not universally, destroyed by the pressure of the womb long before the life of the woman is in danger. Every experienced obstetrician must assent to the truth of this statement, as he must have observed in practice, or read of cases in which the infant was destroyed in a few hours, or has not been felt to move for one or more days, without the powers of the woman having been exhausted, or her life endangered; so that, as a general rule, he may wait the destruction of the infant by uterine action in most cases, and ought not to effect it, if ever, by cutting instruments. Such is the rule I have adopted for nearly twenty years; and I have never lost a woman on whose infant I performed craniotomy after death, for I never did while it was living, nor never shall, because I conscientiously consider it unwarrantable, unscriptural, and unjustifiable to act otherwise, for the reasons above assigned, and for others I shall adduce in the account of the caesarean operation hereafter.

Neither can I agree with the French and other European obstetricians, that the caesarean or other incisive operations already mentioned, ought to be performed on a woman to save the life of her infant, for the purpose of baptizing it, as it would be a barbarous proceeding, [245/246] in my opinion, to subject the mother to dangerous, indeed almost fatal operations, for the extraction perhaps of a dying or dead infant, nor do I believe that there is any scriptural authority to warrant such a proceeding, or that alludes to intra-uterine baptism. But when this is considered necessary, it can be readily accomplished in most cases by passing the tube attached to the modern double injecting, or enema syringe into the womb, and in contact with some part of the infant, and the water be applied freely. There are also many cases in which the water might be applied with the finger, when the infant is low in the pelvis.

I cannot agree to the opinion, that father, mother, or obstetrician, is ever justified in warranting or effecting the destruction of a living infant in the womb under any circumstance, even before or after intra-uterine baptism, and consequently I differ from the position of Professor Hatin of Paris, a most able obstetric author, "that embryotomy ought to be performed when the mother refuses every other operation." I maintain she has no right, or any one else, to destroy the life of the infant in the womb. (See Caesarian operation hereafter).

I also differ from Dr. Blundell and many other eminent British obstetricians on this subject.

This is not the place for theological or religious discussion, but I cannot help observing that the commandment, "Thou shalt not kill," cannot be broken without incurring great culpability. I am well aware that all religionists act differently from the principles of the Bible in modern times, but whether they are justified remains to be determined. It is a point of great importance in all cases requiring cutting operations, either on the infant or mother, to determine whether the former is alive or dead, which can be done, in most cases, by means of modern inventions, *as* the stethoscope and metroscope, (see *Signs and Detection of Pregnancy*, p. 159), which will enable the obstetrician to arrive at correct conclusions in general.

[254] It is now an established obstetric rule in all civilized countries, that no woman should be allowed to die undelivered, and that in all cases of parturition, however difficult, attempts may be made to save the lives of mother and infant, or to save the life of the one, even by destroying that of the other. However laudable may be the first part of this proposition, the last part of it, as to the destruction of the life of either woman or infant, is clearly contrary to the divine precept, "thou shalt not kill;" and I fearlessly maintain, that no case can occur in civilized society, not even homicide, in which any human being is justified in destroying the life of a fellow-being in any rank of society, more especially on account of the disqualification of such being or beings for the performance of a natural function, either parturition, respiration, digestion, &c., under any circumstances whatever.

In rude ages, the husband was suffered to murder his children, but at the present enlightened period, I am unacquainted with any scriptural, civil, statute, or other law, in any civilized country, which allows either husband or wife, or father or mother, to destroy the life of either or of their offspring, whether before or after birth. There is no law, which warrants kings, queens, legislators, judges of any court, criminal, civil, naval, military, ecclesiastical, &c. &c., or physicians, surgeons, or any class of society, to destroy, either directly or indirectly, by whatever means, the life of a fellow-subject, or any member of the human family – not, I repeat, even for the horrible crime of homicide. This is the general opinion of many of the ablest British and other statesmen, judges, theologians, jurisconsults, physicians and medical practitioners, as well as of the leading portion of our free, most powerful, and unequalled public press, and of all enlightened individuals, at the approach of the middle of the nineteenth century, AD 1840. Nevertheless, I am grieved to admit, that there are as yet some eminent members of the medical profession in this and other civilized countries, who maintain that the life of the infant in the womb may be sacrificed to preserve that of the mother, although they have forgotten to specify, in an accurate manner, for the best of all reasons, because they could not, the class of cases which would warrant their inhuman and unjustifiable conclusions. I feel convinced that no well-informed member of the medical profession could attempt such a classification, or precisely describe the [254/255] particular cases, amidst the human family on the face of the globe, in which the destruction of the infant would or could save the life of the mother, or vice versa. It is for this reason that the French, German, and most European and American obstetricians prefer symphyseotomy and gastro-hysterotomy to craniotomy, although preferred to the former operations by a majority of British medical practitioners. But so great is the difficulty in determining cases requiring any of these operations, that a consultation must always be held before resorting to any of them.

It is supposed, very erroneously, by some persons, that the husband possesses the prerogative of deciding upon the preservation of the life or death of his wife or infant, or the father of the life of his victim of seduction or concubine, and infant. I think, however, that were husband or father, to cause the death of either, he would be found guilty of murder, or infanticide, or manslaughter, according to the laws of this country. (See Author's *Medical Jurisprudence,* 1836, *Articles – Abortion and Infanticide*.[)]

In my opinion, a medical practitioner is equally amenable to the laws, and is never justified in consulting a husband or father concerning the preservation of the life of either wife or mother, or of the infant in the womb, as every person endowed with common sense will admit. No one but an educated and experienced medical practitioner can be a competent judge of the nature of any difficult case in midwifery, and no husband or father or any other person has the slightest right, in my opinion, to offer any suggestion as to the treatment, or as to saving of the life either of the woman or her infant – a question which must be entirely decided by medical opinion, and one on which even the most eminent of the faculty are still very much divided.

It was properly referred by the medical faculty of France to the doctors in theology at the Sorbonne, in Paris, A.D. 1648, who decided as follows, and I think correctly, according to the Bible: – "Nous sous signés docteurs en théologie de la faculté de Paris, sommes d'avis, que si l'on ne peut tirer l'enfant sans le tuer, l'on ne peut sans péché mortel le tirer; et qu'en ce cas la, il faut tenir à la maxime de St. Ambroise, – '*Si alteri subveniri non potest, nisi alter lædatur, commodius est neutrum juvare.*' " – Délibré à Paris le 24 Avril 1648. "We the undersigned doctors in theology of the faculty of Paris, are of opinion, that if the infant cannot be extracted without killing it, it cannot be extracted without committing a mortal sin, and that in such a case, it would be best to hold the maxim of St. Ambrose, – 'If one cannot be assisted without seriously injuring the other (by wound, blow, or otherwise, *see lædo*), it is best not to assist either.' " – Delivered at Paris, April 4, 1648. This is still the doctrine of the Roman Catholic Church. Another great objection to embryotomy was, that the infant could not enter Heaven without baptism. But it was contended by Thomas Aquinas, that the infant could not be baptized in the womb, for, according to Scripture, it should first be born, that is, it should be *natus* before it could be *renatus*, reborn by baptism. This difficulty was, however, over-ruled by the Sorbonne doctors in 1773, who declared that baptism was valid if the water touched any part of the infant's coverings. They decided, "*dum modo infans sit vivus, et arte seu industria medicorum possit aqua* [255/256] *ad ejus corpus immediate pervenire.*", "Whilst the infant is alive, and by the skill of medical practitioners, the water is brought in contact with its body." They might have added, or in contact with the membrane which covers it, and lines the womb, which is now considered part of the infant. It was formerly undecided whether the outer surface of the membrane, which encloses the water and infant, belongs to the womb, as it is closely attached to it, or to the foetus.

The decision of the Sorbonne doctors removed one objection to embryotomy, but admitted the former.

In this country, obstetricians are generally in favour of embryotomy, whether the infant be alive or not, as the more valuable life of the mother, they contend, ought to be preserved. It is said, "the tree should be preferred to the fruit." But the French, German, and American obstetricians are in the favour of the caesarean operation.

[260] It is also a strong objection [to embryotomy] that we have no certain and no positive signs indicative of the death of the infant, except those afforded by the stetho-

scope and metroscope, see p. 159; and hence the European and American authors hold that the sacrifice of the infant is murder; and that in cases of extreme deformity, its mutilation will not save the mother. On the whole, the most respectable and eminent of the foreign obstetric authors are unanimous in preferring gastrohysterotomy to embryotomy. They contend, that if the former operation were performed at an early period, the woman's life would not be so much endangered as by the latter.

Professor Lizars, the eminent surgeon of Edinburgh, has frequently performed gastrotomy with success, and the wound in the abdomen healed by the first intention. Hull informs us, in a note in his valuable [260/261] translation of Baudelocque's work, that of two hundred and thirty one women, operated on by gastrohysterotomy in this and foreign countries, one hundred and thirty-nine recovered; and the recent reports of the German practitioners are still more favourable.

Denman observed, that in cases where the infant should be invariably destroyed, a question ought to arise, whether a woman who was warned of this, again becoming pregnant, ought to be relieved by embryotomy. He, as well as Burns, Hull, and Dewees, are advocates for gastrohysterotomy, while a few others despise the "silly theological discussions, concerning the question of saving the life of the mother or infant," and agree as to the destruction of the latter. Henry VIII. was asked this question before the birth of his son Edward; and with that barbarity and cruelty, for which he was so remarkably distinguished, he exclaimed in a rage, "Save the infant, for it is easier to get wives than children." So he found it. The operation was performed on the mother, to whom it proved fatal. The same question was put to Bonaparte, by Dubois, before the birth of his son, and he answered the terrified accoucheur, "treat the Empress as you would a shopkeeper's wife, in the Rue St. Martin; but if one life must be lost, by all means save the mother."– *O'Meara's St. Helena*. Both were saved.

I have already argued in this article, that no private individual either high or low, has any right or power to advise a medical practitioner; nor has the latter any right whatever to consult a husband, or father, or mother, emperor, empress, king, queen, or any other individual, as to the destruction, or as to the destroying the life of a human being under any circumstances whatever. All the human race are equal in this respect, and there is no exception. The world would have gone as well if neither Henry VIII nor Napoleon never existed, so far as religion, morality, and the rights of humanity were or are concerned. They were mere mortals, like the rest of their species, but somewhat differently situated, and they had, however, no more right to sacrifice the life of a human being, in a social point of view, than the humblest of their subjects. (See pp. 254, 255.)

It will, however, appear, in a succeeding part of this article, that prolicide is unhesitatingly advised by one of the most eminent British obstetricians now living, in women who require the Caesarian operation, and also the removal of a portion of the Fallopian tube and other equally indefensible operations, to induce sterility; on the immorality and illegality of which I shall comment hereafter. (See also pp. 241–243). These recommendations are in unison with the principles of modern new lights, who would prevent generation, or destroy its products, contrary to the antiquated dictates of nature and her Divine Master. Such are the new moral world gentry; some march of intellect philosophers, and many others, who are desirous of destroying the effects of licentiousness and the proofs of infamy, – even at the horrible alternative – the destruction of human life. I have exposed the utter fallacy of these infamous doctrines, in my works on Marriage, Prostitution, and Medical Jurisprudence, already quoted; and likewise in the strictures in the concluding part of this section. But I am proud to record it, that an overwhelming majority – indeed there is only one solitary exception amongst the enlightened and learned members of the faculty, [261/262] who do not agree with me in opinion. They never have, never can sanction the destruction of the foetus in the womb, under any pretence whatsoever, not even to prevent the necessity of the Caesarian operation. They will not do evil that good may follow. They will not commit prolicide, which is justly considered homicide according to the divine, civil, and criminal laws of this and all other civilized countries; notwithstanding the falsely imputed infidelity laid to their charge, or the large bribes which they are so frequently offered.

It is with much pain that I indite these strictures, but when I find a doctrine lightly and flippantly proposed, as will appear hereafter, by an individual whose opinion has great influence upon the rising members of the profession, both junior practitioners and students, in this kingdom and elsewhere, although diametrically opposed to that of the medical faculty in all civilized nations at present, and too well calculated to lead to the commission of the most atrocious crimes, the deaths of most women and their offspring, who may be subjected to fruitless attempts to destroy the embryo in the womb, which generally kill both woman and offspring, (see all modern works on Medical Jurisprudence), the laws of nature, humanity, religion, medicine, and of civilized society, compel me to expose and denounce its inhuman and baneful influence. I neither mean nor intend any personal or professional offence, but medical science and practice is a republic in which every member is entitled to his own opinion, and so also is civilized society. I shall also, in candour and truth, add, that there is not a member of the profession in any country, of whom I entertain a higher opinion as a practical obstetrician, than of my opponent, yet I cannot agree with him on the point under consideration.[4] I shall adduce many additional cogent reasons hereafter, and shall therefore dismiss the subject for the present.

But reverting to the Caesarian operation, I have to observe, that there is not as yet, so far as I know, a correct history of the antiquity or fatality of gastrohysterotomy, in any of our ancient or modern works on obstetricy; a knowledge of which is indispensably necessary to form an accurate opinion of the danger of this most formidable and fatal operation. I trust I shall therefore be excused for attempting a summary of its history.

[264] *Indication.* – The operation [Cæsarian section, or gastrohysterotomy] is necessary when the sacro-pubic diameter of the brim is reduced to an inch and a half, and even when this strait is not so much contracted, when we cannot succeed in extracting the infant by any other mode of proceeding. [264/265]

When the antero-posterior diameter of the brim of the pelvis measures from two inches and a half to two inches and three-quarters, the French propose the forceps, version, or section of the pubic bones, as already stated. But in such cases, the attempts made with the hand or forceps, often contuse or lacerate the vagina or uterus. The vulva, vagina, and all parts contained in the pelvis may be inflamed, swollen, indurated, or gangrenous; the womb may be entirely detached from the vagina, and the woman so prostrate or debilitated, that she may die on the slightest attempt being made to save her life. In such a case are we to attempt to relieve her, or leave her in despair? I think with most modern obstetricians, that efforts should be made to save her life; because it is utterly impossible to determine, in most cases, the exact state of the organs in the pelvis, or whether they are so diseased as to destroy life in a short time. But while ever the slightest hope remains, we are, I feel convinced, bound to operate; and M. Velpeau did so, in such a case, in 1833, contrary to the opinion of many, but with the approbation of Maygrier, Moulin, Halma-Grand, and Bientot. The woman died soon after the operation; but he still persists in believing that conscience dictates similar conduct under similar circumstances. In extreme cases, the first incision may destroy life, as was the result of a remarkable example related to the Medical Society of London, a few years since. But if all signs were fatal, and death near at hand, I should not operate; as an incision would most probably only extinguish life, when the patient is moribund. I shall, however, immediately show that women supposed to be dead for two hours were re-animated by the operation; and in one case in which version was performed, the life of the infant was saved; the mother, though apparently dead, recovered, and was alive four years afterwards.

The operation is also indicated when the pelvis is contracted in the manner already stated, and the woman more than seven months pregnant; also in those cases in which the cavity, or inferior aperture of the pelvis is so contracted as not to admit of delivery by any other means, as if there were tumours in the cavity which it is impossible to remove; and likewise when the bodies of twins are united together, and cannot be extracted on account of their size. (*vide ante* p. 223.)

The operation should also be performed soon after the death of the woman in other cases, as the infant has been saved at the lapse of twelve, twenty-four, and forty-eight hours after the mother had expired.

The Princess of Schwartzenberg died at Paris of a burn, and next day the infant was found alive. Gardien relates a similar case, in which the operation was not performed until forty-eight hours after death, and the infant was living. Cangiamila states in his *Embryo1ogia Sacra*, that twenty-one infants were saved in this way in four years; and that the operation was performed twenty times at Syracuse in eighteen months.[5] Numa Pompilius enacted a law, which still exists, in a work entitled *Legregia Diget.* lib. xx., which commanded the physicians to open the bodies of pregnant women after death with the intention of preserving the citizens of the state, 600 AC. The same law prevailed in Venice in 1608 and 1721, which punished medical practitioners severely unless they used the same caution in operating on the dead as [265/266] on living women. In 1749, the King of Sicily punished medical attendants with death who omitted the operation on women soon after they expired, (see *Author's Manual of Medical Jurisprudence, 1836.*)

Ebel states, that an infant was born after the interment of the mother, whose body was exhumed in consequence of a judicial inquiry. (*Burns's Midwifery, 1837.* Several witnesses attested, that a woman who died at seven o'clock a.m., and appeared so in the evening, was found to have given birth, next day, to an infant. (*J. Univ. Med. des Sci. Med. Tom.* 7, p. 149). Sarrois states, that a living infant was extracted two hours after the death of the mother. Jackson, Deleau, Huguier, Jolly, Duparcque, Laaverjat, Reicke, Green, Blundell, &c. have extracted living infants, by the caesarean operation, a few minutes after the deaths of the mothers. Dr. Blundell succeeded in fifteen minutes after the death of the woman. – (See former edition, 1831, of this work). Van Swieten, Baudelocque, and many other celebrated authors, cite cases of women supposed to be dead, who were roused from their lethargic, cataleptic, or hysterical state, by the caesarean section. Peu states, that in a case in which he commenced the first incision, the body trembled, the woman moved her lips and ground her teeth, to his great horror. Trinchinetti relates a case nearly similar. Rigaudeaux was summoned to a woman at Douay, whom he supposed to have been dead for two hours. Before he proceeded to open the abdomen, he judiciously examined the pelvis, found it natural, brought down the infant by the feet, and succeeded in restoring it to life in two hours. The limbs of the mother were still supple, and he properly advised that the body should not be buried while they remained so. The woman finally awoke from her lethargy, and, four years afterwards, told her surgeon she was not dead as yet. I have also recorded many singular cases of premature dissection and burial, well worthy of perusal, in another work, to which I refer the reader, (*Manual qf Medical Jurisprudence, 1836*). Art. *Inhumation – Burial of the Dead,* p. 485).

When called to a woman who has suddenly died in labour, or in the last month of pregnancy, we should always examine the state of the pelvis, and when sufficiently capacious, dilate the mouth of the womb, if possible, and bring down the feet of the infant when there is the slightest chance of saving its life. I have, however, failed in such attempts, in two cases in the last month of pregnancy, and the caesarean section would not be allowed. The operation is much easier made than effected.

If the Caesarian operation is tolerated, it should be performed according to the rules hereafter described, and with the same care and caution as during life.

[269] The [Caesarian] operation was performed at Saltzburgh, but delayed a day, because it was doubtful whether the infant was alive or not; a decision which made all the difference between the caesarean section and embryotomy. The motions of the infant were perceived next day, when the former operation was determined upon. The infant was extracted alive, but died after half an hour, and the woman recovered. A reviewer remarks on this case, "we consider the above procedure on the part of the surgeon as well deserving of condign punishment, whether the woman survived or not. To perform the caesarean operation, in preference to embryotomy (where the latter is practicable) is most unwarrantable, and evinces a lamentable, not to say a culpable want of judgment, as to the proper estimate of the value of human life." (Vide *ante*, p. 245). The French are deserving "of this condign punishment," as well as the Germans; for M. Duges, a late writer, and many others, assert that the crotchet aigu, which is the same as the perforator, should not be employed until after the death of the infant: "nous avons dit, qu'on ne pouvait 1'appliquer qu'apres la mort certaine du fœtus; ne peut être applique que sur un enfant indubitablement mort." He likewise asserts that the caesarean section is the only resource in the excessive deformities of the pelvis. The Americans also deserve this condign punishment. My distinguished corre-

spondent, Professor Dewees, late of Philadelphia, well observes, "from an attentive consideration of both operations, the crotchet and caesarean section, we are free to confess ourselves in favour of the latter, and for the following reasons: First, because the infant must be destroyed by the crotchet; second, because the risks are often very great to the woman; third, because there are cases in which it is impossible to deliver with the crotchet; fourth, because where this instrument is employed, there is a great risk to the [269/270] mother, without a chance of benefit to the infant. These remarks refer to cases in which it is ascertained, or presumed, that the infant is living; if it is dead, then the crotchet may be used under a sufficient diameter of the pelvis. But if the infant is dead, and the delivery impossible by the crotchet, the caesarean operation should be proposed," (p. 594). He further observes, "for what reprehension, what punishment would be sufficiently severe for that practitioner, who after having destroyed the infant, should find it impossible to deliver it; and then, for its accomplishment, subject the poor woman to the caesarean section? He would scarcely merit the plea of *quo animo* in his favour." Par. 1479. (See Craniotomy and Symphyseotomy, as well as Gastrohysterotomy, in this work).

When we contrast the number of unsuccessful and successful authenticated cases in the eighteenth and nineteenth centuries, we shall find the results as follow: 147 deaths, 118 recoveries.

The operation is only performed in the British dominions in extreme cases of deformity, the sacro-pubic diameter being one ich and a half in. and upwards.

In eighty cases the following were the admeasurements in sixty-two:–one inch in 1 case; one inch and a half in 8; same and two lines in 23; two inches and a half and two lines in 25; two inches and a half and two three-quarters in 5=62.

From 1821 to 1830 the operation was performed in sixty-one cases, and only twenty-eight from 1810 to 1820.

In thirty-six operations in lying-in hospitals, eleven were fortunate and twenty-five unfortunate. In one practice, thirty-one succeeded out of sixty.

When the operation was performed before or immediately after the discharge of the amniotic fluid or water, the infants were extracted alive. In such cases, the proportion of women saved to that of those lost, is as 4:3.

Total of infants living 67, dead 29. The general proportion of favourable to unfavourable cases of the caesarean operation is 3:4.

According to M. Velpeau, from whom I quote, the causes of death were the following:

Peritonitis and enteritis, 13; gangrene, 8; hæmorrhage, 7; effusion into the abdomen, 3; meteorism or tympanites, 3; prostration or sinking, 3; shock of operation, 2; convulsions, 2; and colliquative diarrhœa, 1.

Days on which death occurred.— 1st day eight died; 2—six; 3—ten; 4—five; 5—twelve; 6—four; 7—one; 8—three; 18, 20, 27, 30, 45 —one on each day.

Time of recoveries.— In 3 weeks three recovered ; 4 in three ; 5--five; 6—five; 7—three; 8—three; and 10 in two weeks.

It is utterly impossible, in my opinion, in the present state of science, to arrive at any positive conclusion, either as to the mortality of women, caused by the caesarean operation, because no two cases, constitutions, or circumstances are alike, or can be determined.

It is not quite certain, perhaps, indeed it is very improbable, that all the unfavourable cases have been recorded, but there can be no doubt [270/271] as to successful ones. M. Velpeau observes, I think very properly, that up to the present time, the caesarean operation has been fatal at least in one in three, if not, in one in two cases. The great majority of authors are likewise of this opinion, and a great majority agree with the sentiment of Sir Fielding Ould, of Dublin, "that to practise it is a proof of detestable, illegal, and barbarous inhumanity." – I should say, if it can possibly be avoided.

The preceding details must convince obstetricians of the great danger of the operation, and that they should never have recourse to it without absolute necessity.

As the English law now stands, the destruction of the infant is a felony, even to save the life of the mother; yet the contrary opinion is the prevailing one in this country. "In this country and in France, however painful it may be to destroy the life of the child, the mischief is considered less serious in its consequences, than the destruction (or we ought, perhaps, rather to say) the probable destruction of the mother." (*Lancet*, 1828, p. 328). Dr. Kind asserts, that in Germany, the expediency, as in this country, of destroying the child, is always left to the judgment of the medical attendant. (*Op. Cit.* p. 415. *Vide ante*, p. 245). The first quotation is not correct.

The validity of this conclusion, will be further acknowledged, when we consider the number of important parts, divided by the incisions in gastro-hysterotomy – the parietes of the abdomen; the double division of the peritoneum; the incision of the enlarged arteries, veins, and nerves of the gravid or pregnant womb; and of the substance of that organ itself, all of which may be followed by inflammation, gangrene, or haemorrhage; especially when the powers of life are greatly weakened, or almost destroyed. The escape of the amniotic fluid, by tedious labour, as in the cases in this country, and of blood into the cavity of the abdomen, may, unless removed, induce fatal peritonitis.

It is, however, to be borne in mind, that the substance of the womb is not very sensitive, so that it has been often wounded or ruptured, without the supervention of inflammation.

The experiments of Blundell, the operations of gastrotomy performed by Lizars, and the numerous cases of penetrating wounds of the abdomen, even the intestines being transfixed, as recorded in the works on military surgery, and in Professor Cooper's valuable Dictionary of Surgery, in which recoveries took place, though there was effusion of blood and faeces into the cavity of the abdomen, clearly show, that when the powers of life are not weakened, as in the inferior animals, in women of good constitution, who are operated on in France and Germany, but rarely in this country: or in soldiers who are well fed, that gastrotomy is by no means so fatal, as gastro-hysterotomy in women of the worst constitutions, suffering from osteomalaxia, who are greatly exhausted by long continued labour-pains, and who are too often, indeed generally sinking before the operation is commenced, which has almost been hitherto the case in this kingdom, except in Mr. Barlow's solitary example, or perhaps the case of Mary Dunnally, already noticed.[6]

[277] The fatal results are so great, that some modern obstetricians of eminence have gravely proposed to excise a portion of either Fallopian or uterine tube, or the whole uterus, when the abdomen is laid open, and before inserting the sutures, for the purpose of inducing sterility. This proposal requires to be rigidly examined, on moral, social, and medical grounds. (see p. 261).

Two obstetric authors of eminence have proposed operations for the prevention of future conceptions when gastrohysterotomy has been performed. These proposals are, in my opinion, untenable in a moral point of view, and have never been hitherto adopted. Michaelis proposed to prevent future conceptions and the necessity of the caesarean operation, and to avoid a very great reaction, by which, I presume, he means conjugal dissension and infidelity, to extirpate the uterus (*Kilian, Op. Cit.*); and Dr. Blundell not only sanctions this operation, but also proposes a substitute, viz., the removal of a line of the Fallopian tube, right or left, so as to obliterate its calibre – the larger blood-vessels being avoided: mere division of the tube might be sufficient to produce sterility; but the further removal of a portion of the tube appears to be the surer practice. I recommend this precaution, therefore, as an improvement of the operation. (Work, by Lee and Rogers, 1840, already quoted, p. 359).

I cannot admit the correctness of this physiology, because it is on record and well authenticated, that a woman who had but one ovary and uterine tube, and who was found to possess no more on actual dissection, not only was not sterile, but was the mother of infants of the different sexes. No less an authority than M. Velpeau asserts this fact from actual dissection.

It therefore clearly follows, that obliteration, or even absence of one fallopian or uterine tube, or even the absence of one tube and ovary, neither might nor can produce sterility. It is equally well known, that a man who has one testicle destroyed by disease or removed by castra- [277/278] tion, may beget offspring of both sexes, proofs of which will be found in all the standard treatises on Medical Jurisprudence of modern times.

I am perfectly at a loss to perceive upon what moral or physical grounds, women who may or may not be deformed, are to be rendered sterile by surgical operations, or other means, and men exempted, though really the sources from which our race is perpetuated. This is not according to the laws of nature. I shall not dwell further upon this subject, as its various bearings would lead me into a long digression, and more particularly as I have discussed it elsewhere (see *Philosophy of Marriage*), and shall only add, that perhaps at some future period, during the rapid march of intelligence and knowledge, it may be gravely proposed to emasculate or render sterile all husbands and others, whose wives or mistresses may have been subjected or liable to, the performance of the caesarean operation.

The removal of the womb, if justifiable, which I strongly deny, though much more dangerous, and so fatal, that even when the organ is incurably diseased, has been totally abandoned of late years, would be a more effectual preventive; but even this remains to be proved, as every one knows there are extra-uterine conceptions. I cannot here enter upon physiological discussions, which would be out of place, but I refer the interested reader to the account of generation in a preceding chapter.

I am well aware that the womb, when diseased, has been cut away, and also after delivery absolutely torn away by ignorant practitioners, and even separated during the action of the bowels or bladder; but such cases are not in point, and do not relate to the question under consideration. (See p. 184).

That question is this, is it right or moral to use any means or operation to prevent procreation? I boldly answer in the negative – because any such means or operation prove a doubt in the minds of those who employ them, of the omnipotence of Divine Providence, and of his power to preserve the human race. The induction of abortion, to destroy the life of the foetus, is a felony according to the laws of this country, and is considered homicide by the most eminent of the medical faculty in all countries.

The following proposals by Dr. Blundell have not been as yet tried, so far as I know, nor are they ever likely to be sanctioned by the profession, on account of their danger.

"Now, is there any other mode in which, when the obstruction of the pelvis is insuperable, the formation of a foetus may be prevented? In my opinion, there is. If a woman were in that condition, in which delivery could not take place by the natural passage, – provided she distrusted the circumstance in which she was placed, – I would advise an incision (of 1 in. in length) in the linea alba, above the symphysis pubis. I would advise further, that the Fallopian tube (on either side) should be drawn up to this aperture; and, lastly, that a portion of the tube should be removed; – an operation easily performed; when the woman would, for ever afterwards, be sterile. All this may be done after due consideration; – circumstances not forbidding. 'But the abdominal inci- [278/279] sion; – that is bad.' True; but the caesarean incision; – that is worse! Is not that true also?

"*Destruction of the Ovum.* – If a woman, in the earlier months of pregnancy, be known to have a pelvis contracted in a high degree, is there nothing which you may then do to prevent an ultimate need of the caesarean operation? Abortive medicines might, in this case, be thought of; or, these failing or rejected, – if you could feel the os uteri, – you might introduce a female sound, or any other instrument of that kind; and, passing this sound into the uterine cavity, you might completely break up the structure of the ovum, so as to prevent the progress of generation. In doing this, there would always be a risk of haemorrhage; but where you are endeavouring to avoid the necessity of the caesarean incisions, this risk would be justifiable. The substitution of the smaller evil for the greater, is frequently the principle of the healing art. But what if the os uteri be inaccessible? Is there, in such a case, any other expedient to which we may have recourse? In a case like this, were my opinion consulted, I should be inclined to reply, – 'As a substitute for the caesarean operation, let an incision be made as before, above the symphysis pubis; then let some instrument (such as a trocar, or canula) be carried into the cavity of the uterus; let this instrument be sufficiently stiff to enter the cavity, and retain its form there under pressure; and then let it be resolutely moved about in the uterus, so as to break up completely the texture of the ovum. The whole instrument need not be much thicker than a bell-wire. The process is allied to that of acupuncture. The point of the trocar, on entering the uterus, should be withdrawn with the canula; a finger should be carefully placed on the uterus, so as to guide the instrument, and guard against injury of the intestines or the bladder.' Scribblers had better content themselves with sneering at the operation; – surgeons had better perform it! – Artem quisque suam! To produce future sterility, the Fallopian tubes might be rendered impervious." – *Op. Cit.*

All these proposals are liable to the most serious objections. The recommendation of destroying the ovum or fetus in the womb by so high an authority, may so influence many junior members of the profession, as to lead them to attempt to accomplish it in cases of illegitimate pregnancies, and render themselves liable to criminal proceedings for felony, punishable by long imprisonment, flogging, transportation for different periods, and in some cases, by an ignominious death. (See Author's *Manual of Medical Jurisprudence,* second edition, 1836; see also pp. 241–243, 254, 255, 261, 262).

The induction of premature labour is effected for the purpose of saving the life of the infant, preventing the necessity of craniotomy and the caesarean operation, and thus saving the lives of both mother and infant, is a laudable and humane proceeding, the most contrary to that under consideration, which is, in my opinion, properly considered, most reprehensible and penal. (See p. 240).

The mode of inducing premature labour I described in a dead language, to prevent, as far as possible, its being practised for criminal purposes. (See p 240).[7] As to the above quotation, he who runs may read; but, fortunately, the description is so general and vague, that few, [279/280] if any, could succeed in breaking up the ovum without more or less injuring the mother, and most probably destroying her life, as well as that of her innocent and unoffending offspring.

The orifice of the womb is inaccessible in the early months of pregnancy (see p. 155), and no instrument can be passed into it. It is for this reason that we are advised to resort to other operations, nearly as dangerous as the caesarean section, which, though tried on inferior animals, have never been attempted on the human female, nor never can be, by any one who properly values the rights of humanity, and his professional reputation. They may appear very ingenious and bold in a lecture-room, to novices who are incapable of forming a correct opinion on the subject, or to practitioners unacquainted with the histories of gastrotomy and gastrohysterotomy, but certainly not to learned and experienced obstetricians, who will and must avoid them. Here I must add, in conclusion, the old adage, – "Amicus Socrates, amicus Plato, sed magis amica veritas." Socrates is a friend, Plato is a friend, but truth is a greater friend.

Notes

1. *Editors' note*: Dr. William Potts Dewees (1768–1841) was professor of midwifery at the University of Pennsylvania, author of many works on midwifery, and a favorite authority of Ryan's.
2. *Editors' note*: Ryan repeats most of this material in his 1841 edition, but intersperses it with a good deal more technical information on the diagnosis and management of labor and its complications; *Manual of Midwifery* (1841): 166–179.

3. *Editors' note*: Thomas Denman (1733–1815) was the author of an introduction to midwifery which went though numerous editions, and several volumes of "aphorisms" on various topics. Ryan edited a volume of his aphorisms posthumously.
4. *Editors' note*: The "opponent" Ryan here refers to would apparently be Dr. Blundell, whose ideas are discussed below, p. 277 of Ryan's text. This was presumably Dr. James Blundell, of Guy's Hospital (1790–1878), author of several obstetric works.
5. *Editors' note:* Francesco Emmanuele Cangiamila, 1702–1763.
6. *Editors' note:* Ryan had earlier, in some detail, relayed a 1793 case in which Mr. Barlow successfully performed a Caesarian section on a patient who then survived until 1826, although the infant was dead at the time of the operation; in this case the severely contracted pelvis made embryotomy impossible. Ryan also cited the case of Mary Dunnally, an Irish midwife, who extracted a dead infant with a razor in 1738, sewed the wound with tailor's thread, and applied egg whites. The patient recovered in 27 days. This was apparently the first recorded successful caesarian section in Britain. Ryan, *Manual of Midwifery* (1841), 256–260.
7. *Editors' note*: Ryan described the specific features of the "operation" in Latin, pp. 242–243 of this volume.

Bibliography

I Writings by Michael Ryan

Ryan, Michael. *Tentamentum Medico-Physicum Inaugurale, de Genere Humano ejusque Varietatibus...* Edinburgh: P. Neill, 1821.

Ryan, Michael. "Remarks on the Use of Hydrocyanic Acid in Angina Pectoris and other Diseases of the Heart, and in Bronchitis, Phthisis, and Dyspepsia," *London Medical and Physical Journal* 51 (1824): 369.

Ryan, Michael. *A Treatise on the Most Celebrated Mineral Waters of Ireland containing an Account of the Waters of Ballyspellan, Castleconnel, Ballynahinch, Mallow, Lucan, Swadlinbar, Goldenbridgre, Kilmainham, &c., &c.: and of the spa lately discovered at Brownstown, near Kilkenny, with Plain Directions during the Use of Mineral Waters and an Account of some of those Diseases in which they are Most Useful.* Kilkenny: Printed and published by J. Reynolds, for Hodges, M'Arthur, and J. Cummings, Dublin, 1824.

Ryan, Michael. "An Essay on the Natural, Chemical, and Medical History of Water, in its simple and Combined States; including an Account of the Chemical composition, and Medical Effects, of the principal Mineral Waters in the United Kingdom of Great Britain and Ireland, and also the Continent of Europe," *London Medical and Physical Journal* 54 (1825): 442–61.

Ryan, Michael. "Delirium Tremens, Treated as a Nervous Disease," *Lancet* 2 (1827–28): 791–93.

Ryan, Michael. "Introductory Lecture to the Theory and Practice of Midwifery," *Lancet* 2 (1827–28): 394–400.

Ryan, Michael. *A Manual on Midwifery; or a Summary of the Science and Art of Obstetric Medicine, including the Anatomy, Physiology, Pathology, and Therapeutics, Peculiar to Females; Treatment of Parturition, Puerperal, and Infantile Diseases; and an Exposition of Obstetrical-Legal Medicine.* London: Longmans, 1828.

Ryan, Michael. *Remarks on the supply of water to the metropolis: with an account of the natural history of water in its simple and combined states, and of the chemical composition and medical uses of all the known mineral waters being a guide to foreign and British watering places.* London: Longmans & Co., Anderson, and Messrs. Underwood and Highley, 1828.

Ryan, Michael. "To the Editor of the *London Medical Gazette* [reply to report of meeting of the Medical Society of London]," *London Medical Gazette* 3 (1828–29): 684.

Ryan, Michael. "Drs. Ryan and Gordon Smith [letter]," *Lancet* 1 (1830–31): 106–108.

Ryan, Michael. *Lectures on population, marriage, and divorce, as questions of state medicine, comprising an account of the causes and treatment of impotence and sterility, and of the morbid and curative effects of marriage; forming a part of an extended course on medical jurisprudence, delivered at the Medical Theatre, Hatton-Garden.* London: Renshaw and Rush, 1831.

Ryan, Michael. *Manual of Medical Jurisprudence*, 1st ed. London: Renshaw and Rush, 1831. *Note:* Material in this volume was initially published in Ryan's journal, as follows:

Hippocratic ethics, *London Medical and Surgical Journal* 3 (1829): 485–95.
Importance of the study of state medicine, *London Medical and Surgical Journal* 4 (1830): 217 ff.
[Ancient medical ethics], *London Medical and Surgical Journal* 4 (1830): 300 ff.
[Medical ethics of the Middle Ages], *London Medical and Surgical Journal* 4 (1830): 400 ff.
[Medical ethics of the present period], *London Medical and Surgical Journal* 4 (1830): 492 ff.
Ryan, Michael. "To the Editor of *The Lancet*" [in reply to *Lancet* review of *Manual of Medical Jurisprudence*]. *Lancet* 1 (1831–32): 222–24.
Ryan, Michael. "Tweedie v. Ramadge [with Ryan's addendum]," *London Medical and Surgical Journal* 7 (1831); reprinted, *Lancet* 2 (1831–32): 408–9.
Ryan, Michael. "Reason of the Verdict against Ryan and Co." [reply to *Lancet* editorial]. *Lancet* 2 (1831–32): 496–97.
Edwards, Milne and Vavasseur, P. *A New Practical Formulary of Hospitals of England, Scotland, Ireland, France, Germany, Italy, Spain, Portugal, &c., &c.,* [Translated from the French and augmented by Michael Ryan]. Henderson, London, July 1835.
Ryan, Michael. *A Manual of Medical Jurisprudence and State Medicine, Compiled from the Latest Legal and Medical Works, of Beck, Paris, Christison, Fodere, Ofila, etc.* 2nd ed. London: Sherwood, Gilbert, and Piper, 1836. [Available at http://books.google.com/books?id = vToEAAAAQAAJ&printsec = frontcover&dq = michael + ryan&lr =, accessed December 14, 2007.]
Ryan, Michael. "Hyoscyamus and Its Preparations," [letter] *London Medical Gazette* 18 (1836): 609–10.
Hooper, Robert. *Physician's Vademecum*,. Michael Ryan ed. London: H. Renshaw, 1837.
Ryan, Michael. *The Philosophy of Marriage in its Social, Moral, and Physical Relations; with an Account of the Diseases of the Genito-urinary Organs, Which Impair or Destroy the Reproductive Function, and Induce a Variety of Complaints; with the Physiology of Generation in the Vegetable and Animal Kingdoms; Being Part of a Course of Obstetric Lectures Delivered at the North London School of Medicine, Charlotte Street, Bloomsbury, Bedford Square.* London: John Churchill, 1837. [Available at http://books.google.com/books?id = x0kEAAAAQAAJ&printsec = frontcover&dq = michael + ryan&lr =, accessed December 14, 2007.]
Ryan, Michael. *The Medico-Chirurgical Pharmacopœia; or, a Conspectus of the Best Prescriptions in Medicine, Surgery, Obstetricy, and Infantile Medicine; with a Table of the Doses of All Medicines in Use; the Additions in the London Pharmacopoeia, 1836; M. Magendie's Formulary; Aphorisms on the Treatment of Poisoning; Reduction of Dislocations and Fractures; and on Natural and Difficult Parturitions, with Puerpal Diseases,* 2nd ed. London: John Churchill, 1838.
Ryan, Michael. "Dr. Ryan's Reply to the Notice of Mr. Houston's Manual on Diseases of the Eye," *Lancet* 1 (1838–39): 705–6.
Ryan, Michael. *Prostitution in London with a Comparative View of That of Paris and New York, as Illustrative of the Capitals and Large Towns of All Countries; and Proving Moral Depravation to Be the Most Fertile Source of Crime, and of Personal and Social Misery; with an Account of the Nature and Treatment of Various Diseases, Caused by the Abuses of the Reproductive Function.* London: H. Bailliere, 1839. [Available at http://books.google.com/books?id = zHsEAAAAQAAJ&printsec = frontcover&dq = michael + ryan + prostitution + in + london&lr =, accessed December 14, 2007.]
Ryan, Michael. *A Manual of Midwifery and Diseases of Women and Children…,* 4th ed. London: (self-published), 1841.

II Secondary Sources and Other Works

Aikin, John. *Thoughts on Hospitals/with a Letter to the Author, by Thomas Percival*. London: Joseph Johnson, 1771.

[Anonymous]. "Death of Dr. Ryan," *Provincial Medical and Surgical Journal*, 1 (1840–41): 206–7.

[Anonymous]. "Dr. Ramadge and St. John Long," *London Medical Gazette* 8 (1831): 117–20.

[Anonymous]. "[Review of] *A Manual of Medical Jurisprudence* ... by Michael Ryan..." *Lancet* 1 (1831–32): 137–43.

[Anonymous]. [Review of *Medical Ethics...*, by Thomas Percival], *Lancet* 12 (series 2) (1826–27): 696.

[Anonymous]. [Review of Ryan, *A Manual of Midwifery*, 3rd ed.] *London Medical Gazette* 9 (1831–32): 52.

[Anonymous]. [Review of Ryan, *A Manual on Midwifery*,] *London Medical Gazette* 2 (1828): 714–15.

[Anonymous]. [Review of Ryan, *Prostitution in London*], *British and Foreign Medical Review* 7 (1839): 540.

[Anonymous], *Review of Works on Medical Ethics*. Edinburgh: Murray & Gibb, 1850.

[Anonymous]. [Review of S. Little's *A Manual of Diseases of the Eye; or, Treatise on Ophthalmology*,] *Lancet* 1 (1838–39): 494.

[Anonymous]. Subscription for Dr. Ryan, *Lancet* 1 (1832–33): 351–352.

[Anonymous]. "Subscription for the Widow and Children of the late Dr. Ryan," *Lancet* 1 (1840–41): 903.

Bacon, Francis. *The Advancement of Learning*, ed. Michael Kiernan. Oxford, UK: Clarendon Press, 2000.

Baker, Robert B., Arthur L. Caplan, Linda L. Emanuel, and Stephen R. Latham (eds.) *The American Medical Ethics Revolution*. Baltimore, MD: Johns Hopkins University Press, 1999.

Baker, Robert B. "The Discourses of Practitioners in Nineteenth- and Twentieth-Century Britain and America." In *Cambridge World History of Medical Ethics*, eds. Baker, Robert B. and Laurence B. McCullough. Cambridge, UK: Cambridge University Press, 2009:446–464.

Baker, Robert B., Arthur L. Caplan, Linda L. Emanuel, and Stephen R. Latham, "Introduction," In *The American Medical Ethics Revolution: How the AMA's Code of Ethics Has Transformed Physicians' Relationships to Patients, Professionals, and Society*, eds. Baker, Robert B., Arthur L. Caplan, Linda L. Emanuel, and Stephen R. Latham. Baltimore, MD: Johns Hopkins University Press, 1999.

Baker, Robert B. and Laurence B. McCullough. "What Is the History of Medical Ethics?" In: *Cambridge World History of Medical Ethics*, eds. Robert B. Baker and Laurence B. McCullough. Cambridge, UK: Cambridge University Press, 2009:3-15.

Bartrip, Peter. "An Introduction to Jukes Styrap's *A Code of Medical Ethics* (1878)," in *The Codification of Medical Morality*, ed. Robert Baker. Boston, MA: Kluwer, 1995: 145–148.

Beecher, Henry K. *Research and the Individual: Human Studies*, Boston, MA: Little, Brown & Co, 1970.

Berlant, Jeffrey. *Profession and Monopoly: A Study of Medicine in the United States and Great Britain*, Berkeley, CA: University of California Press, 1975.

Burns, Chester R. "Reciprocity in the Development of Anglo-American Medical Ethics," in *The Codification of Medical Morality*, ed. Robert Baker, Boston: Kluwer, 1995: 135–143.

Caldwell, John C. "Malthus and the Less Developed World: The Pivotal Role of India," *Population and Development Review* 1998; 24: 675–96.

Castro, Roderic a (Rodrigo de Castro), *Medicus-politicus: sive de officiis medico-politicis tractatus, quatuor distinctus libris: in quibus non solum bonorum medicorum mores ac virtutes exprimuntur, malorum vero fraudes et imposturae deteguntur ...* (The Politic Physician: or a Treatise on Medico-Political Duties ...), Hamburg, [no publisher identified] 1614.

Celsus, Aulus Cornelius. *De Medicina*. Florence: Nicolaus Laurentii, Alamanus, 1478.

Clarke, James F. *Autobiographical Recollections of the Medical Profession*, London: J. & A. Churchill, 1874.

Cope, Zachary. "The Private Medical Schools of London, 1746–1914." In F. N. L. Poynter, ed. *The Evolution of Medical Education in Britain*. Baltimore, MD: Williams & Wilkins, 1966: 89–109.

Corden, Thompson. *A Letter to the Public on the Necessity of Anatomical Pursuits; with Reference to Popular Prejudices, and to the Principles on Which Legislative Interference in These Matters Ought to Proceed*. London: Taylor, 1830.

D., H.W. "Obituary: John Gordon Smith, M.D., F.R.S.L.," *Gentleman's Magazine*, September 1833: 278–79.

Denman, Thomas. *The Obstetrician's Vademecum; or Aphorisms on Natural and Difficult Parturition; the Application and Use of Instruments in Preternatural Labours; on Labours Complicated with Hemorrhage, Convulsions, etc* (edited and augmented by Michael Ryan). London: Cox, 1836.

Dermott, George Darby. *A Discussion on the Organic Materiality of the Mind, the Immateriality of the Soul, and the Non-Identity of the Two*. London: Callow & Wilson, 1830.

Deslandes, Leopold. *De l'Onanisme et des autres Abus Veneriens*. Paris: A. Lelarge, 1835.

Desmond, Adrian. *The Politics of Evolution: Morphology, Medicine, and Reform in Radical London*. Chicago, IL: University of Chicago Press, 1989.

Epps, Elizabeth, ed. *Diary of the Late John Epps*. London: Kent & Co., 1875.

Epps, John. *Internal Evidences of Christianity Induced from Phrenology*, Edinburgh: J. Anderson, 1827.

Fraser, William. *Queries in Medical Ethics* [read before the Medico-Chirurgical Society of Aberdeen, 5th April 1849]. *London Medical Gazette*, Vol. 2, 1849: 181–7, 227–32.

Godman, John D. *Addresses Delivered on Various Public Occasions*. Philadelphia, PA: Carey, Lea & Carey, 1829.

Godman, John D. *Professional Reputation: An Oration before the Philadelphia Medical Society* [on February 8, 1826]. Philadelphia, PA: Benjamin & Thomas Kill. 1826.

Grant, Alexander. *The Story of the University of Edinburgh During Its First Three Hundred Years*. London: Longmans, Green &Co., 1884.

Gregory, James. *Conspectus Medicinæ Theoreticæ ad Usum Academicum*. Edinburgh: A. Constable, 1818.

[Gregory, John.] *A Comparative View of the State and Faculties of Man with those of the Animal World*, 2nd ed. London: J. Dodsley, 1766.

Gregory, John. *A Father's Legacy to his Daughters*. Edinburgh: A. Strahan and W. Creech, 1788.

Gregory, John. *Elements of the Practice of Physic: For the Use of Students*. 2nd ed. London: W. Strahan and T. Cadell, 1774.

Gregory, John. *Lectures on the Duties and Qualifications of a Physician*. London: W. Strahan and T. Cadell, 1772.

Gregory, John. *On the Method of Prosecuting Enquiries in Philosophy*, 1770.

Haakonssen, Lisbeth. *Medicine and Morals in the Enlightenment: John Gregory, Thomas Percival, and Benjamin Rush*. Atlanta: Rodopi, 1997.

Harris, James. *Hermes: or, a Philosophical Inquiry Concerning Language and Universal Grammar*. London: J. Nourse and P. Vaillan, 1771.

Haughton, Brian. "Feral Children: The Unsolved Mystery of Kaspar Hauser – Wild Child of Europe." Downloaded on August 25, 2004, from http://www.mysteriouspeople.com/Hauser1.htm

Hoffmann, Friedrich. *Medicus Politicus sive regulae prudentiae secundum quas medicus juvenis studia sua et vitae rationem dirigere debet…* Lugduni Batavorum: Apud Philippum Bonk [etc.], 1746.

Knips Macoppe, Alessandro, 1662–1744. *La politica del medico nell'esercizio dell'arte sua esposta in cento aforismi …/Traduzione italiana con note del Dottore Ignazio Lomeni*. Milan: G. Pirotta, 1826.

Leake, Chauncey D. Preface to Percival's *Medical Ethics*, Baltimore, MD: Williams & Wilkins, 1927.

Levine, Robert. "Clarifying the concepts of research ethics," *Hastings Center Report* 9 (3) (1979): 21–26.

Malthus, T. Robert. *An Essay on the Principle of Population as It Affects the Future Improvement of Society with Remarks on the Speculations of Mr. Godwin, M. Condorcet, and Other Writers.* 1798. [accessed from http://www.faculty.rsu.edu/~felwell/Theorists/Malthus/Essay.htm#35 on 16 November 2007]

McCullough, Laurence B. *John Gregory and the Invention of Professional Medical Ethics and the Profession of Medicine,* Hingham, MA: Kluwer, 1998.

McCullough, Laurence B. (ed.), *John Gregory's Writings on Medical Ethics and Philosophy of Medicine,* Boston, MA: Kluwer, 1998.

McCullough, Laurence B. "Virtues, Etiquette, and Anglo-American Medical Ethics in the Eighteenth and Nineteenth Centuries," in *Virtues and Medicine: Explorations in the Character of Medicine*, ed. Earl E. Shelp, Boston, MA: D. Reidel, 1985.

Moore, Theophilus. *Marriage Customs and Modes of Courtship of the Various Nations of the Universe. Remarks on the Condition of Women, Penn's Maxims, and Counsel to the Single and Married, &c. &c.* 2nd ed. London: [printed for John Bumpus, Holborn] Hamblin, 1820.

O'Brien, Eoin (ed.) *The Charitable Infirmary, Jervis Street, 1718–1987: A Farewell Tribute*, Dublin: Anniversary Press, 1987.

O' Meara, Barry Edward. *Napoleon in Exile, or, A Voice from St. Helena: The Opinions and Reflections of Napoleon on the Most Important Events in His Life and Government, in His Own Words*. London: Printed for W. Simpkin and R. Marshall, 1822.

Parent-Duchatelet, Alexandre Jean Baptiste. *De la Prostitution dans la ville de Paris, considérée sous le rapport de l'hygiène publique, de la morale et de l'administration; ouvrage appuyé de documents statistiques puisés dans les Archives de la Préfecture de Police; et précédé d'une Notice sur la vie et les ouvrages de l'auteur* par Fr. Leuret. Paris: J.-B. Baillière, 1837.

Paris, J.A. *Pharmacologia, Corrected and Extended, in Accordance with the London Pharmacopoeia of 1824, and with the Generally Advanced State of Chemical Science*. London: W. Phillips, 1825.

Percival, Thomas. *Essays Medical, Philosophical, and Experimental*. London: Printed by W. Eyres for J. Johnson, London, 1788–1789.

Percival, Thomas. *Medical Ethics; or, a Code of Institutes and Precepts adapted to the Professional Conduct of Physicians and Surgeons…* Manchester: Russell, 1803 (modern reprint by the Classics of Medicine Library, Birmingham, AL: 1985).

Percival, Thomas. *Medical Ethics; or, a Code of Institutes and Precepts, Adapted to the Professional Conduct of Physicians and Surgeons…by the late Thomas Percival…*, London: W. Jackson, 1827.

Pernick, Martin. The Patient's Role in Medical Decisionmaking: A Social History of Informed Consent in Medical Therapy," in: President's Commission for the Study of Ethical Problems in Medicine and Biomedical and Behavioral Research, *Making Health Care Decisions: The Ethical and Legal Implications of Informed Consent in the Patient-Practitioner Relationship; Volume Three: Appendices: Studies on the Foundations of Informed Consent*, Washington, DC: U.S. Government Printing Office, 1982: 1–35.

Phelan, Denis. *A Statistical Inquiry into the Present State of the Medical Charities of Ireland: With Suggestions for a Medical Poor Law, by Which They May be Rendered Much More Extensively Efficient*. Dublin: Hodges and Smith, 1835.

Pickstone, J.V. and Butler, S.V.F. "The Politics of Medicine in Manchester, 1788–1792: Hospital Reform and Public Health Services in the Early Industrial City," *Medical History* 28 (1984): 227–49.

Place, Francis. *Illustrations and Proofs of the Principle of Population: Being the First Work on Population in the English Language Recommending Birth Control* (1822); London: Routledge/Thoemmes, 1994.

Pliny the Elder. *The Natural History* (translated with notes and illustrations by John Bostock and H.T. Riley). London: Taylor & Francis, 1855.

Porter, Roy. *The Greatest Benefit to Mankind,* New York: W.W. Norton, 1997.

Ramadge, Francis H. "London Medical Society: Expulsion of Dr. Ramadge," *Lancet* 2 (1830–31): 251–252; *London Medical and Surgical Journal* 7 (1831).

Ramadge, Francis H. "[John] Long the Quack" *Lancet* 2 (1830–31); No. 398 (16 April 1831): 90–93; No. 400 (30 April 1831): 154–156.

Rivett, Geoffrey. *The Development of the London Hospital System, 1823–1982*. London: King's Fund, 1986. [Section: Voluntary Hospitals" accessed http://www.nhshistory.net/voluntary_hospitals.htm (accessed January 23, 2005.)]

Schleiner, Winfried. *Medical Ethics in the Renaissance*. Washington, DC: Georgetown University Press, 1995.

Smith, John Gordon. *An Analysis of Medical Evidence Comprising Directions for Practitioners in the View of Becoming Witnesses in Courts of Justice, and an appendix of Professional Testimony*. London: T. & G. Underwood, 1825.

Smith, Gordon. "Drs. Gordon Smith, Ryan, and A. Thomson," *Lancet* 1 (1830–31): 72–73.

Styrap, Jukes. *A Code of Medical Ethics*. London: J. & A. Churchill, 1878.

U.S. National Commission for the Protection of Human Subjects of Biomedical and Behavioral Research. *The Belmont Report: Ethical Principles and Guidelines for the Protections of Human Subjects*. Washington, DC: U.S. National Commission, 1978.

Veatch, Robert M. *Disrupted Dialogue: Medical Ethics and the Collapse of Physician-Humanist Communication (1770–1980)*. New York: Oxford University Press, 2005.

Velpeau, Alfred Armand L.M. *Traité complet de l'art des accouchemens*. Paris: Baillière, 1835.

Vinten-Johansen, Peter, Brody, Howard, Paneth, Nigel, Rachman, Stephen and Rip, Michael. *Cholera, Chloroform, and the Science of Medicine: A Life of John Snow*, New York: Oxford University Press, 2003.

[Wakley, Thomas.] [Editorial]. *Lancet* 1 (1838–39): 706.

[Wakley, Thomas.] [Editorial]. *Lancet* 2 (1831–32): 92, 21 April 1832.

[Wakley, Thomas.] "Letter from Dr. Ryan," *Lancet* 1 (1831–32): 222–26.

[Wakley, Thomas.] "Ramadge v. Wakley," *Lancet* 2 (1831–32): 408–409.

[Wakley, Thomas.] "Ramadge v. Ryan et al.," *Lancet* 2 (1831–32): 408–409.

[Wakley, Thomas.] "Tweedie v. Ramadge," *Lancet* 2 (1830–31), no. 419 (10 Sept. 1831).

Watson, Richard. *An Apology for the Bible in A Series of Lectures Addressed to Thomas Paine*. Boston, MA: James White, 1796.

Widdess, J.D.H. *The Royal College of Surgeons in Ireland and its Medical School, 1784–1984*, Dublin: Royal College of Surgeons in Ireland, 1984.

Index

C

R

S

MIX
Papier aus verantwortungsvollen Quellen
Paper from responsible sources
FSC® C105338

If you have any concerns about our products,
you can contact us on
ProductSafety@springernature.com

In case Publisher is established outside the EU,
the EU authorized representative is:
Springer Nature Customer Service Center GmbH
Europaplatz 3, 69115 Heidelberg, Germany

Printed by Libri Plureos GmbH
in Hamburg, Germany

INTERNATIONAL ENVIRONMENTAL LAW AND POLICY IN AFRICA

ENVIRONMENT & POLICY

VOLUME 36

The titles published in this series are listed at the end of this volume.

International Environmental Law and Policy in Africa

by

Beatrice Chaytor

Programme Director,
Foundation for International Environmental Law and Development,
London, U.K.

and

Kevin R. Gray

Fellow in Public International Law,
British Institute of International and Comparative Law,
London, U.K.

Springer Science+Business Media, LLC

A C.I.P. Catalogue record for this book is available from the Library of Congress.

ISBN 978-90-481-6287-1 ISBN 978-94-017-0135-8 (eBook)
DOI 10.1007/978-94-017-0135-8

Printed on acid-free paper

TABLE OF CONTENTS

Development of Law and Policy Instruments

FOREWORD

C.O. OKIDI[1]

I welcome the opportunity to prepare a Foreword to the book on Environmental Policy and Law in Africa, edited by Kevin R. Gray and Beatrice Chaytor. It is a pleasure to do that because the book is a contribution to the cause of capacity building for development and implementation of environmental law in Africa, a goal towards which I have had an undivided focus over the last two decades.

There is still some belief in and outside Africa that for developing countries in general, and Africa in particular, development and implementation of environmental law is not a priority. This belief prevails strongly in many quarters of the industrialised countries. In fact, the view is held either out of blatant ignorance or by some renegade industrialists who fail to appreciate Michael Royston's 1979 thesis that *Pollution Prevention Pays*.[2] That group, for obvious reasons, must have their correspondent counterparts in Africa to provide hope that industries rejected as derelict in the West or inoperable due to rigorous environmental regulation, can find homes to which they can escape and dump their polluting industries.

Moreover, the group forgets or is unaware that the environmental agenda, which is organised around the pre-*Stockholm Declaration* narrow scope of pollution control and management of urban squalor, was firmly rejected by developing countries. Not only were the concepts leading to a narrow approach but it was bound to lead to strengthened command and control legal mechanisms with very little or no consideration for dynamic management and development consideration.

The fast action by the Secretary General of the United Nations Conference on Human Environment in constituting the Founex experts that packaged the management and development concepts on the road to Stockholm saved the Conference. It is arguable, therefore that the Brundtland Commission, and its compelling direction of the Rio Conference, built on the agenda of the developing countries. None of the African countries has rejected the pursuit of the general objective of sustainable development, which became the organizing concept and main theme of Agenda 21 and Rio Principles. The World Summit on Sustainable Development (WSSD) scheduled for Johannesburg in August 2002 *should* properly be on the agenda of the developing countries. What has happened in

---

[1] The views expressed herein are those of the author.

[2] (Oxford: Pergamon Press; 1979).

Africa may be register only on a small scale but its impacts are nevertheless dramatic. Out of the 53 African country members of the OAU, about thirty have already adopted framework environmental legislation, albeit at different levels of sophistication. These countries are discretely endeavouring to establish apex institutions and to bring in line the sectoral and functional agencies. It is also well known that the countries are struggling in the face of limited human and institutional capacity to streamline the institutional arrangements for efficient implementation. The Rio Conference left an expectation that the rich countries would promote availability of resources to support implementation of *Agenda 21*. One may assume that the WSSD process will take stocks of the extent to which that objective has actually been realized.

A decisive lesson is obtainable from the pilot project on Environmental Law and Institutions in Africa that has been implemented systematically in seven African countries from 1995 to 2001,and with funding from The Dutch Government. Through the methodology operationalising the concept of capacity building and broadly based public participation, the African project countries have demonstrated a commitment to development and implementation of environmental law. Built on national motivation and drive, the project has confirmed that African countries need opportunities, information and resources to support human and institutional capacity building. With these forms of support, development and implementation of environmental law in Africa, will doubtlessly match their aspiration commonplace in conference diplomacy.

Public officials as well as the civil society organisations have shown keenness to participate actively in development and implementation of legal norms in such highly technical areas as biodiversity, biotechnology, climate change, protection of ozone layer, in addition to the traditional fields of environmental law. There is also widespread interest in procedural rights in the environmental field and these are increasingly linked to the quest for democratic governance. Use of judicial mechanisms to address environmental ills in Africa has caught some western industrialists by surprise as was in the Titanium case in Kenya.[3] The industrialists found that the old and tired slogans such as the argument that environmental enforcement will hamper development just do not sell well any more. More examples could be cited from other African countries, particularly from Uganda, South Africa and Tanzania. With increased awareness, information and resources, a critical mass of environmental experts may be built in each African country and their cooperation with the civil society organizations will lead to effective environmental management in Africa.

It is in the context of information for capacity building, that the present volume on environmental law and policy in Africa is a timely contribution. First, the range of topics is diverse, and we have submitted that no field of environmental law is beyond the interest of African environmental law experts. Secondly, the background of the authors is wide. About half of the thirteen

---

[3] *Rodgers Muema Nzioka & 2 others* v. *Tiomin Kenya Ltd in the High Court of Kenya at Mombasa*, Civil case No. 97 of 2001.

authors are Africans. The non-Africans are people with significant experience and dedication to the study of environmental management in Africa. This brings together diverse background and comparative perspectives from which African and non-African readers can extrapolate principles applicable to the continent. But each of the papers examines how the issues relate to Africa.

This book should symbolically encourage more students of environmental law to contribute to the pool of information, which will, in turn, support its development, understanding and implementation in the continent. These initiatives will continue to systematically explode the myth that African countries have no interest in environmental law.

Prof. C.O. Okidi
Nairobi, Kenya

25th April, 2002

INTRODUCTION

A few years ago, we discussed the possibility of a publication that would examine the nature and content of environmental law and policy in Africa. It appeared that other developing country regions, such as Latin America and Asia, were given more scholarly attention in this field, perhaps attributable to their levels of development. Analysis of environmental law and policy in individual African countries existed in journal articles and through project case studies, but a comprehensive look at African environmental law and policy was systematised in few scholarly works.Our primary interest in the book was to uncover, mainly through the contributions of African environmental legal specialists, how international environmental law commitments were being implemented in individual countries. The response to our call for contributions was quite enthusiastic. We were impressed with the number of authors who had first hand experience, governmental or otherwise, with implementation. Moreover, they provided empirical insight into the difficulty of meeting international obligations and identifying where environmental management registered on the list of government priorities. This book is divided into 3 sections. Section 1 examines how Multilateral Environmental Agreements (MEAs) are implemented in African countries. Section 2 deals with the challenges facing African environmental authorities in managing the continent's natural resource sectors. In Section 3 developments of specific legal and policy instruments are assessed in meeting environmental protection needs.

The chapters in the first section of the book provide observations on the multilateral environmental agreement (MEA) negotiation processes and the extent to which African countries are making an impact. There appears to be contrasting experiences with each MEA. For instance, the limited participation of the African Group in the Climate Change regime stands in marked contrast to the influence of African States evident in the development of the *Desertification Convention* and the *Bamako Convention.* A basic premise of the book is that African countries will increasingly have to meet more complex international commitments. Treaty obligations will become more inter-related, thereby engaging Africa further into the environmental community. Kevin Gray and Joyeeta Gupta discuss the fledgling Climate Change regime where African countries are not currently subject to emission reduction targets. However, this is expected to change as the treaty matures. African countries will be seen to make such commitments to reduce energy consumption and greenhouse gas production in order to take advantage of the economic opportunities arising from participating in the climate change regime. Moreover, African economic development will require a degree of long-term sustainability that participation in the treaty can help foster. Another MEA with few specific commitments is the *Desertification Convention.* Here, implementation of the Convention's objectives is a matter of pure survival. However, implementation appears to be greatly influenced by the level of financial assistance and technological transfer passed from industrialised countries. Elisabeth Correll's contribution notes that the lofty objectives of the *Desertification Convention* will only be met by increased development assistance in addition to the political will by African countries. Lacking a proscribed financial mechanism, the effect of the *Desertification Convention* may be seriously limited. For transboundary waste management, the chapter by Beatrice Chaytor and Mita Manek traces the development of modern hazardous waste management through customary international law principles and culminating in the adoption of two treaties: *Basel* and *Bamako*. Dissatisfaction with the *Basel Convention* prompted African countries to develop the *Bamako Convention* (under the auspices of the Organisation of African Unity), which prohibits the imports of hazardous wastes into any African Contracting Party, in the first instance and establishes a Basel style management system for wastes generated within Africa. While both Conventions broadly adopt similar approaches to the control of transboundary movement of hazardous wastes, Bamako has a more stringent regime for monitoring and enforcement. Yet with the ban on the movement of wastes for recycling between OECD and non-OECD countries, there may be an argument for developing the synergies between the two treaties and ultimately merging them. The chapter on *CITES (Convention on the International Trade in Endangered Species)* also looks at a well-developed MEA regime where African countries have been principal participants in its formulation and design. Michael Kidd and Michael Cowling focus on the management regime developed under *CITES* for the African Elephant and the regulation of the ivory trade, suggesting that the *CITES* regime has been unsuccessful in managing the elephant. The chapter traces the developments within CITES to delist the elephant, particularly the efforts of key Range States which have been successful in their conservation of the African elephant, to recommence exports of ivory. It is argued that listing African elephants on Appendix II will enable the sale of elephant hides, the export of hunting trophies under programmes like CAMPFIRE and the export of live animals in the place of culling. Such measures could operate alongside a zero quota on ivory.

The section covering natural resources management assesses how governments aim to achieve the optimum balance between exploitation of the resource and sustaining the environmental integrity of the area. What will be important for African countries is exploring mechanisms to take advantage of the commercial opportunities that MEAs present. Rachel Wynberg addresses the question of the inevitability of the commercialisation of biodiversity, arguing that African countries should find ways to procure the financial benefits from the exploitation of indigenous species while balancing sustainable development of the genetic resources and traditional practices behind the agricultural and pharmaceutical products. Kent Nnadozie explores the relationship between the oil and gas industry and the environment in Nigeria. The chapter highlights the fact that the industry is entering a new era of legal regulation as exploration and production activities are increasingly subjected to international law, national requirements as well as civil society pressures. The heightened profile of environmental issues associated with the oil industry in Nigeria has produced a range of voluntary and non-regulatory approaches alongside certain regulatory and monitoring institutions to better enforce existing law and policy. Consideration is also being given to effective economic incentives for the use of advanced technology and adequate infrastructure to reduce environmental damage. Similarly, the chapter on Policing Industrial Pollution in Nigeria by Ali Ahmad examines the previous abuses by the private sector and how these relate to a limited enforcement regime that is generated by a low level of political will to regulate pollution. In the chapter on Tanzanian forest law by John Lovett, consistent international influence on the country's forest law is highlighted, tracing the establishment of forest reserves by colonial powers to the development and implementation of forest law and policy. Each major change in administration in Tanzania left its own legacy of law, policy and land use. Tanzanian forests are among the most valuable biodiversity resources in the world. Approaches to management in forest policy reflect changes in social values. In response to international developments, there is an increased emphasis on conservation of biodiversity. At the same time, there is concern for the equitable inclusion of people affected by forest management in forest policy decision-making. Lovett argues that the conflict over forest policy has revolved around the requirements of resource protection and local needs. The two factors affecting the nature of African fisheries legislation are their diversity and changes in the industry and the institutions regulating them. Ali Mekouar presents the legal and regulatory framework governing fisheries management in Africa through their provisions on management and surveillance and the special problems of small-scale fisheries. Although most African countries have adequate laws and policies in place to ensure the long-term sustainability of their fisheries, implementation remains a problem. The FAO's *Compliance Agreement* and its *Code of Conduct for Responsible Fisheries* and the *UN Straddling Stocks Agreement* has had a positive impact on fisheries management, ensuring that conservation and environmental considerations are gradually incorporated into the legal framework. Regional co-operation in management and surveillance is essential and improving, due to the shared and interdependent fish stocks between countries. Transboundary management of fisheries resources is also treated by the contribution from Cecil Machena. He reviews the efforts of local artisan fishers in cooperation with the Zimbabwean and Zambian government to sustainably manage fish resources. A local level strategy of implementation was employed differing from the previous top down approach to regulation practiced by the national governments. Institutions were created in partnership so that concession areas are jointly managed. Community management institutions enhance the level of public participation, stimulating greater incentives to comply with regulatory requirements. Although bilateral intergovernmental cooperation was essential for the system to work, the integration of local fishers and other resource users rendered the project successful, since commonalities between these participants transcended the natural boundaries.

The contributions in the final section focus on how particular, and sometimes local, ecological problems are addressed by using a range of policy instruments based on national and international commitments. Civil society participation has become a crucial element in environmental policy making in recent years. Kofi Kufuor's chapter on Popular Participation in Environmental Decision Making in Ghana explores the development and design of the framework governing popular participation in environmental decision-making in Ghana. Avenues for citizen and NGO participation became current in Ghana in the mid-1970s after the 1972 UN Conference on the Human Environment, with the establishment of the Environment Protection Council. Those avenues however, appeared to be accessible only by a small number of societal elites. Fundamental change came in the late 1980s with the emergence of an identifiable environmental civil society in Ghana created by the move towards liberal democracy in the country. Links were built between global environmental organisations such as Friends of the Earth and local environmental organisations, enabling the latter to grow. In turn, environmental legislation in Ghana began to accord greater importance to the role of individuals and local communities in providing innovative solutions to key environmental problems. Significantly the 1992 Constitution enshrines provisions on citizen participation in environmental decision-making. Kufuor argues that these constitutional provisions are not mere policy guidelines, but may be used to support the creation of specific rights by citizens to participate in environmental decision-making. It is conceded however, that civil society in Ghana has yet to develop into the environmental movement necessary to activate these rights. Public participation is an essential component of another policy/legal instrument - Environmental Impact Assessment (EIA). Rolph Poyet's chapter on EIA in the Seychelles offers a detailed glimpse on the establishment of a EIA regime. Its ability to address the impact of fledgling tourism development industry on the environment demonstrates that locally based decision-making can result in the best possible sustainable development outcomes. Mohammed Bekhechi provides a comprehensive survey of the national EIA regimes in Africa, demonstrating how it is becoming a critical tool to pre-empt unsustainable patterns of development. However, similar problems of limited enforcement and capacity pervade many African country experiences. The need for harmonisation of EIA legislation in Africa and the universal requirement of public participation are seen as the key ingredients for successful EIA implementation in Africa. This theme is picked up in the chapter by Carl Bruch analysing the potential for an environmental governance treaty for Africa. Using the *Aarhus Convention* as an example, Bruch explores the feasibility of developing a similar instrument for Africa that enshrines the three pillars of environmental democracy; access to information; public participation in decision-making and access to administrative or judicial redress. This is seen as not necessarily providing a panacea for the continent but a conduit towards better informed and therefore improved government planning and decision-making. Alternative policy instruments for environmental management are discussed by Pendo Susswein. The majority of current environmental policies in the Southern African Development Community (SADC) region are based on regulatory command and control instruments.

However decentralisation, the advent of globalisation and free market economics have enabled most countries in the sub-region to consider a broader range and mix of regulatory measures and economic instruments to facilitate their transition to sustainable development. It is argued that institutional failure, lack of enforcement, duplication and fragmentation of authority have contributed to weaknesses in implementation of command and control instruments. The chapter examines the institutional, legal and policy frameworks for the use of such market based instruments in the consumption, management and conservation of water, marine and soil resources and the industrial sector in the SADC region. Ultimately, the benefits to be gained from market-based instruments depend largely on developing coherent laws and policies and strengthening the institutional structures to enforce them.

The book's title, while hinting at breadth, can only cover a limited number of countries and environmental issues. It would have been an enormous task to encompass analyses of the status of law and policy in each African country. However, we hope that the book makes three positive contributions. First, that it continues to support the existing pan-African scholarship in the field of environmental law, secondly that it encourages new scholarship in the field, and thirdly, that such existing and new scholarship will assist in providing the vital underpinning to informed law- and policy-making on environment in African countries. As the economies of Africa develop, there will be resulting development of new environmental laws and regulations. Further study of the evolution of this process is needed in order to determine the extent to which this effects real change in the continent. This book testifies to the amount of knowledge and scholarship available within and outside Africa on this topic.

Beatrice Chaytor
Kevin R. Gray

DRYLAND DEGRADATION - AFRICA'S MAIN ENVIRONMENTAL CHALLENGE:

International activities from the 1970s to the 1990s and the future of The United Nations Convention to Combat Desertification

ELISABETH CORELL[1]

Desertification is arguably Africa's main environmental challenge. In fact, it has provided such a formidable challenge that Africa has not been able to meet it alone. Since the 1970s, numerous international efforts have been devoted to designing responses to this environmental problem. This chapter traces the successes and failures of these international activities.

The 1970s witnessed the United Nation Conference on Desertification (UNCOD), and resulted in the 1977 *Plan of Action to Combat Desertification* (*PACD*). However, as evidence grew that the *PACD* was not being implemented, desertification was placed on the agenda of the 1992 United Nations Conference on Environment and Development (UNCED). Chapter 12 of *Agenda 21* is devoted to desertification and drought. At UNCED, delegates also decided to negotiate a CCD. The 1994 *United Nations Convention to Combat Desertification*[2] was negotiated in a little over a year and is the only legally binding instrument to address dryland management issues at the international, regional, sub-regional, national, and local levels.

The chapter begins by presenting the environmental issue of desertification. It then describes the history of international attempts at addressing desertification, followed by an examination of the negotiations and provisions of the *CCD*. The last part of the chapter discusses the future of the *CCD* and particularly the possibilities for its implementation. The analysis concludes that implementation of the *CCD* in Africa is contingent on three factors: (1) funding; (2) political commitment; and (3) the bottom-up approach.

DESERTIFICATION – AN ENVIRONMENT AND DEVELOPMENT ISSUE

Similar to other major global environmental issues, such as biodiversity loss and climate change, desertification is a complex process that involves numerous physical, social and economic factors. Simply put, desertification is environmental degradation in and around arid lands. More specifically, the *CCD* defines desertification as "land degradation in arid, semi-arid and dry sub-humid areas resulting from various factors, including climatic variations and human activities."[3] These arid, semi-arid and dry sub-humid areas – covering 40 percent of the world's land surface – are referred to as "susceptible drylands".[4]

In the 1992 *World Atlas of Desertification*, the United Nations Environment Programme (UNEP) estimated that desertification affected 70 percent of all drylands used for agricultural purposes in the geographical regions of Africa, Asia, Australia, North, Central and South America, and the area of the former Soviet Union.[5] It threatened about one sixth of the world's population – 900 million people.[6] By 1997, this figure had risen to one billion.[7] Asia suffers most from desertification in terms of total land area affected, while Africa and North America experience the greatest severity of degradation in their drylands.[8]

The impacts of desertification can include: soil erosion; changes in vegetation species and loss of biodiversity; salinisation; reduction in plant growth; reduced grazing areas; reduction in soil fertility leading to loss of agriculturally productive land; and may lead to concerns about food security and migration. However, the extent of the impacts, remedies and costs of desertification continue to be debated and arguments regarding its increase and existence should be approached with caution. While some researchers argue that desertification is a misused term – having become a powerful development narrative for policy-makers – they also caution observers not to dismiss or underestimate the consequences of environmental degradation for the peoples eking out a living in the drylands.[9]

Causes and impacts of desertification

The causes of desertification are contested among scientists. The *CCD* notes "desertification is caused by complex interactions among physical, biological, political, social, cultural and economic factors".[10] While most scientists agree that human activities contribute to desertification, opinions diverge about the causes of those activities. Most scientists group the causes into two categories: natural ("physical, biological") and human ("political, social, cultural and economic"). Some emphasise that desertification is largely the result of natural causes; while others claim that human activity is mainly responsible. A synthesis perspective holds that human factors combined with natural causes lead to desertification.

Natural causes may include: recurrent periods of drought; rainfall variability; seasonal temperature changes; and quantitative and qualitative changes in the vegetation cover, all of which are sometimes loosely sorted under the heading of climatic factors.[11] From this perspective, desertification can be explained, for instance, by the fact that dryland climates are prone to unusually high levels of variability, particularly in precipitation levels.[12] Parts of dryland Africa have suffered a major downwards shift in rainfall since the end of the 1960s, and this lesser rainfall and its greater variability have increased vulnerability to natural resource degradation since it becomes more difficult for ecological and social systems to withstand stress.[13]

Global climate change may also have implications for desertification. For example, some climatologists believe that climate change and global warming are responsible for periods of drought in Sahelian Africa.[14] However, the precise link between desertification and climate change is the subject of ongoing research, as is the impact of human activities on local and regional climate conditions.[15]

The *human activities* commonly believed to cause desertification are overgrazing, over-cultivation, deforestation and poor irrigation management.[16] Overgrazing, being the excessive livestock grazing on the same piece of land for too long, and over-cultivation meaning the intensive cultivation of land without allowing for natural regeneration, both contribute to land degradation. Deforestation is the clearing of natural vegetative cover and woodlands – often to gain firewood. Lastly, poor irrigation management can lead to waterlogging and salinisation, and thus reduce soil productivity.[17]

The driving forces behind these practices derive from economic, social and political structures and realities. These include: population change; underdevelopment; internationalisation of the economy; colonial legacies; inappropriate technology and advice; and, civil unrest and war.[18] Indeed, the *CCD* also considers "the impact of trade and relevant aspects of international economic relations on the ability of affected countries to combat desertification", and that there are "interrelationships with important social problems such as poverty, poor health and nutrition, lack of food security, and those arising from migration, displacement of persons and demographic dynamics".[19]

The impacts of desertification can be extremely serious. They are divided in this chapter into physical, social and economic impacts. The *physical impacts* can be observed in soils and vegetation and include: biodiversity loss; soil erosion and salinisation; loss of arable land; reduction in plant growth; sedimentation of water bodies; and the reduction of biomass.[20] Not surprisingly then, areas experiencing these types of effects also suffer serious *social and economic impacts*. For the pastoralists, nomads and agriculturalists living in affected drylands, this means reduced or eliminated grazing areas and the loss of agriculturally productive land. In 1992, UNEP estimated that every year nearly six million hectares of previously productive land in arid, semi-arid and dry sub-humid areas lost its capacity to produce food.[21] With a growing world population, loss of arable land for food cultivation is a global concern since it ultimately affects world food security.[22] Desertification can also lead to population migration away from degraded areas, resulting in both increased urbanisation and migration across borders, which in turn can lead to the loss of traditional lifestyles.

With such wide-ranging causes and impacts, effective responses to desertification cannot be confined to environmental considerations only. The African desertification problem has an obvious connection to social and economic development. In the formulation of policy to curb desertification, the politics of development have become an influential factor. Discussions of the desertification concept have become battles between conflicting understandings of the problem and about different perceptions of how the term should be used.

Understanding and defining desertification

Scientists have been moving towards a more sophisticated understanding of desertification. The development has been described as beginning with an age of awareness, followed by an age of misconceptions in the 1970s and early 1980s. Then, ten years after *UNCOD*, came an age of doubt, followed by the "dawning of a new realism."[23] For instance, some early desertification scientists assumed desert-like conditions in arid regions to be a constant state and often blamed local people's ignorance of proper land management for the degradation of the land.[24] The understanding of arid ecosystems has since become more advanced – dryland areas are now understood to posses a remarkable resilience. Arid ecosystems are subject to changing conditions and local populations often know how to adequately cope with them.[25]

The difficulties in identifying the precise causes and effects of desertification and the debate over the relative importance of human or climatic factors has had major implications for the way in which desertification is defined. A review of the definitions of desertification used at the international level from 1976 to 1994 reveals a movement from a focus on natural causes in the early 1970s, to prominence given to both natural and human causes in the late 1970s and the early 1980s. In the early 1990s, the focus shifted to human causes. However, the 1992 UNCED and 1993-94 *CCD* negotiations again highlighted *both* natural and human causes.[26] Thus, according to the *CCD*, desertification is the result of "various factors, including climatic variations and human activities".[27]

The inclusion of climatic factors as a partial cause of desertification in the *CCD's* definition may have been intended to soften the political consequences of a definition that places all the responsibility on human activity. It may be argued that if the responsibility for dryland degradation could be placed entirely with local populations and administrators, there would be little incentive for donor countries to contribute financial aid to mitigating desertification. If, on the other hand, the locals were victims of climatic factors, and their land management practices exacerbated the problem, there would be more reason for donor countries to fund projects to change or re-direct such human activities. Scientists involved in developing the early 1990s definition used a formulation stating that desertification resulted from the "mainly adverse human impact", which in their view also allowed for climatic factors as causes of desertification.[28] However, the decision-makers that negotiated and agreed to the legally binding text had to consider the political implications of the definition they chose. As a result, both human activities and climatic variations were expressly mentioned. While the prevailing scientific view finds human activity to be the main cause of desertification, the re-introduction of climatic factors in the *CCD* definition gives the impression that they are given equal weight to human factors as causes of desertification.

Land degradation, not desertification

The debate over definitions emphasises that there are different meanings attached to the notion of desertification. Some use the image of the expanding desert, while others use desertification to describe the more substantial problem of dealing with loss of land productivity.[29] Due to the concept's ambiguity, numerous writers have criticised the term desertification for engendering muddled thinking about the exact problems to be addressed and have argued that the term should be abandoned because it combines too many different processes that should be distinguished and dealt with separately.[30] Many scientists today prefer using the term *land degradation* to describe the results of deteriorating processes. In this view, to address the problems of land degradation, *dryland management* should be implemented in affected areas.

However, despite conflicting interpretations of the concept of desertification and demands to abandon the term for alternatives that better reflect the complexities of dryland problems, the term stubbornly remains in use. The term's persistence may result from efforts to broaden it to include a series of problems in drylands. Some observers suggest that the plethora of definitions and the incidental inclusion of a growing number of geographic areas may be the result of efforts to mobilise extra funding for desertification and to increase the number of potential beneficiary nations.[31]

The term's durability may also stem from the fact that desertification has become a catch phrase. For example, one scientist suggested that it was a "a stroke of political genius" to include desertification in the title of the first international conference – UNCOD – since it "carried the connotation of disaster and required no explanation."[32] The image conveyed by the word has perhaps been regarded as more important than using a term that communicates a more accurate picture of the problem. Desertification conveys an urgent need for action, and has been used to market desertification on the international political stage to attract the attention of donors. In their efforts to make the issue known, international institutions, environmental advocates and the media have perpetuated the powerful image of wandering sand dunes to visualise the problems of dryland degradation. These actors use the term desertification in a dramatic, emotional context to maximise people's attention and focus on humans, rather than the landscape.[33] According to some observers, this image has inflated the dimensions and scope of desertification. They speak of a "myth" created by the United Nations Environment Programme, among others.[34]

To conclude, while scientists wanted to avoid using it as early as the 1970s, the term desertification remains an essential catchword to illustrate dryland degradation. Desertification is "the most publicised form of land degradation" and in the public, and many decision-makers' minds, desertification and advancing sand dunes continue to be confused.[35] While some believe the term is a greater burden than an aid to understanding dryland degradation and has ceased to be useful, others argue that it is still better than the dull "land degradation" –

"desertification" has a political appeal that land degradation does not have. Although desertification is not limited to the developing world or the poorest people, it was the drought crisis in Africa that became the focus for concern and inspired the United Nations to initiate international action.

THE HISTORY OF ADDRESSING DESERTIFICATION

The international community has discussed desertification several times, first at the 1977 United Nations Conference on Desertification (UNCOD), then at the 1992 United Nations Conference on Environment and Development, and finally during the 1993-94 negotiations of the *CCD* and its subsequent meetings. This section examines these events, and their documents and strategies.

In the 1950s and 1960s, a host of international agencies such as UNESCO, FAO, WMO, ICSU and IUCN, conducted research and collected a wealth of information on environmental constraints in arid zones.[36] However, desertification attracted little attention at the 1972 United Nations Convention on the Human Environment (UNCHE) conference. It was only when the Great Sahelian Drought[37] (1968-1973) became an international issue some years later that most of the nations affected by various degrees of land degradation viewed desertification as a process that deserved full international attention.

While the Sahelian drought served as a catalyst to elevate desertification to the status of a "global" problem, there were nevertheless differing views on the desirability of holding an international conference on the subject. Developing countries maintained that desertification could not be solved nationally or regionally but must be approached through a global strategy. They also argued that such a strategy was appropriate since the most affected states lacked the financial and technical resources needed to effectively combat desertification. Industrialised states, on the other hand, did not agree that an international conference was needed or would be effective – even if some did agree on the need for a world strategy for desertification. They argued that creating a new institution would not solve the problem and that the UN system already had adequate capacity to consider all aspects of desertification.[38]

However, the Sahelian drought had galvanised world opinion in favor of the need to "fight desertification" and what one observer called a "global desertification movement" demanded a fully-fledged UN conference to bring political weight to addressing the problem.[39] Some of the rhetoric used to promote this view was extreme. For example, a UN document published in preparation for the UNCOD stated that "if desert spreading is not stopped, two African countries will completely disappear within the next decade".[40]

Within the UN system, member states' concern for desertification and the movement towards convening an international conference was expressed through a number of resolutions and decisions adopted by the United Nations General Assembly (UNGA), the Economic and Social Council (ECOSOC), UNEP and the United Nations Development Programme (UNDP). *UNGA resolution 3337*

(XXIX) of 17 December 1974 called for a UN Conference on Desertification. Delegates from Burkina Faso (then Upper Volta), which had been especially badly affected by the Sahelian Drought, introduced this resolution.[41] The resolution delegated the responsibility for preparing the conference to the newly established UNEP. To ensure that "all available knowledge in this area is fully utilized", UNEP Executive Director Mostafa Tolba – who subsequently served as the Secretary-General for UNCOD – then organised extensive scientific preparatory work, involving numerous representatives from the world scientific community, to form a basis for the Conference.[42]

The 1977 United Nations Conference on Desertification

Delegates from 95 States, 5 UN agencies, 8 intergovernmental organizations, and observers from 65 non-governmental organizations (NGOs) attended the UN Conference on Desertification (UNCOD), held in Nairobi, Kenya from 29 August to 9 September 1977.[43] The Conference had three central aims: (1) to increase global awareness about desertification; (2) to collect together all the available scientific and technical knowledge on the problem and its possible solutions; and (3) to instigate a program to combat desertification.[44] The first two goals were, in part, fulfilled through the scientific preparations for UNCOD and resulting documents distributed to the conference participants. The negotiation of the *Plan of Action to Combat Desertification* (*PACD*) accomplished the third objective and provided the Conference's most important outcome. The immediate goal of the Plan was:

> "to prevent and to arrest the advance of desertification and, where possible, to reclaim desertified land for productive use. The ultimate objective is to sustain and promote, within ecological limits, the productivity of the arid, semi-arid, sub-humid and other areas vulnerable to desertification in order to improve the quality of life of their inhabitants."[45]

The *PACD* was to reinforce, integrate and coordinate national, regional and international actions against desertification outside and inside the UN system, and to encourage the creation of new programmes. Its drafters envisioned an international campaign against desertification and, through the Plan, encouraged developing countries subject to desertification to include such a vision in their national development plans. The *PACD* addressed social problems in land-use management, stating that prevention would cost far less than the cure.[46] A central theme was to immediately apply and adapt "existing knowledge, particularly in the implementation of urgent corrective measures against desertification, in educating the people and the affected communities to an awareness of the problem".[47] The *PACD* contained 28 recommendations for national, regional and international action. These recommendations called for actions that sought to: (1) arrest and reverse the environmental processes of desertification; (2) establish ecologically appropriate, productive and sustainable land uses; and (3) secure the social and economic advancement of the communities affected.[48]

The *PACD* was to be implemented by the year 2000 - the period 1978 to 1984 was chosen for the implementation of immediate action. At the end of that seven-year period, a first general assessment of progress was to be made.[49] The *PACD* assigned responsibility for implementation to governments, and the overall responsibility of following up and coordinating the implementation of the *PACD* to UNEP. The UNCOD decisions also resulted in the formation of two groups: (1) the Interagency Working Group on Desertification, which was responsible for providing guidance to UNEP regarding overall implementation of the *PACD* and coordinating participation of the UN system in the world actions; and (2) the Consultative Group for Desertification Control (DESCON), which assisted in mobilising financial resources for combating desertification.[50]

While the UNCOD Conference successfully placed desertification on the international environmental agenda for the first time, implementation of the *PACD* was insufficient to accomplish its goals. UNEP has noted that several global conferences and studies identified the following reasons for the failure of the *PACD*: (1) low priority by funding agencies; (2) lack of external financial and technical assistance to allow developing countries to cope with the problem; (3) lack of integration of desertification control programs into other socio-developmental programs; (4) failure to include local populations in the solutions; and (5) the search for technical solutions for socio-political and socio-economic problems.[51]

Moreover, the Advisory Panel on Food Security, Agriculture, Forestry and Environment to the Brundtland Commission noted that many national governments had not even developed national action plans to combat desertification as proposed by the *PACD*. While many governments had accepted the regional nature of the problem, regional organisations such as the Permanent Interstate Committee on Drought Control in the Sahel (CILSS) that were created in response to this awareness were ineffective because of "political squabbling and shortage of qualified staff".[52]

In general, the *PACD* triggered debate rather than action and offered more of a statement of good intentions than an international agreement seriously intended to be implemented by its parties. A 1990 evaluation of the *PACD* stated that the main cause of its failure was the lack of political commitment by both affected countries and the donors.[53] Participants had misjudged available resources, based on their overly optimistic assumption that the industrialised world would affect resource and technology transfers to developing countries to address problems the donors did not feel posed a threat.[54] For example, DESCON proved to be ineffective in raising the financial resources hoped for and was closed down in 1990.[55] Additionally, a special account for voluntary donations for the *PACD*, created in 1978 by the UN, had only received USD 166,886 after ten years - the General Assembly closed the account in December 1989.[56] In 1986, UNEP's Executive Director Tolba agreed that little had been achieved in tackling desertification as a result of the recommendations in the *PACD*.[57] The issue then re-emerged on the international agenda six years later at the 1992 UNCED, but this time the intent was to negotiate a binding convention.

Desertification on the 1992 UNCED agenda

Based on UN General Assembly resolutions prior to the negotiations that led up to the Earth Summit, it seemed as if desertification would occupy a prominent place on the UNCED agenda. Even before the adoption in 1989 of the UN General Assembly *Resolution 44/228* that created UNCED, the *1989 Resolution 44/172* invited UNCED to "accord high priority to desertification control and to deploy all means necessary, including financial, scientific and technological resources, to halt and reverse the progress of desertification with a view to preserving the ecological balance of the planct." It also invited UNEP to provide a report on the progress of the implementation of the *PACD*.[58] *Resolution 44/228* subsequently accorded high priority to drought and desertification control.

The first three sessions of the UNCED Preparatory Committee (PrepCom), however, paid little attention to desertification. Decision 1/15 (Soil loss, desertification and drought) of PrepCom I, in August 1990, simply repeated the invitation to UNEP to report on the implementation of the *PACD* and requested the UNCED Secretariat to consult with the specialised agencies on the need for further international cooperation to combat desertification and drought. The discussion on desertification, held in Working Group I and chaired by Sweden's Bo Kjellén, was cut short due to lack of time and unavailability of documents in all working languages. However, at the insistence of the African Group, it was decided that Working Group I would put the issue of desertification at the top of its agenda for PrepCom IV.[59]

In August 1991, the Secretariat presented the background paper "*Protection and Management of Land Resources: Combating Desertification and Drought*" to the PrepCom III.[60] This report contained a review of the issues, a report on current work by the various UN agencies and a discussion of progress made in implementing measures to control desertification. The document also contained a UNEP progress report, a review of drought monitoring and research activities, and reports on reforestation activities to combat land degradation and desertification, and alternative and sustainable systems of production and livelihoods in marginal lands.

The initiative for a global convention developed from an African regional preparatory meeting for UNCED, held in November 1991 between PrepComs III and IV. Over 40 African environment ministers met in Abidjan, Côte d'Ivoire and, among other things, called for a convention to combat desertification as one of the concrete items to be included in *Agenda 21*.[61] The African Group felt that the interests of other regions of the world were being met either by the biodiversity or climate change conventions, both of which were later signed at UNCED.[62] Africa needed something in exchange.

A draft of what eventually became the chapter on desertification in *Agenda 21* was presented at PrepCom IV, in March 1992.[63] . The African Group presented a series of amendments. Subsequently the revised text became the basis for almost the entire *Agenda 21* chapter on combating desertification. There was, however, no consensus the proposal for a future binding convention on desertification. [64]

During the preparations for UNCED, African demands for an agreement became politically linked to the deforestation issue. Industrialised countries were pressing for a convention on forests, while developing countries wanted one for desertification. The European Community and the United States referred to the failed *PACD* and doubted whether desertification, perceived as a cluster of local, or at least regional environmental issues, was a problem suited for a global convention. One solution to the dilemma was suggested in a package deal: developing countries would agree to a forests convention while industrialised countries would support the *CCD*.[65]

However, when delegates arrived at UNCED, this deal proved difficult to put into operation. Tommy Koh of Singapore, Chairman of the Main Committee, held informal consultations on the issue of a convention to combat desertification. African countries, supported by the rest of the developing world, forcefully pushed the desertification issue meeting resistance from industrialised countries. The turning point came when the United States agreed to support the proposal and the pressure became too strong for the European Community to resist.[66] A chapter on desertification and drought was included in *Agenda 21* and, on 10 June 1992, UNCED requested the UN General Assembly to establish an intergovernmental negotiating committee to elaborate on the *CCD*.[67]

As requested by UNCED, the 47th session of the UNGA began discussions on the preparations for a convention. In early October 1992, the African Group and later the G-77 and China, held consultations on a draft resolution establishing an intergovernmental negotiating committee and submitted a final draft on 7 November. In *Resolution 47/188* of 22 December 1992, the UNGA established a negotiating committee for the elaboration of a convention to combat desertification.[68] The resolution, referring to Chapter 12 of Agenda 21, stated that the negotiations needed to be finalised by June 1994 and that the first negotiating session would devote its first week to "information sharing" on drought and desertification. The resolution also established a multidisciplinary panel of experts to assist the Secretariat of the negotiations, a special voluntary fund to assist developing country participation in the negotiations and a trust fund for existing UN funds and contributions to be administered by a secretariat, which would also handle the negotiations. Additionally, NGOs, particularly from developing countries, were invited to contribute to the negotiating process.

THE 1994 UNITED NATIONS CONVENTION TO COMBAT DESERTIFICATION

The *CCD* is the third international environmental convention arising directly from UNCED and it is often conceived of as a "sister" of the other two Rio agreements on biodiversity and climate change.[69] Unlike the majority of environmental conventions negotiated prior to UNCED in 1992, developing countries were the initiators of the *CCD*. It is also regarded as the first sustainable development treaty, since it considers economic and social development needs while also addressing a serious environmental problem.[70] The *CCD* is different from previous attempts to combat desertification, such as the *PACD*, because it is legally binding; countries that accede to it will be obliged to implement it. It includes concrete national commitments for practical action, particularly at the local level, and places emphasis on the machinery needed to implement the *CCD* and monitor its progress. The *CCD* defines desertification as "land degradation in arid, semi-arid, dry sub-humid areas resulting from various factors, including climatic variations and human activities", the definition already agreed to in Chapter 12 of *Agenda 21* on drought and desertification.[71]

The content of the CCD

In addition to the overriding goal of combating desertification and mitigating the effects of drought,[72] the goals of the Convention are: (1) to achieve a more effective international instrument to combat desertification than the 1977 *PACD*; (2) to raise the international community's awareness of the effects of drought and/or desertification on the environment and the possibilities of achieving sustainable development; (3) to acknowledge the overrepresentation of developing countries – and especially least developed countries – among those most affected by drought and/or desertification and to assure that such countries receive resources to enable them to deal with this problem; and (4) to stress that, to mitigate desertification and drought, the human beings living in the affected areas must be at the centre of concern, and that women, NGOs, and local participation have particularly important roles to play. This last goal is what has informally been labeled the "bottom-up approach".

The main text of the *CCD* contains 40 articles, supplemented by four regional implementation annexes for Africa (19 articles), Asia (8 articles), Latin America and the Caribbean (7 articles), and the Northern Mediterranean (10 articles). It provides a code of good practices for dryland management for governments of affected countries and for donors as well as offering a framework for cooperation between local land users, NGOs, governments, international organisations, funding agencies, and Northern donor countries.[73]

To pursue the *CCD*'s objectives, Parties are to undertake several obligations. Parties are divided into "affected country Parties" and "developed country Parties". Three articles outline their key obligations. Article 4 outlines general obligations for all Parties, including the adoption of integrated approaches for combating desertification, giving "due attention" to the situation of affected developing country Parties, and integrating strategies for poverty eradication. Article 5 contains the obligations of affected country Parties. These include: the establishment of strategies to combat desertification; addressing the underlying socio-economic causes that contribute to desertification processes; awareness raising; and ensuring the participation of local populations, with the support of NGOs. Developed country Parties are to undertake additional obligations found in article 6 relating to the provision and mobilisation of financial resources for affected developing country Parties.

The question of available resources is central to the implementation of the *CCD*. While the *CCD* refers to affected country Parties and developed country Parties, there is no list of what countries belong in which category. Parties themselves decide to which category they belong. While it is possible to be both an affected country and a developed country, article 4(3) explicitly notes that only affected *developing* country Parties are eligible for assistance. All Parties are to "promote the use of existing bilateral and multilateral financial mechanisms" and channel "substantial financial resources to affected developing country Parties". To achieve this objective, article 21 creates the Global Mechanism to "promote actions leading to the mobilisation and channeling of substantial financial resources".

The core of the *CCD* is the development of national, subregional and regional action programmes to combat desertification, building on existing environmental frameworks where they exist.[74] National governments are to develop these action programmes in close cooperation with donors, local populations and NGOs. The National Action Programmes (NAPs) may include early warning systems, drought contingency plans, strengthened food security systems, alternative livelihood projects and sustainable irrigation programs. Donors are to commit to achieving better coordination among themselves to prevent duplication of actions and contradictions between their approaches and conditions.

The *CCD* stresses the importance of the participation, knowledge, know-how and experience of local populations, women and NGOs – the bottom-up approach – in its implementation. Article 19 states that Parties shall promote capacity building "through the full participation at all levels of local people, particularly at the local level, especially women and youth, with the cooperation of non-governmental and local organizations". This could take place, for example, "by fostering the use and dissemination of the knowledge, know-how and practices of local people in technical cooperation programmes".

The *CCD* establishes a Conference of the Parties (COP), which serves as its governing body. The COP reviews the implementation of the *CCD*, promotes information exchange, and adopts amendments. The Committee on Science and Technology (CST), a subsidiary body of the COP, provides the COP with information and advice related to combating desertification. It is a multidisciplinary committee open to all Parties, composed of government representatives competent in the relevant fields of expertise. The COP also maintains a roster of independent experts with expertise and experience in the relevant fields that may be called upon by the COP to provide advice on specific issues.

Negotiating the CCD

The *CCD* was negotiated in a little over a year by the Intergovernmental Negotiating Committee for the Elaboration of an International Convention to Combat Desertification in Those Countries Experiencing Serious Drought and/or Desertification, Particularly in Africa (INCD). Between 1993 and 1997, the INCD met twelve times. The organisational session in January 1993 was followed by five sessions during which the *CCD* text was negotiated. It was completed on 17 June 1994 and was opened for signature in October of the same year. While waiting for the *CCD* to enter into force, delegates prepared for the first Conference of the Parties (COP-1) during six interim negotiating sessions. The *CCD* entered into force on 26 December 1996, three months after the fiftieth instrument of ratification had been submitted to the United Nations.

Priority for Africa

A major element in the negotiations, as indicated by the title of the *CCD*, was the special role for Africa. African countries pushed for a start to convention negotiations at UNCED, supported by the other developing countries of the G-77 and China. At the INCD organisational meeting, the Chairman proposed that the special concern for Africa be expressed in the form of an annex or protocol to the *CCD* while other parts of the world could have protocols or annexes in the future.[75] However, this caused a split in the G-77. At INCD-1, negotiators grappled with what type of agreement to negotiate. There was a divide between those who favoured a framework type agreement (a set of overarching principles comparable to those of the *Climate Change Convention*) and supporters of an operative convention that could provide guidance and support for national, regional, and worldwide programmes of action.[76]

The Chairman eventually proposed the compromise of a framework agreement *and* an operative instrument for Africa. However, Latin American and Asian delegates in particular resisted this proposal, insisting that instruments for other regions should be negotiated simultaneously. They argued that the problem was worldwide and that their efforts to combat desertification deserved equal attention to those of Africa. Unless areas other than Africa received attention, the *CCD* would not appear to be global and thus would be unlikely to receive support from countries in other regions. The African countries thus faced a balancing act in trying to uphold their particular role in the *CCD* while retaining the support of the rest of the G-77 countries. This disagreement was so strong in the last hours of INCD-1 that the entire negotiations seemed to be in jeopardy. At INCD-3, negotiators finally settled on three regional implementation annexes for Africa, Asia, and Latin America and the Caribbean. A fourth annex for the Northern Mediterranean region was added at INCD-5.

The greater size and detail of the 19-Article Regional Implementation Annex for Africa illustrates the emphasis placed on Africa throughout the negotiations. The Annex outlines the particular conditions of the African region, including: the high proportion of arid, semi-arid and dry sub-humid areas; the substantial number of countries and populations affected by desertification; and the weak infrastructural base and the insufficient institutional, legal, scientific, technical and educational frameworks, which lead to substantial capacity building needs. The Annex also: specifies the obligations of both African country Parties and developed country Parties; emphasises that the National Action Programmes should be part of the broader process of formulating national sustainable development policies; underscores the priority of affected African country Parties in the allocation of financial resources; and also emphasises the importance of technical assistance and technology transfer.

The African Annex provides detailed descriptions of the content and preparation of the National, Sub-regional, and Regional Action Programmes (NAPs, SRAPs and RAPs). In preparing the NAPs, each country needs to designate a national coordinating body or focal point to function as a catalyst in the preparation, implementation and evaluation of its action programme. NAPs are to use past experiences of combating desertification, identify factors contributing to desertification, and increase the participation of local populations and communities. NAPs should also include measures to: improve the economic environment to eradicate poverty; conserve natural resources; improve institutional organisation; improve knowledge of desertification; and monitor and assess the effects of drought. Sub-regional Action Programmes (SRAPs) focus on issues better addressed at the sub-regional level. Mechanisms for management of shared natural resources can be established to handle transboundary problems associated with desertification and drought and shall provide support for the harmonious implementation of NAPs. Regional Action Programmes are devoted to issues that are best handled at the regional level, including: coordination of SRAPs for building regional consensus on key policy areas; capacity building; seeking solutions with the international community to global economic and social issues that have an impact on affected areas; exchange of information and appropriate techniques; and coordination of early warning systems and drought contingency plans.

In accordance with the *CCD's* title, special attention was also devoted to the African region through the June 1994 adoption of resolutions on urgent action for Africa and interim arrangements for the period between the adoption of the *CCD* and its entry into force.[77] Disappointment with the previous fifteen-year record on desertification activities was the driving force behind this decision; many participants did not want to delay implementation in Africa until it had come into force after a lengthy formal ratification process. Governments of affected countries and their donor partners were thus encouraged to start joint activities on a pilot basis. The interim sessions (INCD-6 to the resumed INCD-10) received progress reports on implementation from numerous African countries, which helped maintain momentum since it reminded participants that activities were already taking place at the national level. Yet, as the two-and-one-half year interim period neared its end, some African delegates worried that the priority for Africa would lose prominence, since the *CCD* would be implemented in all regions.[78]

Prospects for the implementation of the CCD

With the negotiation of the *CCD* and its coming into force at the end of 1996, the goal of creating a legally binding agreement set up by the 1991 African environment meeting in Abidjan was achieved. By the time of COP-3 in November 1999, parties had nearly three years to implement the agreement. In addition, African country Parties were implementing anti-desertification activities since the 1994 adoption of the resolution on urgent action for Africa. Not enough time has passed to allow for a comprehensive evaluation of whether the *CCD* is a successful international instrument for dryland management. However, the level and range of *CCD*-triggered activities initiated since 1994 provide an indication of how implementation is progressing in Africa. This section discusses the prospects for implementing the *CCD* in Africa. The point of departure is a brief comparison between the *PACD* and the *CCD*, followed by a discussion of the roles of funding, political commitment, and the bottom-up approach.

LEARNING FROM PREVIOUS MISTAKES? A COMPARISON OF THE PACD AND THE CCD

One of the main goals of the *CCD* was to avoid earlier mistakes and replace the failed *PACD*. The *PACD* was a non-binding agreement resulting from a single two-week conference and its preparations in the 1970s. The *CCD* is a legally binding agreement resulting from a thirteen-month multilateral negotiating process in the 1990s. Unlike the *PACD*, the *CCD* can be amended in the future. A review of the similarities and differences between the two agreements highlights their different international contexts and changing attitudes towards environment and development issues. A comparison also provides a starting point for an analysis of the prospects for successful implementation of the *CCD*.

The *CCD* and the *PACD* share many similarities. First, industrialised countries raised doubts before both the UNCOD and *CCD* negotiations. They suggested that holding a conference or negotiating a convention was not necessary for primarily a local problem, and that creating new institutions would not make any difference. Second, both agreements resulted from strong lobbying. UNCOD was pushed by a "desertification movement", and the *CCD* was promoted by African governments and UNEP.[79] In both cases, there were strong political motivations to convene these meetings; the challenge became how to create, and fund, internationally workable solutions. A third similarity is that high expectations about the provision of new and substantial financial resources were ultimately not realised. The trouble in financing both the *PACD* and the *CCD* demonstrates the difficulties in mobilising new resources for problems that donors do not feel apply to them.

There are also areas in which the *PACD* and the *CCD* appear to be similar, but prove to differ upon closer examination. First, both the *PACD* and *CCD* definitions say that desertification is caused by natural and human factors. However, the bases for these two definitions are quite different. The *PACD* was more of "a treatise on natural resource management" and the main emphasis was still on natural causes.[80] The *CCD* definition, on the other hand, was based on the idea that desertification was in fact mainly caused by human activities. The main message the *CCD* conveys is that human activities and underlying socio-economic causes are at the centre of the problem that needs to be addressed.

Secondly, while the *CCD* is said to be innovative in its recognition of the need for the involvement of local populations and making use of their knowledge in the development of NAPs, almost identical formulations are used in the following passage from the *PACD*: "Particular attention should be given to the utilization of local experience, knowledge and expertise in the implementation of the recommendations of the Plan at the national level in the countries concerned."[81] References to this issue are few in the *PACD*, particularly in comparison with the numerous references in the *CCD*. There is also reason to believe that, while the *PACD* contained this wording, the drafters still saw the solution in financial and technical resources coming from outside the dryland areas. The *CCD* adopts a bottom-up approach by consistently referring to the role of local populations and the need for their involvement in the development of desertification action programmes.

There are also a number of distinct differences between the *PACD* and the *CCD*. The first is the extent of scientific preparation. UNCOD was preceded by several years of scientific work of four major groups that prepared documents to be considered at the Conference. This is clearly visible in the text of the *PACD* and related UNCOD documents, which contain statements of what is and is not desertification and a detailed description of how desertification takes place. The *CCD* only lists definitions of the central terms used and does not contain any description of the problem. Many *CCD* negotiation participants believed that they already knew what desertification was, or they were at least were willing to accept the definition that was provided in *Agenda 21*, and agreed that legal obligations were what was needed.

A second difference relates to how the *PACD* and the *CCD* view the solutions to the environmental problem of desertification. The *PACD* emphasises the need for "corrective measures" to "reverse" desertification, while the *CCD* promotes a bottom-up approach to address dryland problems.[82] The *PACD* has been criticised for taking a fragmented approach to desertification and overlooking socio-economic aspects by focusing on technological solutions.[83] Whether the *CCD* proves to be an effective instrument or not, it does attempt to address the socio-economic issues that the *PACD* avoided.[84]

These similarities and differences reflect the different political climates in which these instruments were crafted. The negotiations on the *PACD* were deeply affected by the human misery caused by the Sahelian Drought of 1968 to 1973. At UNCOD, the natural environmental problem of desertification was very much at the centre for negotiators and the preparations for the negotiations consisted mainly of trying to understand the problem. Desertification was "bad" because it hampered human development, and in the long run, human progress.

On the other hand, a greater awareness of the complexity of environmental problems preceded the *CCD*, where participants believed that the remedies to a large extent lay in how humans organise their lives, rather than which technologies were used.[85] Lessons from failed development aid projects demonstrated that, even though a technology is available, it may not be used appropriately or be useful in all cases. Additionally, the notion of sustainable development only gained prominence in the 1980s. On a practical level, the *CCD* was also a product of the preparations for UNCED, where predominantly African developing countries wanted something in return for all the attention devoted to what they perceived to be northern interests. In addition, the decision to negotiate the *CCD* had as much to do with generating new technical and financial resources, in a period when development assistance was dwindling, as with curbing the environmental problem of desertification.

The *CCD* negotiations were an attempt to learn from the lessons of the past and provide a new international approach to dryland management. It places desertification in context by regarding activities to prevent dryland degradation as a means to obtaining sustainable development. The *CCD* also addresses the previously underestimated role of socio-economic factors for addressing desertification. Finally, the *CCD* negotiators had learned that the top-down approach assumed in the *PACD*, as well as in the implementation of many development projects, needed to be replaced by a bottom-up approach if projects were to be effective and remain in operation after the departure of the development personnel.

It is also clear, however, that the *CCD* suffers from some of the same weaknesses as the *PACD* regarding the funding of anti-desertification measures for developing country Parties. As the *PACD* experience demonstrated, political commitment is a precondition for action. In addition, the bottom-up approach may prove complex to develop in practice. As discussed further below, NGOs are a significant asset, but their activities alone may not be sufficient to ensure successful implementation. For these reasons, the success of the *CCD* is not assured.

Funding for dryland management under the CCD

Financial assistance for dryland management in affected developing country Parties remained a contentious issue throughout the *CCD* negotiation process. This was partly due to the general trend of diminishing development assistance in the beginning of the 1990s, and partly due to the failure to implement the *PACD*. The disbanding of DESCON as well as the highly insufficient funding provided to the UN special account for voluntary donations for the *PACD* both indicated the difficulties in raising money for dryland management.

To avoid a similar development in the 1990s, developing and particularly African countries pursued several avenues to secure resources. One way of attaining what in the UN context is referred to as "new and additional resources" was through international funding agencies. The Global Environment Facility (GEF) was established in 1990, during preparations for UNCED, to serve as an instrument for multilateral cooperation in addressing global environmental concerns. The GEF's implementing agencies are the World Bank, UNEP and UNDP. The GEF provides funding for investments and technical assistance in four focal areas: global warming, biodiversity, international waters, and ozone depletion.[86]

During the early stages of the desertification negotiations, developing countries hoped for a fifth GEF focal area through which desertification projects could be funded. It soon became clear, however, that desertification projects would only be funded if they could be related to existing GEF priority areas, such as biodiversity.[87] As a result, while the *CCD* contains a reference to the promotion of new and additional funding from GEF, desertification activities cannot receive direct funding and must relate to the GEF's four focal areas.[88]

Another way for the developing countries to raise funding was to turn to the bilateral donors, the industrialised countries. Drawing upon the lessons and recognised shortcomings of the *PACD*, the G-77 and China had demanded the creation of a special desertification fund already during the preparations for UNCED. Many developing countries voiced their concern and explained how many previous national and regional plans of action to combat desertification had not been effective due to a lack of financial resources.[89] However, money proved to be a contentious issue between the North and South.

On the one hand, developing – predominantly African – countries came to the negotiations hoping that the *CCD* could provide new and additional financial resources. In fact, the G-77 and China claimed that the *CCD* would be pointless unless it was backed by new resources and insisted on a new, independent, specialised fund capable of mobilising resources. They pointed to *UNGA resolution 44/288* (establishment of UNCED) and Chapters 12 (combating desertification) and 33 (financial resources and mechanisms) of *Agenda 21*, where commitments had been made for the mobilisation of new and additional financial resources.

The industrialised countries, on the other hand, feared that establishing new institutions would only add to an increasing international bureaucracy. They wanted to rationalise and strengthen existing resources before committing to provide new and additional ones. Specifically, rather than creating a new fund to combat desertification, the donors wanted to coordinate the available national, bilateral and multilateral funding mechanisms. They also noted that special funds had not always been effective, and pointed to the discouraging examples of DESCON and the UN special account for voluntary contributions.[90]

This question – of whether there would be any new and additional funding and an institution to manage it – threatened to topple the entire negotiation process. At the final negotiating session, however, delegates reached an agreement. The United States proposed the establishment of a Global Mechanism that would promote actions leading to the mobilisation and channeling of financial resources. How this was to take place was not specified. As a concession to this compromise proposal, the G-77 and China withdrew its proposal for a special desertification fund, thus making it possible to conclude the negotiations of the *CCD*. However, the nature of the administration and functioning of the Mechanism remained open for debate.

The period after the conclusion of the *CCD* was characterised by disagreements about the Global Mechanism. The G-77 and China unsuccessfully attempted to persuade the industrialised countries to fund the Global Mechanism. This conflict was based on the differing views of the functions of the Global Mechanism. Those who were reluctant to create another ineffective administrative body stressed that it should not be an institution in itself. Others were apprehensive that no new and additional resources would be forthcoming.[91]

Delegates finally agreed that the main function of the Global Mechanism would be to provide guidance on where to find resources for desertification projects (for recipients) and where to allocate such resources (for donors).[92] This decision consequently confirmed that the Global Mechanism would not have any significant resources that it could itself allocate to desertification projects and that it would only help those looking for money to find such funding.

A third way for developing countries to raise new and additional resources was suggested by industrialised countries at the June 1997 special session of the UNGA (UNGASS) that reviewed the implementation of *Agenda 21*. It was proposed that developing countries turn to the private sector for funding, since a considerably larger portion of money flowing into developing countries stems from private investments rather than from government development assistance. This proposal, however, met with disbelief among some Africans. Private financial flows tend to go to a few South East Asian countries rather than to Africa, since private investments are not made where there is serious poverty. Dryland areas are poor and therefore the success of the *CCD* was regarded as depending more on official development assistance than on foreign direct investment.[93]

The result of the struggle for new resources was that the *CCD* established a Global Mechanism to "promote actions leading to the mobilisation and channeling of substantial financial resources".[94] However, while "substantial financial resources" are to be channeled to affected developing country Parties under the *CCD*, this is to be done by using existing bilateral and multilateral financial mechanisms. It is notable that the Global Mechanism only is to *promote* the mobilisation and channeling of financial resources. The Mechanism does not raise or manage any such money itself. The "new and additional" financial resources developing countries kept pleading for simply did not materialise. The result is that industrialised countries can maintain their bilateral development programmes, keeping developing countries dependent on decisions taken abroad. Developing countries are thus mostly unable to change any existing development assistance patterns that may be counterproductive or poorly adjusted to national conditions. The financial message of the *CCD* is business as usual. In that respect, the African environment ministers failed to reach their goal of raising more money to countries suffering from reduced development assistance by mobilising extra funding for anti-desertification activities.

The importance of political commitment for implementation

The disagreement between North and South over funding for dryland management is an expression of an underlying contention about commitment. The *PACD* experience demonstrated what is crucial to implementation: the main cause of the *PACD's* failure was the lack of political commitment by both affected countries and donors. This is the dilemma for African countries that argue they need funding to address desertification - without money it is difficult to achieve results. However, the message is equally clear from the donor community that more money will not be forthcoming unless results are shown.

Donors who are requested to demonstrate political commitment by providing new and additional money for dryland degradation projects are aware of the failure of the *PACD*. They point to the findings made while preparing the *Brundtland Report* in the mid-1980s that many governments in affected countries had not even developed national action plans to combat desertification as was proposed by the *PACD*.

In the *CCD* process, however, urgent action was taken in Africa – even before it had come into force. This allowed African countries to begin their anti-desertification activities at an early stage, thereby asserting the special role the African countries have in the *CCD*. All African countries have also ratified the *CCD* demonstrating their commitment to the process.

At COP-3 in November 1999, African countries reported on the status of their implementation of the *CCD*.[95] All African countries are preparing and implementing their NAPs, including information and awareness-raising, country-wide consultative processes, national forums, and adoption and implementation of the programmes. Progress has been slower than desired, however, due to the complexity of managing cross-sectoral and participatory programming exercises. At the national level, focal points supporting the implementation of the *CCD* are yet to get the full attention of some important ministries. The focal points also strive to encourage and broaden the involvement of civil society and try to mobilise a critical level of resources.

The overall conclusions from African implementation reports were that nearly every African country has set up institutional coordinating bodies with national steering committees, focal point secretariats and technical subcommittees. These organs have multi-disciplinary and multi-sectoral representation from government departments, institutions of higher learning, NGOs and community-based organisations, the private sector and the donor community. Some countries also reported progress in setting up a framework for implementing the *CCD* through scientific and technological support, legislative measures, preferential policies on tax systems, land use and ownership, and support to decentralised management and privatisation. Many countries are also developing mechanisms to synergise the implementation of the three conventions on biological diversity, climate change and desertification.[96]

At COP-3, the African sub-regions also reported on their specific progress. Southern Africa presented measures taken to ensure NAP implementation, including local level capacity building through consultative workshops and national forums on desertification. The sub-region had difficulties gaining support from international partners and reported that benchmarks and indicators were under development at the sub-regional and regional levels. East Africa countries emphasised the need to mainstream desertification issues with other strategies dealing with sustainable development and poverty eradication. The sub-region stressed the importance of stakeholder participation in the NAP process, particularly women and youth. The sub-region also reported that many countries were establishing national desertification funds but that there was a lack of funding for desertification control. Central Africa stated strong political will to implement the *CCD*, but had difficulties because of political instability in the region, inadequate financial resources and the absence of coordinating bodies. West African countries also reported lack of progress in countries with social conflict and political instability. The region experienced difficulties in implementing the participatory approach due to varying levels of preparedness, and inadequate resources. Insufficient data available for planning had constrained the NAP process in some of the region's countries. It was also pointed out that NGO partners lacking the necessary resources had been dependent of the support from northern NGOs. Northern Africa underlined the importance of coordinating key actors and achieving local level participation. The sub-region representatives called for greater cooperation with the donor community.[97]

Another *Brundtland Report* finding on *PACD* implementation was that, while many governments had accepted the regional nature of desertification, regional organisations were ineffective. The political commitment of organisations such as AMU (Arab Maghreb Union), CILSS (Permanent Interstate Committee on Drought Control in the Sahel), ECOWAS (Economic Community of West African States), IGAD (Intergovernmental Authority on Development), and SADC (Southern African Development Community), will be particularly significant for *CCD* implementation. Under the *CCD*, these are key organisations since the sub-regional and regional levels are the highest levels of implementation in the treaty. There are no bodies at the international level responsible for implementation; the Global Mechanism is only to facilitate implementation, which is to take place at the regional, sub-regional, national, and local levels.

The bottom-up approach – NGOs as "partners in development"

Much of the uniqueness of the *CCD* compared to other environmental agreements lies in its bottom-up approach. The *CCD* is in fact a set of targets to be achieved mostly through local action. It seeks to address the degradation of fragile drylands by involving local populations in affected areas in the development of NAPs. The *CCD* bottom-up approach draws on *Agenda 21*, which notes that solutions to environmental problems have to be found and worked out locally. The *CCD* negotiations provided the first chance after UNCED to reflect the agreed centrality of people's participation. There was widespread consensus from the start of the negotiations that the *CCD* should use a bottom-up approach.

The rationale behind the bottom-up approach in the *CCD* – since the agreement addressed human activities in drylands affected by desertification – was that solutions and prevention strategies had to focus on the local level. Experience from decades of failed development aid projects underscored the importance of empowering politically marginalised peoples affected by desertification; local populations had to be involved in plans that were designed to help them in order for the plans to succeed.

Throughout the negotiations, the emphasis on the bottom-up approach and public participation legitimised a high profile role for NGOs. In turn, the NGOs played a central role in ensuring that the bottom-up approach was included – and remained included – in the final text of the *CCD*. In fact, given their observer role in the negotiations, the NGOs exerted significant influence on the outcome of the negotiations.[98] NGOs were regarded by negotiators – particularly those representing donor countries – as future "partners in development", providing the link between decisions made at the international level and the local populations the *CCD* was trying to address. For these purposes, the *CCD* text reflects the importance of NGO participation in the implementation of desertification programmes.[99]

While the NGOs had significant influence during the negotiations of the *CCD*, the challenge for NGOs in the implementation phase is to demonstrate political commitment by following up on decisions made at international and national levels.

The current development trend is to rely heavily on NGOs, which may prove to be a formidable task for them:

> "aid projects, successful and unsuccessful, have barely scratched the surface of the vast acreage's of Africa that are suffering. The only real hope is for there to be grass-roots movements for change, and although these are springing up, poverty and growing populations will make their task very difficult."[100]

The NGOs are expected to function as domestic guarantors of implementation, acting as reviewers of implementation efforts as well as domestic actors who can make the connection between national governments, local administrations and local populations. In many affected developing countries, successful implementation of the *CCD* rests with the NGOs. While this is a great opportunity for NGOs to make an impact, it also provides a challenge for groups that often pursue narrow agendas and may not always be representative of, or accountable to, the local populations they often claim to represent.

CONCLUDING REMARKS ON THE FUTURE OF THE CCD

The *CCD* is truly a sustainable development agreement: it addresses an environmental problem with serious socio-economic components. Successful implementation can therefore not ignore – nor be separated from –North-South development issues. The agreement highlights the responsibility of the rich towards those in less affluent parts of the world. The struggle over additional money for dryland management is part of the North-South tensions that have dominated UN interactions since the end of the Cold War in the early 1990s.

It is important for both donor and recipient countries to prove that the *CCD* – unlike the *PACD* – is not only a statement of good intentions, but also an international agreement that is intended to be implemented. Successful implementation of the *CCD* will depend on the political commitment of donor countries, affected developing countries, regional organisations, NGOs, grassroots organisations and local populations. It will not be possible for actors in affected countries to demonstrate such commitment, however, without adequate resources to do so. The funding of anti-desertification measures is therefore essential.

Africans disappointed with the failure to create a desertification fund as a result of the *CCD* negotiations may find reason to revise their view. There is reason to believe that – while a desertification fund *could* have resulted in additional funding – such a fund would likely have encountered increasing regulation, given donor countries' desire to control funding allocation. That would decrease recipient countries' ability to influence resource allocation, one of the motives behind the need for a desertification fund in the first place. Instead, the Global Mechanism facilitates the identification of financial resources which, given the current constraints in development aid, may be a more realistic option. If the attempts in the 1970s to raise funds failed, why should similar efforts be successful in the early 2000s – at a time when there is even less development aid?

The Global Mechanism strategy – to localise and channel existing resources to projects that need funding – may prove to be more feasible and useful and may in fact allow for more flexibility and recipient country influence than a desertification fund. It will, however, be some time before it is possible to evaluate Global Mechanism activities in this light.

Of the five reasons UNEP lists for the failure of the *PACD*, the *CCD* attempts to address three of them.[1] First, desertification control programs are to be integrated with other socio-developmental programmes – or what we today would call sustainable development programmes. Second, local populations are to be included in dryland management solutions. Third, the *CCD* focuses on the socio-economic problems and discourages attempts to address them only with technical solutions. In this respect, the *CCD* includes lessons from the *PACD* experience and shows a promise of better opportunities for implementation. There was, however, little success in addressing the two remaining reasons for *PACD* failure: (1) low priority by funding agencies, and (2) lack of external financial and technical assistance to allow developing countries to cope with the problem. The future will tell whether addressing three of the five reasons is sufficient to achieve successful implementation, or if additional efforts to create more funding and assistance will be required.

---

[1] Dr. Elisabeth Corell is a Wallenberg Fellow in Environment and Sustainability at the Department of Urban Studies and Planning at the Massachusetts Institute of Technology in Cambridge, USA. In 1999, she completed her PhD dissertation *The Negotiable Desert: Expert Knowledge in the Negotiations of the Convention to Combat Desertification*at Linköping University, Sweden. For this research she attended all CCD negotiating sessions from 1993 to 1997. Her field is international relations and her research interests include the relationship between science and politics in international environmental decision-making, expert knowledge, sustainable development in developing countries, and the translation of global agreements to local action.

[2] *United Nations Convention to Combat Desertification in Those Countries Experiencing Serious Drought and/or Desertification, Particularly in Africa: Text with Annexes*(Nairobi: UNEP, 1994), hereinafter eferred to here as the *CCD*.

[3] *CCD*, Article 1(a).

[4] F. Cardy (1997), "Foreword", pp. v-vi in N. J. Middleton and D. S. G. Thomas (eds),*World Atlas of Desertification* (2nd ed.), (London: Arnold Publishers and UNEP).

[5] M. K. Tolba, (1992), "Preface", p. iv in Middleton & Thomas (1992), *World Atlas of Desertification.*

[6] F. Cardy, (1993), "Global Programme of Action!", 17:1/2 *Ecoforum* 22-23; F. Cardy (1993), *Desertification -- A Fresh Approach*, (undated paper provided at the *CCD* negotiations, May 1993).

[7] E. Dowdeswell, (1997), "Preface", p. iv in Middleton and Thomas, *World Atlas of Desertification.*

[8] Tolba, M. K., "Preface", p. iv in Middleton and Thomas, *World Atlas of Desertification*, 1992.

[9] S. Batterbury, S. and A. Warren (2000), "Desertification",*International Encyclopedia of the Social & Behavioral Sciences* (Oxford: Oxford University Press).

[10] Preamble, paragraph 6.

[11] On natural causes of desertification see, for instance, A. H. Bakhit (1993), "Desertification: Reconciling Intellectual Conceptualization and Intervention Effort", 31:3 *GeoJournal*, 33-40; Helldén, "Desertification - Time for an assessment?", at 372-383; and Ahlcrona, *The Impact of Climate and Man on Land Transformation in Central Sudan.*

---

[1] See section 2.1.

[12] R. C. Balling (1993), "Interrelationships between the Global Climate System, Drought and Desertification, Including the Impact of Desertification on Climate Change and Global Warming", (paper presented at the *CCD* negotiations, May 1993).
[13] C. Toulmin (1993), *Combating Desertification: Setting the Agenda for a Global Convention* (London: International Institute for Environment and Development, paper no.42) at 6.
[14] *Ibid* at 7, Bakhit, (1993).
[15] See, for example, Balling (1993); F. K. Hare (1993), *Climate Variations, Drought and Desertification* (Geneva: World Meteorological Organization, no. 653). If climate change does lead to changed precipitation patterns and surface and water temperatures, then this could have important implications for dryland degradation.
[16] See, for instance, A. Grainger (1990), *The Threatening Desert: Controlling Desertification* (London: Earthscan Publications); see also D. S. G. Thomas and N. J. Middleton (1994), *Desertification: Exploding the Myth* (Chichester: John Wiley & Sons).
[17] For an elaboration of the importance of human activities, see H. E. Dregne,*Desertification of Arid Lands*; and R. D. Graetz (1991), "Desertification: a Tale of Two Feedbacks", in H. A. Mooney, E. Medina, D. W. Schindler, E. D. Schultze and B. H. Walker (eds.), (1991),*Ecosystem Experiments* (Chichester: John Wiley & Sons, SCOPE no. 45).
[18] Adapted from Thomas and Middleton (1994) at 85.
[19] *CCD,* preamble paras. 7 and 9.
[20] M. Kassas, Y. J. Ahmad and B. Rozanov (1991), "Desertification and Drought: An Ecological and Economic Analysis" 20 *Desertification Control Bulletin* 19-29.
[21] M. K. Tolba, "Preface" at iv in Middleton and Thomas (1992).
[22] M. Kassas *et al.* (1991). For further detail about the implications of land management, water resource use, population growth and other factors on food security, see WCED (1987),*Food 2000: Global Policies for Sustainable Agriculture* (London: Zed Books).
[23] Mainguet, *Desertification*, pp. 6-15.
[24] P. Blaikie and H. Brookfield, (1987), "Colonialism, Development and Degradation" at 100-121 in P. Blaikie and H. Brookfield, *Land Degradation and Society* (London: Routledge).
[25] A. Warren (1995), "Changing Understandings of African Pastoralism and the Nature of Environmental Paradigms", 20:2 *Transactions of the Institute of British Geographers* 193-203.
[26] E. Corell (1999), *The Negotiable Desert: Expert Knowledge in the Negotiations of the Convention to Combat Desertification.* (Linköping, Sweden: PhD Dissertation, Linköping Studies in Arts and Science) 53-62. This review covers the definitions used in the international context, such as in multilateral negotiations or in the work of international or intergovernmental expert groups, and does not survey the over a hundred scientific definitions of desertification.
[27] Article 1(a).
[28] Corell (1999).
[29] A. Warren, A (1993), "Desertification as a Global Environmental Issue", (1993) 31:1*GeoJournal* 11-14.
[30] Toulmin (1993) at 2.
[31] O. Greene (1996) "The Development and Implementation of International Environmental Regimes" at 280-312 in R. Blackmore and A. Reddish (eds.) (1996),*Global Environmental Issues* (London: Hodder & Stoughton); M. Simons (1994), "Nations Sign Pact to Stop Desert Growth:Accord Aims to Save Land for Agriculture" (1994) *The New York Times* at 4.
[32] H.E. Dregne, "Reflections on the PACD" (unpublished paper).
[33] Graetz (1991), "Desertification: a Tale of Two Feedbacks", in Mooney *et al.* (1991) at 60.
[34] See, for instance, Thomas and Middleton (1994) at 99-127; and A. Warren and C. Agnew (1988), *An Assessment of Desertification and Land Degradation in Arid and Semi-arid Areas* (London: International Institute for Environment and Development Drylands Paper no. 2). It is importantto note that those who discuss a desertification "myth" do not question whether desertification occurs, but rather they question the inflated image of the problem that has been conveyed to the international public and decision-makers.

[35] P. Harrison (1992), *The Third Revolution: Population, Environment and a Sustainable World* (London: Penguin Books) at 117.

[36] P. Brinck (1976), "Preface", at 7-10 in A. Rapp, H. N. Le Houérou and B. Lundholm (eds.),*Can Desert Encroachment be Stopped? A Study with Emphasis on Africa* (Stockholm: Swedish Natural Science Research Council and United Nations Environment Programme); R. S. Odingo (1990), "Review of UNEP's Definition of Desertification and its Programmatic Implications" at 7-44 in R. S. Odingo (ed.) (1990), *Desertification Revisited: Proceedings of an Ad Hoc Consultative Meeting on the Assessment of Desertification* (Nairobi: UNEP-DC/PAC) at 8.

[37] The protracted drought affected six countries on the southern border of the Sahara– Mauritania, Senegal, Mali, Upper Volta (today Burkina Faso), Niger and Chad. Crops failed, millions of livestock died, and pastoralists had to flee drought-stricken areas to seek help in the urban areas of the more humid south. It is estimated that 200,000 people died. See H. E. Dregne (1997), "Land Degradation Control in the Drylands: Establishing Priorities", at 73-88 in H. N. Barakat and K. Hegazy (eds.) *Reviews in Ecology: Desert Conservation and Development*(Cairo: Metropole); Dregne (1986); and *The United Nations Convention to Combat Desertification: A New Response to an Age-old Problem*, background paper prepared for Earth Summit+5, available at http://www.un.org/ecosocdev/geninfo/sustdev/desert.htm , published by the United Nations Department of Public Information, DPI/1870/SD (February 1997)

[38] M. Glantz, (1977), "The UN and Desertification: Dealing with a Global Problem", at 1-15 in M. H. Glantz (ed.), *Desertification: Environmental Degradation in and around Arid Lands* (Boulder: Westview Press).

[39] Odingo (1990) at 8.

[40] United Nations Conference on Desertification (UNCOD) (1976),*document OPI/NGO/SB/96*, 20 May 1976, p. 1, as cited in Glantz (1977).

[41] Thomas and Middleton (1994) at 29.

[42] Odingo (1999), "Review of UNEP's definition of desertification and its programmatic implications", in Odingo (1999) at 8; UNCOD (1978),*Round-up, Plan of Action and Resolutions*, (New York: United Nations) at 1.

[43] United Nations (1980), *Yearbook of the United Nations 1977* (New York: United Nations Department of Public Information) at 509.

[44] M. Verstraete (1986), "The United Nations Organization and the Issue of Desertification", at 42-51 in F. El Baz and M. H. A. Hassan (eds.) (1986) *Physics of Desertification* (Dordrecht: Martinus Nijhoff), as cited in Thomas and Middleton (1994) at 29.

[45] UNCOD (1977) at 7.

[46] UNCOD (1977) at 7; United Nations (1977) at 509.

[47] UNCOD: *Round-up, Plan of Action and Resolutions*, at 8.

[48] This division of the recommendations into three objectives was made by J. A. Mabutt (1987), "A Review of Progress Since the UN Conference on Desertification", 15*Desertification Control Bulletin* 12-23.

[49] United Nations (1977) at 509.

[50] A. F. El Kassas (1994), "Desertification: A General Overview", 59:1 *Development and Socio-Economic Progress*, 14-27.

[51] UNEP (1991), *Report of the Executive Director - Status of Desertification and Implementation of the United Nations Plan of Action to Combat Desertification* (Nairobi: UNEP/GCSS.III/3) at xiv.

[52] World Commission on Environment and Development (WCED) (1987),*Food 2000: Global Policies for Sustainable Agriculture* (London: Zed Books) at 73-74.

[53] A. Bounajuti (1991), "External Evaluation of the Plan of Action to Combat Desertification", 20 *Desertification Control Bulletin*, at 30-33.

[54] T. Brenton (1994), *The Greening of Machiavelli: The Evolution of International Environmental Politics* (London: Earthscan Publications) at 123; Bounajuti (1991).

[55] Bounajuti (1991) at 30-31; Toulmin (1995) at 455-457.

[56] El Kassas (1994) at 21.

---

[57] See Desertification Control Programme Activity Centre - UNEP (1987), *Rolling Back the Desert: Ten Years After UNCOD* (Nairobi: UNEP) at 17.

[58] United Nations General Assembly (UNGA) resolution A/RES/44/172, "Implementation of the Plan of Action to Combat Desertification", 19 December 1989. The UNEP report was presented to its Governing Council in February 1992 as Report of the Executive Director, *Status of Desertification and Implementation of the United Nations Plan of Action to Combat Desertification.*

[59] J. Bernstein, P. Chasek, L. Goree, R. Jordan and J. McMahan (1997), "UNCED PrepCom III, Third Week Synopsis", 7:2-3, *Earth Negotiations Bulletin*; J. Bernstein, P. Chasek, and L. Goree (1993), "A Summary of the Proceedings for the Organizational Session of the INC forthe Elaboration of an International Convention to Combat Desertification", 4:1 *Earth Negotiations Bulletin* at 5.

[60] UN document A/CONF.151/PC/62.

[61] Bernstein, Chasek, and Goree (1993), "A Summary of the Proceedings for the Organizational Session of the INC for the Elaboration of an International Convention to Combat Desertification", at 5.

[62] P. Chatterjee and M. Finger (1994), *The Earth Brokers: Power, Politics and World Development* (London: Routledge,) at 48.

[63] UN document A/CONF.151/PC/100/Add.17, *Managing Fragile Ecosystems: Combating Desertification and Drought.*

[64] J. Bernstein, P. Chasek and L. Goree (1992), "PrepCom IV Highlights, Tuesday 10 March", 1:8 *Earth Negotiations Bulletin*; and Bernstein, Chasek, and Goree (1992), "A Summary of the Proceedings for the Organizational Session of the INC for the Elaboration of an International Convention to Combat Desertification", at 5.

[65] The subsequent discussion on the forest issue did not in fact result in a convention, but only in a statement of principles. See Brenton (1994) at pp. 216, 229.

[66] J. P. Bernstein, P. Chasek and L. Goree (1992), "UNCED Highlights, 10 June", 2:9 *Earth Negotiations Bulletin*; see also S. P. Johnson (1993), *The Earth Summit: The United Nations Conference on Environment and Development (UNCED)* (London: Graham & Trotman/Martinus Nijhoff). At 243, Johnson argues that the United States changed its position to avoid a negative perception of their role in the Conference and because they did not want to antagonise the Africans. Additionally, the African Group had indicated throughout PrepCom IV and UNCED that it would insist on a convention before approving other UNCED documents. J. Bernstein, P. Chasek and L. Goree (1992) "PrepCom IV Highlights" 1:11 *Earth Negotiations Bulletin.*

[67] The request to begin negotiations is found in paragraph 12.40 of Chapter 12 of *Agenda 21*, "Managing Fragile Ecosystems: Combating Desertification and Drought". See United Nations (1992), *Agenda 21: The United Nations Programme of Action from Rio* (New York: United Nations Department of Public Information) 98-108.

[68] *UNGA resolution A/RES/47/188*, 22 December 1992.

[69] O. Greene (1996), "The Development and Implementation of International Environmental Regimes" at 280-312 in R. Blackmore and A. Reddish (eds.), *Global Environmental Issues* (London: Hodder & Stoughton and the Open University) at 296; S. Briceño, "Synergy Amongst and Between the Rio Conventions" at 183-187 in A. J. Fairclough (ed.) (1998), *World Development Aid & Joint Venture Finance 1998/99* (London: Kensington Publications and World Business Council for Sustainable Development). For a discussion on the biodiversity and climate change conventions see M. Grubb (1993), *The "Earth Summit" Agreements: A Guide and Assessment– An Analysis of the Rio '92 UN Conference on Environment and Development* (London: Earthscan).

[70] P. Chasek (1997), "The Convention to Combat Desertification: Lessons Learned for Sustainable Development", 6:2 *Journal of Environment and Development* 147-169.

[71] UN, *Agenda 21*, Chapter 12, at 98; *CCD*, Art.1(a).

[72] Art. 2(1)

[73] Greene (1996) at 296.

[74] Arts. 10, 11.

[75] Bernstein, Chasek, and Goree (1992) at 2.

[76] El Kassas (1993-1994).

[77] INCD document A/AC.241/L.22/Rev.1,*Resolution on Urgent Action for Africa*; INCD document A/AC.241/L.20, *Interim Arrangements*, 7 June 1994.
[78] Corell, E., W. Mwangi, T. Prather and L. Wagner, "Summary of the Resumed Tenth Session of the INC for the Convention to Combat Desertification", 18-22 August 1997", 4:106*Earth Negotiations Bulletin*, vol. 4 at 9.
[79] Odingo (1999).
[80] Odingo (1999) at 21.
[81] UNCOD: *Round-up, Plan of Action and Resolutions* at. 9, para. 16 (o).
[82] UNCOD: *Round up, Plan of Action and Resolutions* at 8; R. S. Odingo (1990), "The Definition of Desertification: Its Programmatic Consequences for UNEP and the international community", 18 *Desertification Control Bulletin* at. 31-50.
[83] UNEP, *Status of Desertification and Implementation of the United Nations Plan of Action to Combat Desertification*, at xiv-xv.
[84] K.. Danish (1995), "International Environmental Law and the 'Bottom-Up' Approach: A Review of the CCD", 3:1 *Indiana Journal of Global Legal Studies* 133-176.
[85] See, for instance, WCED (1987),*Our Common Future* (Oxford: Oxford University Press); and J. P. MacNeill, P. Winsemius and T. Yakushiji (1991), *Beyond Interdependence: The Meshing of the World's Economy and the Earth's Ecology* (New York: Oxford University Press).
[86] In the first two years of GEF operation, 83 percent of its funds were allocated to global warming and biodiversity. GEF Scientific and Technical Advisory Panel (1992), *Criteria for Eligibility and Priorities for Selection of Global Environmental Facility Projects*; GEF (1992),*Global Environment Facility: The Pilot Phase and Beyond* (Working Paper Series, Number 1).
[87] P. Chasek, L. Goree and W. Mwangi (1993), "INCD Highlights Wednesday 26 May 1993", 4:5 *Earth Negotiations Bulletin* at 2.
[88] *CCD*, art. 20 (2)(b).
[89] Chasek, Goree and Mwangi (1993), "INCD Highlights" at 1.
[90] J. Bernstein, P. Chasek and W. Mwangi (1994), "Summary of the Fourth Session of the INC for the Elaboration of an International Convention to Combat Desertification," 4:44*Earth Negotiations Bulletin* at 7.
[91] See C. Carpenter, P. Doran, A. Gupta and L. Wagner (1997), "UNGASS Highlights", 5:86*Earth Negotiations Bulletin* at 2.
[92] Corell, Mwangi, Prather and Wagner (1997), "Summary of the First Conference of the Parties to the Convention to Combat Desertification" at. 6. It was also agreed that the Global Mechanism would be hosted by a lead institution, the International Fund for Agricultural Development (IFAD), that the UN Development Programme (UNDP) would appoint the head of the institution, and that the World Bank would assist the IFAD and UNDP. See E. S. Howard and E. Corell (1998), "Soil/Desertification", *Yearbook of International Environmental Law* (Oxford: Clarendon Press).
[93] E. Corell (1998), "North-South Financial Tensions: Desertification after UNGASS", 7:1 *Environmental Politics* at 222-226.
[94] *CCD*, art. 21(4).
[95] ICCD/COP(3)/5, 19 August 1999, "Implementation of the Convention: Review of the implementation of the Convention, including support to regional action programmes".
[96] ICCD/COP(3)/5/Add.2 (D) and ICCD/COP(3)/5/Add.2 (E), "Implementation of the Convention: Review of Reports on Implementation of Affected African Country Parties, Including the Participatory Process and on Experiences Gained and Results Achieved in the Preparation and Implementation of National Action Programmes" 23 August 1999.
[97] A. Churie, J-S. Fritz, M. Schulman, and L. Wagner (1999), "Highlights from CCD COP-3", 4:131 *Earth Negotiations Bulletin*, at 2. For details from the 39 country reports submitted by African countries to COP-3, see ICCD/COP(3)/5/Add.2 (A), (B), (C), (D), and (E), "Implementation of the Convention: Review of reports on implementation of affected African country Parties, including on the participatory process and on experiences gained and results achieved in the preparation and implementation of National Action Programmes" (23 August 1999).

---

[98] E. Corell (1999), "Actor Influence in the 1993-97 Negotiations of the Convention to Combat Desertification", 4:2 *International Negotiation* 197-223.

[99] *CCD*, preamble para. 21 and arts. 9(3), 13(1)(b) and 21(3).

[100] S. Batterbury and A. Warren (2000), "Desertification", *International Encyclopaedia of the Social & Behavioral Sciences* (Oxford: Oxford University Press).

RECONCILING BASEL AND BAMAKO: THE FUTURE OF HAZARDOUS WASTE MANAGEMENT IN AFRICA

BEATRICE CHAYTOR AND MITA MANEK

INTRODUCTION – BACKGROUND TO INTERNATIONAL WASTE MANAGEMENT

The Stockholm Declaration and Customary International Law

Prior to 1992 the international aspects of hazardous waste management were addressed in a fragmented and incomplete manner. The roots of the international regime can be traced to Principles 21 and 22 of the *Stockholm Declaration.*[1] Principle 21 establishes that States have the sovereign right to develop their own domestic environmental policies, provided that those policies do not cause environmental damage in areas outside of a State's national jurisdiction. Principle 22 encourages States to cooperate in the development of international law regarding liability and compensation for victims of environmental damage. However, neither principle is of legal force, although Principle 21 has become customary international law and finds more forceful embodiment through rulings of international tribunals. According to the rulings in the *Trail Smelter Arbitration* (1938)[2] and the *Corfu Channel* case (1949)[3], States have a clear obligation to exercise due diligence regarding hazardous materials within their jurisdictions. In both cases, the fundamental conclusion was that States might be held liable for the negative effects outside their jurisdiction of hazardous materials used within their jurisdiction. This liability may exist whether the State is able to undertake mitigating measures or not. Indeed where it cannot, the relevant State has a duty to warn those who may be potentially affected. These responsibilities and liabilities are a close parallel to the more commonly seen theories of negligence in the law of torts.

Drawbacks to the Historical Regime

This rudimentary regime for hazardous waste management created by Principles 21 and 22 had major drawbacks. To enforce a complaint, an aggrieved State is must take its case to either the International Court of Justice (ICJ) or to international arbitration, assuming the infringing State agreed to submit to such action. However, any cause of action would be based on broad principles of customary international law rather than precise rules addressing hazardous wastes. Yet the increasing frequency of abuses in international waste management and the limited capacity among developing countries (which were by and large the targets of such waste materials)[4] to bring cases in the ICJ and arbitration panels, made it imperative for the international community to develop rules specifically addressing hazardous waste management.

The response was an initiative under the auspices of the United Nations Environment Programme (UNEP) beginning in the early 1980s. UNEP convened an Ad Hoc Meeting of senior government officials, expert in environmental law in 1981.[5] The meeting resulted in the *Cairo Guidelines and Principles for the Environmentally sound Management of Hazardous Wastes.*[6] The purpose of the *Cairo Guidelines* is to assist national governments in development of effective policies to handle issues of hazardous waste management.[7] The *Cairo Guidelines* urge States to take steps to "ensure the protection of health and the environment from damage arising from the generation and management of hazardous wastes".[8] The UNEP Governing Council authorised the Ad Hoc Working Group to prepare a global convention on the transboundary movement of hazardous waste.[9] The ultimate result of the working group meetings was the Conference of the Plenipotentiaries on the Global Convention on the Control of Transboundary Movements of Hazardous Wastes, which convened in Basel, Switzerland in March 1989, and which adopted the *Basel Convention on the Control of Transboundary Movements of Hazardous Wastes and their Disposal* ("the *Basel Convention*").[10] This treaty, which entered into force on 5 May, 1992, creates a global regime that takes precedence over the existing principles and that strictly regulates the control and transboundary movement of hazardous wastes and other wastes, obligating Parties to ensure their environmentally sound management, especially during the disposal process. The length of time it took for the *Basel Convention* to enter into force (eleven years) illustrates the highly contentious, and often highly technical, nature of international waste management.

THE BASEL CONVENTION

Goal

> "The overall goal of the Basel Convention is to protect human health and the environment against the adverse effects which may result from the generation, transboundary movements and management of hazardous and other wastes".[11]

The goal of the *Basel Convention* rests on two basic assumptions. The first is that hazardous waste is inherently dangerous. The second is that the generation of waste is inevitable. These two assumptions dictate the basic character of the *Basel Convention in* that it can be described as negative, its intention is clearly to restrict rather than facilitate. The controversy over the two basic assumptions is largely what links the two of them. If waste is indeed inevitable then should it be considered as a liability rather than as an opportunity? By this it is meant that if waste is inevitably generated, there is automatically an opportunity for commercial activity related to its effective disposal – a sort of 'multifunctionality' of hazardous waste. The inevitability is itself sometimes called into question, particularly by environmental advocates, on the basis that if effective recycling programmes and disincentives for waste generation were implemented then the volume of waste would be dramatically cut. The counter argument presented by industry advocates is that the costs and technology related to recycling or abatement measures may be prohibitive, even for developed countries, and that cutting waste generation would in many instances equate to a ceasing of production.

Objectives

With its main goal as a guiding principle, the *Basel Convention* Secretariat has identified six key objectives that provide the context for the Convention's functions. The first of these is: *to reduce transboundary movements of hazardous wastes and other wastes subject to the Basel Convention to a minimum consistent with their environmentally sound management.* There are several implications of this objective. Implicit in the desire to minimise transboundary movement is recognition of the widespread and possibly even systematic movement of waste. Minimising any kind of transportation means that movements of waste are far more easily monitored and at the same time shifts the burden of proof to the transporter to demonstrate why s/he should be allowed to transport instead of national authorities establishing why transport should not be permitted. The references to minimisation, rather than prohibition, and to environmentally sound management establish the principle that in some instances it may be desirable for wastes to be transported elsewhere for disposal. This is based on the idea that there are some locations that are far better suited than others, whether technically, financially or environmentally, to handling hazardous waste. For instance the United Kingdom has made conscious efforts to develop itself as a leader in the reprocessing of spent nuclear fuels. While radioactive waste is specifically excluded from the *Basel Convention* it is true that other countries have developed similar areas of expertise for other forms of waste. A particularly important point here is that developing countries are generally poorly equipped, by a range of criteria, to handle their own wastes let alone that of others. A final overall theme to this objective rests on the inherently dangerous nature of hazardous wastes. If hazardous waste is inherently dangerous then it should be transported as little as possible, since transportation tends to involve a rather high level of unexpected events – effectively a form of the precautionary principle.

This leads one directly to the second key objective, *[t]o dispose of the hazardous wastes and other wastes generated, as close as possible to their source of generation.* If transport of hazardous wastes should either be prohibited or minimised, then the logical conclusion is that it should be disposed of as close to the point of generation as possible. This objective is implicitly linked to the polluter pays principle. This is a method of ensuring waste producers directly address the externalities associated with their activities.

The difficulty with this lies in the question of capacity - given (natural or acquired) comparative advantage, might it be a better idea to acknowledge specialisation? In other words, is the point of origin the most suitable, or most environmentally sound, point at which to handle waste disposal? The apparent simplicity of the third objective: *[t]o minimize the generation of hazardous wastes in terms of quantity and hazardousness,* masks its complex nature. This is essentially the same issue of inevitability, discussed earlier.

The next objective is *[t]o ensure strict control over the movements of hazardous wastes across borders as well as the prevention of illegal traffic..* This ties into the transport question and the question of abuse. With its strict controls, the *Basel Convention* is simultaneously creating a mechanism that will restrict overall volume of movement while also allowing for the application of the precautionary principle to the high risks associated with transport. Nevertheless, abuse may take the form of action, which although not illegal, may be understood to constitute activities, contrary to the spirit of the *Basel Convention.* In such case, strict controls that are properly monitored limit the possibilities for activities such as undue influence in the issuance of documentation and pressure due to economic imbalances between importing and exporting countries. Prevention of illegal traffic addresses the other side of abuse. The level of illegal traffic should not be underestimated: documentation may be falsified and hazardous waste may be illegally dumped without consideration for the receiving State's laws on health or environment.

The fifth objective is *[t]o prohibit shipments of hazardous wastes to countries lacking the legal, administrative and technical capacity to manage and dispose of them in an environmentally sound manner.* This expresses the first form of abuse discussed above, where external imbalances and influences may undermine the control system. This is highlighted in several provisions of the *Basel Convention* in recognition of the fact that such hazardous waste activity probably constitutes the greatest threat to the effectiveness of the *Convention.* Legal, administrative and technical capacity are either limited or lacking in developing countries, leading to a situation where the onus must be on the transporter to establish that any transport is legal and based on prior informed consent.

The final objective, [t]*o assist developing countries and countries with economies in transition in environmentally sound management of the hazardous wastes they generate,* recognises the fact that even if abuses of the system are eliminated, waste will still be generated and constitute a challenge to the capacity of developing countries. UNEP has highlighted the special problems faced by developing countries in handling hazardous wastes. They include: "poor control of pollution, lack of financial resources and staff with technical and managerial skills, a low level of public awareness and lack of infrastructure for management and control."[12] To address this situation, the technology and experience of developed countries needs to be applied to and shared with developing countries. The risk with this is that an overemphasis of such technical assistance could obscure the fact that in some instances transport of waste to a specialised disposal point, may be a more cost effective and environmentally sound option. The focus of treatment and disposal of waste at source stems from fear of abuse of the system should the idea of the inherent value of transport be established.

Overall the six key objectives illustrate inter linkages through themes such as prevention of abuse and suspicion of transport. However, they also illustrate the tensions among the varying perspectives of the Parties to the *Basel Convention*, some of whom advocate the transport of waste, mostly for reasons of flexibility, and others that advocate the banning of waste transportation due to its innate dangers.

System

A key accomplishment of the *Basel Convention* is that it establishes a system for the international regulation of transboundary movement of hazardous wastes. The main operative principles are contained in Article 4, "*General Obligations*". The first of these, 4.1(a), is a reaffirmation of states' sovereign right to prohibit the import of waste, the only qualification being that the implementation of this sovereign right must be communicated to all other parties. Article 4.1(a) does make specific reference to waste "*for disposal*" but this is not expressed as final disposal. From the definitions included in Article 2 and Annex IV recycling and reprocessing are considered to be within the ambit of "disposal". While this definition may be convenient it would seem to contribute to some of the difficulties regarding the status of recycling and reprocessing under other elements of the *Basel Convention*. States should obviously have the sovereign right to reject all imports of waste material but at the same time a clearer division between waste destined for final disposal and recycling and reprocessing is a curious omission. Article 4.2(b) essentially supports 4.2(a) by requiring that Parties notified of a prohibition are required to recognise such by imposing a consequent prohibition on exports to countries with a ban. The purpose of this is largely to deal with the potential for 'leakage' in the *Basel Convention* in continuing recognition of the fact that some developing countries may experience abuses due to their general lack of capacity in enforcement.

The third, and central, operative principle is that any permitted international transporting of waste must be accompanied by written consent from the intended importing country, the principle of prior informed consent (PIC). This paragraph is qualified to state that written consent from a state with a ban is not acceptable. A State may export waste only after the importing and transit States have been provided with enough information to enable them to evaluate the waste and its proposed disposal, and have then consented to its import. States can forbid the import of the waste or can require additional information before consenting to import. The *Basle Convention* further provides that "any transboundary movement of hazardous wastes...without notification, or without consent...or with consent obtained through falsification, misrepresentation or fraud or that does not conform in a material way with the documents...shall be deemed illegal."[13] The *Basel Convention* relies on the Parties to pass appropriate domestic legislation to prevent illegal traffic.[14] In the event illegal traffic does occur, the State at fault is responsible for disposal of the waste, even if that means the exporting State must take back the offending waste.[15] To add weight to these provisions, Article 4.3 establishes that all illegal transport of waste is a criminal offence. This has been explicitly adopted due to the uncertain, but significant, level of illegal activities regarding the dumping of waste, particularly in developing countries and countries in transition. In many ways it is the scale of these and some borderline activities that led to the development of *Basel Convention* in the first place. Article 4.5 prohibits trading of waste between Parties to the *Basel Convention* and non-Parties but curiously this is essentially reversed in Article 11 where such transport is permitted, provided that the standards of the *Basel Convention* are observed in any agreement to transport. It is possible that this duality exists due to tensions during the negotiations for the *Basel Convention* between those who sought a blanket ban on transport and those who sought a more limited regime.

The *Basel Convention* provides for the Secretariat to act as a channel of information for State Parties. It may prepare reports and compile information concerning national sites and facilities, but Parties must specifically request any technical assistance, since the Secretariat does not actively assist Parties in handling hazardous waste movements.[16] One further notable principle exists in Article 4.6, which prohibits all exports to Antarctica. On the one hand this is driven by Antarctica's position as a relatively pristine wilderness but is also a reflection of the preexisting international agreements prohibiting the exploitation of the continent's resources. Article 4 does include further operative principles, but these are generally subsidiary requirements that oblige Parties to employ theories of best practice in areas such as mitigation of environmental impacts and enforcement.[17]

Overall what the principles in Article 4, and the detail provided in subsequent articles create, is a system that allows for the prohibition of international transporting of waste but that rests on the idea that such transport can be operated on a closely monitored system of best practice. As long as a prospective transporter obtains the requisite documentation and promises to abide by the relevant principles, its activities will be permitted. It is this basic idea that so alarmed developing countries and consequently led to the development of the *Bamako Convention* in Africa.[18] In February 1989, the OAU formally recognised in *Resolution 1199* that the *Basel Convention* was inadequate for the level of protection desired by African countries.[19] Developing countries were concerned about a range of issues that the *Basel Convention* failed to resolve, including *inter alia*, the right of non-signatory transit States to prior informed consent, the scope of the powers of the Convention Secretariat, coverage of territorial seas, the definition of 'environmentally sound management',[20] and liability for damage caused by the disposal of hazardous wastes.[21] *Resolution 1199* concluded by calling on African nations to agree on a "common African position for ameliorating the inadequacies contained in the [Basel] Convention."[22]

The reason that the idea of a closely monitored system with further, less enforceable, principles is so alarming to many developing countries comes from the recognition of two facts of life. The first is that the majority of developing countries have limited technical or administrative capacity in any of the areas required for the management or monitoring of hazardous waste transportation. In other words, acknowledgement that industrialised and developing countries do not possess the same capacity for dealing with hazardous wastes. Consequently it is unreasonable to expect that developing countries will be able to derive much realistic protection in the near future through their own efforts from the system that the *Basel Convention* proposes. However, what should provide the requisite protection for developing countries are the provisions of the *Basel Convention* that allow for banning transport of hazardous wastes, particularly the requirement that Parties amend their laws or policies to compliment bans imposed by other Parties.[23] In this manner the limited capacity of developing countries is less of a factor since there is a corresponding obligation on the part of waste exporters. In effect, if Bamako Parties were also Parties to the *Basel Convention*, the latter would have to adopt the ban on the import of wastes into Africa.

The second factor that particularly concerns developing countries is derived from the first, this is that illegal or borderline transport of waste has been such a common occurrence, particularly in Africa. While illegal transport is expressly addressed in the *Basel Convention* it is borderline cases that most disturb African countries. The majority of these are what are known as "sham" or "dirty" recycling incidents. A "sham" deal is, in most cases, likely to be illegal, as it is the transporting of waste under nominally legal documentation, which has been obtained under false pretences or by similar means. Such documentation normally indicates that a shipment is destined for recycling or reprocessing when it is actually destined for final disposal, and unlikely to be the "environmentally sound" final disposal required by the *Basel Convention*. A "dirty" deal is almost always legal as the waste transported is actually destined for some form of recycling but in such a manner that creates adverse environmental impacts at the point of recycling.[24] The lack of specific standards on what is "environmentally sound" allows an importer to argue that their environmental standards are the best possible under the circumstances. However, once again this fear should be abated by the allowance for import and export bans. The levels of evasion in the past have brought into sharp relief the complexities of the hazardous waste industry.

AN AFRICAN RESPONSE TO BASEL – THE BAMAKO CONVENTION

Goal

The goal of the *Bamako Convention* is almost purely a response to the perceived inadequacies of the *Basel Convention* discussed above.[25] It is basically the imposing of a complete ban on the importing of waste into Africa from elsewhere and a Basel like restriction of any intra-African or external transport of hazardous wastes. The system, as with the *Basel Convention*, is largely achieved through the requirements of Article 4. Article 4.1 lays down the blanket ban for imports of "*all hazardous waste, into Africa from non-Contracting Parties.*" The non-Contracting Parties stipulation is a little unusual: according to the terms of Article 22.1, only Member States of the Organization of African Unity are eligible to become Parties meaning that the ban of imports into Africa would seem to be sufficient. However, if one reads the clause as "into Africa *and* from non-Contracting Parties" then it makes a lot more sense. Such a reading excludes the possibility of 'sideways' imports that would be brought into a non-contracting African state and then subsequently transferred. This intention seems to be reinforced by 4.1(b) that requires an import ban on hazardous wastes from non-Parties.

Similar to the *Basel Convention*, the *Bamako Convention* permits bilateral or multilateral agreements on transboundary movement of wastes among Parties and non-Parties. Such agreements are only allowed for hazardous wastes generated in Africa, meaning that even if non-Parties brought hazardous wastes into Africa those wastes may not be exported to the rest of Africa. This rule is of tremendous significance for recycled and reused wastes. Again, it prevents a non-Party from importing hazardous wastes as raw materials for use in industry and then disposing of those wastes after use in other African countries. The goal of the *Bamako Convention* may seem a little one sided in that importing is prohibited while exporting is permitted, but this is founded on two quite reasonable theories. The first is that African exports of waste have not historically been a problem - it is exports from developed countries to Africa that engender great concern. The second concerns capacity, and particularly technical capacity. African countries generally acknowledge that they do not have the capacity to handle waste in most instances, but this does not mean that some countries do not have greater capacities and thus are the most suitable destinations for waste. This position is widely reflected in national legislation that in many instances also prohibits most exports.[26] In recognition of the possibility of changing circumstances, the Parties to the *Bamako Convention* agree in Article 15.4 that the Conference of the Parties will "*keep under continuous review and evaluation the effective implementation of this Convention*". However, this is a broad statement only made effective by the amendment provisions of Article 17, which states that a consensus should be sought, and failing that a two-thirds majority is required. Effectively this means that if and when African States reach a point where they need to address international waste transport through regulation and monitoring, they will be able to do so under the *Bamako Convention*.

Objectives

A survey of the Preamble to the *Bamako Convention* produces eight main objectives. The first addresses the generation of hazardous waste itself. This operates on two levels – the first is to limit any increase in waste generation, and the other to encourage a decrease in current level, either in terms of quantity and/or hazard potential. The flexibility in terms of quantity and/or hazardous potential provided by the Preamble is an interesting concept, since it implicitly recognises situations where responsible handling of waste may be preferable to the prevention of its generation. This could amount to an acknowledgement of the inevitability of waste generation from certain key industries, or that the production and environmentally sound disposal of waste is more economically viable than addressing its generation at source. The second objective is to establish the principle that transboundary movement of waste is an undesirable activity. This is driven by three factors – the first is the proposition that the transport of such waste is inherently dangerous. The second relates to the fear that Africa is a frequent destination for transboundary movement of waste. The final factor is the belief that the prohibition on transboundary movement of waste will lead to more effective and responsible activity at the point of origin. All of these factors present difficulties.

The idea that transboundary movement is inherently dangerous leads to the assumption that it cannot be accomplished competently and effectively. In many instances, transboundary movement of hazardous substances and materials is undertaken without significant threat to either human health or the environment. Given historical events the second factor is completely justified, but should not be the driving force of the management system. The rationale for an effective management system should be to ensure that adequate measures are in place to address the problems of the waste disposal industry. The third factor underscores the motive of regulatory authorities: to change producer behaviour. Yet there may be occasions when either the significant reduction of waste generation or its disposal at the point of origin, are not realistic options. To impose a complete ban creates a straight jacket that may in its turn encourage illegal or borderline behavior. Overall some flexibility in this objective may well be desirable.

Another objective expressed in the Preamble effectively recognises the polluter pays principle. By this principle, the generators of undesirable materials and by-products thereof would be responsible for their safe handling. There is also an internationally accepted concept featured in the Preamble: that states have the sovereign right to ban imports of hazardous waste into their territory. The *Bamako Convention* deviates from this slightly by imposing qualifications, i.e. the right to ban for environmental and human health reasons only, although this is not likely to fundamentally affect the basic sovereign right.

The fifth objective has become well accepted in the field of waste management. It is that local disposal is encouraged. This is practical in that it creates a situation that allows for transport where local disposal is not feasible or desirable. This is in light of the fact that wherever feasible, the waste generator should take full responsibility for disposal, whether locally or otherwise. The next objective provides the underpinning for the whole system; it establishes human rights and conservation as the twin pillars upon which waste disposal decisions should be made. The right to life has become an almost universal element of African constitutions and legal systems. Generally, this is broadly interpreted and understood to include rights such as the right to a healthy environment. [27] It is through these interpretations that the human rights - conservation link is made in the context of waste disposal. The conservation element is normally reinforced through recognition of the need to operate on a theory of sustainable development, where a State's natural resource base is not prejudiced or unduly threatened.

The sixth objective is to employ clean production methods wherever possible.[28] This ties into many of the earlier objectives but particularly the need to minimize the generation of waste wherever possible. The difference here is that methods for the disposal of waste are explicitly considered to fall within the ambit of "clean production methods". This is interesting because again, it is recognition of the inevitability of a certain level of waste generation and the fact that there can be specialised methods of disposal.

The final readily identifiable objective is a need for flexibility when dealing with African generated waste. This is a sensible approach, in recognition of the fact that waste disposal and treatment within Africa may require transport to other countries that may be better equipped to deal with such wastes. This recognition largely derives from the identification of developed countries as the sources of the most problematic waste. Such identification is justified by historical experiences and may even still be valid today. Nevertheless, African States are clearly pursuing an industrialising agenda (with varying degrees of success), and it is often posited that they cannot be held to the environmental standards of the North. The risk in this is that a variety of African States may well soon be involved in the generation of waste similar to levels in developed countries. Where levels of hazardous waste increase in some African countries, it is important that they do not follow former patterns of developed country behaviour by shipping the wastes to other less developed African countries.

System

The other half of the *Bamako Convention* regime - a regulatory system for permissible trade - is extremely similar to the system designed in the *Basel Convention*. However in certain areas, there are different approaches. Under the *Bamako Convention*, all hazardous waste generators have to report to the Secretariat on waste they generate.[29] This is useful from the perspective that the Secretariat can take an overview of the waste management system in Africa. However this raises questions as to its feasibility either in terms of the likelihood of reporting by generators, or of the ability of the Secretariat to cope with the volume of information received. Despite this question of capacity, the principle is still probably a good one - the Secretariat needs information to enable it to correctly assess the picture of the hazardous waste industry in Africa. Parties to the *Bamako Convention* are required to make adequate disposal facilities available.[30]

Mirroring the *Basel Convention*, the *Bamako Convention* does not permit exports to those who have imposed a ban.[31] However, the *Bamako Convention* goes further, prohibiting exports to countries that do not have adequate disposal facilities.[32] The burden of verification falls upon the exporting State. Parties under both the *Basel* and *Bamako Conventions* are thus responsible for waste during the entire transport cycle. This is a useful concept in that it provides for clear allocation of responsibilities, however, it introduces elements of extra-territoriality that may need careful handling. A further element in the heightened responsibilities that the *Bamako Convention* imposes upon its signatories is that any export of waste is only allowed if the state of origin doesn't have the technical capacity and facilities for disposal, including recycling and reprocessing.[33] This provision has obvious advantages with the only potential drawback being the difficulty in assessing what might be considered adequate facilities.

In effect, given the nature of hazardous waste it would be wise for it to be treated at the most suitable site possible, whether that happens to be local or not. For example one might have a low-grade final disposal facility in the state of origin and high grade recycling facility elsewhere. In such circumstances, does it make sense to insist that waste be handled at the local facility? This will inevitably depend on whether the Secretariat is prepared to look at the relative quality of facilities in addition to their simple existence. Such an approach will require careful monitoring leading to great advantages. This flexibility may be found in the second half of the same clause, which allows the Parties to the *Bamako Convention* to set out other export criteria that conform to the objectives of the *Convention.*[34] Since one must assume the good faith of the Parties, adopting a flexible approach is clearly positive. A similar provision exists in the *Basel Convention* but it is accompanied by an allowance for exports for the purpose of recycling or reprocessing, enabling the allowance for good faith exports.[35] However, the question of the relative quality of facilities is also not addressed in the *Basel Convention*. The fundamental issue of the nature of waste being transported is addressed by both conventions in a similar manner. Under the *Bamako Convention*, the documentation covering any waste shipment must declare its potential danger on the environment and human health.[36] In the *Basel Convention*, declaration is dependent on the possible effects of the proposed movements in question.[37] The Bamako approach is superior to that in the *Basel Convention* in that, from a legal perspective, it allows far less opportunity for loopholes. Under the Basel regime, one could argue that there were effects that the exporter was unaware of, or in the extreme that they are not sufficiently probable to be considered or not inherent in the waste itself. The *Basel Convention* only deals with the specific movement at hand rather then the long-term risks of wastes. Bamako's use of "potential danger" means that all possibilities should be considered, probably to some sort of reasonableness standard, but as a general principle nonetheless.

Another significant point relating to documentation is that the *Basel Convention* allows for multiple shipment documents covering a twelve-month period while specific documents for each shipment under the *Bamako Convention*.[38] Although the *Basel Convention* stipulates that any multiple shipment documents provide details of individual shipments, Bamako's approach would seem to be safer despite the likely higher transaction costs. There is less of a risk of documents being confusing or misleading, and specific tracking of each shipment would be a simpler process. Equally, if monitoring is close enough, the Bamako approach should allow authorities to have more flexibility in authorising or refusing shipments depending on previous performance and/or changing circumstances. It is true that corporate interests would prefer the increased certainty of longer-term arrangements but this does not really outweigh the need for flexibility when protecting human health or the environment.

The Basel and Bamako regimes also differ over the modalities for shipment of waste. The *Bamako Convention* requires that Parties nominate specific points of entry that are competent to handle waste imports and exports.[39] The basis for this is that waste transportation involves unusual technical skills and facilities that are not likely to be available at all ports of entry in a given state. Consequently it makes sense for such activity to be channeled to specific points where the requisite skills and facilities can be made available. This has implications for the safe handling of waste but also could be extremely helpful for enforcement as monitoring capacity can be focused along with handling. It is possible that the *Basel Convention* is relying on the assumption that states of import will specify particular points of entry when providing their consent for a shipment but even accepting that premise, the Bamako approach would seem to be more secure.

Overall the fundamentals of the regulatory side of the *Bamako Convention* and the *Basel Convention* are vastly similar. Where they diverge, the differences are almost exclusively tied to Bamako parties' desire to either impose higher standards of care or to require more certain methods of verification. In the subsequent analysis, it would seem that many of the innovations in the *Bamako Convention* could meet future consideration by the Basel Secretariat.

The Basel – Bamako Nexus, Synergies and Conflicts

The most salient themes in the relationship between both conventions relate to the import ban, liability, information disclosure and the role of the Secretariat. Do the similarities between the two conventions justify attempts to merge or reconcile them? What would be the purpose of this 'merger'? Would it serve the interests of both developing and developed countries? There are no inconsistencies in the provisions, but the duplication of effort involved would seem to suggest that a search for synergies would be a rewarding activity.

The *Basel Convention* in effect could be interpreted as prohibiting the import of wastes into Africa through the operation of Article 4.2 (e), which provides for equivalent prohibition measures where the destination State has prohibited imports of hazardous wastes into its territory. The recent decision by the Conference of the Parties to the *Basel Convention* to impose a ban on the transboundary movement of hazardous wastes destined for recycling and recovery from OECD to non-OECD States, brings the *Basel Convention* even closer in ambit to the *Bamako Convention*.[40] Incorporating an import ban into the *Basel Convention* under specified conditions may therefore be feasible.

However, on the treatment of liability for damage resulting from transboundary movement and disposal of hazardous wastes, the two treaties diverge. The *Basel Convention* has postponed addressing this issue while the *Bamako Convention* treats it directly.[41] It imposes "strict, unlimited liability as well as joint and several liability...on hazardous waste generators."[42] Such liability regime only applies to injuries arising out of disposal or hazardous waste; accidents during transboundary movement are not covered.

Both conventions emphasise the transmission of specific information as a means to controlling waste movements.[43] The respective Secretariats play a key role in the access and channeling of information. *Bamako Convention* Parties must provide, even without the request by the Secretariat, information concerning: illegal activities; accidents; breaches of the treaty; and movement notification. Moreover, the *Bamako Convention* Secretariat is required to: conduct hazardous waste audits; investigate breaches of the treaty; provide information about consultants; and communicate with relevant intergovernmental organisations and non-governmental organisations (NGOs) about the implementation of the *Convention*. By contrast, the *Basel Convention* does not require its Secretariat to undertake such activities. In this respect the roles of the respective Secretariats are different with the *Bamako Convention* Secretariat designed to be more proactive.

Although in several respects, Bamako deviates from Basel, could its methods serve to protect the interests of most developing countries and be feasibly incorporated into the *Basel Convention*? The *Bamako Convention* establishes a dual system of regulation that may actually make it easier for developing countries to control hazardous waste movements and import bans plus a comprehensive regulatory framework not unlike what is present under the *Basel Convention*. Given this situation, a combination of the two systems with some added elements would seem to present the way forward in international waste management. If one took the basic Basel system and added in some of the heightened monitoring mechanisms of Bamako, one would go a long way towards stronger regulation of the international trade in waste.

Hazardous Waste – Future Regulatory Structures

The aim of a future regulatory system on hazardous waste management should be on the one hand to limit the generation of waste wherever possible but on the other hand to recognise that the generation of some waste is inevitable. Where this recognition is accepted, initiatives can be made to ensure that the waste generated is handled in the most efficient and creative way possible. To do this waste generators have to be given the right incentives to comply with the system rather than try to evade it, and states need to be given the requisite tools to develop their individual enforcement and monitoring procedures. How can the crucial element of the correct incentives be provided alongside acknowledgement of the fact that developing countries currently have limited capacity for recycling, reprocessing or final disposal? A solution might be the introduction of a certification system to go along with what is essentially the current end-user certificate mechanism in the Basel and Bamako regimes. In such a system the presumption would be that all Parties have imposed a ban on waste imports unless they apply to the Convention Secretariat for registration as a potential importer of waste for whatever purpose. To obtain certification the Secretariat could be required to assess the competence of the applicant State according to criteria set by the Parties. Such criteria could include the quality of facilities, technical capacity and enforcement capacity. Once certification has been granted, waste generators could then proceed with the current system of documentation covering all stages of transport based on prior informed consent of both the importing and transit States.

CONCLUSION

Four main points appear of particular interest when examining the relationship between the *Basel* and *Bamako Conventions*. These are the *raison d'être* of the *Bamako Convention*, the driving forces behind the regimes, the question of liability in the event of problems and finally the relationship between the public and private sectors. All four points are strongly related but the first two and the last two are particularly intertwined.

The *raison d'être* of the *Bamako Convention* stems from the dissatisfaction with the regime that the *Basel Convention* imposes. This is clearly evident from the negotiating process leading up to the *Basel Convention* involveing strong lobbying by developing countries, and particularly Africa, for a complete ban of waste exports.[44] As has been mentioned earlier, once it became clear that this aim would not be realized, African States resorted to drafting the *Bamako Convention*. The problem with this situation is that it may discourage effective African participation in the Basel regime or the commitment of developed country Parties to the *Basel Convention* in providing prescribed assistance. The membership of the two conventions overlaps, yet a substantially higher number of African States are Parties to the *Basel Convention*. The majority of *Bamako Convention* States are parties to both agreements.[45]

Both Conventions have as their underlying premise the subtle discouragement of transport in hazardous waste. In the *Bamako Convention*, even waste destined for recycling is to be treated where it is generated. This ethic of leaving waste where it may not be adequately disposed of rather than transport it, raises concerns about the flexibility of the Conventions to adapt to changing circumstances.

The second set of points relates to the relationship between the public and private sectors under the two regimes. Both, as international instruments, are agreements between governments. On the other hand, both Conventions are almost exclusively concerned with regulating the activities of private actors for the benefit of the general public (also private actors). The question of treaty obligations, which also potentially binds corporations and affects public interests, is mirrored in other multilateral environmental agreements such as *CITES*,[46] the *Montreal Protocol*,[47] the *Biosafety Protocol*,[48] and the *Kyoto Protocol*.[49] International treaty negotiations are increasingly demonstrating that private interests play a greater role in the process than what is seen on the surface. At the same time, the question may be raised as to how such multilateral frameworks can accommodate both the interests of the private sector and the general public.

This leads to the fourth and final point of particular interest in the nexus between the two Conventions - the question of liability. Both treaties assign various responsibilities to the State Parties, but liability is restricted to the private actors; exporters; transporters; and importers. Thus both the government and private sectors have a responsibility to see that waste is handled in an environmentally sound manner. Where waste is not handled according to the regulatory system set out in either Convention, responsibility must be apportioned to the party solely responsible for the resultant damage. In the first instance, the relevant generator of the waste must bear responsibility, yet governments bear a secondary responsibility to regulate the activities of waste generators within their jurisdiction. These latter issues have come to be some of the most controversial ones in the continuing development of both Conventions, where the establishment and enforcement of protocols on liability and compensation are being discussed.

In the final analysis, there are advantages and drawbacks to both the *Bamako* and *Basel Conventions*. The latter agreement has some flexibility in approach to waste management but its monitoring and enforcement mechanisms are not as stringent as they might be. The *Bamako Convention* is more rigid but has some useful ideas on monitoring and enforcement. The basic difficulties are that both Conventions are reactive rather than proactive. The future of the international system for managing the transport of hazardous wastes inevitably depends on the needs and aspirations of the Parties to the respective treaties.

---

[1] *Declaration of the United Nations conference on the Human Environment*, adopted by the UN Conference on the Human Environment, Stockholm, 16 June 1972 (A/CONF.48/14/Rev.1).

[2] (1939) 33 *AJIL* 182 (1941) 35 AJIL 684

[3] *ICJ Rep*. (1949) 4

[4] The primary incentive for targeting developing countries for waste disposal is economic. The disposal costs in developing countries are often a fraction of the costs in developed countries. Another factor is the lack of environmental regulation or enforcement in developing countries.

[5] United Nations Environment Programme (1981), *Report of the Ad Hoc Meeting of Senior Government Officials Expert in Environmental Law*. UN Doc. UNEP/GC.10/5/Add2 (1981), at 10-11.

[6] United Nations Environment Programme (1987), Governing Council, 14th Sess.,*Cairo Guidelines and Principles for the Environmentally Sound Management of Hazardous Wastes*, UN Doc, UNEP/GC.14/17 Annex II (1987) [*Cairo Guidelines*].

[7] *Cairo Guidelines*, at 3.

[8] *Cairo Guidelines*, at 5.

[9] UN GAOR, 42nd Sess., Supp No. 25 at 83; UN Doc. A/42/25 (1987), at 84.

[10] UN Doc UNEP/WG. 190/4. [28 ILM (1989) 657]

[11] *Basel Convention*, Preamble.

[12] UNEP (1991), Governing Council, 16th Sess., Report of the Executive Director, UN Doc, UNEP/GC.16/20 (1991) at 24.

[13] Article 9.1.
[14] Article 9.5. States are therefore empowered to make some types of transboundary movements of hazardous wastes subject to criminal sanction.
[15] Article 9.2.
[16] See Article 16.1.
[17] See Article 4 paragraphs 7 to 13.
[18] *Resolution 1153* passed by the OAU Council of Ministers in 1988 was the first formal response by African countries to the dumping of hazardous wastes in Africa. See OAU (1988),*Resolution on Dumping of Nuclear and Industrial Waste in Africa*, OAU Council of Ministers Res. CM/Res. 1153 (XLVIII) (23 May 1988), reprinted in 28 ILM 567. The Resolution contained key themes: a) a ban on the import of hazardous wastes into Africa; b) the liability of waste generators for damage caused by their wastes; c) assistance from developed countries in monitoring and controlling hazardous waste movements; and d) the use of ecologically rational methods of disposal of hazardous wastes.
[19] OAU (1989), *Resolution on a Global Convention for the Control of Transboundary Movement of Hazardous Wastes*, OAU Council of Ministers Res. CM/Res.1199 (XLIX) (25 Feb. 1989), reprinted in UN GAOR, 44th Sess., UN Doc A/44/291 (1989), at 31-32.
[20] The precise meaning of 'environmentally sound management' was left open in the *Basel Convention.*
[21] (1989), *Western, African Nations Fail to Agree on Transboundary Movement of Toxic Wastes*, 12 International Environmental Reporter (BNA) (8 Feb. 1989) at 49
[22] Ibid. at 32. As a result, no OAU member state signed the *Basel Convention* in 1989. However, Nigeria has since signed and ratified the *Basel Convention.*
[23] See Article 4.2 (e).
[24] See J. Krueger (1998), "The Basel Convention and Transboundary Movements of Hazardous Wastes", (The Royal Institute of International Affairs, Energy and Environmental Programme, Briefing No.45, May 1998).
[25] *Bamako Convention on the Ban of the Import into Africa and the Control of Transboundary Movement of Hazardous Wastes within Africa*, opened for signature 30 January 1991, 30 ILM, 773. The Convention entered into force in February 1996.
[26] See *Law on Toxic and Nuclear Waste,* Cote d'Ivoire Law No. 88651, 7 July 1988, translated in 28 ILM 391.
[27] See C. Bruch (1999), "What's the Constitutional Basis of Africa's Environmental Law?" 6:2 *Innovation;* C. Bruch (2000), *Constitutional Environmental Law: Giving Force to Fundamental Principles in Africa*, Environmental Law Institute, Research Report, May 2000.
[28] See Article 3 (f) and (g).
[29] Article 4.3 (a).
[30] Article 4.3 (d).
[31] Article 4.3 (r).
[32] See Article 4.3 (j).
[33] Article 4.3 (n) (i).
[34] Article 4.3 (n) (ii)
[35] See Article 4.9 (c), (b).
[36] Article 4.3 (u).
[37] Article 4.2 (f).
[38] See *Basel Convention*, Article 6.8; *Bamako Convention*, Article 6.1, 6.6.
[39] *Bamako Convention*, Article 5.
[40] See *Basel Convention Conference of the Parties* Decision III/1.
[41] See *Basel Convention*, Article 12.
[42] *Bamako Convention*, Article 4.3(b).
[43] *Bamako Convention*, Article 13, *Basel Convention*, Article 13.
[44] See *Western, African Nations Fail to Agree on Transboundary Movement of Toxic Wastes*, 12 Int'l Env't Rep. (BNA) (4 Feb. 1989), at 49.
[45] For instance, *inter alia*, Cote d'Ivoire, Senegal, Tanzania and the Democratic Republic of Congo are parties to both Basel and Bamako.
[46] *Convention on International Trade in Endangered Species of Wild Fauna and Flora (Washington), (*1973) 12 ILM 1085.

[47] *Protocol on Substances that Deplete the Ozone Layer (Montreal)* (1987) 26 ILM 1550.
[48] *Protocol on Biosafety (Cartagena)*, (2000) 39 ILM 1027.
[49] *Protocol to the Framework Convention on Climate Change (Kyoto)*, (1997)(1998) 37 ILM 22.

CITES AND THE AFRICAN ELEPHANT

MICHAEL KIDD[*] & MICHAEL COWLING[+]

INTRODUCTION

Probably the most contentious issue that has arisen within the scope of the Convention on the International Trade in Endangered Species of Flora and Fauna (CITES) is that of the ivory trade. This article considers the status of the African elephant (*loxodonta africana)* under CITES and suggests that CITES has failed the elephant. CITES obviously also concerns many other species found in Africa, but the elephant will be focused on in this Chapter as a particular case study that highlights many of the problems faced by CITES and the international political influence on the Convention.

CITES is concerned with the regulation of international trade in endangered species and 'specimens' – any animal, both living or dead and including any readily recognisable part or derivative thereof.[1] This would include elephants' tusks – ivory. CITES does this by means of a system of Appendices. For present purposes, we are concerned with Appendix I and II. Animals listed in Appendix I are considered to be threatened with extinction and hence any form of international commercial trade is prohibited. Appendix II species are regarded as potentially threatened with extinction with the result that international trade is permitted but strictly controlled by means of a permit system. The objective of the latter is to ensure that trade in such species is not detrimental to this survival.

At the outset it is necessary to consider why the elephant is threatened. The ivory trade is certainly a major threat, but so are loss of habitat and, related to this, conflict with human land use, particularly cultivation of crops. Elephants require extensive natural habitat in order to survive. In this regard a UNEP Report has stated that the average daily intake of dry weight of grass is 22kg whereas the annual home range of the African bush elephant is in the region of 3000km.[2] This then raises the issue of competition for resources between elephants on the one hand and the indigenous population on the other. One of the main reasons for this is that, although there is 'some controversy about the preferred diet of the African elephant, there is no doubt that they have a strong predilection for the crops grown by human in their market gardens and their farms'.[3] This has led one observer to note rather wryly that 'human fondness for elephants stands in inverse proportion to the distance at which one lives from them'.[4] Another important contributing factor is the instability (if not war) prevalent in many elephant range states. CITES is concerned only with the trade issues, and this is an important point to bear in mind.

In order to appreciate the way in which the elephant debate has proceeded within CITES, it is necessary to examine how different range states have approached the issue. In South Africa, a relatively healthy elephant population is found only in protected areas. Until recently, the conservation authorities have selectively killed (culled) a number of elephant that it considers to exceed the carrying capacity of the protected area in question. This controversial culling policy is currently in abeyance, but has not been permanently terminated. The culling has generated stocks of ivory that South Africa would like to sell in order to raise funds for conservation.

Other southern African states – Zimbabwe, Botswana, Namibia, Malawi and Zambia – also have healthy elephant populations, but in these countries elephants exist outside of protected areas and consequently come into conflict with people. The argument that has been put forward by all of these countries (not always consistently) is that people will tolerate elephants in their vicinity and the threat that they pose to their crops and to their personal safety, if they see the elephants as having some kind of benefit to them. In Zimbabawe, for example, under the CAMPFIRE project, communities benefit financially from selling the rights to hunt elephants to foreign hunters. If the elephant is seen as having no economic value, then it is treated as just another nuisance.[5]

In other African countries, however, particularly East African countries, elephants have suffered a far more serious toll. The reasons for this are not entirely clear but poaching for purposes of the illegal ivory trade certainly has played a major role in the decline of elephant numbers in East Africa.

CITES AND THE AFRICAN ELEPHANT

The population of African elephants diminished significantly during the 1980s – it has been estimated that the 'illegal killing of elephants for their ivory has reduced Africa's elephant population from 1.5 million to fewer than 500 000 in the [1980-1990] decade'.[6] The international response to this up until 1994, including the measures adopted by CITES, is excellently documented by David Harland.[7] The detail of the responses will not be repeated here, but in summary, the measures adopted by the CITES parties have been as follows:

CITES 7[8] – LAUSANNE –1989

The parties accepted a proposal which essentially required the placing of all African elephant populations onto Appendix I, but to establish a mechanism under which particular populations could be downlisted at a later date if they could prove that they met certain biological and management criteria.[9][10]

The Appendix I listing came into affect on 18 January 1990. Despite the fact that most of the southern African states lodged reservations against the listing, the market essentially collapsed because the importing states (in particular Japan) stood by the CITES 7 decision. The consequence of this was apparently a marked reduction in elephant poaching, especially in East Africa. As Harland suggests, 'the ban was more effective than many of its critics had believed possible'.[11]

CITES 8 – KYOTO –1992

The parties agreed on a proposal recognising that 'commercial trade in wildlife and wildlife products can be beneficial to the conservation of species and ecosystems'.[12] The parties also agreed to overhaul the Berne Criteria (requirements for the listing of species on the Appendices) with the idea that these would be presented at CITES 9. There was no success at CITES 8 for the southern African states' attempts to change the listing status for individual species. There appeared often to be no scientific basis for the decisions made. For example, a southern African proposal to downlist the leopard from Appendix I to II, well supported by 'excellent' population and trade data, was rejected out of hand half an hour after the North American bobcat was down-listed 'without question', despite the absence of 'convincing reason of any sort'.[13] The difference was 'that the leopard had a following in the nongovernmental community that the bobcat did not'.[14]

Despite this lack of success, there were two significant victories. The first was achieving a move towards the idea of sustainable use underpinning the Convention. Secondly, southern Africa had made its case very strongly in the media (at least the 'serious' media) and this is very important in an arena in which public opinion counts for so much.

CITES 9 – FORT LAUDERDALE – 1994

The existing Berne criteria for deciding the grounds for listing species on the appendices, were replaced by the so-called 'Everglades Criteria'. There is no space in this Chapter to consider these requirements in detail,[15] but insofar as the criteria relate to downlisting from Appendix I, the criteria, if strictly applied, appear to be so wide as to make it highly unlikely that the elephant will ever be transferred to Appendix II or, more to the point, that the ivory trade will ever be opened again.

CITES 10 – HARARE – 1997

After heated debate,[16] a limited resumption in the ivory trade was sanctioned by the parties. Specifically, the elephant populations of Botswana, Namibia and Zimbabwe would be transferred from Appendix I to II from 18 September 1997. This would allow the countries, from that date, to export sport hunting trophies for non-commercial purposes and live elephants to appropriate and acceptable destinations. The three countries were allowed to sell, at a once-off sale, stockpiles of ivory and quotas were set for each country as follows. Such trade could only take place after 21 months had elapsed from CITES 10 and then only after certain conditions had been met, as certified by the Standing Committee.

CITES 10 also made certain resolutions concerning the elephant, which related to the downlisting decision but were certainly not as contentious or high-profile. The first of these resolutions[17] dealt with the consideration of proposals for the transfer of African elephant populations from Appendix I to Appendix II, repealing similar criteria adopted at Lausanne eight years earlier. The new criteria,[18] seemingly augmenting the Everglades criteria, constitute together with those criteria, on paper at least, a formidable array of considerations which have to be satisfied before the elephant in particular can be transferred from Appendix I to Appendix II.

CITES 10 also made some important decisions relating to monitoring. The parties recognised the Bad Ivory Database System (BIDS), established by TRAFFIC for the purpose of monitoring illegal trade in ivory and other elephant specimens in 1992. Monitoring of illegal hunting in range states was also raised, with the decision being to establish 'a system through which the impact of CITES decisions with respect to elephants and trade in elephant specimens can be measured … [including] a simple system of international reporting of incidents of illegal hunting as a baseline against which changes in trends can be detected'.[19]

This, of course, raises the issue of whether there *was*, in fact, an increase in illegal hunting of elephants after CITES 10. The anti-trade lobby were predicting 'doom and gloom' after the decision to downlist the African elephant, albeit in a very limited sense. It was described variously as sounding the 'death knell' for Africa's elephants,[20] and 'a new wave of poaching and possible extinction of elephants outside of Southern Africa'.[21] The International Fund for Animal Welfare reported in October 1997 that significant numbers of elephants were being poached in the Democratic Republic of Congo and Central African Republic by Sudanese poachers, poaching figures were increasing in Zimbabwe, and that Kenyan poaching figures, seemingly insignificant within national parks, were increasing 'dramatically' on private land.[22]

Shortly before the Standing Committee met in February 1999 to decide whether the conditions set down at CITES 10 for the one-off trade in stockpiled ivory by Zimbabwe, Namibia and Botswana had been met, reports claimed that 'dozens' of elephants had been killed and their tusks removed in and around Kenyan national parks, seemingly in anticipation of the Standing Committee's positive decision.[23] Earlier, the CITES Secretariat had indicated the importance of gathering information about illegal killing and had mentioned 'unsubstantiated reports' of such killing, but rang no alarm bells to suggest that the animal rightists' gory predictions were coming true.[24]

The Standing Committee met in Geneva in early March 1999, and decided that Namibia and Zimbabwe could sell the quotas of ivory agreed upon at CITES 10 to Japan. Botswana was considered as 'almost' meeting the conditions, and would be allowed to do so once a CITES Team's visit to the country determined that the conditions were satisfied there.[25] Botswana subsequently received approval and on 9 April 1999, Namibia put just over 12 tons of ivory on auction. Four days later, Zimbabwe sold 20 tons and on 17 April, Botswana sold nearly 18 tons.[26]

Not six months later, Kenyan authorities apparently noticed an increase in elephant poaching, prompting the media to state that poaching had 'soared since a ban on ivory was lifted earlier this year'.[27] The figures which prompted this alarm were 58 elephants killed in 1999 up to September, compared to 41 in the whole of 1998,[28] an increase and some cause for concern, but hardly alarming. But more 'evidence' of illegal killings of elephants would be raised in the time leading up to the most recent Conference of the Parties, the eleventh.

CITES 11 – GIGIRI - 2000

Six proposals on African elephants were to be placed on the table at CITES 11, four of them from southern African states proposing trade, outlined in more detail below. Following the pro-trade lobby's success at Harare, South Africa now joined the proposers amidst claims that this was 'irresponsible'.[29] Shortly before the meeting, a Kenya-based organisation called Save the Elephants released a report indicating that there was significant smuggling of ivory in Africa and a 'thriving black market' trade in ivory.[30] Save the Elephants and the Kenya Wildlife Service established in aerial surveys late in 1999 that elephant populations had not declined but that none had increased at predicted rates and some had remained stable. The elephant carcasses numbered more than in previous counts.[31] At the same time, IFAW was claiming that the Democratic Republic of Congo was losing 'as many as four elephants a week'.[32] International conservation groups, the Species Survival Network and Born Free Foundation, claim that 'official' elephant poaching figures for 1998 and 1999 are 'dramatically' under-reported and that elephants poached are as many as 25 times more than the 235 estimated by United Nations reports.[33]

It would be easy, perhaps, to conclude from the observed increase in elephant poaching that the cause was either reaction (albeit misguided) to the 1997 Harare decisions, or the anticipated further relaxation of trade barriers at CITES 11. Yet a more balanced view is presented by TRAFFIC:

> Since COP 10, there is clear evidence that elephant poaching has escalated in some parts of Africa experiencing civil unrest and conflict, but it is difficult to establish a causal link between such killing and elephant developments under CITES. In certain areas, a meat trade to provision soldiers or feed tens of thousands of civilian refugees stands behind large-scale elephant killing. ... [In response to media reports of increased poaching, it] is not possible to establish any meaningful trend at this time ... very few Range States have filed incident or national reports on the killing of elephants in their countries with the CITES Secretariat'.[34]

Moreover, TRAFFIC asserts that the backdrop against which elephant poaching takes place warrants consideration. They indicate that domestic ivory markets, which are not subject to CITES regulation, continue in much of Africa and that, since the 1989 ban on the ivory trade, demand for ivory (raw and worked) has continued in the southeast Asia region (excluding Japan), as evidenced by their seizure data.

The animal-rights groups' claims about 'dramatic' poaching levels, based on unsubstantiated data, were designed to attract support for a proposal tabled by Kenya and India at CITES 11, to the effect that all African elephant populations be moved from Appendix II to Appendix I.[35] On the other hand, South Africa's proposal was to transfer the South African elephant population from Appendix I to Appendix II to allow an experimental trade in a maximum of 30 tonnes of ivory from government stocks, trade in live animals, hides and leather goods, and hunting trophies for non-commercial purposes.[36] South Africa's neighbours, Botswana, Namibia and Zimbabwe, tabled similar proposals. Their proposals were to maintain their elephant populations on Appendix II and allow trade in registered stocks of ivory (with annual quotas of 12, 2 and 10 tonnes respectively), as well as trade in live animals, hunting trophies and hides and leather goods.[37]

Elephants, then, were once again to be the 'top issue' at CITES 11.[38] The southern African states looked to be facing a difficult challenge when it emerged that the European Union member nations (excluding Ireland) as well as the United States would be supporting Kenya and India.[39] In the event, Botswana, Namibia and Zimbabwe withdrew their proposals, as did Kenya and India, and an amended South African proposal was accepted by consensus. The amendment allowed South Africa's elephant population to be transferred to Appendix II but there would be a zero ivory quota,[40] the ivory ban to be reviewed at the twelfth Conference of the Parties in 2002-3, probably to be held in Santiago, Chile.[41] The elephant populations in Botswana, Namibia and Zimbabwe would remain on Appendix II also with no ivory quota.

This is, it is submitted, a sensible decision. It means that these countries, which claim that trade in elephant and elephant specimens is necessary for their conservation in the region, having to 'pay their way' to survive, can trade in live elephants (allowing for relocation of elephants as an alternative to culling, for example) and trade in elephant products other than ivory, hides, for example. Trade in elephant hides would appear not to constitute a threat to elephants as a stimulus to poaching,[42] probably because of the logistical challenges in removing commercial-quality hides from elephant carcasses. This is a significant potential income earner for conservation in southern Africa: between 1985 and 1989, the sale of elephant hides earned the South African National Parks (SANP) an annual income of more than two million rand, and it constituted the second largest income earner for SANP behind ivory.[43]

Another significant outcome of CITES 11 for the African elephant was that the Parties showed support for the two long-term monitoring systems (for illegal killing of elephants and the elephant trade) mandated at CITES 10. In response to Resolution Conf. 10.10, two monitoring systems were designed. The first is MIKE (Monitoring Illegal Killing of Elephants), developed by the Species Survival Commission of the IUCN in order to monitor elephants in the Range States across Africa and Asia. MIKE is still in the process of development.[44] The second system is ETIS (Elephant Trade Information System), which was designed by TRAFFIC as a refined version of BIDS, which was recognised at CITES 10. Kenya initially opposed these two systems, but that opposition was withdrawn at Gigiri and the Conference adopted MIKE and renewed support for ETIS as the official monitoring systems. There were pledges of financial support for the systems from the European Commission, Belgium, United Kingdom, United States and Japan.[45] It is anticipated that the two systems will be in a position to deliver analytical reports to CITES 12,[46] but there is evidently still considerable scope for the development of the two systems

THE FUTURE OF THE AFRICAN ELEPHANT UNDER CITES

Where does CITES appear to be heading after the eleventh Conference of the Parties in April 2000 and what does this mean for the African elephant? Several trends can be observed. First, there would appear to be a movement towards a philosophy of sustainable use within CITES, so there is no reason, in principle, why the ivory trade could not be resumed within the ambit of CITES. Secondly, there is a comprehensive collection of criteria consisting of the Everglades criteria and augmented by Resolution Conf. 10.9 at Harare, which determine the conditions under which the elephant can be transferred from Appendix I to Appendix II. Third, monitoring systems have been established to monitor trade and killing of elephants, although these have some way to go before being fully functional. Fourth, there seems to be consensus that elephant poaching levels have dropped since 1989, when the ivory trade ban came into effect and the ivory price plummeted.

These, however, are not the only criteria to be taken into account in assessing the ivory trade ban and, more generally, the CITES status of the African elephant. There are other factors which are not as clear as those outlined in the previous paragraph. First, there would appear to be evidence to suggest that the limited relaxation of the ivory trade ban in 1997 led to an increase in elephant poaching, although figures are sketchy and much 'evidence' seems to be anecdotal. Second, and perhaps most crucial, there is little, if anything, being done on the international legal plane to address causes of dwindling elephant populations other than trade in elephant specimens.

This last point raises a very important factor, which is often not taken into consideration in the CITES-elephant debate, but which has been stressed by several authors dealing with the topic, and that is that CITES is 'only one piece of a much larger puzzle and must be viewed for what it is: a limited treaty designed to ensure that international trade of fauna and flora is not detrimental to the survival of a species'.[47] Given that 'there is no simple mono-casual link between trade and conservation status of a species according to its IUCN Red List category',[48] a ban in the ivory trade would not necessarily lead to an overnight improvement in the conservation status of the elephant. However, there would appear to be evidence to suggest that the ivory trade ban has had an important positive effect on population levels of the African elephant, at least in the short term. There are certain consequences of this that need careful attention, however, particularly by the same international community which is driving the development of CITES.

The main point to be made recognises the argument that the ivory trade is not the only threat to the African elephant and is the point consistently raised by those Range States wanting to reopen the ivory trade. Probably the most important threat to species generally, and this would be true of the elephant as well, is loss of habitat, due particularly to the growth in human populations and human 'development' of natural areas.[49] As Julienne du Toit indicates,

> 'where elephants have disappeared, in many cases humans have taken up the space. There are over 800 million people in Africa, and the population is growing, and there are just half a million (600 000) elephants, so which species will win the space race?'[50]

This situation is exacerbated in the case of the elephant in that it is a very difficult animal for people to live alongside, as pointed out above. Ever since the ivory trade ban began, there has been hope, particularly from southern African states, that the trade would be reopened, even if only in a limited manner. This has probably influenced the way in which elephants are seen by those people who live with them. If the door is seen as being firmly shut, however, the threat to elephants outside of protected areas is not to be discounted. This is not to suggest that CITES should be used to 'alleviate the social problems of those countries that contain endangered species habitat'[51] but that decisions involving trade bans should take into account as an important factor the consequences such decision may have on how people in the Range States may view elephants and how this may affect the animal's habitat.

Issues such as this, however, were seemingly not considered in the decision to ban the trade in ivory. How, then, are we to assess the ban? Considering that CITES aims at regulating trade in endangered species in order to protect the species in question, one could view the ivory trade ban as essentially an admission of defeat by CITES to the elephant poachers. CITES sought to regulate trade in the African elephant and its numbers in the wild in Africa were halved. The result was that elephants, still numbering approximately half a million animals in the wild,[52] joined the tiger, with only 4 600-7 700 animals left, on Appendix I.[53] It left behind on Appendix II animals such as the polar bear, with 23 000-40 000 left in the wild.[54] With these comparative figures in mind, one would hardly think of the elephant as endangered, but neither the tiger nor the polar bear lost approximately 700 000 animals in about a decade. Nevertheless, banning the ivory trade when there are thousands of elephants left in the wild is a clear admission that the regulation of the ivory trade was a disaster. What were the reasons for this?

Clearly, many Range States were, as *The Economist* somewhat unkindly put it, 'too inept or corrupt to prevent poaching.'[55] CITES relies on its effectiveness for the Parties to carry out certain responsibilities, and many countries are seemingly unwilling or unable to carry out these duties. Sand has pointed out that

> 'A survey of 81 CITES parties carried out by the IUCN Environmental Law Centre in 1993-4 indicated that only 12 of the countries surveyed had completed the full range of legislative and administrative measures needed to give effect to

all aspects of the Convention and related resolutions and decisions of the Conference of the Parties; legislation in at least 26 countries was found not to meet the four minimum requirements set by the Conference;[56] and legislation in 43 other countries was considered incomplete or deficient in some specific aspects'.[57]

Not only have several CITES parties who are elephant Range States failed to control elephant poaching and the illegal ivory trade (and this fact seems to be beyond debate), but there were other contributing factors as well. According to Sand, these reasons were 'partly as a result of civil war in several range states (where both sides used ivory poaching to buy weapons abroad) and the free rider behaviour of non-Party states like Burundi and the UAE (serving as entrepôts for huge stockpiles of smuggled ivory destined for the Far East)'.[58] Civil war and civil unrest in many African countries, including the Democratic Republic of the Congo (DRC) which has one of the largest wild elephant populations in Africa, continues today, indicating that conditions have not changed sufficiently since the end of the 1980s for there to be increased confidence that the resumption of the ivory trade would see improved enforcement and control. If relatively stable countries like Kenya and Tanzania are, by their own admission, unable to control elephant poaching and the illicit ivory trade, what chance is there for the DRC and Angola, for example?

The situation is exacerbated by the fact that the ivory trade ban has led to a decline in poaching in Range States, with the result that efforts aimed at combating poaching in the past have been cut back. For example,

> 'Since ... the late Eighties and early Nineties, the [Kenyan] anti-poaching teams have suffered from a cost-cutting drive. In 1992 the Tsavo parks had 80 teams of 10 men each. Now they have just 40 teams, of five men each'.[59]

This is clearly another major reason why these countries are so vigorously opposed to reopening of the ivory trade.

But what of the argument that those countries which were unable to manage their elephant herds adequately are now benefiting most from the ivory trade ban at the expense of those countries which were, and still are, managing their elephants well and which are strong proponents of the view that elephants should pay their way in order to warrant such management? There is something about this plea which appeals to one's sense of justice, but ultimately it is not compelling. Should the pro-trade camp hold sway, and if the result of this is that their herds remain relatively safe but that their northern neighbours' herds are once again decimated, is this a satisfactory result? It is easy to say something to the effect that "That's their problem," but the whole point behind CITES is that it is not *their* problem but *everybody's* problem. States effectively surrendered sovereignty over their wild animals when they became Parties to CITES and decisions that are in the interest of the global conservation of wild animals should hold sway over regional or domestic interests.

But that is not where the argument ends. So far, there has been no independently verifiable evidence that a partial lifting of the ivory trade has been responsible for a resurgence in elephant poaching, despite extravagant claims that this has been the case. On the other hand, and to be fair to those factions making such claims, neither is there satisfactory evidence establishing the linkages between the ivory trade and the health of all elephant populations that may be potentially affected by such trade. This is a point made by several eminent biologists specialising in the African elephant, who argue that further ivory sales ought not to be authorised in the absence of such information.[60]

It is this consideration which suggests that the monitoring systems approved by CITES - MIKE and ETIS - have a vital role to play. However, before the data produced by these systems is more reliable than that which is on the table at present, it would appear that a resumption in the ivory trade would be premature. As Steven Broad, International Executive Director of TRAFFIC says:

> "A deeper understanding of the illegal trade and the illegal killing of elephants is fundamental before any further legal trade in ivory can be allowed. Until monitoring systems are fully operational, new ivory quotas would have been premature'.[61]

While this would be very much acceptable to western animal rights organisations and East African and Asian Range States, the southern African Range States, who argue that they rely on the income of elephant products to fund their conservation and, moreover, that an elephant outside of a protected area without value to the community is a dead elephant, would be less satisfied (to put it mildly).

Part of the solution to the conflict of positions is already catered for by CITES. An Appendix II listing of the elephant allows the international sale of elephant hides for commercial purposes (a lucrative business), as well as the export of hunting trophies (which also potentially brings in important revenue to rural communities, like the Zimbabwean CAMPFIRE projects) and the export of live animals, which provides an alternative to culling to reduce population pressures in protected areas. There is no reason why this situation cannot continue with a proviso that the zero quota on ivory continues.

The second possible part of the solution is more contentious. Karen Steuer, IFAW Director of Commercial Trade and Exploitation, has been quoted as saying,

> 'IFAW has heard the concerns expressed by the majority of elephant range states, who must bear the financial burden of protecting their elephants when poachers respond to newly opened ivory markets. Have we learnt nothing from the last decade?"[62]

But why should it be the Range States who bear the burden of protecting, not *their* elephants, but the world's elephants, whether through the costs of anti-poaching measures or through loss of income from sale of elephant products? The point has often been made, particularly by Range States, that if the world wants a particular species to be conserved, then the world should pay for it.[63] The plummeting population numbers of elephant in the 1980s suggests that many African states do not have the means to control poaching. The perceived solution, a ban on the trade of ivory, severely cuts back the income of those countries which were being relatively successful in their management of elephant. Ten years ago, *The Economist* argued,

> Logically, if the rest of the world wants the elephant to survive, then it should not only compensate Africa for $60m of lost exports, but help to foot the bill– of perhaps $80m-100m – for an effective war on poachers'.[64]

The figures would surely have increased since then, but the cost of fighting poaching has arguably dropped with the ivory trade ban. The basic message is a valid one, however, and one which should be addressed by the international community, particularly those governments and non-governmental organisations which value the future survival of the elephant. Harland suggests that 'the best system would likely be one in which landowners, public or private, were paid a price – at least equal to the value of each animal's ivory – for every elephant kept alive in its original habitat'.[65] Whatever system is used, this is probably beyond the realm of CITES and therefore will not be pursued further in this chapter, but it is certainly an issue that could usefully be pursued in the future. The International Fund for Animal Welfare (IFAW), for example, has already made a substantial contribution to South Africa to purchase land to augment an already existing protected area, with the quid pro quo being an agreement not to cull elephants in the reserve in question.[66] Payments from the international community as compensation for income foregone as a result of not being able to trade in ivory would, it is submitted, strongly quell the rumblings of discontent from southern Africa and ultimately benefit the elephant.

In short, then, the suggestion is that the ivory trade ban should continue for the foreseeable future, with trade in other elephant products allowed and some efforts being expended to establish a system for compensating Range States for their efforts in conserving the elephant. It is not being suggested that the ivory trade ban should necessarily be permanent, despite the views of people like Garrison, who suggests that, '[g]iven the failed efforts to control poaching and past illegal ivory trade practices, it is questionable if any control measures are sufficient for authorized international ivory trade to ever be a sustainable and properly controlled enterprise'.[67] It is our position that, if the links between trade and poaching can be satisfactorily established for all elephant populations, it may be possible some time in the future to reconsider the ivory trade ban.

It is suggested, however, that a system such as that proposed by Harland would best solve many of the problems which beset international trade before the ban. He proposes an exclusive marketing agreement between exporters who can manage their herds at sustainable levels and importers who were both willing and able to enforce import controls requiring them to accept ivory from approved producers alone.[68] Criteria for deciding which range states (exporters) would participate could be:

> a. a proven record in elephant management over an extended period of time in which elephant populations have been managed at sustainable levels;
>
> b. a proven ability to control poaching over the national range of the elephant;
>
> c. an ability to control the mixing of legal and illegal ivory before it is exported.[69]

> There would be an approved centre (the 'CITES centre') which would accept ivory from states meeting the above criteria. It would then be sold to buyers meeting these criteria:
>
> a. a proven ability to control the flow of ivory into and out of its territory over an extended period of time;
>
> b. a willingness to be legally bound to banning all trade in ivory with the sole exception of direct supervised imports from the CITES centre.[70]

The ultimate system adopted, however, would have to cater for the trends and considerations raised by the data produced by the monitoring systems, which makes discussion of a possible trade regime somewhat speculative. The important point, though, is that there may well be scope for trade in ivory in the future, but that the manner in which it should be done must be carefully thought out and not decided upon in a 'knee-jerk' manner.

Unfortunately, the debate over the past decade over the ivory trade and the African elephant has been characterised by seemingly as much emotion as scientific logic. As one delegate at CITES 10 is reported to have said,

> "There is no rationality here. What should be an intelligent debate on the future of man and animal has been mired in endless human politics that have no relevance to what is actually happening in the bush and the oceans".[71]

If this is to be the pattern in the future, nobody, including the elephant, will win. Decisions on trade must be based on objective criteria, and ought to be supplemented by international measures other than CITES designed to consider the protection of the elephant's habitat from undue human encroachment. The elephant is under threat only partly because of the ivory trade. We do not yet know to what extent the ivory trade is responsible for the decline in the elephant's population, but ignorance of this is no excuse for failing to address those other causes that we are currently able to identify.

---

* B Com LLB LLM (Natal), Associate Professor of Law in the University of Natal, South Africa.

\+ BA(Rhodes), LLB (Natal), LLM Mphil (Cantab), Professor of Law in the University of Natal, South Africa.

1 Article I (a).

2 UNEP *The African Elephant* (1989) at 8.

3 David Harland *Killing Game: International Law and the African Elephant* (1994) at 19.

4 B Killoch *The Shamba Raiders* (1972) at 88.

5

6 Michael J Glennon 'Has international law failed the elephant?' (1990) 84*American Journal of International Law* 1 at 3, quoting 11 June 1989 *New York Times* §1 at 6.

7 David Harland *Killing Game: International Law and the African Elephant (*1994).

8 This refers to the 7th Conference of the Parties to CITES.

99 Harland op cit at 98.

10 CITES 7 Doc 7.43.8, quoted in Harland op cit at 99-100.

11 Harland op cit at 101.

12 CITES 8 Doc 8.48.

13 Ibid.

14 Harland op cit at 153.

15 For detailed discussion in the context of the significance of these criteria for the elephant, see Michael Kidd & Michael Cowling 'CITES – An African Perpsective' (2000) *SA Yearbook of International Law* 189.

16 Christopher Munnion 'Plea to relax ivory ban lost after bitter debate'*Electronic Telegraph* 18 June 1997, described it as 'one of the most emotional and contentious debates in the history of conservation ... characterised by what one conservationist described as "skulduggery and street-fighting politics that have nothing to do with the future of endangered animal species'.

17 Resolution Conf. 10.9

18 Set out in full in Kidd & Cowling op cit.

19 Annex 2 to Conf.10.10.

20 Charles Clover 'Three African states allowed to sell ivory' *Electronic Telegraph* 20 June 1997.

21 Eddie Koch 'All clear for ivory trade' *Weekly Mail & Guardian* 27 June 1997. See also Koch 'Now to prevent the slaughter of the elephants'*Weekly Mail & Guardian* 27 June 1997.

22 SAPA Report 29 October 1997, quoted by the South African Environment Project at http://www.saep.org/subject/natcon/natcites.html (accessed date 23 March 2000).

23 David Harrison 'Elephants in peril as ivory ban is set to be lifted'*Electronic Telegraph* 7 February 1999. Cf 'Tsavo elephant numbers rise, but at a slower rate'*African Wildlife News* Summer 1999, where evidence of poaching is described, but at insignificant levels (http://www.awf.org/news/news11.html [accessed 23 March 2000]).

24 'African elephant range states dialogue meeting' (January 1999) 17 (3)*TRAFFIC Bulletin* at http://www.traffic.org/bulletin/archive/january99/africanelephant-status.html etc (accessed 30 September 2001)

25 'CITES permits one-time sale of African elephant ivory' *Environment News Service* 11 February 1999 at http://ens.lycos.com/ens/feb99/1999L-0211-03.html (accessed 30 September 2001).

26 'Legal ivory sales end 9-year ban on global trade'*African Wildlife News* Summer 1999 at http://www.awf.org/news/news11.html (accessed 23 March 2000).

27 Paul Harris 'Kenya's elephant poaching toll soars' *Electronic Telegraph* 19 September 1999.

28 Ibid.

29 'South Africa announces proposal for commercial trade of elephant hides, leather goods and 30 tonnes of ivory: International Fund for Animal Welfare calls proposal "irresponsible"' *Environment News Service* 10 November 1999 at http://ens.lycos.com/e-wire/Nov99/10Nov9903.html (accessed 30 September 2001).

30 Tervil Okoko 'Thriving black market in ivory demolishing African elephants'*Environment News Service* 29 March 2000 (accessed date 30 September 2001).

31 Ibid.

32 Ibid.

33 Cat Lazaroff 'UN elephant poaching figures dangerously low'*Environment News Service* 3 April 2000.

34 Tom Milliken 'African elephants and the eleventh meeting of the Conference of the Parties to CITES' *TRAFFIC Network Briefing Document* at http://www.traffic.org/briefings/elephants-11thmeeting.html (accessed 21 June 2000).

35 Proposal 11.24.

[36] Proposal 11.20. See also Ministry of Environmental Affairs and Tourism*Briefing Document on the African Elephant Proposal by South Africa to CITES 11*at http://www.environment.gov.za/docs/2000/cites_elephant.htm(accessed 23 June 2000).

[37] Proposals 11.21, 11.22 and 11.23 respectively.

[38] Barrack Otieno 'Elephants are top issue on day one of CITES Conference' *Environment News Service* 11 April 2000 at http://ens.lycos.com/ens/apr2000/2000L-04-1102.html (accessed 13 April 2000).

[39] Barrack Otieno 'Whales, elephants divide endangered species delegates'*Environment News Service* 14 April 2000 at http://ens.lycos.com/ens/apr2000/2000L-04-1402.html (accessed 17 April 2000).

[40] CITES 'Outcome of the 11th meeting of the Conference of the Parties to CITES' at http://www.cites.org/CITES/eng/index.shtml (accessed 25 June 2000). Note that at the time of writing this chapter, the official proceedings of CITES 11 had not yet been published.

[41] Barrack Otieno 'Endangered species win limited protection at CITES' *Environment News Service* 20 April 2000 at http://ens-news.com/ens/apr2000/2000L04-20-04.html (accessed 20 April 2000).

[42] Milliken op cit.

[43] Ministry of Environmental Affairs and Tourism*Briefing Document on the African Elephant Proposal by South Africa to CITES 11*at http://www.environment.gov.za/docs/2000/cites_elephant.htm (accessed 23 June 2000). Two million rand was, in September 2001, equivalent to about US$125 000.

[44] Milliken op cit.

[45] TRAFFIC 'Boost for elephant trade monitoring systems' at http://www.traffic.org/cop11/newsroom/elephant.html (accessed 25 June 2000).

[46] Milliken op cit.

[47] John L Garrison 'The Convention on International Trade in Endangered Species of Wild Fauna and Flora (CITES) and the debate over sustainable use' (1994) 12*Pace Environmental Law Review* 301 at 390. See also Peter H Sand 'Commodity or taboo? International regulation of trade in endangered species' (1997) *Green Globe Yearbook* 19 at 26.

[48] Sand op cit at 19.

[49] See Shawn M Dansky 'The CITES "objective" listing criteria: Are they "objective" enough to protect the African elephant?' (1999) 73 *Tulane LR* 961 at 970-1.

[50] Julienne du Toit 'The big five – Africa's flagship species' in Endangered Wildlife Trust Seventh Annual *Endangered Wildlife: Business, Ecotourism and the Environment* (1999) at 42.

[51] Dansky op cit at 979.

[52] Estimates show that the wild population of the elephant in African is between 400 000 and 500 000 animals. The African Elephant Database, an IUCN/SSC African Elephant Specialist Group and UNEP programme, estimated in May 1997 (the latest avalable figures) that there were 285 246 'definite' elephants, 101 285 'probable', 171 892 'possible' and 22 752 'speculative': http://indaba.iucn.org/external/themes/ssc/sgs/afesg/index.html (accessed 30 September 2001). The sum of the definite and probable elephants is 386541. Cf Cat Lazaroff 'UN elephant poaching figures dangerously low' *Environment News Service* 3 April 2000, who claims that current population estimates are between 301 773 and 487 345. She does not name a source.

[53] See http://www.tigersincrisis.com/status.htm (accessed 30 September 2001).

[54] See http://www.polarbearsalive.org/facts2.htm(accessed 30 September 2001).

[55] Anon 'Saving the elephant: Nature's great masterpiece' *The Economist* 1 July 1989 at 16.

[56] CITES Resolution Conf. 8.4 (1992), which defines the minimum measures as: (i) designation of at least one Management Authority and one Scientific Authority; (ii) prohibition of trade in specimens in violation of the Convention; (iii) penalization of such trade; and (iv) confiscation of specimens illegally traded or possessed.

[57] Sand op cit at 25.

[58] Sand op cit at 23. *The Economist* op cit, indicates that 'for many years tiny Burundi ... had just one elephant. In 1986 this prolific beast produced 23 000 tusks, all carefully documented as originating in the country' (at 17).

[59] Paul Harris 'Kenya's elephant poaching toll soars' *Electronic Telegraph* 19 September 1999.

---

[60] Katy Payne, Iain Douglas-Hamilton, Cynthia Moss & Joyce Poole 'Concern over ivory sales' letter to the *Electronic Telegraph* 24 March 1999.

[61] TRAFFIC 'Wildlife trade conference ends on mixed note' 20 April 2000 at http://www.traffic.org/cop11/newsroom/finalday.html(accessed 22 June 2000).

[62] 'South Africa announces proposal for commercial trade of elephant hides, leather goods and 30 tonnes of ivory: International Fund for Animal Welfare calls proposal "irresponsible"' *Environment News Service* 10 November 1999 at http://ens.lycos.com/e-wire/Nov99/10Nov9903.html (accessed 30 September 2001)..

[63] See Dansky op cit at 978-9.

[64] Anon 'Saving the elephant: Nature's great masterpiece' *The Economist* 1 July 1989 at 17.

[65] Harland op cit at 170.

[66] SAPA Report 29 October 1997 quoted by the South African Environment Project at http://www.saep.org/subject/natcon/natcites.html (accessed 20 June 2000).

[67] John L Garrison 'The Convention on International Trade in Endangered Species of Wild Fauna and Flora (CITES) and the debate over sustainable use' (1994) 12*Pace Environmental Law Review* 301 at 373.

[68] Harland op cit at 171.

[69] Ibid.

[70] Harland op cit at 172.

[71] Christopher Munnion 'Plea to relax ivory ban lost after bitter debate'*Electronic Telegraph* 18 June 1997.

THE UNITED NATIONS CLIMATE CHANGE REGIME AND AFRICA

KEVIN R. GRAY & JOYEETA GUPTA[1]

INTRODUCTION

Kofi Annan, Secretary General to the United Nations, has argued that: '"The recognition that the industrialised countries should take the lead in tackling climate change is one of the political cornerstones of the Convention."[2] He was referring to fact that the *United Nations Framework Convention on Climate Change* explicitly states that the developed countries should take the lead[3] and includes articles that implicitly emphasise that developed countries should take the lead.[4] This was urged during the *Climate Change Convention* negotiations, where the developing countries argued that the primary burden of action lies with the North and the developed countries (initially) acknowledged this responsibility.[5]

While there is no doubt that the problem of climate change can be historically attributed to unchecked economic development in industrialised countries, there is increasing realisation that the economic consequences of reducing emissions may form a political barrier for action in the developed countries. Furthermore, time is running out and developing countries can neither afford to wait for the developed countries to take action - nor is it wise for them to invest in development paths that might lock their countries into unsustainable technological trajectories. The climate change regime is global in scope, thereby necessitating a partnership of all nations to adopt measures and take action.

The *Climate Change Convention* as well as the *Kyoto Protocol*[6] highlight the common but differentiated responsibility principle by exempting developing countries from being subject to any quantified emission related requirements.[7] Varying commitments are a true manifestation of the principle of common but differentiated responsibility. The principle is a mainstay in international environmental law.[8] It recognises that although the international community shares a burden to address an environmental problem, the countries that have predominantly contributed to the problem, and have the best capacity to take action, are subject to stricter obligations than other states that have not. This is especially poignant for climate change since although more than three-fourths of the global population live in developing countries, their energy consumption is still less than one tenth of that of the North.[9]

The principle can be applied in numerous ways such as the setting of differential standards or transition periods as seen in the *Montreal Protocol.*[10] In some agreements, they can arguably include a provision guaranteeing technical and financial assistance for the purposes of implementation.[11] The *Climate Change Convention* adheres to the principle in various ways. Overall, the specific needs and special circumstances of developing countries are to be given full consideration.[12] There is a differential standard in that industrialised countries, including those from the transition economies of Central and Eastern Europe, are subject to emission related targets. Differentiation has also been exercised among the group of developed countries, with some countries allowed to increase their emissions,[13] others allowed to bring back emissions to 1990 levels,[14] while some have to reduce their emissions.[15] The developing countries are not yet subject to any quantitative restrictions under the *Kyoto Protocol.*

There is no greater need for a genuine application of the common but differentiated responsibility tan in Africa. African countries are scarcely responsible for the problem of climate change. Africa, consisting of 53 countries, contributes merely 7% of the global emissions on a per capita basis.[16] Yet, it arguably represents the area that will disproportionately suffer its greatest effects. The continent has several coastal nations and islands, and arid and semi-arid regions, all which are highly vulnerable to climate change.

Not only are there moral questions concerning African emission reduction commitments, but also there are severe resource limitations for African countries hindering the possibility of taking effective action. The effects of climate change will be difficult to mitigate for states facing a wide range of developmental problems. It is anticipated that the climate change regime will eventually mature so that developing countries will take on quantitative greenhouse gas reduction commitments. This compounds the problem for African countries that may have to reduce their emissions and therefore adopt greener technology while dedicating limited resources to combat the effects of climate change induced predominantly by industrialised country development.

The purpose of this chapter is to examine the implementation of the *Climate Change Convention* in Africa. Overall, there are no quantitative commitments for developing countries beyond their national reporting obligations to the Climate Change Secretariat.[17] The developing countries are, however, under an obligation to prepare national inventories of greenhouse gases, national policies to deal with their emissions, to take climate change into account in related planning.[18] As a result, an analysis of how developing countries, and in this case, African countries are implementing their commitments may not be appropriate or possible. However, this paper looks at implementation in a different light. It views implementation as something that African countries must address as their economies, and corresponding energy needs, develop. Ensuring that the industrialised countries meet their obligations is as much dependent on African States as it is with other parties participating in the climate change regime.

Moreover, implementation in Africa will only deepen in necessity as ecological conditions worsen.

African countries, numerically the strongest negotiating bloc in the climate change regime, are severely underrepresented. Their voice is overwhelmed by industrialised countries that command the negotiations agenda. Africa has been unable to effectively: mount a common position internally; develop coalitions with other like-minded G77 countries; and negotiate its position in relation with the rest of the G77 countries. This paper examines some of these problems and attempts to assess the impact of the negotiation process on Africa.

THE CLIMATE CHANGE REGIME

The *Climate Change Convention* was signed in 1992, being one of the environmental instruments negotiated at the United Nations Convention on Environment and Development (UNCED). It is essentially a framework agreement, requiring in weak legal language the industrialised countries (listed in Annex I), including OECD countries,[19] to reduce their greenhouse gas emission levels recorded in 1990 by the year 2000. The ultimate objective of the *Climate Change Agreement* is to stablisise GHG concentrations in the atmosphere.[20] Article 3(1) recognises that industrialised countries are to take the lead in combating climate change. Developing countries (or countries not listed in Annex I), are expected to follow, subsequently given their low emissions per capita and their ability to take action. Developed countries listed in Annex II (i.e. excluding countries with economies in transition) also have the responsibility of transferring financial and technological resources to developing countries.[21]

In 1997, the parties adopted the *Kyoto Protocol,* establishing a timetable for developed countries listed in Annex B countries to reduce their emissions. Non-Annex I countries include countries that are particularly vulnerable to the adverse effects of climate change and the parties that are expected to bear a disproportionate or abnormal burden under the *Climate Change Convention.* Based on an aggregate target of emission reductions of 5.2 % in 2008-2012 in relation to emission levels in 1990, each developed state is subject to reduction targets, with some countries permitted to increase their emissions.[22] The first commitment period commences in 2008-2012. The common but differentiated principle, as applied, reflects the special needs and concerns of developing countries,[23] and specific needs and special situations of Least Developed Countries (LDCs).[24] Consequently, developing countries are not held to any emission related targets, but are still to be given new and additional financial resources to meet the incremental costs of preparing their National Communications and to help them adopt modern technologies.[25]

In addition to the national reporting requirements mentioned above, there are some general obligations for developing country parties under the climate change regime. Under Article 4(1((b), all parties are to formulate and implement national programmes including measures to mitigate climate change by addressing emissions and measures for adaptation to the effects of climate change. Parties are also required to integrate climate change considerations into relevant social, economic and environmental policies and actions as well as deploy appropriate assessment of projects.[26] The objectives of achieving sustainable development for non-Annex I countries, a requirement for all parties, is a cornerstone of the climate change regime.[27] However, economic and social development and poverty eradication are considered to be the first and overriding priorities for developing countries.[28]

FUTURE OBLIGATIONS

As the climate change regime develops, it is anticipated that developing countries that graduate into high economic categories will eventually be subject to emission related commitments. In fact, consultations in the subsidiary bodies are investigating on how to broaden emission targets to include some developing countries.[29] Rapidly industrialising countries will be under growing pressure to take on such obligations. We have already seen developing countries themselves initiating their own sense of "meaningful participation" in the climate change regime by voluntarily taking on emission reduction commitments. Argentina announced its intention at the Fourth Conference of the Parties, to be bound by emission reduction requirements, although there has been little movement on this. This is not surprising given the national economic collapse since then. Kazakhstan successfully moved to Annex I, perhaps in order to become eligible for participation in emissions trading schemes under the *Kyoto Protocol,* which can engender technology transfer and the infusion of other know how and resources.[30] Irrespective of what transpires at the international level, some constituencies in developing countries may push their own governments to adopt more substantive measures on climate change including binding emission reduction schedules.[31] It will be difficult for some developing countries to maintain their opposition to quantitative obligations for a long period of time.[32]

Clearly emission related objectives for non-Annex 1 countries may become a reality. From a political perspective, this might be the only way that Annex I countries will agree to further reductions, beyond the first and perhaps second emission reduction commitment. The prospect of Annex I countries taking on further emission reduction targets beyond the first period will depend on some or all developing countries agreeing to a fixed reduction timetable. Although, it is envisaged that the newly industrialising countries will be among the first to be subject to quantitative commitments, it is possible that the qualitative obligations for the remaining developing countries will become more and more specific over time. This could impose an enormous burden for resource-deprived African parties, which might, in comparison, render current difficulties by Annex I countries negligible.[33] African emissions are growing, from 3% in 1990, to 7% in

2000.[34] Nevertheless emissions from Africa will continue to be low in global terms with the exception of South Africa, which is highly dependent on coal and contributes 1.4% to global GHG emissions.[35]

In any event, future commitments that may be taken by developing countries will ultimately rest on the successful implementation of commitments by industrialised countries. What takes the principle of common but differentiated responsibility to a different dimension is the notion of contingency. Developing country commitments, despite being soft obligations, are dependent on the fulfillment of industrialised country obligations. Article 4(7) stipulates that effective developing country implementation depends on how industrialised countries meet their commitments relating to financial resources and technology transfer. This is particularly problematic for an agreement such as the *Climate Change Convention* where industrialised country implementation has been slow on many fronts. Not many Annex I countries have reportedly reduced their emissions. Nor have they transferred significant quantities of technologies under Article 4.5 of the Convention, or made substantial sums of financial resources available under Article 11 of the Convention.

An issue that is highly relevant for African countries, considering their dependency on industrialised country implementation, is compliance. New provisions dealing with compliance, that were agreed to at the 7th Conference of the Parties, will not be in operation until the *Kyoto Protocol* is in force.[36] The architecture of the compliance system was developed at the 6th bis Conference of the Parties, allowing for developing country participation in decisions relating to compliance by all parties. [37] Although negotiating a compliance regime is a monumental process, the legal form of the procedures and mechanisms relating to non-compliance were not specified and therefore subject to decision-making at the Conference/Meeting of the Parties instead of the enforcement branch of the Compliance Committee. Developing country commitments relating to technology transfer and financial assistance are not under the compliance regime. Implementation relating to Article 3.14 commitments regarding Annex I country commitments on funding, insurance and technological transfer for adaptation to the adverse effects, as well as supplementarity (how much Annex I countries can use the flexible mechanisms to meet their emission reduction targets), are reviewed only by the facilitative branch of the Compliance Committee whose powers are merely advisory or recommendatory, in addition to offering financial and technical assistance in the event of non-compliance. Coupled with the non-specificity of the requirements for all of these obligations, developing country complaints against lack of compliance in these matters will be subject to the general negotiation processes in the Climate Change regime. In effect, there is no definitive consequence such as inability to use the mechanisms, and therefore undertake more domestic action, as a result of a Compliance Committee determination of noncompliance.

Although developing countries may benefit from not having to make the necessary economic sacrifices in order to achieve emission reductions, this does not assist those countries that encounter the impacts of climate change on a daily level. There is an imperative need in some countries for early-warning systems for climate related disasters and coastal adaptation technologies. The slow progress on implementing Articles 3 and 4 of the *Climate Change Convention,* has caused serious delays in taking concrete actions to counter the adverse effects of climate change in Africa, potentially leading to "serious drawbacks in laying the foundations for sustainable economic growth and development on the Continent".[38] Any investment that is driven by the regime's provisions, such as the Clean Development Mechanism,[39] may facilitate the infusion of technology and other sustainable development investments as well as the delivery of assistance for adaptation. Active engagement in the regime will be of direct benefit to African countries.

Insisting on the application of common but differentiated responsibilities does not provide a panacea for developing countries. Regardless of the presence of any strict emission reduction requirements, it is arguably in the developing world's best interests to curtail the growth of GHG emissions. Global contribution to climate change does a disservice to countries already highly vulnerable to the impacts. By taking domestic action and engaging in the CDM and other processes, this not only capitalises on using more energy efficient and greener technology but also leads these countries to more sustainable development paths that can be commenced at a much earlier stage than their industrialised country counterparts. Economic growth does not preclude this option.

THE AFRICAN CONTEXT

Africa, as a whole has several coastal nations and islands, and arid and semi-arid regions, and as such has much to lose from the potential impacts of climate change. The consequences of climate change are expected to be high for the continent. The Intergovernmental Panel on Climate Change report from 1998 states:

> "Several climate regimes characterise the African continent; the wet tropical, dry tropical, and alternating wet and dry climates are the most common. Many countries on the continent are prone to recurrent droughts; some drought episodes particularly in Southeast Africa are associated with ENSO phenomena. Deterioration in terms of trade, inappropriate policies, high population growth rates, and lack of significant investment- coupled with a highly variable climate - have made it difficult for several countries to develop patterns of livelihood that would reduce pressure on the natural resource base. Under the assumption that access to adequate financing is not provided, Africa is the continent most vulnerable to the impacts of projected climate change because widespread poverty limits adaptation capabilities."[40]

In the 20th century, an average warming of 0.05 degrees Celsius has been experienced in the Southern African region. The six warmest years occurred since 1980. There is a trend towards reduced rainfall and increased droughts.[41]

The *IPCC Report* points out that the temperature of the continent may rise by about 0.2 degrees Celsius per decade up to 2020, with the coastal zones warming more slowly than the interior.[42] This warming can lead to increased open water and soil/plant evaporation with an average rise in sea level of around 25 cm expected by 2050. This will have profound effects on terrestrial ecosystems, hydrology and water resources, and agriculture and food security in Africa due to the decrease in both precipitation and water infiltration into the soil. Forest resources will be lost while there should be reductions in rangelands and livestock numbers.

Africa, as a whole, is most likely to suffer from the impacts of climate change. This may be most severe in terms of drought and desertification since at least 36 countries are affected by desertification. The low capacity and high dependence on natural systems for livelihood of citizens weaken the ability of these African countries to respond to adverse impacts of climate change.[43] The possible impacts of climate change in terms of sea level rise are likely to affect both island and coastal states. Food and water insecurity will play an increasing role in Africa as well, contributing to vector migration and disease.

At an individual country level, the situation does not appear to be much brighter. Although significant data on climate change in particular African countries is not abundant, some work has been done. A study by Abira & Awuor[44] argues that the limited surface and ground water in Kenya will come under increasing pressure as a result of climate change. Onyango *et al*[45] anticipates that increased CO2 levels may lead to extra water consumption and that while some species may thrive under warmer circumstances, corals may get bleached and there may be stagnant growth. In addition, the minor changes in precipitation levels may have major impacts on flora and fauna, food supply and water availability, which may affect wild life and humans.

Ironically, the African contribution to the climate change problem is fairly negligible. The African countries are recorded to have the lowest contribution to the climate change problem in terms of emitting greenhouse gasses.[46] It amounts in total to about 7% of global emissions coming from 1/4th of the planet's countries having a combined population of 750 million. A significant source is from South Africa's bustling economy, which has the third highest emissions in the developing world after China and India. It has the highest per capita emissions in Africa, followed by Egypt, Nigeria and Algeria. The per capita contributions for the rest of Africa remain extremely small. However, this does not negate the importance of effective implementation of the *Climate Change Agreement* or the *Kyoto Protocol*. African countries have the most at stake due to the failure of industrialised countries to meet their obligations. The disproportionate vulnerability to the adverse effects of climate change and the further exposure to more natural disasters might marginalise Africa even further. The existing slow rate of implementation has already engendered skepticism about whether African countries, and particularly LDCs, will ever be in a position to address the effects of climate change and achieve sustainable economic growth and development.[47]

ADAPTATION

A key concern for African States is the availability of assistance for climate change adaptation. The *Climate Change Convention* acknowledges this by stating that the agreed full incremental costs of adaptation[48] especially for Africa[49] and other countries that are particularly vulnerable to the impacts of climate change should be provided.[50] In 1995, the Conference of the Parties set in motion a process to undertake studies to identify countries that are vulnerable and possible measures. However, decisions on actual projects under the Global Environment Facility, aimed to increase capacity and deliver assistance to adaptation projects were delayed. At the 7th Conference of the Parties, it was agreed that the special climate change fund, in addition to the adaptation fund mentioned in the *Kyoto Protocol*, would finance activities on adaptation, improving and monitoring of diseases and vectors, and capacity building.

The adaptation fund in the *Kyoto Protocol* is financed by CDM projects where the parties must dedicate a share of the proceeds from certified project activities to partly assisting in meeting the adaptation costs for countries that are particularly vulnerable to the adverse effects of climate change.[51] Ultimately, the cooperation between North and South through the CDM is to be taxed to finance adaptation. The other two flexible mechanisms have not been made similarly subject to such a tax. Whatever funds may eventually be generated on adaptation from CDM projects, will have to be divided over 153 non-Annex I developing country parties. Special attention to African countries is not mandated.

In the *Marrakesh Accords,* the Parties established an adaptation fund "to finance concrete adaptation projects and programmes" in developing countries. This fund is to be financed from the share of the proceeds garnered under CDM activities and other sources of funding. Annex I parties are invited to provide additional resources to CDM share of the proceeds for the fund. Shares of the proceeds to meet adaptation costs referred to in Art. 12(8) of the *Kyoto Protocol,* are to be 2% of the certified emissions reductions issued for a CDM project activity.

TECHNOLOGY TRANSFER AND THE CDM

As referred to above, technology transfer requirements are essential provisions for African countries. Implementation of the *Climate Change Convention* depends on effective industrialised country implementation of commitments relating to financial resources and technology transfer.[52] It is hoped that technology can be sent to developing countries to assist in the pursuit of sustainable energy projects. In turn, African countries can generate markets for CDM projects.

Under the Clean Development Mechanism, Annex I parties can undertake project activities in industrialised countries, and gain certified emission reductions to be used to assist in achieving compliance in the first commitment period.[53] A party can acquire the Certified Emission Reduction (CER's) units if it has demonstrated to a Compliance Committee that it has met the eligibility requirements. In turn, a non-Annex I party can host a CDM project if they have ratified the *Kyoto Protocol.* The operational entity would be in charge of verification and certification, with CERs ultimately being issued by an Executive Board to be established under the *Kyoto Protocol.* The Executive Board would make recommendations to the Conference of the Parties to the *Climate Change Convention*/Meeting of the Parties to the *Kyoto Protocol,* on entities certifying CERs in accordance with Article 12.5, which includes validating and registering the projects. An independent entity is justified to combat incentives for developing countries to overstate the CERs generated from the project.[54] The CDM is seen by numerous developing countries as a way to access clean technology and attract new investments to their economies.

Annex I parties are to ensure that there is financial and environmental additionality for CDM projects.[55] The *Bonn Agreements on the Implementation of the Buenos Aires Plan of Action* requires that CDM projects are not to result in a diversion of overseas development assistance and not computed in the amount affecting Annex I countries' financial obligations of the parties under both the *Convention* and the *Kyoto Protocol.* However, the Clean Development Mechanism will be principally a private sector driven process. Investors will be concerned with generating an economic return for their activities. In addition to generating easily quantifiable CERs, there will be a need for a strong economic climate to attract investors. Africa suffers from attracting the lowest percentage of foreign direct investment.[56] Technology transfer accompanies such investment, widening the incoming FDI rate disparities amongst developing countries. If the Activities Implemented Jointly (AIJ) projects are any measure, very little investment will flow to Africa.[57] Countries likely to benefit from CDM investment are larger rapidly industrialising countries that have low marginal abatement costs and (mainly coal users) are generally successful in attracting FDI, such as China, India or South Africa.[58]

Another difficulty in attracting CDM projects relates to the state of African economic development. Generally, African needs are still focussed on infrastructure development whereas investors may have interests in other sectors. Although Africa is expanding its manufacturing sector, there is a gradual shift of investment attention to software sectors such as information technology and capital goods where Africans have not made an impact.[59] African markets are not stable enough for an investor to adopt the inherent risks associated with investment. Basic economic development may be a prerequisite for substantial energy saving projects.

In order for the CDM to achieve emission reduction and sustainable development objectives in Africa, it must include socio-economic and development indicators.[60] Suitable criteria for projects include: sustained economic growth; poverty eradication; social development; meeting basic needs; reduction or elimination of foreign debt; real and meaningful technological transfer; awareness and capacity building; and intra and intergenerational equity.[61] Another possibility, for the purpose of efficiency and avoidance of duplication, could be to account for the development of regional energy cooperation mechanisms that are guided by clear views on how to incorporate sustainable development, such as ECOWAS and SADC.[62] However, the development of criteria for CDM projects may not be region-specific. There may be factors to be considered that are misplaced in Africa or not appropriate for all countries on the continent. Moreover, it may be difficult to achieve a consensus on the continent concerning the proper criteria, especially since African countries may be pitted against each other in competing for CDM investor interest. This can prove to be counterproductive since regional cooperation can assist African countries in developing the necessary technical expertise and compiling data needed to undertake the accounting of emissions.[63] Regionally based projects may bring about a greater return on investment, in terms of emission reductions, while better meeting the energy needs of a group of countries.[64]

One area where some African countries are well endowed is in forests. Carbon sinks will be permitted as CDM projects so that an afforestation or reforestation project can earn CERs for western investors.[65] However, this might pose difficulties in light of the energy dependency on fuel wood in sub-Saharan Africa. Forestry and land use change is one of the most significant contributors of GHG emissions on the continent. In many rural areas, there is a strong reliance on timber and charcoal for energy use.[66] The uses of forestry resources form a critical part of African development and planning, in addition to being a chief source of energy from fuelwood.[67]

CAPACITY BUILDING

A prerequisite for implementing any existing or future commitments in Africa is capacity building. All African countries would certainly face difficulties in implementing any emission reduction commitments. They lack the necessary technology and financial resources for economic development let alone greening their energy consumption. Whether the Climate Change regime will encourage economic development through technology transfer or delivery of other resources remains to be seen.

As mentioned, the only current obligation for African countries is the submission of their national communications. This is essentially a reporting requirement, consisting of information on implementation; national inventories of anthropogenic emissions by sources and removals by sinks of all greenhouse gases; a general description of steps taken or envisaged to implement the *Climate Change Convention*; in addition to other relevant information.[68] Effective implementation of these commitments is connected to how industrialised country parties meet their financial commitments to assist developing country efforts.[69] As a manifestation of common but differentiated responsibilities, non-Annex I countries have been given more time to prepare their communications and even received financial and technical assistance for such purpose. Due to the limited resources that have been forwarded to African countries, African countries have historically had difficulties meeting their reporting deadlines. Extensions have not resulted in overall compliance. Only 21 African countries have thus far submitted their National Communications third reports at the last (7th) Conference of the Parties. Furthermore, the communications have been submitted without the inclusion of adequate policies and measures to meet Article 4 commitments.

AFRICA IN UNFCCC NEGOTIATIONS

One area where capacity building is certainly needed is in negotiations. African countries are at a massive disadvantage during negotiations, dwarfed by the extensive manpower and resources assisting their inudstrialised country counterparts at the negotiations. Africa is the largest negotiating bloc consisting of 53 countries - representing more than one fourth of the total number of parties to the *Climate Change Convention*.[70] By virtue of this, it has the largest number of votes in the negotiating process. Africa's negotiating history in the climate change issue can be divided into two phases. In the pre-1996 period, Africa was more or less swept into the negotiation process. In the post-1996 period, perhaps coinciding with the Chairmanship of Tanzania, there was an increased awareness with the existence of a larger number of sub-regional meetings and the beginning of preparatory work for the negotiations done in advance of the Conference of the Parties.[71] However, most of this history has seen African government participation led by climatologists, who form the focal points for IPCC communications and in cases, have not always been able to pass on incoming information and activities to other relevant departments or other economic sectors in a usable manner.[72] This has led to the effecting of little impact on the text of the negotiations and a muted voice during negotiations.

These developments stem from a relative deficit in negotiating skills, technical capacity, human resources and preparatory work. There is little southern-based knowledge for developing country negotiators to use and therefore developing countries assume that the western knowledge of climate change impacts and mitigation is correct. When African countries advance novel positions, they are rooted in principles and ideology without having any context to them. The lack of resources available to African countries compounds their inability to participate in every meeting and cover all the issues. There are a great deal of endogenous African resources available to tap into, including those present in universities and NGOs, however, many African countries have not attempted to consult with these actors in order to better inform itself in advance of negotiations. This in itself results from limited manpower at ministries and an organisational culture that is not based on consulting stakeholders.

Another impeding factor for African negotiators is their limited mandate. Similar to other countries, the negotiators are confined and cannot take positions contrary to the official government position. The problem is that the official national position often does not go much further than a statement of principles to be adopted. Many negotiators are new to the process, having been rotated into their current position, and therefore lack the courage to deviate from instructions. Institutional memory of what was advanced at previous meetings disappears with the entry of new government personnel. Absence of a comprehensive negotiating strategy or even the requisite policy coordination to develop this leads to a hollow mandate.[73] As a result, the necessary flexibility for engaging in negotiations does not exist.

The interaction of all of these factors leads to what can be referred to as a defensive negotiating strategy based on pseudo factors of legitimacy. Positions taken by the countries are often oppositional without any injection of new ideas. African parties are also vulnerable to vacillating between various positions and can be influenced by bilateral discussions with a powerful donor government. In light of the common problems in many African countries, it is difficult to coordinate their position as a regional negotiating block.

TOWARDS A COMMON POSITION IN AFRICA

As far back as in 1992, a proposal was made regarding the content of Africa's strategy on climate change. This included the preparation of inventories and scenarios, inventory of the impacts and hence the development of priorities for Africa.[74] There are sufficient similarities shared by African countries to prepare a common negotiating position. An "African" position is achievable while being consistent with: the ultimate objectives of the *Climate Change Convention*; its guiding principles with special reference to common but differentiated' principles; and the need for the industrialised countries to take the lead (and comply with their obligations).[75] Africa has a common position in terms of calling on the industrialised countries to take serious measures to reduce their greenhouse gasses and to take on board serious commitments. At the same time, Africans believe that the special needs and circumstances of developing countries need to be taken into account. The Africans are also willing to accept their responsibilities under Article 4.1 para. (a) to para. (j) and Article 12.1 of the *UNFCCC,* implementing these commitments in accordance with their limits and capabilities.

Despite the common ground, there have been historic difficulties for Africa to unite on a position. This is due to general mistrust between the countries themselves, which might be aggravated by an implicit competition between the countries to attract CDM and other climate change related ODA. This disjointed regional position is alarming, when compared to other groups of countries in the G77: AOSIS; Latin America; and OPEC; that present a stronger united front despite differences on key issues.[76] The African continent as a whole represents a plurality of interests and issues. There is a handful of oil producing countries in Africa that are concerned about any commitments that reduce the need for their exported oil resources. Other countries, especially those in the Sub-Sahelian region of Africa, experience the primary consequences of climate change such as desertification. Coastal countries are concerned about their shrinking coastlines as well as the diminishing fisheries stocks that can be linked to climate change. However, all countries share a severe vulnerability to climate change impacts and a dearth of real resources needed to combat climate change.

CONCLUSION

Despite pressing need for action on climate change, implementation of the *Kyoto Protocol* still is pending ratification. It may come into force when the World Summit on Sustainable Development convenes in South Africa in August 2002. The significance of this event occurring on African soil elevates its importance, considering Africa may be the continent that benefits the most from its implementation. It is notable that the 7th Conference of the Parties, where the *Agreement on the Implementation of the Buenos Aires Plan of Action,* otherwise known as the *Marrakech Accords,* was finalised, also occurred on the continent in Morocco.

The African Group at the 7th Conference of the Parties welcomed the *Marrakech Accords*, perceiving that it could engender prompt implementation of the *Kyoto Protocol* and also a quick start to CDM projects. At the 7th COP, the membership of the CDM Executive Board was created, with a mandate to develop the procedures for fast-track CDM projects.[77] There is no guarantee that parties will quickly implement their commitments nor channel CDM projects and corresponding technology to Africa. Despite the urgent need for technology transfer and financial assistance, it has been argued that African countries should not rely on any hard commitments obligating them under the regime. Action to combat climate change can include both efforts to mitigate the impacts as well as preventative action that does not increase greenhouse gas emissions. Moreover, more integrating environmental concerns into energy expansion does not act as an inhibitor to economic development and can even act as an agent for longer and more sustainable growth. In effect, proactive activities may ease the burden of subsequent emission reduction commitments that will eventually apply to African States.

[1] Kevin R. Gray is the Research Fellow in Public International Law and British Institute of International and Comparative Law in London, UK. Joyeeta Gupta is head of the Programme on International Environmental Governance, and interim head of the Department of Environmental Policy Analysis at the Institute for Environmental Studies at the Vrije Universiteit in Amsterdam.
[2] Kofi Annan (1997), 'We need decisive action in Kyoto to limit Greenhouse Gas Emissions,*Climate Change Bulletin*, 15 4th Quarterly at 4.
[3] *United Nations Framework Convention on Climate Change,* (1992) 32 I.L.M. 849, (herinafter *Climate Change Convention*); see Article 3(1) and 4(1).
[4] See, for example, Articles 4.3 and 4.7 of the *Climate Change Convention* and the Preamble, See for more details on leadership in the *Climate Change Convention*, Joyeeta Gupta, 'Leadership in the Climate Change Regime', *RECEIL* (1998), Vol. 7, Issue 2.
[5] See, for example, pre-negotiation Declarations on Climate Change, such as the (1989)*Noordwijk Declaration on Climate Change* and the (1990) *Declaration of the Second World Climate Conference.*
[6] (1998) 37 I.L.M. 22.
[7] This is stated in Article 3 of the *Climate Change Convention.*
[8] Principle 7 of the *Rio Declaration.* (*Rio Declaration on Environment and Development*, (1992) 31 I.L.M. 874).
[9] C. Alzamora (1987), *Environment and Development in the Context of the North-South Debate".*
[10] *Protocol on Substances that Deplete the Ozone Layer,* (1987) 26 I.L.M. 154.
[11] For an interesting discussion, see D. French (2000),"Developing States and International Environmental Law; The Importance of Differentiated Responsibilities", 49 *ICLQ* 35.
[12] Art. 3.2.
[13] Notably, Iceland, Australia and Norway who are allowed to increase their emissions by 10%, 8% and 1% respectively in the period 2008-2012 with respect to levels in 1990; see Annex B of the Kyoto Protocol.
[14] For example, Ukraine and Russia are, in effect, allowed to increase their emissions to 1990 levels. Given that both economies have collapsed considerably, their emissions have decreased substantially since 1990.
[15] This includes the European Union countries that have to reduce their emissions by 8%, USA by 7% and Japan and Canada by 6% respectively; see Annex B of the Kyoto Protocol.
[16] South Africa, Nigeria and Egypt are not included in this calculation.
[17] Arts. 4(1)(a), 12, *Climate Change Convention.* This would include a national inventory of emissions, steps taken to implement the *Climate Change Convention and* other information relevant to the achievement of the *Convention's* objectives.
[18] See for example, Article 4.1 of the *Climate Change Convention*, and Art 10 of the *Kyoto Protocol.*
[19] The countries subject to emission reduction requirements are listed in Annex I. These include both industrialised countries as well as the countries of eastern and central Europe.
[20] Art. 2.
[21] See, for example, Article 4.3 and 4.7 of the *Climate Change Convention.*
[22] Under the *Kyoto Protocol,* Australia is permitted to increase their emissions by 8% while Iceland is permitted a 10 % increase.
[23] Art. 4.8
[24] Art. 4.9
[25] Funding is to be additional to what is given through both bilateral ODA (Overseas Development Assistance) and multilateral assistance, as well as what is allocated to the Global Environmental Facility (GEF) climate change focal area. *Bonn Agreements on the Implementation of the Buenos Aires Plan of Action*-Decision 5/CP.6, July 27, 2001). At the 6th Conference of the Parties*bis* of the parties, the creation of a special climate change fund was to be established and funded by both Annex I and Annex II countries, to finance activities, programmes and measures related to climate change in the following areas: adaptation; technology transfer; energy; transport; agriculture; industry; forestry; waste management; and activities to diversify the economies of developing countries.
[26] Art. 4(2)(e).
[27] *Climate Change Convention,* Art. 3(4), *Kyoto Protocol,* Arts. 2(1), 4(1)(d), 10, 12(2).
[28] *Kyoto Protocol,* Article 4(7).

[29] The Umbrella Group Countries (Canada, Australia, Japan, Russia) wanted this issue to be put on the formal agenda of the 8th Conference of the Parties, which was opposed by the G77 and the EU.
[30] However, Kazakhstan still remains an Annex I party under the *Climate Change Agreement.*
[31] Such measures may be the result of internal transformation and the need for local companies to become competitive; see, for example, Gupta, J., J. Vlasblom and C. Kroeze with contributions from C. Boudri, K. Blok and K. Dorland (2001),*An Asian Dilemma: Modernising the Electricity Sector in China and India in the context of Rapid Economic Growth and the Concern for Climate Change*, Institute for Environmental Studies, Report number E-01/04, Amsterdam.
[32] J. Gupta, (1997) *The Climate Change Convention and Developing Countries- From Conflict to Consensus?* (Environment and Policy Series, Kluwer Academic Publishers: Dodrecht) at 24.
[33] R.N. Cooper (1998), "Towards a Real Global Warming Treaty" 77:2 *Foreign Affairs* 66.
[34] R. Spalding-Fecher, K. Matibe & G. Simmonds, (1998), "The Clean Development Mechanism: Energy Project for Africa" in UNEP/DANIDA. The WRI estimates this figure to be 7.6% World Resources Institute, (1996-1997) *World Resources: A Guide to the Global Environment 1996-97,* (WRI/UNEP/UNDP/World Bank; Oxford University Press, New York USA)
[35] O. R. Davidson & Y. Sokona (1998), "African and the Clean Development Mechanism: Perspectives for Growth" at 11 in UNEP/DANIDA,*African Perspectives on the Clean Development Mechanism*, Regional Workshop New Partnerships for Sustainable Development: The Clean Development Mechanism under the Kyoto Protocol, Accra, Ghana, 21-24 September 1998).
[36] As of March 6, 2002, 49 parties have either ratified or acceded to the *Kyoto Protocol.* Article 25(1) states that 55 Parties, incorporating Parties included in Annex I accounting in total for at least 55 per cent of the total carbon dioxide emissions for 1990. Although reaching the number of required parties will probably be met by the end of 2002, it will take all of the OECD country ratifications to meet the 55 % target in light of the US current policy of withdrawing from the climate change process.
[37] *Marrakech Accords,* www.unfccc.org, doc. FCCC/CP/2001/13.
[38] P. A. Cham (2000),*African Interests and Stakes Under the United Nations Framework Convention on Climate Change,* (Paper delivered at the Climate Change Capacity Project-Africa Workshop in Dakar, Senegal, July 17-21, 2000) at 5.
[39] For discussion, see *infra.*
[40] R.T. Watson, M.C. Zinyowera, R.H. Moss (eds.)(1988),*The Regional Impacts of Climate Change: An Assessment of Vulnerability*, (Special Report of IPCC Working Group II: Cambridge University Press) at 31.
[41] Mike Hume (ed) (1996), *Climate Change and Southern Africa: an Exploration of Some Potential Impacts and Implications in the SADC Region.*,(CRU/WWF).
[42] *Ibid.* Hume explains that it is expected that regional temperatures may rise by between 1 degree and 2 degrees Centigrade in the 21st century, and about 15-20% of the nature reserves will experience a change in biome.
[43] Cham (2000) at 5.
[44] Abira, M.A. (1997). "Water Resources" in M. A. Abira V. O. Awuor (eds.) (1997),*Potential Impacts of Climate Change in Kenya*, (Climate Network Africa: Nairobi) at 52.
[45] J.C.O. Onyango, J.C.O., E.O. Massawa and M.A. Abira (1997), "Biodiversity and Climate Change" in J. S. Ogala, M.A.Abira, V.O.Awuor (eds.) (1997),*Potential Impacts of Climate Change in Kenya*, Climate Network Africa, Nairobi at 60-61.
[46] IPCC (1998), See also World Resources Institute (1996-1997),*The Regional Impacts of Climate Change: An Assessment of Vulnerability* at 33. (World Resources Institute: Washington, D.C.).
[47] Cham (2000) at 14. Under the*Bonn Agreement on the Implementation of the Buenos Aires Plan of Action* (2001), a fund for least developing countries was established to support a work programme for LDCs.
[48] Art. 4.3.
[49] Art. 4.1 (e).
[50] See Arts. 4.3, 4.1(e) and 4.4. Under Article 4.8, assistance should be provided to those developing countries that are small island countries, have low-lying coastal areas, arid areas, areas prone to natural disasters, desertification, high urban atmospheric pollution; countries with fragile ecosystems,

economies that are highly dependent on fossil fuels and land-locked countries.
[51] Art. 12.8. of the *Kyoto Protocol*
[52] Article 4(7).
[53] *Kyoto Protocol,* Art. 12(10).
[54] Yamin et al. (2001) at 203. The authors note that the same incentive does not exist for JI since overstatement by the CIS countries can result in lower emission limits for themselves.
[55] Art. 12.5(c)), *Kyoto Protocol.*
[56] Estimates are that Africa attracts only 3% of total FDI. See UNCTAD (1999), *World Investment Report 1999: Foreign Direct Investment and the Challenge of Development,* (UNCTAD: Geneva).
[57] AIJ projects resulted from the pilot phase of Joint Implementation projects under the *Kyoto Protocol.* Sokona *et al.* (1998) notes that Africa was excluded from the AIJ pilot phase due to its focus on emissions reductions and its reliance largely on market forces. See Y. Sokona, S. Humphreys and J-P Thomas (1998) *The Clean Development Mechanism: What Prospects for Africa?,* (ENDA: Dakar, Senegal).
[58] Yamin (2000).
[59] R. S. Maya and J. Turkson "CDM Baseline and Additionality in the African Context- The Issues" at 19 at 33 in UNEP/DANIDA, *African Perspectives on the Clean Development Mechanism*, Regional Workshop New Partnerships for Sustainable Development: The Clean Development Mechanism under the Kyoto Protocol, (Accra, Ghana, 21-24 September 1998).
[60] Sokona *et al.* (1998).
[61] Climate Network Africa, (1998) *Preliminary Report of the Regional Workshop for Eastern and Southern Africa on Clean Development Mechanism in the Context of the Kyoto Protocol to the United Nations Framework Convention on Climate Change,* (Nairobi, Kenya).
[62] Spalding-Fecher (1998) at 67.
[63] B. V. Braatz *et al* (1995) "African Greenhouse Gas Emission Inventories and Mitigation Options: Forestry, Land-Use Change, and Agriculture" in J. F. Fitzgerald *et al African Greenhouse Gas Emissions Inventories and Mitigation Options: Forestry, Land-Use Change, and Agriculture,* (Kluwer Academic Publishers; Dordrecht) at 10.
[64] A natural gas pipeline is being built that will start from Nigeria and go through Benin, Togo and Ghana to Cote D'Ivoire, displacing the fuel oil use at thermal generation stations with significant local cost savings. Discussion has surfaced about extending the pipeline to include Burkina Faso, Mali and Niger.
[65] Annex I countries are restricted to use sink projects only up to 1% of their base year emissions.
[66] In Kenya for instance, 71% of the energy supply comes from the forests. See Onyango (1997). For all of Southern Africa, this statistic has been recorded as high as 80%. See S. Moyo (1996), "Improving Access to Natural Resource Potentials and Regional Cooperation in Southern Africa", at 167 in T. Abdul-Raheem, *Pan Africanism: The Politics, Economy and Social Change in the Twenty-first Century,* (London: Pluto Press) at 171.
[67] Over 80% of Southern African People rely on wood for fuel – due to inaccessibility of available electricity resources and technologies that depend on energy. The poor and working class rely on charcoal, which consumes word in its production, and coal, which is unsound for health. S. Moyo, (1996) "Improving Access to Natural Resource Potentials and Regional Cooperation in Southern Africa", at 167 in T. Abdul-Raheem, *Pan Africanism: The Politics, Economy and social Change in the Twenty-first Century,* (London: Pluto Press) at 171.
[68] Arts. 4(1)(j) and 12(1).
[69] Arts. 4(3)(7).
[70] Of these only Liberia and Somalia have not ratified the *Convention.* Twelve African countries have thus far ratified the Kyoto Protocol (as of March 28, 2002).
[71] See Gupta (1997); R.S. Maya and A. Churie, "Critique of the African Approach to Negotiating JI- Bargaining or Posturing?" in R. S. Maya & J. Gupta (1996), *Joint Implementation: Carbon Colonies or Business Opportunities?"* (Southern Centre for Energy and Environment; Harare).
[72] Maya & Churie (1996) at 35.
[73] Gupta (1997) at 117.
[74] H.W.O. Okoth –Ogendo, J.B. Ojwang, A. Churie (1992), *Global Climate Change and the*

Environment: Setting an African Agenda for Policy Responses to the United Nations Framework Convention on Climate Change, (African Centre for Technology Studies, Nairobi).

[75] Cham (2000).

[76] For instance, Latin American countries have been divided about whether CDM projects should include all types of forestry projects.

[77][77] These are prescribed as renewables up to 15 MW, energy efficiency up to 15 GWh, and small projects up to 14 ktCO2e.

BIOTECHNOLOGY AND THE COMMERCIALISATION OF BIODIVERSITY IN AFRICA

RACHEL WYNBERG[1]

AFRICAN RICHES, AFRICAN WOES

African Riches

Far from being the proverbial basket case, Africa is a continent rich in natural and cultural diversity, and remarkable in the innovations her people have developed to use and conserve biological resources. The region contains a quarter of the world's biodiversity, encompassing a wide range of habitats from vast savannah plains and tropical forests, through to coral reefs, the world's largest desert region, and some of the most spectacular freshwater systems in the world.[2] These habitats are repositories for an extraordinary range of animal and plant species. Up to 60 000 plant species occur on the continent, including 35 000 endemics that occur nowhere else on Earth.[3] The region also contains three of the most biologically diverse countries in the world (South Africa, Madagascar and the Democratic Republic of the Congo),[4] three of the world's biodiversity "hotspots" – or high biodiversity ecosystems under the greatest threat of destruction,[5] and one of the world's eight major centres of crop plant diversity (Ethiopia)[6]. Many thousands of useful plants and animals are found in the region, with plants such as sorghum, millet, palm oil, cotton and coffee, as well as several timber species, making key contributions to world agriculture. Additionally, many ornamental plants and medicinal products have their origins in Africa.

A Matter of Survival

While important for the international community and for global environmental sustainability, for the majority of Africans, biodiversity is a matter of survival.

More than any other region in the world, biological resources constitute the basis of national economies and livelihoods for the 700 million people who inhabit the continent.[7] The vast majority of these people rely on biodiversity for food, medicine, low-cost building materials, fuelwood, craft materials and income.[8] Its use, abundance and variety represent an indispensable buffer against poverty and the seemingly endemic occurrences of drought, environmental change and war. Unlike many other parts of the world, where traditional knowledge about biodiversity is held by geographically distinct indigenous groupings, in Africa such knowledge is intrinsic to the daily functioning of virtually all rural households. This finds expression in the enormously diverse cultures of the more than 2000 ethnic groups that inhabit the continent, and in the central role played by plants and animals in African indigenous systems of medicine and agriculture.

In addition to subsistence use, the bulk of employment, economic output, and export earnings in Africa is generated through the exploitation of biological resources. For example, agriculture accounts for between 30% and 60% of GDP in sub-Saharan Africa,[9] and employs more than 60% of the labour force - a dependency equalled only by South Asia.[10] Forestry and fisheries likewise play key roles in many national African economies, contributing over 60% to foreign exchange in countries such as Equatorial Guinea and Mauritania respectively.[11] Tourism, largely dependent on the extensive wildlife found in parts of Africa, is a significant contributor to GDP in countries such as Kenya, where it generates about 10% of total employment and ranks second only to agriculture in foreign exchange earnings.[12] Together, these activities illustrate the heavy dependency of Africa upon a productive, diverse and readily available biological resource base.

Vulnerability and Control

Dependency however, brings with it vulnerability and Africa is indeed vulnerable. The continent is characterised by crippling levels of foreign debt,[13] stagnant or declining economic performance, and the highest levels of HIV/AIDS in the world.[14] Currently 76% of the population lives on less than $2 per day, and the number of people living in poverty is rising.[15] Biodiversity loss is accelerating – and millions of livelihoods are threatened - through the clearing of forests for commercial agriculture, monoculture cropping and forestry, overfishing, invasive alien species, mining, and the overharvesting of natural resources. Accompanying these activities is the erosion and loss of traditional knowledge about biodiversity. Perhaps most pervasively however, people's livelihoods are in danger of being profoundly impacted by the intensifying control over food, agriculture and healthcare by a diminishing number of large profit-driven corporations.

The most powerful players in the global economy, the multinational corporations, aided by new technologies and a world trade regime that ensures open access to biological resources through liberalisation and legal appropriation through intellectual property rights, are encroaching upon African biodiversity on an unprecedented scale. Increasingly, the agrochemical, seed, and pharmaceutical industries are converging with one another through takeovers, mergers and alliances. Large investments in biotechnology have accelerated these trends, together with the granting of patents on genetically engineered organisms, and associated pressures to commercialise new products. Today, just a handful of 'Gene Giants' – including Aventis, DuPont, Monsanto (now owned by Pharmacia), Dow and Syngenta dominate the market. Between them, the top ten 'Gene Giants' account for 84% of the $30[16] billion global agrochemicals market, one-third of the $24 billion commercial seed market,[17] virtually the entire market for genetically engineered seeds,[18] and are increasingly converging with the $300 billion pharmaceutical industry.[19]

Sales of this magnitude have ensured such companies complete dominance over smaller enterprises and national institutions. In Africa, just ten companies account for 88% of the agrochemical market.[20] Three of the biggest pesticide companies – Syngenta, Monsanto, and DuPont – also dominate the African market in genetically engineered seeds[21], and increasingly, the local supply and marketing of seeds. Similarly, multinationals hold the majority of local markets for pharmaceuticals in sub-Saharan Africa, which together with the Middle East was estimated to be over $8 billion in 1997.[22]

Box 1. Pesticide Use in Africa[1]

Since the companies controlling seed supply are the same as those providing agrichemicals, important questions must be asked of the implications for fertiliser, herbicide and pesticide use. Biotechnology products in particular are often sold together as a package by the same company – an example being crops which are engineered to be resistant to a particular herbicide.

With less than 4% of the global market, Sub-Saharan Africa is the region of the world which uses the least pesticides[1, 1]. Can these footnotes be combined?? However, at a national level pesticide use is significant, especially for cash crops, and in some countries, for vegetables. Insufficient capacity, information and infrastructure mean that safe application of hazardous products is virtually impossible, and the impacts on the environment and human health go unrecorded[1]. This is compounded by the growing stockpiles on the continent of hazardous, obsolete and unidentified pesticides.

Products of the leading agrichemical companies are all available in Africa, the most active being Syngenta (then Novartis), which in 1991 estimated it held half the market share of sub-Saharan Africa[1]. Other industries marketing pesticides in Africa include American Cyanamid (previously Shell and now part of the American Home Products group); AstraZeneca / Syngenta in Anglophone Africa; Rhone Poulenc in Francophone countries; as well as Bayer, particularly for domestic pest control. Pesticides are also imported from cheaper generic producers, particularly in China and India. With the exception of South Africa (the largest user of pesticides in sub-Saharan Africa) there is no production of the active ingredients in African countries.

Although government subsidies for pesticides are gradually being phased out, a significant proportion of pesticides supplied to Africa are donated under aid programmes. The Japanese KR2 programme, for example, accounts for 25% of all pesticides used[1]. Donors also supply large quantities of pesticides for migratory pest control, as well as chemical fertilisers. In many cases there are insufficient studies of the environmental impacts of application or follow up studies on biodiversity and efficacy. Additionally, unanswered questions remain about the wisdom of creating donor-driven dependencies on external chemical inputs.

TRADE, NOT AID

Globalising Africa

Trade liberalisation and the new regimes ushered in by the World Trade Organisation (WTO) and other trade agreements underpin many threats to biodiversity and the loss of control people face over their livelihoods. Africa, as an untapped market and emerging trade partner, is a strategic target for producers and investors. The continent's integration into the global economy is considered crucial to attain this goal. "Trade not aid", is the new mantra of donors, investors and lenders, despite Africa's astonishing debt burden of some $230 billion, repayments of which amount to double that spent on education and healthcare combined.[23] It is a difficult pressure to resist, given that Africa's share in world trade is a mere 2%, and that conditionality imposed by the World Bank and IMF in many African countries go hand in glove with trade liberalisation and privatisation. However, deeper integration into the global economy is likely to further undermine African public interest goals of food security, healthcare and environmental conservation.

Through the *TRIPS (Trade-Related Aspects of Intellectual Property Rights)* Agreement of the WTO, a global regime has been created that facilitates stricter intellectual property protection over biodiversity and opens the door for the patenting of life forms. Driven by the multi-billion dollar biotechnology and pharmaceutical industries, it raises profound questions about the ethics of commercialising life forms, about the concentration of knowledge in the hands of few, and about the rights of holders of traditional knowledge. As members of the WTO, most countries of the world are now obliged to allow patents and other forms of intellectual property protection to enter the realm of agriculture, food production and healthcare.[24] From being a common heritage of humanity, biodiversity is increasingly becoming a privately held commodity to be used for individual or corporate benefit.

Biotrade

Trade in biodiversity forms an important part of this picture. The past decade has witnessed a surge of interest in the commercial use of wild species and genetic resources, on a scale unprecedented in history, and linking markets in virtually every corner of the globe. Bioprospecting – or the exploration of biodiversity for commercially valuable genetic and biochemical resources,[25] has become a boom industry, as has the rapidly growing botanical medicine industry based on plant materials. Natural product-derived pharmaceuticals alone contributed an estimated $120 billion – or 40% - of global pharmaceutical sales in 1997,[26] with global trade in raw botanical materials approximating $8 billion in the same year.[27] New applications for genetic resources in the biotechnology industry have led to new and varied demands for biodiversity, in forms previously unimagined.

How does Africa stand to benefit from such developments? If one considers that Africa is the repository for 25% of the world's biodiversity, and at a crude estimate the combined total value of all products derived from the world's genetic resources lies between $500 and $800 billion annually,[28] it is clear that Africa is a rich, profitable seam of raw material and knowledge for the development of new medicines, foods, cosmetics and other products from biodiversity. Historically, however, benefits derived from the commercialisation of these resources have amounted to nothing for the region, while colonial powers have gained substantial economic advantage through their use.[29] This situation continues today.

Redressing these inequalities is a key objective of the *Convention on Biological Diversity* (*CBD*),[30] which 47 African countries are party to. In what has been referred to as the "Grand Bargain"[31], countries providing genetic resources will receive a suite of benefits from those commercialising the resources, including a fair share of the profits generated, as well as non-monetary benefits such as technology and the opportunity to participate in research. In exchange, provider countries are to facilitate access to their genetic resources and associated knowledge.[32] However, access is to be granted on "mutually acceptable terms" and subject to the prior informed consent of the provider country.[33] In Article 8(j) of the *CBD,* recognition is given to holders of traditional knowledge, and to the importance of fairly sharing benefits derived from the use of this knowledge. Importantly, the Convention asserts that custodians of biological resources have control over these resources. Comparison between these provisions and those of the WTO agreements present some disturbing contradictions that are now becoming increasingly apparent in Africa and elsewhere, between the priorities of sustainable and equitable biodiversity use, and the tenets of the global marketplace.

TAKING STOCK: FOOD, AGRICULTURE AND HEALTHCARE

Agriculture in Africa

Nowhere are polarities between the WTO and the CBD more apparent than in agriculture, and in the two systems of production that co-exist in African agriculture. The first system, and that practised by the vast majority of African farmers, is typically based on multiple cropping and relies on farm-saved seeds, low chemical inputs, and rainfall. On-farm crop selection remains vital in these traditional systems, with ownership of resources, seeds, and knowledge often being of a collective nature and indeed "often shared with pride and given away as a great honour".[34] Traditional farming produces over 90% of food in sub-Saharan Africa although situations obviously vary from country to country.[35]

The second system mirrors that of industrial nations and is based on purchased seed, high chemical inputs, irrigation, mechanisation, and mono-cropping. This sector contributes significantly to the export earnings of many African countries and produces mainly cash crops such as coffee, cotton, cocoa, tobacco, tea and sugar. With the exception of maize production, this takes place largely at the expense of producing food crops for local consumption.[36] Most seeds are imported, high-yielding hybrids and cultivars, which need to be bought anew each season from the corporations.[37] Although the seeds may have their origins in Africa, ownership of the seeds – usually through monopoly patents - resides with the corporations responsible for their development.

Traditional farming conflicts directly with the objectives of agribusiness. It limits the market for agrochemicals and, with farm-saved seeds representing about 90% of planted seeds on the continent,[38] constitutes a considerable constraint to the expansion of the commercial seed market.[39] There are aggressive efforts to change this on many fronts as part of the globalisation of agriculture and its markets.

Increasingly, African governments are under pressure to "modernise" their agricultural systems through the use of high-yielding hybrids, transgenic varieties, fertilisers, herbicides and pesticides. This is most apparent in the supply and distribution of seeds. Through structural adjustment programmes, virtually every government in Africa is under pressure to privatise state-owned seed supply systems. In Malawi for example, a World Bank and IMF programme aims to reform the agricultural sector, a result of which is the privatisation of the National Seed Company of Malawi.[40] In Senegal and other West African countries, structural adjustment has similarly resulted in the privatisation of the seed sector.[41] Specific guidelines issued by the World Bank's Sub-Saharan Africa Seed Initiative make compliance with seed reforms part of any negotiations with governments for loans or investments.[42]

The recent establishment of the African Seed Trade Association (AFSTA) is part of this trend. Its stated mission is to "represent the interests of African seed industries and to promote the development of seed industries for the betterment of crop production in Africa".[43] In theory an indigenous seed industry could be developed to meet the needs of African farmers. However, recent indications show that financial pressures are driving indigenous seed companies into relationships with large monopolies, or forcing their takeover. Sensako, a local seed producer in South Africa, has sold a majority stake to Monsanto "in order to compete with other multinational seed companies",[44] and another South African seed company, Carnia, has witnessed a similar fate. Together, it is estimated that the mergers will result in job losses for a quarter of the staff.[45]

What do these developments mean for the millions of small farmers dependent on farm-saved seed for their livelihoods? Arguments for agricultural reform in Africa emphasise the need to provide farmers with access to quality seed and new research developments, and to improve export earnings. Yet few farmers in Africa have the capital to purchase high-yielding seeds, much less transgenic varieties, which are often engineered to lock the farmer into buying the associated packages of herbicides and pesticides made by the same company. Moreover, the tighter intellectual property protection that accompanies the use of new varieties will prevent farmers from saving and exchanging "proprietary" seeds.

This is likely to be intensified with obligations under *TRIPS* for member countries of the WTO to extend intellectual property protection over plant varieties. Although some African countries are adopting innovative approaches towards meeting this requirement,[46] an increasing number are taking up a model of plant variety protection that conforms to the highly restrictive *UPOV Convention* of 1991 (*UPOV*).[47] This form of Plant Breeder's Rights is widely regarded to promote genetic uniformity in agriculture and be biased towards the commercial interests of industrial breeders in the North. In Francophone Africa, member states of the Organisation Africaine de la Propriété Intellectuelle (OAPI) have agreed to a revision of their 1999 *Bangui Convention*, and the provision of a PVP regime that conforms with *UPOV*. With the rights of some 20 million farmers affected by this decision, it would seem that no holds are barred in pursuit of the vision - espoused by governments and industry alike - to globalise Africa.

Biotechnology – Greed or Need?

As part of the vision, modern biotechnology is being aggressively pushed as a solution to "counter famine, environmental degradation, and poverty".[48] Proponents of the technology allege that without it, Africa will not be able to feed her growing population, and are vigorously promoting its use and development in the continent.[49] Life science giants such as Monsanto or Syngenta that combine agriculture, pharmaceutical, and nutritional arms justify many of their practices with the assertion that by integrating all aspects of food production, health and nutrition they alone are posed to resolve the world's related problems (ignoring equity and distribution issues altogether). The increasing rejection of genetically engineered products by Northern consumers has accelerated the vigorous promotion of their use and development in Africa, and the United States in particular is keen to find new markets for its $60 billion agricultural industry.

Evidence of this is found in the US-led "Partnership for Growth and Opportunity for Africa", a scientific exchange programme to enhance crop biotechnology research in sub-Saharan Africa.[50] The programme stems directly from the *African Growth and Opportunity Act* (*AGOA*)[51] recently passed by the US Congress and which has been widely condemned for the extensive rights and benefits it grants to transnational corporations operating in Africa, and the lack of protection for workers and the environment.[52] Several other initiatives also aim to promote biotechnology in Africa. In Algiers, an African Agency for Biotechnology has recently been established to help develop biotechnology across the continent.[53] In Kenya, the International Service for the Acquisition of Agri-Biotech aims to "facilitate the transfer of biotechnology to developing countries for the primary benefit of the rural poor and small scale farmers."[54] In South Africa, biotechnology receives prominence in the US-South Africa Binational Commission and is also identified as a major growth area warranting the support of the taxpayer in the government's Innovation Fund. Recently, some 47 groups, including Monsanto, Delta and Pine Lands, AgrEvo, Novartis/Syngenta, Pioneer Hi-Bred, as well as several research institutions and producers joined in a major consortium to promote biotechnology in South Africa and to "provide one strong voice for lobbying government on biotechnology ... and ensuring that unjustified trade barriers are not established which restrict its members".[55]

What do these trends imply for food security and agriculture in Africa? Many question the relevance of biotechnology in Africa,[56] and suggest the technology and its value system to be a new form of colonialism, imposed by Northern countries and their industries and inappropriate to African needs.[57] Ever since the inception of biotechnology, Africa has remained a mere supplier of raw materials, used by agribusiness and research institutions in the North. Most African countries do not have basic facilities for the simplest of tissue culture methods, let alone for recombinant DNA techniques.[58] Those countries, which are developing or applying modern biotechnology - South Africa, Egypt, Kenya, Nigeria and Zimbabwe - do so without capacities for risk assessment and management.[59] Although transgenic crops have been released in Morocco, Zimbabwe, Egypt, South Africa and Kenya, no environmental impact assessments have been undertaken, nor appraisal for the need and desirability of the crop.[60] This despite recent findings that illuminate the potential ecological risks of biotechnology, and concerns about the impact of genetic engineering on Africa's biodiversity.[61] Moreover, the large majority of commercial releases and field trials have been for herbicide tolerant and pest resistant crops rather than for developments that could make a real impact on African food production, such as improvements in nitrogen fixation, or drought and viral resistance. Similarly, industry places emphasis on high volume crops, which offer most opportunities for sales, rather than food crops used traditionally by Africans, such as sorghum or millet. In South Africa, where approximately 200 field trials are underway, and 5 commercial and commodity releases have been approved, over 90% of applications for transgenic crop testing were for insect resistant and herbicide tolerant strains. Seventy per cent of these applications were received from multinational "gene giants", including Monsanto, Pioneer Hi-Bred, AgrEvo, Delta and Pine, Syngenta and Du Pont.

Far from being the panacea for Africa, genetic engineering brings with it a host of socio-economic and environmental problems for the continent. The destruction of livelihoods for small farmers is one such problem, with the International Labour Organisation suggesting employment losses of up to 50% in developing countries due to biotechnology.[62] Genetic engineering means it will become possible to produce in the laboratory or in temperate zones crops that have hitherto been grown exclusively in the tropics. In Madagascar, some 70 000 small farmers growing vanilla could be threatened by the laboratory production of vanilla aroma using biotechnology.[63] Similarly, synthetic production of thaumatin – derived from a West African plant – threatens the livelihoods of thousands of people collecting the resource from the wild in Cote d'Ivoire and other West African countries. Higher yields from genetically engineered cocoa varieties could reduce prices and jeopardise smallhold production of the crop in West Africa.[64] Commercialisation of the so-called Terminator Technology,[65] designed to prevent seed reproduction and thus assure repeated sales, as well as other coercive technologies that enforce loyalty to seed companies, would have untold effects on the millions of small-scale farmers throughout Africa that depend on replanting farm-saved seeds. Patent applications for the technology in up to 90 developing countries further fuel these concerns and public commitments from Monsanto and AstraZeneca not to commercialise seed sterilisation technology have been short-lived. In 1999 AstraZeneca conducted its first field trial on genetic trait control technology in the United Kingdom, and the technology is now broadly acknowledged to be on a 'fast track' to commercialisation.[66]

The need to improve African food security and agricultural productivity is an obvious concern shared by all. However, is genetic engineering the answer? Many would contend that the problem is not insufficient food, but rather issues around its distribution and access, including the difficulties of poor farmers obtaining credit, lack of storage facilities and inadequate infrastructure.[67] These problems are unlikely to be resolved by biotechnology, especially under current conditions, in which research and development is almost exclusively the domain of the private sector.[68] In an industry with such massive and increasing research costs, the emphasis is on products, which will generate sales large enough to recoup research costs and generate profits. Food security, human development and environmental sustainability simply do not factor in this value system.

Healthcare in Africa

A similar story is unfolding for Africa's healthcare. Privatisation, and the distortion of intellectual property and trade measures are being pursued to pursue profit at the expense of meeting the primary healthcare needs of the majority of the population. As with agriculture, two systems of medicine co-exist in Africa - one based on traditional medicine and the use of biodiversity, and the other on western approaches to healthcare and the use of pharmaceutical products. In general, western approaches to medicine have dominated in most national health systems in the region, but slowly this is changing, and numerous countries are now integrating traditional medicine into their official healthcare programmes.[69]

For the estimated 70-80% of African people who rely on traditional medicine and the variety of plants and animals upon which it is based,[70] the conservation of biodiversity and its sustainable use are vital. Exploitation for domestic use constitutes the bulk of medicines harvested, and with most medicines originating from the wild, an increasing number of species are becoming scarce and vulnerable.[71] With Western markets for African herbal products such as the bark of *Prunus africana* used to treat men's prostitis, growing at 10% per annum, added to rapid population growth in Africa, pressure on these resources is likely to increase. Thus there is a growing realisation that actions must be taken soon to secure a sustainable supply of medicinal products.

Conservation is critical not only to protect biodiversity, but also to meet the health needs of the continent. For both rural and urban Africans, traditional medicine is a form of healthcare that is more available and often cheaper than western medicine. More than this, it embodies an integral part of a spiritual healing process, environmental ethic and ancestral belief system that is uniquely African.[72] Knowledge, experiences and observations are handed down from one generation to another, resulting in a health system that is tribally and culturally specific, but also continuously changing and strongly influenced by social, economic and political factors.[73]

Use of this knowledge and experience is big business, and most of the top 150 plant-derived prescription drugs correlate with knowledge about traditional medicine that has been derived from communities throughout the world.[74] Although Africa has yet to yield a "block-buster" drug, traditional African knowledge has been used to develop numerous commercial products, or potential products with medical, cosmetic, food or agricultural value, and the raw material for their manufacturing continues to be sourced from farms and forests today. In many instances, such knowledge has been patented, usually by researchers or companies in industrialised countries, with no compensation to the original holders of the knowledge. Examples abound of patents - from sweeteners through to appetite suppressants, cardiotonics, and treatments for cancer, diabetes, rheumatism and mental disorders - all of which have their origins in African species and traditional use by African communities.[75]

The great irony is that such products are unlikely to assist with the health crises that plague the African region. The same giants that control the agrochemical and seed industries dominate the pharmaceutical industry, and within five years are predicted to own 75% of pharmaceutical companies worldwide.[76] They too are following the trend towards an ever more sophisticated set of molecular techniques, centred on genomics[77] and biotechnology. At an estimated $300 million USD to develop and introduce a new drug, the focus is on drugs that pay. Even if a compound proves effective and safe, most pharmaceutical companies will simply not bother to develop and market it unless potential sales exceed $100 million a year.[78] With the average African spending less than $10 per person each year on healthcare,[79] and debt-strapped governments unable or unwilling to adequately subsidise medical care, these are clearly not products aimed at the African market. Research and development is focussed on commodities for Western markets from which a financial return is guaranteed – such as anti-obesity drugs, anti-depressants, cardiotonics and cancer treatments – rather than on medicines that could make a difference to the millions of Africans suffering from malaria, tuberculosis and malnutrition. Where treatments are appropriate, such as those developed for AIDS, drugs are prohibitively expensive because of their patenting. Accessibility to essential drugs by Africans is likely to continue to diminish with the patenting requirements of *TRIPS*, which will further increase the price of drugs and concentrate research and development in industrialised countries.[80] Revision of the WTO *General Agreement on Trade in Services* is predicted to accelerate these trends and open up foreign healthcare markets to creeping privatisation and foreign commercial interests. This has alarming implications for the majority of people in the world who simply cannot afford to pay for health services.

Biodiversity Commercialisation - Problem or Panacea?

How can this scenario – and the tremendous biological resources and human innovations of the African continent - be turned around to benefit the region? A much-touted solution, embraced by players as diverse as the World Bank, national governments, UN agencies, NGOs and the private sector, is to actively commercialise biodiversity and traditional knowledge, within the context of the benefit-sharing and conservation provisions of the *CBD*. Biodiversity, it is argued, cannot be adequately conserved without economic return; through commercialisation, the biological riches of developing countries will bring economic opportunities and much needed technology transfer and capacity building. While few would argue with this approach, its interpretation and practice in Africa requires careful scrutiny.

On a daily basis, biological resources are collected throughout the continent for commercial purposes, to be screened for potential therapeutic or other benefits, or to be packaged and marketed as herbal drugs, cosmetics or other natural products. Without doubt, such collections are on the increase – as reports from as far afield as Namibia, Senegal, Uganda, Kenya, South Africa, Mauritius, Zimbabwe, Cameroon and Ethiopia indicate.[81] Local interest is also peaked, with universities, museums, botanic gardens, and other research institutions often collaborating in bioprospecting projects. They provide foreign companies or research institutions with help in field collections, the provision of biological material or information, or – in the case of more advanced developing countries – through direct participation in the process of new product discovery. Individual biologists, chemists, or healers from developing countries may be approached by companies or foreign research institutions wishing to investigate a country's biodiversity. Generally through a lack of awareness they will accept an *ad hoc* payment or enter into an agreement out of line with the access and benefit-sharing provisions of the *CBD*. Pressures to supplement dwindling research budgets have made developing country universities and other research institutions especially vulnerable to the lure of Western scientists and companies.

A major problem is that most African countries lack the technological and scientific capacity to capitalise on commercial collaborations, and to make use of the opportunities created by the *CBD*. The necessary business and legal skills to ensure the development of a fair deal are also often lacking, a constraint heightened by the absence of legislation in most African countries to regulate access to genetic resources and to set parameters for benefit-sharing. Thus bioprospecting for the majority of African countries simply reinforces their role as mere suppliers of raw material and results in trivial benefits for their struggling economies.

A few African countries – notably South Africa, Kenya and Nigeria – have engaged at a higher level, through adding value to information and resources supplied, undertaking in-country research and development and in some instances gaining access to screening technologies and product development. While such projects undoubtedly serve to strengthen local institutions, scientific capacities, and biodiversity inventories, there is still little indication of how socio-economic uplift for African people is achieved, how incentives are created for biodiversity conservation, how holders of traditional knowledge are compensated, or how the vexed questions of patenting of life forms are resolved. In short, commercialisation seems ultimately to serve rather than address the economic imbalances and inequities of the global trading system.

A question that must be asked is whether the high risks and levels of sophistication that accompany new drug discovery are indeed appropriate routes for African countries wishing to commercialise their biological resources. Even countries as advanced as South Africa are doubting the ability of bioprospecting to pay dividends, and are increasingly recognising the importance of developing and supporting industries based on phytomedicines, personal care products and food supplements.[82] This option offers fewer risks and delays,[83] the use of technologies more appropriate to developing countries, and also a far greater chance of benefits reaching the ground. Thus in Namibia, a women's cooperative is involved in the commercialisation of a resource long used and nurtured by them; in South Africa, a job creation programme for retrenched mineworkers is spearheading the commercialisation of products made from the indigenous marula tree; and in Botswana and Zimbabwe numerous projects are being launched to develop products, owned and managed by custodians of indigenous biological resources. While countries obviously need to develop strategies to respond to bioprospecting and the search for new drugs and other products, Africa is perhaps best positioned to invest its energies in more locally driven solutions to economic and environmental crises.

Ultimately it would seem that the commercialisation of biodiversity is unlikely to achieve revolutionary changes in the ways in which resources are used and conserved, nor in the way that business is conducted. Although the *CBD* has undoubtedly advanced approaches to equity and benefit sharing, the commercial use of biodiversity cannot in itself answer broader questions around fair trade, social justice, the commodification of resources and economic reform. Yet the links between these issues are increasingly becoming entangled and inseparable, as responses from the continent demonstrate.

WAKING UP TO THE CHALLENGE

A New Philosophy: Global For a and Regional Strength

From Addis Ababa to Cape Town – from government officials through to NGOs, small-scale farmers and local communities - the voice from the continent resonates loud and clear: Africa stands united in its commitment to protect the rights and interests of communities and holders of traditional knowledge, to reject the patenting of life forms, and to develop indigenous technologies and innovations that are appropriate to local conditions and needs. Despite the enormous obstacles facing the continent, at the turn of the new millennium a new and dynamic political phase is underway in Africa, identified as the "second independence struggle" or, according to South Africa's President Thabo Mbeki, an "African Renaissance". This is a philosophy driven by the need to find "African solutions for African problems", to mobilise the people of Africa to take their destiny into their own hands, to break neo-colonial relations between Africa and the world's economic powers, and to vigorously pursue the economic recovery of the continent.[84] It is a call that represents a rejection of Afro-pessimism and a realisation of the ability of Africa to confidently take control of her own future, on equal terms with the rest of the world. Building on this sentiment, the New Partnership for Africa's Development (NEPAD)[85] aims to place countries on a path of "sustainable growth and development" whilst being competitive in the global economy.

Tremendous energies are being mobilised to ensure these perspectives are reflected at all levels – from the halls of the United Nations and its suited diplomats, to communities and farmers. Within international fora, these sentiments are illustrated by the increasingly influential role played by Africa in the various initiatives stemming from the *CBD* and the WTO. A united stance from the continent has been key to ensuring that the environmental, ethical, social and economic consequences of modern biotechnology have not been neglected in negotiations to develop a *Biosafety Protocol.*[86] A determination from Africa to protect the rights of farmers to save seeds and continue their traditional practices has contributed to the successful conclusion of an International Treaty on Plant Genetic Resources for Food and Agriculture. Emerging issues in the Agreements of the World Trade Organisation, and the revision of the *TRIPS Agreement* in particular, are receiving priority attention by African governments, regional bodies such as the Organisation for African Unity (OAU)[87] and SADC[88] and diplomatic groups such as the African Group of Ambassadors. All of these groups argue for the right of states to introduce systems of their own choice in order to ensure national food security, livelihoods, healthcare and the development of sustainable agriculture. These factors, among others, catalysed African countries to play a pivotal role in resisting the massive political and economic pressures from industrialised countries to reach consensus at the 1999 WTO meeting in Seattle.[89] Indeed, the fact that sub-Saharan Africa was better prepared and more active than ever before was a key contributor to the infamous breakdown of these trade talks.[90]

At the regional level, initiatives are well underway to prepare the ground for compliance with *TRIPS* and other international agreements such as the *CBD* and *Biosafety Protocol*. Within the OAU, an *African Model Legislation for the Protection of the Rights of Local Communities, Farmers and Breeders, and for the Regulation of Access to Biological Resources*,[91] has been developed. This model expands on an earlier version developed by the OAU and aims "to ensure the conservation, evaluation, and sustainable use of biological resources, including agricultural genetic resources, and knowledge and technologies in order to maintain and improve their diversity as a means of sustaining the life support systems." The model law aims to assist countries to fulfil obligations under *TRIPS* and the *CBD* while protecting the rights of communities to control access to resources and knowledge, and of farmers to save, use, multiply and sell seed.

In cooperation with the OAU, SADC is similarly drafting a common legislative framework for *sui generis* rights, recommending an approach to cater for different sectoral activities within a country – from horticultural products through to open-pollinated crops and medicinal plants. Another recent OAU initiative has resulted in the drafting of biosafety model legislation, which would make it illegal for a country to export genetically modified food without first seeking permission from the importing country.[92] Through such undertakings, and the bringing together of countries to debate options, it is intended that governments expedite the establishment of domestic regimes that regulate biotechnology, control access to genetic resources and ensure equitable benefit-sharing, and protect the rights of farmers, communities and plant breeders.

National Legislation and Grassroots Initiatives

Whether or not countries can translate these complex and often cumbersome proposals into workable solutions remains the acid test. While it is still too early to assess the impacts of OAU processes, awareness is undoubtedly growing among governments as to the urgency of putting measures in place to deal with these issues. In many countries, *TRIPS* requirements for countries to develop a regime providing intellectual property protection over plant varieties have provided a major impetus for action. In countries, which have significant plant breeding activities, such as South Africa, Zimbabwe, and Kenya, legislation has long existed for plant variety protection. In Zimbabwe, such legislation is coming under increasing scrutiny and efforts are underway to develop a *sui generis* legislation, in line with the proposal described above for SADC. Other countries such as Zambia are committed to protecting community rights by means of a *Plant Breeder's Act* presently being formulated through wide consultation.[93] In Uganda, draft legislation to protect community rights is similarly on the table, based upon the *OAU Model Law*. In South Africa, there are ongoing policy processes to ensure the protection of traditional knowledge and Farmers Rights, although powerful lobbies are arguing against such measures.

Slowly, governments are realising the importance of legislative measures to control access to genetic resources and to regulate biotechnology, although in most countries there exists a policy vacuum with respect to these issues. Exceptions include South Africa, which has recently promulgated a *Genetically Modified Organisms Act;*[94] Namibia, which is presently drafting regulations to control access to genetic resources and has published a policy on biosafety; Cameroon, which has broad provisions to regulate access to genetic resources that are incorporated into forestry[95] and environmental framework laws[96]; and Malawi, which is presently developing a policy on access to genetic resources and benefit-sharing. Perennial constraints of capacity, resources, vested interests, corruption, and lack of political will are critical obstacles to government action being successful and taken more broadly. A major problem, and one that is not unique to Africa, is that institutional links between biodiversity issues - access to genetic resources, community and farmer rights, biotechnology, patenting and plant variety protection, and biodiversity conservation – are often not made within and across government departments.

All too frequently, policy dialogues and international relations remain peripheral to people on the ground. While this remains true for the bulk of Africa's populace, there is a growing movement of civil society organisations, farmers, scientists, and citizens throughout the region that are aware of and concerned about the continent's threatened biodiversity, and about the changing face of agriculture and healthcare on the continent. Traditionally, biodiversity in Africa has been the domain of a few conservation NGOs, typically involved with maintaining a network of protected areas and preserving endangered species, without consideration of broader social issues.[97] While this important lobby remains strong, the "voice of biodiversity" is gradually being replaced by one that is more mindful of social and economic realities, and of the need for clear political commitment towards ensuring implementation of the *CBD's three* objectives: conservation, sustainable use and the equitable sharing of benefits derived from the use of biological resources.[98]

Growing coalitions of traditionally disparate organisations similarly mirror the increasing recognition being given to the links between trade, structural adjustment, biodiversity conservation, and basic development issues. The African Trade Network, for example, comprising over twenty NGOs and civil society groups from ten African countries, has called on African political leaders to prohibit the patenting of life forms and ensure the protection of traditional knowledge of biological resources.[99] In Zimbabwe, a range of scientific and consumer groups are calling on the government to tighten legislation controlling the import of genetically engineered seeds, plants and food, and to delay the importation of genetically engineered crops and food until the health risks can be scientifically evaluated. Similar calls are being made in Kenya and Uganda, backed up by a strong voice from farmer groups to reject the patenting of life forms and to assert the importance of collective ownership of genetic resources and associated innovations. In South Africa, a flourishing coalition of trade unions, political parties, NGOs, consumer bodies, and farmer groups is demanding a five year freeze on the use and release of genetically engineered crops.[100] Similar movements are springing up throughout the continent with remarkably similar calls.

A NEW MILLENNIUM

These trends are highly significant for Africa as the continent commences the new Millennium with renewed hope for a better future. Maintaining the optimism and momentum to drive an African Renaissance – in the face of seemingly insurmountable obstacles of poverty and debt – presents perhaps the greatest challenge of all. Likewise the need for the international community at large to take responsibility for alleviating these problems, and to put in place measures that substantially reduce the continent's debt burden. Enormous interests are at stake – from the faceless transnational corporations pushing biotechnology products at all costs, through to government officials already reaping significant benefits through hefty handouts and iniquitous deals. Continued networking and monitoring by civil society groups are fundamental to counter these interests, to conserve biodiversity, and to protect the rights of ordinary Africans to safe food, appropriate and adequate healthcare, and sustainable livelihoods. Full and open disclosure of information is integral to these objectives being met, as is the development of mechanisms to enable different sectors to learn from one another. Capacity building, a much vaunted but poorly developed activity, is essential to empower people from all levels of society to make informed and locally-driven choices about these issues. Technology transfer from industrialised countries remains a critical component to enable African countries to add value to raw biological material, but as important is the need for Africans to consider using indigenous technologies and experiences as starting points for technological improvement, economic development, and environmental sustainability. Africa's future and that of her unborn generations is at a crossroads. The moment is now but can the moment be seized fast enough?

---

[1] Graduate School of Environmental Studies, University of Strathclyde. Current address: PO Box 69, St James, 7946, South Africa. tel: +27 21 788 7677; fax: +27 21 788 9169; e-mail: rachel@iafrica.com, This paper has formed part of a broader research initiative, led by GRAIN and the Gaia Foundation, to investigate the linkages between trade and biodiversity. These organisations are gratefully acknowledged for allowing these research findings to be used in this publication. Particular thanks are due to the following people for their assistance with providing information, advice and comment: Renee Vellve, Katharine Ainger, Henk Hobbelink, and Anna-Rosa Martinez (GRAIN); Liz Hosken and Tanya Green (GAIA Foundation); Sarah Laird; Robert Lettington (African Centre for Technology Studies, Kenya); Phoebe Barnard (Namibian Directorate of Environmental Affairs); Herta Kolberg (National Plant Genetic Resources Centre of Namibia); Jean Marie Fondoun (Programme

Conservation des Ressources Genetiques, Cameroon); Athman Mgumia (Mtandao wa Vikundi wa Wakulima Tanzania); Thomas Kentos Bakyatita (Joint Energy and Environment Projects Uganda); Paul Therence Senghor (Department of Environment, Senegal); Dr JC Autrey (Mauritius Sugar Industry Research Institute); Doreen Mnyulwa and Julius Mugwagwa (Biotechnology Trust of Zimbabwe); Dr Abebe (Ethiopia Biodiversity Institute); Barbara Dinham (Pesticide Action Network); R. Guyer (Novartis); Robert Maybury and Charles Weiss (International Organisation for Chemical Sciences in Development); Dr Hans Herren and Dr Wilber Lwande (International Centre of Insect Physiology and Ecology, Nairobi, Kenya); Cyril Lombard (CRIAA SA-DC); Carl van der Lingen.

[2] H. Beentje *et al.* (1994), 'Regional Overview: Africa', in S.D. Davis *et al, Centres of Plant Diversity, Volume 1 Europe, Africa, South West Asia and the Middle East,*(WWF, IUCN, European Commission, ODA, 1994), 101.

[3] Beentjie (1994).

[4] Conservation International (1999). http://www.conservation.org/web/fieldact/megadiv/tables/divrank.htm

[5] While unique and threatened ecosystems are found in all African countries, identified 'global hotspots' include Madagascar and the Indian Oceans, the Cape Floristic Province and Succulent Karoo of South Africa, and the Guinean Forests of West Africa.

[6] Conservation International (1999). http://www.conservation.org/web/fieldact/hotspots/Priority.htm

[7] See for example A.B. Cunningham (1992), 'Local Initiatives and Institutional Roles', in L.A. Bennun *et al, Proceedings of a Conference held at the National Museum of Kenya, 30 August– 3 September 1992.* (National Museums of Kenya, Nairobi, Kenya); B.M. Campbell (1986), 'The importance of wild fruits for peasant households in Zimbabwe', 12 *Food and Nutrition* 38.

[8] Biodiversity Support Programme (1993), *African Biodiversity: Foundation for the Future. A Framework for Integrating Biodiversity Conservation and Sustainable Development* (Biodiversity Support Programme, Maryland).

[9] World Bank (1992), *World Development Report* (Oxford: Oxford University Press).

[10] World Bank (1992).

[11] R. Fraser (1991), *The World Trade System,* (Harlow, Essex, U.K. Longman Group). These figures are however likely to have changed over the last few years, especially given Mauritania's recent exploitation of oil reserves.

[12] J. Mugabe (1998), 'Biodiversity and Sustainable Development in Africa', in J. Mugabe and N. Clark, *Managing Biodiversity. National Systems of Conservation and Innovation in Africa,*(ACTS Press, Nairobi) at 5.

[13] In Sub-Sarahan Africa, foreign debt approximated to 66% of GNP in 1997. However, for least developed countries, many of which occur in Africa, this figure was as high as 92,3% of GNP in 1997. See UNDP (1999), *Human Development Report* (New York: United Nations).

[14] Global statistics for HIV/AIDS indicate an average 39,7 people per 100 000 are afflicted by the disease. In Africa, 111,1 people per 100 000 are affected by HIV/AIDS, nearly three times the world average. See UNDP (1999).

[15] World Bank (1999), *Global Economic Prospects for Developing Countries 2000.* (Washington, D.C.; World Bank).

[16] $ refer to US$ throughout the document.

[17] Annual figures based on estimates from P. Heffer, (1999), "Basic Figures on International Seed Trade" (*Preparatory Meeting for the Establishment of an African Seed Trade Association*, Lilongwe, Malawi, 8-10 April 1999) at 29-32. See also ETC Group (2001). "Globalization, Inc. Concentration in Corporate Power: The Unmentioned Agenda". Communique July/August 2001, Issue 71. www.rafi.org

[18] ETC Group (2001).

[19] S. A. Laird and K. ten Kate (1999), "Natural Products and the Pharmaceutical Industry", in K. ten Kate & S.A. Laird, *The Commercial Use of Biodiversity. Access to Genetic Resources and Benefit-Sharing.* (London: Earthscan) at 34-77.

[20] B. Dinham (1999), 'Pesticide Use in Sub-Saharan Africa', (Unpublished report, Pesticide Action Network).

[21] Dinham (1999).

[22] Laird and ten Kate (1999).

[23] C. Landsberg and C. Kabemba (1998), 'The US and Africa. A New Beginning?', 15:2*Indicator South Africa* 36.
[24] The UNDP Human Development Report (1999) reports that industrial countries hold 97% of all patents worldwide. More than 80% of patents granted in developing countries are held by residents of industrial countries.
[25] Reid, W.V. *et al,* (1993), *Biodiversity Prospecting. Using Genetic Resources for Sustainable Development,* (World Resources Institute, Instituto Nacional de Biodiversidad, Rainforest Alliance, and African Centre for Technology Studies).
[26] Laird and ten Kate (1999).
[27] S.A. Laird (1999), 'The botanical medicine industry', in Laird and ten Kate (1999).
[28] Laird (1999).
[29] C. Juma (1989), *The Gene Hunters,* (London: Zed Books, London).
[30] (1992) 31 ILM 851.
[31] M. Gollin, 'An intellectual property rights framework for biodiversity prospecting', in Reid (1993).
[32] Article 15, *CBD.*
[33] Art. 15.
[34] W. Mundaka, 'Farmers' Rights: The Crisis Every Ugandan Farmer Needs to Know About', 8*Rural News*. http://www.nic.ug/IRDI/html/rights.html.
[35] F. Dakora (1997), 'Using indigenous knowledge to increase agricultural productivity in Africa', in H, Normann *et al, Indigenous Knowledge and its Uses in Sauthern Africa,* (HSRC Cooperative Programme and the Institute for Indigenous Theory and Practice).
[36] A.K. Banda (1999), Keynote Address on the Official Opening of Proceedings of the meeting. International Seed Trade Federation, (FIS; See also GRAIN, 'Plantvariety protection to feed Africa? Rhetoric versus reality', (GRAIN, 1999), http://www.grain.org/publications/reports/variety.htm
[37] *Ibid.*
[38] J. Shah (1999), 'The Seed Industry in Kenya: An Overview'. (Proceedings of the meeting. International Seed Trade Federation).
[39] P. Heffer (1999), (Proceedings of the meeting. International Seed Trade Federation); A.J.G. van Gastel *et al,* 'Promotion of Seed Marketing in West Africa', 113, (Proceedings of the meeting of the International Seed Trade Federation)
[40] Banda (1999).
[41] D. Ndoye (1999), 'Seed Industry in Senegal', 68, (Proceedings of the Meeting of the International Seed Trade Federation; van Gastel (1999).
[42] SSASI Team, World Bank, "Initiatives for Sustainable Seed Systems in Africa", http://www.fao.org/ag/AGP/Abidjan/Paper12.htm
[43] Proceedings of the meeting of the International Seed Trade Federation.
[44] M. Mlangeni (1999), 'US Company Buys Majority Stake of South Africa Seed Producer Sensako', Business Day 5 March 1999, http://www.bday.co.za/99/0305/company/c3.htm.
[45] L. Cook (1999), "Seed Firm to Lose Staff". Business Day 25 August 1999, http://www.bday.co.za/99/0825
[46] GRAIN (1999), 'Beyond UPOV: Examples of developing countries preparing non-UPOV "sui generis" plant variety protection schemes for complance with TRIPS', http://www.grain.org/publications/reports/nonupov.htm.
[47] (1991) *The Union for the Protection of New Varieties of Plants* an intergovernmental convention managed by the UN World Intellectual Property Organisation. Its original *Plant Breeder's Rights Convention* was adopted in Paris in 1961. Since then the Convention has been amended several times and two forms of Plant Breeder Rights are now in common use. Most UPOV members adhere to its 1978 treaty, which is widely interpreted by governments to allow farmers to save and exchange seed. UPOV's 1991 *Convention*, however, assumes that farmers cannot save seed unless governments permit specific exemptions.
[48] F. Wambugu (1999), 'Why Africa Needs Agricultural Biotech', 400:15-16 *Nature* (1 July 1999).
[49] *Ibid;* See also J.A. Thompson and Zhang-Liang (2000), 'Biotechnology Fear is a Rich Nation's Luxury', No. 3212 *Sunday Advocate* (16 April 2000). http://www.biotechknowledge.com/showlib us.php3?3212.
[50] USDA (1999), 'USDA launches biotech research project for SubSaharan Africa', July 27 1999, USDA Press Release. <http://www.gene.ch/>.

[51] (H.R. 434).

[52] J. Dube (1999), 'African Ministers Prepare for the Millennium Round', (IPS: 14 July 1999); J. Lobe (1999), 'Trade-US: Political Wrangling Derails Africa, CBI Trade Bills', (IPS: 29 October 1999).

[53] 'Africa moves to strengthen biotechnology', 399:6 *Nature* (6 May 1999).

[54] ISAAA (1997), 'ISAAA Launches New Intellectual Property / Technology Transfer Initative', (12 November 1997), <http://www.isaaa.org/new1.htmo>.

[55] N. Opperman (1999), 'Plans for biotechnology association for food, feed and fibre sectors',*The Farmer*, 25 (October 1999).

[56] For example, in response to a Monsanto advert promoting biotechnology as the solution to the world's food crisis, 24 African countries and 30 community, environmental and farmer organisations attending a United Nations meeting in August 1998 published a joint statement entitled "Let the Harvest Continue", objecting strongly that the image of the poor and hungry is being used to push a technology they do not perceive to be safe, environmentally friendly, or economically beneficial.

[57] D. Sharma (1999), 'USDA Pushing Gene Foods on Third World', *The Pakistan Observer, Islamabad* (27 June 1999); J. Otieno (1999), 'Biotechnology Controversy Rages',*Daily Nation, Nairobi* (16 September 1999).

[58] A. De Kathcn, (1999), 'Transgenic Crops in Developing Countries. A Report on Field Releases, Biosafety Regulations and Environmental Impact Assessment, UBA-TEXTE 58/99 (1999).

[59] *Ibid.*

[60] See for example, M. Mayet (1999), 'Critical Review of Existing Legislative Framework for Genetic Engineering in South Africa', Biowatch South Africa (August 1999). Available from Biowatch South Africa: nicci@biowatch.org.za

[61] F. Huang *et al.* (1999), ' Inheritance of resistance to *Bacillus thuringiensis* toxin (Dipel ES) in the European Corn Borer', 284 *Science* (1999), 965; E.L. Losey *et al.* (1999), 'Transgenic pollen harms monarch larvae', 399 *Nature* (1999), 214; T. Spears (2000), 'Birds Do It Bees Do It. Nature Frustrates Efforts to Confine the Spread of GM Plants', *The Ottawa Citizen* (4 January 2000), http://www.ottawacitizen.com/national/000104/3385145.html.; "Killer virus",*New Scientist*, January 10 2001.

[62] A. Seiler (1996), 'Impacts of Biotechnology on the Third World', 17 *IPR info* (Institute for Agriculture and Trade Policy 1996).

[63] Novartis Foundation for Sustainable Development, 'The Socio-Political Impact of Biotechnology in Developing Countries', (1999), http://www.foundation.novartis.com.

[64] *Ibid.*

[65] RAFI (2000), 'Suicide Seeds on the Fast Track: Terminator 2 Years Later', (February 2000).

[66] *Ibid.*

[67] H.R. Herren (1999), 'Potentials and Threats of the Genetic Engineering Technology: Quest for an African Strategy at the Dawn of a New Millennium', (Unpublished paper, International Centre of Insect Physiology and Ecology, 1999).

[68] Cassava provides a good example of this. Despite the fact that cassava serves as a staple food for at least 300 million Africans, the biotechnology industry made no research efforts to improve crop yield and production until the crop was shown to be a feed substitute for pig industry in States. See Sharma (1999).

[69] N. Marshall (1998), *Searching for a Cure. Conservation of Medicinal Wildlife Resources in Eastern and Southern Africa*, (Cambridge: TRAFFIC International).

[70] L. Ake-Assi (1998), 'Plantes medicinales: quelques La Gumineuses utilises dans la medecine de tradition Africaine en Côte d'Ivoire', 25 *Monogr. Syst. Bot. Gard;* 309; O. Akerele (1993), 'Summary of WHO Guidelines for the Assessment of Herbal Medicines', 28 *Herbalgram* 13.

[71] Marshall (1998).

[72] M.M. Iwu (1993), *Handbook of African Medicinal Plants,* (Boca Raton: CRC Press).

[73] Iwu (1993).

[74] F. Grifo *et al* (1996), 'The Origins of Prescription Drugs', in F. Grifo and J. Rosenthal,*Biodiversity and Human Health*, (Washington D.C.: Island Press).

[75] See R. Wynberg (2000), *Privatising the Means for Survival. The Commercialisation of Biodiversity in Africa,* (Gaia/GRAIN Briefing).

[76] J. Barton (1999), cited in the *International Conference on Biotechnology in the Global Economy*, 2-3 September 1999, http://www.iisd.ca/sustainable development/har/index.html.

[77] Genomics refers to technology for the automated, high-throughput sequencing of DNA for the study for the 120 000 genes estimated to form the human genome.
[78] L. Glowka (1998), *et al*, 'Best Practices for Access to Genetic Resources. Background Paper for the Correspondent Worlshop on "Best Practices for Access to Genetic Resources"', (14-17 January 1998, Cordoba, Spain).
[79] 'Drugs for AIDS in Africa', (*New York Times* Editorial, August 23 1999).
[80] O. Ombaka (1999), 'Trade-Related Aspects of Intellectual Property Rights (TRIPs) and Pharmaeuticals', 15 *Echoes* 19.
[81] Reponses to a questionnaire administered in January 1999 by GAIA and GRAIN.
[82] M. Mander (1998), *Marketing of Indigenous Medicinal Plants in South Africa. A Case Study in KwaZulu-Natal,* (Food and Agriculture Organisation, Rome).
[83] C. Weiss, and T. Eisner (1998), 'Partnerships for Value-Added through Bioprospecting', 20 *Technology in Society* 481.
[84] R. Cornwell (1998), 'The African Renaissance, The Art of the State', 15 *IndicatorSA*, 9.
[85] The New Partnership for Africa's Development (2001). http://www.nepad.org
[86] *Protocol on Biosafety (Cartagena),* (2000) 39 ILM 1027. T.B.G. Egziabher, African spokesperson for negotiations to develop a Biosafety Protocol and to revise the International Undertaking on Plant Genetic Resources.
[87] "Elements for a Positive Agenda in the New Trade Negotiations Under the WTO from an African Perspective", Council of Ministers, 70th Ordinary Session / 5th Ordinary Session of the AEC, 6-10 July, Algiers, Algeria. CM.2110 (LXX) Annex IV.
[88] "SADC Ministers Agreed Negotiating Objectives for the Third WTO Ministerial Conference", WT/L/317. 1 October 1999.
[89] D. Sharma (1999), "The Message from the Seattle Fiasco. Wrong Trade Organisation. Deccan Herald, Bangalore, India Dec 18, 1999.
[90] L. Machipisa and R. Oyo (1999), 'Failure of WTO Talks Applauded, Africa Feels Vindicated', IPS (Lagos, 10 December 1999).
[91] See www.oau-oua.org. Hereinafter referred to as the *OAU Model Law*. See also J.A. Ekpere (2001), 'The OAU's Model Law: The Protection of the Rights of Local Communities, Farmers and Breeders, and for the Regulation of Access to Biological Resources, An Explanatory Booklet', Organisation of African Unity, Scientific, Technical and Research Commission, Lagos, Nigeria.
[92] E. Masood (1999), 'Africa seeks laws on GM food exports', *Nature*, (5 August 1999).
[93] E. D. Zulu, R. M. Makano and A. Banda (1999), *National experiences and plans to implement a sui generis system of protection in Zambia*. Paper presented at the UPOV-WIPO-WTO Joint Regional Workshop on "The Protection of Plant Varieties Under Article 27.3(b) of the TRIPS Agreement". Nairobi, 6-7 May 1999.
[94] *Act 15* of 1997.
[95] *Law No. 94/01*, Articles 11 and 12.
[96] *Law No. 96/12*, Articles 64 (1c) and 65 (1) and (2).
[97] R. Bonner (1993), *At the Hand of Man. Peril and Hope for Africa's Wildlife,* (Vintage Books, New York).
[98] Article 1.
[99] Third World Network (1999), 'Joint NGO statement of support for the African Group Proposals on Reviewing the WTO TRIPs Agreement (Article 27.3b)', (August 1999), 'NGOs ask governments to reject new trade round', 4505 *SUNS* (7 September 1999).
[100] http://www.safeage@mweb.co.za

ENVIRONMENTAL REGULATION OF THE OIL AND GAS INDUSTRY IN NIGERIA.

KENT NNADOZIE

INTRODUCTION

The industrial development and societal practices of the 20th century have undoubtedly had a disturbing impact on the environment of the planet earth. The wasteful use of natural resources, development of energy sources and improper disposal of energy by-products poses adverse impact on, not only the air, water and land, but increasingly the entire ecosystem and human health. The causes may be diverse and numerous, but the problems created by the production and consumption of fossil fuel in general and petroleum in particular represent some of the greatest concerns shared by the entire international community.[1]

For nearly three decades, petroleum production and consumption has probably brought out both the best and worst of modern civilisation in Nigeria. It has contributed enormously to the country's economic growth and, on the other hand, has left profound adverse impact on the natural environment and has generated a number of other socio-economic concerns including human rights issues. Nigeria is currently the largest producer of crude petroleum in Africa and sixth largest in the world. Estimates of Nigeria's oil reserves range from 16 billion to 22 billion barrels and almost all of these reserves are found in relatively simple geological structures along the country's coastal Niger Delta. This makes it highly susceptible to more marine pollution through spillage and blowouts especially in the exploration and production (E & P) stages coupled with the continuous flaring of associated gas. The World Bank estimates that gas flaring in the Niger Delta releases some 35 million tonnes of carbon dioxide annually into the air. The increasing profile of climate change as an issue of growing global concern has drawn tremendous attention to the role of Nigeria's gas flaring in the build-up of greenhouse gases while incessant communal conflicts have highlighted the socio-economic dimensions of petroleum E & P.

In the light of the profound dilemma facing the country, the task of reconciling the demands of economic development, social justice and environmental protection has become a challenge facing the entire spectrum of the Nigerian society, from the government and its agencies, the oil industry, local communities and the entire citizenry. Clearly, the environment versus petroleum extraction balance has tended to tilt more in favour of production in Nigeria.

However, the industry is entering a new era of legal regulation as E&P activities and products are increasingly subjected to international law, national and institutional regulations and requirements as well as civil society pressures.

This chapter seeks to examine the relationship between the oil and gas industry and the environment in Nigeria and the legal framework governing this relationship. It gives a brief overview of the balance between these aspects and explores the legal regimes used to maintain or account for that balance. It also seeks to examine the current state of the regulatory approaches and management tools, and their effectiveness and effect on the industry. It will further attempt to highlight the new themes for petroleum environmental regulation and the possible direction for future legal development especially within the regional and international context.

OIL AND GAS RELATED ENVIRONMENTAL RISKS

Exploration, Production and Consumption

The location of oil facilities especially in the upstream sector is not determined by free choice, nor conditioned by proximity to markets, but is subject to the imperatives of geology and geography. Production facilities are typically designed for specific sites based on the location of the oil, so the environmental effects tend to be governed by site-specific factors (nature of land: plain, mountains, hills; location of rivers, access or nearness to the sea; climate; transportation infrastructure; population and land use).[2] The environmental effects of the petroleum are, to a large extent, not predetermined and change both in geographical location, scope and intensity as the phases of exploration, production processing and utilisation advance.[3] However, substantial proportion of environmental disasters involving the oil industry in Nigeria have been as a result of oil spills through either accidents or ageing pipelines and sometimes sabotage. Oil spills, especially in the Niger Delta area, have had devastating effects on wild life and its habitats and have seriously affected the livelihood and health of people in the immediate or surrounding communities.

Although with respect to the environmental consequences of petroleum E & P, the major focus in Nigeria is on the upstream operations of the industry, the entire spectrum of activities in the industry, from exploration, to production through conversion, transportation and to consumption, raise serious environmental concerns. Environmental problems start right away with exploration activities such as seismic surveys and geological prospecting, although the environmental interference and disturbance at this stage are limited and localised. In the upstream operations, environmental problems continue from initial access and acoustic activities at the exploration phase, accidental spills and blow-out at the development stage, and to operational discharges, spills, wastes and emissions during the production period and transportation stage. Further environmental damage can be caused at the production stage by the flaring of the associated gas. This process can create considerable air pollution (carbon monoxide, oxides of nitrogen and sulphur) particulate fallout and heat that scorches the earth and withers all vegetation in the vicinity of the flares.

Petroleum environmental problems do not end with energy consumption, rather they go much further beyond the petroleum energy cycle, as evidenced by the phenomena of global warming, Ozone layer depletion and acid rain. In fact, global climate change has effectively become the greatest environmental concern of our time and is mainly caused by the emission of green house gases from use of fossil fuels.[4]

Apart from the traditional environmental problems, other problems that arise as a result of the operations of the industry include economic issues, ethics and human rights. Oil and gas exploration and exploitation also have social impacts on the culture and heritage of the people in the operating localities

Gas Flaring

Quite often, oil is found in reservoirs containing natural gas, which is produced with the oil. This associated gas is separated from the oil at flow stations and, in Nigeria, more than 80 per cent of this gas is flared and is estimated to be about a quarter of the gas the world flares and vents. This represents a substantial loss of income to the nation as well as a significant contribution to global greenhouse gas emissions. The principal problem in this regard is the low level of natural gas utilisation that results from technical, commercial and policy factors.

Gas flaring has also been blamed for causing acid rain and all sorts of health problems (especially respiratory ailments) in the Niger Delta - though this is hotly contested by the industry. But, burning gas produces carbon dioxide which is a greenhouse gas and a major contributor to global warming. Flaring also releases carbon monoxide and other particulate matter which are hazardous to human health and constitute visible nuisance as the resultant soot settles on rooftops, vegetation and other surfaces. The elimination or significant reduction in flaring of associated gas will, no doubt, benefit the environment substantially and conserve resources for the nation.

INTERNATIONAL ENVIRONMENTAL REGIMES

Background

Although environmental problems have always been associated with industrial and societal practices, pollution, especially in the past, mostly occurred at local levels and their effects were limited while nature's apparent abundance and resilience made them seem remediable or negligible. This paradigm is reflected in the early laws (or lack thereof) governing the protection of the environment from the industrial process. Petroleum production and consumption were no exception to this perception. A general survey of international laws and systems a few decades ago reveals that there were few provisions governing the environmental aspects of oil and gas activities. This situation was basically informed as much by the dearth of thorough scientific knowledge as by the general belief that nature had the capacity to absorb and mitigate the adverse impacts of the relevant operations and fuel utilisation as well as the overarching drive for industrialisation and rapid economic growth.

In this milieu, modern international environmental law has a short history. However, early international legal development did have significant, albeit, limited relevance to the environmental issues in oil and gas activities. This can be found in customary international law and founded in legal principles like the *sic utero tuo* rule which was clearly enunciated in the famous case of the *Trail Smelter* in 1941[5]. The dispute arose between the United States and Canada over the emission of sulphur fumes from a smelter in Canada which caused environmental damages to its neighbouring State of Washington. In the arbitral award, the Tribunal held that, under international law, "no state has the right to use or permit the use of its territory in such a manner as to cause injury by fumes in or to the territory of another or the properties or persons therein . . ." The principles of neighbourhood and liability for pollution damage have a close bearing, even to date, on oil and gas operations. Indeed, it has been argued that this customary law should be applied to offshore petroleum operations.[6]

With the rising profile of related environmental problems and the recognition that the environmental issues and policies relating to the industry invariably have international dimensions, subsequent treaties, conventions and national legislation started incorporating environmental considerations as vital components and stipulations. Further, industry practices and commercial imperatives helped reinforce this trend as trans-national corporations also incorporated environmental considerations into their internal policies and practices as well as part of concession agreements. Over the years, several international agreements imposing environmental obligations on nations have been reached and several have a direct bearing on the industry and form the basis for more comprehensive national or local regulations.

Emerging International Environmental Law.

Over the years Multilateral environmental agreements (MEAs) have increasingly become important in international environmental management evoking trade and sustainable development principles and forming the basis for domestic national laws as effective means for the protection of environmental integrity. Even though the provisions of some of the agreements are sometimes so open-ended that they that they have little or no effect at all on the operations of the industry, they do, however, give some impetus to the formulation - though no guarantee for effective enforcement - of environmental legislation in the signatory countries.

The United Nations Conference on Environment and Development (UNCED) -- better known as the **"Earth Summit"** -- produced an unprecedented vision for guiding the world community towards a more secure, equitable and sustainable future. That vision, and the agreements that resulted from it, called for a new global commitment to sustainable development -- economic growth that meets human needs while protecting nature's ability to renew itself. In spite of the Earth Summit's initial success, much more needs to be done to translate the Earth Summit commitments into realities. Appropriate governance, management and delivery systems as well as legal mechanisms are still urgently needed, especially at the national and local levels, to develop the necessary social, political, and financial capacity required to operationalise sustainable development. Several years later, it is apparent that despite these commitments and the accompanying public affirmations, the basic concept of sustainable development is still not well understood and the policies and legal structures required to implement the Earth Summit agreements, in particular the Convention on Climate Change, are still not fully in place.

Some of the international Conventions that Nigeria has ratified and which have a direct bearing on the industry include:

(a) The United Nations Convention on the Law of the Sea (UNCLOS)

The Convention aims to establish basic environmental protection principles and rules on global and regional co-operation, technical assistance, monitoring and environmental assessment, and adoption & enforcement of international rules and standards and national legislation with respect to all sources of marine pollution.

(b) Continental Shelf

Although UNCLOS also deals with matters relating to the Continental Shelf, the main treaty governing the use of the continental shelf is the 1958 ***Geneva Convention on the Continental Shelf.*** It recognises that the rights of coastal states to exploit its natural resources are exclusive and inherent and no other State may undertake activities without the express consent of the coastal State. The exercise of these inherent rights must not, however, infringe or result in any unjustifiable interference with navigation or other rights and freedoms of other States. Article 5.5 requires that any installations which are abandoned or disused must be removed.

(c) International Convention for the Prevention of Pollution from Ships (MARPOL)[7]

Safe shipping not only reduces the risk to life and property but also to the environment and this constitutes the central theme of the Convention. The Convention contains regulations and requirements with which all ships must comply in order to prevent the discharge of oil and oily wastes into the sea. The Oil in Navigable Waters Act of 1968 implements the Convention in Nigeria, and the Nigerian Maritime Authority (NMA) is authorised to check incoming ships' compliance with MARPOL and is empowered to detain and sanction any vessel found to be in contravention until the breach is rectified.

(d) The Civil Liability and Fund Conventions

Four International Maritime Organisation (IMO) conventions govern compensation for damage caused by persistent oil spills from laden tankers - the 1969 and 1992 International Conventions on Civil Liability for Oil Pollution Damage (1969 CLC and the 1992 CLC) and the 1971 and 1992 International Conventions on the Establishment of an International Fund for Compensation for Oil Pollution Damage (1971 and 1992 Fund Conventions).

The 1969 CLC entered into force in 1975 and lays down the principle of strict liability (i.e. liability even in the absence of fault) for tanker owners and creates a system of compulsory liability insurance. Claims for compensation for oil pollution damage (including clean-up costs) may be brought against the owner of the tanker which caused the damage or directly against the owner's insurer. The tanker owner is normally entitled to limit his liability to an amount which is linked to the tonnage of the tanker causing the pollution.

(e) The Framework Convention on Climate Change (UNFCCC).

One of the major agreements reached at the UNCED, the convention is a binding international agreement that seeks to limit or reduce emissions of gases – greenhouse gases (GHG), mainly carbon dioxide and methane – associated with the potential for global warming. The third session of the Conference of Parties to the Convention, held in Japan in 1997, through the Kyoto Protocol attempted to set definite limits to developed countries' emission levels by committing them to reduce those levels by specific percentages below 1990 levels. Nigeria, along with other developing countries, were exempt from making any immediate commitments, though encouraged to take definite steps to supplement global efforts. Nigeria's position and situation with respect to climate change is paradoxical and delicate - with respect to her choices as how to address the issue as well as the choices and extent of the international community action. Apart from the fact that it suffers similar vulnerabilities as do other developing countries, it is in a peculiar dilemma. Concerted action to reduce GHG emissions, expectedly, which hinge on cutting the consumption of fossil fuel will adversely affect the national economy almost entirely dependent on revenues from oil and gas. The matter is complicated further by the fact that concrete steps within the country to cut GHG emissions is being financed by revenues generated from the oil industry. Practically all approaches to cut emissions do, to a greater or lesser extent, involve substantial reductions in consumption or demand for fossil fuels, whether through direct reductions, fuel efficiency, alternative/renewable energy sources or the use of fiscal measures such as energy or carbon taxes. The choices are, therefore, not easy nor the solutions simple.[8] The Federal Government has, however, issued a national directive to oil companies to achieve zero flaring by the year 2008, though some operators in the industry as well as the Ministry of Environment claim that achieving that target by 2004 is actually feasible. There are currently concerted efforts by the government and the industry to promote the use of natural gas to further facilitate the realisation of this target as well as capture the concomitant economic and environmental benefits.

The **Kyoto Protocol** to the convention was negotiated in furtherance of the objectives of the UNFCCC to stabilise the atmospheric concentration of GHGs at a level that would prevent anthropogenic interferences with the climate system, and in recognition of the need to ensure more commitment from the industrialised countries to these objectives. The main goal of the protocol is to achieve a reduction in global emissions of GHGs in 2010 that are approximately 5% of the 1990 levels. The Kyoto Protocol provides for three flexible international market-driven mechanisms which nations may use to achieve GHG reduction targets and the private sector is expected to participate in implementing all three. They are: ***Emission Trading*** – Annex 1 Countries, i.e. developed countries, can trade their surplus emission rights with one another to meet their commitments; ***Joint Implementation*** – Annex 1 countries can generate Emission Reduction Units (ERUs) through cross-border investment in projects that reduce emissions or enhance GHG sinks. The ERUs can be credited towards the investing country's commitment and debited from the host country; and ***Clean Development Mechanism (CDM)*** – Projects which contribute to sustainable development in a developing country (non-Annex 1 country) and which result in real, measurable and long-term GHG reductions can earn Certified Emissions Reductions (CERs). CERs created in the pre-commitment period of 2000 – 2007 can be banked for use by the investing Annex 1 country in fulfilling its commitments. The Kyoto Protocol, yet to come into force, is currently beset by numerous controversies topmost amongst which is the recent withdrawal of the US, while the modalities for the actual implementation of its several components are yet to be fully worked out and agreed upon. Nigeria, though party to the UNFCCC, has not signed or ratified the Protocol yet and does not appear in a hurry to do so in the light of the implications to national development objectives and inherent dilemma it faces in respect of the related issues.

REGULATION UNDER NATIONAL LAW

There are several approaches and legal instruments used to bring environmental objectives to bear on the industry at the national level.

Legislative Instruments

The predominant approach towards environmental regulation in Nigeria is command and control which, as in most other developing countries, seeks to prescribe specific regimes or standards. Statutory instruments comprising general laws and subsidiary regulations that stipulate standards or procedures for the protection of the environment and prescribe limits for air and water pollution with stipulations for waste disposal, in particular, hazardous waste

From a general survey of early legislation touching upon the oil sector one thing is clear – that they were not particularly environment-oriented. Even those that were subsequently directed at protecting the environment displayed certain fundamental inadequacies by their vagueness and ostensible unenforceability or their practical irrelevance in addressing the critical environmental issues that emanate from the oil and gas sector. On a scale, therefore, the environment versus petroleum extraction balance have tended to tilt more in favour of production in Nigeria because there is generally an overriding policy to commercially maximise exploitation often to the detriment of the environment.[9]

Until recently, the sector had been characterised by low commitment on the part of the operators to implement adequate environmental standards and it has been argued that, because of the lax provisions of the laws, international oil companies, irrespective of their official policies, in developing countries apply standards of operation different from the ones complied with in developed countries. Also, the often-proclaimed principle of "adaptation to local laws and conditions" often leads to adaptation to domestic standards (legal or de facto) of lower environmental quality.

The principal legislation and formal instruments directly governing the oil industry and having environmental components include:

(a) The Oil Pipelines Act, 1956 as amended in 1965[10]
(b) The Mineral Oil (Safety) Regulations, 1963
(c) The Petroleum Regulations, 1967
(d) Oil in Navigable Waters Act, 1968[11]
(e) The Petroleum Act, 1968.[12]
(f) The Petroleum (Drilling and Production) Regulations, 1969 - made pursuant to (e) above. Section 25 requires the adoption of "all practicable precautions" to avoid pollution.
(g) Petroleum (Refining) Regulations, 1974
(h) Associated Gas Re-injection Act, 1979.[13]

To a large extent, the provisions of these instruments stipulate general principles or require subsidiary instruments by competent authorities in order to become operational. It has, however, been argued, and rightfully so, that in the hands of the relevant government agencies with the principal mandate to promote petroleum development, such general principles inevitably meant that environmental objectives had to be subjugated to investment and production priorities.

For instance, the ***Oil in Navigable Waters Act*** of 1968 was largely devoid of environmental provisions, and apparently enacted in observance of the International Convention for the Prevention of Pollution of the Sea by Oil, 1954 (as amended in 1962) and not necessarily as a deliberate and considered attempt to curb environmental pollution connected with the industry. However, that is not to say that it was not a significant legal development in the history of environmental regulation in the industry. Also, the 1956 *Oil Pipelines Act*, which was one of the earliest deliberate attempts at legislating against environmental pollution by oil, in section 31(c) provides that the Governor General (now the President and Head of State) may by regulation prescribe "measures in respect of public safety, the avoidance of interference with works of public utility in, over and under any land and the prevention of pollution of any land or water." Further, the open-endedness of the provisions typically found in the petroleum or related laws provided convenient justification for inaction.

Although the ***Petroleum Act*** of 1969 tended to make fairly comprehensive provisions with respect to the general operations of the industry, it incidentally lacked adequate considerations of their environmental impacts. It merely empowers the relevant authority to make regulations for "the prevention of pollution of water courses and atmosphere".[14] This power was very discretionary and provided for no institutional framework for enforcement and implementation nor any tangible sanctions for any violations.[15] While the *Oil in Navigable Waters Act* suffers certain limitations, it did have a few fairly specific provisions with respect to pollution prevention. Section 1 makes it an offence for any ship to discharge oil or any mixture containing not less than 100 parts into a port or "the prohibited sea area". It further prohibits the discharge of oil, or mixture containing oil, into waters from "any vessel, or from any place on land, or from any apparatus used for transferring oil from or to any vessel".

One of the most prominent characteristics common to existing laws is the emphasis on criminal sanctions as a means of deterring environmentally undesirable behaviour - which leaves the enforcement on compliance in the hands of designated or stipulated officials or authorities. Performance apart, most of the relevant bodies either lack the capacity or are handicapped in many other ways that they are largely incapable of thorough monitoring and effective prosecution and punishment of offenders. On the other hand, there are few or no provisions for individual or interest group action which, most often, provide a more vigorous and enthusiastic enforcement of environmental responsibility and liability.

Apart from the fact that the Oil in Navigable Waters Act merely stipulates the imposition of a fine for violation, albeit without prescribing the minimum or maximum amount, it provides and affords some sweeping defences which have the effect of negating its very intent and, therefore, making enforcement virtually ineffectual. Available defences include discharge of oil for the purpose of securing the safety of any vessel, or preventing damage to any vessel or cargo or saving life; discharge due to an escape in consequence of damage to the vessel; escape of oil without want of reasonable care on the part of the occupier or due to the Act of God or a third party trespasser; or if contained in refinery effluents which was not reasonably practical to dispose of otherwise than by discharging it into water. "Surely, by the time all these defences are pleaded, it is hardly feasible to convict anybody under the provisions of the enactments."[16] Considering the enormity and severity of damage usually involved in pollution incidents, liability should be strict as in most other jurisdictions backed probably with mandatory requirements for insurance. Another severe limitation to the effectiveness of the Act as an anti-pollution legislation is the stipulation contained in Section 12 that no action could be brought under the Act except with the consent of the Attorney General of the Federation. Considered in the light of the nature of the current political environment, the policy thrust of the government and the existing legal system, substantial delays might be engendered, or possibly outright reluctance on the part of the authorities, in enforcing compliance or prosecuting offenders.

The Petroleum Regulations 1967 made pursuant to the Petroleum Act equally lack extensive environmental considerations and stipulations. The central theme was the control of the storage, transportation and marketing of petroleum products while the vast areas prospecting, production and refining are left untouched.[17]

Pursuant to the Petroleum Act enacted in 1969, the Federal Commissioner for Mines and Power issued the Petroleum (Drilling & Production) Regulations, 1969. Just as the principal Act did not adopt environmental considerations as part of its central theme, the regulations also treated the issue as merely incidental to its major concern, which is the regulation of drilling and production of petroleum to maximise revenue for the government.[18] Regulation 25 merely stipulates that an operator "shall adopt all practicable precautions, including the provision of up-to-date equipment ... to prevent the pollution of inland waters, rivers, water courses, the territorial waters of Nigeria or the high seas by oil, mud or other fluids or substances which might contaminate the water, banks or shoreline or which might cause harm or destruction to fresh water or marine life, and where any such pollution occurs or has occurred, shall take prompt steps to control and, if possible, end it."

For all practical purposes, these stipulations are mere exhortations or, at best, directives, as they do not prescribe process or the methods for determining how they are to be carried out. This general lack of provisions to compel or stimulate compliance renders them virtually unenforceable.[19]

A subsequent relevant instrument was the Petroleum Refining Regulations issued in 1974, and like its predecessors lacked adequate environmental protection stipulations. It merely required the manager of a refinery to adopt all practicable precautions including the provision of up-to-date equipment ... to prevent the pollution of the environment by petroleum or petroleum products. And in the disposal of refinery waste and effluent, the Manager shall conform "to good refining practices." Penalty for default is a mere 100 naira (approximately

$1) or imprisonment for six months. Such amorphous terms as "good refining practices" were also not defined nor did the Regulations stipulate any criteria for establishing the standard of such "practices".

A further reference point for enforcement and possible sanction for pollution could be the ***Criminal Code*** [20] which provides, in section 245, for a six-month jail term for "any person who corrupts or fouls the water of any spring, stream, well, tank or place so as to render it less fit for the purpose for which it is ordinarily used." It also prohibits acts "that vitiate the atmosphere or any place so as to make it noxious to the health of persons generally within the neighbourhood or passing along the public way, or any act which is likely to spread infection of any disease dangerous to life, whether human or animal, with a sanction of six months' imprisonment."[21] Technically these provisions may be invoked with respect to oil pollution, but they are feeble and palpably inadequate to effectively address the problems of environmental pollution from oil and gas operations.

Existing environmental legislation in the country has been interpreted subjectively and the record on enforcement is patchy. The general conclusion then is that most of the legislation and instruments governing the oil industry do not go far enough to make provisions for the protection of the environment and where they did make attempts, did not provide for adequate administrative and enforcement machinery to make them meaningful.

Environmental Policies

The current Federal Government's policy on energy and environment is embodied in the 1989 National Policy on the Environment (currently under review). According to the National Policy on the Environment, Nigeria should ensure a balanced mix of various energy types that will be compatible with sound environmental practice, and the reductions of negative impacts of energy production and use on the environment.

Among the strategies proposed in the policy include:

i. encouraging the use of energy forms that are environmentally benign and sustainable;
ii. promoting safe and pollution-free operations in energy production and use;
iii. monitoring and controlling the level of noxious by-product in energy production and use, such as the emissions of CO, NO_X, SO_2, CO_2, and none methane hydrocarbons, thereby reducing the "greenhouse" effect;
iv. establishing stringent standards of safety in all energy production processes;
v. monitoring oil spill contingency plans, including national, co-operative and company-level plans;

vi. ensuring effective monitoring and assessment of environmental protection programmes in upstream and downstream (exploration, production, refining, petrochemicals, transportation and marketing) activities in the petroleum industry;
vii. establishing standards for the control of fuel additive with respect to trace elements especially Pb, S, Va and Ni, Cr and Zn;
viii. encouraging re-injection and utilisation of produced gases to prevent the adverse environmental impacts of gas flares;
ix. encouraging research and development programmes that promote environmentally sound application of coal as a domestic energy source through the reduction of the ash and noxious chemical content;
x. licensing of energy waste disposal sites;
xi. encouraging a multi-sectoral approach to the monitoring and control of environmental problems associated with energy production and use; and
xii. promoting and encouraging research in the development of energy resources.

Environmental Impact Assessment

Both the Department of Petroleum Resources (DPR) and the Federal Environmental Protection Agency (FEPA)(now Federal Ministry of Environment) acting within their respective mandates prohibit operators from embarking on major projects or activities, without due consideration, at an early stage, of any environmental effects. Both agencies have laid down Environmental Impact Assessment (EIA) procedures and set out mandatory study activity lists which spell out offences and penalties for non-compliance. However, the possibilities of discrepancies and overlaps between the respective mandates of FEPA and DPR on EIA procedures have raised jurisdictional issues and have on occasions confused operators. It also has the potential to cause delays and increase costs. There have, however, been recent meetings amongst the Regulatory Agencies with a view to harmonising EIA procedures.

Prior to 1992, only the petroleum sector in Nigeria had any sort of formal regulations issued by DPR requiring prior consideration of environmental impact before the commencement of a project. The regulations were rather loose and enforcement weak while compliance was principally founded on the voluntary initiatives of the oil companies most of which had reasonably high internally set standards.

In December 1992, the Federal Military Government enacted the Environmental Impact Assessment Decree, No. 86 of 1992 (EIA Decree). Although displaying some awkward draughtsmanship, it was a fundamental legal development in the environmental protection efforts and goals of the nation. Section 1 of the EIA

Decree sets out the objectives of any EIA as being, inter alia, "to establish that before a decision is taken by any person or authority intending to undertake or authorise the undertaking of any activity that may likely, or to a significant extent, affect the environment, the environmental effects of that activity shall have been first taken into account".

The stipulated rules are an important lever for environmental protection since they could provide an avenue for allowing communities that might possibly be affected and environmental groups raise objections against operator-sponsored impact statement. The EIA Decree is, in context, geared towards facilitating the assessment of the physical, biological, socio-economic, etc. effects so as to allow logical and rational decisions to be made on the predicted impacts of the proposed activity.[22]

FEPA is presently implementing the set of guidelines developed by it pursuant to the Decree and contains explanations and detailed descriptions of the procedural activities to be followed in the preparation and submission of an EIA to the Agency for review and decision. They also contain checklists, advice on how to prepare an EIA, the format and content, presentation techniques and the monitoring and post-audit procedures.

Common Law

Common Law principles provide avenues for individual actions against polluters and civil remedies for compensation for damages and remediation of injury to the person or damage to property. The importance of common law lies in the fact that it is basically under some of the principles that one can found civil liability in pollution cases, because statutory law is lacking significantly in that regard. The common law torts of Negligence, Nuisance and the rule in ***Rylands v Fletcher*** [23] provide legal principles through which individuals can seek redress or enforce their rights. However, these principles are limited in scope as they do not, per se, recognise any general obligation to protect the environment nor do they provide redress for general environmental pollution. The relevance of the remedies they provide in pollution control cases is with reference to and in the context of damages or injury suffered in respect of the individual's right to enjoy his property and immediate environment free from interference or disturbance.

4.4 (a)

Under the tort of **Negligence**, of paramount importance and primary consideration in establishing liability is the concept of 'duty of care' which a complainant must prove is owed him personally by the polluter. He must show that the latter or his servant or servants were so negligent or omitted to take reasonable care in handling a situation or thing which is under their control as to actuate some injury to him, the complainant. Alternatively he can show, *res ipsa* *loquitor,* that the polluting consequences of the act or omission complained about were such that they could not have occurred in the natural course of events if the defendant polluter or his servant(s) had acted diligently or had taken reasonable care.

4.4 (b)

Nuisance, by definition, involves some form of pollution or inconvenience caused by the wrongful interference with the ordinary physical comfort of human existence, and may be public or private. Public nuisance relates to the interference with the enjoyment of a right which all members of the community are entitled to such as fresh air, free movement through public highways etc. while private nuisance consists of any wrongful interference with a person's use or enjoyment of land say, by allowing the escape of deleterious things into his land, e.g. smoke, noise, smell, fumes, heat, disease-germs etc. Remedies include damages for injury, injunction to restrain continuance, or mandamus to compel abatement, of polluting act or consequences.

Limitations to the principle lie in the fact that, in the case of private nuisance, action can only be brought by the person in possession of land injuriously affected by it, and only by the Attorney General or in relation to him in the case of public nuisance. Furthermore, the person complaining must establish damage or injury which is foreseeable and, in respect of public nuisance, that which is also special and exclusive to him. He must also show a causal link between the act of the polluter and the consequences complained of. So, where for any reason the Attorney General fails or refuses to act, or is not involved in the case of public nuisance, pollution affecting a major section of the public would have to be endured without legal remedy. [24]

4.4 (c)

Under the rule in ***Rylands v. Fletcher***, "the person who for his own purpose brings on his land and collects and keeps there anything likely to do mischief if it escapes, must keep it in at his peril, and, if he does so, is prima facie answerable for all the damage which is the natural consequences of its escape". The principle of strict liability applies here so that a polluter may be held responsible for the harm caused irrespective of wrongful intent or negligence. However, in order to establish liability, a plaintiff must prove that there was an "escape" which was caused by the "non-natural user" of land by the defendant-polluter - a requirement to which no objective or universal test can really be applied. The rule is further excepted where there is statutory authority to perform certain acts, say, where a polluter has a licence to carry on a certain project the consequence of which is being complained of.[25]

Contractual Environmental Regulation

Most contemporary concession contracts contain obligations, formulated as general principles, to minimise environmental degradation and to adhere to good environmental practices and sometimes a general formulation on reclamation, exploitation and utilisation of associated natural gas and decommissioning or abandonment of installations. Increasingly, licences and contract terms are playing a major role in regulating environmental behaviour by petroleum operators in the absence of comprehensive and appropriate legislation. Apart from defining the 'property right' inherent in mineral title, they have been taking on a more significant role in expressing regulatory - as contrasted with mere property right - concerns of public policy.[26]

In addition to the general clauses affirming environmental responsiveness usually found in these agreements, a number of specific issues now tend to be incorporated whether or not already provided for by statutes or specialised regulations, such as:

- Obligation to submit a separate environmental impact statement and a comprehensive environmental management plan;
- Obligation to submit bonds/guarantees for environmental liability (in particular offshore oil spillage) and to include environmental liability in mandatory insurance cover;
- Obligation to carry out periodic environmental audits or to allow the government agency to arrange for an independent environmental audit evaluating compliance with environmental obligations, best environmental practices and standards, mitigation and disaster management programmes.
- Obligations to restore areas used for exploration and extraction to a safe state, in particular to comply with current international guidelines for abandonment of offshore installations.

Further, the growing concern about the effects of gas flaring as well as the need for the commercial utilisation of natural gas has made the Government introduce associated gas utilisation clauses in recent production sharing contracts (PSCs) signed with oil prospectors. This is in line with current practice in other major oil producing countries and in compliance with international environmental standards. This provision generally requires that the prospector must, prior to the commencement of oil production, make arrangements for the commercial utilisation of associated gas or otherwise, its safe management through re-injection.

Voluntary and Non-Regulation approaches

Several past events, beginning from the Koko toxic waste saga[27], the execution of Environmentalist, Ken Saro-Wiwa, to some of the widely reported spills and blowouts, have contributed to heightening the profile of environmental issues in

Nigeria. These have largely resulted in greater public pressures on the operators as well as an increasing militancy on the part of some of the local communities and Non-Governmental Organisations (NGOs). Operators in the industry now, more and more, are adopting a wider range of strategic environmental protection approaches that embrace the essential components of sustainable development: economic prosperity, environmental health, and social equity and well-being.

Following also the growing global trend away from prescriptive legal controls towards co-operation, negotiation and self-regulation, voluntary actions by the operators in the industry are now being vigorously encouraged. This is further justified by the fact that the regulating authorities do not have the capacity to effectively monitor or fully enforce existing standards. Voluntary approaches have been found to have tremendous advantages and can be used to achieve significant but measurable environmental results either alone or in combination with other approaches and are, therefore, fast gaining popularity. The social and economic benefits of improved environmental performance and the anticipation of regulation are also driving corporations to get ahead of legislation and to demonstrate to governments and to the public that self-regulation is a viable option.

Further, beyond the traditional arguments of environmentalists, the fears of local communities and the government's concern for the protection of the eco-system, other very practical aspects bordering on the future survival and profitability of the operators themselves also inform these developments. They have come to realise that sound environmental strategy and practice is good for business, whether in the oil sector or in the banking industry, and that environmental progress can be part of the continuous corporate effort to improve effectiveness and quality and at the same time enhance profitability.

Regulatory and Monitoring Institutions

The Department of Petroleum Resources (DPR) of the Ministry of Petroleum Resources is the unit directly empowered to make, monitor and ensure compliance with regulations in the operations of all upstream oil companies. Despite this, all oil exploration and production companies operating in Nigeria have to report to several government agencies on environmental matters. These include:

- The Federal Environmental Protection Agency (FEPA) or now the Federal Ministry of Environment.
- The States' Environmental Protection Agencies, and
- The Local Government Environmental departments and other "task forces".
- The Nigerian Maritime Authority.
- The Energy Commission of Nigeria.

(a) The Department of Petroleum Resources

The Department of Petroleum Resources (DPR), of the Petroleum Resources Ministry, took over from the Petroleum Inspectorate as the regulator and monitor of environmental issues in the oil industry. Its principal functions include giving due and timely consideration to the needs for environmental protection of the public and the working population from hazards arising from the exploitation, conversion and utilisation of energy from petroleum. It also is in charge of setting environmental and safety standards, ensuring its enforcement and monitoring compliance. DPR has an established set of guidelines for operators but which is currently under review.

Its concurrent mandates with FEPA/Ministry of Environment have raised the issue of possible inter-departmental rivalry and conflict in the course of carrying out their duties. However, this conflict has not been expressly manifest because, in practice, they appear to stay out of each other's way or otherwise co-operate in the fulfilment of their respective mandates. Currently, there are moves to relocate the Environment monitoring and compliance unit of the DPR to the newly created Ministry of Environment. The rationale is that it is reasonable to have this unit under the Ministry of Petroleum Resources whose primary mandate is to maximise production and an active participant in E&P activities and, therefore, a contributor to the concomitant environmental problems.

(b) Energy Commission of Nigeria

The Energy Commission of Nigeria was established by Decree No. 62, 1992 and is charged with the responsibility of the strategic planning and co-ordination of national policies in the field of energy in all its ramifications. The Energy Policy prepared by the Commission contains policy guidelines on the relationship between energy production, utilisation and the Nigerian environment. Its status is not regulatory per se, but mostly advisory and functions through co-operation with the industry and related institutions in the promotion of research and development of sustainable energy sources and utilisation.

(c) Federal Environmental Protection Agency (FEPA)[28]

The Koko toxic wastes dumping incident[29] precipitated and catalysed the establishment of FEPA by Decree 58 of 1988, which charged it with the responsibility for the protection and development of the environment, including initiation of policy in relation to environmental regulation, research and technology.

FEPA is charged with the responsibility for setting, monitoring and enforcing national environmental standards. It was authorised to handle all facets of environmental problems and formulate policies in dealing with them. FEPA Officers are given significant inspection and enforcement powers, including powers of seizure and arrest, although its current capacity is grossly inadequate to contend with the environmental challenges that the country now faces. The

Decree also contains penal provisions against discharging hazardous substances in harmful quantities into the air, land or waters and against obstructing a FEPA official or failing to comply with environmental standards.

In the exercise of its powers and functions FEPA has instituted a number of Regulations, Guidelines and other publications aimed at controlling pollution and protecting the environment. Notable among them are:

(i) National Policy on the Environment, 1989 - which aims to provide a rational, practicable, coherent, comprehensive and sustainable approach to the pursuit of economic and social development. It sets policy goals, provides the concepts and stipulates the strategies which will lead to the procedures and other concrete actions required for launching the country into an era of sustainable development.

(ii) Guidelines and Standards for Environmental Pollution Control in Nigeria, 1991 - which was an interim order, or rather prescription, directing industries to improve the quality of the environment and free it from pollutants and other environmental hazards. They serve, more or less, as recommended standards of environmentally good behaviour for industries.

(iii) The National Environmental Protection (Effluent Limitations) Regulations, 1991 - which enjoin every industry, among others things, to install anti-pollution equipment for detoxification of its effluent and chemical discharges and treat the same to stipulated levels before discharging.

(iv) The National Environmental Protection (Pollution Abatement in Industries and Facilities Generating Wastes) Regulations, 1991 - which prohibit the release of hazardous or toxic substances into the environment beyond the limits approved by the Agency and provide for other requirements and procedures e.g. in cases of accidental or unusual discharges, and also require industries to obtain permits for discharges that are beyond permissible limits.

(v) Waste Management and Hazardous Wastes Regulations, 1991 - which regulate the collection, treatment and disposal of solid and hazardous wastes from municipal and industrial sources, and provide a comprehensive list of chemicals and chemical wastes by toxicity categories.

FEPA is now a body under the Ministry of Environment which was created by the government of President Olusegun Obasanjo on assumption of office in May 1999 and is, expectedly, undergoing major reorganisation to streamline its mandates and activities within the context of its new position.

ENVIRONMENTAL PROTECTION AND SOCIO-ECONOMIC IMPACT

The Nigerian society vis-à-vis the oil industry is a tremendous paradox. The country is blessed with enormous reserves, as the 6th largest producer in the world, but insecurity and insufficiency of commercial energy supplies, high energy import bills, and a depressed economy have contributed to constrain major development efforts and threaten the maintenance of sound ecological integrity.[30] The interplay between the current breakdown in economic development, social conflicts, energy crises and an over-stressed environment has tended to reinforce each other and has exacerbated the decline in human conditions especially among the communities in the oil producing areas.

Apart from the traditional environmental problems already mentioned, a number of other issues emerge from the operations of the industry, including socio-economic problems, cultural impacts, indigenous peoples' marginalisation, and human rights violations – all inexorably linked to the environmental issues. Also, the handling of the demands and protests of local communities by successive governments in the country, especially in the past, has also raised a lot of local and international furore with respect to issues of their fundamental rights and social equity.

Community Relations

Local communities are typically affected as much by cultural diffusion as by noise, fumes and water pollution from oil & gas operations, or by pipeline and other transportation, the unattractive sights and by the risks of accidents (blow-outs; spills) which occur during the extraction and transportation of products. Since the boom of the seventies, when oil exports earned over 100 billion dollars, people from the oil-producing areas have seen money pouring from their land into Lagos, and later Abuja, where it has been concentrated in the hands of only the ruling groups and the multinationals. Communities in the Delta area now feel unfairly treated in the entire process of oil prospecting and production, and see themselves as being at the receiving end of only the adverse consequences. They believe that they have not received an equitable share of the tremendous oil revenues, which is being derived from their land and territories, especially in the light of the disruptive consequences on their health and sources of livelihood.

Past insensitivity to local communities' rights and claims - by both governments and companies - has resulted in subsequent tensions imperilling production operations and human life. Therefore, related to physical environmental concerns, are the conflicts between the technical imperatives of petroleum operations and the existence and interests of local communities affected by such operations. Although there are obligations to prepare a socio-cultural impact statement along with the EIA, it is rare to see procedures in practice that give a formal role to local community representatives in the decision making and approval process. The centralist - and undemocratic - character of previous governments in the country had militated against the notion of community and citizen participation in such administrative procedures. The situation appears to be improving, however, because in recent times several proposed projects subjected to the EIA process by Ministry of Environment have had the Environmental Impact Statements published or displayed for public examination and comments. It is, however, submitted that despite these developments, at critical moments of decision, the government can override any public (even where valid), objections to a proposed project, largely because of its paternal nature and the power it exerts.

Compensation and Remediation

Laws, contracts and licenses generally take into account local community interest by providing for compensation (not necessarily appropriate or adequate) to landowners and others affected by the operation of the industry. Compensation for environmental damage and loss of use of land, undoubtedly, is about the single most critical issue in community relations for the industry. There are four basic levels to which the issue of compensation concerns - the individual, family/communal, the State and National - depending on who possesses, has custody or ownership of the land/territory within which E & P activities are undertaken. There is, therefore, a direct link between property law regimes, and land tenure in particular, and the payment of compensation generally for environmental damage or other loss.

By Section 1 of the Land Use Act of 1978, all land comprised in the territory of each State is vested in the Governor of that State and such land is expected to "be held in trust and administered for the use and common benefit of all Nigerians". In accordance with the provisions of the Decree, the Governor can grant either a certificate of occupancy or rights of occupancy to individuals or groups for use and enjoyment. Although the act removes absolute ownership of land from the citizens, it does recognise communal and family forms of tenure which are derived from customary rights of occupancy, albeit subject to the Governor's title. And because tenure is vested in the Governor, compensation for land acquisition is due and paid to him "in trust" for the occupants. But in most cases, the payments never reach the people for whom they are meant and who bear the direct brunt of the E & P operations. It is fairly typical that the current fiscal regime - designed, negotiated and implemented by the Federal and State Governments exercising sovereignty and ownership over minerals and land - generates revenues exclusively for the central government. In this situation, local dissatisfaction over suffering damage without access to the revenues usually target the company as the conspicuous and present perpetrator of the damage making it in fact the victim of constitutional and political tensions existing in the country. [31]

Ostensibly in response to the inherent inequities in the system, the Oil Mineral Producing Areas Development Commission (OMPADEC) was established by decree in 1992 by the Federal Government for the development of oil-producing areas. It received as a special fund monthly allocation from the Federation account for this purpose and was mandated to administer the sums in accordance with confirmed ratio of oil production in each State. Its roles included creating understanding and co-operation between oil companies and the communities. It was also mandated, inter alia, to embark on the development of projects properly agreed upon with the local communities of the oil mineral producing areas and mitigate environmental damage arising from E & P activities. But it was alleged that a very minute proportion of the designated funds reached the target communities necessitating its dissolution and the review of its activities. An unrepresentative arm of the federal bureaucracy, and a hostage to the relentless struggles for personal and ethnic advantage which bedevil Nigerian governmental institutions, OMPADEC was never able to win the confidence of the people in the oil producing areas and never delivered meaningful development."[32] It represented but another futile and failed attempt to fully and properly address the fundamental issues that underlie the continued ecological disaster, conflict and ethnic tensions in the oil producing areas. The current Government has, however, passed into law, the Niger Delta Development Commission (Establishment etc.) Act, No. 6 of 2000, which is a comprehensive law dealing with revenue allocation and administration in the Niger Delta and major oil producing areas for the development projects, compensation and general welfare of the indigenes as well as the remediation of environmental damage. It is devised to address some of the fundamental gaps and errors which previous efforts have suffered in the search for solutions, especially with respect to participation and equity. Specifically, section 6 of the Act stipulates that the powers and functions of the Commission are, inter alia, to formulate policies and guidelines for the development of the Niger Delta area, tackle ecological and environmental problems that arise from the exploration of oil mineral in the Niger Delta area and advise the Federal Government and the member states on the prevention and control of oil spillages, gas flaring and environmental pollution; liase with the various oil mineral and gas prospecting and producing companies on all matters of pollution prevention and control. The Act further repealed the OMPADEC Decree 1998 and stipulated the provision of funds through the allocation of specified sums from the Federation Account and other contributions for the fulfilment of its mandates.

FUTURE TRENDS AND NEW REGULATORY APPROACHES

Legislative Means

Despite the growing global trend away from prescriptive legal controls towards co-operation, negotiation and self-regulation, it is expected that legislation and regulations are still going to change and evolve significantly in the direction of prescribing higher and more exacting standards. This is particularly critical because of the peculiarities of the polity in which laws take time to adjust to practical realities and trends, and when they do finally change are often sudden, drastic and seemingly cumbersome.[33] And one can also expect more vigorous and aggressive enforcement and implementation of existing laws and standards as a result of increasing public awareness and pressures. In this situation also, only firms that have made progressive technological advances enabling them to voluntarily exceed legislative standards, will find compliance with stringent environmental standards easier.

Pollution reduction, effluent improvement, waste reduction, resource conservation and energy efficiency, all deserve continued implementation and support but will ultimately not be enough to prevent regulation. Legislative measures will continue to play a major and dominant role in environmental issues concerning the industry. For instance, the Federal Ministry of Environment is in the process of developing a law on response, compensation and liability for environmental damage in Nigeria. The overall objective of the proposed law is to put in place a legal framework capable of delivering an acceptable compensation formula and provide for response and remediation regimes to cater for the interest of the environment and particularly to ensure that:

- the polluted environment is adequately compensated by way of remediation and restoration;
- the victim is also equitably compensated; and
- the offender or saboteur is made fully liable.

The law is also expected to set up a fund, akin to the Superfund in the United States, which would be used to remedy and respond to environmental damages and disasters, especially in the oil industry. It will also be utilised for the purpose of rehabilitating areas where operations have ceased, pursuing conservation programmes and carrying out continuous data collection and studies. Corporations could be held liable for negligent acts committed in the past with present adverse effects. The law will also require that certain categories of companies listed in it shall take out a specified minimum environmental risk insurance coverage in addition to a comprehensive general liability insurance. The proposed law will greatly expand the role of the Federal Government through the Ministry of Environment in steering the management of pollution arising from improper waste disposal activities, hazardous substance emissions, and in cases of oil spills. While there is still an inclination in the country in favour of production maximisation as against environmental effects, there is no doubt that companies operating within the country will in time have to adapt - by technology and management - to the advanced levels of environmental regulation.

POLICY REFORMS

One of the greatest challenges facing the management of the petroleum-environment nexus in Nigeria is the dearth of a comprehensive energy policy and regulation for the country. There is no doubt that the country needs an energy policy and a set of regulations that takes into account the inter-relatedness of the nations various energy systems, the economic and social implications of the industry as well as the environmental sensitivity of its operation.[34]

There is substantial room for improvement in energy resource utilisation, domestic environmental protection and enhanced contribution to global efforts in combating greenhouse effect. Policy instruments like prices, efficiency achieved through demand side management, and infrastructural maintenance can greatly enhance the achievement of energy security for sustainable development and contribute to reduce the country's role in increasing the greenhouse effect.[35]

Nigeria's total lack of demand management has promoted energy wastes and intensified environmental degradation. The Petroleum Resources Ministry has traditionally focused only on energy production and supplies and minimally on energy efficiency. The present legislative framework and pricing mechanisms do not leave a lot of room for policy innovations capable of benefiting both economic and environmental considerations in the country. In view of the huge benefits to both the private sector and the national environmental policy, a comprehensive energy management strategy will be in the interest of the country as a whole to design and implement. It is, however, believed that the environment will increasingly play a more dominant role in future government policies as a result of increasing public awareness as well as international trends and pressures.

One of the major characteristics of the industry in Nigeria has been the steady growth in consumption in the past 8-10 years, which, even under peak bouts of recession characterising this period, has shown a phenomenal inelasticity to economic decline.[36] From all indications, continued future growth seems inevitable. Meeting this growing energy need on a sustainable basis without its concomitant environmental implications will require large-scale investments, coherent policies, disciplined implementation and robust enforcement of standards and regulations. The prospect of meeting these energy needs sustainably will require allowing a greater role for natural gas. Current projects to develop flared natural gas, especially through the Nigeria Liquefied Natural Gas (NLNG) project, stand out as the most economically and ecologically profitable future energy investments in the country. They will further substantially cut down on the country's current contribution to global warming while providing huge revenues for socio-economic development.

Command-Control versus Market-based Approaches and Voluntary Initiatives

One of the significant issues for the future concerns the choice between the now traditional command-control approach prescribing specific standards and solutions to reduce environmental impact on one hand and a more market-oriented approach relying rather on economic incentives, operational autonomy and the self-interest of firms participating in a competitive market.[37] The question, however, is whether the "market" approach can be effectively introduced into the existing regulatory system for petroleum development which is currently subject to significant government controls and subsidies. Certainly, more sophisticated market-based environmental policies will definitely need more understanding, fiscal discipline, political will and an effective state machinery than exists in the country at the moment.

In addition to direct regulation and contractual stipulations, adverse environmental impact can be minimised by imposing, or providing effective economic incentives for, the use of state-of-the-art technology and installing adequate infrastructure.[38] The government currently imposes monetary penalties on operators flaring gas, though the sums are not substantial. The rates are calculated in accordance with the estimated volume of gas flared and are aimed at discouraging flaring and at the same time encourage its utilisation. Although penalties and taxes can go a long way in discouraging operators from flaring gas, adequate infrastructural adaptations need to be made to encourage demand and consumption.

Without doubt, there is room for improvement in the current environmental legislative arrangements for oil and gas exploration and development, but maintaining the current level of economic contribution of the industry still necessitates an ongoing exploration effort in prospective areas. Nevertheless, maintaining the integrity of the country's environment is essential to maintaining the exploration and production investment in Nigeria and efforts will need to be intensified to harmonise these interests.

The industry, on its own, is increasingly assuming responsibility and now accepts that it has a definite role to play in environmental management through its operations, and ongoing research and monitoring activities. For this reason, the Niger Delta Environmental Survey (NDES) a private sector initiative is highly commendable and should be encouraged. The independent NDES, funded by the Nigerian oil industry, looked at flaring in the Delta and addresses concerns about air quality and soil and water contamination, with reference to international standards. This is a unique initiative by the industry that has been accused of being traditionally insensitive to the effects of its operations on the environment. The NDES, estimated to cost more than $4.5 million, was initiated in February, 1995, by Shell Petroleum Development Company (SPDC) on behalf of its joint venture partners (Nigerian National Petroleum Company, ELF and AGIP) to undertake an environmental study of the delta region and to provide the required database for a more complete understanding of the environmental pressures on the Niger Delta. Although initiated by Shell Nigeria and partners, the Oil Producers Trade Section (OPTS) of the Lagos Chamber of Commerce and Industry, is providing further funding for the survey, significantly broadening its base of support.

On another level, companies or local subsidiaries of multi-nationals are increasingly reporting on their environmental performance in line with international trends and the current trend of internal company environmental audits could continue and move towards independent environmental audits. Much like independent financial accounting to protect the financial constituencies of companies, the practice of independent environmental auditing, as a complement to government regulations, could help to ensure compliance with environmental obligations.

CONCLUSIONS.

Understandably, the industry by its very nature has the potential to endanger human health and safety as well as the physical environment. It is the responsibility of Government to establish a regulatory framework that minimises such dangers without imposing excessive cost burdens on the industry and thereby jeopardising its economic viability. On the other hand, the operators have a duty to adopt such measures as complement any government efforts to minimise the adverse effects of their activities. In the present circumstances, making choices and decisions based on reliable and up-to-date data of economic and environmental implications is therefore imperative. However, the energy/environment policy-makers are yet to enjoy the benefit of informed choices while political and economic obstacles continue to frustrate the prospects of appropriate energy and co-ordinated environmental policies despite the obvious developmental yields such initiatives could have brought to the country.[39] Further, capacity gaps, especially at the institutional level and human resources which have been a major obstacle to progress in this regard need to be addressed, through training and provision of adequate resources to the regulating and monitoring bodies and agencies.

Despite progress made on many fronts in the country, Nigeria has still not made the fundamental transition to the legal structures and a development pathway that will provide the entire populace with a sustainable and secure future as environmental deterioration continues and the forces which drive it persist.

The policy trend whereby the Government sacrifices the needs of environmental imperatives on the altar of political expediency and economic interests needs to be revised. Rather, the Government needs to be more committed to improving environmental protection through the use of legal devices and innovative market-based approaches combined with the latest in energy-efficiency technologies. That way the country need not necessarily have to choose between a healthy economy and a healthy environment.

---

[1] Z. Gao, "Environmental Regulation of the Oil and Gas Industries" in *CEPMLP On-line Journal on Natural Resources, Energy and International Business Transactions* http://www.dundee.ac.uk/cepmlp/main/html/journal/vol2-11.htm

[22] T. Walde, "Environmental Policies Towards Mining in Developing Countries" in *Journal of Energy and Natural Resources Law*, Vol. 10 No.4 (1992), 327-357.

[3] Ibid.

[4] Z. Gao, n. 1above.

[5] 3 UN Arb. Awards 1905 (1941) in Z. Gao, n. 1 above.

[6] Z. Gao, n. 1above.

[7] 1973, substantially amended in 1998.

[8] This dilemma and ensuing contradictions was recently brought to the fore in a recent public debate during the Workshop on Building Capacity to Facilitate Clean Development Mechanism Projects in the Oil and Gas Industry Sectors in Nigeria (Abuja, 10 –12 September 2001). There, the Honourable Rilwanu Lukman, Presidential Adviser on Petroleum and Energy (equivalent of a cabinet Minister), stated: "I am constantly on the look-out for opportunities to attract investments that will contribute to the maintenance, improvement, and growth of Nigeria's oil and gas industry... We are talking about improving and expanding our production capacity of these fuels that are central to our nation's economic well-being, as well as modernising and increasing our domestic refining capacity and our oil pipeline and gas transmission system... I need not emphasise that ***Nigeria is a poor, developing country that simply has too many other items on what could be called its "priority to do list"*** (emphasis mine) ... we cannot afford to indulge ourselves..." He preferred the extended period of 2008 for gas flaring elimination. On the other hand the Minster of State for Environment, Dr. Imeh Okopido, while insisting 2004 is and should be the target date, called for more rigorous efforts by the oil and gas industry in respect of environment standards. According to him, "there is a significant impact of the high level of gas flaring on the ecosystem, environmental health and on global climate. This bears on our national image vis-à-vis the global environmental agenda to which we are signatories... On our part as a responsible government and the conscience of the people, we commit all our efforts as flag bearers of the African environment to foster innovative partnership between the public and private sectors, with a view to terminating gas flaring in Nigeria by 2004".

[9] See note 8 above.

[10] Cap 339 Laws of the Federation of Nigeria (hereinafter LFN), 1990.

[11] Cap 337 LFN, 1990.

[12] Cap 350 LFN, 1990.

[13] Cap 26 LFN, 1990.

[14] Section 8(1)(b)(iii).

[15] O. Akanle, *Pollution Control Regulation in the Nigerian Oil Industry*, (Nigerian Institute of Advanced Legal Studies, Lagos 1991) at 6. See also note 21, at 6. "It is generally assumed, however, that government through the Petroleum Inspectorate may penalise the operator who has polluted the environment. But it is submitted that there is no legal basis for this in any of the enactments. Again, though the Inspectorate may withdraw an operator's licence for not operating in accordance with "good oil-field practice" or "good refining practice", or that the operator did not employ up-to-date equipment, it is hardly feasible how the sanction can be evoked in practice... Also, the

NNPC, of which the Inspectorate is an integral part, action on behalf of the government is either the owner or substantial part-owner of most of the petroleum operations in the country".

[16] O. Akanle, n. 13 above, at p9.

[17] ibid. at p10.

[18] ibid. at p11.

[19] ibid.

[20] Cap 77 LFN, 1990.

[21] S. 247, ibid.

[22] K. Nnadozie, "Legal and Administrative Requirements of Environmental Impact Assessment in Nigeria", January 1998 *Nigerian Petroleum News*.

[23] [1866] L.R. 3 H.L.300

[24] *Jimoh Lawani v. West African Portland Cement Company Limited* [1971] Abeokuta High Court Suit No. AB/82/71; see also *Amos v. Shell BP Petroleum Development Corporation* [1974] 4 U.I.L.R. 345.

[25] See also *Sam Ikpede v. Shell-BP Petroleum Development Company Nigeria* [1973] M.W.S.J. 61.

[26] T. Waelde, "The Current Status of International Petroleum Investment: Regulating, Licensing & Taxing and Contracting" in *CEPMLP On-line Journal on Natural Resources, Energy and International Business Transactions*, http://www.dundee.ac.uk/cepmlp/main/html/journal/vol1-5.htm

[27] In September 1987, an Italian businessman acting on behalf of some Italian companies, with the collaboration of some local officials and citizens, imported and deposited about 4,000 tons of highly toxic industrial waste on a farmland near the local port at Koko town in Delta State.

[28] The Federal Environment Protection Agency Decree 1978 establishing the Agency is yet to be repealed despite the creation the Ministry of Environment. So, the Agency is still legally in existence though it no longer operates as such. A comprehensive Framework Environmental Act is currently being drawn up to take into account the new realities and policy direction of the government.

[29] Ibid.

[30] E. O. Eleri, "Sub-Sahara Africa: Energy and Environmental Challenges" *Working Paper 1992:6*, Centre for International Climate and Energy Research, Oslo, at 1.

[31] ibid.

[32] I. O. Oguine, "Nigeria's Oil Revenues and the Oil Producing Areas" in *CEPMLP On-line Journal on Natural Resources, Energy and International Business Transactions*; http://www.dundee.ac.uk/cepmlp/main/html/journal/article4-10.htm

[33] K. Nnadozie, "From Black to Green: Environmentalism in the Oil Industry" February1998 *Nigerian Petroleum News*.

[34] See also E. O. Eleri, n. 28 at 43.

[35] ibid. at p66.

[36] ibid. at p57.

[37] T. Waelde, n. 26 above.

[38] Gaëtan Verhoosel (1988); "Foreign Investment And Environmental Regulatory Change In Developing And Transition Economies: How To Reconcile The Tension For The Benefit Of Technology Transfer?" in *CEPMLP On-line Journal on Natural Resources, Energy and International Business Transactions*; http://www.dundee.ac.uk/cepmlp/main/html/journal/vol1-1.htm

[3939] E. O. Eleri, n. 28 above at 71.

POLICING INDUSTRIAL POLLUTION IN NIGERIA

ALI AHMAD[1*]

Environmental harm is caused by a host of human activities, each differing in intensity and effect. Pollution by industries is one of the major contributors to environmental degradation and biodiversity loss in Africa, especially the West African sub region.[2] The level of development in Africa and insistence of developing countries on complete sovereign right with regards to exploitation of their natural resources, which unhappily is still the norm in international environmental law, clearly suggests that this trend will continue.[3] However, two factors make industrial pollution in Africa especially worrisome: first, the meteoric rise in the level of hazardous wastes compared to dwindling mechanization and industrial processes due to downward economic trends, and secondly, the character and legal treatment of the polluters. Major actors in industrial pollution are made up of entities that by implication are either protected from the general theme of environmental laws because of government interest in their activities, or whose real decisionmakers are fairly protected from the consequences of the decisions they make because their entities are *persona ficta*.

Nigeria is one of the first African countries to design and adopt a National Environmental Policy.[4] The country is a party to major international environmental agreements.[5] To implement these agreements, it has put in place enabling acts and institutions such as the federal enforcement agency, the Federal Environmental Protection Agency (FEPA) or its successor, the Federal Ministry of Environment.[6] State governments and specialised sectors such as petroleum, food and drug industries also have specific laws or state environmental ministries regulating relevant activities. Furthermore, the Federal Ministry has adopted FEPA's rules, standards, and procedures established for various industrial sectors. These regulatory expressions and institutions notwithstanding, pollution still continues unabated. The consistency and frequency of environmental infraction by industries belie the existence of established environmental laws and standards.[7] The task of this paper is to demonstrate the need for shifting the focus of environmental policy in Africa in general, and Nigeria in particular, from the wrongful assumption of voluntary compliance by industries to a proactive strategy by agencies of active enforcement and implementation of existing rules.

It is believed that continuing deterioration of the environment in Africa is more to do with enforcement and compliance rather than the lack of applicable rules, regulations, or institutions. A scheme that is geared toward addressing industrial pollution on a sustainable basis must not consider enacting an endless body of laws as an end in itself. Rather, environmental compliance is a process that begins with creating awareness, building consensus, enacting rules, and more importantly providing public access to vital information. By examining the major polluting industries in Nigeria and identifying the factors responsible for noncompliance, to the chapter offers an understanding of the problem of growing industrial pollution in the face of the current drive toward liberalisation and privatisation of the economy in comparison with the failings of an uncoordinated enforcement, the need for a policy objective that aims at eliciting compliance through a strategy that is consistent, industry-specific, and enlists support of the citizens is emphasised.[8]

LEVEL OF INDUSTRIAL POLLUTION IN NIGERIA: CAUSES AND EFFECTS

A suitable method for determining the level of industrial pollution in Nigeria is to classify the industries into identifiable categories. For this purpose, major polluting industries may be limited to five. There are, however, a host of other enterprises whose activities also contribute in no small measures to environmental degradation. All polluting industries in Nigeria are not listed even if it were possible. Indeed, medium- and small-scale industries are by nature scattered and not amenable to easy monitoring. The amount of pollution caused by sawmills, automobile repair shops, petrol stations, and printers cannot be overemphasised. Nevertheless, the examination is restricted to five large-scale industries in the country.

A. Steel Work Sector

There are four steel facilities operating in Nigeria - the fifth at Ajaokuta is under construction and may be completed soon. All are federal government-owned. Sixteen other private companies operate both crude steel-making facilities and rolling mills.[9] Principal air emissions by steel industry include particulate matter, sulfur dioxide, fume, and alkaline oxide emissions from blast furnace. Others include suspended solids, phenols, amonia and cyanide. These steel plants and rolling mills seem to operate without adherence to national and international environmental standards. For instance, the Delta Steel Company regularly deposited metals in downtown areas.[10] Indeed, pollution from the Delta plant continues to impose significant health risks on local communities, despite the Environmental Ministry's elaborate regulations.[11]

B. Food Processing

The food processing industry brazenly pollutes all the environmental media--water, land, and air. Significant pollution results from dairy products, bakeries, sugar products, confectioneries, beverages, and edible oil. This diverse group unites in disposing of large amounts of untreated organic wastes that contribute to high chlorine levels of watercourses. The industry is not known to comply with effluent standards. Wastes from the dairy sector are largely liquid and these disrupt aquatic systems because they decrease oxygen and increase suspended solids.[12] A study of waste discharge from a juice company into a stream in Ibadan observed that oxygen demand, temperature, and dissolved solid levels were so high that the water was no longer usable for irrigation. This imposed economic losses on downstream users for irrigation.[13] Biochemical Oxygen Demand (B.O.D.), a measure of the amount of oxygen in water, has always been extremely limited in almost all the rivers in the country.[14]

C. *Tanneries*

Tanneries mostly pollute watercourses, although they also generate large quantities of solid wastes, acids, nitrogen compounds, sulfides, vegetable tannins, oil, and grease. Their wastes raise oxygen demand in aquatic life. They also pollute air through emissions of particulate matter as well as nitrogen oxides, ammonia and hydrogen sulfide.[15] Non-compliance with standards is also rampant. Studies involving many of the tanneries showed that B.O.D. level was found to be over six times above the standard. Suspended solids were 130 times more concentrated than the acceptable limit. Total chromium levels measured 920 times. It was found that more than any single factor, tanneries contributed most of the water quality degradation of the Kaduna River as it flew past the Kaduna South Industrial Estate.[16]

D. *Textiles*

Until recently, the textiles industry was one of the largest in Nigeria. It contributed N8.5 billion or 30 percent of non-oil Gross Domestic Product in 1991.[17] There is little government ownership in the industry and foreign investment holds sway. The industry produces large quantities of fabric and fibre solid wastes. Much of the wastes are recycled and the rest are sent to landfills, burned, or buried in shallow ditches. Although the industry does not contribute much to air pollution, its degradation of watercourses is alarming. Residues of fibre make textile wastewater high in B.O.D. and suspended solids. It contains a wide variety of chemicals like dyes, surfactants, oxidixing and bleaching agents, reducing agents, silicates, and inorganic salts. Decreasing dissolved oxygen, caused by high oxygen demand, and the large number of different chemicals used by the industry greatly stress aquatic environments.[18]

It has been observed that textile companies do not generally treat their effluents before release. For instance, the General Cotton Mill in Onitsha normally discharged its untreated effluents into a settling pond that overflows into the Niger River.[19] Almost all textile factories have failed to comply with water quality guidelines set by the environmental ministry.[20]

E. *Petroleum*

Oil pollution from wells, pipelines, ships and refineries has become a scourge in the country. From three oil wells in 1954 that produced 1,876,000 barrels annually, production has increased to 3,930 wells to a total number of 11,935,449,000 barrels of crude oil or 32.699 million barrels per day.[21] The current production rate is about 1.836 billion barrels per day.[22] The main sources of pollution have been leaks and accidental spills.[23] Studies have shown that these discharges are preventable because malfunctioning of equipment often causes them.[24] There are also four refineries and two petrochemical plants, all owned by the government. The remarkable aspect about their operation is that they all contribute to pollution through inefficient use of energy. The Warri refinery uses up 11 percent and the Kaduna refinery 16 percent of crude throughput just to operate. The operating costs for the Warri refinery in 1995 were US$22 per ton instead of the world average costs of US$13. Marginal pollution output would decline dramatically simply by operating more energy-efficient refineries.[25] The impact of oil contamination includes surface and groundwater pollution as well as marine organisms. Some fishing communities have lost much of their source of income. Agricultural crops are sometimes damaged by spills. Oil leaks at the Kaduna Refinery have penetrated into wells in Rido and Romi residential areas.[26] The metals, air pollutants, and wastewater that these refineries generate have adversely impacted the ecosystems near their areas of operation.[27]

THE LEGAL AND INSTITUTIONAL FRAMEWORK

Most of the polluting industries have been in operation in Nigeria for a long time and laws that were aimed at regulating pollution, especially of watercourses, were enacted to facilitate the country's developmental goals through exploitation of natural resources or in response to some international agreements do you mean in response to the requirements under such agreements. A perusal of those laws demonstrates that few of them were enacted as a result of a commitment to the preservation of the natural environment.[28] Subsequently, environmental protection in Nigeria amounted to no more than the monthly National Environmental Sanitation Day, during which citizens were required to remove solid wastes abutting their surroundings.[29] Lack of commitment to the environment was at its height when a Court held, in an action alleging damage by an oil company to land and pollution to a fish pond and creek, that nothing should be done to disturb the operations of a trade that was the main source of the country's revenue.[30] Due to this lack of commitment, redress for environmental infractions followed the traditional common law approach with its substantive and procedural bottlenecks concerning requirement of formal proof of causation and, where this was satisfied, the absence of adequate remedies.[31] All the environmental treaties that Nigeria signed did not alter this lack of commitment to environmental protection until after the Koko incident.

This incident happened in 1988 when an Italian company dumped 3,800 tons of extremely toxic wastes encased in several drums and containers in a remote Koko Village, populated by about 15,000 people.[32] The wastes, which lay there for the next year, were secretly shipped into Nigeria as ordinary items of harmless merchandise from various European countries and the United States, in a transaction arranged by an Italian merchant. It was part of an ongoing trend whereby Pacific, Caribbean, and especially African countries became dumping centers for hazardous wastes from industrialised countries. The incident radically elevated environmental awareness not only in the country but also in Africa as a whole. Shortly thereafter when the government of Benin Republic also granted a "concession" to a European firm to dump toxic waste in that country in exchange for $10m, the contract was rescinded following the adverse publicity the Nigerian experience generated. Similarly, President Stevens of Sierra Leone, under pressure generated by the Nigerian experience, rejected a $25 million deal from a Colorado firm for permission to dispose its hazardous waste in that country.[33] The existing federal and state sanitation laws of Nigeria, which were then the expression of government's concern for the environment, was not adequate in addressing the incident that came to mark a true beginning of grassroots environmental awareness. To address the issue of transboundary movement of hazardous wastes, African States were further ahead than what was agreed to in multilateral international negotiations, by reaching a consensus in the *Bamako Agreement* on the preventive effect of the precautionary principle under international environmental law.[34] The parties quickly agreed to ban the import of all hazardous wastes into Africa.[35] The implementation action in Nigeria included the establishment of the Federal Environment Protection Agency (FEPA), which has since metamorphosed into a full Ministry of Environment to strengthen environmental regulation in the country.[36]

The *FEPA Act* gives the Agency, now the Ministry of Environment, the responsibility of protecting and developing the environment as well as conservation of biological diversity and sustainable development of Nigeria's natural resources.[37] The Ministry is to establish criteria, guidelines, specifications, and standards for water, air, and noise pollution.[38] The Ministry has power to inspect without warrant any facility it suspects of polluting the environment, as well as power to search, seize, and cause arrest to be made for violating environmental laws and subsidiary rules.[39] It is remarkable that within a short period, the old FEPA succeeded in generating immense awareness on environmental issues in Nigeria. It has been involved with reputable research institutes both locally and internationally. However, minimum compliance with established standards is still illusory and the environment remains largely abused by corporate polluters. It is disheartening that over a decade after the initial establishment of FEPA, most residents living near rivers are unable to put the downstream water to any purposeful use due to industrial discharges. Those in oil producing areas fare even worse.

In implementing laws and standards, the Ministry has been sensitive to social and developmental realities of the country in order not to cause adverse economic repercussions for the industrial sector.[40] Its choice of emission-standard strategy, also known as the best-practicable-means strategy, attempts to reduce pollution to the greatest extent possible with the practical methods available.[41]

In its early years, FEPA was involved with jurisdictional overlap with other allied enforcement agencies. It was thought that FEPA had no authority over matters regarding conservation in view of the power vested in the Natural

Resources Conservation Council to coordinate matters concerning conservation of natural resources.[42] This unnecessary separation and the resulting overlap were resolved by placing all environmental issues under coordination and oversight of FEPA.[43] However, the history of inter-agency conflicts between FEPA and the petroleum industry has not been clearly resolved. Before the establishment of FEPA in 1988, the Department of Petroleum Resources (DPR) exclusively implemented the petroleum laws, set standards and guidelines and enforced them.[44] Under its enabling laws, which predate FEPA, it has three regional offices and claims to be the highest authority on environmental degradation stemming from petroleum activities throughout the country. It has competence to make and enforce provisions for, among others, "the prevention of pollution of water courses and the atmosphere".[45]

The *FEPA Act* or its subsequent amendments did not resolve this inter-agency conflict, and it is arguable whether the resulting confusion has not been deliberate. It merely obligates it to *cooperate* with the DPR for the removal of oil related pollutants and "play such supportive role as the Ministry of Petroleum Resources may from time to time request from the Agency."[46] But the former Director-General of FEPA, relying on the same statutory provision, asserted FEPA's control over DPR on environmental matters:[47]

> "The oil companies. Let me start with them. I've put them on notice that most of the areas they degrade...they would have to revisit and clean up...If they don't of course; we have the wherewithal to sanction, to deal with them. DPR has control over the safety and the workings of the industry, not the environment. Environment is my jurisdiction. It is FEPA's jurisdiction. There is no environment for oil industry separately from environment for the gas industry, or environment for the water industry or environment for the land industry. Environment is one."

Despite the Ministry's enabling laws and its establishment of standards and regulations as well as its sensitivity to the country's socio-economic development, there are no indications that industrial pollution has ebbed. This is because existing regulations and standards are not being adequately enforced. On their part, industries have not shown readiness for voluntary compliance with these standards. With the level of pollution that is taking place without rescue from the expected quarters, it is obvious that both Ministry of Environment and the regulated community are aware of the overwhelming violations of environmental standards that occur daily.

Even a five-year moratorium that the then FEPA granted corporations, during which it suspended enforcement of its standards and rules to allow industries adequate opportunity to install pollution abatement facilities and which expired in 1994, did little in encouraging industries to adopt a voluntary compliance method that is expected to follow any such moratorium.[48] This moratorium proved to be overbroad, highly unnecessary, and sent wrong signals to many corporations that were not ready to abide by any standards that would require any shift in their standard operation procedures.

Moratorium has been used as a strategy to induce targeted polluters to comply with rules. An offer of moratorium needs to be specific with a promise given by an industry or enterprise to fully comply with specific requirements of the rules at the end of the allotted time. However, selective use of moratorium has been the successful strategy.[49] The following account illustrates the point more vividly.

The Continental Iron and Steel Company Plc. (CISCO), a steel factory located in a residential area in Lagos, disposed its sludge and liquid wastes without treatment. Long before FEPA was established, the Lagos State Ministry of Environment and Physical Planning had issued directives to the company since 1980 for requiring a more acceptable operation. The company did not yield its operations. Later, FEPA carried out several inspections of the factory but this produced no result. In 1995, one year after the expiration of the general moratorium offered to all industries, FEPA sent series of letters and carried out further inspections but the company did not abide by FEPA's instructions.

In 1996, FEPA issued a directive closing down the factory instantly, until the Agency was satisfied with the company's installations and abatement measures for its furnace emission, solid wastes spent oil, and lubricants.[50] Apparently, CISCO benefited from the general amnesty that FEPA granted in 1989 but refused to change its detrimental practices. CISCO should have been closed since 1989 without having the moratorium. The country has gained nothing from the amnesty from which CISCO and many others like it across the country have benefited from. It would have been consistent with FEPA's subsequent enforcement drive to make the company pay for the devastation of the environment it had caused in violation of the law. Shutting the company down after it had enjoyed an unsolicited moratorium was hardly a commendable enforcement measure. Perhaps, if citizens were aware of the intensity and level of noncompliance of this company, they would have taken appropriate judicial action against it, seeking that eh company desist and carry out clean-up action.

Therefore, one critical area where the law may be enhanced is in the generation, preparation, and provision of information. The absence of legal requirements for self-reporting or for preparing and safekeeping of environmental and compliance information by corporations has rendered non-compliance easy and has increased the difficulty of enforcement. Except for petroleum industry where records are well maintained for purposes unrelated to the environmental but scrupulously withheld from the public, a lot of the firms are unconcerned about tracking their pollution monitoring programs. The Ministry cannot generate this information or work without it.

As observed earlier, non-compliance with environmental standards has been caused mainly by lack of resources of the Environmental Ministry and by the cost consciousness of corporations. Although FEPA derived its funding from the Ecological Fund, which is a reserve of one-percent of the total annual federal budget, it has only been entitled to 25 % of that amount [51] It is not yet clear whether the Ministry is still entitled to draw directly from the Fund now that it is a full-fledged ministry. FEPA was not known to have or maintain laboratories even in the four most polluted cities in the country: Port-Harcourt; Ibadan; Kano; and Kaduna. As a result, it must rely on laboratories of the corporations it is expected to regulate.

However, given the widespread environmental degradation and lack of readiness of industries to comply, the Ministry can generate enough funds from fines and levies to run its operations for a number of years.[52] The highest pollution fee in Nigeria has been N60, 000 or approximately $600.[53] What is more concerning is that it has been assessed that the country loses about $5 billion annually to environmental degradation.[54] In China and India, the majority of agency revenue comes from penalties and fines, offering a strong incentive to apprehend polluters.

There is little doubt that despite established rules and regulations, lack of consistent enforcement has provided incentives for pollution, making infringement of these rules more attractive to a rational cost-conscious corporate decision maker. The Ministry of Environment has not deemed it fit to pursue enforcement through imposition of fines and penalties despite its power to do so under its enabling law. Perhaps it has been heeding the advice of the Manufacturers' Association of Nigeria handed down by its erstwhile chairman who ironically became Nigeria's first Minister of Environment that, "any approach taken to tackle a very important issue like environmental pollution should not be seen as a revenue-generating exercise."[55]

Closely connected with the Ministry's incapacity is the issue of corruption. Where a regulatory agency is corruptible, all its operations are adversely affected. The Ministry must have an internal modality for detecting corrupt inspectors and punishing them. This is possible though only where the inspectors are motivated and adequately remunerated. The issue of corruption is however an unwholesome phenomenon that has been pervasive in all segments of the country, and is not unique to the Ministry. It is still remembered that during the Second Republic (1979-1984), many state governors were found to have misappropriated their states' share of the Ecological Fund.[56]

Inadequacy of skilled personnel also affects the Ministry's enforcement ability. It is highly unlikely that an inspector with average skill will technically prevail in detecting a violation and win over a highly knowledgeable head of the pollution unit of a corporation who manipulates state-of-the-art technology. The impact of unskilled personnel on detection and enforcement of rules is perhaps most evident in the failure of the customs officials to detect the deleterious substance of the Koko incident. The customs officials were told that the 3,800 tons of toxic wastes were chemicals used to manufacture fertiliser.[57]

III. FACTORS RESPONSIBLE FOR NON-COMPLIANCE

Apart from the aforementioned agency-induced causes of non-compliance, the breakdown of environmental law and order may be traced to sensitivity of industries to what they perceive as high cost of environmental compliance. A typical board of directors does not authorise a more polluting activity because it takes pleasure in fouling the environment. Cost of achieving environmentally friendly operation is a critical factor that results in non-compliance by industries. Actual cost of providing for cleaner facilities or operations is hard for industries to internalise. In the oil industry for instance, oil spills are the major cause of environmental degradation by oil companies, most of whose pipelines were laid as far back as 1959.[58] There is not doubt that a lot of money is required to upgrade or repair those pipelines, most of which are rusted and fretted and therefore susceptible to constant damage. It has been demonstrated that corporate polluters will characteristically avoid pollution abatement measures for the sole reason of keeping down the cost of their operations.[59] This corporate decision-making is not likely to change unless the legal framework is able to provide enough motivation to make the decision to pollute economically irrational to corporations.

A typical illustration of cost consideration in corporate environmental decision-making is offered by an incident involving Mobil. Mobil Producing Nigeria has one of its platforms at Ubit, 34 kilometers offshore Akwa Ibom State. As at 1995, the platform produced 85,000 barrels of oil per day that was worth US$1.3 million at the then price of US$15 per barrel. The company decided to expand the platform. As engineering work involved welding, its in-house safety experts were reported to have advised the company to shut down for the duration of the expansion.[60] Apparently concerned about potential loss of earnings during the expansion period, the company chose to carry out the expansion simultaneously with normal production. A leak of oil from one of the pipes ignited fire at the platform causing massive explosion that ruptured the pipeline linking it with the Qua Iboe Terminal. This also spewed out gas, oil, and water at high pressure. Ten of the workers were reportedly killed and nineteen others were seriously injured.[61] More than 200,000 barrels of oil and polluted water were estimated to have spilled into the Atlantic Ocean. Ultimately the platform was shut down for more than a month. Most corporations may not behave differently.[62]

The international consensus as expressed in the *polluter-pays* principle has addressed cost consideration by trying to shift the legal treatment of environmental accountability from victims of pollution to the perpetrators.[63] The statement may be supported that the effect of the principle is that states are required to promote policies that bind polluters to internalise cost of pollution abatement so that the polluters reflect it in the cost of their product for the market to provide an efficient level of environmental management. To the extent the principle urges states to adopt laws for such purpose, an overwhelming number of states have adopted the principle, although many states qualify it with a few exceptions where its strict application would lead to serious economic disturbances.[64] The version of the principle that the Ministry has adopted mandates new industries to incorporate pollution control facilities at all new points but encourages old industries to install such facilities as may be practicable and reasonable from technical and economic standpoints.[65] However, nothing substantial has been done to ensure compliance with this modest requirement for both old and new industries.

Thus, since it is still cheaper to pollute and not install pollution abatement facilities mainly due to lack of enforcement of the standards, the rational corporate decision is to not install those facilities or otherwise adopt any significant procedures that would increase cost of production and reduce profit margin. It is only logical that an enforcement strategy that makes the decision to pollute more expensive than it is to comply with environmental requirements is needed in light of the limited expectations surrounding voluntary compliance from cost-conscious, profit-oriented entities. The United States has chosen an enforcement strategy through administrative fines or the judicial award of damages.[66]

A popular response to any inadequacy in enforcement and compliance with rules has been to strengthen the laws and increase fines that are payable by polluters. While these may be desirable, experience has shown that mere enactment of strict laws and stiff penalties will more likely suffer from a similar problem of random and inefficient enforcement, thereby legitimising acts of environmental degradation through ostensible but ill-enforced rules. The efforts of the Environmental Ministry will continue to fail until it has been able to develop and harmonise various enforcement resources with different categories of actors in the business of industrial pollution.

LEGAL CHARACTER OF ACTORS

Part of the failure of regulatory scheme against corporate polluters in Nigeria is due to a unidirectional strategy with respect to corporations of all sorts. It appears that the Ministry of Environment has chosen a uniform enforcement approach with respect to all corporate polluters, making no principled distinction between persistent and casual polluters on the one hand and between government owned and other corporate polluters on the other. An examination of past enforcement strategy reveals that the Ministry has pursued enforcement with equal vigor or apathy against all corporations and granted reprieve to all without discrimination, thereby offering no incentive to corporations that voluntarily complied while indirectly encouraging other potential polluters.

Given the objective of a strategy that solicits compliance, polluting industries may be classified into three categories based on ownership and control. These are wholly domestic corporations, multinational corporations, and municipal or government owned corporations. This classification is not inflexible as the government may have an interest in the shareholding or management of a multinational corporation. Most of the failures of environmental enforcement in African countries may be traced to the symmetry of approach by regulators with respect to disparate polluting entities. It is believed that an enforcement policy that is industry-specific promises to be a more beneficial mode of enhancing regulatory compliance by industries in Nigeria.

Domestic Corporations

A great number of corporate polluters are constituted by wholly domestic companies, meaning companies that are incorporated under general company laws of the country although different nationals may hold their shares. They are referred to as wholly domestic companies because they are subject to the ordinary rules of corporate laws. This category of corporations presents no special problems to the enforcement machinery of Nigeria beyond normal factors attributable to their corporate character. If we understand the factors that are at play when this category of corporate actors takes decisions as well as the reasons behind those decisions, the government is better able to fashion out an appropriate enforcement mechanism and improve compliance.

A central factor in corporate decision-making is the legal protection that its decision-makers enjoy. A corporation is a distinct entity and it alone is generally responsible for its corporate actions.[67] The Ministry is empowered to apply administrative, civil, and criminal sanctions. It seems that regulators in Nigeria have failed to appreciate the fact that corporate entities are different from humans with respect to their objectives and the effects of sanctions. A human-based approach to enforcement and to imposition of sanctions will tend either to assume strict enforcement for inconsequential and technical breaches of corporate violations or to neglect enforcement of regular and continuing ones. Both extremes are counterproductive and have been responsible for non-compliance.[68] Unlike a criminal offender, even the most egregious corporation may be willing to be law-abiding because it cannot run away. Apparently, shutting down a corporation is no substitute for imprisonment nor is it in the interests of the society.

A more effective approach is to utilise administrative sanctions to the fullest because it is cost-effective and able to respond quickly to corporate maneuvers. Judicial enforcement, especially the criminal process, is to be used sparingly as the objective of deterrence has little concern for or ability to remediate a damaged environment.[69] An enforcement policy is to be preferable where it supports corporations' efforts to initiate the most effective internal organisational changes amenable to the extent of what doing so saves the society.[70] One will agree with Professor Mitchell's view that 'pre-monitory' control measures, -that is, efforts to inspect and survey behavior before violations occur, rather than to detect and investigate them afterwards -will definitely raise the level of corporate compliance.[71]

Numerous studies suggest that regulation is typified both by a positive inducement, which seeks to accommodate the regulated community, and negative sanction, which deters persistent offenders.[72] Inducement consists of restraint or forbearance from enforcement. This sociological perspective makes distinction between persistent failure to comply and a one-off polluter. Although it involves discriminatory enforcement strategy to different offenders, it is a relationship of reciprocity that an environmental inspector gets for not enforcing a violation. This entails an undertaking by the polluter to comply strictly in accordance with the bargain. Where the polluter subsequently does not comply with the terms of the bargain, the inspector resorts to stricter enforcement of the rules. Forbearance entails overlooking violations that do not pose serious risk or not enforcing regulatory requirements that would be especially costly or disruptive to a corporation in relation to the additional degree of protection they would provide. It may also entail the granting of a reasonable time to come into compliance, accepting measures that would provide substantial compliance and making allowance for good faith efforts on the part of the polluter.[73]

In some cases, formal and strict enforcement emphasises a confrontational style of eliciting compliance and sanctioning against rule breaking. It operates under an assumption that if violators are detected and adequately punished, they and other violators will be deterred in future. When violation becomes a trend and beyond consent of the regulatory agency or above tolerable limits to the law, the uncooperative attitude may only be addressed by invoking the law's full sanctions.[74] A reasonable agency is likely to use some mixture of both inducement and sanction in different degrees for varying circumstances. The degree to which one strategy predominates over the other in a given situation and time marks an agency's understanding of an optimum compliance strategy. In all cases, non-compliance must be made less profitable to corporate polluters. Unfortunately, the situation in environmental enforcement in Nigeria, where oil corporations pay a meager penalty of ten cents for flaring one thousand cubic feet of gas, points to the opposite direction.[75]

Multinational Corporations

The second category comprises of multinational corporations. In the present context, these are corporations whose shareholding, management, and operation span many countries.[76] They are different from domestic corporations mainly because of their allegiance to foreign parent entities and to operations in multiple jurisdictions. By their character, multinational corporations have immense resources and a superior technological base. In some cases government is not merely interested in the activities of these corporations, but is involved in a partnership, joint venture, or profit sharing relationship with them.[77] Furthermore, when the business of such corporations accounts for a sizable amount of a country's foreign exchange earnings, as is the petroleum industry in Nigeria that is the main source of foreign exchange earnings, their influence becomes glaring.

It is obvious from earlier discussion that in Nigeria, most of the corporations whose operations greatly impact on the environment are involved in the exploitation of natural resources, chemical industry, food processing, and the textiles business.[78] Their multi-jurisdictional character and involvement with foreign exchange earnings places multinational corporations under enforcement regime distinct from the domestic corporations in developing countries. These sophisticated corporations are formidable and it is questionable that the Ministry simply adopts an enforcement method that is similar in all respects with other less formidable polluters.

In most cases, multinational corporations are responsible corporations.[79] However, the overall trend in Nigeria and other developing countries shows that multinational corporations do not hesitate to take advantage of lower environmental standards or laxity in enforcement.[80] The desire of these multinational corporations to uphold their reputation and their immense resource base will dictate an enforcement policy that seeks a bargain and that provides for a high penalty in the event of deliberate infringement. An enforcement agency may improve voluntary compliance by merging the interests of cost-consciousness of multinational corporations with adverse publicity of their environmental profiles and enforcement responses.

Further, it is more desirable that a strategy establishes a mandatory requirement of keeping environmental records. Multinational corporations should be made to keep records of their effluents and other discharges and to carry out environmental audits whereby internal environmentalists or third-party professionals determine their level of compliance with regulatory standards. More importantly, a process needs to be established whereby information on compliance or discharges is made available to citizens within a reasonable time and for a reasonable fee. There is no doubt that public access to vital corporate information is one of the most effective ways to elicit environmental compliance from multinational corporations. The *Rio Declaration* acknowledges the importance of public participation in facilitating accountability and efficacy.[81] Since governments have neither time nor resources to provide for complete enforcement, availability of information will provide the tool for concerned citizens to seek judicial enforcement of existing laws.[82]

If more enactment were needed for the abatement of industrial pollution, it would create the specific situation of engaging powerful entities that enjoy government support. While environmental enforcement agencies in developing countries may have acknowledged this fact, they have failed to articulate a response to the overwhelming challenge to enforcement by powerful multinational corporations that enjoy government benefaction. Environmental law reform in those countries need not take the form of prescription of further rules or penalties. Instead, it should be directed toward compliance enhancement through requirement of provision of information, involvement of citizens, and clearly articulating pollution levels that are acceptable or permissible through defined or clearly select entities to satisfy any desirable government policy objective.

Municipal Corporations

Finally, municipal corporations also present a distinct problem that is yet to be fully rationalised with the general enforcement scheme. Municipal corporations are those incorporated under special federal or state acts or similar charters. Mostly, they provide essential or public services, such as those of various ministries but some of them, like state mass transit corporations, are profit-oriented. Ordinarily, these corporations and their subcontractors are not immune from environmental laws and standards, but not many records are found relating to federal or state environmental agencies seeking enforcement against these entities. More transparency is required to determine the extent of their environmental obligations.

A desirable framework for effective enforcement against this category of polluters is for the authorities to possibly exclude those entities as are required under the general regulation of environmental law, if that is desirable and is not in conflict with any obligations under international law. Such exception should be made explicit and concise as to whether the exemption or immunity covers the whole or some aspect of the entity. With regard to all other non-excluded municipal entities, citizens should be able to seek injunctive relief restraining environmental pollution. However, there is no good reason for excepting from environmental laws municipal entities that operate for profit, because then their activities will be unreasonably subsided.

The foregoing classification of polluting industries in Nigeria indicates that government participation in the affairs of corporations is a critical factor for non-compliance with environmental standards. The Nigerian government owns substantial interests in the concessions of major petroleum producing companies. It wholly owns all the refineries, fertiliser companies, and steel rolling mills.[83] Available data shows that compliance is low in the corporations where government holds certain interests while those multinational corporations that are free of such interest have been more compliant.[84] Thus, government participation in economic activities presents one of the greatest challenges to any enforcement policy in Nigeria. Perhaps, current privatisation and divestment efforts of various governments from major economic activities will have the salutary effect of significantly enhancing environmental enforcement.

CONCLUSION

The success of a regulatory policy is measured not by copious display of laws or duplication of bureaucracies but by delimiting the acceptable level of pollution and effectively implementing it. Nigeria is a party to numerous environmental treaties and yet contributes more than a fair share to global environmental degradation. Although corporations have kept their costs of operations low by disregarding environmental standards, the country loses billions of dollars annually through failure to enforce effectively the established environmental standards. This is in addition to endangering the country's pristine ecosystems to the disadvantage of future generations and aggravating public health concerns of its citizens.

African environmental enforcement ministries or agencies will become more relevant and justify their continued existence when they develop an enforcement policy that acknowledges from the outset the challenges posed by different entities that generate industrial pollution. The agencies need to give preference to an enforcement strategy that excludes identifiable activities or enterprises from the general requirement of environmental standards rather than the current pretentious appearance of a overall applicability of such standards. Corporations should keep and provide information to the interested members of the public to enable citizens to assert their rights to a healthy environment. Rather than expending efforts at churning out more rules or stiffer penalties, Nigeria needs to articulate enforcement strategies that incorporate the objectives of transparency, public participation and accountability.

* Dr. Ali Ahmad (LL.B. Hons., M.B.A., LL.M, S.J.D.) is a Lecturer at Faculty of Law Bayero University, Kano, Nigeria. He was a Visiting Scholar at the Environmental Law Institute, Washington, D.C. 2000-2001. This chapter is based partly on an essay submitted to Harvard Law School, Cambridge, USA, in partial fulfillment of the award of degree of LL.M.

[2] See Federal Republic of Nigeria.*(1991) Achieving Sustainable Development In Nigeria: National Report to UNCED,* (hereafter *National Report to UNCED*). Most of the 2,000 industrial establishments in Lagos State, with a population of over six million, discharge effluents directly into the Lagos lagoon, which is suspected to be one of the worst polluted inland fresh-water bodies in Africa. *See* "Nigeria Five Others Adopt Pollution Monitoring Manual", *The Guardian*, Mar. 15, 1999 (all references to "The Guardian" are to the Nigerian Guardian newspaper). Further, oil producing companies in Nigeria flare about 75 percent of total gas production and about 95 percent of associated gas, a by-product of oil extraction, thereby releasing about 35 million tons of the global warming carbon dioxide. This ranks Nigeria as one of the worstgas-flaring nations in the world. *See* World Bank (1990), *Towards the Preparation of Nigerian Environmental Action Plan,* at 23.

[3]Principle 21 of the *Stockholm Declaration on the Human Environment* provides: "States have, in accordance with the Charter of the United Nations and the principles of international law, the sovereign right to exploit their own resources pursuant to their own environmental policies..." U.N. Doc. A/CONF. 48/14, at 2-65, and Corr.1(1972), reprinted in 11 I.L.M. 1416 (1972). Commenting on the principle, Professor Louis B. Sohn showed his preference for the Holy See's proposal on the draft, which had proposed that in the exploitation of national resources, states should follow a "just environmental policy". He lamented: "While this provision does not go as far as to assert that a state has unlimited sovereignty over its environment, it comes quite close to such an assertion." Louis Sohn (1973), "The Stockholm Declaration on the Human Environment",14 *Harv. Int'l L. J.* 423 at 492. There has not been an improvement in outlook regarding sovereignty and resource exploitation. In reality, the clamor for complete sovereignty got some impetus by stressing that states have sovereign right to exploitation not merely in accordance with theirenvironmental policies however inauspicious, but also their developmental goals. *Rio Declaration on Environment and Development* (1992) U.N. Doc. A/CONF. 151/26, princ. 2 (vol. I) (1992) reprinted in 31 I.L.M. 874.

[4] *National Report to UNCED*, (1991) at 14.

[5] These include the 1968 *African Convention on the Conservation of Nature and Natural Resources* (1968) 1001 U.N.T.S. 3; *Convention on the Prevention of Marine Pollution by Dumping of Wastes and other Matter*, (1972) 1046 U.N.T.S. 120 *Convention on International Trade in Endangered Species of Wild Fauna and Flora*, (1973) 993 U.N.T.S. 243, Vienna Convention for the Protection of the Ozone Layer (1985) 26 I.L.M. 1520;*Montreal Protocol on Substances that Deplete the Ozone Layer* (1987) 26 I.L.M. 1550.

[6] Reference to FEPA necessarily incorporates the newly established Federal Ministry of Environment, especially its Enforcement Department. The Ministry has effectively taken over the activities of FEPA. It may be observed that there are more than fifty federal laws enacted to ensure protection of the environment. These include: *Associated Gas Re-Injection Act*, 26 Laws of the Federation of Nigeria (LFN) (1990), *Energy Commission of Nigeria Act*, 109 LFN (1990);*Factories Act*, 126 LFN (1990), *Federal Environmental Protection Act*, 131 LFN (1990),*Harmful Waste* (Special Criminal Provision, etc.) Act, 165 LFN (1990), *Hydrocarbon Oil Refineries Act*, 170 LFN (1990), *Minerals Act*, 226 LFN (1990), *Mines and Quarries (Control of Buildings Act)*, 227 LFN (1990), *Oil in Navigable Water Act*, 337 LFN (1990), *Petroleum Act*, 350 LFN (1990), *Petroleum (Drilling and Production) Regulations* 350 LFN (1990), *and Environmental Impact Assessment Decree No. 86* (1992).

[7] In an editorial, *The Guardian* chronicles the brazen disregard for law and environmental standards by oil corporations. It rightly stated that the 1969*Petroleum (Drilling and Production) Regulations,* directed oil corporations to commence use of the associated gas from their operations within 5 years. Then the *Associate Gas Re-Injection Act* (1979) went further to provide sanctions that included forfeiture of oil concessions for non-compliance. The laws were ignored. Thereafter, government introduced a fine of 2 kobo or 0.02 cents, and later 10 Naira or 10 cents per 1,000 standard cubic feet of gas flared. The companies willingly pay the nominal fine and continue with the gas flaring. *See The Guardian*, (Dec. 30, 1999).

[8]It is to be noted that the paper reflects the pre-August 1999 posture of enforcement, when FEPA was transformed to form the nucleus of a larger, full-fledged Federal Ministry of Environment. *See* Emmanuel Onwumbiko (1999), "Government Scraps FEPA, Creates Departments Agencies," *The Guardian*, Aug. 4, 1999. The structure of the Ministry is still being workd out but it is believed that a new unit of the Ministry, the Compliance and Enforcement Department, will carry out FEPA's functions. Recently, international conventions have strengthened the role of the public in the development, implementation, and enforcement of international obligations. The most notable declaration crystallising public involvement norms is Principle 10 of the 1992 *Rio Declaration*:

> Environmental issues are best handled with the participation of all concerned citizens, at the relevant level. At the national level, each individual shall have appropriate access to information concerning the environment that is held by public authorities, including information on hazardous materials and activities in their communities, and the opportunity to participate in decision-making processes. States shall facilitate and encourage public awareness and participation by making information widely available. Effective access to judicial and administrative proceedings, including redress and remedy, shall be provided.

See also UNDP (1995) *United Nations Development Programme on Sustainable Human Development, The United Nations System-Wide Special Initiative on Africa Booklet* available at http://www.undp.org/news/unsia00.htm, seeking to stimulate social and economic development in Africa by placing reliance on public participation.

[9] Nigeria Federal Ministry of Mines, Power and Steel (1991), Report on Market Survey for Rolled and Flat Steel Products in Nigeria *Federal Ministry of Mines, Power and Steel.*

[10]World Bank (1995) Nigeria*: Strategic Options for Redressing Industrial Pollution*3. (World Bank: Washington, D.C.). (hereinafter, *Strategic Options*). The deposits were done with abandon and were not unusual occurrences. Lead from the pellet plant was about 7.5 times above the acceptable background level. Metal concentrations of nearby cultivated crops were also elevated and nickel concentrations were over the background level.

[11]*Ibid.*

[12]*Ibid.* at 6.

[13]FEPA (1991), *Guidelines and Standards for Environmental Pollution Control in Nigeria* 37 (hereinafter *FEPA Guidelines*)).

[14]Biochemical Oxygen Demand is a measure of the amount of oxygen consumed in the biological processes that break down organic matter in water. The greater the BOD, the greater the degree of pollution. On the other hand, Chemical Oxygen Demand (COD) is a measure of the oxygen required to oxidise all compounds in water, both organic and inorganic. Due to the discharges by the juice company in Ibadan, the BOD levels of the nearby stream were approximately 60 times the limit of 15 milligrams per litre. FEPA Guidelines, (1991) at 87.

[15] Strategic Options(1995) at 8.

[16] *Ibid.* p 7.

[17] World Bank (1993), *Nigerian Industry: An Assessment,* (World Bank Country Operations Division: Washington, D.C.).

[18]Strategic Options, (1995) at 9.

[19]*Ibid.*

[20]Suspended solids, for instance, were measured to be 1,930 milligrams per liter instead of 5mpl. BOD levels were 27 times and COD 63 times higher. Although effluents must be colourless, FEPA inspections have always found them to be yellow, green or deep purple depending on the factory. The effluents also exceed their limit of manganese, turbidity, and alkalinity. FEPA Guidelines (1995) at 88.

[21] Nigeria (1988) Nigeria National Petroleum Corporation, the Department of Petroleum Resources: The Regulatory Arm of the Oil Industry at 10 (hereafter Department of Petroleum Resources). There were in 1987, 27,520 of petroleum product storage and sale, out of which 4,710 are petrol filling stations.

[22] Joseph Sesebo (1999) "N470b Budget to Relieve Poverty", *The Guardian*, 25 Nov., 1999.

[23] Most prominently are the Shell-BP Bomu 11 blowout of 1970 and the Satrap (now Elf) Obagi 21 blowout of 1972. The Funiwa 5 oil well blowout of January 17th, 1980 released 400,000 barrels of crude oil into the marine environment, with N12m of compensation provided. *See* B.A. Osuno, (1982) "Impact of Oil Industry on the Environment", in *Environmental Awareness Seminar for National Policy Makers,* (Federal Ministry of Housing and Environment) at 53.

[24]Equipment failure accounted for 92% of the net volume of crude oil spills in Nigeria between 1976 and 1979; oil spills as a result of natural causes are almost non existent: A. Guobadia (1991), "Defining Corporate Social Responsibility forNigeria's Oil and Gas Section", 13 *Af J. Int'l. & Comp L.* 472, at 478 quoting Awotayo, (1980) *An Analysis of Oil Spills in Nigeria 19791980.*

[25]Strategic Options, (1993) at 10.

[26]Other refineries have not operated differently. The Warri refinery was also non-compliant with environmental standards. Concentrations of chromium was three times higher, lead was four times higher than the normal standard. Zinc was also four times above the background while copper was six times above its normal measure. *See* Strategic Options (1993) at 11.

[27]Other major polluting industries are pharmaceuticals, industrial chemicals, paint, pulp and paper, breweries, cement and fertiliser. All of them disposed their wastes without much regard to the environmental rules and standards.

[28]*See The Guardian* (1999) at 5.

[29] A. Okediran, (1989) "An Appraisal of Environmental Sanitation Edicts in Nigeria" in F. Shylon, *The Law and the Environment in Nigeria,* 23.

[30] *The Allar Irou* v. *Shell B. P. (Nigeria) Ltd.* Suit No.W/89/71, Warri High Court 11/26/1973(unreported) cited in M.A. Ajomo and O Adewale (eds.) (1994),*Environmental Law and Sustainable Development in Nigeria* (Lagos: Nigerian Institute of Advanced Legal Studies: British Council).

[31] Victims of environmental pollution were often caught up between the wedge of the apparent damage they have suffered and the inadequacy of common law relief, which was not easy to obtain in Court. For instance, the Supreme Court refused to find for plaintiff in*Seismograph Services Ltd.* v *Akpruovo* because he failed to prove damages in an action alleging nuisance. The plaintiff had asserted that defendant's operation caused damage to his building and household (6 S.C. 119 (1974). See also *Amos* v. *Shell B. P. Nigeria Ltd.* (1974) 4 E.C.S.L.R. 486, where a representative action alleging damage to a creek through flooding from a large earth dam was dismissed because the creek was held to be a public water way and a representative action for special damages was improper as plaintiffs' losses were distinct. But see *Umudje* v. *Shell B. P. Nigeria Ltd.*(1975) 11 S.C. 155, where defendant was held to be strictly liable for allowing escape of oil from its fields.

[32] Sylvia F. Liu, "The Koko Incident: Developing International Norms for the Transboundary Movement of Hazardous Waste," 8 *J. Nat. Resources & Envt'l L.* 121 at 131.

[33]*Ibid.* at 124.

[34] "Each Party shall strive to adopt and implement the preventive, precautionary approach to pollution problems which entails, inter alia, preventing the release into the environment of substances which may cause harm to humans or the environment without waiting for scientific proof regarding such harm." Organization Of African Unity: Bamako Convention On The Ban Of The Import Into Africa And The Control Of Transboundary Movement And Management Of Hazardous Wastes Within Africa, (1991) 30 I.L.M. 775.

[35] 'Art. 4.

[36] Cap. 131 LFN (1990) as amended by *Decree No. 59* (1992). The Agency first operated under the Federal Ministry of Works and Housing but was later transferred to the Presidency, specifically under the Office of the Secretary to the Government. Presently, it is part of the Ministry of Environment.

[37]Section 4.

[38]Section 17(1)(d).

[39] Sections 25 and 26.

[40] 2 FEPA (1992)*Monograph* 13, at 1818,

[41] *Ibid.* Other alternatives involving use of economic instruments such as emission charges or the use of permit-based system may be highly complex to operate in the absence of adequate technical and information database.

[42]*Decree No. 50* (1989).

[43]See *Decree No 5* (1992) s. 11 (1992) as amended. The former Natural Resources Conservation Council (NARESCON), the Remote Sensing and Cartography Unit, the Wildlife Management Unit, the Drought and Desertification Control Unit of the Federal Ministry of Agriculture, the Soil Erosion and Flood Control Division and the environmental schools of the Federal Ministry of Water Resources were to operate under FEPA. *See* further, (1994) 6 *The Nigerian Environment*, Mar./June.

[44] *Petroleum (Amendment Decree) NO. 33* (1977). Enforcement of petroleum laws used to be the responsibility of The Petroleum Inspectorate Division of the Nigerian National Petroleum Corporation (NNPC), a wholly owned government company that sets policy on petroleum and also took part in commercial oil activities. In 1985 the policy arm was separated and named the Ministry of Petroleum Resources. But the Petroleum Inspectorate still remained with the NNPC until 1988 when the Inspectorate was transferred to the Ministry and renamed Department of Petroleum Resources. *See* Department of Petroleum Resources (1988)at 11.

[45] S. 9(1)(b)(iii) *Petroleum Act 1969*, 350 LFN (1990).

[46] S. 23 *FEPA Act*, 131 LFN (1990).

[47]*The Guardian*, May 4, 1996.

[48] In 1989, FEPA granted a five-year moratorium to all industries in the country with a view to allowing them time to comply with various regulations, especially the installation of pollution abatement facilities. See 6 *The Nigerian Environment*, Sept./Dec. (1994).

[49] For instance, selective corporate amnesty by the antitrust division of the Department of Justice allowing complete corporate amnesty in certain circumstances has been highly successful in eliciting compliance and preventing violation. To qualify for amnesty, a corporation must terminate its involvement in the regulated activity promptly, report its violation before investigation begins, cooperate completely with the investigation, repay any injured parties, and not be the leader or instigator of the activity. See "Growing the Carrot: Encouraging Effective Corporate Compliance" (1996) 109 *Harvard L. Rev* 1783, at 1795.

[50] *The New Nigerian* (1996) Jan. 24.

[51] Section 7, *FEPA Act* (as amended by *Decree No. 59 1992*). It has been a policy of the Federal Government since 1981 to set aside 1% of the Federation Account for the amelioration of ecological problems. Before FEPA was made the adviser on its disbursement, 7 ministries performed this function. From 1985-1992 about N1.48 billion ($180 million) was disbursed to Federal, state and local government agencies for control of environmental degradation. By 1992, the Federal Government had reportedly spent about $500 million a year on environmental protection and resource management programs. This represented roughly 1.5% of the country's GDP. *See National Report to UNCED*, (1991) at 17 and 44.

[52]In China, enforcement is based on fees for discharges up to the concentration limit and fines for any amount above the limit. 20% of the revenues fund the agencies while 80% is put into a revolving fund, which is loaned out to corporations to partially subsidise the financing of pollution abatement facilities. In India, the environmental boards are financed mainly by fines and penalties. *See* Strategic Options, (1995) at 27.

[53] *Lagos State Environmental Pollution Edict No. 13* (1989). In Kano, the second most polluted city in the country, the highest fines for unlawful discharges were N1, 000 or $10. *Refuse Disposal Agency Edict No. 1*, s. 16 (1994).

[54] "As mentioned earlier, quantitative estimates of the long-term losses to Nigeria from environmental degradation have been calculated to be in the region of US $5 billion annually in the absence of near-term mitigative measures." World Bank (1990), *Towards the Preparation of Nigerian Environmental Action Plan*, xiii. (World Bank: Washington, D.C.).

[55]Hassan Adamu, in *FEPA Monograph* (1992) at 7.

[56] The case of *W.S. Kirkpatrick & Co. Inc. et al. v. Environmental Tectonics Corp. International*, 110 S. Ct. 701 (1990), demonstrates the prevalence of corruption in Nigeria. In that case, the United States Supreme Court unanimously held that act of state doctrine was not available as a defense for an American corporation that was involved in bribing officials of the Nigerian government. The corporation had agreed with Akindele that in the event Akindele secured a contract for the corporation, a commission equal to 20% of the contract price would be paid to him and he would in turn give the money as a bribe to officials of the government.

[57] *Newswatch* (1988), Jul. 4, at 16.

[58] Most pollution caused by oil industry are clearly preventable: K. O. Oyefolu, & O. A. Awobajo (1979), "Environmental Aspect of the Petroleum Industry in the Niger Delta: Problems and Solutions" in *Proceedings of Petroleum Industry and the Environment of the Niger Delta* 145.

[59] In an interesting case study of an asbestos company that willfully neglected to comply with environmental regulations, secrete correspondence between the management revealed that "the modifications which had been suggested on the basis of those counts were of such an expensive nature that might require closure of the mine." N. Gunningham (1987), "Negotiated Non-Compliance: A Case Study of Regulatory Failure", 9 *L & Pol'y* 69. The corporation, therefore, decided against installing the necessary dust-control technology, which resulted in loss ofmany lives as a result of the asbestosis.

[60] (1996), 23:2 *Newswatch* (Jan. 8).

[61] *Ibid.*

[62] Shell Producing Development Company, for example, has some of its installations closer than one kilometer to residential areas, contrary to environmental regulations. Indeed one of such is located 10 meters from a residential bungalow in Erhoike, Delta State and it produced 18,000 bpd. Resolution of this lies in relocation of the residents or the oil installations, whomever came later. Shell claims earlier settlement. In any event, it has been estimated that relocation of either of them would cost Shell at least N2 billion (US$25million), a cost reportedly considered by a Shell official to be "prohibitive". *Ibid.*

[63] "National authorities should endeavor to promote the internalisation of environmental costs and the use of economic instruments, taking into account the approach that the polluter should, in principle, bear the cost of pollution, with due regard to the public interest and without distorting international trade and investment." *Rio Declaration on Environment and Development*, (1992),principle 16.

[64] *EEC's Recommendation 75/436*, cited in J. Mcloughlin and E. Bellinger (1993)*Environmental Pollution Control: An Introduction to Principles and Practice of Administration*(Boston: Graham & Trotman).

[65] *FEPA Mongraph* (1992) at 18.

[66] Judicial awards of damages or fines in the United States are substantial enough to deter potential polluters. For instance, knowing violations of the provisions of the*Resource Conservation and Recovery Act* (42 U.S.C.S. § 6928 (e) and (g) (1994)) could attract a criminal penalty of $50,000 for each day of violation, $1 million where the offender is an organisation, and $25,000 per day for a civil penalty. Liability under the statute is strict and the court narrowly construes it to give effect to the legislative intent. Thus, a penalty of $1,000 per day was appropriate for operation of land disposal facility for 194 days without interim status or permit, where owner/operator had already paid penalty of $19,500 to state and was proceeding to close that facility. *U.S.* v. *T & S Brass and Bronze Works, Inc.*, 865 F.2d 1261 (1988).

[67] However, directors or officers are increasingly losing their common law protection in modern statutory frameworks. The *Harmful Waste Act* (1990) 165 LFN s. 7 provides that where a criminal act has been committed by a body corporate and it is proved that it was committed with the consent or connivance or neglect of an officer, he as well as the body corporate, shall be guilty of the crime and shall be liable to be proceeded against and punished accordingly.

[68]Gunningham (1987) at 30.

[69] R. A. Kagan,(1984) "On Regulatory Inspectorates and Police", at 37 in K. Hawkins and J. M. Thomas, eds., *Enforcing Regulation* (Boston: Kluwer-Nijhoff Hingham) at 37.

[70]C.D. Stone (1991),, , *Where the Law Ends: The Social Control of Corporate Behaviour*(Illinois: Waveland) at 35.

[71] R.B. Mitchell (1996), "Compliance Theory: An Overview"in J. Cameron, J. Werksman *et al.* (eds), *Improving Compliance with International Law* 22 (London: Earthscan).

[72] N. Gunninham (1974), *Pollution, Social Interest and the Law* (London: Robertson); R. Cranston (1979), *Regulating Business: Law and Consumer Agencies* (London: Macmillan, Social Science Research Council) J. Scholz, (1984), "Voluntary Compliance and Regulatory Enforcement", *6L & Pol'y* 385; and C. S. Russel (1990),,*Monitoring and Enforcement*, at 243 in P. Portney (ed.), *Public Policies for Environmental Protection* (Washington D.C.; Resources for the Future).

[73] E. Bardach & R. A. Kagan (1982), *Going by the Book: The Problem of Regulatory Unreasonableness* (Philadelphia: Temple University Press)at 123.

[74] K. Hawkins (1983), "Bargain and Bluff: Compliance Strategy and Deterrence in the Enforcement of Regulation", 5 *L. & Pol'y. Q.* 35.
[75]*See The Guardian* (1999) at 6.
[76] Multinational corporations are better defined by their attributes and functions. The attributes include size and the kinds of business activities in which the corporations have specialised; another is the pattern of their management and control. *See* R. Vernon, (1977), *Storm Over the Multinationals: The Real Issues,* (London: MacMillan) at 19.
[77] Y. Omorogbe (1987) "The Legal Framework for the Production of Petroleum in Nigeria",5 *of Energy and Natural Resources L.* 273 at 278.
[78] Strategic Options (1995).
[79]For instance, multinational corporations dominated FEPA's list of honour for ten corporations that were adjudged environmentally friendly in Nigeria. No oil corporation or other corporations where government holds equity or any other interests however made the list. FEPA used five criteria which were: effective effluent treatment; emission pollution control systems; environmental auditing,; self monitoring and reporting system,; effective solid waste management systems,; and good house keeping and sound contingency plans. See6 *The Nigerian Environment* (1995) Jul./Oct.
[80]A representative case on the issue of utilizing lower standards in developing countries is the on-going case of *Maria Aguinda v. Texaco,* Inc. 175 F.R.D. 50 (1997). In that case certain citizens of Ecuador brought an action in the U.S. against Texaco, Inc., the parent corporation of Petroecuador. The claim was that for the past twenty years ending in 1990, Texaco, Inc., through Petroecuador, owned and operated a number of oilfields in Ecuador and allegedly abused the environment through careless disposal of inadequately treated oil wastes, contrary to prevailing standards. The case is mired in complex procedural issues, the latest of which was inadequate waiver of immunity by one of the parties, the government of Ecuador. Likewise, Shell Petroleum has been widely accused of setting substantially lower operating standards in Nigeria than in developing countries. It is stated that Shell operates in 100 countries but that 40 % of its oil spills have occurred in Nigeria. Accordingly, in its 1997 annual shareholders' meeting, Shell defeated a resolution, filed by a group of its minority shareholders, which called for heightened environmental standards regarding the company's operations in Nigeria. *See* David Weinraub (1997), "Shell Defeats the Pirc Group Resolution", *Colo J. Int'l Envt'l L & Pol'y* 136.
[81]"Environmental issues are best handled with the participation of all concerned citizens, at the relevant level. At the national level, each individual shall have appropriate access to information concerning the environment that is held by public authorities, including information on hazardous materials and activities in their communities, and the opportunity to participate in decision-making processes. States shall facilitate and encourage public awareness and participation by making information widely available. Effective access to judicial and administrative proceedings, including redress and remedy, shall be provided." *Rio Declaration* (1992), Principle 10. In the U.S. a number of legislative instruments have created enforceable rights of access even to governmentally held documents, e.g. *Freedom of Information Act,* 5 U.S.C.S. § 552 (1988); *Emergency Planning and Community Right-to-know Act,* 42 U.S.C.S. §§ 11001-11050 (1988).
[82]The United States *Clean Air Act* (1970) 42 U U.S.C.S. § 7604 (1988) (as amended) is one of the first statutory instruments to provide that any person may commence a civil action against any person including the United States government that is alleged to be in violation of emission standards as stipulated by the law. There is yet no clear authority for citizens' suits in Nigeria remotely comparable to the *Clean Air Act.*
[83]*See* texts accompanying *supra* note 8..
[84]*See* Nigerian Environment (1995).

TANZANIAN FOREST LAW

JON C. LOVETT[1]

INTRODUCTION

There has been a substantial international influence in Tanzanian Forest law. Firstly, State forest reserves were established by colonial powers, demarcating and alienating forests from customary control. Secondly, forest laws have been developed and implemented by reference to international treaties, conventions and agreements; and thirdly, international donor organisations have assisted in supporting both the creation and implementation of forest policy. Historical contingency has also played a major role in development of Tanzanian forest law. In the last one hundred years or so Tanzania has undergone three major changes of administration, each leaving a legacy of law, policy and land use. Acquisition of the State forest estate has been progressive. Forest reserves gazetted under the German administration of 1891-1919 were retained by the British from 1919-1961 and most are still recognised today by the Tanzanian government. Over time, each administration continued to add to the State controlled area with more forest reserves, game reserves and national parks.

While the physical boundaries of these forest reserves have usually remained the same over time, the policies under which they were managed have changed. Under customary law forest was preserved for traditional rituals, resource conservation and acted as a source of famine foods or shelter during inclement times. The eastern African climate is unpredictable and maintaining a mosaic of vegetation types with a range of resources was a well regulated part of traditional land use management that enabled people to survive droughts or excessive rains. The German and British administration were concerned about over-exploitation of the forest resource by timber extraction, replacement by cash-crop plantations and degradation by fire. After independence the Government of Tanzania continued to use the 1957 British forest ordinance, but this was replaced in 1998 by a new forest policy and in June 2002 by a completely revised Forest Act. The new policy and law integrate the sentiments of the international 1992 Biodiversity Convention with a recognition that people living near to forest reserves should be involved in forest management.

Changes in Tanzanian forest policy reflect changes in social values. Firstly, there is an increased emphasis on conservation of biodiversity, and secondly there is concern for an equitable inclusion of people affected by forest management in determination of how the forests should be managed. Conservation of forests has always been an important part of forest law in Tanzania, but only recently has forest biological diversity been explicitly part of forest policy and management. The forests of eastern Tanzania are one of the top twenty-five global biodiversity hotspots and so are internationally as well as nationally important for biodiversity conservation. The focus on conservation can conflict with the desire for equity. What is of importance to the global community may not be compatible with needs of the State and local community. These tensions can be seen in the Convention on Biological Diversity (CBD)[2]. The convention recognises sovereign rights of the State over its own resources under Article 3[3], but also requires the State to both form a system of protected areas and respect traditional lifestyles under Article 8[4]. It is in these latter two requirements, protection and local needs that there has historically been the most conflict.

This chapter first presents an overview of forests in Tanzania, focusing on the mountains in the east of the country which are regarded as being the most important for biodiversity conservation, and then reviews the history of Tanzanian forest management and law. In conclusion the problem of biological conservation and social equity is discussed.

THE FORESTS OF TANZANIA

Tanzania is a predominately dry country, with about half of the land area receiving less than 760 mm of rain a year in four years out of five[5]. Only about 4% of the country receives greater than 1270 mm of reliable rainfall a year. This simple meteorological fact has enormous ramifications for the type of vegetation that the land can support and the use to which the land can be put. In the absence of groundwater, closed canopy forest requires greater than about 1200 mm of rain a year to develop. If the rainfall is less than this, or is markedly seasonal, then open woodland, thorn thicket or grassland prevails. This chapter deals specifically with closed canopy, predominately evergreen forest as it is in this vegetation type that a high proportion of Tanzania's rare and endemic plants and animals are found. Much of the higher rainfall areas are on mountains, where lower temperatures at higher elevations mean that precipitation exceeds evaporation. Thus the naturally forested areas are also catchments where water can be harvested. These catchment forests feed rivers which can be used to irrigate crops, supply cities and generate hydroelectric power. Although forests use water through evapo-transpiration, the consensus is that natural forest cover is an appropriate land-use for catchments as natural forests regulate micro-climate, ameliorate flooding and capture fog and mist through occult precipitation[6]. Maintenance of forest cover as a State priority is in preference to agriculture or grazing as an alternative land-use for the high rainfall areas, and this in turn has historically led to conflict between national needs of the State forestry authority and other land users.

More recently, the role of forests for biodiversity conservation has been incorporated into forest management. The forests in Tanzania vary greatly in terms of both the number and type of plant and animal species they contain. There are marked differences in species composition associated with changes in rainfall, elevation and location[7]. The forests can be divided on the basis of geology and climate (Figure 1).

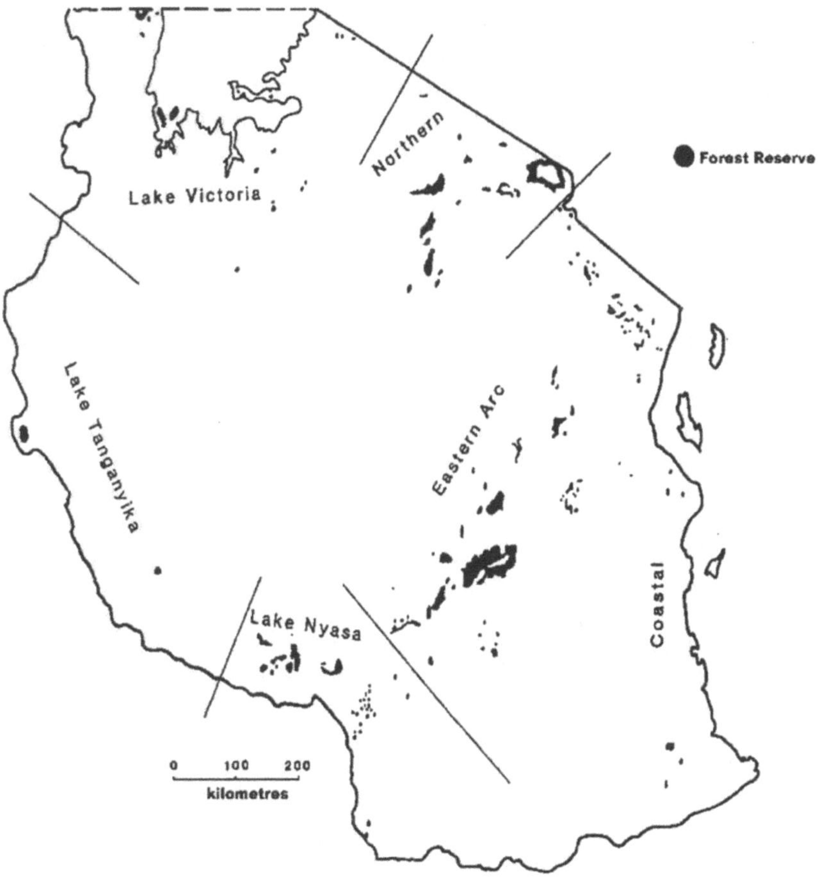

Figure 1. Division of Tanzanian forests on the basis of geology and climate. Forest distribution is based on forest reserves containing closed forest formations. Coastal, Eastern Arc and Northern forests are under the direct climatic influence of the Indian Ocean, but Coastal forests are predominantly on sedimentary rocks, the Eastern Arc are on igneous and metamorphic rocks, and Northern forests are predominantly on volcanic areas (with the exception of the Mbulu highlands). Forests associated with the great lakes of Victoria, Tanganyika and Nyasa are subject to environmental fluctuations associated with variation in the local climates associated with these lakes.

In western Tanzania the forests are under the climatic influence of the Atlantic Ocean and local climates generated by the great lakes of Victoria and Tanganyika. In eastern Tanzania, the forests are under the climatic influence of the Indian Ocean and Lake Nyasa. The forests of Kilimanjaro and Meru in the north, and Rungwe in the south are on volcanoes, where-as those on the arc of mountains from the Pare to the Udzungwa are on crystalline gneiss. It is the forests on this latter range of mountains, termed the Eastern Arc[8], that are so remarkable from a biodiversity perspective. Together with forests on the tropical African Indian Ocean coastal plain they constitute a fragmented series of forest islands that contain an exceptional number of plant and animal species found nowhere else in the world.

The eastern Tanzanian forests have been recommended many times as a priority for conservation and research. Some 25-30 % of the plant species are endemic to the Eastern Arc[9], and endemism for some groups of organisms, such as millipedes is greater than 80%[10]. Some of the forest plants are of actual or potential economic importance as genetic resources[11]. Eastern Arc populations of more widespread timber trees may be of importance as provenances for breeding as they have been isolated from the main population for some time. Oil containing plants such as the oil palm, a widespread Guineo-Congolian species with a distinctive tall slender lowland Eastern Arc population[12] and *Allanblackia stuhlmannii* the fruits of which contain a high molecular weight fat[13] may also be of economic interest. Other genetic resources include endemic species of wild coffee[14], and the horticulturally important African Violet, *Saintpaulia*[15], African Primrose, *Streptocarpus*[16] and Bizzy-Lizzy, *Impatiens*[17].

In a paper presented at the 1966 meeting of the Association for the Taxonomic Study of the Flora of Tropical Africa (AETFAT) Polhill[18] gave lists of endemic species and areas as priorities for conservation, including the Usambara, Nguru, Uluguru, Rubeho and Udzungwa mountains among other sites of botanical importance in Tanzania. The National Research Council[19] of the United States of America listed the montane forests of eastern Tanzania as one of the eleven tropical areas of highest priority for biological inventory. In the early 1980's the World Wildlife Fund (WWF, now known as the World Wide Fund for Nature) listed them as a priority area in both their Tropical Forests and Plants campaigns, using the African Violet (the whole genus is endemic to the Eastern Arc and Coastal forests) as the logo for their Plants campaign. The International Union for Conservation of Nature and Natural Resources (IUCN) selected the Usambara and Udzungwa mountains as some of the 150 areas world-wide containing important ecosystems for conservation[20], and together with the Coastal forests, the Eastern Arc mountains have been identified as a globally important "hot-spot" area of biodiversity[21]. In 1989 the Tanzania Tropical Forest Action Plan called for protection of the Eastern Arc forests and regarded the Eastern Arc Forest Management Plan as the central project for ecosystems conservation[22]. This latter plan has been particularly influential in Tanzanian forest policy as it provided a focal point for the co-ordination of development-aid donor activity.

Most forest areas are under some form of protected status and in the last two decades there has been substantial involvement of international development agencies in their management. All the Eastern Arc forests in central government forest reserves are considered catchment reserves and so are nominally set aside for watershed protection[23]. Those in Kilimanjaro Region (Pare mountains), Tanga Region (Usambara and Northern Nguru mountains) and Morogoro region (Ukaguru, Rubeho, Uluguru, Mahenge and eastern Udzungwa mountains) are under the Catchment Forest Project, which has received support from the Norwegian development agency NORAD[24]. Those in the East Usambara are under the East Usambara Catchment Forest Project which has received support from the Finnish agency FINNIDA[25] with additional inputs outside the forest with the EEC[26] funded, IUCN administered, East Usambara project. The German GTZ[27] supported a Soil Erosion Control and Agroforestry project in the West Usambara and Pare mountains which included natural forest management. In Iringa District the Hifadhi ya Mazingira (HIMA) project, supported by the Danish agency DANIDA[28] includes a forest management component[29].

In addition to forests administered in forest reserves, some forests are under strict protection for nature conservation. A 450 ha patch of submontane forest on Malundwe Hill is in Mikumi National Park[30], and the 450 ha University forest reserve at Mazumbai in the West Usambara was protected because it was on private land donated to the University of Dar es Salaam for conservation and research[31]. A 2000 km^2 National Park covering the escarpment and upland forests and grasslands of the north-east Udzungwa mountains was gazetted in February 1992 following discovery of high biodiversity values[32] and has received support from the World Wide Fund for Nature. A riverine strip containing a few hectares of montane forest along the Luisenga Stream on private land in the southern Udzungwa has been declared a nature reserve by its owners, Brooke Bond Tanzania Ltd.[33], and an 8380 ha nature reserve was established in 1997 at Amani in the East Usambara mountains[34].

Protection of forests for water catchment is compatible with biodiversity conservation, but conflict does arise when natural resources are extracted or the forests are converted to another form of land-use. Traditionally the forests are of considerable social importance, both as a primary source for many locally utilised forest products and as a culture base. Forest bamboo is locally of economic importance for manufacturing baskets and for building. The Brooke Bond Tea Estates in the southern Udzungwa use bamboo for making plucking baskets, and Imalutwa village near Iringa produces bamboo baskets (*tenga*) for transport of agricultural produce. Reeds (*milulu*) are gathered to make mats and baskets as are fronds of the montane forest palm, *Phoenix,* which is also used to make winnowing baskets[35]. Many other forest plant species are used for knife and hoe handles, pestles and mortars, spoons and other household implements. No Eastern Arc forest species which yield edible fruit offers commercial possibilities, but they are used locally, though they are not a major source of food except under exceptional circumstances. For example, *Parinari excelsa* fruits were formerly an important food source during famine. Building poles are an important forest product as the forests produce strong rot and termite resistant poles used in house building. Two types of pole are cut: large straight poles for house main frame construction (*nguzo*); and slender poles used for wattle and daub and to hold thatching (*fito*). In easily accessed areas close to high populations, up to 50% of the available poles can be cut[36]. Many plant species are used medicinally, with some sold on a commercial basis. For example the stimulant, *Mirunge* (Ghat, *Catha edulis*), is gathered from wild trees in the West Usambara mountains for sale. Honey is regarded as an important forest product by villagers. It is extracted either from wild nests, or from traditional log hives hung in the forest. Forests are also a source of meat. The large bush pig is hunted with dogs and pitfall traps; and the tree hyrax and small forest antelopes, duikers, are trapped with snares. Black and White Colobus are also trapped, reputably eaten, and the skins used for ceremonial purposes by the Wahehe in the Udzungwa mountains. Colobus skins are also used ceremonially by the Wazigua who live from the Nguru and East Usambara mountains to the coast. The forests are of some importance as a local culture base. On Nyumba Ng'itu mountain in the West Kilombero forest reserve traditional ceremonies are held annually by the Wahehe. In many other forest areas, particularly those which were formerly inhabited prior to gazettement, similar traditional ceremonies for *tambiko* are conducted. In other parts of the Eastern Arc the forests are associated with rain and fertility, such as the rain-shrines for Kolelo in the Nguru and Uluguru mountains[37]. Local cultural interaction with the forest also includes many superstitions and legends.

The Tanzanian forests contain a number of tree species which yield valuable timber. The better known of these are: Mvule (Iroko, *Milicia excelsa*), East African Camphor, Podo (Yellow wood, *Podocarpus* spp.) and the East African Mahoganies (*Entandrophragma* and *Khaya* spp.). Many other species are used for lower quality timber, and the less valuable species are being increasingly extracted as stocks of more sought after timbers are depleted by harvesting. Some of the timber species have been heavily exploited. For example, Brenan & Greenway[38] wrote of *Beilschmeidia kweo* (Mfimbo or Mkanta which also occurs in the Udzungwa) in the East Usambara "An economic timber before 1914, exported to Germany where it was used for wagon and ship building, but supplies now exhausted". In most forests supplies of quality East African Mahogany, Mvule and East African Camphor are finished. Some research has been done on regeneration of valuable timber species, both for enrichment of natural forests and for the use of native trees in plantations[39]. Although some trees, such as East African Camphor can be fast growing, and a number of species have been successfully introduced into experimental plantations, commercial exploitation of timber in natural forests has relied on old growth and exotic species such as pines from Central America and teak from Asia have been used in larger plantations

The management of the forests for conservation and extraction of forest products clashed when the FINNIDA supported mechanised logging of old growth forests in the biodiversity rich East Usambara mountains during the mid to late 1980s caused an international outcry from conservationists[40]. At this time mechanised logging was also taking place in the lowland forests of the Udzungwa mountains with Chinese support, and extraction of timber by organised pit sawing affected practically all forest areas containing timber of any value. The rampant logging not only raised issues of forest protection, but also equity. Over the years forests had been sequestered by the State, and access to forest resources denied to local villagers, because of fears of forest loss and degradation. During the 1980's unsustainable extraction of forest resources was not by local villagers, but by logging operators who were usually not local and who often brought in skilled labour from outside the area. The historical contingency of forest policy meant that the villagers were not directly involved in forest management, received little or no benefit from commercial exploitation of the forest resource, but were punished if they were caught exercising traditional rights without permits. To understand how this situation arose the next section reviews the history of management of the Tanzanian forests in four time periods[41], precolonial (before 1891), German (1891-1919), British (1919-1961), and post-independence (1961 to date).

HISTORY OF MANAGEMENT

Precolonial (before 1891)

Man has lived in eastern Africa since the earliest times, true hominids emerged in the area by 5 Myr BP[42]. Controlled use of fire developed over 1 million years ago[43] has probably had a major effect on African vegetation for at least 400000 years particularly when global climate was changing from glacial to interglacial conditions[44]. Farming probably developed in two forms, systems based on root crops in West Africa from the fourth millennium BC and a system based on grain and herding in north-east Africa from the sixth millennium BC[45]. The practise of cultivation is believed to have been introduced into the interlacustrine region of central Africa by Bantu speaking iron-workers before 2000 yrs BP, and a pollen core from south-west Uganda suggests cultivation in the area as early as 4800 years BP[46]; though the charcoal deposits in the core could also be due to natural fires following climatic changes[47]. Agriculture and iron smelting in the Kagara region of western Tanzania during the last 2500 years has had a major impact on the closed forests. By AD 400 iron producers were exploiting secondary growth and in the late fifth century AD sedimentary evidence suggests severe erosion, possibly caused by deforestation that in turn led to collapse of the iron industry[48]. Cyclical forest regrowth and exploitation in Kagara means that the forests today are a product of a long history of social use of natural resources. In the East Usambara mountains of north-east Tanzania, pottery in surface soil layers under diverse tall natural forest has been ascribed to Early Iron Age settlements from 2000 yrs BP[49]. On the coast there has been trade for thousands of years[50] between the indigenous Swahili culture[51] and other western Indian Ocean civilisations which would have resulted in exploitation of forests for natural resources or agricultural land[52]. In coastal south-east Kenya and north-east Tanzania sacred forest patches enclose *kaya* clearings which are the home of ancestral spirits of the Mijikenda[53] and which have archaeological remains dating from late in the first millennium AD[54]. Islamic influence on the tropical eastern African coastal strip dates from at least the late eighth century[55], and prior to European influence the Sultan of Zanzibar applied Islamic law to settlement of land disputes[56].

Many forest areas in Tanzania have been disturbed by human influence and show signs of previous habitation. Around Lake Tanganyika at Gombe[57] and on the Mahale mountain[58], the forests appear to be secondary and some contain remains such as pot-shards. Both areas are now National Parks. On the Ufipa plateau the Mbisi forest is thought to be a relict of more widespread forests[59]. On southern Kilimanjaro the German administration found the edge of the closed forest at a lower level than it had been in the past with a wide zone of secondary forest containing the remains of old habitations and fortifications[60]. In the Eastern Arc in the Ukaguru mountains[61], Rubeho mountains[62] and southern Udzungwa mountains[63], many montane forests are secondary and show signs of cultivation ridges, houses and pot-shards. Forest adjacent to Maasai pastoral lands in northern Tanzania, such as on Mt. Hanang and Monduli, show evidence that fire was used to open up the forest for cattle grazing[64]. Access to wetter forested sites may be an important part of traditional seasonal herding movements and it is likely that such forest management practices date back to the domestication of stock in the area.

Some of the Eastern Arc secondary forests are at least 100 years old and probably much older. The Ukaguru forests were gazetted as forest reserves during the German administration prior to 1914 and formerly cultivated areas under the Ukwiva forest in the Rubeho mountains are said by the local people to have been forest when they arrived in the area from Iringa shortly after 1894[65]. In the West Usambara mountains the forests were places of refuge during periods of hostility and sources of food during times of famine[66]. Clearings deep in the forest were maintained for these purposes. Old inhabited areas within forest reserves are still of importance for spiritual reasons. Forests were traditionally protected on the Pare mountains for ceremonial purposes, and traditional forest reserves are also found on the Nguru, Uluguru and Udzungwa mountains. For example the forests of Kanga mountain in the northern Nguru are held scared by the Wazigua, and a ceremony is still needed if visitors to the forest are to enter with the blessing of the local community[67], though migrants from other areas such as Kilimanjaro no longer respect traditional norms.

Customary rights to land and natural resources in Tanzania are complex and reflect a multi-faceted relationship between society and the environment[68]. For example, in the late seventeenth century access to forest natural resources in Kagara was regulated by the Hinda dynasty to centralise control over the iron industry[69]. Rights to forest and forest land in Kagara would be very different to those practised by the Maasai for dry season grazing in moist mountain forests. In turn these systems would not necessarily have the same priorities as those of the Wahehe cultivating forest land in the Udzungwa mountains, or of the Mijikenda in stockaded forest patches on the coast. Towards the nineteenth century the arrival of a new and pervasive administrative hegemony was to find established customary rights both a difficulty and a boon. A difficulty because conflict was inevitable when communities were denied their traditional rights. A boon because the colonists regarded customary rights to be usufructory rights in which the community allocated deemed right of occupancy and access[70]. Land and natural resources where thus considered to be held in trust for the whole community. The new State simply assumed that mantle of trust and the heterogeneous indigenous forest management systems were lost, giving way to national priorities for protection of water catchment and commercial exploitation.

German administration (1891-1919)

Extraordinary circumstances surrounded establishment of German colonial authority over the vast tract of land that covers the current area of mainland Tanzania, Rwanda and Burundi. In 1884 a young man, Dr Carl Peters, formed a *Gesellschaft für deutsche Kolonisation* (Association for German Colonisation)[71]. The same year he travelled to Tanzania and collected signatures on treaties that apparently ceded 140000 square kilometres of territory to the Association. In 1885 the German *Reichskanzler* (Imperial Chancellor), Otto von Bismarck, decided to effectively create a German protectorate in East Africa by granting protection to the Association with a *Kaiserlicher Schutzbrief* (Imperial Charter), and the Association became a chartered company, the *Deutsch-Ostafrickanische Gesellschaft* (DOAG). The DOAG controlled the coastal hinterland because of treaties with Africans and the Imperial Charter, and the coastal strip through a treaty with the Sultan of Zanzibar in 1888. Despite these gains of land, the DOAG was not financially successful and unable to control the Arab revolt of 1888-1890. The German Empire sent German soldiers and several hundreds of mercenaries from Sudan to suppress the revolt, and on 1 January 1891 the Empire took control of the DOAG's protectorates. The development of legislative infrastructure and conflict that were to follow in the next few years are still reflected in the current pattern of forest reserves in Tanzania.

The first professional forester arrived in 1892, one year after the start of colonial administration (forestry under the German administration is reviewed by Grant, Lundgren and Schabel[72]). During the early years there was little control over forest exploitation by settlers, and forests were logged, especially near agricultural plantations in the eastern Usambara. In 1895 the Usambara Forest Ordinance attempted to control destructive practices with fines, imprisonment and corporal punishment. However early ordinances were difficult to enforce because of staff shortages and the large remote areas they had to cover. Despite numerous forest ordinances and other legislation, the German administration did not draw up a comprehensive forest law. The State had however assumed control of most of the land in the German Imperial Ordinance of 26 November 1895 which deemed all land without clearly specified ownership rights (*herrenlos*) to be Crown Land (*Kronland*) with right of occupation vested in the government[73]. This had the advantage of preventing land speculation by Europeans seeking to acquire vast estates and extending State control over apparently unoccupied land, but the disadvantage of not recognising customary land rights beyond land specifically under cultivation of crops. Access to natural resources such as seasonal grazing land, forest produce or sites of cultural importance was now from Crown Land.

Prior to 1896 the dry *Juniperus* forests near Kwai in the West Usambara were largely unsettled due to Maasai raids. However, following the arrival of rinderpest in northern Tanzania around 1891, the Maasai population declined drastically due to loss of their cattle herds[74]. Influence of the Maasai in dry forest areas waned and in the Usambara cultivators moved into the *Juniperus* forest and practised shifting cultivation, grazing and stripping of *Juniperus* bark for roofing. This area was to be later reserved as the Shume-Magamba forest and subsequently a large portion of the reserve was excised for cultivation after independence in 1963.

In 1898 the first trial plantations of teak *Tectona*, *Eucalyptus*, and *Casurina* were established. In 1902 experimental plantations were being made in Lushoto using *Eucalyptus* and Australian Wattle (*Acacia mearnsii*) in an attempt to redress deforestation. In 1902 the Biological-Agricultural station at Amani was set up and planted a large number of tropical crops and timbers, which exist today as the Amani arboretum. In 1904, passing of the Forest Conservation Ordinance initiated conversion of 7500 sq. km of crown land into 231 forest reserves where no settlement, farming, grazing and other unauthorised use was allowed. Local people were moved and compensated. The importance of forests in conserving water supplies was a key reason for alienation of this land from local control, and forest reserves were created as protection reserves in the northern volcanic mountains at Nou (314 sq. km), Marang (356 sq. km), Ngorongoro-Olmalassin (894 sq. km), Monduli (60 sq. km), Mt. Meru (442 sq. km), Mt. Kilimanjaro (1875 sq. km, though the forest was not legally reserved until 1929 as a portion of the plan was lost); in the Eastern Arc mountains at Chomme (143 sq. km) in the Pare, Shume-Magamba (247 sq. km) and Shagayu (83 sq. km) in the West Usambara, North Nguru (88 sq. km), Nguru (187 sq. km), Uluguru (278 sq. km), Ukaguru (140 sq. km); and in the Lake Nyasa forests on Rungwe (138 sq. km); plus many other smaller reserves. This network of reserves forms a large proportion of the present day pattern of central government administered forestry estate.

In contrast with some parts of Tanzania, little forestry work was carried out in districts under military rule such as the Udzungwa and Mahenge mountains following resistance from the Wahehe and the Maji Maji uprising. Famine and disruption in these southern highlands from the late nineteenth century to the start of the British administration resulted in depopulation of large areas. Shifting tribal alliances, suppression of Chief Mkwawa's Wahehe stand against German colonial authority, the widespread Maji Maji rebellion and commandeering of supplies during World War I were all contributing factors to disruption in the region[75]. Much of the Udzungwa mountain forests remained unreserved until the late 1950's when the British administration alienated a huge, almost unpopulated, area to forest department control. A large part of which was to become the Udzungwa Mountains National Park in 1992.

Despite protection being the primary aim of State reservation of forests, timber was needed for development of the new colony and revenue required to run the forest department. After 1906 in the mountains of the north and north-east, stands of *Juniperus* and *Podocarpus* were subject to large scale exploitation and foresters became interested in regenerating these species. A concession was granted to the Wilkins & Weiss company to extract 3000 hectares of *Juniperus*, *Podocarpus* and other hardwoods from Shume-Magamba in the West Usambara, and by 1909 they were operating a 9 km funicular railway from their sawmill at Shume (1870 m in altitude) to Mkumbura (450 m). Similarly the German East Africa Lumber Company built 24 km of narrow gauge tramway to open up 12000 ha of forest in the East Usambara mountains. In the Usambara mountains, sawmills were established at Sigi, Kwamkoro, Derema, Ambangulu, and Shume. Up to 1912, some 710 ha had been planted by the forest service, mainly of indigenous species such as *Juniperus* and *Podocarpus* in cut-over montane forest. In many respects large-scale commercial timber extraction contradicted management for catchment protection, but was an inevitable result of colonial development giving rise to a need for both timber and income to pay for establishment of the forest department. Hence alienation of the forests to protect them, led ultimately to what is now known to be unsustainable timber harvesting with benefits bypassing the local people who had also lost access to resources that traditionally would have been theirs.

Economically, forestry operations during the German period were loss making. Only *Juniperus*, mangrove products, the sale of maize from agroforestry lots and rubber latex proved profitable. Annual forest income remained consistently at around half the administrative expenses, primarily due to the cost of establishing the many forest reserves[76]. The overall effects of German administration of the forest estate were to exploit many of the timber resources, whilst creating a system of protective reserves, establishing plantations of exotics and trials of exotics. For local communities, the impact of German laws was denial of access to forest resources, bans on field burning to prevent fires damaging forests and prohibitions on hunting leading to an increase in crop predators harboured by forests[77].

British administration (1919-1961)

During the First World War a fluid campaign was fought through Tanzania and into northern Mozambique between German and British led troops. British Martial Law covered some parts of the country in 1917 and with capitulation of German forces in 1918, Germany ceded German East Africa under the Peace Treaty of Versailles in 1919[78]. The British administered the then Tanganyika (covering the area of present mainland Tanzania) under a League of Nations mandate from 1922 to 1946, and under a United Nations Trusteeship until independence in 1961. The 1922 mandatory agreement specifically referred to respect of the rights and interests of the native population in framing laws relating to holding land (Article 6). This mandate is reflected in the Land Ordinance, 1923, which recognised the importance of customary rights and "natural fruits", but as with the previous German administration and the subsequent independent Government of Tanzania, took the precaution of declaring all land to be public and so under state control:

> "WHEREAS it is expedient that the existing customary rights of the native of the Tanganyika Territory to use and enjoy the land of the Territory and the natural fruits thereof in sufficient quantity to enable them to provide for the sustenance of themselves and their families and their posterity should be assured protected and preserved;
>
> AND WHEREAS it is expedient that the existing native customs with regard to the use and occupation of land as far as possible be preserved;
>
> AND WHEREAS it is expedient that the rights and obligations of the Government in regard to the whole of the lands within the Territory and also the rights and obligations of cultivators or other persons claiming to have an interest in such lands should be defined by law.
>
> BE IT THEREFORE ENACTED by the Governor and Commander-inChief of the Tanganyika Territory as follows:-
> 1. This Ordinance may be cited as the Land Ordinance, 1923.
> 2. The whole of the lands of the Territory, whether occupied or unoccupied, on the date of the commencement of this Ordinance are hereby declared to be public lands...."

Forestry continued where the German administration had left off (forestry in Tanganyika is reviewed by Grant, Troup, Parry and Lundgren[79]). During the military campaign of 1914-1918 the lack of any effective control resulted in encroachment into many of the forest reserves established during the German administration. The previous German forest law was followed during early British days, with Government Notice No. 15 of 1st February 1920 invoking the Ordinance of 27th February 1909 to protect trees on the Mindu Hills near Morogoro. On the 18th December 1920 a Forest Department was instituted by appointment of an officer to be the Conservator of Forests and in 1921 a Forest Ordinance (GN 32)[80] was passed, based on the Ordinance used in adjacent Kenya colony with inclusion of provision for control of forests on private lands. The ordinance was aimed at protection and development of forest reserves gazetted by the Germans, plus reservation of further areas, formation of plantations, control of cutting and collection of forest revenue. The German reserves were all regazetted in GN 32, and prohibition of cultivation, habitation, grazing and entering within forest reserves placed at the discretion of the Governor (Articles 5 and 10). Access to forest produce was strictly controlled (Article 14):

> "Any person found within a forest area, or in its vicinity, and having in his possession any forest produce, who, on being thereunto required by any Forest Officer, refuses to give satisfactory account of the manner in which he became possessed of any such produce may be taken by the officer interrogating him before a Magistrate..."

For the purposes of the Ordinance, forest produce was defined as "... earth, trees, timber, plants, grass, reeds, rushes, peat, creepers, fibres, leaves, moss, fruit, seeds, galls, spices, bark, rubber, gum, resin, sap, charcoal, honey, wax and such other things as the Governor may, by notice in the Gazette, declare to be included in the said term." However, the initial strict nature of the 1921 Ordinance was modified in 1926 (GN 86)[81] to specify that "no royalty was payable by natives in respect of any forest produce taken by them for their own use only" and in a later modification (GN 163)[82] it was made possible for a forest officer to issue a written permit for a specified quantity of forest produce to be extracted without fee or royalty. This brought the Ordinance more in line with the spirit of the 1923 Land Ordinance and League of Nations Mandate, though this was later to change again and revert to tighter control of "natural fruits".

The Tanganyika Territory Forest Department's report to the 1923 British Empire Forestry Conference[83] focuses on timber production as an important product from reserved forest land, though 40% of the total forest area was regarded as being unprofitable or inaccessible and valuable for water catchment and prevention of erosion. The report raises concern about degradation of the reserves as a result of encroachment and setting of fires. The forests most at risk from fire were considered to be those of Arusha and Kilimanjaro where pastoralists set fires to obtain fresh grazing on the forest edge and where honey and hyrax hunters set fires inside the forests. The department initially had a small staff of eleven Europeans and 106 local forest guards, but embarked on an ambitious programme of planting, protection and creation of new reserves.

The *Juniperus* plantations established by the Germans continued to be monitored and used as trials for thinning practices and growth measurements. Planting of indigenous species continued, though this was set back in 1925 when a drought destroyed many thousands of half a million planted seedlings principally of *Juniperus, Milicia, Commiphora, Prunus, Ocotea usambarensis* and *Syzygium cordatum*[84]. Timber concessions continued to be granted in the early 1920's in both the West Usambara and Kilimanjaro forests, though the latter were not formerly gazetted in a reserve until 1929[85].

Forest Policy increasingly became concerned with protection of water supplies and sustainable utilisation. On 14th April 1925 the Forest Policy submitted to, and accepted by, the Governor included[86]: "Examination of existing forest resources on both provisionally reserved and unreserved areas and determination of the areas suitable for permanent reservation, for forest production as well as for regulation of stream-flow, conservation of water supplies and maintenance of soil fertility. Regulation of forest exploitation within reserves to the permissible quantities having in view a sustained yield.... Protection of the forest estate from damage by fire, man and beast." This policy was reflected in the government gazette of 1927[87]: "If in the opinion of the Conservator of Forests felling of forest will jeopardise any natural water supply, the Conservator of Forests may as regards the catchment area of that water supply (1) prohibit any and all felling, (2) limit felling by area or quantity." Forests on private lands could be protected and managed by the Governor under the 1921 Forest Ordinance and in 1927 an amendment to the Ordinance enabled such forests to be scheduled in the same way as reserves on public lands[88]. The private forests of the Eastern Usambara were scheduled under this rule[89] to prevent excessive felling in order to protect water catchments. The Forest of Ordinance of 1921 was replaced in 1933 by a new set of rules that included both protection of water supplies and control of deforestation on private land[90]. The 1933

Ordinance continued to restrict access and extraction of forest produce from reserves (Article 3):

> 3." No person shall, in any forest reserve do any of the acts or things following:-
> (1) Fell, cut, take, work, burn, injure or remove any tree or forest produce;
> (2) Squat or reside or build any hut or cattle enclosure;
> (3) Fire any grass or undergrowth;
> (4) Graze or depasture cattle;
> (5) Clear, cultivate, or break up land for cultivation or any other purpose;
> (6) Enter any part of a forest reserve, which, by order of the Governor, may be closed to trespassers; or
> (7) Deface, injure or remove any forest boundary mark."

Importantly, in contrast to recognition of customary rights and access to "natural fruits" embodied in the 1923 Land Ordinance, provision for obligatory free issue of forest products from forest reserves to natives was omitted. Collection of timber and other forest produce, cultivation, residence and grazing of stock were permissible only by permit.

In the mid-1930's, choice of plantation species shifted from indigenous trees to exotics for several reasons, notably because indigenous species were slow growing in comparison to pines, cypresses, eucalypts and black wattle; and reforestation projects with these exotics in South Africa and Kenya had proved very successful. The shift reflected a focus on timber as the most important forest product to the exclusion of local needs as plantations of exotics do not contain the many different non-timber forest products used by local people. In 1935 an arboretum was established at Kigogo in the southern Udzungwa where 55 species (all but 4 exotics) were planted up to 1939. In 1936 a school for training local people in forestry was started at Arusha. The forest policy at this time was aimed at bringing local people into closer contact with forestry and encouraging them to protect forest on public lands and to reforest denuded areas[91].

Forest reserves continued to be gazetted for the purposes of watershed protection, and by 1942 most of the gaps left in the pattern of German reserves as a result of the Wahehe resistance and Maji-Maji rebellion had been filled. The most important being: Uzungwa scarp (208 sq. km), Kisinga-Rugaro (143 sq. km), Ugwama (260 sq. km) and Nyumbanito (62 sq. km) on the eastern escarpment of the Udzungwa mountains (later to become the West Kilombero Scarp reserve); and the Livingstone escarpment (265 sq. km) and Poroto ridge (94 sq. km) in the southern highlands.

The exigencies of the Second World War resulted in a large increase of timber extraction[92], emphasising the role of forest reserves as a resource for the State rather than a benign form of protection for posterity. Timber consumption doubled and the volume of exports more than quadrupled. Exports of minor forest products, such as wild rubber, tanning bark and gums, more than doubled in value. Half of the increase in timber production was for railway sleepers and came from woodland outside reserved forests, but extraction in closed moist forests also increased. For example in early 1941 the Shume concession in the West Usambara mountains was worked for four months by the 39th Forestry

Company of the South African Engineering Corps. Increased forestry activity in the war resulted in the unusual state of affairs of forestry revenue exceeding expenditure[93], but the forest department remained committed to protection, with the forest policy redefined in the 1945 Forest Department annual report:

> "So to protect, to preserve and where necessary to restore the forest cover as to provide the maximum insurance against depletion of water supplies, degradation of the soil and deterioration of the local climate and by correct land usage reducing to a minimum damage by dangerous and unnecessary use of fire and, while bearing in mind the needs of the rural economy, preventing whenever possible the cultivation of unsuitable sites."

In a Government White Paper presented to the Legislative Council in 1953 the importance of forests for water catchment and sustainable supplies of forest produce became part of a Government Forest Policy[94]. The policy explicitly recognises the importance of forests in supplying benefits for local people, but the means of achieving this policy was paradoxical as it required further alienation of forest resources:

> "To demarcate and reserve in perpetuity, for the benefit of present and future inhabitants of the country, sufficient forested land or land capable of afforestation to preserve or improve local climates and water supplies, stabilise land which is liable to deterioration, and provide a sustained yield of forest produce of all kinds for internal use and also for export."

The result was an increase in forest reservation with an aim of creating a permanent forest estate of at least 8% of the land area[95]. Large areas of catchment reserves were gazetted in the 1950's. The West Kilombero group of reserves in the Udzungwa totalled 2930 sq. km, a great swathe of land covering areas that were affected by social disruption caused by the Maji Maji rebellion and other events during the German administration. Other large areas included: Ukwiva (546 sq. km) in the Rubeho, and Kipengere range (340 sq. km) in the southern highlands. In Maasai-land a reserve was gazetted on the mountain of Essimingor (60 sq. km) but although plans were drawn up for Gelai (245 Sq km), Kitumbeine (130 sq. km), Longido (20 sq km), Loliondo, Ngelai and others, they remain ungazetted to the present day.

In 1951 a central silvicultural research station was set up in Amani and moved to Lushoto in 1952 where an arboretum of 180 species, mostly exotics, was established. Further arboreta were established at Mombo and Rondo. Research on natural forest regeneration continued with studies on *Ocotea* regeneration on Kilimanjaro, *Milicia* on the Rondo Plateau, *Olea* on Meru, *Juniperus* on the West Usambara, and *Podocarpus* at Minziro[96].

Large scale planting was resumed after the war, again with timber as the main forest product, mainly using fast-growing pines (*Pinus patula, P. radiata*) and cypress (*Cupressus lusitanica*). It was considered that commercial, non-destructive, harvesting of natural forests gave yields too low to pay for their protection and so logged natural forests on Meru, Kilimanjaro and the West Usambara were replanted with exotics[97], ignoring importance of non-timber forest products to local communities. It was also thought that logging practises at this time left much to be desired. There was nothing to stop sawmills removing all good quality trees of valuable species whilst leaving inferior species; and there was no requirement to protect natural regeneration of these species. Thus the proportion of inferior species increased whilst the quantity and quality of better species decreased[98]. Nonetheless there was still interest in regenerating natural forests with indigenous species[99].

In summary, the British administration continued exploitation of timber in the West Usambara and Kilimanjaro, but also expanded the system of protective reserves started by the Germans. It was considered that reservation and protection of natural forests was the first and most important duty of the Forest Department, and this was the main aim of forest policy throughout the British administration[100]. The second important achievement of the British administration was to substantially expand plantations of fast-growing exotics aiming to meet future timber demands, though in some areas this was done at the expense of natural forest. In lands outside the reserves, "public lands", the forests continued to be managed according to native law and custom with the exception of certain species of valuable timber which were reserved by law and for which a licence was required if they were to be felled[101]. In 1957 a Forest Ordinance became law[102] which was broadly similar to earlier forest law under the British administration, and which was to remain in force until June 2002.

After Independence (1961 to date)

On the first of May 1961 the British administration gave way to internal self-government with independence following on December 9, 1961 and formation of a republic one year later. The United Republic of Tanzania was formed in 1964 by a union with Zanzibar. The move to independence entailed many changes, but there were some important significant constitutional and legal features[103]. Firstly the intention was that Tanganyika would be an African nation, later embodied in the notion of self-reliance in the 1967 Arusha Declaration. Secondly the President of the Republic had exceptional powers, unfettered by the balance of an independent civil service or opposition political parties. The President was also head of the Tanganyika African National Union (T.A.N.U.) which was to become the sole political party in 1965, the opposition African National Congress being declared an unlawful society and its president arrested and imprisoned. Later events have shown the principle of self-reliance to be paradoxical as Tanzania became heavily dependent on donor aid-funding and mired in external debt. By the mid-1980s political changes led the way to multi-party elections in 1995 following pressure from foreign donors. Not only was the international community a major influence in Tanzanian politics, it was also heavily involved in natural resource management. Sometimes these influences were contradictory, for example one donor supported logging whilst in the same area another supported conservation; or one might support strict protection of nature reserves whilst another recommended access of local communities to forest products.

Independence brought some dramatic changes, but surprisingly many laws remained the same. As in the colonial period, all lands continued to be regarded as state owned and the 1923 Land Ordinance remained in force until May 2001,

40 years after initiation of self-government. In contrast, during the late 1960s and early 1970s there was extensive forced movement of people in rural areas under a villagization policy aimed to improve access to services. Although a well intentioned part of socialist Ujamaa ideals, the policy ignored customary rights and caused immense rural disruption. Moreover it was not supported by legislation until the Rural Lands (Planning and Utilization) Act in 1973 and the Villages and Ujamaa Villages (Registration, Designation and Administration) Act, 1975[104]. Customary rights were further downgraded with the enactment of The Regulation of Land Tenure (Established Villages Act in 1992), which has been considered unconstitutional[105].

Given the major political changes following independence it is perhaps surprising that, as with the Land Ordinance, forest legislation and reservation remained unchanged until June 2002. Until then forest law was determined by the 1957 Forest Ordinance (Cap 389) in which rights of access to reserves were broadly similar to the 1933 Forest Ordinance. In a few places there were changes to extent of the forest estate, and perhaps it is the resulting deforestation that prevented more far-reaching revision of forest regulations. In 1963 popular pressure in the densely populated West Usambara mountains resulted in 12000 hectares of the Shume-Magamba forest reserve being degazetted[106]. By 1978 this area was totally deforested and a smaller area of 500 hectares already planted with Cypress and Pines had to be given up in 1971-72 after taungya farmers refused to leave the area[107]. When public pressure for release of the privately owned Mazumbai forest in the West Usambara for clearance and cultivation became too great in the 1960's, the owner donated it to the University of Dar es Salaam for research and teaching[108]. In the Uluguru mountains the area of forest has declined from 300 sq km in 1955 to 230 sq km today, with deforestation being on land under village control and the remaining forest being now almost completely confined to government catchment forest reserves[109].

The forest development plan for 1960-1964 allocated less than 10% of expenditure on field work to natural regeneration, though about 6000 ha of high forest were classified as regenerating or under treatment compared to the 10500 ha which were under state plantation[110]. *Ocotea* regeneration experiments were established at Shume-Magamba in the West Usambara mountains in 1962[111] and *Cephalosphaera* regeneration plots in the East Usambara[112]. In the late 1960's research policy showed renewed interest in the regeneration of natural forests[113], with reassessment of natural regeneration plots established in the 1950's[114]and a re-examination of the potential of indigenous species[115]. The emphasis of forest policy in the 1960's remained on protection[116]:

> (a) To set aside as permanent reserves sufficient land throughout the Republic to ensure:
> (i) Conservation of water supplies.
> (ii) Preservation of local climate and stabilisation of land.
> (iii) Maintenance of adequate supplies of forest produce, both for the present and for the future.

In 1969 the centrally controlled forest department was placed under regional and district authorities whose short term aims to meet financial targets by exploiting forest reserves were in many cases not compatible with the long term protective and sustainable management aims of the foresters. The Catchment Forest Projects, where reserves important for catchment purposes were brought back under central government control, were initiated following the Iringa Declaration on Agriculture (Siasa ni Kilimo - Politics is Agriculture). It was directed that moist forests be managed and improved as catchments and a national project with this aim was established in 1976 covering Morogoro, Tanga, Kilimanjaro and Arusha regions[117]. Iringa Region was not included in the Catchment Project due to lack of funds, despite discussion and recommendations that it should be done[118]. In 1984, the Norwegian donor agency NORAD were approached to assist in establishing management plans for the most important catchment forests.

During the 1980's extraction of Eastern Arc moist forest timber increased to the point where very few areas were not exploited. Large scale mechanical logging was carried out by Sikh sawmills in the submontane forests of the East Usambara mountains with assistance from the Finnish donor agency FINNIDA[119] and lowland forests of the eastern Udzungwa mountains were logged by a joint Chinese-Tanzanian timber company UWOP[120]. Pit sawyers operated in even the most remote forests where planks were head-loaded over a period of up to two days out of the forest to where they could be loaded onto lorries. Pit sawyers were also working ahead of both the East Usambara and Udzungwa mechanised logging operations, taking out the best trees and then using logging operation access roads to remove planks.

The NORAD supported Catchment Forestry Project was initiated in 1988 to administer 110 Catchment Forest reserves covering 685000 ha in Morogoro, Tanga, Kilimanjaro and Arusha regions. It aimed to improve management of the reserves and initially focused on improving performance of the existing field organisation in addition to carrying out mapping and surveying using remote sensing. A pilot project was also started to investigate social approaches to forest management[121]. A number of other donor supported projects aimed at improving management in catchment forests were established in the late 1980's. The Danish donor agency DANIDA supported HIMA project in Iringa District was initiated 1989 and included a moist forest management component, and supported a Regional Catchment Forest Officer from 1991. Following an international outcry by conservationists over logging activities in the East Usambara forests, logging was stopped and instead FINNIDA supported the East Usambara Catchment Forest project which commenced in 1990 with four components: nature conservation, catchment forestry, plantation forestry, and institutional support[122]. Also in the East Usambara an EEC supported, IUCN administered, East Usambara Agricultural Development and Environmental Conservation Project (EUADEC) was established in 1987 under the Ministry of Agriculture and Livestock Development, together with the Ministry of Lands, Natural Resources and Tourism. The objective of the EUADEC was to improve villager's living conditions through improving agriculture and so assist in forest conservation by developing sustainable cultivation[123]. Until the FINNIDA supported logging operation in the East Usambara was halted there was a remarkable contradiction in that one donor funded project was trying to conserve forests by working with villagers and a few kilometres away another was logging them with sophisticated imported machinery. Catchment forests were also covered by EEC supported Forestry and Agroforestry projects in: Mbeya Region, Songea District and Mbinga District, under the Ministry of Agriculture and Livestock Development. Other projects with a catchment forest component were KIDEP[124] in Kigoma, RUDEP[125] in Rukwa (both answerable to the Prime Minister's Office), and the German GTZ supported SECAP[126] in the West Usambara mountains which was primarily a soil erosion control and agroforestry project. GTZ also had activities in the Pare mountains, where the Japanese JICA[127] have also been working.

General concern over unsustainable tropical forest management in the 1980's led the UN and World Bank to launch a global Tropical Forest Action Plan in 1985 overseen by the FAO[128]. The action plan was criticised for being inconsistent, failing to align activities in different sectors, not having broad participation, and not integrating issues such as land tenure, agricultural development and rural poverty[129]. In Tanzania this planning process was implemented as the Tanzania Forestry Action Plan (TFAP) in 1988. It was a broad-based review, and so did not suffer from some of the problems of the original FAO action plan, and was completed and adopted by the Tanzanian government in 1989. The TFAP was a multi-sectoral review concentrating on forest resources, but also including wildlife and conservation, with support from FINNIDA (who also supported a revision of the TFAP in 2001[130]). The Conservation of Ecosystems and Genetic Resources sector review emphasised conservation of moist forests with the Eastern Arc as the central project[131]. The 1989 TFAP stated that conservation of biological resources should be considered at three levels:

- conservation of the various ecosystems
- conservation of species
- conservation of genetic variability within species

In emphasising that conservation implies sustainable utilisation, the TFAP gave the following criteria for assessing sustainability, which contain both conservation and fulfilment of basic needs:

- The maintenance of the harvest of all products at sustainable levels.
- The maintenance of essential ecological processes, such as watershed regulation and nutrient cycling.
- The maintenance of biological diversity at all three levels listed above.
- The satisfaction of the basic needs of people living in and around the forests.

As recommended in the 1989 TFAP, the importance of biodiversity conservation in the forests was emphasised in 1992 with the creation of the Udzungwa Mountains National Park[132]. Covering the areas in whole or in part of the Mwanihana, Matundu, Iwonde, Nyanganje and West Kilombero Scarp forest reserves, the national park consolidated protection of forests initiated in the 1950s through gazettement of the West Kilombero group of reserves; which themselves had followed on from depopulation as a result of German military rule at the end of the 19th century. Inevitably there were conflicts of interest. In the same year that the Udzungwa National Park was gazetted, a study on forest management in Iringa District near the park discovered that villagers considered the central problem in forest management to be the lack of consensus between forest conservation regulations and villager's activities[133]. The villagers view of restrictive forest regulations embodied in the 1957 Forest Ordinance, and still in place in the 1990s, was that constraints of the law resulted in long distances being walked to collect firewood and water, prevented collection of honey, prevented traditional healers from collecting herbs and prevented traditional rituals being performed appropriately.

In the late 1980s the Forest Division started revising its forest policy. The policy enunciated in 1953, and reviewed in 1963, was still in place despite major social, economic and political changes since that time. Shortcomings of the earlier policy were considered to be: highly generalised statements without guidelines on level of implementation; no emphasis on inter-dependence with other sectors; no emphasis on the need for community forestry; no emphasis on the role of forestry as a rural employer; and no emphasis of people's participation[134]. The new policy was approved in 1998 and represented a significant change from the earlier British policy, notably for its specific recognition of biodiversity and the involvement of local communities in forest management[135]:

> POLICY STATEMENT (15)
>
> New forest reserves for biodiversity conservation will be established in areas of high biodiversity value. Forest reserves with protection objectives of national strategic importance may be declared as nature reserves.
>
> POLICY STATEMEMT (16):
>
> Biodiversity conservation and management will be included in the management plans for all protection forests. Involvement of local communities and other stakeholders in conservation and management will be encouraged through joint management agreements.
>
> POLICY STATEMENT (17):
>
> Biodiversity research and information dissemination will be strengthened in order to improve biodiversity conservation and management.

The 1998 Forest Policy was supported by enactment of a new Forest Act in 1992[136]. The new act retains some features of the earlier British Forest Ordinances, in that it is highly restrictive of what activities can be carried out in forest reserves, for example Article 26 is essentially the same in spirit to Article 3 of the 1933 Ordinance and Article 15 of the 1957 ordinance in that no cutting of vegetation, cultivation, building, grazing, or hunting is allowed. Restrictions have been updated to cover both tourism and research, neither of which were an important issue in 1933 and 1957. However, the new forest act differs in some important respects, particularly with respect to conservation of biodiversity, community involvement in management, and capturing value that the international community places on the forests. Conservation of biodiversity is mentioned throughout the act, but is explicitly covered in Part 8 "Conservation of Trees, Wild Plants and Wild Animals" (Articles 65 to 69). However, the conservation component of the Act is not restricted to lists of protected fauna and flora as property rights are also asserted. Influence of Articles 3 and 15 the Convention on Biological Diversity, concerning sovereignty over biodiversity is apparent in Article 69 of the Forest Act:

> 69-(1) All biological resources and their intangible products, whether naturally occurring or naturalised within forests including genetic resources belongs to the government in accordance with Article 27 of the Constitution and shall be conserved and utilised for the people of Tanzania in accordance with the provisions of this Act and any other written law on biological resources.
> (2) The transfer of any biological resources, their derivative products or intangible components from forest shall not operate so as to extinguish the sovereignty of Tanzania over those resources.
> (3) Without prejudice to the provisions in subsection (1), the right to determine and regulate access to genetic resources is vested in the Government and may be exercised by the competent authority in consultation with relevant organisations in accordance with the provisions of this Act or any other written law on biological resources.

Influence of the Convention on Biological Diversity is apparent in other sections of the Forest Act, for example a requirement for Environmental Impact Assessment, as requested by Article 14 of the CBD is covered in Article 18 of the new Act:

> 18.-(1) In any proposed development in a forest reserve, private forest or sensitive forest area including watersheds to which this section applies, whether that development is proposed by, or is to be implemented by, a person or organisation in the public or private sector, the proposer of the development shall prepare and submit to the Director an Environmental Impact Assessment of the proposed development from independent consultants selected from a list approved by the Government of the United Republic of Tanzania and, notwithstanding any other law to the contrary, no development to which this section applies shall be commenced unless and until an Environmental Impact Assessment has been completed to the satisfaction of the Director.

Community involvement in forest management is regulated in sections on village land forest reserves (Articles 32 to 41) and community forest reserves (Articles 42 to 48), but most significantly under Article 27, the functions of managing a national forest reserve can be undertaken by a village council or community group. The article also allows individuals, private sector organisations and non-governmental organisations to manage forest reserves, but only if cooperation with stakeholders within the forest reserve is established through consultation. However, if a community takes over management of a reserve, then it will have to observe all the other provisions in the Act which includes restrictions on harvesting forest products and protecting listed plants and animals.

Capturing value placed on Tanzania's biodiversity rich forests is covered in Part 10 "Financial Provisions and Establishment of a Fund" (Articles 77 to 83) which regulates establishment of a fund which is to be used for promoting awareness of the importance of protection, development and sustainable use of forest resources; promoting community forestry; funding research; assisting

Tanzania to benefit from international funds directed towards conservation; and assist participation in public debates on forestry. The fund is supported by a levy on fees and royalties payable under the Act and:

> 79 (2) (c) grants, donations, bequests or such sums contributed by any private individuals, corporate bodies, foundations, or international organisations or funds within or outside the Country.

Development of the new Forest Act has taken many years and the international community has been influential in determining Tanzania's forest policy and in supporting forestry activities. For example, in implementing the Convention on Biological Diversity, Tanzania has drawn up a national biodiversity strategy with assistance from NORAD[137], and guidelines for community based forest management have been supported by SIDA[138]. The Global Environment Facility are also involved in Tanzania forest conservation activities through an active project on mountain forests near the borders of Tanzania, Kenya and Uganda, and have an Eastern Arc Forest Conservation and Management Project in the pipeline[139]. In the Uluguru mountains, DANIDA is supporting the Uluguru Mountains Biodiversity Conservation Project (UMBCP) which was initiated in 1999. The project is managed by an NGO, the Wildlife Conservation Society of Tanzania in collaboration with the Regional and District Natural Resources Offices of the Tanzania Government, the Catchment Forest Project of the Ministry of Natural Resources and Tourism, and the Uluguru Mountains Agriculture Development Project[140]. DANIDA is also supporting the Udzungwa Mountains Forest Management and Biodiversity Conservation (UMFM) which started in 1999 and aims to develop joint forest management plans in line with the 1998 forest policy.

In summary, independent Tanzania retained the pattern of forest reserves built up during the German and British administrations, and also continued to use the British 1957 Forest Ordinance until June 2002. This lead to an unusual situation in that on one side of a forest reserve boundary it was possible to stand in under the rules of post-independence African Socialist Ujamaa Tanzania, and on the other side in the law of the British Empire administration. Continued alienation of forest resources from villagers living near the reserves has lead to conflict between the needs of local people and the national desire for forest protection. Inclusion of villagers in formulation of management plans as laid out in the new 1998 forest policy should enable forest conservation and community development go hand in hand, and this policy is now supported by the 2002 Forest Act. But although there have been some small scale attempts to create village forest reserves[141], deforestation in the past on village lands in the Usambara and Uluguru mountains, and the need to bring catchment forests back under central government in 1976 following decentralisation in 1969, suggests that there will be conflicts of interests in preparation of joint forest management agreements. The State and international community wish to conserve forests in perpetuity, where-as local communities want more agricultural land now.

CONCLUSION

Throughout the management history of Tanzanian forests there has been a tension between State protection for exploitation and local community exploitation for consumption or conversion. Prior to establishment of national colonial governments, regional authorities limited forest loss to cultivation and fire by restricting access through social norms. International involvement in Tanzanian forest management started with development of a forestry department under German administration, bringing with it further social norms on forest conservation. This resulted in alienation of large areas of land into a national forestry estate, a process which has continued to the present day. The "natural fruits" (in the sense of the 1923 Land Ordinance) in the form of timber from the forest reserves, have been gathered by State licensed enterprises under German, British and Independent governments. Traditional rights of local communities to forest products are emphasised in the 1923 Land Ordinance, the 1992 Biodiversity Convention and the 1998 Tanzanian Forest Policy, but benefits have by-passed people who live adjacent to the reserves. With recognition now of internationally recognised biodiversity values in addition to national priorities for watershed protection, finding a balance between conservation and equitable access has become more difficult, particularly when past experience has shown that when villagers are given control of forested land it is converted to agriculture.

Is it possible to have both ecology and equity? Tanzanian forests are extra-ordinarily rich in plant and animal species of restricted distribution. This unique ecological nature is irreplaceable and makes the forests globally important. But does this mean that customary rights of access to forest resources should be denied to those who traditionally utilised forest products or had access to forest land? Laws introduced in the 20th century were not initially put in place to deny local access, but to prevent over-exploitation by incoming colonists. However, the laws consolidated State power and gradually local rights were eroded to better protect natural resources. Initially this was for direct economic national interests, such as maintenance of water catchments and timber exploitation. More recently it has been for biodiversity conservation as a "common concern of humankind"[142]. Now that community-based management of the forests is again coming to the fore, a problem with returning to customary management systems is that social values of the forests have changed from the time when local ownership ensured conservation. Forests were once part of a broad-based natural resource management system under communal ownership that benefited from richness in habitat diversity. An unpredictable and changeable climate meant that it was an advantage for societies to have a range of ecological options, so it made sense to conserve forests under customary law. In the present day, it is only the most remote communities that still lead this type of lifestyle. For most areas, especially those with a high population density such as the Usambara and Uluguru mountains, value of the forest lies not in the vegetation but in the land on which it stands. Thus, when forests are returned to customary ownership, as in the case of the Shume-Magamba forest in 1963, the forest is removed and the land used for agriculture. Traditional forest protection from over-exploitation, such as historically in Kagara, in the coastal *kaya* forests or some of the Eastern Arc forests important for *tambiko*, often has a spiritual or religious aspect. When people migrate, these rituals become less important, for example Wachagga moving from the highly populated slopes of Mt Kilimanjaro to the less populated Kanga mountain in the southern Nguru are not concerned about the indigenous Wazigua respect for the forest and are keen to plant coffee on forest land. Without the social constraint of taboo or legal constraint of forest reserves, the incentive is to convert forests to alternative land uses. This is straight-forward economic sense to the individuals concerned, returns from agriculture are greater than the individual's share of social benefits from existence of the forests.

The principles of cost-benefit analysis suggest that if values accruing from forest conservation exceed benefits derived from alternative land use and the costs of community forest management, then sustainable forest management as envisaged in the 1998 Forest Policy and 2002 Forest Act should be possible. There has been much international involvement in development of Tanzania's forest policy, helping to cover costs of implementation and provide benefits to local communities is an important future role, and one that is also covered in the Convention on Biological Diversity under Article 20 (2):

> "The developed country Parties shall provide new and additional financial resources to enable developing country Parties to meet the agreed full incremental costs to them of implementing measures which fulfil the obligations of this Convention...".

The 2002 Forest Act contains provision for establishment of a fund by which these financial resources can be channelled. It remains to be seen if, firstly these resources will be made available, and secondly, whether they will be passed on to the villagers who historically would have had control over the forests and the natural fruits they contain.

Acknowledgements

A shorter version of this chapter was originally published in the 1992 Udzungwa Forest Management Project report. I am grateful to support from Danida during preparation of the Udzungwa report and permission to use material contained in the report. I am grateful for comments on an earlier draft of the chapter from Neil Burgess, Kevin Gray, David Ockwell and Harald Sippel.

---

[1] Centre for Ecology Law and Policy, Environment Department, University of York, York YO10 5DD, England
[2] Convention on Biological Diversity, 8 June 1992. Tanzania became a signatory in 1992 and ratified the convention in 1996.
[3] Article 3 recognises sovereign rights of States "States have, in accordance with the Charter of the United Nations and the principles of international law, the sovereign right to exploit their own resources pursuant to their own environmental policies, and the responsibility to ensure that activities within their jurisdiction or control do not cause damage to the environment of other States or of areas beyond the limits of national jurisdiction".
[4] Article 8 on *in situ* conservation requires the signatories to "8(a) Establish a system of protected areas or areas where special measures need to be taken to conserve biological diversity", whilst at the same time under "8(j) Subject to its national legislation, respect, preserve and maintain knowledge, innovations and practices of indigenous and local communities embodying traditional lifestyles relevant for the conservation and sustainable use of biological diversity and promote their wider application with the approval and involvement of the holders of such knowledge, innovations and practices and encourage the equitable sharing of the benefits arising from the utilization of such knowledge, innovations and practices."
[5] Griffiths, J.F. 1972. 'Climate'. In W.T.W. Morgan (ed.) *East Africa: its people and resources*. (Oxford University Press, Nairobi). Pp. 106-125
[6] Kerfoot, O. 1968. 'Mist precipitation on vegetation'. *For. Abstr.* 29: 8-20.
Pócs, T. 1976. 'The role of the epiphytic vegetation in the water balance and humus production of the rain forests of the Uluguru mountains, East Africa'. *Boissiera* 24: pp. 499-503
Smiet, F. 1987. 'Tropical watershed forestry under attack'. *Ambio* 16: pp. 156-158
[7] Lovett, J.C. 1990. 'Classification and status of the moist forests of Tanzania' *Mitt. Inst. Allg. Bot. Hamburg*,. 23a: pp. 287-300
Lovett, J.C. 1996. 'Elevational and latitudinal changes in tree associations and diversity in the Eastern Arc mountains of Tanzania'. *Journal of Tropical Ecology* 12(5): pp. 629-650.
Lovett, J.C. 1998. 'Continuous change in Tanzanian moist forest tree communities with elevation.' *Journal of Tropical Ecology* 14:7: pp. 19-722.
[8] Lovett, J.C. 1988. 'Endemism and affinities of the Tanzanian montane forest flora'. *Monogr. Syst. Bot. Missouri Bot. Gard.* 25: pp. 591-598.
Lovett, J.C. 1990. Op. Cit.
[9] Lovett, J.C. 1988. Op. Cit.
[10] Rodgers W.A. and K.M. Homewood. 1982. 'Species richness and endemism in the Usambara mountain forests, Tanzania'. *Biol. J. Linn. Soc.* 18: pp. 197-242.
Lovett, J. C. and Wasser, S. K. (Eds.). (1993). *'The biogeography and ecology of the forests of eastern Africa'*. (Cambridge University Press).
[11] Lovett, J.C. 1988a. 'Practical aspects of moist forest conservation in Tanzania'. *Monogr. Syst. Bot. Missouri Bot. Gard.* 25: pp. 491-496.
[12] Brenan, J.P.M. and P. Greenway. 1949. 'Check-lists ofthe forest trees and shrubs of the British Empire. No. 5. Tanganyika Territory. Part II'. (Imperial Forestry Institute, Oxford).
[13] FAO, 1983. Food and fruit-bearing forest species. 1: Examples from eastern Africa. FAO Forestry Paper 44/1: 1-172.
[14] Bridson, D.M. and B. Verdcourt. 1988. 'Rubiaceae (Part 2)'. In R.M. Polhill (ed.) *Flora of Tropical East Africa.* (Balkema, Rotterdam).
[15] Baatvik, S.T. 1993. 'The genus Saintpaulia (Gesneriaceae) 100 years: History taxonomy, ecology, distribution and conservation'. 2(1):*Fragm. Flor. Geobot. Suppl.* 97-112.
Johansson, D.R. 1978. '*Saintpaulias* in their natural environment with notes on their present status in Tanzania and Kenya'. 14: *Biol. Conserv.* 45-62.
[16] Hilliard, O.M. and B.L. Burtt. 1971. '*Streptocarpus an African Plant Study*'. (University of Natal Press, Pietermaritzberg).
[17] Grey-Wilson, C. 1980. '*Impatiens* of Africa'. (Balkema, Rotterdam).
[18] Polhill, R.M. 1968. 'Tanzania', In I. Hedberg and O. Hedberg (eds.) *Conservation of vegetation south of the Sahara. Acta Phytogeogr. Suec.* 54: pp. 166-178.
[19] National Research Council. 1980. '*Research priorities in tropical biology*.' National Academy of Sciences, Washington, D.C.

[20] Droop, S.J.M. 1988. 'The work of the IUCN Conservation Monitoring Centre at Kew, and the forthcoming centres for plant diversity book'. *Monogr. Syst. Bot. Missouri Bot. Gard.* 25: pp. 465-469.
Davis S.D., V.H. Heywood et al. (editors) (1994). 'Centres of Plant Diversity. A guide and strategy for their conservation'. Volume 1. Europe, Africa, South West Asia and the Middle East. (IUCN, Cambridge).
[21] Mittermeier. R. A. and N. Myers, et al. 1998. 'Biodiversity hotspots and major tropical wilderness areas: approaches to setting conservation priorities'. *Conservation Biology* 12: pp. 516-520.
Myers, N. 1990. 'The biodiversity challenge: Expanded Hot-Spots analysis'. 10: *The Environmentalist* 243-256.
Myers, N., R.A. Mittermeier, et al. 2000. ' Biodiversity hotspots for conservation priorities'. 403: *Nature* 853-858.
[22] Bensted-Smith, R. and T.H. Msangi. 1989. 'Report on the conservation of ecosystems and genetic resources. Tanzania Forestry Action Plan', United Republic of Tanzania.
[23] Lovett, J.C. and T. Pócs. 1993. 'Assessment of the condition of the Catchment Forest Reserves, a botanical appraisal'. Catchment Forestry Report 93.3. Forestry and Beekeeping Division, Ministry of Tourism, Natural Resources and Environment, Dar es Salaam.
[24] Norweign Agency for Development Co-operation.
[25] Finnish International Development Agency.
[26] European Economic Community.
[27] Gesellschaft für Technische Zusammenarbeit.
[28] Danish International Development Agency.
[29] Lovett, J.C. 1992. 'Main report of the Udzungwa Forest Management Project Identification Mission'. Pp. 108 plus 16 technical annexes. Forest and Beekeeping Division, Ministry of Tourism, Natural Resources and Environment, and DANIDA. Dar es Salaam.
[30] Lovett, J.C. and G.W. Norton. 1989. 'Afromontane rain forest on Malundwe Hill, Mikumi National Park, Tanzania'. *Biol. Conserv.* 48: pp.13-19.
[31] Redhead, J.F. 1981. 'The Mazumbai Forest: an island of lower montane rain forest in the West Usambara'. *Afr. J. Ecol.* 19: pp. 195-199.
[32] Rodgers W.A. and K.M. Homewood. 1982a. 'Biological values and conservation prospects for the forests and primate populations of the Uzungwa mountains', Tanzania. *Biol. Conserv.* 24: pp. 285-304.
[33] Lovett, J.C. and D.C. Moyer. 1992. 'A new nature reserve in the Eastern Arc mountains', Tanzania. *Oyrx* 26: pp. 115-118.
[34] Schulman, L., L., Junikka, A. Mndolwa, and I. Rajabu, 'Trees of the Amani Nature Reserve, NE Tanzania'. Ministry of Natural Resources and Tourism, Tanzania.
[35] Kabuye, C.H.S. 1988. 'East African plants used in basketry'. *Monogr. Syst. Bot. Missouri Bot. Gard.* 25: pp. 351-361.
[36] Hall, J.B. and W.A. Rodgers. 1986. 'Pole cutting pressure in a Tanzanian Forest'. *For. Ecol. & Manage.* 14: pp. 133-140.
[37] Iliffe, J. 1979. *A modern history of Tanganyika.* African Studies Series, 25. (Cambridge University Press).
[38] Brenan and Greenway 1949 Op. Cit.
[39] Mugasha, A.G. 1978. ' Tanzania Natural Forests Silvicultural Research: Review Report'. *Tanzania Silviculture Technical Note (New Series)* 39: pp. 1-41.
Mugasha, A.G. 1983. 'The silviculture of Tanzanian indigenous tree species. IV Cephalosphaera usambarensis'. *Tanzania Silviculture Technical Note (New Series):* 58.
Willan, R.L. 1965. 'Natural regeneration of high forest in Tanganyika'. *E. Afr. Agric. For. J.* 31: pp. 43-53.
[40] Hamilton, A.C. 1988. 'Conservation of the East Usambara forests with emphasis on biological conservation'. 28: *Symb, Bot. Ups.* Pp. 244-254.
Hamilton and R. Bensted-Smith (eds.). *Forest Conservation in the East Usambara mountains, Tanzania.* (IUCN, Gland). pp.75-78.
[41] Rodgers, W.A. 1993. 'The conservation of the forest resources of eastern Africa: Past influences, present practices and future needs'. In J.C. Lovett and S.K. Wasser (eds.) *Biogeography and ecology of the forests of eastern Africa.* (Cambridge University Press). pp. 283-331.

Burgess, N.D. and S.B. Mbwana. 2000. Forestry, In *Coastal Forests of eastern Africa*. (IUCN, Cambridge and Gland). pp. 263-279.
[42] Pilbeam, D. 1985. 'Patterns of hominoid evolution'. In E. Delson (ed.) *Ancestors: The hard evidence*. (Alan R. Liss), New York. . pp. 51-59
[43] Brain, C.K. 1993. 'Swartkrans: A cave's chronicle of early man'. Transvaal Museum Monograph No. 8. Transvaal Museum, Pretoria.
Brain, C.K. and A. Sillen, 1988. 'Evidence from the Swartkrans cave for the earliest use of fire', *Nature*, 336: pp. 464-466.
[44] Bird, M.I. and J.A. Cali, 1998. 'A million-year record of fire in sub-Saharan Africa'. *Nature* 394: pp. 767-769.
[45] Vansina, J. 1995. 'A slow revolution: farming in subequatorial Africa'. *Azania* 29-30: pp. 15-26.
[46] Hamilton, A.C., D. Taylor and J.C. Vogel. 1986. 'Early forest clearance and environmental degradation in south-west Uganda'. *Nature* 320: pp. 164-167.
[47] Perrott, R.A. 1987. 'Early forest clearance and the environment in southwest Uganda'. *Nature* 325: pp. 89-90.
[48] Schmidt, P.R. 1997. 'Archaeological views on a history of landscape change in East Africa'. *Journal of African History* 38: pp. 393-421
[49] Schmidt, P.R. 1990. 'Early exploitation and settlement in the Usambara mountains', In A.C. Hamilton and R. Bensted-Smith (eds.) *Forest Conservation in the East Usambara mountains, Tanzania* (IUCN, Gland) pp. 75-78
[50] Iliffe 1979 Op. Cit.
[51] Pouwels, R.L. 1999. 'Review article: East African coastal history'. *Journal of African History* 40: pp. 285-296.
[52] Clarke, G.P. and N.J. Karoma. 2000. 'History of anthropic disturbance'. In Burgess, N.D. and Clarke, G.P. (eds.) *Coastal Forests of Eastern Africa*. (IUCN, Cambridge & Gland). pp. 251-261
[53] Hawthorne, W.D. 1993. 'East African coastal forest botany', in Lovett, J.C. and Wasser, S.K. (eds.) *Biogeography and ecology of the rain forests of eastern Africa*. (Cambridge University Press, Cambridge). pp.57-99
[54] Mutoro, H.W. 1995. 'Tana ware and the kaya settlements of the coastal hinterland of Kenya'. *Azania* 29-30: pp. 257-260.
[55] Horton, M. 1996. 'Shanga. The archaeology of a Muslim trading community on the coast of East Africa'. Memoirs of the British Institute of East Africa 14. London: The British Institute in Eastern Africa
[56] Sippel, H. 1996. 'Aspects of colonial land law in German East Africa'. In R. Debusmann and S. Arnold (eds.) *Land law and land ownership in Africa*. (Bayreuth African Studies 41, Bayreuth University). pp.3-38
[57] Clutton-Brock, T.H. and J.B. Gillett. 1979. 'A survey of forest composition in Gombe National Park, Tanzania'. *Afr. J. Ecol.* 17: pp. 131-158.
[58] Kielland, J. 1978. 'A provisional checklist of the Rhopalocera of the eastern side of Lake Tanganyika'. *Tijdschrift voor Entomologies, Deel 121, Afl.* 4: pp. 147-237.
Nishida, T. and S. Uehara. 1981. ' Kitongwe name of plants: A preliminary listing'. *African Study Monographs.* 1: pp.109-131.
[59] Rodgers, W.A., T.T. Struhsaker and C.C. West. 1984. 'Observations on the red colobus (Colobus badius tephrosceles) of Mbizi forest, south-west Tanzania'. *Afr. J. Ecol* 22: pp. 187-194
[60] Parry, M.S. 1962. 'Progress in the protection of stream-source areas in Tanganyika'. *East African Agricultural and Forestry Journal* 27: pp. 104-106.
[61] Parry, M.S. 1962 Op. Cit.
[62] Lovett J.C. and T.R.A. Minja. 1990. 'Notes on a visit to Ukwiva forest, Tanzania.' *East Africa Natural History Society Bulletin*, 20: pp. 4-5.
[63] Greenway, P.J. 1973. 'A classification of the vegetation of East Africa'. *Kirkia,* 9: pp.1-68.
Lovett. J.C. and T.C.E. Congdon. 1989. 'Notes on the Ihangana forest and Luhega forest near Uhafiwa, Uzungwa mountains, Tanzania'. *East Africa Natural History Society Bulletin*, 19: pp.30-31.
[64] Lovett and Pócs 1993 Op. Cit.
[65] Lovett and Minja, 1990 Op. Cit.
[66] Mgaa Sabuni, pers. comm.

[67] Lovett, J.C., and D.W. Thomas. 1988. A report on a visit to the Nguru Mountains, Tanzania.*East Africa Natural History Society Bulletin*, 18: pp.19-22.
[68] James, R.W. and G.M. Fimbo, 1973. 'Customary land laws of Tanzania'. A source book. East Africa Literature Bureau, Nairobi.
[69] Schmidt 1997 Op. Cit.
[70] Twaib, F. 1996. 'The dilemma of the customary landholder: The conflict between customary and statutory rights of occupancy in Tanzania'. In R. Debusmann and S. Arnold (eds.) *Land law and land ownership in Africa*. (Bayreuth African Studies 41, Bayreuth University), pp. 81-112
[71] Iliffe 1979, Op. Cit.,
Sippel, 1996, Op. Cit.
[72] Grant, D.K.S. 1924. 'Forestry in Tanganyika'. *Empire Forestry,* 3: pp. 33-38.
Parry, 1962. Op. Cit.
Lundgren, B. 1975. *Landuse in Kenya and Tanzania*. Royal College of Forestry, Stockholm, Sweden
Schabel, H.G. 1990. Tanganyika forestry under German colonial administration, 1891-1919'. *Forest and Conservation History*34: pp. 130-141.
[73] James and Fimbo, 1973 Op. Cit.
Sippel, 1996 Op. Cit.
[74] Iliffe, 1979. Op. Cit.
[75] Monson, J. 1998. 'Relocating Maji Maji: the politics of alliance and authority in the southern highlands of Tanzania 1870-1918'. *Journal of African History,* 39: pp. 95-120.
[76] Schabel 1990 Op. Cit.
[77] Sunseri, T. 1997. 'Famine and wild pigs: gender struggles and the outbreak of the Maji Maji war in Uzaramo (Tanzania)'. *Journal of African History*, 38: pp. 235-259.
[78] Richter, R.E. 1996. 'Land law in Tanganyika since the British military occupation and under the British mandate of the League of Nations, 1916-1946'. In R. Debusmann and S. Arnold (eds.) *Land law and land ownership in Africa*. (Bayreuth African Studies 41, Bayreuth University). pp. 39-80.
[79] Grant, 1924 Op. Cit.
Troup, R.S. 1940. *Colonial forest administration*. (Oxford University Press).
Parry, 1962 Op. Cit.
Lundgren, B. 1978. 'Soil conditions and nutrient cycling under natural and plantation forests in Tanzanian highlands'. 31: *Department of forest soils, Swedish University of Agricultural Science, Reports in forest ecology and forest soils.* Pp. 1-429.
[80] GN No. 32 of 17th August 1921.
[81] GN No. 86 of 20th August 1926.
[82] GN No. 163 of 9th October 1930.
[83] Forest Department. 1923. Tanganyika Territory. British Empire Forestry Conference, 1923. Forest Department, Lushoto.
[84] Grant, D.K.S. 1925. 'Tanganyika Notes'. *Empire Forestry*. 4: pp. 290.
[85] GN 119 of 25th May 1929.
[86] Forest Department. 1928. Additions to the Statement on Tanganyika Territory (Forests). Made for Forestry Conference, Canada, 1923. Proceedings of the Third British Empire Forestry Conference, Canberra.
[87] GN No. 14 of 29th January 1927.
[88] GN 18 of 15th February 1927.
[89] GN No. 99 of 31st July 1931, GN No. 141 of 12th October 1931
[90] GN No. 119 of 4th August 1933.
[91] Troup 1940 Op. Cit.
[92] Forest Department. 1947. Empire forests and the war: Tanganyika Territory. Appendix in the Proceedings of the fifth British Empire forestry conference. London: His Majesty's Stationay Office.
[93] Burgess and Mbwana 2000, Op. Cit.
[94] Legislative Council of Tanganyika; Sessional Paper No. 1 of 1953.
[95] Forest Department. 1957. Forestry in Tanganyika. Statement prepared by the Forest Department for presentation to the British Commonwealth Forestry Conference, 1957. Government Printer, Dar es Salaam.
[96] Willam, 1965 Op. Cit.

[97] Lundgren, 1975. Op. Cit.
[98] Champion, F.W. 1951. 'Subsidiary silvicultural operations in Tanganyika'. *The East African Agricultural Journal*, July 1951: pp. 19-23.
[99] Champion, 1951. Op. Cit.
[100] Parry, 1962. Op. Cit.
[101] Forest Department. 1962. Progress report 1955-1960 by the Forest Department of Tanganyika. Prepared for the British Commonwealth Forestry Conference 1962. Forest Department, Dar es Salaam.
[102] The Forest Ordinance Cap 389 of 1957.
[103] Cole, J.S.R. and W.N. Denison, 1964. '*Tanganyika. The Development of its laws and constitution. Vol. 12. The British Commonwealth. The development of its laws and constitution*', General editor G.W. Keeton. (London: Stevens & Son).
[104] Twaib 1996 Op. Cit.
[105] Twaib 1996 Op. Cit.
[106] Lovett, J.C. and S.N. Stuart. 2001. Avifauna and vegetation of the Shume*Juniperus* forest of the West Usambara mountains, Tanzania. *Scopus.* 21: pp. 1-14.
[107] Lundgren 1978 Op. Cit.
[108] Lundgren 1978 Op. Cit.
[109] Burgess, N., Doggart, N., Lovett, J.C. 2002. 'The Uluguru Mountains of eastern Tanzania: could we lose their globally significant biodiversity values?' *Oryx* 36: pp. 140-152.
[110] Willan, 1965 Op. Cit.
[111] Kimariyo, P.E. 1971. 'Regeneration of *Ocotea usambarensis* Engl. at Sungwi, West Usambaras'. *Tanzania Silvicultural Research Note,* 21: pp.1-3.
[112] Willan, 1965 Op. Cit.
[113] Mugasha, 1978. Op. Cit.
[114] Kimariyo, 1971. Op. Cit.
Kimariyo, P.E. 1972. 'Initial intensive and medium thinnings increase DBH growth in second growth Camphor regeneration'. *Tanzania Silviculture Research Note.* 26: pp. 1-8
Borota, J. 1969. 'The result of gap planting in natural forests'. *Tanzania Silviculture Research Note.* 13: pp. 1-4
[115] Mugasha 1978 Op. Cit.
[116] Forest Division. 1968. Progress Report 1960-65. Prepared for the Commonwealth Forestry Conference 1968. Forest Division, Ministry of Agriculture and Co-operatives, Dar es Salaam.
[117] Hermansen, J.E., F. Benedict, et al. 1985. *Catchment Forestry in Tanzania, Status and Management.* (Institute for Environmental Analysis, Bo, Norway).
[118] Hermansen et al. 1985. Op. Cit.
[119] Hamilton, A.C. and I.V. Mwasha. 1989. 'History of Resource Utilisation and Management'. In A.C. Hamilton and R. Bensted-Smith (eds.) *Forest Conservation in the East Usambara Mountains, Tanzania.* (IUCN Gland), pp. 45-56.
[120] Ushirika (Partnership) Wood Products; Tanzania held a 51% stake and China a 49% stake.
[121] Mitzlaff, U. 1991. 'Managing the Catchment Forest Reserve. Who is to be involved?' Catchment Forestry Project, Forest and Beekeeping Division. Dar es Salaam
[122] East Usambara Catchment Forest Project. 1990. Project Document 1990-1993. Forestry and Beekeeping Division, MLNRT and FINNIDA.
[123] Stocking, M. and Perkin, S. 1992. 'Conservation-with-development: an application of the concept in the Usambara Mountains, Tanzania'. *Trans. Inst. Br. Geogr. N.S.* 17: pp. 337-349.
[124] Kigoma Development Project.
[125] Rukwa Development Project.
[126] Soil Erosion Control and Agroforestry Project.
[127] Japan International Co-operation Agency.
[128] Food and Agriculture Organisation.
[129] Colchester, M. and Lohmann, L. 1990. The Tropical Forestry Action Plan at the Crossroads. World Rainforest Movement.
Winterbottom, R. 1990. The Tropical Forestry Action Plan after five years. World Resources Institute, Washington, D.C.
WWF. 1990. Reforming the Tropical Forestry Action Plan, a WWF position. WWF, Gland.

[130] Government of Tanzania. 2001. National Forest Programme in Tanzania. Ministry of Natural Resources and Tourism, Forest and Beekeeping Division, Dar es Salaam.
[131] Bensted-Smith, R. and T.H. Msangi. 1989 Op. Cit.
[132] Government Notice No. 39 published on 20/3/92.
[133] Meshack, M.V. and Kaijage, A.S. 1992. 'Logical Framework Approach Workshop Report of 3/4-12/4/1992. Annex 3'. In Lovett, J.C. Udzungwa Forest Management Project. Danida, Dar es Salaam..
[134] Rodgers 1993. Op. Cit.
[135] Ministry of Natural Resources and Tourism. 1998. National Forest Policy. Government Printer, Dar es Salaam
[136] Gazette of the United Republic of Tanzania No. 23 Vol. 83, 7 June 2002.
[137] Government of Tanzania. 1999. National Strategy and Action Plan for Biodiversity. Vice President's Office, Dar es Salaam.
[138] Swedish International Development Agency.
Government of Tanzania. 2001. Community based forest management guidelines. Ministry of Natural Resources and Tourism, Forest and Beekeeping Division, Dar es Salaam.
[139] http://www.gefweb.org/index.html
[140] http://www.africanconservation.com/uluguru/
[141] Wily, L. 1996. Collaborative Forest Management Villagers & Government: The Case of Mgori Forest, Tanzania.
[142] Convention on Biological Diversity Preamable "Affirming that the conservation of biodiversity is a common concern of humankind".

COMMON FEATURES AND TRENDS OF FISHERIES LEGISLATION IN AFRICA

ANTONIO TAVARES

INTRODUCTION

Leaving aside considerations regarding its socio-economic importance, the main "common" characteristics of fisheries in Africa may well be their heterogeneity and changing nature[1]. This is primarily due to the uneven distribution of resources[2]. The fishing industries and management systems based on those resources vary accordingly. Thus, Africa has a few countries with highly sophisticated and efficient fisheries management systems, benefiting from particularly favourable natural conditions, such as Namibia, which compare to the management systems of developed countries. Similarly, the administration of South African fisheries is fairly efficient, although the multiplicity of landing sites was seen as raising some difficulties for the proper implementation of a quota system. Other countries, such as Mozambique, have also efficient segments of specialized fisheries, such as the shrimp fishery[3]. There are countries like Mauritania, Morocco or Senegal endowed with abundant resources and fast growing fishing sectors. The fisheries of the countries boarding the Mediterranean Sea are particular, insofar as their environmental and social importance prevail over their economic relevance. Until the establishment of exclusive economic zones, the countries bordering the Gulf of Guinea had active shipowners who used to send their vessels to operate in Northwest and Southwest Africa, but significant fishing activities in those countries are now almost limited to the artisanal sub-sector. There are a number of archipelagic States, such as Cape Verde, Mauritius or Seychelles, endowed with substantial migratory resources, but confronted with obvious difficulties to manage and control the exploitation of resources lying in vast maritime areas.

Change, in its widest sense, has also been a common feature of African fisheries. Thus, some countries succeeded in developing their fisheries and marketing systems and made major strides in enhancing their institutional capabilities to deal with fisheries management problems. In other countries, notably in the Gulf of Guinea, promising fishing industries, based on active markets, which had to rely on access to other countries' waters, fell into crisis and virtually disappeared. Foreign fishing remains an important feature in the African fisheries scenery, but is now confined to high value, specialized fisheries, and tends to diminish as local fishing capacity expands according to a scheme that would have looked excessively simplistic some fifteen years ago. Many attempts to promote international cooperation at the regional, sub-regional and even bilateral levels did not hold their promises, but a more balanced assessment of the various experiences would need to be carried out. In some sub-regions, especially in Eastern and Southern Africa, inland fisheries landings have shown a steady increase over the past twenty years. Production from the Sahelian region continues to show great variability[4].

A few major distinctions conditioning the approach to fisheries legislation, and the very nature of applicable law, must be effected[5]. One of such distinctions is between marine and inland fisheries if only because each of these types of fisheries poses substantially different problems. The regions where significant inland fisheries activities take place are the Sahelian region[6] - where they reflect sometimes very old traditions - the Great Lakes Region and Southern and Eastern Africa. With the exception of some Lakes of Southern and Eastern Africa where modern fisheries have been developed, the management of inland fisheries is sometimes more akin to land use than to what one might see as being designated by fisheries management. In other central African countries, fishing remains, by and large, a mere traditional harvesting activity.

Another crucial distinction must be made in Africa between the so-called industrial fisheries and artisanal fisheries. Thus, without any exception, all basic laws and regulations dealing with marine fisheries lay down criteria for the paramount differentiation between industrial and artisanal fisheries. Industrial fisheries are usually based on the operation of fishing vessels of greater size and equipment of a more sophisticated nature. This sub-sector is subject to modern law. Artisanal fisheries, also referred to from a socio-economic point of view as small-scale fisheries, rely on smaller or traditional boats, normally with mobile engines or no engines and rudimentary gear and tend to be subject, at least partly, to a legal regime different from that governing industrial fisheries. Often, this legal regime combines elements of modern and traditional law and is, therefore, difficult to apprehend. Almost everywhere, these two sub-sectors can be distinguished easily. They target different species, operate for distinct markets and present their own specific socio-economic characteristics[7]. In many countries, the artisanal sub-sector plays a fundamental role and accounts for a growing percentage of landings but the lawyers and the administrators are faced sometimes with intractable difficulties to address management issues and challenges raised by this sub-sector.

The legal and regulatory framework governing fisheries management is, therefore, conditioned by general conditions which may differ radically from country to country, or at least from sub-region to sub-region. It presents also distinct features depending on whether it deals with marine or inland fisheries or industrial or artisanal fisheries. In the same vein, countries with more developed fisheries have also improved legislation. Consequently, it is essential that the existing diversity of situations, the differentiated nature of African fisheries and the variety of frameworks be kept in mind when considering this attempt to write about common features and trends of fisheries laws and regulations in Africa. This chapter seeks to present those laws and regulations through their provisions regarding management and surveillance of fisheries, and the special problems of small-scale fisheries.

MANAGEMENT OF FISHERIES

In general, African countries were among the most active supporters of the extension of the limits of national jurisdiction and, more generally, played a fundamental role in the definition of the new law of the sea[8]. One of the arguments which they put forward staunchly in support of their claims was that a regime of free access did not create conditions for a proper management of fisheries. However, substantial experience acquired since the extension of limits of national jurisdiction shows that many countries have not yet been able to make the most of their opportunities.

Fisheries management planning

Many African countries have adopted laws and regulations laying down general management objectives, or providing for the preparation, adoption and up-dating of fisheries management plans establishing the main management objectives and options and authorized levels of fishing effort for different fisheries. For instance, in Gambia, one of the first countries to adjust its fisheries legislation to the new law of the sea, the Fisheries Act provides that fisheries management plans are to be prepared. Each fisheries management plan is required to contain an identification of the major fisheries or fisheries management areas concerned and an assessment of the state of their exploitation; a statement of the objectives to be achieved in the management and development of the fisheries of fisheries management areas; the specification of any management and development measures to be taken and policies to be adopted with respect to the fisheries or fisheries management areas; an indication of the main requirements for statistical information on fisheries and the means to be used to obtain such information; the specification of any licensing programmes to be followed for any fishery or fisheries management area, any limitations to be applied to local fishing operations and the amount of fishing if any to be allocated to foreign fishing vessels and such other matters as may be necessary to provide for the proper management of fisheries[9]. Comparable provisions are found in many fisheries laws of African countries. In some sub-regions, laws, as implemented by regulations and ministerial orders, include detailed procedures for the preparation of those plans, including procedures for the full involvement of the profession and cooperation with fisheries research institutions. Various countries have set up consultative mechanisms that contribute to the preparation of the management plans[10].

Thus, many African countries, with particular reference to those countries with important fishing industries, have now a modern policy framework allowing them to implement, in an efficient manner, adequate policies likely to ensure the long-term sustainability of their fisheries[11].

However, progress still needs to be made in the implementation of existing frameworks. Current legal frameworks and provisions dealing with fisheries management planning have seldom been fully implemented. During the course of recent years, it has also appeared in some countries that the administrations are reluctant to define in legally-binding documents the main management objectives and options, as they fear that that could be a source of disputes with fishermen or be used as a basis for complaints against the administration[12]. While, again, this situation varies from country to country, much has still to be done towards ensuring the complete and meaningful involvement of research institutions in the fisheries management planning process.

That said, the fact that, in general, management systems need to be improved is not peculiar to Africa and it is well known that one of the reasons which accounted for the adoption of the Code of Conduct for Responsible Fisheries was the realization that new approaches to fisheries management, embracing conservation and environmental considerations, were urgently needed[13].

In this connection, mention should be made of the fact that of late some new concepts introduced by this Code of Conduct have been, or are in the process of being incorporated into new laws and regulations. Thus some countries have adopted legislation providing that in managing their resources they will apply a "precautionary approach". Recently, a number of countries, have reconsidered and improved the provisions of their laws dealing with the procedures for the elaboration of management plans. In doing so they made an attempt to involve the industry more closely in the management process, thus giving effect to the requirement that "states should seek to identify relevant domestic parties having a legitimate interest in the use and management of fisheries resources and establish arrangements for consulting them to gain their collaboration in achieving responsible fisheries". An effort to ensure transparency in the mechanisms for fisheries management and in the related decision-making process has been made in reconsidering those procedures.

An issue that ranks high among the priorities reflected in the Code of Conduct for Responsible Fisheries is that of the control of fishing capacity. Many provisions of this Code deal with the need to institute procedures and mechanisms ensuring that fishing capacity should remain commensurate with the sustainable use of fisheries resources so that operators work under economic conditions that promote responsible fisheries. Since the mid-eighties several African administrations were fully aware that this problem would acquire, sooner or later, a critical nature. At that time, various countries had adopted legislation calling for tight controls over the construction or importation of fishing vessels[14]. However, they were not able to apply effectively this preventive approach[15] and are now faced with the hard reality that once overcapacity is build up, it is difficult for a wide range of reasons of a political, social and economic nature, to eliminate excess fishing capacity. The problem has now a critical dimension in a number of countries. It is possible that preventive controls will be applied in future in a stricter manner than in the past, at least as far as industrial vessels are concerned, so that the exploitation of resources could remain economically viable. The problem is far more complex as regards the increase in fishing capacity in the artisanal sub-sector. In this sub-sector it is politically more difficult to institute and apply preventive controls over fishing capacity. Control of fishing capacity will remain a major future problem in many African countries, especially as concerns artisanal fisheries and some specific fisheries of an industrial nature.

Management regimes and licensing schemes

By and large, despite nuances, there are two methods for the control of fishing effort[16], i.e. a direct method based on the determination of total allowable catches and their allocation through quotas, and an indirect system based on the control of fishing capacity, relying usually on the limitation of the number of vessels authorized to operate. The quota system is largely used in predominantly single species fisheries with particular reference to developed countries. However, this system implies a tight follow-up of fisheries operations and strict controls over catches, transhipments and landings. It implies, of course, the determination of total allowable catches for all the fisheries, which many African countries are not in a position to effect. It may tend to exacerbate negative trends towards over-capitalization. It is also difficult to implement a system of quotas in contexts of multispecific fisheries, which prevail in most African countries. However, a few African countries do implement quota systems at least with regard to some of their fisheries[17]. In general, African countries have developed indirect systems for the control of fishing effort based on controls over the number of vessels authorized to operate. This system is considered to be more appropriate for the majority of African countries in view of the overall biological and socio-economic characteristics of their fisheries[18] and the generally limited capacities of the administrations. As it is normally supplemented by various other conservation measures, a system of control of fishing units may be, or may evolve towards, a fairly efficient and sophisticated one.

Usually, licences cover all fishing activities of a commercial nature. They concern both industrial and artisanal fisheries, although there are some notable exceptions to this rule in the case of artisanal fisheries[19]. The licences are normally granted to the owner with respect to a given vessel or canoe, in the case of artisanal fisheries. As a general rule, logistic support and transshipment operations at sea, which increase the ability of vessels to remain at sea and, consequently, increase the levels of fishing effort, are subject to licensing. In general, these operations are tightly regulated.

The licensing schemes instituted in order to give effect to mechanisms of control of fishing effort, mainly those being implemented in countries with important commercial fisheries such as Angola, Guinea, Mauritania, Morocco, Mozambique and Senegal, have been considerably tightened up. In addition, differentiated fishing licenses tend to be established with respect to specific fisheries. Actual fishing licenses may also become, through conditions set out therein or established with respect to categories of license holders, an instrument for the regulation of specific fisheries.

Regulatory technical measures

In general, the fisheries laws and regulations that a number of countries adopted in the years following their independence used to contain very many technical conservation measures, such as closed areas and seasons, differentiated fishing areas established sometimes on the basis of bathymetric lines, minimum catching or landing sizes for fishes, regulation of authorized gear including minimum mesh sizes. In many cases, those regulatory measures were adopted on the basis of what the scientists considered ideal scientific requirements, to the exclusion of other concerns, including administrative convenience. A common feature of many laws and regulations was perhaps that they were hardly enforceable. A broader approach to the conservation of resources emerged progressively in some countries as the administrations were confronted with the need to manage efficiently their resources under the new law of the sea. These administrations tended to realize that what was essential for the management of the resources was an appropriate general framework and adequate legal and institutional tools to manage and thus assuring the sustainability of the fisheries, rather than sets of complex technical measures, based on questionable or changing scientific justification.

Thus, technical measures are being increasingly seen as important complements to systems for the control of fishing effort, rather than central management tools *per se*, as they may have been sometimes considered wrongly. Thus, in a number of recent processes of revision of fisheries laws and regulations a careful assessment of the actual need for a number of regulatory technical measures in light of the enforcement capabilities of the countries was effected. There has been a general awareness on the part of those administrations that it is not always easy to ensure compliance with minimum catching or landing sizes or weight of fishes or minimum mesh sizes or tight regulation of authorized fishing gear. Instead, there has been a growing tendency to privilege differentiated fishing zones and closed areas or closed seasons that are generally easy to control. Thus Senegal has developed over the past ten years a reasonably sophisticated management system which is essentially based on the determination of fishing zones depending on the type of fishing. Similarly, Morocco has instituted a closed season ("période de repos biologique") which is considered cost effective and might be generalized in other countries of the region[20]. Both countries have made an assessment of conservation measures available to them and decided to adopt those more cost-effective measures, while they could sometimes be regarded as too simplistic from a purely scientific perspective.

Efforts were carried out in some regions with a view to assuring harmonization of conservation measures. Thus, in the late seventies, the FAO Fishery Committee for the Eastern Central Atlantic (CECAF) recommended to its Coastal Member States the adoption of a single, minimum mesh size for all demersal trawling and proposed uniform methods for the measurement of mesh sizes. In July 1993 an agreement was concluded under the aegis of the Sub-Regional Fisheries Commission providing, *inter alia*, for the harmonization of a number of conservation measures[21]. At the time when these instruments were adopted their provisions were generally implemented by the concerned countries[22]. More recently, there has been a growing tendency on the part of the most important fishing countries to keep the efficacy of conservation measures under continuous review, which has led to increased differentiation of those measures. In general, regional organizations and commissions were not always able to accompany this evolution although some matters of a very important nature in Africa, such as discards, the need to minimize waste, catch of non-targeted species or selectivity of fishing gear, have been the subject of a number of preliminary reflections.

Foreign fishing

In general substantial foreign fishing activities, which would be affected by the establishment of Exclusive Economic Zones, used to be carried out in African countries. Thus Portuguese and Spanish boats used to operate for various centuries in some maritime areas of Northwest Africa. Following the establishment of such zones, coastal countries continued to allow access by foreigners, as required by the Law of the Sea Convention, which provided that in giving access to other States to the surpluses[23] of resources of their exclusive economic zones the coastal States should take into account "the need to minimize economic dislocation in States whose nationals (had) habitually fished in the zone or which (had) made substantial efforts in research and identification of stocks"[24]. Foreign fishing continues to play a major role in some countries and to contribute to foreign exchange earnings, but these activities tend to focus now on specific, high value fisheries[25]. These activities are normally conducted under the terms of fishing agreements between the coastal state and the flag countries or the European Community.

Intergovernmental access agreements were instrumental in developing the concept of flag state responsibility for compliance control. To varying degrees, Governments have always been held responsible for the conduct of their vessels operating under bilateral access agreements, a responsibility which flows naturally from the principle of international law *pacta sund servanda*. African countries were able to benefit from active support from flag States in assuring that their vessels complied with the terms of access agreements and with coastal state fisheries laws and regulations and that violators were brought to task. In practice, the matter lost most of its interest in Africa, as it turned out progressively that foreign operators tended to practice fewer offences than local operators.

Many fishing laws and regulations dealing with marine fisheries used to provide, and continue to provide, for the establishment of stricter access conditions for foreign fishing vessels, especially as concerns monitoring and surveillance. However, while the possibility of adopting differentiated access conditions is still provided for in most laws, this is clearly a situation that might disappear progressively and, except for the level of fishing fees and other payments, access conditions tend to be the same for local and foreign fishing vessels[26].

A related question is that of the determination of nationality criteria for fishing vessels. A majority of African countries instituted tight nationality criteria for fishing vessels whereby, in order for a fishing vessel to benefit from local access regime, it had to belong effectively to nationals of the coastal State. In the case of vessels belonging to companies, nationals of the coastal State had to hold a controlling interest in that company, which was normally set at 51% of the shares. Other countries were less concerned with ensuring that their vessels were effectively owned by nationals[27]. They tended to consider that if a fishing boat was effectively based in the country and if their owners had a long-term commitment to that country, the question of the actual nationality of those owners was of subsidiary importance. The prevailing trend now seems to be clearly in favour of this "open" approach[28].

Regional cooperation

Africa is a region where there are many shared or interdependent stocks making it necessary for the countries to carry out appropriate cooperation activities for the management and surveillance of those resources. This contributes to explain why there are in Africa a high number of organizations and agreements dealing with regional or sub-regional cooperation in fisheries. A presentation of such organizations and agreements would, in itself, form the basis of a special study[29].

The need for cooperation for the management and surveillance of fisheries is reflected in several fishing laws. Thus the Gambian Fisheries Act provides that the Minister may, on the advice of the Director of Fisheries, enter into arrangements or agreements with other countries in the region or with other competent regional organizations to provide, *inter alia*, for the harmonization of systems for the collection of statistics and stock assessment surveys; the harmonization of licensing procedures and conditions in respect of foreign fishing vessels; the adoption of joint or harmonized enforcement measures in respect of foreign vessels contravening fisheries laws in the region; the establishment and operation of fisheries management bodies; the establishment of a regional register of fishing vessels and other cooperative measures, including measures for the promotion of the welfare of fishermen. Comparable provisions are found in the laws of other countries.

Some countries have recently refined the provisions of their laws regarding cooperation in light of the requirements of the Code of Conduct for Responsible Fisheries calling for compatibility of management measures for transboundary fish stocks[30].

Of more direct interest for the present review is the question of whether the activities of regional organizations have had an impact upon the management and surveillance regimes of coastal countries. Substantial efforts towards sub-regional harmonization of access conditions have also been carried out in some areas[31]. Some multilateral agreements calling for the harmonization of specific access conditions have been adopted in some sub-regions. As already mentioned, a convention on the harmonization of conservation measures was adopted in July 1993 among the Members of the Sub-Regional Fisheries Commission. Later, in

September of that year, a very ambitious multilateral agreement on cooperation in the exercise of surveillance operations, to be implemented through protocols, was also adopted in 1993 among the members of this Sub-Regional Fisheries Commission. As a consequence of this process, many provisions of fisheries laws in Northwest Africa are similar, especially in the areas of monitoring, control and surveillance of fisheries. However, despite progress in the harmonization of monitoring and surveillance schemes[32], significant results have not yet been attained in fisheries management, neither as concerns general policies, nor with regard to such fundamental specific issues as, for instance, coordinated determination of levels of authorized fishing effort[33]. It is also of interest to note that exploratory studies were carried out with a view to establishing a comprehensive bilateral framework for the management and surveillance of transboundary stocks between Angola and Namibia[34].

SPECIAL PROBLEMS CONCERNING SMALL-SCALE FISHERIES

With some notable exceptions[35], small-scale fisheries, carried out from canoes, play a fundamental role in most African countries. The artisanal sub-sector is extremely active in some countries. For instance, in Senegal it accounted for 78% of landings in 1997 and a substantial proportion of those landings are exported. In most countries of the Gulf of Guinea, artisanal fisheries also account for a majority of the catches. The characteristics of this sub-sector, such as its versatility, low-cost and efficiency, suggest that it will be a lasting feature of the fisheries scene in Africa. However, its informal and dispersed character poses obvious management problems, whose consideration would go beyond the scope of this chapter. Two series of questions seem of particular importance. One concerns present trends towards co-management of artisanal fisheries by both the central administrations and the fishing communities. Another concerns the role of customary law in fisheries management.

Co-management of fisheries resources

A complex question, if only from a conceptual point of view, is that of the participation by fishing communities and communities in the management of fisheries, an emerging general trend in natural resources management. Various laws and regulations of West African countries have set up local fishing councils entrusted with responsibilities with regard to management and surveillance of fisheries of interest to artisanal fishing communities[36]. Thus, the new fisheries law of Senegal provides for the establishment of such "local advisory councils" that consist of representatives of concerned fishing associations and local communities, traditional authorities and representatives of the administration. The councils' functions include the provision of advice on all matters pertaining to artisanal fisheries within their area of competence, the prevention and settlement of disputes arising out of conflicts among fishermen using different types of gear and the organization of fishermen so that they can second the administration in the conduct of surveillance and enforcement of fishing activities[37]. Comparable provisions are under consideration in the Gulf of Guinea, in particular in Côte d'Ivoire in the context of the management of a number of specific lagunes[38]. Drawing on experience gained in the implementation of projects carried out by the International Union for the Conservation of Nature (IUCN) in the Bijagos archipel, the African State which seems to have gone furthest in establishing legal mechanisms for co-management of fisheries is Guinea-Bissau. Thus the maritime, estuarine and river waters of the country are to be divided progressively into various management areas taking into account their characteristics. For each of these areas a regional consultative council for artisanal fisheries is to be created. The councils must be consulted by the administration regarding any proposed management measures envisaged for their area of competence and the licensing programme for that zone. It is expected that the councils would be involved progressively in the surveillance of fishing activities[39].

The general objective sought has been to lay down some minimal conditions allowing the councils to take over increased responsibilities in a pragmatic, progressive, manner, thus allowing for some degree of self-regulation of artisanal fisheries. However, central administrations have never entrusted local councils with direct management authority. The State remains the institution responsible for the regulation of natural resources management. Current constitutional and administrative regimes do not reflect a desire of sharing with, or transferring to, local communities that responsibility. Legal systems remain centralized and seek an objective of uniformity of legal rules. In general, local communities are not given competence under general legislation to manage fisheries resources. While local communities carry out a number of management and development activities, they do not have a legal status commensurate with those activities. This is largely due to reasons of a political nature, insofar as a policy involving the association of fishing communities to the management of resources implies the delimitation of zones where "land-based rights" are to be exercised, thereby giving to those communities a "right of exclusion" of third parties. The political authorities of African States are aware of the risks inherent in such prerogatives, which explains why they give an ambiguous support to experiences of decentralization of fisheries management[40]. This may also explain why African experiences of land-based rights are seldom statutorily recognized and remain largely the result of empirical, de facto, rather than legal recognition[41]. It can be anticipated that current efforts towards recognizing a measure of management authority in favour of local communities will continue, but the central administrations are reluctant to go as far as granting wide management and surveillance prerogatives to those communities. Thus, the so-called co-management approach, which reflects a compromise between conflicting central and local concerns, is likely to remain a long-term trend of fisheries policies in Africa. Similar observations can be made in connection with the question of the recognition of customary rules and practices.

Customary systems for the regulation of fisheries

In various geographical areas, with particular reference to the Sahelian region, customary systems of regulation of fisheries are found, especially in the context of traditional and subsistence fishing. Substantial scientific literature is available on the matter[42]. A common feature of these systems is the recognition of exclusive rights of exploitation to the benefit of fishing communities. Management and use rights are normally entrusted to a "maître d'eau" or a traditional chief; these rights are usually derived from conquest or first occupancy. Practices such as prohibition of fishing on certain dates or days, prohibition of various types of gear and other traditional practices are considered to provide a set of coherent customary rules allowing for effective management systems. Sometimes, modern laws have tried to give a measure of recognition to those rights, provided that they did not conflict with fundamental principles of modern law, as is demonstrated by the example of Mali or Niger[43]. In the case of Mali, various conventions were concluded between the central or regional administrations and fishing communities recognizing their exclusive rights and traditional practices over some areas[44].

Again, the recognition of customary practices by modern law in natural resources management has been limited and ambiguous. Modern law has been unable to integrate customary practices or create conditions for their evolution. Traditional regulatory systems tend to be now confined to remote areas or to subsistence fishing and one can say that, elsewhere, they sustain a process of erosion as all other customary systems for the management of natural resources. In this regard this process is perhaps the reflection of the crisis of traditional society as a result of demographic growth, increasing urbanization, wars, natural disasters such as drought, the introduction of new techniques, interference by modern administration, the construction of water infrastructures, the loss of prestige of traditional groups and authorities. This phenomenon has probably been exacerbated by some features of customary rules such as their diversity and lack of uniformity, their imprecise character and sometimes the fact that they ignore fundamental aspects of modern ethics[45]. The "co-management" approach could perhaps contribute to a progressive evolution of traditional customary systems towards modern systems.

SURVEILLANCE OF FISHERIES

It is not exaggerated to say that African countries, especially those endowed with abundant resources, made considerable efforts to improve their monitoring, surveillance and enforcement systems. Sometimes these countries have vast maritime areas under national jurisdiction and control over such areas was a deep concern of the administrations. As early as in the 1970s and early 1980s various reflections and initiatives took place intended to assist the countries to improve their monitoring and surveillance capabilities[46]. The efforts of the countries concerned both the monitoring of fishing activities and administrative and judicial proceedings.

Monitoring of fishing activities

In general, African countries tried to develop or improve practical and legal mechanisms allowing them to follow closely fishing operations and to obtain full information on all aspects of those operations.

Thus, many countries have included in their laws and regulations provisions requiring that statements of catches or fishing logbooks be supplied to the local authorities. The effectiveness of those provisions has been enhanced by the imposition of comparatively high penalties in case of non-compliance with those provisions. As part of their efforts to closely monitor fishing operations, some countries have also imposed requirements of self-reporting by radio of the vessels' positions and/or catches. The implementation of those provisions with regard to foreign vessels has not always been successful, and could cause diplomatic disputes[47].

In general, fisheries laws and regulations of African countries allow for the placing of observers on board fishing vessels. In many countries all industrial fishing vessels were required to have an observer on board who would report on the fishing activities being conducted to its administration. Observers are not authorized officers empowered to make records of possible offences but their findings could be used on the occasion of administrative or judicial proceedings. Much debate has surrounded what has been both a traditional and a controversial institution[48]. Nowadays, one can witness a tendency in many coastal states towards not imposing strictly the requirement of an observer on board each vessel. Attempts have also been made, whenever vessels are licensed to fish in the waters of neighbouring countries, to designate "sub-regional observers" but this proposal did not materialize[49].

In the same spirit of seeking to design procedures aimed at improving the monitoring of fishing operations, attempts have been made to develop a harmonized system for the marking of fishing vessels. The need for common criteria and specifications for the marking of fishing vessels was highlighted by the fact that many fishing vessels in Africa operate successively in waters under the jurisdiction of various countries. It was thus important that marking criteria and requirements be common. As a result of a series of technical consultations organized by FAO with support from the Government of Canada it was proposed that the international radio call sign allocated by the International Telecommunications Union should form the basis of a system for the marking of fishing vessels[50]. The technical consultations proposed also dimensions for the marking signs. Some countries incorporated in their fisheries regulations those criteria and requirements[51]. In practice, however, although they are perhaps in the process of being generalized in some sub-regions, these provisions are not always implemented strictly[52].

Other provisions, likely to improve the ability of the countries to monitor and follow activities in their zones under national jurisdiction, are controls over the transit of unlicensed fishing vessels or requirements in terms of stowage of fishing gear[53].

The terms of the debate on the effectiveness and use of "soft" means of surveillance for industrial fisheries may be undergoing radical changes as a consequence of the possibility of using satellite based vessel monitoring systems (VMS)[54], which are attracting considerable attention worldwide in general, and in Africa, in particular, where a few experimental projects are being carried out. A VMS consists of several components, namely a transmitter or receiver, capable of fixing a position, the most common of which is a Global Positioning System, and an automated reporting system that controls the transmission of the position data and possibly other information via a communication system to a monitoring station. The communications system conveys data between the transceiver or transmitter on the vessel and the monitoring authority. Owing to scientific and technical progress VMS can now produce, directly or indirectly, a wide range of increasingly accurate information about activities of fishing vessels. Costs have also been considerably reduced and VMS are affordable both by coastal countries and shipowners. Various preparatory reflections are being carried out. Preliminary conclusions show that the implementation and increased reliance by States on VMS requires the establishment of appropriate national legal regimes involving, *inter alia*, definition of terms peculiar to VMS and the usage thereof to avoid interpretation problems, specification of the national monitoring agency or department, determination of minimum performance standards, approval and verification procedures, registration of some components to prevent their removal and transfer and the establishment of offences and penalties that would have a deterrent effect and ensure the credibility of VMS systems[55].

Administrative and judicial proceedings

Much could be said about administrative and judicial proceedings for the surveillance of fisheries and it is not easy to make a selection of issues deserving consideration. Undoubtedly, the issues involved in the surveillance of fisheries are among the most interesting for the practitioner.

To begin with, the question of the statutory responsibility for the overview and conduct of surveillance operations has been the subject of much debate in many African countries. In normal circumstances, fisheries surveillance should be an administrative task of a civilian nature, as a civilian department should be better prepared to deal with matters such as control of mesh sizes, composition of catches, compliance with fishing areas and so on. However, fisheries surveillance implies recourse to aerial and naval means of surveillance and communication mechanisms over which military authorities always seek to retain control, or which they are sometimes the only authorities in a position to control. As a result of this, all surveillance systems in African countries, especially in countries in which the fisheries sector is important, are based on collaborative mechanisms between the fisheries administration and the Army with a great involvement of military services in all logistic and operational aspects. Several fisheries laws reflect this situation and have set up joint supervisory arrangements[56].

The duality of supervisory authorities has had implications with respect to the identification of authorized officers. In general, both officers of the fisheries administration and officials of the Navy or the Airforce are authorized officers empowered to record fishing offences. The actual powers of those officers are defined in different manners depending on the legal system to which their countries belong. Common law countries tend to define their powers in great detail, whereas countries with a civil law tradition define those powers in general, broader terms. But there has been a recent tendency to define surveillance procedures in great detail, especially when carried out at sea. Similarly, procedures such as boarding, visit, inspection, arrest and seizure of vessels, gear and catches are normally defined in detail in existing laws and regulations. This is in fact an area in which, at least in some sub-regions, harmonization of procedures has almost reached the state of uniformity. An effort to harmonize surveillance practices, especially as concerns the provisional seizure of fishing vessels and their taking to a local port, is being carried out in some sub-regions[57]. In line with the relevant provisions of the Convention on the Law of the Sea, many laws make provision for the release of crew and vessel that were arrested pending the completion of proceedings, upon posting of a reasonable bond[58].

A related question concerns the exercise of the right of hot pursuit of fishing vessels. In a region with many maritime boundaries, it is normal that the exercise of this practice, provided for in article 111 of the Law of the Sea Convention and in many fisheries laws, should be seen as providing a useful tool for the control of zones under national jurisdiction. However, in practice, African coastal States have refrained from exercising the right of hot pursuit from their exclusive economic zones and treated maritime boundaries as a limit beyond which they would, under no circumstance, assert their rights. Instead, some coastal States have concluded agreements allowing the exercise of the right of pursuit into the waters of their neighbours. Thus on 1 September 1993 a Sub-Regional Convention on the Exercise of the Right of Hot Pursuit Among Member States of the Sub-Regional Fisheries Commission was signed. This framework Convention was supplemented by a Protocol laying down some modalities for cooperation among surveillance units of the concerned countries[59].

In general fishing offences are defined in great detail in fisheries laws. Normally, the offence that is considered to be of the most serious gravity is unauthorized fishing. In some countries unauthorized fishing is punished with the forfeiture of the fishing vessels, of the gear and the catch on board. By and large, the gravity of fishing fines is sometimes closely associated with the overall objectives of the policy and the fisheries management system of a country. Thus, in the very few countries that implement a quota system, offences which involve eluding declarations of catches (under-reporting, unauthorized transshipment) are sanctioned with heavy fines as is the case in Namibia. In Mauritania, where all catches had to be landed, unauthorized transshipment was severely punished. Senegal has developed progressively a sophisticated management system for industrial vessels based on the establishment of differentiated fishing zones. Consequently, fishing in an unauthorized area is repressed strictly.

In line with the provisions of article 73, paragraph 3 of the Convention on the Law of the Sea, coastal state penalties for violations of fisheries laws and regulations in the exclusive economic zone do not include imprisonment or any other form of corporal punishment. But in Africa the level of fines for fishing offences used to be among the highest in the world. This was a reflection of a situation where the coastal countries had limited means of surveillance[60]. Progressive readjustments took place and the level of fines are comparable to that existing elsewhere in the world.

An important issue, which raised, on occasion, intractable debates, is that of the authority competent to impose fishing penalties. Normally, authority to impose penalties is vested in the tribunals. However, traditionally in a number of countries the possibility for administrative authorities to compound offences has been recognized, both in common law and civil law countries. But there were cases where it turned out that a fisheries administration was not adequately equipped to deal with situations of this nature, given, in particular, the high level of fines being imposed and political pressures exercised upon the concerned administrators. As a result, there was in some countries an evolution towards the recognition of exclusive competence of the tribunals for the imposition of penalties. This situation is not deemed to be fully satisfactory either, insofar as judicial proceedings are lengthy and many offences, of a minor and routine nature, may either go unpunished or lead to costly and deleterious proceedings. A more balanced evolution that seems to have taken place over the past ten years could be summarized by distinguishing two situations. In a number of countries, fishing offences, however their gravity, are punished through a fine imposed by, or on behalf of, the head of the fisheries department. However, this possibility is always subject to judicial review upon the initiative of the author or the presumed author of the offence. In some cases, given the delicate nature of this procedure, advisory committees, that propose a fine to the Minister of Fisheries, have been set up[61]. In other countries, penalties for offences of more gravity - normally unauthorized fishing - are subject to the exclusive competence of the tribunals while penalties for other offences can be imposed by the administration subject to judicial review.

There are significant differences between civil law and common law systems as concerns the question of the burden of proof. Civil law countries tend to accept reports from authorized officers, as unquestionable evidence, until the contrary is demonstrated. This is in line with underlying principles of civil law countries where authorized officers and the so-called "Ministere public" both have, in accordance with the inquisitorial procedure, already established facts beyond any reasonable doubt when a case is brought before a court. Common law jurisdictions usually require proof of the facts contained in records of offences. This traditional approach was changed in many common law countries, especially in areas of natural resources management, which gave rise to some debate. In general, countries tended to change traditional evidentiary rules as part of their strategy to improve the enforcement of their fisheries laws, in particular by imposing the burden of proof on the fishermen and establishing various presumptions[62]. A related issue that attracted considerable controversy in some African countries was that, as a condition to their support to a number of management and surveillance initiatives, some foreign partners pressed in favour of greater recognition of evidence supplied by modern tools such as aerial photographs and video and satellite provided information[63]. As indicated earlier the situation in this respect may be changing radically.

PRELIMINARY CONCLUSIONS

This brief overview of the main trends and features of fisheries legislation in Africa, remains little more than an outline of the most important characteristics of the legal and regulatory frameworks in force, the practical problems raised by their implementation and the prospects for their future evolution. A broader, more in-depth study or specific studies on particular subject matters covered by this chapter would present great interest and fill in existing gaps.

The situation continues and will continue to undergo major changes. It is likely that the great differentiation that exists already among the main sub-regions of Africa will continue to increase. Countries with more resources and important and more sophisticated fisheries sectors might go on improving their institutional and legal frameworks. Small-scale fisheries and the development of management approaches more adequate to the nature of these fisheries might also continue to evolve, if only because a number of external partners have decided to actively support this sub-sector. Some countries are also assessing their national laws and regulations in light of some initiatives of a broader nature and scope, carried out under the aegis of the United Nations or FAO, such as the Agreement for the Implementation of the Provisions of the United Nations Convention on the Law of the Sea of 10 December 1982 Relating to the Conservation and Management of Straddling Fish Stocks and Highly Migratory Stocks, the Code of Conduct on Responsible Fisheries and the so-called "Compliance Agreement"[64]. The progressive development by FAO in the course of the next few years of a policy and regulatory framework for illegal, unauthorized and unregulated fishing, may also impact on African laws dealing with fisheries.

---

[1] The following Internet site of FAO contains exhaustive information and fisheries statistics: http://www.fao.org/fi/default.asp. A number of important legal instruments, with particular reference to the Code of Conduct for Responsible Fisheries, are also available from this Website.

[2]2 For general information on the distribution of fisheries resources see *The state of world fisheries and aquaculture* published annually by FAO. The latest edition concerns 1998 and was published in 1999. More specific information regarding Africa is found in *Fisheries and Aquaculture in Sub-Saharan Africa: Situation and Outlook in 1996*, FAO Fisheries Circular N° 922 and *Fisheries and Aquaculture in the Near East and North Africa: Situation and Outlook in 1996*, FAO Fisheries Circular N° 919.

[3] The shrimp fishery of Mozambique was developed towards the late fifties and has since then been implemented in an efficient manner. It is based on a system of allocation of individual quotas to companies and to vessels which catch would be landed and tightly controlled in the South African port of Durban and recourse to a wide range of measures such as observers on board, prohibition of transshipments at sea, sealed holds while the boats were not operating in the fishing grounds, systematic weighting of all catches. The fishery, to which all concerned parties were much attached, was not affected by a difficult political environment conditioning the relations between Mozambique and South Africa from 1975 to 1993.

[4] Historically, Sahelian fisheries have shown great variability. See A. Bonzon and C. Breuil, Stratégies pour l'aménagement et la planification des pêches continentales dans la région du Sahel, *Rapport sur les pêches N° 481*, (FAO, Rome, 1992).

[5] In the present chapter only limited references to actual laws and regulations of the concerned countries are made. These are available from the FAO LEX database found at the Internet site http:\\www.fao.org.legal. Many laws and regulation are available in full or partly in English. A substantial effort is made to keep this database up to date. Since, 1979 FAO has also published occasionally regional compendia of fisheries legislation of African countries bordering the Atlantic Ocean. The latest version of this compendium was published in three volumes in 1995 by project FAO GCP/RAF/302/EEC, Document N° 15. An interesting study, that allows for comparative analysis of legislation is Coastal State Requirements for Foreign Fishing, (*FAO Legislative Study N° 56*, Rome). The study is available on line from the above site.

[6] The inland catches of Mali, expressed in metric tonnes, were for many years comparable to those of Senegal or of Mauritania, two important fishing countries of West Africa.

[7] But this traditional distinction is increasingly blurred in some countries where artisanal fisheries tend to target high value demersal species for export. There is clearly a phenomenon of industrialization of artisanal fisheries in Northwest Africa for example.

[8] There are two fundamental, extremely well-documented, articles on the African States and the new law of the sea: D. Bardonnet, La dénonciation par le Gouvernement sénégalais de la Convention sur la mer territorialc et la zone contigüe et de la Convention sur la pêche et la conservation des ressources biologiques de la haute mer, en date du 29 avril 1959 à Genève,*AFDI*, 1972, p. 123 and D. Bardonnet and J. Carroz., Les Etats d'Afrique de l'ouest et le droit international des pêches maritimes,*AFDI*, 1973, p. 837.

[9] Fisheries Act 1991, 4. These provisions are similar to parallel provisions of the Gambian Fisheries Act 1977.

[10] In general, the functions of "National Advisory Councils" that have been set up in many countries (Gabon, Guinea, Mauritania, Morocco, Mozambique, Senegal) go beyond their contribution to the preparation of management plans. A question that will arise in future is whether they could provide a vehicle for progressive self-regulation of specifc aspects of fisheries by the industry.

[11] Thus many concepts that are today enshrined in Articles 6 and 7 of the Code of Conduct for Responsible Fisheries are found in many fisheries laws of African countries.

[12] In a few African countries administrative tribunals have, on a number of occasions, handed down judgements against the administrations. Some administrations are either loath to set out in legal documents a line of conduct or wish to assess the possible implications of planning documents before issuing them. This may be part of an emerging trend in some African political and constitutional systems where tribunals tend to constitute a counterweight to the authorities.

[13] G. Moore, "The Code of Conduct for Responsible Fisheries", in E. Hey (ed.),*Developments in International Fisheries Law*, (Kluwer Law International), Netherlands, 85.

[14] Preventive legal systems to control local fishing capacity do exist in many African countries in West, South and East Africa. These include Morocco, Mauritania, Senegal, Guinea-Bissau, Guinea, Gabon, Angola, Namibia, South Africa, Mozambique.

[15] One could assume that countries, which for many years could not exploite, their resources because they did not have a fishing fleet, would not be very strict in controlling the increase of national fishing capacity.

[16] J.R. Beddington and R.B. Retig, *Methods for the regulation of fishing effort*, FAO, Fisheries Technical Report N° 243, 1984.

[17] There are three countries which apply quota systems in Africa, i.e. and Namibia and South Africa (for all fisheries of commercial interest) and Mozambique (with respect to high value fisheries, notably that of shrimp). See L. Clark, Le système de gestion des pêches de la Namibie,*Espaces et Ressources Maritimes*, (1999).

[18] See for a discussion of this system J.-P. Troadec, "Practices and prospects for fisheries development and management; the case of Northwest African Fisheries", in B.J. Rotschild,*Global Fisheries Perspectives for the 1980's*, (Springer-Verlag, New York, 1983). The FAO Committee for Central Eastern Atlantic Fisheries (CECAF) kept the matter under review for many years. Eventually, it recommended that the concerned countries give effect to a system of control of fishing effort as opposed to a quota system. See *Report of the VI Session of the CECAF Sub-Committee on the Management of Resources within the Limits of National Jurisdiction*, Lome, Togo, 2-5 December 1985, paragraphs 40-43.

[19] Artisanal fishing in Senegal, which accounts for some 70% of the landings, is not subject to prior licensing but to a regime of prior declaration by the fishermen to the administration of their intention to fish for commercial purposes. On the occasion of a recent revision of the Fisheries Act and its implementing Regulations, the Government decided for major political reasons not to impose a system of licensing for artisanal fisheries.

[20] This measure is considered to be effective from a biological perspective. It is also thought to be highly effective from a surveillance point of view insofar as any boat spotted fishing during the closed season is operating illegally.

[21] The members of the Commission are Cape Verde, The Gambia, Guinea, Guinea-Bissau, Mauritania and Senegal.

[22] In the case of the minimum mesh size of 70 mm for trawling for all demersal species proposed by CECAF in 1980, a follow-up mechanism, through periodical questionnaires, was instituted by the CECAF secretariat.

[23] In the context of most African fisheries, the concept of surplus– the difference between the total allowable catch and the harvesting capacity of the coastal State- may have been little more than a political expedient, an element of the compromise that allowed for the recognition of Exclusive Economic Zones under the new Law of the Sea. See, mainly from a scientific point of view, S. Garcia, J.A. Gulland and E. Miles, The new Law of the Sea and the access to surplus resources, *Marine Policy*, (1986) 192; J. Gulland, "Conditions of access to fisheries: Some resource considerations", in *Report of the Expert Consultation on the Conditions of Access to the Fish Resources of the Exclusive Economic Zones*, (FAO, Rome, 1983), 15.

[24] A fundamental study, today mainly of "historical" interest as the information it contains is now outdated, must be quoted in this context: J. Carroz and M. Savini, Les accords de pêche conclus par les Etats africains riverains de l'Atlantique, *Annuaire français de droit international,* 1983, p. 674.

[25] Notwithstanding occasional official statements to the contrary, coastal countries are generally favourable to the continuation of foreign fishing activities under access agreements. This is particularly the case of activities of European vessels under agreements with the European Community which are generally very advantageous for the administrations of the coastal countries given the level of financial payments effected.

[26] At a certain moment foreign fishing vessels used to make more offences to the fisheries laws than local fishing vessels. This is no longer true; in general the level of compliance with laws and regulations is now higher with foreign operators than with local shipowners.

[27] That was the case of Côte d'Ivoire, Mozambique, Namibia. For a systematic comparative analysis of nationality criteria for fishing vessels see Coastal State Requirements for Foreign Fishing, (FAO *Legislative Study N° 57*, Rome, 1996).

[28] In some countries in order for the foreigners to have access to local resources they had to associate themselves with local shipowners, under joint venture arrangements. As many of these operations were not successful, coastal countries tended to prefer the straight licensing of foreign vessels or the conclusion of access agreements with the Flag State. Access agreements are now criticized by the public opinion in some countries, both African and European.

[29] Some organizations have carried out substantial cooperation activities in the course of the past twenty years. One of them is the FAO Fishery Committee for Eastern Atlantic Fisheries (CECAF), already mentioned, established in 1967, competent over an area lying between the Gibraltar Strait and the Congo River. This organization has provided a forum for a dialogue on fisheries management issues and recommended to its members the adoption of various harmonized conservation measures. Another is the Sub-Regional Fisheries Commission consisting of Cape Verde, Guinea, Guinea-Bissau, Mauritania and Senegal set up in the late seventies. The International Commission for the Conservation of Atlantic Tunas (ICCAT), set up in 1966, has also been very active. Some inland fisheries organizations competent with respect to specific lakes or rivers basis have also been active. Finally, a process was launched for the establishment of the South East Atlantic Fisheries Organization, competent over high seas fisheries, under the United Nations Convention on the Law of the Sea on straddling fish stocks and highly migratory fish stocks. The concerned countries are Angola, Namibia, South Africa and the United Kingdom with respect to Ascension, St. Helena and Tristan da Cunha. The organization would deal mainly with hake as a straddling stock and certain high seas stocks such as red crab; it would not be concerned with whales, dealt with by the International Whaling Commission; highly migratory species, covered by ICCAT and species falling

under the purview of the Commission for the Conservation of Antarctic Marine Living Resources. The Organization would not be concerned with species within areas of national jurisdiction. Agreement has not yet been reached on all issues under discussion. On the proposal to set up this organization, see W. Edeson, "Towards Long-Term Sustainable Use: Some Recent Development in the Legal Regime of Fisheries", in *International Law and Sustainable Development, Past Achievements and Future Challenges*, (Oxford University Press), 1999, 194.

[30] Namibia, South Africa, Senegal and Guinea have included in their fisheries legislation provisions on regional cooperation based on the Code of Conduct for Responsible Fisheries.

[31] Extensive work on the harmonization of fisheries legislation has also been carried out in Africa. On the scope of purpose of those initiatives, including the general meaning of the concept of harmonization in the fisheries practice, see *Report of the first seminar on the harmonization of the fisheries legislation of the States members of the Sub-Regional Fisheries Commission*, Dakar, Senegal, 29 November-2 December 1994, (Project FAO GCP/RAF/302/EEC, Document N° 7).

[32] However, an ambitious attempt to establish a sub-regional register of fishing vessels to be used both for management and surveillance purposes under the aegis of the SubRegional Fisheries Commission, supported by important foreign partners, was the subject of much debate from 1986 up to 1997 in Northwest Africa, but did not materialize.

[33] Even in areas where most, if not all, of the species are shared, interrelated or interdependent, the centre of gravity of fisheries management policies remain essentially at the national level. The benefits of regional cooperation imply necessarily long-term, protracted efforts. Those benefits are not visible in the short run and the countries tend thus to neglect cooperation. Experience also shows that regional cooperation is very much affected by internal political changes or instability and, obviously, by the general political climate among the concerned countries. For obvious reasons, countries, including external partners, tend to privilege visible initiatives rather than day-to-day activities of a scientific and technical nature. Thus, the record of regional cooperation in fisheries in Africa remains ambivalent. Realistic legal and institutional tools are required to overcome the current dramatic gap between the requirements of a proper management of fisheries on a eco-system basis and the essentially local administration of those fisheries. Progress is also affected by the fact that, as is the case in Eastern Africa and in the Indian Ocean, many maritime boundaries have not yet been agreed upon.

[34] R. Rackowe, K. Roberts and M. Tandstad, *Cooperation between Angola and Namibia on the management of transboundary stocks*, (Project FAO GCP/RAF/302/EEC, 1996, Document N° 47).

[35] Two exceptions are Morocco and Namibia. In Namibia there is no artisanal sub-sector. In Morocco there is the so-called "pêche cotiere" (inshore fisheries) consisting of small vessels.

[36] The question is not limited to the management of small-scale fisheries. Some countries, with particular reference to Mauritania and Senegal have established national fisheries councils entrusted with general advisory functions, but it is unlikely that these bodies will constitute an institutional scenario allowing for an evolution towards self-regulation of fisheries by theprofession, at least in the near future.

[37] Law N° 98-32 of 14 April 1998 (Loi des pêches).

[38] C. Breuil, E. Charles-Dominique, A. Van Houtte, *Assistance à la préparation d'un plan d'aménagement des pêches en Lagune Aby-Tendo-Ehy*, (Project FAO GCP/RAF/302/EEC, Document N° 20, 1995).

[39] J.-L. Kromer, *Propositions pour l'aménagement et la réglementation de la pêche artisanale en Guinée-Bissau*, (Project FAO GCP/RAF/302/EEC, Document N° 18, 1995) and *Relatório do seminário sobre a regulamentação da pesca artesanal na Guiné-Bissau* (Project FAO GCP/RAF/302/EEC, Document N° 45, 1996).

[40] While the authorities are reluctant to state their position officially, there are a few examples of reservations expressed in unconditional terms with respect to transfer of competence to decentralized administrations. See articles in the French newspaper *Le Monde*, 29 January 1998, page 5 and 3 February 1998, page 3. See also *Marchés Tropicaux et méditerranéens*, N° 2725, 30 January 1998, page 235.

[41] Many reviews have published an impressively high number of articles on co-management or community-based management. Some of these reviews have published special issues; see, inter alia, *Marine Policy*, Special issue: UNCED Agenda 21 and the role of legal authorities, Vol. 18, no 2, 1994; Ocean and Coastal Management, Special Issue: Property Rights and Fisheries Management,

vol. 28, Nos 1-3, 1995. Thus studies dealing with community approaches to management of resources, including fishery resources, can be counted today by the hundreds. Some of them are of exceptionally good quality as the study by Marie Christine Cormier-Salem, *Gestion et évolution des espaces aquatiques: la Casamance*, (Paris, éditions de l'ORSTOM, Série études et thèses, 1991). The main conclusions of this study are published in M.C. Cormier-Salem, Environmental changes, agricultural crisis and small-scale fishing development in the Casamance region, Senegal, *Ocean and Coastal Management*, vol. 24, N°2, 1994, 109. However, studies dealing with the legal aspects of the matter are scarce. In a recent article dealing with community-based management in forestry one can read observations that could be applied to fisheries: "Many, if not most, community-based management efforts around the world lack the legal support needed to provide a way for local people to establish enforceable legal rights to the resources on which they depend, or to play a meaningful part in planning and manage those resources". J.M. Lindsay, Creating a legal framework for community-based management: principles and dilemmas, *Unasylva*, 1999/4, p. 30.

[42] An extensive bibliography on the matter is available. A legislative study published by the Legal Office contains abundant basic information on customary rules of interest to fisheries. M. Ramazzotti, *Readings in African customary law. Anthologie du droit coutumier de l'eau en Afrique.* (FAO, Legislative Study N° 58). When working in a number of countries, the author found two publications of great interest. T. Scuddet and T. Conelly, *Management Systems for Riverine Fisheries,* (FAO Fisheries Technical Paper N° 263, 1985) and R. Welcomme and E. Henderson, *Aspects of the Management of Inland Water for Fisheries*, (FAO Fisheries Technical Paper N° 161, 1976).

[43] In the case of Mali, "Code de la pêche" of 1986; in the case of Niger "Loi des pêches" of 1971. Conventions were inter *alia* concluded with respect to Tenenkou, Youvarou, Djenne and Mopti.

[44] Article 12.12 of the Code of Conduct on Responsible Fisheries reads as follows: "States should investigate and document traditional fisheries knowledge and technologies, in particular those applied to small scale-fisheries, in order to assess their application to sustainable fisheries conservation, management and development".

[45] Another problem affecting inland fisheries is that posed by the fact that inland waters are used for many purposes, and fisheries concerns receive often low priority. A. Tavares de Pinho, La réglementation des pêcheries continentales dans le Sahel, *Revue générale de droit international public*, 1996-2, 391

[46] See *Report of the workshop on the evolution of the law of the Sea and West African fisheries*, Banjul, The Gambia, 19-27 September 1977, (CIDA/FAO/CECAF, Rome, 1978). There are also some more recent studies and reports on the matter. K. Roberts, *Legal and institutional aspects of monitoring, control and surveillance of fisheries and prosecution of offences in West Africa*, (Project FAO GCP/RAF/302/EEC, 1995, Document N° 23) and *Report of a Regional Workshop on Monitoring, Control and Surveillance of Fisheries of African States Bordering the Atlantic Ocean*, (FAO Field Report 92/22, 1992).

[47] Basically, shipowners are reluctant to disclose their positions and catches, as they fear competition on fishing grounds. The situation is likelyto evolve. It should be noted that the development of these procedures was part of an effort to rely on "soft" means of surveillance, based on increased cooperation between the coastal states and the shipowners and involving various mechanisms of strict self reporting, which were expected to alleviate, or even obviate, the deployment of material means of aerial or maritime surveillance. This usefulness of this approach was underlined on the occasion of preparations for the FAO World Conference on Fisheries Management and Development in 1984. The Conference itself called for the design and establishment of "practical mechanisms of compliance control in exclusive economic zones at the national, bilateral and regional levels, that reduce the need for costly enforcement and do not hamper fishing activities more than necessary". The most convincing presentation of this approach is found in G. Moore, Enforcement without force: new concepts in compliance control for foreign fishing operations, in *Essays in memory of Jean Carroz*, (FAO, Rome, 1987), 159. However, experience has clearly shown that those concepts are not of a self-standing nature, and must be seen within the context of a system including both soft and hard means of control.

[48] In the late seventies and early eighties various studies were prepared by FAO on the system of observers. The shortcomings of the system are associated with the fact that life on board a vessel seems to be incompatible with a presence, over a long period of time, of what is orcan be perceived, as a hostile person. In general, with limited exceptions, the system of observers in Africa has been

ineffective. There are a few interesting studies and reports on the system of observers: M.G. Jennings, *Enforcement of fisheries regulations*, (CECAF/TECH/80/22, Dakar, July 1980), P. Flewwelling,*An introduction to monitoring, control and surveillance for capture fisheries*, (FAO Fisheries Technical Paper N° 338, Rome, FAO, 1995) and *Report of a technical consultation on fisheries observers in West Africa, Dakar, Senegal, 11-13 December 1995*, (Project FAO GCP/RAF/302/EEC, Document N° 34). See also a broader *Report of a regional workshop on fisheries monitoring, control and surveillance*, Albion, Mauritius, 16-20 December 1996 (FAO, Rome, May 1997) which contains comprehensive information on countries bordering the Indian Ocean.

[49] These observers would remain on board for limited periods of time only. This is a tendency, favoured by some large distant water fishing countries, but seems to be opposed by coastal states.

[50] Report of the technical consultation on technical specifications for the marking of fishing vessels, Rome, 16-10 June 1986, (FAO, Fisheries Report N° 367).

[51] The interest and importance of this system is not only linked to the fact that it allows for the identification of the fishing vessels. The identification of a vessel through its radio call number is particularly suitable for use in computerized systems.

[52] Developments on the practical difficulties to enforce vessel marking requirements are found in *Workshop on the harmonization of fisheries legislation of member States of the Sub-Regional Fisheries Commission, Saly, Senegal, 17-18 August 1995*, (Project FAO GCP/RAF/302/EEC, 1995, Document N° 26).

[53] While provided for in various laws, these requirements have not been enforced in Africa. See W. Burke, Exclusive Fisheries Zones and Freedom of Navigation, *San Diego Law Review*, 1983, 596.

[54] The main developments contained in this paragraph are taken from a note prepared by the FAO Legal Office entitled "Legal issues relating to the use of vessel monitoring systems in fisheries monitoring control and surveillance".

[55] A number of legal issues relating to the implementation of VMS are the subject of some debate. These issues include, first of all, confidentiality. Ensuring confidentiality of VMS information, particularly vessel positions and catch reports is a sensitive matter to the fishing industry and therefore is fundamental for the success of VMS implementation. Ensuring confidentiality entails, inter alia, limiting its use substantially to fisheries management purposes; restricting disclosure and limiting access to authorized personnel only in the VMS monitoring agency and protecting intellectual property. Difficulties arising out of the uncertainty regarding the maritime boundaries limit the effectiveness of VMS in some areas.

[56] It should be noted that the heavy involvement of military authorities in fisheries surveillance far from precluding the development of international bilateral and sub-regional cooperation seems to have in fact greatly favoured such cooperation. Military authorities tend to share a common system of values, a similar approach to the need to protect resources and a more realistic view of what is politically feasible.

[57] The Law of the Sea Convention provides in its article 73 that the coastal State may take such measures including arrest as may be necessary to ensure compliance with the laws and regulations that it has adopted. It provides further that arrested vessels and crews should be promptly released upon the posting of reasonable bond or other security. In fact, the provisions allowing boats to be brought to port were abused insofar as boats, based in and operating from the coastal States were brought to ports for minor offences. Various reflections were carried out in some countries which have adopted appropriate criteria to deal with such situations. The provisional seizure of a boat fishing for high value species (shrimp for example) is a very serious act involving high costs. There were a few cases where administrations were held liable for undue arrest of boats and had to give compensation.

[58] Several laws impose tight time limits for the release of arrested crews and fishing vessels. For instance article 60 of the Fisheries Law of Angola. Article 62 of this Law provides that the State of Angola shall be liable for damages caused to shipowners as a result of unjustified arrest of a vessel on the occasion of surveillance operations.

[59] B. Ndiaye and A. Tavares de Pinho, Une expérience africaine de coopération halieutique: la Commission sous-régionale des pêches, *Espaces et ressources maritimes* (1994, N° 8), 237.

[60] The high level of fines tended to erode the credibility of the overall surveillance systems. In the case of foreign vessels, flag State authorities used to intervene with the coastal States which was a source of political and diplomatic embarrassment.

[61] This is the case, for instance, in Mauritania, Senegal, Guinea, the Gambia and in Angola. In Gambia, an inter-ministerial committee consisting of the Permanent Secretaries of the Ministries of Natural Resources and Environment (in charge of Fisheries) and Defence, a representative of the Attorney-General, a senior Naval Officer and the Director of Fisheries convenes to compound offences when a vessel owner opts for an out-of-court settlement. It recommends a fine to the Minister in charge of fisheries who imposes the fine. Similar procedures are in the process of being generalized in some sub-regions.

[62] A typical and simple example of this is the provision whereby any fish found on board a boat without a licence is deemed to have been caught by that boat and, usually, forfeited, unless the master or shipowner can provide evidence that that was not so. See D. Freestone, *The Burden of Proof in Natural Resources Legislation: Some Critical Issues for Fisheries Law*, (Legislative Study N° 63, FAO, Rome).

[63] K. Roberts, Legal and Institutional Aspects of Fisheries in West Africa, *The African Journal of International and Comparative Law*, 1998, 88, at 112.

[64] G. Moore, "The Food and Agriculture Organization of the United Nations Compliance Agreement", *The International Journal of Marine and Coastal Law*, 199510, 412.

SETTING UP INSTITUTIONS TO MANAGE TRANSBOUNDARY NATURAL RESOURCES: LAKE KARIBA A CASE IN POINT.

CECIL MACHENA[1]

INTRODUCTION

Lake Kariba[2] was formed by damming the Zambezi River between 1955 and 1958. It was finally filled in 1963. It is the third largest man-made lake today with a length of 277 km, a mean depth of 29m and a surface area of 5 400km^2. The lake is shared between Zambia and Zimbabwe, and has a catchment area of 1 193 500 km^2 which extends over Angola, Zambia, Namibia and Zimbabwe. The lake has an average annual draw down of 3m. This is because rainfall is seasonal, falling between December and February. The lake level rises to reach a maximum in May – June, although it has experienced massive fluctuations of late, following severe droughts. The lowest level (11m below maximum level) was experienced in 1982.[3] This shows that climatic changes will impact the lake and its ecology. The lake is divided into 5 basins on the basis of hydrobiological factors.

There are two types of fisheries on the lake, the inshore and the pelagic fisheries. The pelagic fishery is for commercial exploitation and is based on a clupeid (*Limnothrissa miodon*). The inshore fishery exploits mainly ten species belonging to the cichlid, characid and mormyrid groups. The inshore fishery is artisanal and is operated by individuals and co-operatives that use multi-filament gill-nets and canoes. This fishery is limited to shallow inshore waters of less than 10m depth,[4]

Large-scale artisanal fishing on Lake Kariba commenced on the Zambian shore in 1961 and in 1962 on the Zimbabwean shore. The inshore fisheries from the two countries are effectively separate, as the lake is deep and wide and the inshore fish species are limited to shoreline areas. The inshore fishery potential is limited.[5] Annual inshore yields on the Zimbabwean shore amount to 1 500 tonnes. The fishery is exploited by well over 800 fishers.[6] Annual yields in Zambia amount to 2 000 tonnes.[7]

Lack of effective state control in the Lake Kariba inshore fisheries by both countries put the management of the fisheries into a crisis. The fishery has effectively been vulnerable to open access amongst resource users without accountability. There are a number of reasons for this.

Legislation and policies designed to manage and control fishing operations have historically been centralised. The top-down approach essentially disregarded the needs of the fishers and placed emphasis on strict enforcement. Fishers did not embrace this approach well and resisted the measures. They are legally required to have a license to fish on the lake as a means of controlling access and effort, but this has proved difficult to enforce and monitor. The worsening economic climate in the two countries has not helped either. The people around the lake are basically agrarian, despite the fact that agricultural production frequently fails because the climate is hot and dry. Crops frequently fail making the situation around the lake desperate particularly during dry years. During these periods many people from around the lake go fishing without licenses. Entry into the gill-net fishery is attractive compared to other economic ventures because of the low capital input requirements, making fishing an "occupation of last resort".

Government monitoring of the illegal fishing activity is difficult due to the poor access to most parts of the lake. The Department of National Parks and Wildlife Management, which has the responsibility for fisheries management, lacks sufficient resources to monitor and control fishing activities. The lake is in a remote part of the two countries with poorly developed communications infrastructure. People are able to fish in the remote areas of the lake without being detected.[8] Fishers also operate in prohibited areas (reserve areas, rivers feeding the lake and their mouths).

There are other problems that compound the situation. The fishers operate independently as there are no institutions the fishers can interact with. In the past chiefs had control over fishing villages in sections of the lakeshore in which only their subjects fished. This system has broken down. There are some fishers who continue to observe regulations but they have no 'right' nor 'power' to control the majority who ignore them.

Sanyanga et.al. (1995) and Sanyanga (1996) have shown that some shores of the lake are heavily fished and in these areas, predictable changes have occurred in fish assemblages. Larger species, which are in higher demand, are disappearing from the assemblages and are being replaced by smaller counterparts that have a higher turnover particularly *Synodontis zambezensis*. The mean length of the target species is declining. This pattern shows that some individual species are overfished and are being replaced by smaller species of less value.

The situation above reflects a scenario in which the fishers have failed to self-regulate the exploitation of the resource in their own interest. It is this scenario that strongly persuaded the fisheries managers, being the Department of National Parks and Wild Life Management in Zimbabwe and the Department of Fisheries in Zambia, to introduce a community based management approach and the development of a Protocol as an instrument for a joint approach to the management of the fishery by the two countries.

This Chapter presents an outline the philosophy behind the new management approaches. A joint project between Zambia and Zimbabwe, (the Zambia

Zimbabwe SADC Fisheries Project),[9] provided the opportunity for the two countries to design and launch the new approach.

The Zambia Zimbabwe SADC Fisheries Project and the Formation of the Protocol for the Joint Management of their Fisheries of Lake Kariba

In 1982, the Southern African Development Community (SADC) presented to donors a proposal for a joint Zambia Zimbabwe Lake Kariba Fisheries Project. The Danish and Norwegian International Aid Agencies, DANIDA and NORAD respectively, expressed interest in jointly funding the Project. The two donors funded a feasibility study in 1984, which recommended the need for the Project.

The Zambia Zimbabwe SADC Fisheries Project (ZZSFP), as the Project was subsequently called started with the pre-project phase in March 1988. The aim of this phase was to clarify conditions in the project area, to ascertain precisely the mode of cooperation between the recipient countries (Zambia and Zimbabwe) and to prepare a more detailed Project design plan.

A joint project management system was established at two levels. A Steering Committee was set up. This comprised of the Director of Fisheries (Department of Fisheries, Zambia), the Director of the Department of National Parks and Wildlife Management (Zimbabwe), the Officer-in-Charge (Lake Kariba Fisheries Research Institute), the Chief Fisheries Officer (Department of Fisheries, Zambia) and the Director for the Centre for Applied Social Sciences (University of Zimbabwe), and any other organisation that may be recruited from time to time. NORAD sat on the Steering Committee as observers. The Centre for Applied Social Sciences (CASS) at the University of Zimbabwe had a special role to play. CASS was contracted to be overall responsible for supervising the socio-economic component of the Project (in both countries) and for advising the Steering Committee on integrating the results from socio-economic research into fisheries management plans.

The Steering Committee had the responsibility for providing overall guidance to the Project, approving work plans, budgets and contracts. A Project Co-ordinator served as Secretariat to the Steering Committee. The Steering Committee met at least twice a year. Although the Project was based at the Lake Kariba Fisheries Research Institute (LKFRI), the Chair of the Steering Committee rotated between the two countries annually. The venue for the meetings also alternate between the two countries.

Below the Steering Committee served the Project Co-Managers (the Office-in-Charge, Lake Kariba Fisheries Research Institute, and the Chief Fisheries Officer, Department of Fisheries, Zambia). The Project Co-Managers were responsible for running the Project on a daily basis. Project activities were to be based on an annual implementation plan. Once a year, Project staff was to carry out a review of activities of the previous year and planned activities and budgets for the following year. Project Co-Managers were to have joint meetings at least once every three months. The Co-Managers together with the Coordinator were to prepare for Steering Committee meetings.

Collaboration and co-operation between the two countries at both levels was strong. This critical component contributed undoubtedly to the success of the Project. External Project evaluations also acknowledged that Project management functioned well.

Importantly, the above joint administrative structures were set up as temporal institutions, to get the Project up ultimately to transform them into legally binding permanent institutions. This was to be effected though the "*Protocol on Economic and Technical Co-operation Between the Government of the Republic of Zambia and the Government of the Republic of Zimbabwe Concerning the Management and Development of Fisheries on Lake Kariba and the Transboundary Waters of the Zambezi River*". The *Protocol* is one of the important outcomes of the Zambia Zimbabwe SADC Fisheries Project (ZZSFP). It forms the legal basis for setting up a joint institution between Zambia and Zimbabwe and a framework for collaborative research and fisheries management on Lake Kariba, providing a tool for fisheries management.

It is one of the few protocols developed within SADC to manage transboundary natural resources, despite the natural resources that all SADC members share. Protocols that facilitate of the establishment of inter-state institutions to manage transboundary natural resources offer a number of opportunities. These can be seen at both the community and government levels. Communities along national borders often share common values and experience similar problems calling for cooperation. In colonial Africa, political boundaries were arbitrary and cut across nations and tribes. As an example, the Zambezi River forms the political boundary between Zambia and Zimbabwe. The area that is now covered by Lake Kariba, was occupied by the Tonga tribe. The communities were now divided by the political boundary and were distanced further apart by the Lake.

At the government level, opportunities exist for transfrontier cooperation and joint management of transboundary resources. This forms the basis for setting common standards and accountability e.g. in terms of information sharing, policy development, and investments across the boundary, etc.

The *Protocol* was developed at the end of a long (10 year) project, which had witnessed successful collaboration between Zambia and Zimbabwe. Some of the experience acquired in the project management process may be transcended to both the regional and international levels. The *Protocol* has been signed by high-level government representatives from the two countries. The Protocol authorises the establishment of the Joint Fisheries Committee of Management (JFCM) that has the mandate for joint collaboration on hydrobiology and fisheries research and management on Lake Kariba, regardless of donor funding. At the time of writing this Chapter, the JFCM has not yet been established.

It is envisaged that the JFCM will operate as one of the Committees of the Zambia Zimbabwe Joint Permanent Commission. The operations of the Zambia Zimbabwe Joint Permanent Commission are funded by both countries. It is expected that both countries will also fund the JFCM, and hence ensure its longevity, reducing dependence on Donor funding.

Defining the Context for Sustainable Ecological Production

In the research and management of exploited ecological systems, one overarching issue invariably surfaces - the linkage between Sociology, Ecology and Economics. These three disciplines are connected in a complicated way in exploited ecological systems including fisheries such as Lake Kariba. In an exploited fishery, research focuses on the human impact on the ecological system. The human impact is manifested at the level of the fishing effort. The fishing effort is a balance of many forces invariably determined by the sociology of the fishers (technological capacity, level of or lack of basic needs, poverty, etc.), the economics (the level at which the fishery is perceived as a resource, the macro-economic environment, etc.), and the politics (public policy issues promulgated to manage the environment or lack of these, the importance of the resource at national level, etc.) (Fig.1). In Fig.1, Negative feed backs reflect the response of the ecological system to human impact. These reflect an integrated ecological response to the various anthropological forces (culture, social behaviour, technology, economy, population, information, property rights issues, economic and social incentives, etc.).

Both research and management are a means to an end and not an end in themselves. The desired outcome of the human intervention is the sustained ecological production in functions (e.g. production of raw materials), services (e.g. nutrient cycling) and amenities (recreation, perception of beauty). Research and management attempt to define the parameters for sustained production, and then take these parameters as a management objective.

Sustainable ecological production, which is taken as the basis for sustainable development, is a difficult concept and it is equally difficult to set indicators to signal constancy of behaviour.[10] Sustainability is not a fixed condition as there are many variables that affect or influence it. Uses can either be sustainable within different limits or not. It is difficult to prescribe the condition of sustainable use before the implementation of a particular management regime.

Sustainable production is perceived differently in different environments, societies and cultures. Furthermore, the ecological system is in a dynamic state – forces that support a particular state are continuously changing in composition and strength.[11] A drought year will present a different scenario from a previously wet one. The level of production fluctuates with the variations in the dynamic state of the ecological system. For example. Karenge & Kolding (1994)[12] found highly significant correlations between lake level as well as the Zambezi River flow and fisheries yield in Lake Kariba. There are also spatial and temporal contexts and the levels of standards desired. In the spatial context the distribution of a particular species may vary.

Given these complexities, the context for sustainable use may be defined by answering the following questions; Sustainable how?; Sustainable for what?; Sustainable for whom?; Sustainable, under which conditions?[13]

Due to the influence of many variables, the concept of sustainable production is not an absolute frame, but a state of production perceived by a given community or society depending on their cultural, socio-economic, political

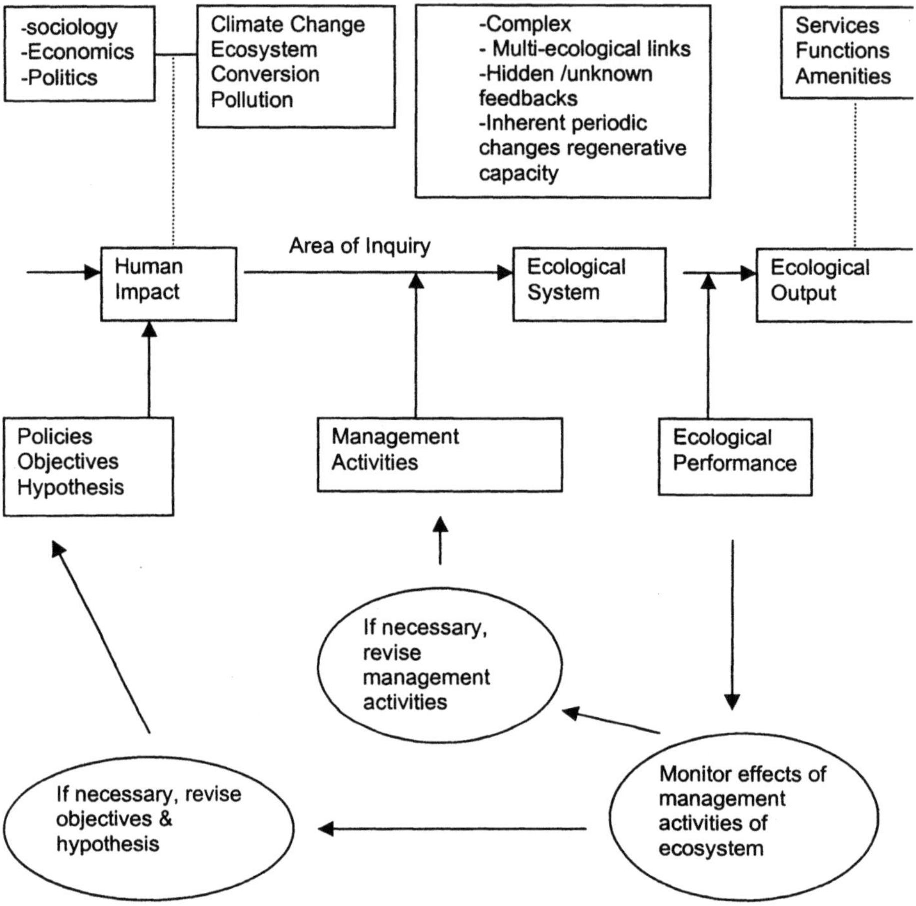

Fig. 1 The Interface Between Ecology and Economics in Natural Resources Management and the Negative Feedback Loops

status, etc. Different cultures may have different off take levels or understandings of sustainable levels because they have different needs potentially varying over time. Environmental degradation for example is associated with poverty or lack of alternate means of survival by a particular community, which will lead the community to exploit the environment unsustainably.

The new system that is being promoted in the management of the Lake Kariba fisheries offers a holistic approach in which fishers' impact on the ecosystem can be perceived the negative feed backs and the ecological forces behind the feed backs can be understood, using certain indicators and respond appropriately. (Fig. 1). The key element is the ability of the fishers to understand the response of the system to their fishing effort and to monitor these changes using a number of ecological and socio-economic indicators. On the basis of this, the fishers will decide whether to increase or reduce fishing effort or change fishing gear, etc. Ability to achieve this at the fishers level is a critical stage in the human/ecological interaction at which self restraint or reduction in exploitation effort (in the case of an overexploited situation) by an empowered community will lead to sustained use or recovery of a particular species. It is therefore important that communities are fully involved in the management of the resources. Incentives are needed for the communities to adopt the above approach as a conservation and development strategy.

The report of the Subsidiary Body on Scientific, Technical and Technological Advice of the *United Nations Convention on Biological Diversity*[14] describes the above approach as "an ecosystem approach" and defines it as "a strategy for the integrated management of land, water and living resources that promotes the conservation and sustainable use in an equitable way". The 5th Conference of the Parties to the *Convention on Biological Diversity* adopted the use of the ecosystem approach in the implementation of its work programme.

Fisheries Co-Management Approaches

Given the complex issues outlined above, the new strategy that has been adopted in the management of the Lake Kariba artisanal fisheries is a bottom up - needs driven approach focusing on local level factors. The lowest institution of fisheries management on Zimbabwe side of the lake is at the level of the Exclusive Fishing Zone or Concession Area while on the Zambia side it is at the level of the fishing camp. A fishing camp comprises of a group of fishing villages within a defined area. The "fishing camp" concept is a new development fully supported by Headmen and Chiefs. Before this, fishers were scattered all along the lakeshore on the Zambian side.

The lakeshore on the Zimbabwean side has been subdivided into Exclusive Fishing Zones (EFZ), which are henceforth taken as "concession areas". The sub-division was carried out following a study that analysed catch per unit effort along the shoreline.

Under the new management strategy, the concession area was taken as the unit for granting proprietorship over inshore resources to fishing communities. It demarcates the water base that defines the natural resource over which proprietorship is granted. Each concession area will host an identifiable group of fishers who will have collective proprietorship over the water base. It is a form of communal property regime in which users for the resource are controlled by an identifiable group and are not privately owned or managed by government. A communal resource management regime is enhanced if it is small enough (in membership size) for all members to be in regular contact, to enforce conformity to rules through peer pressure and to have a collective identity.

The number of fishers within a concession area is based on a Catch Per Unit Effort study using data collected over a long time. Hence the size of the unit is based on the ecological constraints within the concession. It is this identifiable group of fishers within the concession which is used to calculate the lowest community unit of production, management and benefit. Ultimately, the community plays a critical role of resource management through self-regulation.

Benefits and responsibilities are to be conferred through a "lease". The lease will act as a contract between the fishers and government. Each unit of proprietorship will have one lease. This marks a departure from the current practice where fishers are given individual permits. The idea is that this joint responsibility amongst the fishers in each concession area will enhance dialogue and direct contact within the group. Peer pressure is applied, which may help in improving compliance with regulatory measures necessary for sustainable utilisation. In terms or responsibility, the fishers within a concession area will need to set up a system of self-governance (rule making, conflict management, power sharing, dialogue, decision-making, knowledge generation, resource utilisation regulations and how to monitor observance to this, etc.).

The lease given to fishers within a concession area will exclude non-members from fishing activities. The sub-division of the shoreline on the Zimbabwe side and the lease are proscribed in the *Parks and Wild Life Act.*[15] In terms of Section 82 (1), the relevant Minister may, by notice in the gazette, declare any person to be the appropriate authority for the whole or portions of that water. This may be revoked under section 83 (1) of the same *Act.*

The changes being effected on the Zambian side of the lake are also provided for in the *Fisheries Act of Zambia.* What this means is that fishers from both sides of the lake are being given legal powers to make some key decisions as far as the use and management of the fisheries are concerned. However, the fisheries managers still retain some responsibilities. This is seen mainly in giving technical advice on such areas as optimal fishing levels, fishing techniques, fish processing techniques, advice on the impact of fishing operations, etc.

Fishery managers are expected to derive a number of advantages from the set up. As the fishers are able to monitor and control their own operations, this reduces cost to government through savings in enforcement. Empowered fishers will be in a position to record their own catch data. This will reduce enumeration costs to government. At the moment, managers rely on fisheries catch data sampled from a few parts of the lake. As fishers do their own recording, more lake coverage may mean increased reliability of the data collected.

The provision of the lease is a key development in the whole scenario. It signifies sharing of responsibility between government and the fishers. The new management approach developed in the ZZSFP, is a co-management system, which focuses on resource management at the community level. It is a partnership between government and local communities. In this case, the government provides the enabling atmosphere. The legislation guarantees communities a stake in local resources use[16] and powers of exclusion, and other assistance. Local communities assume management responsibility using indigenous capacities driven by the interest generated from the appropriation of benefits accruing to them. According to this management philosophy, the fishers become part of the resource management team, balancing rights and responsibilities and working cooperatively with government managers. Acceptance of this system by both fishers and government provides the necessary legitimacy for the fishers

The co-management approach is being developed as a process of resource management that has flexibility, and capacity to adjust and mature to changing circumstances, particularly as social empowerment is enhanced.

Experience from initiatives such as the Communal Areas Management Programme for Indigenous Resources (CAMPFIRE) show that conservation programmes that begin at the community level have a good chance of working. This is possible provided that positive social and economic incentives that allow the meeting of the interests and needs of the communities are built in. CAMPFIRE is a Zimbabwean programme that confers benefits and responsibilities or proprietorship over wildlife to the inhabitants of communal areas.[17]

Focus in fisheries management has been at the level of an identifiable group of fishers within a concession area or at the level of the fishing camp. Success at this level will mean incremental success between several community management units and finally, success at national level.

ESTABLISHMENT OF INSTITUTIONS FOR MANGEMENT AND THE ROLE OF THE PROTOCOL

As outlined earlier, the group of fishers within the concession area on the Zimbabwe shoreline constitute the lowest community management unit and the basis for creating higher management institutions units. Fishers in each Concession Area come together to form a Concession Area Fishers Association (CAFA) (Fig. 2).

CAFAs within each lake basin (Fig. 2) unite to form a Sub-Area Fishers Association (SAFA). SAFAs will in turn form a District Fishers Association (DFA). There will be two of these DFAs representing the two Rural District Councils, Nyami nyami and Binga that share the lake. These will ultimately come together to form the Zimbabwe Lake Kariba Fishers Association (ZLKFA).

On the Zambian shoreline the lowest unit of fisheries management is a Village Management Committee (VMC). At a higher level, a number of villages constitute a Zone and each Zone has a Zonal Management Committee (ZMC). At this level the ZMC also incorporates commercial fishing companies, which are located within the respective zone. The integration of the artisanal fishers and commercial companies makes a distinct departure from the Zimbabwean regime where the artisanal fishers have a different organisation from the commercial companies.

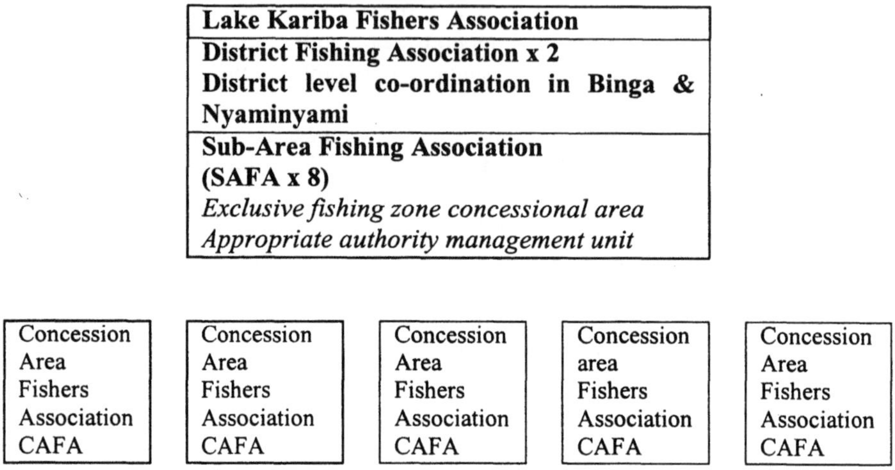

Fig 2. The formation of the Zimbabwe side inshore fishers institutions, showing horizontal and vertical linkages.

The system on both sides of the lake shows horizontal and vertical linkages. The vertical linkage shows a two-way information flow system (CAFA to SAFA to DFA to LKFA and *mutatis mutandis* on the Zimbabwe shoreline and VMC to ZMC to ZFMB and *mutatis mutandis* on the Zambian side). It means that the views of the fishers in a particular concession area can be channelled to the top and decisions made at the top can be communicated to the grass roots. Horizontal linkages i.e. communication from one CAFA to the next and from one VMC to the next is effected through the SAFA and ZMC respectively.

The *Protocol* links the two national fisheries associations at the inter-state level. It calls for the formation of a Joint Fisheries Committee of Management (ZFMC) between Zambia and Zimbabwe, as explained earlier. The two fishers associations, the Zimbabwe Lake Kariba Fishers Association and the Zambia Fisheries Management Board will be linked to each other at the inter-state level through the Joint Fisheries Management Committee. At this point in time, the JFMC has not yet been formed. It is not yet clear whether the ZFMB and the LKFA will have members sitting in the Joint Fisheries Management Committee or whether the two associations will express their views through their respective government fisheries managers sitting in the JFMC.

Through representation at the JFMC, decisions made at the inter-state level will be channelled to fishers at the village level and at the Concession Area level through the vertical and horizontal linkages explained above. Equally the concerns of the fishers at the low levels will be channelled to managers at the national and inter-state levels. Inter-state management allows for the formulation of responses to both regional and global issues (e.g. climate change leading to excessive fluctuations in lake level and transboundary pollution). Addressing the changes at the regional and global levels should be able to filter down to lower communities.

The ecosystem approach as defined above has a number of advantages. By focussing at the community level, it provides the following opportunities: the involvement of as many members of the community as possible in terms of decision - making and participation allowing transparency, and accountability; the use of traditional structures; and the use of local level indicators in monitoring key ecological factors and resources. This means that the governance over natural resources and their protection is at the level of the community.

One important element of the ecosystem approach is that it recognises and takes into account the interaction between man (taking into account cultural, economic and social factors) and the environment in the conservation process. By allowing communities to monitor their own impact and respond appropriately, this leads to the process of adaptive management, which is also outlined in fig.1. Adaptive management introduces another dimension of flexibility, which is critical in taking corrective measures through learning by doing.

Conservation approaches are now changing. In the past, focus was on setting up protected areas. Unfortunately, protected areas comprise about 10% of the world's surface area.[18] This means that the bulk of biodiversity lies outside protected areas. It is outside the protected areas where the majority of the people, who effectively serve as custodians of the ecological resources, live particularly in developing countries. In addition, dependence on biodiversity runs up to 90% of human survival requirements (food, medicines, energy, construction materials, clothing, etc.). Rural people live in impoverished communities fighting to survive on a daily basis. The outcome is that conservation is not only a priority but also a key to survival, creating a cycle between poverty and degradation.

The most persistent political realities of this decade are degradation of biodiversity and poverty. In most cases, biodiversity is the only natural capital rural people have. Biodiversity conservation must therefore be primarily active at the level of the rural community. Biodiversity or natural capital must be used as a vehicle for development and poverty alleviation in the ecosystems under focus. Biological resources have one marked feature - they are renewable. This forms the basis for sustainable development as long as this renewable capacity is not overtaxed. Sustainable development is a operative concept when attempting to meet the conservation objectives and the needs and aspirations of people.

CONCLUSION

A number of key elements in the ecosystem approach have been outlined. These focus on resource rights and decision-making powers governing the use of the resources. Understandably, participation and equity considerations are central issues. These processes have been enhanced by the devolution of authority. As the community management institutions are small, thereby enhancing public participation, this potentially promotes exclusion. Hence institutional design is critical.

The vision of the *Protoco*l is to link conservation efforts on the Zimbabwean side and the Zambian side of Lake Kariba. The pelagic species Limnothrissa miodon represents one fish stock in the Lake. Mismanagement by one country will affect the entire stock. Although the inshore stocks are separate, the *Protocol* provides an opportunity for dialogue between fishers and fisheries managers facilitating common approaches to management. In this case the two countries have adopted a common ecosystem approach in which fishing communities are fully involved in fisheries management. This is one case of a transfrontier conservation area that may work as a model for the management of natural resources shared by countries.

---

[1] I thank my colleagues in the Zambia Zimbabwe SADC Fisheries Project for sharing many ideas that led to conceptualising the fisheries conservation approaches described in this Chapter. I enjoyed sharing many ideas over the ecosystem adaptive management approaches with Rowen B. Martin and my colleagues at the Africa Resources Trust particularly Rob Monro. Finally, I wish to thank Kevin R. Gray who reviewed this paper.

[2] (16^{0} 28' to 18^{0} 06' S; 26^{0} 40' to 29^{0} 03' E)

[3] R.A. Sanyanga (1995), "Variations in the abundance of Synodontis zambezensis (picses: Mochokidae) in the inshore fishery of Lake Kariba", in R. A. Sanyanga,*The inshore fish populations of Lake Kariba with reference to the biology of*Synodontis zambezensis *Peters. 1852*, (PhD thesis. Dept. Systems Ecology, University of Stockholm. 1996).

[4] M. Coke, (1968). "Depth distribution of fish on a bush cleared area of Lake Karib Central Africa" 97.4 *Trans. Am. Fish. Soc.* 460 at 465; R.A. Sanyanga et al (1995), "Abundance and distribution of inshore fish in fished and protected areas in Lake Kariba, Zimbabwe", 306 *Hydrobiologia* 67 at 78.

[5] Sanyanga (1995).

[6] M.W.Murphree et al. (1989) "Pre-Project socio-economic study of the artisanal fishery of Lake Kariba", 1 *Zambia Zimbabwe SADC Fisheries Project Report*, Lake Kariba Fisheries Research Institute, Zimbabwe.

[7] *Ibid.*

[8] C. Machena & H. Moinuddin (1993), "Management of the Lake Kariba inshore fisheries (Zimbabwe), a proposal" 19 *Zambia Zimbabwe SADC Fisheries Project Report*, Lake Kariba Fisheries Research Institute, Zimbabwe at 19.

[9] Hereinafter referred to as the Project.

[10] U. Svedin, (1991)"The Contextual Features of the Economy –Ecology Dialogue" , in C. Folke & T. Kaberger, *Linking the Natural Environment and the Economy: Essays from the Eco. Eco. Group, (* Kluwer Academic Publishers. London, Boston, Dordrecht, 1991) , 3 at 29. A.M. Jansson (1991), " Ecological consequences on long-term landscape transformations in relation to energy and economic development", in C. Folke & T. Kaberger, *Linking the Natural Environment and the Economy: Essays from the Eco Group, (*Kluwer Academic Publishers. London, Boston, Dordrecht, 1991) , 97 at 101.

[11] U. Svedin.

---

[12] L. Karenge & J. Koldingm (1994), "On the relationship between hydrology and fisheries in Lake Kariba, Central Africa", in J. Kolding, *On the Ecology and Exploitation of Fluctuating Tropical Freshwater Systems*, (PhD thesis, University of Bergen, Norway, 1994)

[13] Svedin (1991).

[14] UNEP/CBD/COP/5/3, 25 February 2 000.

[15] Chapter 20:14 as amended

[16] R.B. Martin (1994), "The influence of Governance on conservation and wildlife utilisation" Paper presented at a Conference – Conservation through Sustainable Use of Wildlife, University of Queensland, Brisbane, Australia. 8-11 February 1994.

[17]R.B. Martin. (1986), "Communal Areas Management Programme for Indigenous Resources" (Department of National Parks and Wildlife Management, Harare, Zimbabwe. 1986); R.B. Martin (1994), "Alternative approaches to sustainable use: What does and doesn't work" Paper presented at a Conference – Conservation through sustainable use of wildlife, University of Queensland, Brisbane, Australia. 8-11 February 1994.

[18][18] D.C. Zbicz & M.B.J. Green, M.B.J. 1997. "Status of the World's Transfrontier Protected Areas", 7:3 *Parks* 5 at 10.

AFRICAN ENVIRONMENTAL GOVERNANCE:

Opportunities at the Regional, Subregional, and National Levels

CARL E. BRUCH*

INTRODUCTION

In recent years, the critical role of civil society and the public in protecting the environment has become clear. International declarations and agreements such as the 1992 *Rio Declaration* increasingly recognise the role that individuals (including citizens and non-citizens), non-governmental organisations, and local governments have in sustainable management of natural resources and in protecting critically important areas. Countries are starting to give life to environmental governance principles by establishing institutions that guarantee public access to the decision-making process, as well as by creating and clarifying substantive rights to information, participation, and justice. In fact, many African nations have enshrined these guarantees in their constitutions.[1]

Regional Organisations also are advancing environmental governance in their member countries. Thus, the United Nations Economic Commission for Europe (UN/ECE) convened a process that led to the creation of the 1998 *UN/ECE Convention on Access to Information, Public Participation in Decision-making and Access to Justice in Environmental Matters* (*Aarhus Convention*).[2] Similarly, in April 2000, the Organisation of American States adopted the *Inter-American Strategy for the Promotion of Public Participation in Decision-Making for Sustainable Development* (*ISP*).[3] Still ongoing, the Asia-Europe Meeting (ASEM) is developing Elements of Good Practice in the draft document *Towards Good Practices for Public Involvement in Environmental Policies* (*Draft ASEM Elements*), which is expected to be adopted in 2002.[4] Other regions of the world have implemented similar initiatives.[5]

These regional initiatives are significant both for their substantive provisions promoting public participation and for the process by which they were developed, which saw civil society playing an influential role in the development of these international agreements. Traditionally, international agreements have been the realm of nations to negotiate. Experience with the *Aarhus Convention*, the *ISP*, and the *Draft ASEM Elements* has proven that civil society can contribute meaningfully to the development of international norms, and the newly won respect for civil society has fostered a more collaborative relationship between governments and non-governmental organisations.

While none of these three regional initiatives specifically mentions the term "environmental governance," these initiatives – and others described in this Chapter – seek to advance environmental governance processes. They create or modify institutions and laws to improve public access to decision-making processes that can affect the environment. While some of the international instruments mentioned below broadly advance democracy and governance (such as the *Universal Declaration on Human Rights*[6]), most instruments specifically apply to environmental matters.

As in many cases, practice is catching up with theory as governments and civil society increasingly explore the links between environmental issues on the one hand and democracy and governance on the other. Thus, while environmental campaigns continue to emphasize traditional substantive concerns such as wildlife and hazardous waste, environmental organisations and agencies increasingly are re-examining the decision-making processes that affect the environment.[7] Similarly, democracy and governance organisations have taken note of the groundbreaking experiences of the environmental community in opening governmental decision-making processes to the public. Since the environment and natural resources are crucial to human life, health, and happiness, governments frequently are more amenable to democratic provisions in the environmental context; thus, for example, some countries have adopted laws recognising a public right of access to *environmental* information, but not yet to information generally.[8] As such, the environment can be a wedge issue in opening governmental processes to the public that would otherwise be closed. As governments become accustomed to transparency, accessibility, and accountability, these principles can be extended beyond the environmental context.

African environmental advocates increasingly are considering the possibility of developing an African initiative that would parallel the *Aarhus Convention*, the *ISP*, and the *Draft ASEM Elements*. Such an initiative may be pan-African or subregional. It could be organised under the auspices of the United Nations Economic Commission for Africa, the new African Union, or another international or regional organisation. It may address the same themes as the *Aarhus Convention*, the *ISP*, and the *Draft ASEM Elements*; focus on a subset of those themes; or incorporate other themes necessary for environmental advocacy, such as freedom of association and laws governing the incorporation, operation, and dissolution of NGOs. In fact, starting in late 1999, African advocates and governments have started to explore initiatives at different levels, efforts that are highlighted in this chapter.

This chapter surveys the experiences of ongoing initiatives, focusing on experiences that could be relevant to a possible African initiative. Section 2 briefly discusses the benefits of public access to environmental decision-making and then highlights the global context for promoting such access. Section 3 analyses the three primary regional initiatives. Section 4 surveys the current African context for a regional initiative, and Section 5 outlines some options available to African environmental advocates in pursuing a regional initiative on public access to environmental decision-making.

THE GLOBAL CONTEXT

Over the past few decades, there has been a growing recognition, both domestically and internationally, that environmental protection and management must involve all sectors of civil society. Most environmental impacts are local, and people living in an area frequently know their particular local area more intimately than the government. These individuals and the NGOs that represent them can bring their experience, resources, and energy to bear on environmental problems, supplementing and improving governmental actions. Such public involvement enhances environmental decision-making, encourages sustainable business practices, and strengthens civil society. Furthermore, including civil society in the development, implementation, and enforcement of environmental laws and policies educates the public and the regulated community and builds support for governmental actions. This broad-based popular support can be particularly important for evaluating potentially controversial governmental actions affecting the environment.

As experience has developed at the national level, international declarations, agreements, and conventions increasingly have promoted access to information, participation, and justice.[9] Access to information – usually defined broadly and with a presumption in favor of disclosure – ensures that people can know the actual and threatened environmental impacts of particular actions, and respond appropriately. Public access to governmental decision-making processes allows the public to participate in the development of policies and the implementation of environmental provisions. Access to justice guarantees that the other rights are not just paper rights, but concrete requirements that people can enforce in courts and administrative bodies.[10] Additionally, access to justice may include associational rights (e.g., freedom of assembly), which protect the right of people to come together to advocate their interests through public interest and community-based organisations.[11] Associational rights are particularly important where organisations are in danger of being decertified or disbanded for challenging governmental actions.

As early as 1948, the *Universal Declaration on Human Rights* (1948) provided the kernels for generalized rights of access to information (Article 19) and justice (Articles 8 and 10), as well as the right to associate (Article 20). Similarly, the 1966 *International Covenant on Civil and Political Rights* guarantees the "freedom to seek, receive and impart information and ideas of all kinds."[12] The 1982 *World Charter for Nature* requires public disclosure of conservation information "in time to permit effective consultation and participation" as well as "the opportunity [for all persons] to participate, individually or with others, in the formulation of decisions of direct concern to their environment, and [to] have access to means of redress when their environment has suffered damage or degradation."[13] Following the elaboration of various regional instruments, the 1992 *Rio Declaration* crystallised the emerging public involvement norms in Principle 10:

> Environmental issues are best handled with the participation of all concerned citizens, at the relevant level. At the national level, each individual shall have appropriate access to information concerning the environment that is held by public authorities, including information on hazardous materials and activities in their communities, and the opportunity to participate in decision-making processes. States shall facilitate and encourage public awareness and participation by making information widely available. Effective access to judicial and administrative proceedings, including redress and remedy, shall be provided.

Since the *Rio Declaration*, various international conventions addressing specific environmental issues have incorporated public involvement principles. For example, the *Desertification Convention* (1994) rejected the previous centralised approach, which had proven ineffective in addressing desertification, and instead adopted a model that emphasized "the participation of populations and local communities" in developing and implementing environmental programs.[14] The primary mechanism adopted is the national action programme that provides a framework for identifying the causes of desertification, as well as the steps to be taken to combat and mitigate its effects. In addition to "specify[ing] the respective roles of government, local communities and land users," the national action programmes are required to "facilitate access by local populations to appropriate information and technology" and "provide for effective participation at the local, national and regional levels of non-governmental organisations and local populations . . . in policy planning, decision-making, and implementation and review of national action programmes."[15] The *Biodiversity Convention* (*CBD*) incorporated similar public participation principles.[16]

The global effort to include civil society in environmental decision-making extends to many international institutions that affect the environment. While examining public participation in international institutions is beyond the scope of this Chapter, it is worth noting that a number of these institutions now provide mechanisms for public access to information, for civil society participation in the decision-making process, and for concerned parties to appeal decisions. World Bank Institutions such as the International Bank for Reconstruction and Development (IBRD) and the International Finance Corporation (IFC) make many of their documents (including environmental impact assessments of projects) publicly available, and the IBRD has developed Inspection Panels to investigate claims by citizens and non-governmental organisations that a Bank-funded project harmed the environment or human rights. Likewise, although the World Trade Organisation's Appellate Body recently allowed environmental NGOs to submit *amicus curae* briefs in a case with environmental implications.[17] One of the primary objections voiced by activists at the aborted 1999 WTO negotiations in Seattle – as well as subsequent meetings of international institutions in Washington, DC, Montreal, Prague, and Genoa – was that the WTO operates largely in secrecy and without the opportunity for public involvement. At the same time, other international organizations such as the United Nations and its organs, the Organisation of American States, the European Union, and the North American Commission for Environmental Cooperation have increasingly provided for public access to information and participation in their deliberations.[18]

The global development of public involvement norms has interacted synergistically with regional and national initiatives to promote public participation. Global declarations and conventions lend authority to, and in some cases mandate, regional and national efforts to clarify and implement their norms. In turn, the experiences in the different countries and regional bodies provide lessons learned that then may be incorporated – or avoided – in subsequent international documents. Through this iterative approach, governments are able to learn from the experiences in various countries and contexts, become comfortable with the principles, and gradually implement the principles in their own countries. In fact, since the general statement of *Rio Principle* 10, regional initiatives have developed different constructions of the general principles, and some government and NGO representatives have advocated globalising these regional initiatives. In fact, Kofi Annan, Secretary-General of the United Nations, has noted, "The 2002 Special Session of the United Nations General Assembly marking the 10th anniversary of the Earth Summit would be a timely occasion to examine the relevance of the *Aarhus Convention* as a possible model for strengthening the application of principle 10 in other regions of the world."[19]

Regional Initiatives

Regional initiatives promoting environmental governance have complemented the development of global principles, and continue to be important in clarifying and implementing those principles. In early cases, the regional initiatives were primarily declarations and other "soft law" documents, with little substantive import for the signatories. Nevertheless, these general pronouncements promoting environmental governance principles[20] paved the way for more specific pronouncements, such as the *Rio Declaration* (1992).

The 1981 *African Charter on Human and Peoples' Rights (African Charter)* is particularly relevant, although portions of it may be considered hard law.[21] Article 9 guarantees to every individual the right to receive and disseminate information. Article 13 generally provides that citizens have the "right to participate freely in the government of [their] country." Article 7(1) ensures the right of every person to "have his cause heard," and Article 10 protects the right of free association. Many African constitutions have incorporated these provisions, either explicitly or by reference.[22] These general provisions provide a starting point for elaborating more detailed and concrete rights for environmental governance in Africa. In fact, Article 24 of the *African Charter* guarantees to "all peoples . . . the right to a generally satisfactory environment favorable to their development," a right which is reflected in many African constitutions.[23]

While global instruments have promoted environmental governance through general principles, they have provided little guidance on how to implement those principles. Yet, clear and concrete domestic implementation is essential to the realization of any environmental governance protections. In Europe and the former Soviet Union, the Americas, and Asia regional initiatives are fleshing out the basic principles and providing guidance to states in promoting access to information, participation, and justice in environmental matters.

The Aarhus Convention

The *Aarhus Convention* provides a clear example of how governments and civil society can jointly develop and implement regional norms for environmental governance. In addition to elaborating the details of how to strengthen environmental governance through national laws and institutions, the Convention also saw an unprecedented level of NGO involvement in its conceptualization, drafting, signing, ratification, and implementation.

The Substance of the Aarhus Convention

The *Aarhus Convention* is the first international instrument dedicated to codifying environmental governance principles. As such, it is likely to influence the development of other regional initiatives. In fact, the organisation of American States and the Asia-Europe Meeting considered the *Aarhus Convention* as they developed regional instruments for the Americas and Asia, respectively. These regional initiatives have prompted similar discussions about their relevance as possible models for an African initiative.

The *Aarhus Convention* emphasises three related areas or "pillars" of environmental governance: access to information, public participation in decision-making, and access to administrative and judicial redress. In these three areas, the *Aarhus Convention* provides minimum standards for the state parties to adopt in their domestic legislation.[24] The *Aarhus Convention* relies on enforceable rights of individuals in each of the three areas and also makes reference to a human right to a healthy environment. In contrast to many conventions that establish norms that do not necessarily vest rights in anyone, this rights-based approach underpins many of the specific provisions expanding public access, as well as providing a basis for individuals and NGOs to enforce their procedural and substantive environmental rights in Court. The *Aarhus Convention* also focuses on the obligations of governmental bodies and individuals at the national, sub-national, and local levels, as well as natural and legal persons "performing public administrative functions"[25] and European Union institutions; wholly private bodies, such as for-profit corporations, are generally beyond the scope of the *Aarhus Convention*.

The *Aarhus Convention* specifically prohibits nations from discriminating against natural or legal persons on the basis of "citizenship, nationality or domicile."[26] In fact, this provision seems to apply to all members of the "public," and not just reciprocally to other member states. Thus, to the extent that a citizen has a right to obtain information from the government, participate in decision-making processes, or appeal to independent bodies for recourse, anyone from any country appears to have the same rights under the *Aarhus Convention*.

Access to Information

Access to information forms the basis for public awareness of and response to threats to public health and the environment, and is essential to effective public participation in governmental decision-making. The *Aarhus Convention's* access to information pillar (Articles 4 and 5) establishes public rights to obtain environmental information and governmental responsibilities to collect, organise, and disseminate environmental information. Thus, the government must provide information on request (passive dissemination under Article 4) and collect, analyse, and disseminate certain information widely (active dissemination under Article 5).

Environmental information is broadly defined to include information in any material form (written, electronic, visual, etc.) on the "state of the elements of the environment . . ."; institutional and natural factors "affecting or likely to affect the elements of the environment"; and the "state of human health and safety, conditions of human life, cultural sites and built structures . . .."[27] This definition is non-exhaustive, and extends well beyond the typical definition of "environment" that is primarily linked to human health concerns. It also extends to information held by any public authority within the scope of the *Aarhus Convention*,[28] not merely to information held by environmental authorities.

Under Article 4, which deals with passive dissemination of information, civil society is guaranteed broad rights to obtain environmental information that the government holds. There is a presumption in favour of access, and any person may request the information without demonstrating an interest in the information or stating the purpose for which it will be used.[29] The government must provide the information "as soon as possible and at the latest within one month after the request has been submitted, unless the volume and the complexity of the information justify an extension of this period up to two months after the request."[30] The *Aarhus Convention* allows public authorities to withhold requested information only if its release would harm specific enumerated interests, such as: national defence, commercial confidentiality, intellectual property rights, and personal privacy.[31] The exemptions "shall be interpreted in a restrictive way, taking into account the public interest served by disclosure and taking into account whether the information requested relates to emissions into the environment."[32] Thus, the *Aarhus Convention* seeks to maintain the public's right to information regarding the quantity, nature, and sources of pollutants that could affect them.

If the authority does not have the requested information, it must either forward the request to the authority that does have the information, or it must inform the applicant where it believes the information may be obtained.[33] If the authority refuses to release the information, the refusal must be in writing if the original request was in writing or the applicant requested a written response. The decision must state the reasons for the refusal and provide information on how to administratively appeal or judicially review the refusal, under Article 9 of the *Aarhus Convention*.[34] Finally, when releasing information public authorities may choose to impose a "reasonable" fee for supplying the requested information,[35] as well as the circumstances under which the charges may be waived (for instance, for non-profit institutions).

Article 5 of the *Aarhus Convention*, dealing with active dissemination of information, requires public authorities to collect, update, and actively disseminate environmental information that is "relevant to their functions."[36] When there is an imminent threat to public health or the environment, the authority must immediately disseminate all information that could help the affected public prevent or mitigate the effects.[37] To promote public access to the information, states must inform the public about the nature and extent of information held by the government and how the public may access it.[38] The *Aarhus Convention* requires each state to "progressively" provide information via electronic databases that the public may access through the internet.[39]

Each country must prepare, publish, and disseminate national state-of-the-environment reports.[40] These reports provide the public with information on environmental quality and on the various pressures on the environment. Under the *Aarhus Convention*, reports must be produced at least once every "three or four" years.

The *Aarhus Convention* also requires nations to establish national pollution inventories or registers.[41] These databases must be computerised and publicly accessible, and they may track the inputs, releases, and transfers of specified chemicals. Frequently, such systems take the form of a pollutant release and transfer register (PRTR).[42] The PRTR tracks the releases and transfers of specific chemicals to the different environmental media: air, water, and soil. Generally, any private or public facility that uses or emits more than threshold levels of these chemicals in a given year must report its releases of that chemical.

PRTR systems illustrate the power of public access to environmental information. In determining the amount of chemicals released to the environment, industrial facilities are often surprised to discover the raw materials that they waste.[43] Through the PRTR and materials accounting, companies can reduce the amount of pollution they are emitting, saving money by conserving raw materials and reducing disposal costs. Since PRTR information has facility-specific information, individuals are able to determine the sources, types, and amounts of pollution to which they are exposed. In many cases, this knowledge has empowered people to apply direct pressure on the facility to reduce its emissions, particularly when it releases significantly more than comparable facilities located elsewhere.[44] As a result, a number of facilities reduced the amount of chemicals used and released, or even replaced them with non-toxic substances.[45]

Finally, the information generated through the PRTR process can be valuable in shaping environmental policy and legislation. The chemicals tracked by the PRTR often are not regulated, and the information gathered on their releases and transfers can help prioritise government actions by determining which chemicals are being released in sufficiently large quantities that suggests that further investigation and action may be required.[46] As for the state-of-the-environment reports, harmonisation of PRTR standards could provide an effective tool for monitoring pollutants released throughout a region, as well as providing insight into their sources. The *Aarhus Convention's* PRTR provisions remain primarily hortatory, with few concrete obligations. The UN/ECE established a task force to investigate opportunities for developing PRTR in the region, and a PRTR Protocol is expected to be adopted at the 2003 Kyiv Ministerial Meeting.[47]

Finally, the *Aarhus Convention* requires dissemination of legal and policy documents, as well as publication of the facts and analyses relied upon in developing major environmental policies.[48] The *Aarhus Convention* also promotes mechanisms such as ecolabeling and environmental auditing which provide the public, governments, and industry with specific environmental information.[49]

Public Participation in Decision-making

Public participation in decision-making improves the quality of the decision by widening the knowledge base. Individuals frequently know particular areas – including their unique environmental values and the impacts of environmental threats – more intimately than public authorities do, and civil society can bring added energy, expertise, and innovation to bear on environmental issues. Additionally, by providing people with an opportunity to be heard, public participation builds the public support for the ultimate decision.

For purposes of the *Aarhus Convention*, the "public concerned" has a right to participate in a wide range of governmental decision-making processes under Articles 6, 7, and 8. The public concerned is defined as the "public affected or likely to be affected by, or having an interest in, the environmental decision-making."[50] Furthermore, non-governmental organisations "promoting environmental protection and meeting any requirements under national law shall be deemed to have an interest,"[51] and thus have a right to participate.

Article 6 establishes requirements for public participation in decision-making on licensing and permitting certain kinds of proposed activities, which are listed in Annex I to the *Aarhus Convention*.[52] The activities are similar to those for which an environmental impact assessment is typically required.[53] Additionally, public participation is required "in accordance with . . . national law" for decisions regarding proposed activities that are not listed in Annex I if they may have a "significant effect on the environment."[54] For both sets of proposed activities, there is an exemption for proposed activities that serve national defence purposes if public participation would adversely affect national defence.[55]

For activities for which the *Aarhus Convention* mandates public participation, public authorities must provide adequate, timely, and effective notice to the public of the proposed activity, the nature of the possible decision, the responsible public authority, the process for making the decision (including opportunities for public comment), and other relevant information.[56] The *Aarhus Convention* requires public participation to occur early in the decision-making process, when options remain open and public input can still be incorporated into the final decision.[57] To ensure informed and effective public participation, public authorities must provide all relevant information to the public free of charge and as soon as it is available.[58] The public is guaranteed an opportunity to submit written or oral comments.[59] The *Aarhus Convention* seeks to ensure that the participation is meaningful by requiring parties to "ensure that in the decision due account is taken of the outcome of the public participation" and that public authorities issue their decisions in writing, include the bases for the decisions made, and make the decisions publicly available.[60]

While Article 6 mandates significant opportunities for public participation in the permitting and development of projects, both Articles 7 and 8 are more circumscribed when it comes to involving the public in the development of the overarching norms or the general plans and programmes that implement the norms. Article 7 requires nations to "make appropriate practical and/or other provisions for the public to participate during the preparation of plans and programmes relating to the environment . . . ."[61] The *Aarhus Convention* requires plans and programmes to follow the Article 6 provisions relating to reasonable time frames for public participation, opportunity for early participation, and taking "due account" of the participation. In hortatory language, Article 7 encourages parties "to provide opportunities for public participation in the preparation of policies relating to the environment." Similarly, Article 8 generally encourages nations to promote public participation in the preparation of "executive regulations and other generally applicable legally binding rules that may have a significant effect on the environment."

Access to Justice

The third pillar of the Aarhus Convention – access to justice – provides the enforcement mechanism necessary to ensure that individuals and NGOs can realise their procedural rights under the *Aarhus Convention*, as well as empowering individuals and NGOs to enforce environmental law. Article 9 requires that the judicial and administrative review procedures be "fair, equitable, timely and not prohibitively expensive."[62] The decisions must be in writing. Court decisions must be publicly accessible, and the *Aarhus Convention* encourages other authorities to make their decisions public whenever possible. Additionally, the *Aarhus Convention* encourages parties to consider mechanisms for removing or reducing financial and other barriers to access to information.[63]

Article 9.1 provides that any person who believes that his or her request for information under Article 4 was not dealt with according to the prescribed requirements must have "access to a review procedure before a court of law or another independent and impartial body established by law."[64] Alternatively, the aggrieved person can seek independent reconsideration of the request. Final decisions under either option bind the public authority holding the information; and if the refusal to release information is upheld, the reasons must be in writing.

Article 9.2 provides a right to review the decisions made on projects or activities addressed by Article 6. Members of the public can appeal either the substantive or the procedural legality of any decision, act, or omission. In order to invoke this provision, the appellant must have a "sufficient interest" or claim "impairment of a right," although the Article also provides that these requirements must be interpreted "consistently with the objective of giving the public concerned wide access to justice."

Finally, Article 9.3 requires nations to provide public "access to administrative or judicial procedures to challenge acts and omissions by private persons and public authorities, which contravene provisions of its national law relating to the environment." This allows the public to enforce a broad range of environmental laws against both the government and against private people and corporations.

The Process Leading to the Aarhus Convention

The *Aarhus Convention* grew out of the "Environment for Europe" process, which started in 1991.[65] The Environment for Europe process seeks to pursue a coherent strategy for addressing Europe's environmental problems, with particular emphasis on restoring the environments of countries in transition from Communism. An integral part of this international effort has been to define the role of public participation in implementing sustainable development. It is noteworthy that officials in many of the countries in transition had formerly been in NGOs, and consequently supported efforts to improve transparency, access, and accountability.

The third Ministerial Conference of the process – held in Sofia, Bulgaria, in 1995 – included environment ministers from 49 countries from Europe, North America, and Central Asia (comprising most of the UN/ECE countries), as well as from Australia, Japan, and Mexico, and representatives of the European Community and non-governmental organisations also attended. The Conference considered the *UN/ECE Guidelines on Access to Environmental Information and Public Participation in Environmental Decision-making*. Members of Global Legislators Organisation for a Balanced Environment (GLOBE) and NGOs asserted that the non-binding guidelines were too weak and that a stronger, binding convention was necessary. Ultimately, the Ministerial Declaration highlighted public participation as a key issue, and it recommended that the UN/ECE would take the lead in developing a regional public participation convention with appropriate NGO involvement.

One of the unique features of the Convention is the unprecedented role that NGOs have had in conceptualising, negotiating, drafting, ratifying, and implementing the convention. The benefits associated with the public participation include: added technical and personnel resources in the negotiating process, a stronger agreement, increased public support for the outcome, and a broad-spectrum of advocates for implementing the *Aarhus Convention*. In addition to participating in drafting the Convention, NGO representatives served on official national delegations. The NGOs established a coalition called the "European ECO-Forum," which fielded a delegation that participated in the Environment for Europe conferences and drew upon a resource group of more than 200 public interest environmental experts who reviewed drafts, provided input, and suggested strategies.

In 1993, 1995, and 1998, NGOs held conferences that paralleled the ministerial conferences. NGOs also hosted country roundtables in 15 Central and East European nations, as well as some West European nations, to educate the public about the potential agreement. An NGO coalition – including the Regional Environmental Center for Central and Eastern Europe (REC) – conducted extensive research on the state of environmental governance in the UN/ECE countries. In 1994, the REC published a *Manual on Public Participation*, which assessed public participation practices and opportunities in Central and Eastern Europe, and was essentially a guide of best practices. Later, the NGOs, through the REC, published the *Doors to Democracy*, a series of country-by-country assessments of trends and practices in Western Europe, Central and Eastern Europe, and the Newly Independent States.

Considering the high-profile role of NGOs in this Convention, it may be said that this was the first civil society convention, drafted largely for civil society and with substantial participation by the non-governmental community.

Future Directions

As of March 2002, twenty-two countries have ratified or acceded to the *Aarhus Convention*.[66] It entered into force on October 30, 2001. Some countries, such as Ukraine, have ratified it and now must implement its provisions by modifying existing laws and institutions, as well as adopting new ones. Other countries have opted to reform their laws before formally ratifying or acceding to the *Aarhus Convention*.

While focusing on ratification and implementation by its member countries, the UN/ECE has expressed some interest in expanding the *Aarhus Convention* beyond the UN/ECE region. Under Article 19.3, any country that is a member of the United Nations can accede to the *Aarhus Convention*.

A few unresolved issues remain in implementing the *Aarhus Convention*. Due to a lack of consensus on certain topics during the negotiations, the parties drafted framework provisions that need more elaboration. Consequently, the UN/ECE has established task forces on PRTR, compliance mechanisms, genetically modified organisms, access to justice, and electronic tools. The UN/ECE is also seeking to coordinate its efforts with other convention secretariats in the region, such as the 1991 *Espoo Convention on Transboundary Environmental Impact Assessments*.[67] For example, the 1999 *Protocol on Water and Health* became the first international instrument to expressly incorporate the *Aarhus Convention* provisions, this time in the context of environmental health.[68] The *Strategic Environmental Assessment (SEA) Protocol* to the *Espoo Convention*, that is currently being negotiated, seeks to strengthen the requirements for nations to include the public in the assessments made of plans or programs that will affect the environment.[69]

The UN/ECE is also reaching out to other organisations to support development of similar regional environmental governance initiatives. Thus, the UN/ECE and the Organization for American States (OAS) communicated during the development of the *ISP* in the Americas, and the UN/ECE has expressed interest in the ideas circulating about a possible African initiative. As a step toward developing the principles globally (and particularly in Africa and Asia), the UN/ECE and UNEP/INFOTERRA hosted a consultation on "Promoting the Application of Principle 10 of the Rio Declaration." Held in May 2000 in Rome, this meeting brought together environmental lawyers and experts from Africa, the Americas, Asia, and Europe to discuss ways to link and support ongoing and planned activities to implement Principle 10 in different regions and to chart future actions at the national, regional, and global levels.[70]

The ISP

The *Inter-American Strategy for the Promotion of Public Participation in Decision-Making for Sustainable Development* (*ISP*) is an independent regional initiative of the Organizaation of American States (OAS) promoting environmental governance. While the *Aarhus Convention* is a binding treaty with concrete obligations, the *ISP* is deemed a "Strategy." This Strategy encourages, but does not require, signatories to undertake similar legal and institutional reforms. Thus, while the *ISP* is technically soft law, with no binding obligations or institutional framework, it establishes regional principles and standards that the OAS and its member states are now seeking to implement.

The *ISP* rests on the commitments that OAS members had already made to implementing Principle 10 of the *Rio Declaration* and *Agenda 21*.[71] The OAS initially drafted the *ISP*, which was then revised to incorporate comments from consultations by national focal points, from hemispheric conferences that debated the draft language in detail, and from public comments received (including those from the internet).

The *ISP* consists of two documents: a short, general "Policy Framework" and a longer, detailed "Recommendations for Action". The Policy Framework is the essential document, which the OAS member states adopted in April 2000. It contains the principles, objectives, and policy recommendations of the *ISP*. The recommendations are hortatory, urging member states to take action to: (1) improve communication mechanisms to share information; (2) establish legal and regulatory frameworks to ensure public access to information, decision-making, and justice; (3) promote institutional structures, policies, and procedures for expanding public participation; (4) advance education and training programs; (5) dedicate funding for public participation in decision-making; and (6) promote fora for consultation.

The Policy Framework is intentionally general, establishing the basic elements that the nations are encouraged to implement. By contrast, the Recommendations for Action – appended to the adopted Policy Framework (and originally not intended to be signed by the parties) – provides many different, specific actions that the nations can adopt.

The *ISP* represents a landmark initiative in the Americas.[72] The *ISP* has advanced international law by clarifying the scope of environmental governance principles as they apply in the Americas, even if it is in the form of soft law. The OAS also has included NGOs, industry, and citizens in the development of the *ISP* to an extent that was unprecedented in Inter-American governmental efforts. In fact, the OAS also has been expanding the participation of civil society organisations by developing guidelines for their participation in the OAS.[73]

Draft ASEM Elements

The Asia Europe Meeting (ASEM) was first convened in Bangkok in March 1996, and its participants included ten Asian nations, the European Commission, and all fifteen EU nations.[74] ASEM seeks to establish a cooperative process to promote economic, political, and cultural development, as well as environmental protection. In 1999, ASEM established the Asia-Europe Environmental Technology Centre (AEETC) to provide policy guidance, exchange technologies, and promote cooperation in the areas of environment, agriculture, and technology.[75]

In 2000, AEETC (with the Finnish Environment Institute and the Thailand Environment Institute) initiated a project to promote public involvement in Asia and Europe through guidelines for public involvement. Following four rounds of public comments on draft documents, the guidelines have become elements of good practice for environmental governance.[76] The non-binding elements are still in draft form, and will undergo further discussion at the First ASEM International Conference on Public Participation, to be held June 2002 in Bangkok.

The *Draft ASEM Elements of Good Practice* are rooted in Principle 10 of the *Rio Declaration* and emphasise access to environmental information, public participation, and access to justice, along with implementation. The information provisions (Articles 1-17)[77] provide detailed guidance on collecting, disseminating, and making available environmental information. Environmental information is broadly defined to include information on the state of the environment and human health, and any plans or actions taken that may affect or protect them (Appendix 1). Upon request, any "natural or legal person will have free access" to such information.[78] Information, in general, must be made available for review free of charge, but a reasonable, cost-based fee may be assessed for copies of information.[79]

Public authorities at all levels are responsible for making information available to the public, and requests for information must be responded to "within a specified time limits." [80] Finally, information may be denied only on specific grounds.[81]

Governments are required to collect, update, and "periodically disseminate" environmental information, including "social, health, and cultural data".[82] Information must be "regularly published." and may be disseminated through a State-of-the-Environment report, or through a pollutant release and transfer register (PRTR).[83] Additionally, environmental auditing and ecolabeling programs are "encouraged".[84]

Public participation in decision-making processes is envisioned for all activities – including permitting of projects, as well as preparing policies, plans, and programs – that could have significant environmental implications.[85] Public opinion must be solicited before taking any actions that may affect the environment,[86] and input should be sought at an early stage, "when options are still open."[87] Actors must allow "sufficient time for public discussions"; and comments from the public must be considered before making any decisions.[88] Moreover, states must ensure that people who participate "are not penalized in any way for activities that are lawful".[89]

The *Guiding Principles* require member states to "promote an approach" that ensures people "fair, open, transparent and equitable" access to administrative and judicial proceedings to enforce their rights to information and participation.[90] These proceedings are also available when a person (or NGO) believes that his or her input was "ignored," and where the government's response or decision was "inadequate".[91] To implement these principles, nations should develop clear regulations, promote environmental education for the general public, and regularly evaluate the application of the principles.[92]

The *Draft Elements* are notable for guaranteeing that the cultural diversity and literacy of persons seeking to utilise the environmental governance mechanisms will be "taken into account," even suggesting specific methods such as radio programs, posters, and "helpers" to assist the illiterate.[93] Participation of the public is to be advanced "regardless of gender, cultural or ethnic identity, language, citizenship, nationality or domicile".[94] Such assurances are necessary if the protections afforded by these mechanisms are to be applied uniformly.

As with the other regional initiatives, the Draft ASEM Elements of Good Practice constitute an important regional initiative that starts to promote good environmental governance. At the same time, there remain areas for future development. For example, while the Draft Elements provide judicial or administrative avenues for those claiming failure in the information or participation procedures, they do not explicitly offer access to justice for violations of environmental laws. Also, like the ISP, the Draft ASEM Elements are non-binding. In fact, the level of member nation commitment to environmental governance norms may be waning: the November 2000 draft was on "Guidelines on Public Participation in Environmental Aspects in the ASEM Countries"; the January 5, 2001 draft referred to "Guiding Principles on Public Involvement in Environmental Aspects in the ASEM Countries"; and the June 28, 2001 draft is titled "Towards

Good Practices for Public Involvement in Environmental Policies." Overall, though, the Draft Elements are comprehensive and offer member nations a strong foundation upon which domestic policy may be built. The document clearly and concisely lays out mechanisms for access to information, participation, and justice.

The African Context for a Regional Public Participation Initiative

For the past three decades and more, Africa has struggled with the legacy of colonialism.[95] The current national boundaries were largely drawn at the convenience of the colonial powers, often joining traditional adversaries in a single nation while splitting a single tribe between two nations. African colonies adopted legal systems based on English common law, Dutch Roman law, or French or Portuguese civil law – all equally alien to the diverse traditions of customary law followed by indigenous peoples around the continent. Indigenous institutions were disbanded, subjugated, or subsumed into the colonial institutions. Ownership of the country's natural resources, frequently including land, were consolidated and vested in the colonial crown. The whole system – politico-geographic, legal, institutional, and resource ownership – was designed to maximize profits of the colonial power and to disempower indigenous Africans. When African nations won their independence, they frequently maintained the old laws and institutions for the benefit of the elites.[96]

Over the last few decades, however, increased international trade and communications has contributed to a vibrant exchange of ideas. International and bilateral aid organisations play an increasing role in governance. Domestic and grassroots organisations have sought to improve government institutions – particularly those with authority over natural resources – to be more transparent, inclusive, and accountable.[97]

African NGOs have grown in number, diversity, size, and sophistication. In pursuing their particular missions – human rights, environment, rural empowerment, and so forth – these NGOs frequently encounter vestiges of the old colonial system that discourage transparency, access, or accountability. Consequently, these organisations – particularly environmental NGOs – have turned their attention and energies to changing the governance procedures and institutions. In this context, African environmental advocates have been examining the various regional public participation initiatives and considering options for similar endeavours in Africa.[98]

Ultimately, only Africans can determine whether there is sufficient demand for a regional environmental governance initiative, and what form that initiative should take. As one UNEP official observed, "If the need to elaborate such a framework is recognised, the instrument should be conceived as a specific legal response to the requirements, practices and national conditions of the African region."[99]

In exploring options for promoting environmental governance in Africa, it will be worth considering the particular African context. For example, in many countries, there remains a love-hate relationship with the former colonial power, and there is a particular sensitivity to the presumed superiority of Western ideas. Care should be exercised so that initiatives are not viewed as an export from Europe or the United States. Rather, an African initiative could be placed in the global context as Africa's interpretation of the universally agreed-upon norms enshrined in Principle 10 of the *Rio Declaration* and other global instruments.

An African Convention may choose to have two, three, four, or more "pillars": it need not emphasise the same three basic principles of the *Aarhus Convention*, the *ISP*, or the *Draft ASEM Elements*. For instance, in many African countries, public interest organisations must register with the government in order to operate. Notwithstanding widespread constitutional provisions ensuring the right of free association,[100] non-governmental organisations frequently operate in daily fear of being decertified if they challenge governmental actions. Thus, African advocates may seek to add a pillar guaranteeing and elaborating on the right of free association.[101]

The colonial amalgamation of kingdoms has led in many cases to incredible linguistic diversity in some nations. For example, Tanzania has over 130 different recognised languages and dialects,[102] and some countries have even more. Even in countries with a single official language (usually a relic of colonial days), rural people frequently have at best a rudimentary grasp of that language. On a continent where the vast majority of people are farmers and herders whose first language is their tribal language, the numerous languages pose special challenges for governments that seek to provide effective access to environmental information that they hold. Few governments – particularly African governments – have the requisite time, money, or resources to translate relevant documents into all the necessary languages. There are also technological difficulties, as the option of establishing on-line, publicly accessible electronic databases is not realistic despite their lower maintenance costs. For that matter, in some countries, where literacy rates are very low (Niger has a literacy rate of approximately 14%)[103] the methods of disseminating information through newspapers and government publications would be ineffective. This is compounded by a frequent absence of newspapers and other written media outside the cities. It then becomes necessary to explore other avenues, such as dissemination by radio.

African countries also face overwhelming pressures to develop. Eight of the ten poorest countries in the world are in Africa, as are 29 of the 34 countries that the United Nations ranks lowest in terms of human development, accounting for income, life expectancy, and education.[104] Thus, people frequently are more concerned about locating firewood, food, water, and forage than they are about participatory principles. Environmental education will be critical at the individual level in making the link between today's practices and tomorrow's harm.

Some African governments may be reluctant to allow public access to governmental decision-making processes that relate to natural resources, which are often perceived to be a quick source of necessary hard currency. A similar constraint on promoting environmental governance in many African countries is a culture of secrecy that views information as a source of power and something to be hoarded. In fact, it has been asserted that the Kiswahili word for government –

serikali – comes from *siri kali* meaning "top secret."[105] This reflects a general challenge faced by environmental advocates: how to convince governments to open up the governance processes to civil society when there has been a lack of public participation traditions at the national level, often for decades.

There are a number of ways of addressing these impediments to transparency, access, and accountability. For example, such principles in many cases are traditionally African. Before colonization, many African cultures involved more people in the decision-making processes, and there was usually the opportunity for affected individuals to have their grievances heard. To some extent, these pre-colonial traditions persist in the customary law of the country, although customary law may focus more on familial matters while the colonial and post-colonial laws have attempted to preempt customary norms in the realm of natural resources.[106] Drawing on customary institutions and practices at the local level, it may be possible to build specifically African models for public participation in environmental matters at the national level.

Improving environmental governance at the national level can borrow from experiences at the local level. Countries throughout Africa increasingly are devolving powers to local authorities to manage natural resources. The increased transparency, access, and accountability – as well as the greater share of revenues flowing to the local communities – have established valuable precedents for environmental governance at the local level.[107] Further research is necessary to explore the possible links between existing practices at the local level and developing environmental governance at the national and regional levels in Africa.

Options for Advancing Environmental Governance in Africa

In the event that advocates, individual nations, or groups of nations decide to promote environmental governance in Africa, there is a wide range of options available. These include domestic incorporation of participatory principles, accession to an existing convention, and development of an African convention or other regional initiative.

The first option would be to incorporate participatory principles into domestic legal and institutional frameworks. In doing so, a nation could look to the other regional initiatives as reflective of emerging norms in international law. Indeed, many African countries have already incorporated different participatory provisions in their constitutions, environmental laws, decentralization policies, and institutions.[108]

In practice, though, there frequently remains a need to implement the participatory principles, to expand them, and to extend them to other areas. Some nations still lack the necessary laws and practices, or the political will to advance them. While incorporation of these principles into national laws and institutions will continue to be essential to on-the-ground implementation of public involvement, there remains a need for broader, more regional initiatives that could encourage and foster the development of necessary laws, policies, and institutional practices.

This approach has the benefit of not conditioning action on the participation of any other nation. Additionally, if the nation subsequently participates in developing a convention or decides that it should accede to an agreement, many of the principles will have already been implemented through laws and regulations. From a broader geographic perspective, the experiences of particular African countries in implementing environmental governance principles at the national level would be particularly relevant to the development of a regional or subregional instrument.

African nations also could opt to accede to the *Aarhus Convention*. This approach would take advantage of an instrument that includes concrete obligations and that may reflect growing international consensus on the topic. There may be funding available to assist nations in accession, along with technical assistance from the UN/ECE. However, African environmental advocates have expressed reservations about acceding to a public participation convention in which their governments have not participated.[109] Not only does the *Aarhus Convention* fail to incorporate the particular African context, but the advocates also expressed concern that accession could bypass the collaborative relationship between governments and NGOs that a regional negotiating process could engender.

Pan-Africa

Another option would be for a group of African nations or an intergovernmental organisation to undertake a new initiative on environmental governance. A regional initiative – whether it is couched as a convention, guiding principles, or something else – could be undertaken by the United Nations Environment Programme in Nairobi, which currently has a programme to develop environmental laws in countries throughout Africa. Additionally, INFOTERRA, the Global Environmental Information Exchange of the United Nations has extensive experience in promoting access to information in Africa (and globally) through technology transfer, development of policy and legal frameworks, and general capacity building. Other bodies that could undertake a regional process include the new African Union, the UN Economic Commission for Africa (UN/ECA), and the UN African Ministerial Conference on the Environment (AMCEN). If this option is pursued, then one supporting component – which was useful in developing the *Aarhus Convention* – could be a country-by-country survey of the status of the relevant laws and institutions in the countries likely to sign the convention, as well as a manual of best practices.[110] Similarly, on-the-ground pilot projects, utilised effectively in the ISP process, could be useful in providing African experiences upon which to build.

Already, Africa has taken significant steps toward articulating a regional perspective on the involving civil society in decisions. The earliest pan-African initiatives focused on the need to improve public participation in development. The 1989 African Alternative Framework for Structural Adjustment Programs[111] and the 1990 African Charter for Popular Participation in Development and Transformation[112] responded to the traditional mechanisms for development that did not adequately involve civil society or local governments in the decision-making processes.[113]

The *African Charter* was unanimously adopted by 500 participants attending a UN Economic Commission for African Conference, which was held in Arusha, Tanzania from 12 to 16 February 1990. Participants included representatives from 23 African governments, NGOs throughout Africa, NGOs working in Africa, UN organisations, donors, and development specialists. While the *African Charter* is not legally binding, many commentators have resorted to its declarations,[114] and it appears to have influenced regional and national initiatives.[115]

Starting from the "fact that people have a fundamental right to participate fully in the making of policy decisions which affect their lives at all levels,"[116] the *African Charter* included detailed mechanisms to promote and monitor public participation. It set forth the roles and responsibilities of the people (Articles 13-15, 23(B)), African governments (Articles 16-20, 23(A)), the international community (Articles 21-22, 23(C)), and NGOs (Article 23(D)), as well as the media, women's organisations, organised labor, and youth (Articles 23(E)-(H)). The Charter specifically highlighted the need for the "full support and participation of the people" to address environmental challenges.[117]

To monitor the implementation of the goals and mechanisms of the *African Charter*, it proposed a set of indicators to be used at the national and regional levels. The non-exhaustive list of indicators includes: "the literacy rate ... freedom of association ... representation of the people and their organisations in national bodies ... the rule of law and social and economic justice ... protection of the ecological, human and legal environment ... press and media freedom ... number and scope of grass-roods organisations with effective participation ... extent of implementation of the *Abuja Declaration on Women* (1989) ... political accountability of leadership at all levels ... [and] decentralisation of decision-making processes and institutions."[118]

Following the *African Charter*, African heads of state met with more than 100 African environmental and development NGOs to prepare a common position on the environment and development that could be advanced at the 1992 United Nations Conference on Environment and Development in Rio de Janeiro. Ultimately, this meeting led to the African Common Position on Environment and Development,[119] which was endorsed by the Organisation for African Unity and presented at the Earth Summit. One commentator noted that this process "would appear to be the first international meeting for which African governments and African non-government organisations have sought to coordinate their strategies."[120] The Common Position articulated the African needs, constraints, and opportunities for sustainable development, and called for a convention to combat desertification – ultimately, one of the key instruments developed following the Earth Summit. The *African Common Position on Environment and Development* continues to shape the environmental agenda in Africa, as nations and international organisations invoke the *Common Position* to support and provide historical context for particular endeavours.[121]

In light of the experiences with the *African Common Position on Environment and Development* as well as other African common positions,[122] African nations could develop a Common Position on environmental governance.[123] Such an initiative could build on earlier declarations (such as the *African Charter*) and national experiences and constraints to advance an African voice on environmental governance. There is a risk that a common position might represent the least common denominator and thus have little practical effect.[124] Still, it has been a decade since the clear articulation of the need and mechanisms for public participation in Africa in the Charter, and African nations have significantly more experience with public participation now, there are more democratic governments, and civil society (including both citizens and NGOs, as well as other actors) are becoming integral partners in protecting the environment and promoting development. A contemporary African initiative to articulate the current needs and opportunities for strengthening transparency, participation, and accountability in environmental matters would seem to be timely, particularly in light of other regional initiatives around the world. The preparatory processes leading up to the 2002 World Summit on Sustainable Development ("Rio+10"), focusing on the implementation of sustainable development, offers a timely context in which to develop and advance a common position on environmental governance.

One variation on such a regional initiative could be to incorporate environmental governance elements into an existing international regime. The *Desertification Convention*, which already incorporates the principles to some degree, would be a promising candidate for Africa. In fact, the *Desertification Convention* relies on a bottom-up approach to resource management, emphasizing transparency and community participation.

Incorporation into existing MEAs might be one way to move forward if the political will and necessary funds are lacking to negotiate, draft, sign, ratify, and implement an entirely new convention. The problem with this approach is that it could be difficult to mobilise the broader environmental community if the principles would only apply to desertification and not other sectors. Moreover, there is the challenge of convincing the signatories to the *Desertification Convention* – including a wide range of African and non-African nations – to accept the modifications. Nevertheless, as with the option of domestic incorporation of the principles, the experiences of incorporating them into an existing sectoral convention may be valuable in the development of a broader instrument.

Convention secretariats may revisit the text of their original documents to ascertain whether they adequately incorporate the principles, as happened in the case of the 1992 *UN/ECE Convention on the Protection and Use of the Transboundary Watercourses and International Lakes*, which led to the 1999 (*London) Water and Health Protocol*. Currently, there are discussions about amending the 1968 *African Convention on the Conservation of Nature and Natural Resources*[125] to reflect developments in environmental law over the decades subsequent to its adoption. This agreement predates the international development of environmental impact assessment, and aside from general provisions on conservation education set out in Article XIII, it does not address access to information, process, or courts. Thus a protocol to the 1968 *Convention* or other revisions could emphasise transparency, participation, and rule of law in environmental matters, as well as other governance issues such as decentralisation, as integral approaches to improve the implementation of its substantive goals.[126]

East Africa

If a pan-African convention is not yet feasible, another possibility would be to pursue a smaller sub-regional convention through any of the sub-regional bodies. For example, the *Treaty for the Establishment of the East African Community*[127] between Kenya, Tanzania, and Uganda provides for the "Co-operation in Environment and Natural Resources Management" (Chapter 19), "Co-operation in Tourism and Wildlife Management" (Chapter 20), and harmonization of "Legal and Judicial Affairs" (Chapter 24). These chapters contain the kernels for developing access to information, public participation, and rule of law in East African environmental governance.[128]

A *Memorandum of Understanding for Cooperation on Environmental Matters* (MOU)[129] between the three East African countries is even more explicit about the need for a subregional instrument incorporating public participation. The MOU anticipates the development of a protocol on environmental management under the *Treaty for the Establishment of the East African Community*.[130] Such a protocol would assist in harmonizing domestic environmental and natural resource laws and institutions in East Africa, and promote cooperation in the implementation of these laws. Throughout the MOU, the Partner States committed to promoting access to information, public participation, and even access to courts. The Partner States agreed to ensure "the right of the people of the Partner States to a clean, decent and healthy environment" through their laws.[131] In order to guarantee the right of their people to a clean, decent, and healthy environment, the States will develop and harmonise their laws on environmental impact assessment (EIA) and the "full involvement of their people in the sustainable use and management of [the] environment and natural resources".[132]

As the MOU recognises, environmental impact assessment (EIA) is an essential tool of environmental management. When done properly, EIA highlights the possible environmental and social impacts of a proposed project (including a range of alternatives), allowing the government and project proponents to make an informed decision about whether and how to proceed with the project. By making the EIA and underlying data available to the public at an early stage and providing an opportunity for public comment, potential errors or oversights can be corrected and the project modified accordingly while modification is still feasible. Thus, the Partner States "agree[d] to enact legislation to regulate environmental impact assessment including public participation at all stages of the process related to environmental impact assessment".[133] The States "further agree[d] to develop common programmes and procedures for the dissemination of information on the operation and use of environmental impact assessment as a necessary regulatory measure for sustainable development".[134] The MOU also sets forth the basis for developing transboundary environmental impact assessments.[135]

The MOU contemplates much broader transparency, participation, and accountability in environmental matters, above and beyond the project-specific context of EIA. For example, the Partner States have committed to exchanging information on forest resources, wildlife, the marine and coastal environment and wastes and hazardous waste.[136] While most of the provisions speak generally of information exchange, the MOU specifically mentions "regional and international organisations" in research, monitoring, and information exchange relating to the marine and coastal environment.[137] At the same time, the Partner States agreed to "promote public awareness programmes and access to information" on environmental issues and environmental management generally,[138] and to establish resource centres on environmental information, EIA, and other aspects of environmental law and management.[139] The broad scope of these commitments applies to all environmental issues, and could be a key factor in removing a number of the hurdles to access to information (such as national secrecy laws).

Through the MOU, the Partner States have agreed to adopt approaches to enhance public participation in environmental management generally,[140] as well as in particular contexts. These include EIA (where public participation is to be at "all stages") as well as forest management.[141]

The MOU takes specific steps to improve access to environmental justice in East Africa. The Partner States have agreed to:

> Develop measures, policies and laws, which will grant access, due process and equal treatment in administrative and judicial proceedings to all persons who are or may be affected by environmentally harmful activities in the territory of any of the Partner States.[142]

By granting access to courts and administrative agencies, this provision greatly reduces the burden on plaintiffs to show that they have standing or "locus standi," the legal right to be in court. For example, in Kenya, a number of environmental cases had been dismissed because the plaintiff could not show that she had been injured in a way that was different from the other people who were also injured (and thus it was the province of the state to act, not a private individual).[143] The *MOU* seeks to harmonize their definitions of environmental offences in the three countries as well as the judicial and administrative procedures, which will "promote consistency in compliance and enforcement."[144] Finally, with respect to transboundary environmental damage, the Partner States agreed to grant access to their courts and administrative agencies equally to all nationals and residents of East Africa.[145]

The *MOU* could provide a basis for developing a sub-regional compact on environmental governance generally, or tied to a particular topic and include access to information, public participation and EIA, and access to justice. In fact, the African Centre for Technology Studies, with assistance the U.S. Agency for International Development, is promoting the development and implementation of a common regime of environmental impact assessment in East Africa, centred around Lake Victoria.[146] Finally, the *East African Environmental Protocol* would likely expand upon the environmental governance provisions outlined in the *MOU*.

Southern Africa

The Southern African Development Community (SADC) has developed a number of binding protocols on natural resources for its twelve member states to adopt, including one on shared watercourses and one on mining.[147] These protocols incorporate environmental governance principles to varying degrees. For example, Article 5 of the *Watercourse Protocol* charges River Basin Management Institutions with the "collecting, analysing, storing, retrieving, disseminating, exchanging and utilising data relevant to the integrated development of the resources within shared watercourse systems . . ." and "stimulating public awareness and participation in the sound management and development of the environment including human resources development."[148] These institutions also "promot[e] environmental impact assessment of development projects within the shared water-course systems" and "monitor the effects on the environment and on water quality arising from navigational activities," as well as promoting hydrometeorological monitoring.[149] When considered together, these provisions anticipate the collection of a wide range of information on natural and man-made conditions, making the information publicly available, and promoting public participation in the management of international watercourses in Southern Africa.[150]

The SADC Mining Protocol also mandates public access to information and public participation. It compels member states to "encourage private sector participation in the exploitation of mineral resources" and, with SADC, to "improve the availability of public information to the private sector . . . ."[151] This includes the establishment of a database on mineral resources and means for disseminating information (including through networking) on mineral resources, as well as technical, investment, and marketing information.[152] Article 6 includes a strong mandate for encouraging and sustaining private sector participation in the mining sector. The evident emphasis of this provision is on promoting private mining companies, and the extent that this promotes public participation depends on whether the "private sector" includes individuals and other non-governmental actors. Article 8 on "Environmental Protection" compels states to pursue a "regional approach in conducting environmental impact assessments especially in relation to shared systems and cross border environmental effects."[153] As noted above, EIA processes emphasises transparency and public participation in the planning and evaluation process, as well as an opportunity for citizens to appeal decisions to courts or administrative agencies.[154] In light of this and other environmental priorities (such as sustainable development, environmental protection, and sharing information on environmental protection and rehabilitation),[155] private sector participation could be read broadly to include all of civil society.

The SADC Ministers of Environment agreed in October 1999 to establish a binding environmental protocol.. This protocol could incorporate procedural aspects of environmental management (such as public participation) that are essential to implementing substantive environmental norms and standards. In fact, a December 1998 UNEP/INFOTERRA meeting on "Building Bridges for the *Aarhus Convention*" in Gaborone, Botswana examined the potential relevance of the *Aarhus Convention* to the SADC region.[156] Building on this workshop and Tanzania's experiences with the East African *MOU*, SADC could incorporate environmental governance into the ongoing development of an environmental protocol or develop a separate protocol on environmental governance for Southern Africa.

Other Options

Other subregional organisations present opportunities for promoting environmental governance on a sub-regional basis. The Economic Community of West African States (ECOWAS) includes 16 member states, and has thus far focused on economic integration and peace and security issues in the region. However, its mission includes regional integration of energy, agriculture, and natural resource sectors, and environmental governance would seem to be within its purview. The Economic Community of Central African States (ECCAS, with 11 member state), the Inter-Governmental Authority on Development (IGAD, with 6 member states in the Greater Horn of Africa), and the Council of Arab Ministers Responsible for the Environment (CAMRE, with 11 North African states and 12 West Asian states) also present opportunities for promoting environmental governance at the subregional levels.

Another possibility would be to push for a commitment to developing a global, multilateral framework on environmental governance at the 2002 World Summit on Sustainable Development. A global framework, whether in the form of a convention or guidelines, would harmonize the principles, and it has the added benefit that a large part of the world – indeed Asia, the Americas, Europe, the Newly Independent States of the former Soviet Union, Australia, Israel, Japan, and East Africa – has already been exploring possible international instruments for promoting environmental governance. Unfortunately, some of the most powerful nations – including the United States and Russia – show no inclination toward signing the *Aarhus Convention*, and others such as China, are unlikely to be strong proponents of an agreement. Instead, they appear to prefer to advance public involvement through national regimes and declaratory soft law instruments that are less likely to subject their practices to international scrutiny. Furthermore, a global convention or guidelines will not necessarily lead to an upward harmonisation of standards; instead, it could yield the least common denominator of the principles that the different parties accept, and lead to no improvement.

There are some difficulties associated with developing a new global public participation convention. It is unclear to what extent African governments would be inclined to include African NGOs in the process of negotiating, drafting, and ratifying an environmental governance convention. The recent demonstrations associated with meetings of trade and multilateral financing institutions are likely to increase the caution with which developing countries regard environmental NGOs, and the potential power that they can wield.

In the absence of NGOs to advocate for strong public participation provisions, governments may adopt an instrument that is so vague that it is unenforceable. If NGOs expect to be blocked from participating effectively, they may lobby for accession to the *Aarhus Convention* or the drafting of a global convention.

A combination of different approaches may be the best option. Parallel activities at the national, subregional, regional, and/or global levels can complement one another. At the national level, there is the most flexibility to experiment with different mechanisms. Experiences from countries with similar cultural values, legal systems, and states of economic development can provide invaluable political capital for adopting similar mechanisms, let alone negotiating an international instrument. At the same time, supranational declarations can – even if they are soft law – provide legitimacy for domestic initiatives. And sub-regional initiatives can lead to an overarching regional or global initiative.

A combination of approaches may be undertaken piecemeal, as a series of independent actions, or as part of a coherent saturation strategy. Availability of funding could well dictate which approach is feasible.

CONCLUSION

This Chapter has highlighted various regional initiatives that advance transparency, participation, and accountability in environmental matters, with an eye toward the possibility of an African initiative on environmental governance. The successes and challenges of these initiatives illustrate experiences that African advocates and governments may wish to consider in determining whether to advance a public participation initiative and what form the initiative may assume. This Chapter has specifically avoided making any recommendations as to what path might be most appropriate for African advocates. Indeed, it is likely that advocates, governments, and others will have different needs. In the end, Africa must decide what is best for her, and this should be undertaken through a transparent and participatory process.

---

* Carl Bruch is a Senior Attorney and Director of the Africa Program at the Environmental Law Institute in Washington, D.C. The author expresses his appreciation to Sara van de Kappelle, Stefannie Bernstein, and Jessica Yarnall for their research assistance and to Jay Austin and Kevin Gray for their thoughtful comments on earlier drafts. Resources for research conducted under this project wereprovided by the United States Agency for International Development and the John D. and Catherine T. MacArthur Foundation. A more detailed version of this paper was published as Carl E. Bruch & Roman Czebiniak, Globalising (2002), "Environmental Governance: Making the Leap from Regional Initiatives on Transparency, Participation, and Accountability in Environmental Matters", 32 *Envt'l L Rep.* 10428.

1. See generally C. Bruch (2000)*et al.*, "Constitutional Environmental Law: Giving Force to Fundamental Principles in Africa" 26 *Colum. J. Envtl. L.* 131 (2001) (also noting the nascent state of implementation of these provisions in most African nations).

2. UN/ECE Convention on Access to Information, Public Participation in Decisionmaking and Access to Justice in Environmental Matters, adopted at Aarhus, Denmark on 25 June 1998, entered into force 30 October 2001, ECE/CEP/43. The 55 Member States of the UN/ECE include the nations of West, Central,

and Eastern Europe, the Newly Independent States of the Former Soviet Union, Israel, Turkey, Canada, and the United States. http://www.unece.org/oes/eceintro.htm (visited 13 February 2002).

3. Organisation of American States Inter-American Council for Integral Development (OAS CIDI), Inter American Strategy for the Promotion of Public Participation in Decision Making for Sustainable Development, CIDI/RES. 98 (V-O/00), OEA/Ser.W/II.5, CIDI/doc.25/00 (20 April 2000), adopting Organisation of American States, Unit for Sustainable Development and Environment, Inter-American Strategy for the Promotion of Public Participation in Decision-Making for Sustainable Development, December 1999.

4. *Towards Good Practices for Public Involvement in Environmental Policies* (Draft 28 June 2001), (*Towards Good Practices* (2001)) produced by AEETC for the Consideration of the Environment Minister's Meeting, *available at* http://www.vyh.fi/eng/intcoop/regional/asian/asem/junedraft1.RTF and http://www.aeetc.org (visited 20 February 2002). See generally ASEM Official Home Page http://asem.inter.net.th/; Asian Cooperation: The ASEM Process http://www.vyh.fi/eng/intcoop/regional/asian/asem/asia.htm (visited 25 January 2002); personal communication from Eeva Furman, Finland Environment Institute, to Carl Bruch, Environmental Law Institute (7 August 2001).

5. These include the *North American Agreement on Environmental Cooperation* (NAAEC), (1993) (entered into force 1 January 1994),*reprinted at* 32 I.L.M. 1480;*Memorandum of Understanding between the Republic of Kenya and the United Republic of Tanzania and the Republic of Uganda for Cooperation on Environmental Management* (1998) done at Nairobi, 22 October 1998*Charter of Civil Society for the Caribbean Community, done at St. Johns, Antigua and Barbuda* (1997) (*Caricom Civil Society Charter* 19 February 1997, CARICOM Governments, available at www.caricom.org/chartercivilsoc.html. The Mediterranean Commission on Sustainable Development has also considered the importance of public involvement, but has yet to develop a more formal initiative. See*Recommendations and Proposals for Action on the Theme of: Information, Public Awareness, Environmental Education and Participation,* U.N. Doc. UNEP(OCA)/MED IG.12/9, Annex IV, Appendix I (adopted in Malta, 27-30 October 1999). For more information on these initiatives.

6 UNGA Res. 217A (III).

7. To be certain, some countries have extensive experience in incorporating environmental concerns into the decision-making processes. *E.g.*, *National Environmental Policy Act* of 1969, 42 U.S.C. §§ 4321-4370d (creating environmental impact assessment and incorporating environmental priorities into governmental decision-making processes). Nevertheless, environmental agencies and organisations worldwide increasingly recognize the importance of the procedural aspects of governing the environment.

8. *E.g.*, (1990) *Directive 90/313/EEC of 7 June 1990 on the Freedom of Access to Information on the Environment.*

9. For a good history of international instruments incorporating public participation principles, see S. Stec & S. Casey-Lefkowitz (2000), *The Aarhus Convention: An Implementation Guide* (Regional Environment Center, 2000), at 10-14, *available at* http://www.unece.org/env/pp/acig.htm.

10. *E.g.*, A. Babich (1995), "Citizen Suits: The Teeth in Public Participation", 25*Envt'l L Rep* 10141; S. Casey-Lefkowitz (1997), "A Comparative Look at the Role of Citizens in Environmental Enforcement", (June 1997) *Nat'l Envtl Enforcement J* 29.

11. *See* C. Bruch *et al.*, n. 1 above, at 41-42; T.A. Lissu (2000) "Repackaging Authoritarianism: NGO Policy Reform and the Freedom of Association and Expression in Tanzania,"*LEAT Research Report Series No. 1* (May 2000).

12. *International Covenant on Civil and Political Rights*, done at New York on 16 December 1966, entered into force 23 March 1976, art. 19(2), U.N.G.A. Res. 2200 (XXI), 21 U.N. GAOR, Supp. (No. 16) 52, U.N. Doc. A/6316, reprinted in 6 *I.L.M.* 368 (1967).

13. *World Charter for Nature*, adopted by the U.N. General Assembly, 28 October 1982, arts. 18 & 23, U.N.G.A. Res. 37/7, U.N. GAOR Supp. (No. 51) 21, U.N. Doc. A/37/L.4 and Add. 1 (1982);*see also ibid.*, arts. 15, 18.

14 *Convention to Combat Desertification,* (1994) 33 ILM 1016.

15 *Ibid.*, art. 10(2)(e), (f).

16. *United Nations Framework Convention on Biological Diversity*, done 2 June 1992, entered into force 29 December 1993, U.N. Doc. DPI/130/7, reprinted in 31*I.L.M.* 818 (1992), at art. 14(1)(a) (encouraging

public participation in "environmental impact assessment of proposed projects that are likely to have significant adverse effects on biological diversity"), art. 17 (promoting the exchange of publicly available information). The *Biosafety Protocol* to the *CBD* also relies on access to information (Articles 20, 23(1), and 23(3)) and public participation (Articles 23(2) and 29(8))*See, e.g.*, R.M. Saines (2001), "Rotterdam Treaty on PIC and Biosafety Protocol: Examples of Increased Transparency, Technology Sharing, and Accountability in International Law," 24 *(BNA) Int'l Envtl. Rep*. 623 (2001).

17. WTO AB 12 October 1998, *United States - Import Prohibition of Certain Shrimp and Shrimp Products*, WT/DS58/AB/R (AB-1998-4), at 31; but see*ibid.*, at 35-36 (noting that amicus curiae– or "friend of the court"– briefs not submitted as attachments to a party's filings should not be considered); *see also* S. Pyatt (1999), "The WTO Sea Turtle Decision," 26 *Envtl. L.Q*. 815-837.

18. *E.g.*, North American Commission for Environmental Cooperation Communication of 18 May 1999 on the Draft Framework for Public Participation in Commission for Environmental Cooperation Activities, C/DIR/01/Rev.6; ICTSD (1999), "Accreditation Schemes and Arrangements for Public Participation in Accreditation Schemes and Other Arrangements For Public Participation in International For a; M. Prieur (1998), "Environmental Agreements: Legal Aspects," 7*Rev. Eur. Community & Int'l Envtl. L*. 301 (1998); C. Ciorgetti (1998), "The Role of Nongovernmental Organisations in the Climate Change Negotiations," 9 *Colo. J. Int'l Envtl. L. & Pol'y* 115; *see also* C. Bruch (2001), "Public Involvement in the Management of International Watercourses," 31 *Envtl. L. Rep*. 11389.

19 K. Annan, *Foreword*, in S. Stec & S. Casey-Lefkowitz (2000), *The Aarhus Convention: An Implementation Guide* (Sventandre: Regional Environment Center) at v.

20. For example, in Europe, consider the *Salzburg Declaration on the Protection of the Right of Information and of Participation* (1980), adopted by the Second European Conference on the Environment and Human Rights, at Salzburg, 3 December 1980.

21 *African Charter on Human andPeoples' Rights*, done at Banjul on 26 June 1981, entered into force 21 October 1986, O.A.U. Doc. CAB/LEG/67/3 Rev. 5, reprinted in 21 *I.L.M.* 59 (1982).

22 *See* C. Bruch (2000) at 41-62 (surveying constitutional provisions in African nations governing freedom of association (all nations), access to information (21 nations), public participation (5 nations), and access to justice (more than 35 nations).

23. *See ibid.*, at 11-28.

24 Art. 3.5.

25 Art. 2.2.

26 Art. 3.9.

27. Art. 2.3.

28. Art. 2(2) (defining "public authority").

29. Art. 4.1.

30. Art. 4.2.

31 Art. 4.4; see also art. 4.3.

32 Art. 4.4.

33. Art. 4.5.

34. Art. 4.7.

35 Art.. 4.8.

36 *Ibid.*, art. 5.1(a).

37 Art. 5.1(c)

38. *Ibid.*, art. 5.2.

39 Art. 5.3.

40 Art. 5.4.

41 Art. 5(9).

42 For more information on PRTR, see Global Chemical Right-to-Know Resources, http://www/mapcruzin.com/globalchem.htm (with links to right-to-know initiatives around the world); OECD PRTR Documents, http://www.oecd.org/ehs/ehsmono/#PRTRS (with documents in English, Russian, and French); and the UNEP webpage on PRTR, http://irptc.unep.ch/prtr/.

43 R. Percival *et al.*(1992), *Environmental Regulation: Law, Science, and Policy*, at 624-26.

44 *Ibid.*.

45 *Ibid.*, at 626 (noting how Unocal substituted a solvent derived from citrus peel for methyl chloroform,

an ozone-depleting chemical).
46 *See ibid.*, at 626-27 (providing examples of how the U.S. Congress and Environmental Protection Agency used PRTR information in shaping environmental law and policy).
47 *See generally* "Aarhus PRTR Task Force," available at http://www.ecn.cz/prtr-tf/ (visited 25 January 2002).
48 Arts. 5.5, 5.7.
49 Arts. 5.6, 5.8.
50. *Ibid.*, art. 2.5; *see also* art. 2.4 (defining "the public" to mean "one or more natural or legal persons, and, in accordance with national legislation or practice, their associations, ORGANISations or groups.").
51 Art. 2.5.
52 Art 6.1(a).
53 J Wates (1999) "Introducing the Åarhus Convention: A New International Law on Citizens' Environmental Rights," Background Paper distributed at the Pan-European ECO Forum Conference on Public Participation, Chisinau, Moldova, 16-18 April 1999, at 4.
54. Aarhus Convention, n. 2 above, art. 6.1(b).
55 Art. 6.1(c).
56 Art. 6.2; see also art. 6.3.
57 Art. 6.4.
58 Art. 6.6.
59. Art. 6.7.
60 Arts. 6.8, 6.9.
61. See Wates (1999) at 5 (noting that the scope of this provision is potentially vast, as "relating to the environment" could include broad sectoral activities such as energy, transportation, and farming).
62 Art. 9.4.
63 Art. 9.5.
64 See Wates (1999) at 5 (noting that "the latter option [was] included to accommodate those countries which have a well-functioning office of Ombudsperson.").
65. For more information on the history leading up to the *Aarhus Convention*, see the UN/ECE web page (www.unece.org/env/pp), the web page for the European environmental citizens' organisations (ECOs; at www.participate.org), and Wates (1999) above.
66 *See* http://www.unece.org/env/pp/ctreaty.htm (visited 2 August 2002).
67 *UN/ECE Convention on Environmental Impact Assessment in a Transboundary Context*, (1991) 30 ILM 802.
68 *Water and Health Protocol to the Convention on the Protection and Use of International Watercourses and Transboundary Lakes*, done at London on 18 June 1999, arts. 10 (access to public information), 5i (access to information and public participation), and 15 (public participation in compliance).
69 United Nations Economic Commission for Europe (UN/ECE), *Draft Elements for a Protocol on Strategic Environmental Assessment*, 8 March 2001, MP.EIA/AC.1/2001/3, *available at* http://www.unece.org/env/documents/2001/eia/ac1/mp.eia.ac.1.2001.3.e.pdf.
70. *See* UN/ECE/UNEP *Consultation on Promoting the Application of Principle 10 of the Rio Declaration* (Rome, 20 May 2000) (draft meeting report, on file with author).
71. ISP, n. 3 above, at 2.
72. The *North American Agreement on Environmental Cooperation* (1993) is another important regional instrument, which was established to promote compliance with environmental laws and advance transparency and public participation. See also *CARICOM Civil Society Charter* (1997).
73. *E.g.*, OAS G.A. Res. 1707 (2000) (endorsing the Guidelines for the Participation of Civil Society Organisations in OAS Activities, approved by the Permanent Council, Dec. 15, 1999, CP/RES. 759 (1217/99)); see generally Permanent Council of the Organization of American States, Committee on Civil Society Participation in OAS Activities, *Background Information on Civil Society Participation in OAS Activities*, OEA/Ser.G, CP/CSC-3/99, 26 August 1999, available at http://www.civilsociety.oas.org/CSC-3-99.htm; AG/RES.1661, The Organization of American States and Civil Society," XXIX General Assembly of the OAS, Guatemala City, Guatemala, 7 June 1999; AG/RES.1668, "Strengthening Cooperation Between Governments and Civil Society," XXIX General Assembly of the OAS, Guatemala City, Guatemala, 7 June 1999.

74. The 10 Asian nations are Brunei, China, Indonesia, Japan, Malaysia, the Philippines, Singapore, South Korea, Thailand, and Vietnam.
75. *See* AEETC Strategy on Involvement of the Public in Environmental Aspects, Adopted by the AEETC Pilot Phase Guidance Group 00/02000; http://www.aeetc.org.
76 See *Towards Good Practice* (2001).
77. All citations to the Draft ASEM Elements of Good Practice are to the June 28, 2001 Draft.
78 Art. 1.
79 Art. 8.
80 Art. 7.
81 Arts. 4-6.
82 Art. 4.
83 Arts. 4,9.
84 Art. 14.
85 Arts. 18, 24.
86 Arts. 21, 24.
87 Art. 23.
88 Arts. 23, 21.
89 Art. 26.
90 Arts. 27-29.
91 Article 28, Appendix 3.
92 Arts. 30-36.
93 Art. 9.
94 Art. 19.
95 In the context of environment, see generally B.J. Richardson (2000), "Environmental Law in Postcolonial Societies: Straddling the Local-Global Institutional Spectrum," 11*Colo. J. Int'l Envtl. L. & Pol'u* 1;; A.P. Kameri-Mbote & P. Cullet, (1997) "Law, Colonialism and Environmental Management in Africa," 6 *Rev. Euro. Com. & Int'l L.* 23.
96 *See* B.J. Richardson, (2000) at 30 & 32.
97 *See generally* Richardson (2000) & Kameri-Mbote & Cullet (1997).
98. *See, e.g.*, A.A. Ibrahim & B. Rohan (1999), "The Aarhus Convention as a Model for Advancing Citizens' Rights in Africa," 6(2) *Innovation* 29 (emphasizing the particular application in Niger).
99 A. Timoshenko (1999)*Briefing Note on The 1998 Aarhus Convention on Access to information, Public Participation in Decision-Making and Access to Justice in Environmental Matters and its Relevance to Other Regions*, distributed at the April 1999 AMCEN meeting, para. 18.
100 *See* C. Bruch (2000).
101 For example, see,Lissu, (2000).
102. *See* http://www.sil.org/ethnologue/countries/Tanz.html (visited 17 Aug. 2000).
103 Central Intelligence Agency (1999), "Niger,"*The World Fact Book 1999*, 1 January 1999, available at http://www.odci.gov/cia/publications/factbook/ng.html.
104 D.G. Morris & D. Smith (1999), "The Poorest Getting Poorer," *Washington Post*, 17 July 1999, at A13 (citing the 1999 U.N. Human Development Report).
105 *See* M.O. Odhiambo (1999), "Top Secret: Legal and Institutional Framework for Access to Environmental Information in Kenya," Paper presented at the Workshop on Freedom and Access to Environmental Information in Eastern and Southern Africa (Arusha, Tanzania, 6-7 May 1999) (on file with author).
106 Kameri-Mbote & Cullet (1997) at 25-26.
107 J.C. Ribot, (2000),"Representation and Accountability in Decentralized Sahelian Forestry: Legal Instruments of Political-Administrative Control," 12 *Geo. Int'l Envtl. L. Rev.* 447 (2000).
108. See Bruch (2000) at 41-62 (reviewing the incorporation and nascent implementation oparticipatory principles in African constitutions);*Compendium of Environmental Laws of African Countries, Volume I: Framework Laws and EIA Regulations* (UNEP/UNDP Joint Project on Environmental Law and Institutions in Africa 1996) and 1998 Supplement (containing national environmental laws with public participation provisions for most African nations, as well as environmental impact assessment laws of some African nations – including Algeria, Congo, Niger, Nigeria, Tunisia, and Uganda– which typically

require transparency, an opportunity for the public to participate in the decision-making process, and an opportunity for citizens and NGOs to appeal decisions to administrative and judicial bodies).
109. On this point, the author is particularly indebted to the participants of a May 1999 workshop on Access to Environmental Information in East Africa, held in Arusha, Tanzania.
110 *Regional Environmental Center for Central and Eastern Europe, Doors to Democracy: Current Trends and Practices in Public Participation in Environmental Decision-making in Central and Eastern Europe* (1998). The REC also published *Doors to Democracy* volumes for Western Europe and the Newly Independent States of the Former Soviet Union. The Access Initiative is an ongoing effort to establish criteria for assessing the access to information, participation, and justice in nations around the world. *See* http://www.accessinitiative.org.
111 UN Economic Commission for Africa, *African Alternative Framework for Structural Adjustment Programs*, UN Doc. E/ECA/CM.15/16 Rev. 3 (1989), para. 35 ("development has to be engineered and sustained by the people themselves through their full and active participation"); *see also UN Declaration on the Right to Development*, UNGA Res. 41/128, Annex, UN GAOR Supp. No. 53 at 186, UN Doc. A/41/53 (1986), pmbl. ("development is a comprehensive economic, social, cultural and political process, which aims at the constant improvement of the well-being of the entire population and of all individuals on the basis of their active, free and meaningful participation in development and in the fair distribution of benefits resulting therefrom.").
112 *African Charter for Popular Participation in Development*, done at Arusha, 16 February 1990, UN Doc. E/ECA/CM.16/11; *see also Note Verbale*, UN GAOR, 45th Sess., Agenda Items 12, 82, U.N. Doc. A/45/427 (1990) (including as Appendix I, UN Economic Commission for Africa Conference of Ministers responsible for Economic Planning and Development, 25th Sess., Res. 691 (XXV) (19 May 1990) (adopting the African Charter for Popular Participation in Development and Transformation (Arusha 1990))), available at www.un.org/partners/civil_society/documents/a-45-427.htm
113 UN GAOR, 45th Sess., Agenda Items 12, 82, U.N. Doc. A/45/427 (1990) (noting the "lack of full appreciation of the role popular participation plays in the process of recovery and development.").
114 R.L. Barsh (1991), "Current Development: A Special Session of the UN General Assembly Rethinks the Economic Rights and Duties of States," 85 *Am. J. Int'l L.* 192, at 197; D.D. Bradlow (1991), "Regional Developments: Africa," 25 *Int'l Law.* 1053, 1054-58; J. Cahn (1993), "Challenging the New Imperial Authority: The World Bank and the Democratization of Development," 6 *Harv. Hum. Rts. J.* 159 at 186-87; C. Grossman & D.D. Bradlow (1993), "Are We Being Propelled Towards a People-Centered Transnational Legal Order?," 9 *Am. U.J. Int'l L. & Pol'y* 1, 23; C.R. Taylor (1994), "The Right of Participation in Development Projects," 13 *Dick. J. Int'l L.* 69, 98-99; J. Oloka-Onyango (1995), "Beyond the Rhetoric: Reinvigorating the Struggle for Economic and Social Rights in Africa," 26 *Cal. W. Int'l L.J.* 1, 59-60; J.C.N. Paul (1995), "The United Nations and the Creation of an International Law of Development," 36 *Harv. Int'l L.J.* 307 at 316.
115. Paul (1995) at 316 (noting the influence of the Charter in the subsequent development of the *African Common Position on Human and Social Development in Africa*, which was prepared for the 1994 World Summit on Social Development); D.D. Bradlow (1991), at 1055.
116 *African Charter*, third recitation.
117 Art. 9.
118 Art. 24.
119 UN Economic Commission for Africa, *African Common Position on Environment and Development* (June 1992).
120 D.D. Bradlow, (1991) at 1058.
121 See *Resolution on the Conclusions of the Pan-African Conference on Sustainable Integrated Coastal Management* (PACSICOM), done 24 July 1998 in Maputo, Mozambique, *available at* www.sdnp.org.mz/entidade/pacsicom/conclus3.htm (on the protection of the marine and coastal environment).
122 See African Ministerial Conference on the Environment (AMCEN) (1994), *African Common Perspectives and Position on the Convention on Biological Diversity*, Doc. No. AMCEN/Conventions/CBD/1 (26 October 1994); *African Common Position on the Clean Development Mechanism*, Paper No. 1: Uganda (On Behalf of the African Group), UN Doc. No. FCCC/CP/MISC.7/Add.2, ECE.98-135.

123 In fact, at the Eighth Session of the African Ministerial Conference on Environment (AMCEN), held in Abuja, Nigeria in April 2000, African environment ministers called for a common African position on the environment that could be advanced at the 2002 Earth Summit.
124 A. Mumma (2000), *The Poverty of Africa's Position at the Climate Change Convention Negotiations* www.meteo.go.ke/can/africap.html, (criticising the African Common Position on the Clean Development Mechanism for failing to articulate any significant African position).
125 (1969), 8 ILM 689.
126. A recent draft revision of the African Convention on the Conservation of Nature and Natural Resources in Africa has a new article on procedural rights. African Convention on the Conservation of Nature and Natural Resources, Jan. 18, 2002 Draft, art. XVI (on file with author). This article would require Parties to

> adopt legislative and regulatory measures necessary to ensure timely and appropriate a) dissemination of environmental information; b) access of the public to environmental information; c) participation of the public in decision-making with a potentially significant environmental impact; and d) access to justice in matters related to protection of environment and natural resources.

Ibid. art. XVI.1. Moreover, Parties would commit to ensuring non-discriminatory access to justice for persons in other nations who are affected by a transboundary environmental harm. *Ibid.* art. XVI.2. Article XX also provides for capacity building, education, and training, which will be necessary for people to effectively utilize their procedural and substantive rights and to contribute meaningfully to decision-making processes. These provisions could provide a basis for elaborating a more detailed set of guidelines and tools for ensuring public involvement.
127 Done at Arusha, Tanzania, 30 Nov. 1999.
128 See Arts. 112(2)(b) (environmental impact assessment), 112(2)(d) (public awareness and education regarding chemicals), 112(2)(i) (exchange information on pollution and conservation technology), 114(2)(c) (establish databases and networks to exchange information on mining), 116(b) (exchange information on wildlife management).
129 *Memorandum of Understanding between the Republic of Kenya and the United Republic of Tanzania and the Republic of Uganda for Cooperation on Environmental Matters*, done at Nairobi, 22 October 1998.
130 Art. 3
131 Article 7(1)(a)).
132 Art. 7(1)(b); 7(1)(i).
133 Art. 14(2).
134 Art. 14(3).
135 Art. 16(4)(a) ("The Partner States agree not to carry out any activity that may have significant transboundary environmental impact or harm without prior consultations, timely notification and provision of adequate technical information and data.")
136 Arts. 9(a),(d), 10(1)(c), 11(b), 12(d), (e).
137 Art. 11(b).
138 Art. 16(2)(a).
139 Art. 16(2)(b).
140 Arts. 16(2)(a), 7(1)(i).
141 Arts. 14(2) (EIA), 9(f) (pledging to "institute new techniques or measures which enhance . . . partnerships with local people" in the management of forest resources).
142 Art. 16(2)(d).
143 See *Wangari Maathai* v. *Kenya Times Media Trust*, High Court Civil Case No. 5402 (Nairobi 1989); *Wangari Maathai v. City Council of Nairobi*, High Court Civil Case No. 72 (Nairobi 1994); but see *Environmental Management and Co-ordination Act*, 1999, at art. 3(3) (allowing any person who alleges that their right to a clean and healthy environment, guaranteed by Article 3(1) "is being or is likely to be contravened in relation to him, then . . . that person may apply to the High Court for redress ...").
144. MOU, n. 5 above, art. 16(2)(e).

145 Art. 16(3).
146 This initiative also relates to Article 8 of the MOU, which seeks to "initiate, develop, implement and harmonize policies, laws and programmes to strengthen regional coordination in the management of the resources of the lake Victoria ecosystem . . .."
147 *Protocol on Shared Watercourse Systems in the Southern African Development Community (SADC) Region* (done at Johannesburg, Aug. 1995; entry into force Sept. 1998);*Protocol on Mining in the Southern African Development Community (SADC)* (done at Blantyre, Malawi, 5 Sept. 1997); *see also Draft Protocol on Wildlife Conservation and Law Enforcement in the Southern African Development Community* (done at Blantyre, Malawi, Dec. 1998).
148 *Protocol on Shared Watercourse Systems in the Southern African Development Community (SADC) Region*, at art. 5(b)(i), (iv). While Article 2(5) requires member states of a shared watercourse to "exchange available information and data regarding the hydrological, hydrogeological, water quality, meteorological and ecological condition of such watercourse system," this requirement appears to apply only between states and not necessarily to public access to information.
149 Arts. 5(d)(iii), 5(d)(iv), and 5(e).
150 *See* C.E. Bruch (200), "Role of Public Participation and Information Disclosure" *in* N. Makayama (ed.), *International Waters in Southern Afirca,* (forthcoming), reprinted as (2001), "Charting New Waters: Public Involvement in the Management of International Watercourses," 31 *Envtl. L. Rep.* 11389 (2001).
151. Protocol on Mining in the Southern African Development Community (SADC), at art. 2(7), (9).
152. *Ibid.*, art. 3 ("Information Exchange").
153 Art. 8(2).
154 See D.R. Mandelker (1999), *NEPA Law and Litigation*, 2ND ED; J. Teel (2001), "International Environmental Impact Assessment: A Case Study in Implementation," 31 *Envtl. L. Rep.* 10291 (2001).
155. *Protocol on Mining in the Southern African Development Community (SADC),* arts. 8(1), (3), (4).
156 See B. Ingraham (1998), "Southern African Countries Discuss *Aarhus Convention on Access to Information, Public Participation in Decision-making and Access to Justice in Environmental Matters*," Infoterra listserv, 16 December 1998 (on file with author).

THE EVOLUTION AND STRUCTURE OF POPULAR PARTICIPATION IN ENVIRONMENTAL DECISION-MAKING: THE CASE OF GHANA

KOFI OTENG KUFUOR*

INTRODUCTION

At the United Nations Conference on Environment and Development (UNCED) the significance of citizen participation in environmental decision-making was set out in the Rio Declaration[1] and Agenda 21[2] the two key UNCED instruments. The Rio Declaration recognizes the contribution individuals can make towards tackling environmental issues. It therefore enunciates the right of individual access to environmental information held by public authorities.[3] The Rio Declaration also urges States to facilitate public participation by making information on the environment widely available and, by ensuring individual access to judicial and administrative proceedings.[4] In addition it underscores the importance of women,[5] the youth,[6] and indigenous people[7] in decision-making on matters relating to the environment.

As noted above, Agenda 21 also highlights the significance of popular participation in environmental decision-making. For instance, Agenda 21 states that in the more specific context of environment and development, the need for new forms of participation has emerged.[8] These measures include facilitating the establishment of Non-Governmental Organizations (NGOs), and ensuring the right of NGOs to protect the public interest through legal action.[9]

Given the above stated importance of citizen involvement in environmental issues, this chapter examines the development and design of the framework for popular participation in environmental decision-making in Ghana. Part one outlines the function of popular participation in environmental decision-making. Part two discusses avenues for citizen and NGO participation in environmental decision-making in Ghana prior to the emergence of international concern for global environmental problems. This part therefore covers developments in Ghana from 1957 to 1974. Part three details provision for participation at a time when the international community began to show increasing concern for the environment. Central to part three therefore is the point that the mid-1970s saw a change in attitudes, albeit modest, towards citizen and NGO participation in Ghana. Part four's focus is on developments in participation in Ghana as a result of fundamental economic and political change starting in the 1980. Part five examines aspects of the actual participatory process. The conclusion gives an overview of the paper and examines possible new approaches to help make popular participation more effective.

RATIONALIZING CITIZEN PARTICIPATION IN ENVIRONMENTAL DECISION-MAKING

A number of reasons justify involving ordinary citizens in decisions on the environment. One key reason is that achieving participatory democracy normally allows marginalised groups to help shape decisions about resource use. Thus citizen participation can be used as a vehicle for social justice.[10] Environmental activists argue that politics in liberal democracies can easily lead to a situation where local communities are denied the opportunity to take part in decision-making. The point made in this regard is that liberal democracies tend to rely on experts or elected elites to make decisions concerning the environment.[11] Ordinary persons usually only elect their representatives but the essence of decisions are made within elite circles.[12]

Moreover, it is argued that by its very nature liberal, representative, democracy cannot deal with environmental problems. This is because representative democracies are committed to competitive elections, individual liberty and private property. This commitment leads to myopic environmental policies, policies that are skewed in favour of wealthy interests to the disadvantage of the environment.[13] Thus, an alternative to this is a participatory approach to tackling environmental problems.[14]

Another role that citizen participation plays is that it reduces or eliminates conflicts between local communities and industry. When the State excludes local people from formulating environmental policy, this has the effect of alienating them from any final outcomes. A consequence therefore is that tensions tend to emerge when large industrial concerns are seen to exploit local natural resources and communities without reinvesting part of their profits in the area that has supported their operations.[15] Thus, to avoid this, it is useful to include local people in designing environmental policy thus assuring them that they will be compensated for any environmental degradation caused by industrial activity. In effect when environmental law reflects the concerns of local communities, disputes between them and industry is minimized. Participation is therefore a confidence-building measure for local communities.[16]

The reduction of transaction costs,[17] including the cost of negotiating agreements is another reason why citizen participation is important. As communities realize that there are hidden costs of local commercial activity they are likely to insist on the re-negotiation of the agreements that permit industry to exploit the environment. However, the likelihood of this diminishes if the agreement is drawn up with the full participation of communities that are most affected by economic activity. In effect avoiding the repeated renegotiation of agreements cuts transaction costs not only to the industry concerned but to local citizens as well.[18]

Other rationalizations for citizen participation in environmental policy-formulation are as follows: the view that participation enables decisions to be made rapidly when there is an immediate environmental threat;[19] participation serves to educate citizens about threats to the environment and thus transforms them from parochial and narrow-minded persons to people with a concern for wider society;[20] and it is also seen as a part of the `good life' and thus enhances progress towards wider forms of participatory democracy.[21]

POPULAR PARTICIPATION IN ENVIRONMENTAL DECISION-MAKING IN GHANA FROM 1957-1974

Protection of the environment in Ghana became a national issue beginning in the 1970s. One result of this concern was the establishment of the Environmental Protection Council (EPC).[22] Although promoting citizen participation was not expressly included in the EPC's mandate, the legislation establishing it did provide a framework that could be exploited for the purpose of popular participation. The EPC's powers included conducting and promoting investigations, studies, surveys, research and analyses for improving the environment and maintaining ecological systems;[23] undertaking studies and submitting reports and recommendations on environmental matters.[24] It is these powers granted to the EPC, which implicitly amounted to involving citizens in decisions on the environment.

Prior to the establishment of the EPC there was virtually no Ghanaian legislation on the environment that had provisions for popular participation. Three probable explanations exist for this. First, society's understanding of environmental management at the time did not allow for a link between mankind and the laws of nature. Known as the exclusionist model of environmental thought,[25] the theoretical underpinnings of this model were that there the world's natural resources were unlimited. Also there was an infinite supply of sinks. These sinks would dispose of any waste generated as a consequence of environmental activity.[26] Thus environmental degradation was to be tackled after the given activity had taken place. In effect, environmental concerns should not be allowed to hinder the evolution of the free market.[27] Applied to Ghana, this argument had considerable validity. Poor and underdeveloped, Ghana's aim was to develop as rapidly as possible. This implied, *inter alia*, the use of its natural resources. Obviously, environmental concerns were second to this desire for socio-economic development. As such there was no pressing need for popular participation in environmental decision-making as this could result in non-economic considerations influencing and limiting the exploitation of the environment.[28]

The second probable factor was that popular participation in Ghana was essentially non-existent at the time. Soon after the end of the colonial system in the 1950s and 1960s, and as a result of a complex set of factors, the State in Africa moved from liberalism towards authoritarianism. The sum of the reasons for this was that democracy would be expensive for Africa;[29] the desire for national unity - an objective that could not be achieved if dissenting and disruptive political pressures were allowed to persist;[30] and the view that Africans lacked the degree of sophistication to grasp the complexities of politics.[31] This move towards authoritarianism led to a situation where decision-making was largely elite-driven with no avenues for participation by society's lower strata. This perspective on African politics was enthusiastically applied in Ghana.

Third, Public Choice theory that seeks to explain the `sale' of legislative favours to interest groups in society also probably shaped official attitude towards popular participation in Ghana. Central to Public Choice theory is the argument that laws are in effect commodities to be paid for by interest groups or coalitions. Legislative favours are purchased with, for example, electoral campaign contributions, the guarantee of other favours in the future or bribes.[32] The Public Choice school of thought thus contends that legislation shifts society's wealth to those well-organized groups that have relatively easier access to the corridors of power. In effect, government will pass legislation that reduces society's wealth and economic efficiency in order to benefit such special interest groups.[33] Applied to environmental regulation Public Choice theory argues that interests bent on exploiting the commons may take control of institutions that are supposed to protect the commons.[34]

It should be stated here that interest groups can obtain favours from the State not only through favourable legislation, but also by persuading governments to refrain from enacting any statutes at all.[35] If it is possible that a lobbying group can impose its own agenda on the public then it is also possible that such groups can block a regulatory scheme that would benefit the general public. There is no reason then to automatically conclude that the failure of the State to intervene in some aspect of the economy represents the defeat of a special interest group. Thus, special interest groups are sometimes wary of an activist government.[36] This perspective on government inaction is another probable explanation for the lack of pathways for citizen participation in environmental decision-making in Ghana from independence to the early 1970s.

CITIZEN PARTICIPATION IN GHANA IN THE POST UNCHE ERA

In the 1970s the assumptions of exclusionism were challenged. Sustainable development began to emerge as a counter-paradigm to the exclusionist model.[37] One manifestation of this emerging concept was in the United Nations Conference on the Human Environment, (UNCHE) held at Stockholm in 1972, which resulted in the adoption by the participants of the Stockholm Declaration.[38] The importance of citizen participation in environmental decision-making was recognized here. The Stockholm Declaration noted that protection of the environment demands that citizens, communities, enterprises and institutions assume responsibility for environmental protection.[39] In addition local government was also seen as crucial for shaping environmental policy.[40] The Declaration also acknowledged the need to educate individuals and communities so as to enlighten opinion and forge responsible conduct on the environment.[41]

Ostensibly, the Stockholm Declaration had some impact on policy-makers in Ghana. This claim is supported by the fact that in the mid-1970s there was a shift towards public participation in environmental decision-making resulting, as noted above, in the establishment of the EPC in 1974. This, however, seems to be the only positive development as apart from the establishment of the EPC, there was very little in specific legislation providing paths for citizen participation. For instance, economic legislation with implications for a possible impact on the environment also contained nothing that expressly permitted citizen participation.

Examples of this were the Industrial Free Zone Authority Decree[42] and the 1981 Investment Code.[43] Both laws contained provisions that only when interpreted liberally meant citizens could influence environmental issues relating to their implementation. So there was no express provision for citizen participation in the Industrial Free Zone Authority Decree. However, the Authority established to manage the Zones was empowered to engage in any activity conducive to the performance of its functions[44] and this could be construed to mean citizens could contribute to decisions relating to environmental issues arising out of economic activity in Zones. Similarly, under the provisions of the 1981 Investment Code the approval of foreign investments in Ghana depended, *inter alia*, on a firm's commitment to environmental protection[45] although there was nothing specific here to support citizen participation. Even legislation with a direct bearing on the environment had little if anything at all by way of citizen participation. Thus, for instance, the Forest Protection Decree[46] passed to regulate the use of forests[47] made no mention of the role that local communities could play in enforcing its provisions.

Why therefore, despite the acknowledgement at UNCHE about the importance of the role of individuals in environmental decision-making did policy-makers in Ghana fail to give greater commitment to citizen participation in environmental policy-formulation? The answer to this question lies in two main factors. First, even this gradual movement towards citizen participation was not as a result of pressure from society at large. Rather citizen participation was elite-driven and thus was limited in its scope. This is not surprising given the nature of the State in the developing countries at the time. With very few exceptions, the so-called soft State was emerging in most developing countries in the 1970s and Ghana was no exception. A prime attribute of the soft State was that ruling elites handed down legislation without any pressure from below. This was one reason why socio-economic reform through legislation was flawed and failed to engender the expected degree of social change.[48] Almost all legislative initiatives were the result of elitist perceptions of socio-economic problems. Applied to Ghana there is no evidence of any grassroots movement exploiting the growing awareness of environmental degradation in the 1970s. Consequently, there was no pressure from society to push for legislation that would open up space for society to participate in the determination of environmental policies.

An additional reason that explains the failure to quickly open up pathways for citizen participation is that changes in perceptions of socio-economic reality do not occur immediately or even uniformly within societies. While the essence of accepted policies is challenged the significance of the new assumptions may not yet be fully accepted. This stems from the fact that society's comprehension of reality resists the new way of thinking.[49]

Arguably, this is what happened in Ghana. Uncertain about sustainable development, the general impact of greater environmental regulation on economic efficiency, and cautious about increased empowerment of citizens, the State at the time only gave a tepid response to sustainable development and its call for citizen participation. Hence, the full impact of the paradigm shift probably occurred only when the concept of sustainable development gained widespread currency - something that did not happen until the late 1980s.

THE DEVELOPMENT OF LEGISLATION AND AGENCY REGULATIONS REGARDING CITIZEN PARTICIPATION IN ENVIRONMENTAL DECISION-MAKING

A change of government in 1981 heralded a new era of political representation in Ghana. Dissatisfied with the conventional model of political representation, the new government established novel structures for political representation in the form of Defence Committees.[50] Participatory democracy was now to be the norm as this was supposedly a more effective way of ensuring that the majority of Ghanaians influenced the making and implementation of policy. Presumably, by implication, this meant that any legislation connected with the environment would be designed or influenced considerably by ordinary citizens.

How did the new political dispensation impact upon the structure and process of popular participation in environmental decision-making? There is very little evidence to support an argument that there was significant provision for citizen participation in environmental affairs between 1981-1984, when the concept of participation under the umbrella of the Defence Committees was pursued with the most intense revolutionary fervour. Legislation with a direct bearing on the environment[51] included the following: Timber Concessions (Revesting) Law;[52] Timber Concessions (Revesting) Amendments Law;[53] Road Traffic (Amendment) Law;[54] Gold Mining Products Protection Ordnance (Amendment) Law;[55] Wild Animals Preservation (Amendment) Law;[56] National Energy Board Law;[57] Ghana National Petroleum Corporation Law;[58] Factories, Offices and Shops (Amendment) Law[59]; Mining Rights, Licenses, and Certificates (Imposition of Fees) (Consolidated Amendments) Law, 1983;[60] and the Control of Bush Fires Law.[61] Despite the rhetoric of popular participation however, only the Control of Bush Fires Law expressly specified that Bush Fire Control Sub-Committees, could rely on inputs from ordinary citizens[62] to help enforce and observe its implementation.[63]

Second, the Defence Committees focused mainly on economic crimes such as tackling corruption, enforcing price controls and serving as a conduit for citizen's views on the running of state-owned enterprises and related economic issues. Although not excluded from environmental issues, at the time the economic and political crisis in Ghana was so severe that environmental concerns were hardly ever discussed and, there is no evidence that the Defence Committees accorded any real importance to it.

Beginning in 1984 however, there were greater efforts to secure citizen inputs into environmental policy-formulation in Ghana. Why was this the case? What was it about this period that caused a swing towards a relatively stronger framework for citizen participation? External influences probably played a part just as they had influenced notions of citizen participation from the 1950s to 1970s.

First there was the emergence of an environmental civil society in Ghana. As we have noted above, the lack of environmental activism in the developing world was due, in part, to the presence of authoritarian regimes. However, the latter part of the 1980s saw a worldwide trend towards liberal democracy in countries that had previously scorned the idea. Ghana was caught up in this new swing away from authoritarian rule. In the trail of this democratic re-awakening a number of NGOs began to emerge and play an active role in Ghanaian politics. Among these NGOs were environmental groups.

Moreover, Ghana's environmental civil society built links with organisations such as Friends of the Earth and Conservation International. This is an emerging trend in the developing countries and has helped local environmental NGOs to grow.[64] Environmental NGOs based in the developed world are well endowed and well organized. They are therefore capable of supporting NGOs in developing countries in the pursuit of their objectives. Furthermore, the emerging civil society made information about the environment more easily accessible. Now no longer does the State have a monopoly over the supply environmental information and the education of the public. This access to information and other resources has helped developing country NGOs to influence environmental politics and policy.

Additionally, the severe socio-economic crisis in Ghana in the early 1980s had environmental implications. Environmental problems such as the fall in the water level of the Akosombo dam and the consequent electricity rationing,[65] and rapid deforestation[66] were characteristic of a pattern of environmental problems that required innovative solutions. Experience had indicated that these solutions included involving ordinary citizens in decision-making.

Besides, it was in the latter part of the 1980s that the concept of sustainable development became popular. 1987 saw the publication of the Brundtland Report.[67] It was this Report that made the term sustainable development popular and as such the Report articulated arguments needed to displace the exclusionist model.[68] One of the Brundtland Report's arguments was that reversing unsustainable development policies at the national level would require efforts to inform and secure public support. This role can be fulfilled by, inter alia, private and community groups and NGOs.[69]

Also, as noted above, UNCED emphasized the importance of citizen participation in decisions affecting the environment. The conventions and declarations adopted by the participants at UNCED underscore this fact. Thus, for instance, the statement on forests[70] endorses popular participation as being necessary for sustainable management of forests.[71] So too does the Biodiversity

Convention[72] note that biodiversity loss can be prevented by involving the public in general,[73] and women[74] in particular, in conservation efforts. Ghana was represented at UNCED and therefore national policy-makers were most probably influenced by the UNCED's outcomes on popular participation.

Given these developments it was not surprising that legislation in Ghana began to give greater importance to the role of individuals and local communities in decisions bearing on the environment. So, for instance, in 1988 the EPC began the process of drafting an Environmental Action Plan (EAP). In its final form, the EAP[75] recommended shaping law and institutions to facilitate participation by citizens and their representatives in environmental decision-making.[76]

In addition new local government legislation made District Assemblies responsible for the development, improvement and management of human settlements and environment in the District.[77] This responsibility has implications for citizen participation for the following reasons: two-thirds of the members of District Assemblies are to be elected by universal adult suffrage.[78] These representatives can serve as conduits for citizen participation. Also, up to 30 percent of the members of a District Assembly are to be appointed by the President after consultation with traditional authorities and other interest groups in the district[79] and thus it is possible for environmental interest groups to influence this process so as to have their views heard by members sympathetic to their cause.

1992 marked the return to constitutional rule in Ghana. This period has been marked by a move towards stronger forms of citizen participation in environmental decision-making. This is reflected in constitutional clauses and legislative provisions that aim to strengthen citizen input into decisions on the environment. The 1992 constitution contains provisions bearing on citizen participation in environmental decision-making. Under the Directive Principles of State Policy individuals have a duty to protect and safeguard the environment.[80] Locating citizen participation in the chapter on the Directive Principles of State Policy renders this duty, ostensibly, non-justiciable. The Committee of Experts established to draft the 1992 Constitution made this clear in the proposals they submitted.[81] Thus, seemingly, even if individuals sought to enforce this provision of the constitution and demand the opening-up of avenues for their participation in environmental decision-making, this would limit any claims so made.

However, the Ghana Supreme Court has elaborated on the importance of the Directive Principles. In *New Patriotic Party v Attorney-General*[82] Justice Sophia Akuffo asserted that:

> "..The Directive Principles serve as the ultimate litmus paper for testing the constitutional propriety or legality of every enactment or the conduct of every person in Ghana.."[83]

In the *New Patriotic Party* case, Justice Bamford Addo also noted that the Directive Principles are not justiciable "of and by themselves".[84] However, the words "of and by themselves" implies that where those principles are read in conjunction with other enforceable provisions of the Constitution, by reason of the fact that the courts are mandated to apply them, they are justiciable.[85] Justice Bamford Addo added further that where any provision under the Directive Principles can be interpreted to mean the creation of a legal right, i.e. a guaranteed human right, they become justiciable and protected by the Constitution.[86]

The significance of this statement is fully appreciated when juxtaposed against recent developments in the field of international environmental law. Scholarship in international law has identified and discussed the right to a clean environment.[87] Moreover, the international community has adopted declarations supporting the individual's right to a clean environment.[88] At the regional level, the African Charter on Human and Peoples' Rights states that all peoples shall have the right to a general satisfactory environment favourable to their development.[89] This therefore is one of the rights that the parties, including Ghana, undertake to give effect to through legislative or other measures.[90] These developments lend weight to the contention that the Directive Principles are not mere policy guidelines but can be used to support the creation of rights to participate in environmental decision-making in Ghana.

The post-1992 period has also seen mining and timber firms being subjected to guidelines and legislation on mandating citizen participation in their activities that have significance for the environment. For instance guidelines prepared for mining firms state that when an environmental impact assessment is undertaken, its scope should include liaison with the community to be affected by the proposed mining activity.[91]

Parliament and the Environmental Protection Agency[92] (the succesor to the EPC) have also contributed to the development of citizen participation by the adoption of Environmental Assessment Regulations[93] (the Regulations) and Environmental Impact (EIA) requirements respectively. Under the Regulations the Environmental Protection Agency (EPA) is obliged to hold a public hearing when it receives an application to commence 'undertakings'[94] listed in the Regulations' second schedule.[95]

The preconditions for public hearings are where there appears to be adverse public reaction to the commencement of the proposed undertaking, or the undertaking will involve the dislocation, relocation or resettlement of communities, or the EPA considers that the undertaking could have extensive and far-reaching effect on the environment.[96] Panels charged with the responsibility of conducting public hearings are to be constituted by the EPA.[97] At least a third of a Panel's members are to be residents of the geographical area of the proposed undertaking.[98] The Panel's are supposed to make recommendations in writing to the EPA within 15 days of the commencement of any hearing.[99]

There are other provisions in the Regulations that create avenues for citizen participation. First there is the duty on the EPA, after an application to commence an undertaking is submitted to it, to consider in its initial assessment any concerns expressed by the public in general and immediate residents in particular.[100] Second, the public at large has a right to be heard during the consideration and review of environmental impact statements.[101] Third, individuals, NGOs, and local communities have the right to make any comments and suggestions on any matter on which notices are issued under the Regulations.[102] A fourth avenue for citizen participation is the provision that EPA decisions or actions under the Regulations can be challenged by any aggrieved person[103] with the right to a fair hearing in this regard to be held by a Panel also guaranteed.[104]

CITIZEN PARTICIPATION IN ENVIRONMENTAL DECISION-MAKING IN PRACTICE

Officials at the EPA state that public hearings are well attended by local residents. They assert that local residents are enthusiastic about participating in public hearings and they use such hearings to point out the impact of activities such as mining on local water supplies, and the general health of local residents.[105] Public hearings are also held on-site and in the local language thus affording local residents the ability to participate effectively in proceedings.[106] However, EPA officials were aware of the fact that local citizens hardly ever request hearings. Instead, they are initiated from the top down. Thus participation is largely reactive, i.e. it is normally when the EPA announces a hearing that residents make the effort to have their views on the environment heard.[107]

However, staff of environmental NGOs in Ghana did not view participation in environmental decision-making so favourably. It was pointed out to me that in the first instance there is a considerable degree of ignorance about environmental problems in Ghana. In their experience, people tend to see sanitation as their key environmental problem[108] and critical problems such as air pollution are not taken that seriously. This severely limits environmental activism as participation is usually in the form of clean-up campaigns.[109]

Moreover, NGOs were also aware of the reactive nature of citizen participation in environmental decision-making. They noted that citizen interest in the environment is usually only demonstrated when official workshops on a particular matter are organized and thus there is yet to be any real widespread environmental movement to which large numbers of Ghanaians belong.[110] Besides, the lack of legislation on the right to information severely restricts the extent to which citizens and NGOs can participate in and thus influence government policy on the environment.[111]

Furthermore, reviews of reports submitted pursuant to obligations under the Regulations (noted above) indicate that very little space is devoted to popular participation. For instance, in the Proposed O' Sullivan Housing Estate Report[112] citizen participation is only mentioned in one paragraph. The Report only notes that individuals and organizations were consulted in connection with the study.[113] The EIA Report on the Manso Nkran Air Strip[114] also only mentions that there was a public forum with a cross section of opinion leaders of the various communities to inform them of the project, to listen to any objections or comments that they might have on the project,[115] and it just provides a summary of what was said during the consultations.[116] The same can be said of the report on a Proposed Site and Service Scheme at East Legon[117] where there is only a mention of consultations held with the Municipal Assembly and residents of the area.[118]

CONCLUSION

It is obvious from the analysis above that participation in environmental decision-making in Ghana has seen the practice gradually move away from control by narrow interests towards a situation where the wider public, community groups, and NGOs are beginning to have their voices heard. This evolution has been as a result of internal and external pressures. However, problems still remain. For instance, the evolution of the NGO movement is still in its early stage. Thus environmental civil society will have to continue to expand if it is to have a meaningful impact on policy-formulation.

Environmental NGOs and other community groups will also have to organize their own public forums for discussing environmental issues, attend and influence other public hearings, engage in follow-up monitoring of decisions taken with regard to the environment, and regularly issue and distribute their own reports on the state of the environment. A favourable democratic climate that protects the right to information,[119] the right of free speech and the right of freedom of association in Ghana can only help to strengthen the hand of social and political forces whose aim is a greater say in environmental issues.

The use of citizen suits to enforce statutory environmental obligations will also probably be an important part in the further development of citizen participation in environmental decision-making. The Rio Declaration acknowledges the role of citizen suits by stressing that States should provide individuals effective access to judicial and administrative proceedings.[120] The importance of citizen suits is that they serve to fill in the gap left by public authorities that are unable to ensure compliance with environmental law because of a lack of resources or commitment.[121] As the costs of litigation are rather high, it will most probably fall to NGOs to initiate citizen suits.

In addition, environmental issues are hardly at the core of party political campaigns in Ghana. However, civil society can use such organizations to advance their goals and place participation in environmental decision-making on the political agenda. They can do so by pressing for links with political groupings that have a good record of environmental protection or are willing to develop such a record. Through such links, civil society can ensure that it is consulted on a regular basis before key decisions concerning the environment are taken.

Moreover, another avenue for citizen suits can be through the duties and enforcement powers of the EPA. Article 2 of the EPA Act outlines its functions. Included in these is the duty to "secure in collaboration with such persons as it may determine the control and prevention of discharge of waste into the environment and the protection and improvement of the quality of the environment".[122] The EPA Act also charges the EPA to "ensure compliance with any laid down environmental impact assessment procedures in the planning and execution of development projects, including compliance in respect of existing projects.[123] To secure compliance with these provisions the EPA is granted enforcement powers.[124] This amounts to EPA rule making, a situation where the EPA determines that administrative action is required in the interests of the environment. It is possible that over time, the enforcement notices will crystallize into a body of EPA `case law'. Besides, the EPA will probably develop its own procedures for due process in a manner similar to the courts. If the EPA relies on public consultations and complaints as a trigger for instituting enforcement procedures then through this approach, it will act as a path for quasi citizen-suits to ensure the protection of the environment.

Finally, EIA reports submitted under the Regulations should expand on citizen participation by indicating whether any problems were encountered in this process by the organization that has conducted the impact assessment; how regularly were the consultations held; and whether the applicant gave the local community sufficient prior notice of an intention to carry out consultations.

* Kofi Oteng Kufuor, Research Fellow, School of Law University of East London and, Associate Fellow, Royal Institute of International Affairs (Chatham House). Research for this chapter has benefited from interviews with staff at the following organizations in Ghana: the Environmental Protection Agency, Friends of the Earth, and Green Earth Organization. A generous grant from the University of East London's Law School made the trip to Ghana possible.

[1] Declaration of the UN Conference on Environment and Development, 1992, reprinted in P.W. Birnie and A. Boyle (eds.), *Basic Documents on International Law and the Environment*, (Clarendon Press, 1995), 9 (text begins at 10).

[2] *Programme of Action for Sustainable Development: Report of the United Nations Conference on Environment and Development*, Rio de Janeiro 3-14 June 1992, United Nations: New York (1993), 9 (hereinafter Agenda 21).

[3] Ibid. Principle 10.

[4] Ibid. Principle 10.

[5] Ibid. Principle 20.

[6] Ibid. Principle 21.

[7] Ibid. Principle 22.

[8] Agenda 21, n. 2, above, para. 23.2.

[9] Ibid. paras. 27.10 and 27.13.

[10] B. M. Hayward, "The Greening of Participatory Democracy: A Reconsideration of Theory", 4:4 *Environmental Politics*, (1995), 215, 221.

[11] Ibid. 220.

[12] T. Webler and O. Renn, "A Brief Primer on Participation: Philosophy and Practice", in Fairness and *Competence in Citizen Participation: Evaluating Modules for* Environmental Discourse, O. Renn, T. Webler and P. Wiedemann (eds.), Kluwer, 1995) 17, 21.

[13] B.M. Hayward, n.10 above, 216.

[14] Ibid.

[15] H. G. White, "Including Local Communities in the Negotiation of Mining Agreements: The Ok Tedi Example", 8:2. *The Transnational Lawyer* (1995), 303.

[16] E. Rehbinder, "Ecological Contracts Agreements between Polluters and Local Communities", in *Environmental Law and Ecological Responsibility: The Concept and Practice of Ecological Self Organization*, G. Tuebner, L. Farmer and D. Murphy, (eds.) (John Wiley & Sons, 1994), 147, 15354.

[17] R. Coase, "The Problem of Social Cost", 3:1 *Journal of Law and Economics* (1960), 1.

[18] H.G. White, n.15 above, at 336.

[19] B.M. Hayward, n.10 above, at 220.

[20] Ibid.221.

[21] Ibid.

[22] Environmental Protection Council Decree, 239 (1974).

[23] Ibid. Article 2 (1) (b).

[24] Ibid. 2 (1) (e).

[25] G. Porter and J.W. Brown, *Global Environmental Politics* 2nd ed, (Westview Press, 1996), 23.

[26] Ibid.

[27] Ibid.

[28] For attitudes by national authorities in Ghana towards the protection of the environment see, J.R.A Ayee, "The Formulation and Implementation of Environmental Policy in Ghana", XXIII: 2 *Africa Development* (1998), 99, 114.

[29] I. Shivji, *The Concept of Human Rights in Africa* (CODESRIA, 1989), 83-84.

[30] H. Gouldbourne, "The State, Development and the Need for Participatory Democracy in Africa", in P. A. Nyong'o, *Popular Struggles for Democracy in Africa* (Zed Books, 1987) 26, 35.

[31] A. Seidman and R. Seidman, *State and Law in the Development Process: Problem-Solving and Institutional Change in the Third World* (MacMillan, 1994), 186.

[32], J. R. Macey, "Promoting Public-Regarding Legislation Through Statutory Interpretation: An Interest Group Model", 86:2 *Columbia Law Review*, (1986) 223. On Public Choice theory in general see, J. Buchanan & G. Tullock, The Calculus of Consent: Logical Foundations of Constitutional Democracy, (University of Michigan Press, 1962).

[33], J.R. Macey, n.32 above, 230.

[34] D. A. Dana, "Overcoming the Political Tragedy of the Commons: Lessons Learned from the Reauthorization of the Magnuson Act", 24:4 *Ecology Law Quarterly*, (1997), 823, 834.

[35] See, P. L. Khan, "The Politics of Unregulation: Public Choice and Limits on Government", 75:2 *Cornell Law Review*, (1990), 280, 289.

[36] Ibid. 291-292.

[37] G. Porter and J.W. Brown, n.25 above, 28. This attack began in the United States, where it had its most articulate advocates, and then spreading to Europe and other regions. The main forerunners of this paradigm were the Limits to Growth study by the Club of Rome in 1972 and the Global 2000 Report released by the US Council of Environmental Quality and the Department of State in 1980. Both studies predicted that the earth's natural resources faced depletion and also the degradation of ecosystems. Ibid.

[38] Reprinted in P.W. Birnie and A. Boyle, n.1 above, 1, (text begins at 2).

[39] Ibid. Proclamation, para. 7.

[40] Ibid.

[41] Ibid. Principle 19.

[42] Ghana Industrial Free Zone Authority Decree, 157, (1978).

[43] Investment Code, Act 437, (1981).

[44] Industrial Free Zone Authority Decree, n.42 above, Article 3 (1) (j).

[45] Investment Code, n.43 above, Article 11 (2).

[46] Forest Protection Decree, 243 (1974).

[47] Ibid. Article 1.

[48] See, G. Myrdal, "The "Soft State" in Underdeveloped Countries", 15:4 *UCLA Law Review*, (1968) 1118, 127. Although Myrdal's analysis is with respect to the underdeveloped countries of South Asia, he notes that his analysis has validity for other parts of the developing world including Africa. Ibid. 1121. The shortcomings of legislative and policy initiatives in developing countries are highlighted when compared with similar developments in the developed world. Myrdal points out how greater equality of women in civil law matters was gradually attained after a struggle stretching over a long period. These rights-based struggles implied that when rights were finally granted that

there were eager and mostly organized groups of citizens prepared to make use of them. Ibid.

[49] G. Porter and J.W. Brown, n.25 above, 24-25, describing how the rise of an alternative social paradigm in environmental thought (sustainable development) did not occur overnight.

[50] These organs of popular participation were to replace the middle class alliances that had consistenly dominated Ghanaian politics since independence with a different coalition made up of the radical intelligentsia, the lower ranks of the petty bourgeoisie, the urban workers and the peasants. The revolutionary government claimed that the traditional poltical models had failed the country on 3 counts including reducing Ghanaians to the role of passive spectators while most meaningful decisions were taken in Accra (the capital) by politicians who were beyond public scrutiny once they had secured their election. P. Nugent, Big Men, *Small Boys and Politics in Ghana: Power, Ideology and the Burden of History*, 1982-1994, (Mansell, 1995), 47-48.

[51] G.A. Sarpong, "From Stockholm to Rio: Some Ghanaian Responses to the Problems of the Environment", XIX University of Ghana Law Journal, (1993-1995), 53, 81-82. The list of laws I cite below are drawn from Sarpong's observations, ibid.

[52] Timber Concessions (Revesting Law), 1982, Law 17.

[53] Timber Concessions (Revesting) Amendments Law, 1982, Law 35.

[54] Road Traffic (Amendment) Law, 1983, Law 43.

[55] Gold Mining Products Protection Ordnance (Amendment) Law, 1983, Law 54.

[56] Wild Animals Preservation (Amendment) Law 1983, Law 55.

[57] National Energy Board Law, 1983, Law 62.

[58] Ghana National Petroleum Corporation Law, 1983, Law 64.

[59] Factories, Offices and Shops Act (Amendment) Law, 1983, Law 66.

[60] Mining Rights, Licenses and Certificates (Imposition of Fees) (Consolidated Amendments) Law, 1983, Law 67.

[61] Control of Bush Fires Law, 1983, PNDC Law 46.

[62] Ibid. Article 5 (2-3).

[63] Ibid. Article 6.

[64] J. Haynes, "Power Politics and Environmental Movements in the Third World", 8:1*Environmental Politics*, (1999), 222, 225.

[65] "Darkness in Accra & Abidjan", *New African* (February 1994), 37.

[66] Friends of the Earth, *Plunder in Ghana's Rainforest for Illegal Profit: An Expose of Corruption, Fraud and Other Malpractices in the Timber Trade, Vol. 1*, (Friends of the Earth, 1992), 4.

[67] World Commission on Environment and Development, *Our Common Future*, (Oxford University Press, 1987).

[68] See G. Porter and J.W. Brown, n.25 above, 24-25.

[69] The Brundtland Report noted how NGOs and citizen's groups pioneered in the creation of public awareness and political pressures that stimulated governments to act on environmental issues. NGOs also identify risks, assess environmental impacts, publish "State of the Environment Reports, and provide an efficient and effective alternative to public agencies in the delivery of programmes and projects. See, World Commission on Environment and Development, n.67 above, 326329.

[70] Non-Legally Binding Authoritative Statement of Principles for a Global Consensus on the Management, Conservation and Sustainable Development of All Types of Forests, 31*ILM* (1992), 881.

[71] Ibid.

[72] Convention on Biological Diversity, 31*ILM*, (1992), 822.

[73] Ibid. Preamble

[74] Ibid. Preamble.

[75] *Ghana Environmental Action Plan* (Environmental Protection Council, 1991).

[76] Ibid. 25.

[77] Local Government Act, 462 (1993) Article 10 (3) (e).

[78] Ibid. Article 5 (1) (b).

[79] Ibid. para. (d).

[80] *Constitution of the Republic of Ghana*, 1992 (Ghana Publishing Corporation, 1992), Article 41 (k).

[81] Report of the Committee of Experts (Constitution), *Proposals for a Draft Constitution of Ghana*, (Ghana Publishing Corporation, 1991). The Directive Principles of State Policy are for the guidance of Parliament, the President, the Council of Ministers, Political Parties and other bodies and persons in making and applying public policy for the establishmentof a just and free society. The Principles should not of and by themselves be legally enforceable by any Court. The Courts should, however, have regard to them in interpreting any laws based on them. Ibid. 49.

[82] See, *New Patriotic Party v Attorney General*, (1997), The Supreme Court of Ghana Law Reports (hereinafter *New Patriotic Party case*), 729.

[83] Ibid. 802-803. See also the dicta of Justice Bamford Addo, ibid, 745.

[84] Ibid. p.747.

[85] Ibid. See, also, S.Y. Bimpong-Buta, *The Law of Interpretation in Ghana*, (Advanced Legal Publications, 1995), 128-133.

[86] Ibid.

[87] See the following: N. A.F. Popovic, "In Pursuit of Environmental Human Rights: Commentary on the Draft Declaration of Principles on Human Rights and the Environment", 27 *Columbia Human Rights Law Review*, (1996), 487; C. E. Akello, "The Right to Environment and Generational Equity" 4:2 *East Africa Journal of Peace and Human Rights*, (1998), 125-138; E. Kasimbazi, "The Environment as a Human Right: Lessons from Uganda", *Proceedings of the Tenth Annual Conference of the African Society of International and Comparative Law*, (1998), 145; G. Alfredsson and A. Ovsiouk, "Human Rights and the Environment", 60:1 *Nordic Journal of International Law*, (1991) 19; D. Shelton, "Human Rights, Environmental Rights and the Right to the Environment", 28:1 *Stanford Journal of International Law*, (1991), 103.

[88] For instance see, Declaration of The Hague on the Environment, 11 March 1989, 28 *ILM*, (1989), 1308; UN General Assembly Resolution 45/94 (1990). The Rio Declaration also asserts that "human beings are entitled to a healthy and productive life in harmony with nature". See Rio Declaration, in P.W. Birnie and A. Boyle, n.1 above, Principle 1. The participants at UNCHE also declared that "Man has the fundamental right to freedom, equality and adequate conditions of life, in an environment of a quality that permits a life of dignity and well-being, and he bears a solemn responsibility to protect and improve the environment for present and future generations. See Stockholm Declaration, ibid. Principle 1.

[89] African Charter on Human and Peoples' Rights, 21:1 *ILM*, (1982) 59, Article 24.

[90] Ibid. Article 1.

[91] *Ghana's Mining and Environmental Guidelines* (1994), Prepared by the Minerals Commission and the Environmental Protection Council, 21, and 32.

[92] Environmental Protection Agency Act, 1994, Act 490.

[93] Environmental Assessment Regulations, 1999, Legislative Instrument 1652

[94] Undertakings are defined under the Regulations as "any enterprise, activity, scheme of development, construction, project building, work, investment, plan, programme and any modification, extension, abandonment, demolition, rehabilitation or decommissioning of such undertaking, the implementation of which may have a significant impact". Ibid. Regulation 30 (1).

[95] Ibid. Schedule 2.

[96] Ibid. Regulation 17 (1) (a)-(c).

[97] Ibid. 17 (2).

[98] Ibid. 17 (3).

[99] Ibid. 17 (5)

[100] Ibid. Regulation 5 (1) (c).

[101] Ibid. Regulation 16 (5).

[102] Ibid.

[103] Ibid. Regulation 27 (1).

[104] Ibid. Regulation 27 (5).

[105] This was pointed out to me during interviews with EPA officials.

[106] Interview at EPA.
[107] Interview at EPA
[108] Interview at Green Earth Organization.
[109] Interview at Green Earth Organization.
[110] Interview at Green Earth Organization.
[111] Interview at Friends of the Earth
[112] Proposed O' Sullivan Housing Estate, *Environmental Impact Assessment Report* (January 1999).
[113] Ibid. 4.
[114] Resolute Amansie Limited, *Environmental Impact Assessment: Manso Nkran Air Strip* (no date).
[115] Ibid. 32.
[116] Ibid. Appendix IV.
[117] NTHC Properties Ltd., *Environmental Impact Assessment for the Proposed Site –and – Service Scheme at East Legon*, Adjirigano, Greater Accra (December 1998).
[118] Ibid. 21.
[119] There are measures underway to pass a right to information act in Ghana. See, The Right to Information Bill, 1999 (Draft), reprinted in Justice P.D. Anin, *The Right to Information Bill (Draft)* (Institute of Economic Affairs, 1999).
[120] Rio Declaration, in P.W. Birnie and A. Boyle, n.1, above, Principle 10.
[121] Phillipe Sands, *Principles of International Environmental Law, vol. I: Frameworks Standards and Implementation*, (Manchester University Press, 1995), 144.
[122] Environmental Protection Agency Act, n.94 above Article 2 (a).
[123] Ibid. Article 2 (f).
[124] The EPA has the power to issue an enforcement notice where it appears to it that the activities of any undertaking threaten the environment or public health. Ibid. Article 13.

LEGAL AND REGULATORY FRAMEWORK FOR ENVIRONMENTAL IMPACT ASSESSMENT IN AFRICAN COUNTRIES

MOHAMMED A. BEKHECHI[1]

Laws and regulations suited to country-specific conditions are among the most important instruments for transforming environment and development policies into action" Agenda 21, Chapter 8

INTRODUCTION: IMPORTANCE OF EIA IN THE DEVELOPMENT PROCESS

Environmental Impact Assessment (EIA) is in general defined as a process and a planning tool to identify the likely consequences for the bio-physical environment and the human health and welfare, of implementing particular projects and programs and conveying this information at a stage when it can materially affect their decision, to those responsible for sanctioning the proposals for these projects and programs[2]. It is undisputed that EIA, like other planning and environmental management instruments, plays a critical role in ensuring the sustainable development of a country's economy.

Beyond economy, the potential benefits of EIA include effective compliance with sound environmental standards, sustainable use of natural resources, better decision-making processes by involving stakeholders, improvement in the design of infrastructures, and potential reduction of operating costs. To be effective however, an EIA mechanism needs to be grounded on well-defined legislation and procedural rules where the rights and obligations of all stakeholders are clearly defined and their enforcement ensured through appropriate implementation and compliance monitoring procedures and other instruments.

EIA AND LEGAL DEVELOPMENT AND REFORM: GENERAL CONSIDERATIONS

Environmental laws and regulations constitute critical instruments for environmental protection and management in most countries. Environmental standards provide the context and measure for broader economic means of promoting environmental protection and long-term productivity, the twin goals of sustainable development.[3] In less than three decades a vast majority of countries have adopted policies, guidelines, regulations and laws dealing with environmental assessment. These are the cases not only in industrialized countries, but also in developing countries across all continents. Major factors that spurred this trend include: (1) the adoption of the *National Environmental Policy Act* (NEPA)[4] in the United States (US) which introduced environmental assessment in the realm of public policy, (2) the publication of the *Brundtland Report: Our Common Future* which called for improved environmental impact assessment procedures to be set out at all levels of government[5], (3) the endorsement of the US *NEPA* and the *Brundtland Report* by a growing number of countries, businesses and international institutions,[6] and; (4) the adoption in industrialized and developing countries of formal EIA principles and practices to govern a wide variety of activities that impact the environment.

In Africa, the development of environmental law was and is still often perceived as the direct consequence of the requirements of international donor agencies rather than the outcome of a national environmental willingness to strengthen the environmental agenda. The introduction and acceptance of EIA, as a policy and legislative tool, has been challenged by a general misconception, according to which the EIA will hamper the economic and technological development.[7] In addition, in many cases, opinions have been expressed that the EIA is unnecessary, costly and may duplicate existing monitoring procedures and other planning tools.[8]

However, the recent increase in domestic EIA legislation in Africa has also been brought through; (1) a general increase in awareness of environmental problems and issues; (2) the development of important pieces of environmental legislation, including EIA regulations without any involvement of donor agencies in many Sub-Saharan African countries; (3) the inclusion of environmental legislation reform as part of an overall legal reform agenda; and (4) the need to comply with the newly developed environmental requirements of international trade.

This paper describes the EIA-related legislation of a number of African countries[9] in order to assess where these countries stand in the establishment of their EIA system by using the resources of the law, and to compare the different approaches they have developed for EIA procedure and content.[10] Analysing environmental impact assessment legislation in African countries is an important step to measure what has been achieved so far by these countries with regard to the development of an effective and operational EIA system, but also and ultimately to define future steps to: (1) enhance the legal and regulatory framework; (2) strengthen the institutional framework; (3) develop an enforcement regime and other economic instruments in environmental management; (4) increase public participation in the development decision making process; and (5) develop the capacity and role of stakeholders in environmental protection.

In addition, the paper discusses the potential harmonisation of domestic EIA laws and statutes of the respective countries among them and with other international rules and standards where needed and feasible. In fact, at the international level, environmental impact assessment is increasingly considered as a general principle of international environmental law[11] and as such affect States' behavior. In fact, the mainstay to promote sound economic development practices also lies in the use of environmental laws and regulations, the purpose of which may be to introduce the needed and relevant instruments, mechanisms and other planning tools.[12]

EIA AND INTERNATIONAL LAW

The EIA principle has increasingly become a feature of modern international environmental law and is referred to in major international environmental conventions[13] and other international legal instruments of a "soft law" nature. In addition, many authors are referring to this principle as a general principle of sustainable development law, and it is likely that in a few years from now, all countries will formally recognize it as such and introduce it into their domestic legal order.

EIA and International Conventions:

Four international environmental conventions may be mentioned here to highlight the extent to which EIA is recognized and discussed with regards to domestic compliance with international law. The choice of these four conventions is based on the fact that: (a) thirty-two Sub-Saharan Africa countries have ratified the *United Nations Convention on the Law of the Sea - UNCLOS,*[14] forty-two the *Convention on Biological Diversity Convention*[15], thirty-eight the *UNFCCC*[16] and forty-five the *Convention on Desertification,*[17] and that (b) the relevant provisions of these four conventions illustrate the types of provisions indicating the extent to which States have now accepted EIAs as established principle. The recognition of this principle as legally binding and its transformation to enforceable laws and regulations may appropriately address the need to have a predictable regulatory framework, including clear norms and regulations, to improve the business environment in the African countries,[18] and to establish safeguard policies to mitigate and compensate negative impacts resulting from development projects.

However, the issue remains how to monitor compliance with the provisions of the above-mentioned conventions? This is a very complex task given the fact that these conventions do not deal with the details of how the EIA procedures are to be established, implemented and conducted, but merely establish the obligation for States to undertake EIAs when planned activities have potential detrimental effects.[19] The challenge for these conventions is to enhance their impact and effectiveness, even though they often lack the appropriate mechanisms to do so. In practice, there are no direct references to international conventions in EIA-related legislation of African countries except in few cases, such as in Nigeria.[20] This does not mean, however, that African countries are not willing to implement these conventions. Quite to the contrary, it is easily observable that African Governments: (1) are increasingly involved in international treaty-making process; (2) adopt domestic rules that are consistent with international treaties and conventions;[21] and (3) develop processes to support the implementation of international conventions.

African countries must undertake and should also take strong steps to comply with these conventions by empowering all the actors involved in the EIA processes, and to sanction non-compliance behavior. However, their efforts to implement international law may be hindered by the lack of human and financial capacities. Experience shows that a way to help developing countries to build capacity and to implement their obligations under international conventions is to incorporate adequate financial and technical assistance provisions in the conventions themselves.

Other International Legal Instrument of "Soft Law" Nature

Beside these international conventions, it is worth mentioning that during the Rio Summit, in 1992, the United Nations Conference on Environment and Development (UNCED) has recognized EIA as an essential tool to promote sustainable development and protect the integrity of the global environment. Principle 17 of the *Rio Declaration* provides that "environmental impact assessment, as a national instrument, shall be undertaken for proposed activities that are likely to have significant adverse impact on the environment and are subject to a decision of a competent national authority". It is also worth mentioning that IUCN's Commission of Environmental Law has developed and proposed to the General Assembly of the United Nations a draft *International Covenant on Environment and Development*, which recognizes EIA as a tool for complying with the precautionary principle and promoting sustainable development. Article 37 of the draft *Covenant* provides for the establishment or enhancement of EIA systems and procedures.[22] Finally UNEP has developed and enacted a set of guidelines and principles dealing with EIAs, which are of a particular interest[23] for this paper.

Although all these instruments are not legally binding instruments, it is undisputed that the *Rio Declaration* is viewed as an international instrument that manifests a consensus to certain general principles of law that are viewed and accepted by a large number of countries from both the North and the South. To this extent, if such principles are followed by state practice, they can provide evidence of the existence of the *opinio juris* from which customary rules develop and become binding. It is important to note that States have supplemented their effort to adopt treaties and conventions ("hard law") by the development of international "soft law" instruments and materials to foster their cooperation in the field of environment and develop their environmental management capacity. In many cases, EIA-related laws and regulations in African countries are inspired by the UNEP guidelines or other similar guidelines such as those used by multilateral development institutions or donor agencies.

Some of these laws refer explicitly to the international conventions to which the concerned country is party,[24] but in other instances, compliance with international conventions can be measured in substance, i.e. where the domestic legislation is consistent with international conventions even though there is no explicit reference to the convention in the legislation. Environmental impact assessment in countries that have ratified international conventions that contain EIA-related provisions should include the reference to the relevant provisions of these conventions. Practice reveals some discrepancies between international conventions ratified by countries and domestic law of such countries. This happens when it comes to sharing information with other countries, especially neighboring countries interested in the outcomes of EIAs for projects close to their territory or in adjoining areas (seas, lakes, international waterways). In fact, the EIA report should clearly discuss and analyze the extent and nature of environmental impacts taking into consideration the international obligations to which a country subscribed in becoming party to international conventions. Some international conventions provide international mechanisms for the declaration of national wetlands of international significance and which, once declared, carry state obligations.[25]

This is the policy and legal background in which African countries developed EIA systems and procedures pursuant to their specific national environmental action plans and relevant framework environmental legislation. In this effort to equip them with an enabling environmental legal framework, many Sub-Saharan Africa countries have developed an EIA system including laws, regulations and guidelines that are commensurate to their needs and local specificities. African countries have enacted some form of environmental legislation,[26] which invariably includes, among other elements, statements on policies, objectives, institutional arrangements, content of EIA report, enforcement mechanisms and rules, definition of the scope and functions of the central environmental agency, and a statutory environmental review procedure.

ANALYSIS OF THE CURRENT AFRICAN COUNTRIES EIA-RELATED LAWS, STATUTES AND REGULATIONS: CONTENT AND PERSPECTIVES

By the beginning of 2000, twenty-four (24) out of forty-eight (48) African countries enacted specific or some form of environmental legislation dealing with environmental impact assessment issues.[27] Others have some form of administrative procedure or policy in place with respect to EIA, but without background legislation, while others are considering adopting environmental impact legislation, or are in the process of doing so.[28] The remaining countries do not have formalized procedures and responsibility for EIA is diffuse or EIAs are conducted on an *ad hoc* basis by the project's proponent. An interesting example is that of Benin where a full set of EIA guidelines for various categories of projects have been drafted by the Benin Environmental Agency (ABE) without any legal ground, making the implementation of these guidelines a voluntary exercise by project's proponents. How far such guidelines are effectively followed and implemented remains to be discussed and verified, especially because no incentive is provided for such implementation and no sanction could be enforced against project proponents who refuse to implement them. Current EIA requirements are embodied in framework environmental laws,[29] planning-oriented laws, in statutes and regulations specifically devoted to EIAs,[30] or policy documents that require and describe an EIAs' procedural and substantial aspects.[31] Specific cases are those of South Africa, Zambia and Côte d'Ivoire where framework legislation has introduced EIA requirements and where such framework legislation is implemented through specific regulations and guidelines that describe the content, procedure and sanctions of the EIA process in these countries.[32]

The case of Kenya is somehow unique because, even though it has developed an important body of environmental legislation, it has not yet enacted a statute on EIAs and is still in the process of developing a comprehensive environmental law whose draft was circulated to stakeholders for comments in 1998.[33] It should be noted however, that certain sectoral laws provide for EIAs with regard to specific natural resources sectors such as water (under the *Water Act*) and land use (under the *Land Planning Act* and the *Pest Control* Act).[34] Other countries have a framework law on environmental management and are currently developing implementing decrees or regulations on EIA. Cameroon is doing so by developing a set of EIA regulations to be applied for all development projects to be undertaken in protected areas' buffer zones[35] and another set for all other potentially polluting projects. Algeria is in the process of updating the existing *Decree 90-78* of July 1990 on EIA, which was adopted under the framework environmental law of 1983.[36]

Without any exception, all laws, statutes and regulations examined in this chapter have basic provisions related to EIA definition, categories of projects to be subjected to EIA and within certain cases, the requirement of an initial environmental evaluation or screening before a decision is made on subjecting a given project to a full EIA. Others specify the various mandates and responsibilities of agencies involved in the EIA process, the status of public participation and disclosure and finally few provide for implementation of environmental management plans (EMP). Monitoring and compliance with EIA report recommendations and/or mitigation measures or environmental management plans are sometimes required.

DEFINITION OF EIA

African countries' laws, regulations and statutes refer, without any single exception, to EIA. It should be mentioned here that EIA is a sub-set, or, presented differently, a specific tool of Environmental Assessment (EA), which is defined as a process. For example, in the World Bank language; "EA is a process whose breadth, depth, and type of analysis depend on the nature, scale, and potential environmental impact of the proposed project. EA evaluates a project's potential environmental risks and impacts in its area of influence."[37] While EA covers a project "from cradle to grave", EIA is usually seen as the pre-decision making study which looks into the likely environmental impacts of the project as submitted to assessment and into the alternatives to that particular design. EA, in addition to including an analytical phase before decision-making, also includes the supervision of the project and the actual monitoring of its impacts.

Due to the traditional emphasis on EIA in OECD country laws and regulations, many of the countries which enacted the assessment of environmental impacts of investments prior to decision-making have referred to EIA. The present paper has taken a very pragmatic view of the relevance of EIA to EA and has described the national requirements in individual African countries in terms of EIA.[38]

PROJECTS AND ACTIVITIES SUBJECTED TO EIA:

There is no comprehensive or correct answer to the question;[39] "what projects and activities that must be subjected to EIA? Each country decides on the type and scale of activities to be subject to EIAs. Except few cases where the law imposes a general obligation to undertake EIAs for all projects likely to have significant impacts,[40] African countries' legal instruments define the activities to be subject to EIAs in different ways. Some of the countries have considered that an initial environmental evaluation may be a useful tool to help decide what project would be subjected to an EIA.

Prescribed Projects

Almost all the existing EIA laws and/or regulations provide a list of prescribed projects to be subjected to EIA.[41] In practice there are three different categories of legal approaches to address the definition of projects likely to have harmful impacts and their specifications: (a) a first category of legislation refers to "activities which will be listed by implementing regulations";[42] (close to this category are the laws that provide for the authority mandated to protect the environment to publish through official notices the "types and sizes of projects, which shall not be implemented unless an environmental impact assessment is carried out"[43]); (b) a second category of laws have adopted a broad definition of the activities including "physical" development projects, programs, policies and development plans as well as other planning documents,[44] but no list of such activities is provided; and finally (c) a third category of legislation, which combines parts of the two first categories, establishes lists of activities subject to a full EIA, and activities subject to a more simple environmental analysis.[45] It is worth noting also that some statutes include a list of geographical areas where any development project, whatever its size and sector, is to be subjected to a full EIA.[46] Such a list generally comprised of all environmentally sensitive areas of the country concerned including forest and marine protected areas, cultural heritage sites and other significantly important landscapes. One interesting innovation has been introduced by the Republic of South Africa, which subjects to an EIA all activities leading to the modification or alteration of the biological diversity resources of South Africa,[47] to an EIA .

It is a matter of fact that many laws, regulations and statutes that provide lists of activities to be subject to EIAs are too ambitious,[48] in comparison to similar lists provided by other countries in other continents or development agencies. The Schedule related to the mandatory study activities under the Nigerian legislation on EIA is quite symptomatic of such ambition to subject as many activities as possible to a prior environmental impact assessment.[49] However, it should be noted that activities may be prescribed in terms of type, not size and that the EIA regulatory agency retains the power to establish thresholds for prescribed activities.[50] Finally, although EIAs may be required for new development projects and activities only, some statutes extend them to modifications of existing facilities which involve new or increased release of pollutants, more consumption of natural resources or more production of waste as a consequence of the increased volume of activity.[51]

Requirement of EIA Based on Specific Considerations of Projects Likely to Have Significant Impacts

This project specific approach is not favored by African countries. Some of them, however, retain it and empower their regulatory agencies to screen all development projects in order to require an adequately defined EIA. It should be mentioned that the use of a system requiring an EIA for projects likely to have significant environmental impacts is interesting and may prevent harmful projects to be commenced before a detailed environmental assessment is made. Two consequences of this option are to be considered: (a) no matter what the size of a proposed project is, an EIA may be required after a screening of such a project is carried out where the effects are likely-the problem is that the screening should be sufficiently detailed to ensure that the final decision could be made on sound grounds; and (b) the degree of significance of the potential harmful effect of a proposed project could be assessed in different ways by a government agency and the proponent) and may challenge both of them. This is why there is a clear need for an indisputable, and transparent definition of the size, nature and scale of projects, and definition of criteria to assess "significant impacts" in the legislation to avoid conflict and misinterpretation among stakeholders involved in EIA procedures.

Initial Environmental Evaluation

Many African countries laws and statutes require the project proponent to provide an Initial Environmental Evaluation (IEE) in order to determine whether a full EIA would be required for the proposed project. This requirement is not always clearly spelled out in those laws and statutes, except where they define a prescribed list of projects to be subjected to EIA. A number of African countries require an IEE as part of the mandatory EIA procedure, or sometimes as part of the planning and/or construction law.[52] The remaining questions are whether it is appropriate to subject prescribed categories of projects to an IEE and how to determine projects likely to have a "significant harmful impact", which should be subjected to IEE where no list of prescribed projects exist. No answer is provided in the current laws and regulations examined.

What Impacts Are Dealt With in the EIA Process

These are not defined in a uniform way in all countries. They vary from country to country. The usual answer given by all EIA-related laws and regulations refer in a generic manner to the terms "environmental impacts" followed by a listing of environmental media.[53] Generally, most of the statutes focus on the biological and physical environment while few of them are very specific and describe the likely environmental impacts of the proposed activity in detail.[54] The basic requirements relate to impacts on: water quality and quantity, air quality, noise level, ecosystems and ecological processes, landscapes and cultural resources, including historical sites and cultural heritage. More specifically, some laws and statutes require evaluation of the disposal of wastewater, solid wastes and emissions into the atmosphere and may require identification of the positive environmental impacts of the proposed activity.[55]
The human environment is also taken into account in various laws, regulations and statutes under provisions relating to the "social impact", the "impact on populations", "heath impact" or the "impact on communities". Generally, it includes local economies, health and education issues as well as social relationships. This wider definition, which includes social impacts, seems to be favored by the most recent laws and statutes, except the case of Mali where the 1999 *Decree* requires an EIA to simply identify and evaluate, positive and negative potential impacts, whether direct or indirect, in the short or long-term on the environment.[56] Recently prepared draft EIA-related statutes that are being discussed among stakeholders in various Sub-Saharan Africa countries are considering the social and economic impacts of projects to be subjected to an EIA.[57] In fact, a greater concern is being expressed by a new legislative trend, which considers the analysis of the socio-economic implications as an integral part of an EIA study, as well as the internal work environment and health.[58] Its rationale, in the case of African countries may be grounded on the fact that the vast majority of the African populations have closer and complex relationships to their natural environment. In view of future legal and regulatory developments, it seems appropriate to suggest that EIA-related statutes, regulations and guidelines deal more closely with the important issues of public health and the possible external implications of deterioration in the physical and mental well-being of workers and dweller communities.[59]

One issue of importance is the weakness or in some cases the lack of a definition of the applicable quality standards and norms against which the environmental assessment is done in various African countries. Standards are an indispensable tool for the evaluation of the environmental significance and impacts of certain actions, but in no case should the decision to carry out a project solely depend on the compliance or non-compliance with standards. In almost all of African countries, the impact assessment is made by reference to international standards, business guidelines or foreign country norms and standards including those of international aid agencies.[60] One example is Ghana, which has recently enacted environmental assessment regulations[61] to strengthen the powers of the Environmental Protection Agency and clarify procedures and other matters pertaining to EIA including public hearings. However, as regards to standards to be enforced, there are gaps in the Ghanaian environmental regime. In fact, there is no legislation on discharge of wastes into water, the river systems or the marine environment.[62] There is also no legislation governing air quality standards, product quality standards and control of emissions from industrial plants, motor vehicles and other sources. Throughout the industrial sector, including mining,[63] environmental protection is provided only within the context of the safety, health and welfare of employees. However a new trend is taking place in African countries which are starting to develop specific sets of environmental quality standards that fit into their economic, social, environmental and legal framework consistent with those quality standards adopted by relevant international organisations and institutions.

It is also worth noting that a new trend is developing including reference to the precautionary principle as a key reference principle in environmental impact assessment. However, where the EIA process reveals a weakness or lack of definition of environmental quality standards, it is important to consider developing, in the concerned country, sets of sound environmental quality norms and standards to support EIA implementation.

Public Participation and Consultation during the EIA Process: General Considerations

Public participation in development decision-making process helps meet public needs, enhances access to information, leads to better development decisions, results in fewer court challenges and ultimately reduces conflict between developers and the affected public.[64] Most specifically, public participation strengthens the credibility of the entire EIA process by: (1) improving the understanding of issues among all parties involved; (2) finding common ground and determining whether an agreement can be reached on some of the issues; (3) highlighting tradeoffs that must be addressed during the decision-making process, (4) generating higher quality information about potential environmental impacts, and (5) leading to a decision that avoids public hearings where these are discretionary.[65] Recent laws on EIA emphasise the role of public participation in EIA procedures with a view of ensuring quality of its results and outcomes. Public participation requirements, as they are expressed in EIA-related laws and statutes, may have different aspects. How people contribute to the EIA decision-making process may be organised in different ways.[66]

Provisions related to public Participation and Consultation During the EIA Process in EIA-Related Laws of African countries

By analysing all relevant laws and statutes, one can assess exactly who decides what level of participation is mandated. In general, the complexity of issues at stake determines the scope of participation required. However, in order to fully understand how public participation is conducted, it is necessary to study all pertaining laws and statutes (environmental, administrative, and civil) in all jurisdictions (local, national, regional),[67] including constitutional laws.

Many African constitutions solely mention the right of citizens to a clean and healthy environment, and the right to participate in its protection. For example, Kenya's Constitution does not contain explicit environmental provisions.[68] It does, however, place importance on the right to life, and experts argue that this right to life encompasses the right to a clean and healthy environment.[69] The Constitution of Kenya makes provision for the protection of individual fundamental rights and freedoms. Some of these provisions are of direct relevance to public participation in environmental decision-making, such as those addressing the freedoms of speech, assembly, and association; the right to life; and the right to protection of the law. These fundamental individual rights and freedoms are protected under Chapter V of the Constitution. Of particular significance is section 80, which guarantees every person the right to assemble freely and associate with other persons, and includes the right to form or belong to associations. The Constitution includes the right of access to the High Court for redress with respect to enforcement of fundamental individual rights and freedoms.

It is worth considering the case of Article 14 of chapter III of the Constitution of Tanzania, which establishes the right to life and its protection by society. The High Court of Tanzania has on two occasions ruled that this right includes the right to live in a clean and healthy environment.[70] In one of the cases the presiding Justice critiqued the Dar es Salaam City Council for applying for a stay of execution of a High Court decision stopping the Council from dumping waste in a suburb of Dar es Salaam.[71]

This right is considered a constitutional declaration of a fundamental right but still needs additional legislation to be enforced. When this additional legislation exists, it is not always helpful. In fact, public participation is mentioned in many EIA-related legislation, but it is not always appropriately defined and the procedure for its implementation is not always spelled out with the necessary details.[72] Barring a few exceptions, current statutes in African countries are weak and sometime loosely defined. Few provisions deal with the way public concerns, opinions and views are taken into account in the final decision on the EIA report. The scope and purpose of the public participation are sometimes widely defined and do not allow to measure the real impact of public participation in the EIA decision-making process. Interestingly, no legislation in African countries provides for judicial review to enforce the right to participate in EIAs as a means to enforce the constitutional right to a clean and healthy environment. Few African cases reveal examples of good practice, which deserve to be developed further and implemented.

In Zambia, public participation is provided for in various provisions of the EIA-related legislation, and its process is spelled out in strong language. The public consultation starts at an early stage and follows through the entire EIA process, including the drafting of terms of reference for the EIA, the EIA report preparation, and the review of the final EIA report. It is the developer's obligation "to organise a public consultation process, involving Government agencies, local authorities, non-governmental and community-based organisations and interested and affected parties, to help determine the scope of the work to be done in the conduct of the environmental impact assessment",[73] and "take all measures necessary to seek the views of the people in the communities which will be affected by the project."[74] The Environmental Council will make public the final EIA statement and "may organise, or cause to be organised, public meetings in the locality of the proposed project"[75]. Under some conditions, it may "hold a public hearing on the environmental impact statement".[76] The Zambia case illustrates good practice that can be easily followed by other countries when the legal system and the social conditions are similar.

Another good example is the legislation of the Republic of South Africa, which imposes on the proponent the obligation to provide for "a public participation process to ensure that all interested parties are given the opportunity to participate in all the [EIA] relevant procedures."[77] After its completion, an EIA report becomes a public document.[78] Other environmental statutes have more sophisticated procedures to ensure that public views are properly weighted and taken into account when based on strong evidence. In the case of Nigeria, the applicable statute provides for the Environmental Agency to ensure that "government agencies, members of the public, experts in relevant discipline and interested groups [are given the opportunity] to make comment on environmental impact assessment of the [proposed] activity." In various countries, the legislation recognizes a right to appeal EIA-related decisions for the public at large and more specifically in favour of affected people.[79] This feature gives more strength to the consultation requirement in the EIA procedure and may make it very effective.

These sound examples are, however, balanced by a majority of cases where laws and statutes are vague and weak on the public consultation and participation requirement in the EIA procedure. This is, for example, the case of Côte d'Ivoire where EIA procedure lacks an effective legal ground that provides for effective public participation. Although the *Environmental Code* of 1996 states in article 35.6 that every person has the right to participate and to be informed on the state of the environment and to participate in environmental decision-making process,[80] the implementing decree on EIA fails to mention public participation as a means to ensure the soundness of the EIA. The implementing decree states that the Ministry of Environment may undertake a public inquiry after the EIA is submitted by a proponent without detailing the scope, contents and modalities of such inquiry.[81] In fact, legislation in many African countries has made a public inquiry part of the review process in certain situations or for a certain category of projects. However this public inquiry is limited to the release of the EIA Report in some public places and opportunity for public comments lasting only one to two months.[82]

It is slightly a similar scheme that is followed by the *1997 Environmental Code* of Burkina Faso, which states that the EIA process will be completed by a public inquiry. An implementing decree of the *1997 Environmental Code* is being prepared to provide for a detailed EIA procedure that includes public participation.[83] Other examples of weak provisions on public consultation and participation include the laws and statutes of Comoros, Namibia, Gabon and Togo, all which do not provide for any kind of public participation in the EIA process.

African governments may take stock and benefit from good international practices, which define sound EIA mechanisms that imply that the project proponent is responsible for meeting environmental legal requirements and therefore organizing public participation. All the laws and statutes that recognize and ORGANISE a public participation scheme designate a responsible agency empowered to review, manage, and document the public consultation and participation process. In addition, this Agency, which is also responsible to oversee and monitor the EIA process, should be legally authorized to undertake additional public consultation when required by the nature of the project, and when it deems it fit or necessary. It is important for African countries to follow this positive trend to ensure that the development project proponent fully internalises the EIA requirements and that affected peoples have an opportunity to discuss the project and express their comments on its impacts on them and their living environment. Where an EIA reveals that any proposed project will affect people, these affected people should be notified by appropriate means to give them an effective opportunity to become involved through formal meetings that should document their inputs, views, concerns and opinions and ultimately give them an opportunity to help enforce environmental law and standards. This is not always clearly spelled out in the applicable statutes even though it appears that this was intended.

Public Participation, Consultation and Legal Enforcement Issues in African Countries

If public participation is to become effective in African countries, it is necessary not only to recognize the rights of citizens in an environmental context, but also to ensure that these rights can in some meaningful way be enforced in national courts. In fact, in environmental matters, people may wish to help enforce a rule where they have suffered no legally recognizable damage. Therefore, clear procedures should be defined to allow them to bring administrative law proceedings by way of judicial review or statutory appeal.

It is a matter of fact that communities and environmental groups, for example, may wish to help advance the environmental protection agenda in their countries whether they suffer damage directly or not, but they should have a clearly defined right to seek protection for their community rights. In some of the Sub-Saharan countries the law related to associations (NGOs) recognizes a right to standing in favor of representing NGOs under more or less flexible conditions. In order to be effective, this right should help overcome four barriers in administrative law proceedings that NGOs and communities face: (a) gaining access under the relevant domestic law (question of standing), (b) persuading a court or a relevant administrative body of the substance of their claim, (c) convincing the Court to grant an appropriate remedy after persuading it or the relevant administrative body of the merits of the case, y, and (d) being able to pay for the costs of the proceedings. This fourth condition may constitute a real hindrance to an effective enforcement of a meaningful community rights to participate in an EIA process.[84]

THE EIA REPORT

Two issues are dealt with in relation with the EIA Report. The first one relates to the format and content of the Report requested from the developer or project proponent. The second encompasses the entire EIA report review process, including issues of implementation, monitoring and compliance with the EIA recommendations.

Format and Content of the EIA Report

The EIA procedure provides for the preparation and review of an EIA Report (the Report), sometimes called Environmental Impact Study or Statement (EIS). The Report is a document designed to guide the eventual decision-making process concerning the feasibility and implementation of the project. It is generally prepared by the project's proponent, or on his behalf, by a competent consultant (preferably independent).[85] Few African countries have developed some kind of rules and procedure to certify EIA consultants by the regulatory agency, as it is the case in other countries.[86] The expenses incurred from EIA preparation are borne by the proponent, including in some cases, the costs involved by the EIA report review.[87]

In general, the laws, regulations and statutes define the content of the Report in different ways.[88] The required content of the EIA Report is usually described in African laws or statutes, but the specifics differ from one law or statute to another.[89] Some of them state that the Report should describe the proposed project, its specific purposes, the affected environment and environmental and health impacts, including the impacts on the human and cultural environment in some cases. The Report also is required to provide an examination and evaluation of alternative solutions that might avoid or at least reduce and mitigate some or all of the adverse environmental impacts identified.[90]

However, besides the statutes of the Republic of South Africa which allow some flexibility in defining the content and depth of the Report in accordance with the scale of the proposed project, almost all African countries' statutes and laws do not link the content and depth of the Report to the scale of the proposed project. The South African legislation provides for a four-step procedure that requests the proponent to prepare: (1) a plan of study for scoping; (2) a scoping report; (3) a plan of study for environmental impact assessment and finally; (4) an environmental impact report. At each stage of the preparation of the final report, a real "dialogue" is established between the proponent and the regulatory authority on how to conduct and bring about the next phase of the final EIA Report preparation. The Report is developed by the proponent after the relevant authority agrees successively on the study for scoping, the scoping report and the plan for the preparation of the EIA.[91] This method is somehow sophisticated, and gives the proponent a real opportunity to develop a sound and comprehensive EIA. It also allows the relevant authority to follow step by step and adapt the EIA procedure to the proposed activity and to the specific situation of the potentially impacted environment.

Generally, all the statutes reviewed have a provision requiring that the EIA Report must provide a list of items and analysis The following items are usually required under all the laws and regulations of the countries concerned: (a) a description of the project site and the reasons for rejecting other alternative sites; (b) a description of the proposed activity, the technology and raw materials to be used and their likely impacts on the environment; (c) an identification of the likely impacts; (d) comments expressed in the public participation process (where the public consultation and participation is mandated in the statutes or laws); (e) suggestions of mitigation measures, including restoration if necessary; (f) a plan for monitoring or managing the activity in compliance with the existing environmental laws and regulations; (g) the uncertainties of information; and (h) a non-technical summary to be published. In some cases, the laws and statutes do not provide for submitting a detailed description of the EIA report[92] or refers to specific guidelines to be enacted by the relevant regulatory agency[93] for the preparation of the final EIA report

Review of the EIA Report - Monitoring and Auditing

The purpose of the review is to obtain an impartial judgment of the particular and often conflicting interests of the various parties involved, and to avoid unnecessary costs and delays.[94] This is a very important issue as there is a need to proceed as quickly as possible with development projects and to reduce their costs in developing countries. It is therefore a good practice that a review be held before the final EIA report is submitted for consideration for licensing purposes.[95] One sensitive issue is whether the relevant agency grants an authorization to proceed with the project or just to approve the EIA as complying with the relevant laws and regulations. In many countries, the environmental agency reviews the EIA report and grants a clearance that will allow the line ministry to grant the authorization to proceed with the project or activity. Whatever the formal outcome of the EIA reviews, provisions for guiding the review process of the Report are critical pieces of the quality control of the whole EIA. A good and sound example is provided in the case of the Republic of South Africa where statutes provide for a "dialogue" between the proponent and the relevant authority throughout the EIA process[96] until the review of the final EIA Report. Laws and statutes of other African countries provide for the review of the Report as a final administrative step in the EIA preparation process. According to these laws and statutes, different institutions may be mandated to review the Report: (a) a competent licensing authority; (b) an inter-agency committee; or (c) an independent expert panel.[97] Environmental laws and statutes in Africa insist on the timeframe for the review, the sharing of information between the review agency and other agencies that may be involved in the EIA review process and of the follow up of the decision of the review agency, whether it is a clearance for the project to be implemented or more assessment and data to be provided before such implementation

Two major problems face an effective review of an EIA Report: The first deals with the conflicting situation that may exist between various government agencies involved in the licensing process for the development project. Such conflicts are not always adequately addressed and may hinder the credibility of the EIA review; the status of the responsible agency within the government may greatly influence its level of power to effectively review and enforce decisions related to the EIA Report. The second problem is the weakness of the human and professional capacity of almost all the review agencies in African countries.[98] These problems need to be addressed urgently in order to make EIA mechanism implementation and review effective.

The relevant authority should also have qualified staff at its disposal, including expert consultants from a wide range of skills to ensure that the review of the EIA report is adequately and professionally undertaken.[99] The review is the first stage of a comprehensive monitoring program that would result in real improvement of project management. It is also essential for any further environmental impact auditingof the project.[100]

Finally, the review of the EIA Report should be clearly viewed as part of a wider process in scope and timing, which includes the EIA implementation monitoring and auditing functions. Currently, in the best cases, the EIA report review for major projects with significant environmental impacts is done before the project is implemented. However, since the life cycle of these projects may cover a very long period of time, it is necessary to ensure that the EIA Report review would not stop at the decision to implement the project but rather to obtain a sound environmental management for the whole life of the project. This is why the EIA Report review is to be considered as part of a wider process, which includes the monitoring and auditing functions into the EIA process. Unfortunately, almost all environmental laws and regulations in African countries are missing this point.

CONCLUSION

Two major issues arise when we analyze environmental law in Africa. The first relates to the absence of national capacity at all levels of government and society to bring about a comprehensive and sound environmental management including legal requirements implementation and enforcement. The second is to work together to harmonise EIA-related laws and regulations with those of neighboring countries and international technical and financial institutions.[101] National capacity building and harmonisation may be instrumental in facilitating the EIA requirements concerning implementation especially for those investors and developers who need a standardised framework to help them implement their project on time and at the best cost possible.

Capacity Building: Two issues

Enacting legislation is not enough, but it is an important step to foster environmental protection and sustainable development. However, where the national capacity to implement the EIA requirement is lacking, legislation is just a useless tool. By national capacity, we mean capacity at all the levels where EIA is to be performed, reviewed, discussed and monitored. It includes capacity at central government level, local government and other decentralised agencies, private sector, NGOs and local communities as well.

As of now, laws and regulations in African countries describe EIAs mainly as a mandate of the central government (whether it is a ministry of environment, and/or other sectoral ministries) and national environmental agencies. All the statutes in force recognize that these ministries and agencies are responsible for the regulation and supervision of development activities, as well as for the impact assessment of such activities on the public and the environment. Countries such as Nigeria and South Africa, because of their specific nature as a federal system,[102] for the former, and as a decentralised government for the latter[103] adopted rules to involve their local administrations into the EIA processes. Other statutes of African countries mention local government involvement in the public participation provisions.[104] It should be mentioned that decentralisation is being promoted in many African countries as a means to; (a) improve governance; (b) increase economic growth of the rural areas; and (c) promote institutional efficiency. Therefore it seems appropriate to elaborate a regulatory framework to enable local governments to fully participate in the EIA implementation and review processes. Its absence in the statutes may be compensated by the enhancement of both environmental and decentralisation legislation.

Outside the government scope, EIA increasingly mobilises a whole range of social groups, from the small national group of technical and scientific experts to the grassroots organisations and their representatives. As the role of local governments[105] and local communities in planning and delivering development services increases,[106] so does the involvement of the most advanced members of these local groups in the preparation, review and monitoring of EIA. Without this continuum, the initial consensus built during the EIA cannot be sustained in the longer term, thus wasting the opportunity of getting the actual environmental impacts monitored. Involving local communities, especially the potentially affected ones, started historically with the mainstreaming of public participation in EIA. For the African countries, the implication is that; (a) rules, regulations and capacities should not limit themselves to the capital cities, but should go as far down close to the field as development activities do; (b) training programmes that include government staff, grassroots EIA specialists including the private sector, possibly through the training of trainers, in a "cascade" arrangement,[107] and; (c) technical and financial resources put at the disposal of all these groups of EIA specialists to allow them to fully implement their mandates and responsibilities.[108]

As mentioned above, the EIA report review is part of a comprehensive monitoring and auditing process. In practice, many significant hurdles exist that limit the use of monitoring and auditing to date in almost all African countries. The need to ensure a credible EIA review is still to be correctly and fully taken into account in EIA related laws and regulations in these African countries. Further legal developments would have to define the process and responsibility for monitoring and auditing in order to contribute to a general enhancing of the credibility of the regulatory agencies, private sector, project proponents and EIA processes as such.

Towards Harmonisation of EIA Legislation in Sub-Saharan Countries

In 1995, a World Bank paper stated the importance of EA harmonisation for African countries.[109] Harmonisation of EIA procedures among these countries and with the donor community members is crucial and can produce significant cost-savings. In fact, by resolving discrepancies between EIA procedures, African countries can ease their own effort to develop sub-regional and regional cooperation and facilitate development project funding by the donor community.

Several sub-regional organisations in Africa, notably SADCC and IGADE, have expressed an interest in taking a sub-regional perspective to EIA rules and requirements. This makes perfect sense in the context of sub-regional and regional economic integration to make trade and investment attractive in these countries. None of the sub-regional organisations, though, has succeeded in harmonising such rules and regulations. The World Bank is helping in facilitating such a harmonisation process through the Africa Region's Strategy to assist with EA capacity building in SSA.[110] It will be interesting to see how proactive the sub-regional organisations are to promote common approaches to EIA and how responsive the governments involved will be. Other multilateral institutions and bilateral donors, such as UNDP, UNEP, FAO, the Dutch government are playing an important role to develop and enhance EIA systems in African countries while looking at ways to harmonise them. The Africa Environmental Law Partnership Project[111] could play a critical role in these efforts and help develop common legal approaches to EIA.

EA harmonisation includes among other elements (1) the adoption of sound environmental standards[112], (2) the development of effective enforcement mechanisms including effective judiciary control over EIA processes, and (3) the promotion of public participation as a means to promote sustainable development through the enhancement of the quality and sustainability of development projects. Obviously, each country will have to adapt these elements to its own cultural and social situation to ensure effectiveness and possible implementation. An on-going project under the sponsorship of UNEP and UNDP[113] is currently assisting African countries in a genuine effort of harmonisation by providing African lawyers and officials involved in environmental management, the opportunity to work together and cooperate in the development of modern and sound EIA statutes.[114]

LIST OF ENVIRONEMNTAL IMPACT ASSESSMENT-RELATED LAWS, STATUTES AND REGULATIONS OF SELECTED SUB-SHARAN AGRICAN COUNTRIES[115]

I- ***Burkina Faso***

1. Loi No 005/97/ADP du 30 Janvier 1997, portant Code de l'Environnement au Burkina-Faso.

II- ***Cameroon***

1. Loi 96-12 du 5 Aout 1996 portant loi-cadre relative la gestion de l'environnement

III- ***Comoros***

1. Loi-cadre No. 94-018 du 12 Juin 1994 relative à l'Environnement

IV- ***Congo (R.P)***

1. Loi No. 003/91\du 23 Avril 1991 sur la protection de l'Environnement
2. Décret No. 86/775 du 7/06/86 Rendant obligatoires les Studes d'Impact sur l'Environnement en République Populaire du Congo

V- ***Côte d'Ivoire***

1. Loi 96-766 du 3 Octobre 1996 portant code de l'environnement
2. Decret 96-894 du 8 Novembre 1996 portant etablissment d'une procedure d'etude d'impact sur l'environnement

VI- ***Ethiopia***

1. 1992 National Environmental protection Authority Act

VII- ***Gabon:***

1. Loi no 16/93 du 26 Août 1993 relative à la protection et à l'amélioration

VIII- ***Gambia:***

1. National Environment Management Act No. 13 of 1994.

IX- ***Ghana:***

1. The Environmental Protection Agency Act, 1994 No.490
2. LI 1652 on Procedures and Other matters Pertaining to EIA; date of Gazette Notification 26 February, 1999

X- ***Guinea:***

1. Ordonance No. 045/PRG/87 - Portant Code de L'Environnement de la République de Guinée

XI- ***Kenya***

1. Physical Planning Act, 1996

XII- ***Madagascar***

1. Loi No. 90-033 relative à la Charte de l'Environnement malagasy + Annexe, 1991
2. Decret 99-954 du 15 Decembre 1999 relatif la mise en compatibilit des investissements avec l'environnement (modifiant le Decret portant refonte du decret 92-926 du 21 Octobre 1992)

XIII- ***Malawi***

1. Act No. 23 of 5 August 1995 (The Malawi Gazette Supplement, dated 16th August 1996, containing (No.7c) (Published 16th August, 1996)

XIV- ***Mali:***

1. Loi No. 91-47/AN-RM Relative à la Protection de l'Environnement et du Cadre de Vie
2. Decret 99-189 du 5 Juillet 1999 portant institution de la procedure d'etude d'impact sur l'environnement

XV- ***Mauritius:***

The Environment Protection Act 1991 (Incorporating 1993 Amendments)

XVI- ***Namibia***

1. Namibia's Environmental Assessment Policy Cabinet Resolution, No. 002 of the 16th of August, 1994

XVII- <u>Nigeria:</u>

1. Federal Environmental Agency Decree, 1988
2. Decree No. 86 of 10 December 1992 (Supplement to Official Gazette Extraordinary No. 73 Vol 79, 31st of December, 1992 - Part A A979)
3. Federal Environmental Protection Act, Cap 131, LFN, 1990 (as amended) by Decree No. 59 of 1992

XVIII-<u>Seychelles: (Republic of)</u>

The Environment Protection Act , 1994 (Act of 1994)

XIX- <u>South Africa</u>:

1. Environment Conservation Act No. 73 of 1989
2. National Environmental Management Act No. 107 (NEMA), 1998
3. R. No. 1182 and 1183 Identification under Section 21 of Activities Which May have Substantial Detrimental Effect on the Environment, Date of Regulation Gazette, 5, September 1997

XX- <u>Swaziland</u>

1. Swaziland Environment Authority Act, No. 15, 1992

XXI- <u>Togo</u>

1. Loi No. 88-14 du 3 novembre 1988, insitutant code de l'environnement

XXII- <u>Uganda</u>

1. The National Environmental Statutes of 17 May 1995 (Statutes Supplement to the Uganda Gazette No. 21 Volume LXXXVIII Dated 19th May, 1995 Printed by UPPC,
Entebbe by Order of Government)
2. The Environmental Impact Assessment Regulations, 1998 (Under section 108 of the National Environmental Statutes, No. 4 of 1995)

XXIII-<u>Zambia</u>

1. Environmental Protection and Pollution Control Act (EPPC) No. 12, 1990
2. EPPC (Environmental Impact Assessment) Regulations, 1997

XXIV-<u>Zimbabwe</u>

1. Environmental Impact Assessment Policy, 1994

---

[1] Senior Counsel- LEGEN, The World Bank. Previously Professor of Law, Faculty of Law (University of Oran. Algeria). Currently Senior Counsel, Environmental and International Law Unit of the Legal Department of the World Bank. The opinions expressed are those of the author and do not express those of the World Bank.

2 See for example World Bank (1995), *Environmental Assessment in Africa. A World Bank Commitment*. (Proceedings of the Durban Workshop, World Bank, June 25, 1995) (Washington, D.C.; 1995). Many African countries are aware of the costs of environmental degradation and the effect of development planning, and, in some of them, there are now many years of experience in attempting to assess, mitigate, and monitor adverse environmental impacts of development projects and programs. However, in others, formative steps are only now being taken towards the introduction of an environmental impact assessment mechanism, as the potential benefits of the implementation of EIA are more widely understood.

3 See Faith Halter (1990), "Towards More Effective Environmental Regulation in Developing Countries" (Paper presented at OECD Development Center Seminar (August 27, 1990) (Paris: OECD).

[4] *National Environmental Policy Act* (1969) § 101, 42 U.S.C. § 4331 as amended (1994).

[5] World Commission on Environment and Development (1987)

6 Since then, International Financial Institutions (IFI), including the World Bank have tied the provisions of development funding to the assessment of the environmental impacts of proposed projects.

[7] The best example is Algeria where the *Decret 90-78* of July,10 1990 on environmental impact assessment provides that large development projects such as highways, dams or other infrastructure are not explicitly defined as being subject to EIA. A list of defined project does not include them although article 2 of the decree states that the Decree 90-78 which states that «Shall be submitted to environmental impact assessment: all works (travaux, amenagements et ouvrages in French) which by their size and impacts, may directly or indirectly affect the environment and notably public health, agriculture, natural landscapes, fauna, flora, sites and historical monuments conservation and neighborhood». It is only recently that the Government of Algeria decided to solve this contradiction between the general objective of article 2 and the list of defined projects and activities. The Government is preparing a new decret more in line with the objectives of a modern vision of EIA, including subjecting large development projects to prior environmental impact assessment. Another specific case is in the Kingdom of Morocco where an environmental framework law is in the process of being prepared since 1987 and a draft decree on EIA since 1991. Up to now, both texts are not yet adopted. EIA is done as part of an international financing requirement or as a pro-bono work by the developer, but not as a legal requirement under domestic law.

8 In fact, in French-speaking Africa, an old legislation known as "Reglementation des etablissements classées" whose introduction by the then French colonial administration goes as far back as 1926 empowers the government at local and central level to closely monitor pollutions emanating from polluting plants and/or harmful activities. In some cases, this pollution control monitoring involves a public consultation process called "enquête commodo-incommodo". However, this legislation is far less sophisticated than the EIA mechanism and does not cover a wide range of projects and programs.

[9] These countries and the laws and statutes considered in this study are listed in an Annex to this paper.

10 A good attempt to summarize such legal development on EIA has been made by MMM for the Commission for Environmental Impact Assessment of Netherlands (1998) in a document called "EIA Profiles of Developing Countries", Leiden. Nineteen summaries for African countries are reported.

11 In the ICJ Decision concerning the *GabčicoboNagymoros (Gabčikovo-Nagymaros Project (Hungary/Slovakia)* ICJ Rep. 1997, 3.)), Judge Weeramantry, in a separate opinion, considered that the principle of prior environmental assessment is part of the customary law. For further

elements on customary environmental law see D. Bodanski (1995), "Customary (and not so Customary) International Environmental Law", 3 *Indiana J. of Global Legal Studies* 105,

12 Among these, economic instruments, incentives and quality standards are of the greatest importance to complement environmental regulations.

13 See especially the *Convention on Environmental Impact Assessment in a Transboundary Context* (1991), 30 I.L.M. 802 (*Espoo Convention*), which is the most comprehensive international instrument on EIAs. In addition, other instruments make reference to EIAs. See, for e.g., Article 14 (1) (a) of the *United Nations Convention on Biological Diversity*, June 5, 1992, 31 I.L.M. 818; Article 4 (1) f. of the *United Framework Convention on Climate Change*, 31 I.L.M. 849 (entered into force Mar. 21, 1992) that suggests means for complying with the provision; Article 206 of the *United Nations Convention on the Law of the Sea* (*UNCLOS*), Dec. 10, 1982, UN Doc. A/CONF. 62/122 (1982), 1295 U.N.T.S. 211, 21 I.L.M. 1261 (entered into force on Nov. 16, 1994), and Article 13 of the *West and Central African Marine Environment Convention* (1981).

14 *United Nations Convention on the Law of the Sea* (*UNCLOS*), adopted in Montego Bay in December 1992, especially its article 206 which states that:"When States have reasonable grounds for believing that planned activities under their jurisdiction and control may cause substantial pollution of or significant and harmful changes to the marine environment, they shall, as far as practicable, assess the potential effects of such activities on the marine environment and shall communicate reports of the results of such assessments in the manner provided in article 205."

15 Adopted at Rio during the UNCED, 1992. Article 14 states that: "Each contracting party, as far as possible and as appropriate, shall: (a) introduce appropriate procedures requiring environmental impact assessment of its proposed projects, that are likely to have significant adverse effects on biological diversity...".

16 *United Nations Framework Convention on Climate Change* (*UNFCCC*), adopted in Rio during UNCED, 1992, which stipulates in article 4 that: " All parties...shall (f) take climate change considerations into account, in their relevant social, economic and environmental policies and actions, and employ appropriate methods, for example impact assessment, formulated and determined nationally, with a view of minimizing adverse environmental effects..."

17 *United Nations Convention to Combat Desertification in Those Countries Experiencing Serious Drought and/or Desertification, Particularly in Africa (1994)* 13 ILM 3328. See for example article 10

18 It is clear that the private sector especially in relation to large development projects is willing to undertake EIAs to respond to the growing demand for better environmental management and sustainable development of natural resources. The business community is seeking governments to develop regulatory frameworks including sound EIA procedures and review processes, as well as pollution control guidelines to avoid environmental risks and high cost insurance for their development projects. For the private sector, EIA procedures may also be seen as an instrument to: (1) prevent delays in obtaining authorizations; (2) identify mitigation measures involving recycling and recovery of components of waste streams; (3) create a cleaner working environment; and, (4) identify cheaper alternatives.

19 Shihata, I.F.I (1996), *Implementation, Enforcement and Compliance with International Environmental Agreements: Practical Suggestions in Light of the World Bank's Experience* (Paper presented before the Expert Meeting with international Environmental Agreements, held at Georgetown University Law Center, in Washington D.C, on May 20-21, 1996).

20 Nigeria, *Decree # 86* of 1992 Section 56 "International Agreement".

21 See for example article 77.2 of the *Code of Forestry of Burkina Faso* (Loi 005/97/ADP portant Code Forestier) which states "En application des conventions internationales dûment ratifiées par le Burkina Faso et selon les besoins, il peut être créé d'autres types d'aires de protection faunique".

22 N. Robinson (1995) "IUCN's Proposed Covenant on Environment and Development", 13 (1995) *Pace Environmental Law Review* 133.

23 UNEP, *Council decision 14/25* of June 17, 1987, "Principles and Goals of EIA" defined by a group of experts. See *UNEP Environmental Law Series* No. 9.

24 Nigeria, *EIA Decree*, section 36.

25 See for example the (1971) *Convention on Wetlands of International Importance Especially as waterfowl habitat* (1972) ILM 698.*(Ramsar Conventon)*.

26 World Bank (1995).

27 This includes all sub-Saharan countries. Information about current environmental impact assessment legislation was gathered from governmental agencies, academia and local NGOs. Information has also been gathered from the UNEP/UNDP Joint Project on Environmental Law and Institutions in Africa, which published a Four-Volume Compendium of Environmental Law of African Countries (including North Africa).

28 This is the case of Burkina Faso, Benin, Gabon, Kenya, Lesotho, Senegal, Uganda and Cameroon, Niger where draft legislation on EIA was circulated for comment and discussion among stakeholders including ministries, NGOs, development agencies, academia, etc. Many of these are being prepared with the assistance of bilateral donors, or in the framework of Bankfinanced projects.

29 This is the case for Burkina Faso (*Loi 97/005 du 17 Mars 1997 portant Code de l'environnement*), Comoros (*Loi-cadre 94-018 du 22 Juin 1994 relative à l'environnement*), of Gabon (*Loi 16/93 du 26 Août 1993 relative à la protection et à l'amélioration de l'environnement*), Gambia (*Environmental Management Act # 13 of 1994*), Ghana (*Environmental Protection Agency Act No. 490 of 1994*), Malawi (*Environmental Management Act of 1996*), Mauritius (*Environmental Protection Act # 34 as amended in 1993*), Seychelles (*Environmental Protection Act of 1994*), Swaziland (*Swaziland Environment Authority Act # 15 of 1992*), Togo (*Loi 88-14 du 3 Novembre 1988 relative au Code de l'environnement*), and Uganda (*National Environmental Statutes of May, 17, 1995*).

30This is the case for Congo (Rep. Pop. du) with *Décret 86-775 du 7 Juin 1986 rendant obligatoires les études d'impact sur l'environnement*, Côte d'Ivoire with *Décret 96-894 du 8 Novembre 1996 déterminant les règles et procédures applicables aux études relatives à l'impact environnemental des projets de développement*, Madagascar with *Décret 92-926 du 21 Octobre 1992 relatif à la mise en compatibilité des investissements avec l'environnement*, Nigeria, *Decree #86 of December 10, 1992 on EIA*, South Africa, *Environmental Conservation Act* (regulations R.1182, Identification under section 21 of activities which may have a substantial detrimental effect on the environment) and Regulations regarding activities identified under section 21 (1). See also for South Africa the *Explanatory Memorandum regarding Government Notice 1750, 1752 and 1751* of November 1, 1996, on proposed actions in terms of section 21, 22 and 26 of the *Environmental Conservation Act*, 1989. See also Zambia, *The Environmental Protection and Pollution Control (Environmental Impact Assessment) Regulations*, 1997.

31 This is the case for Namibia. See the Environmental Assessment Policy, Cabinet resolution 002 of August, 16, 1994. For Zimbabwe, see Environmental Impact assessment Policy of July 1994.

32 South Africa, *Environmental Conservation Act #73* of 1989, Zambia, *Environmental Protection and Pollution Control Act of 1990*, and Côte d'Ivoire, *Loi 96-766* du 3 Ooctobre 1996 portant *Code de l'environnement* (See reference to implementing decrees in *supra* note 2).

33 This has now been adopted.

34 In Kenya, at the cross-sectoral level, an EIA process is administered by an inter-ministerial committee, which has developed administrative guidelines and questionnaires. The guidelines' coverage is adequate to facilitate informed decision-making by relevant authorities on the environmental impact of the proposed projects. However, this procedure is purely administrative with no legal ground and it is implemented in the context of wide discretionary powers. It is clear that the desire to attract investments may override ecological considerations. See Bondi D. Ogolla (1992), "Kenya. Environmental Management Policy and Law", 3 *EPL* 22, 1992 pages 164-174.

35 *Projet de decret sur les etudes d'impact sur l'environnement dans les zones adjacentes des aires protégées*, Oct. 1999

36 Law 83-03 of February 1983 related to the protection of environment

37 World Bank (1991), *Operational Policy OP 4.01 Environmental Assessment*. See also, William Sheate (1996), "From Environmental Impact Assessment to Strategic Environmental Assessment: Sustainability and Decision-Making" in Jane Holder, *Impact of EC Environmental Law in the UK*, (New York: J. Wiley & Sons).. 268-289. The precise language of World Bank's *OP 4.01 on Environmental Assessment and EIA* is "the EA report for a Category A project is normally an

environmental impact assessment, with elements of other instruments included as appropriate. Any report for a Category A operation uses the components described in this annex, but Category A sectoral and regional EAs require a different perspective and emphasis among the components".

38 The World Bank views adopting an EIA legislation as the first, major and indispensable step towards putting in place an EA system.

39 See for example, the World Bank *Operational Policy on Environmental Assessment*, which defines different categories of projects: (1) Category A for projects "likely to have significant adverse environmental impacts that are sensitive, diverse or unprecedented"; (2) Category B, for projects that "potential adverse environmental impacts on human populations or environmentally important areas ... are less than those of category A projects"; (3) Category C for projects that are "likely to have minimal or no adverse environmental impacts", and finally (4) Category FI for projects that involve "investment of Bank funds through a financial intermediary, in sub-projects that may result in environmental impacts".

40 The Mauritius *Environmental Protection Act of 1991*, Section 13 (1) mentions that "notwithstanding any license, permit or approval granted under any other enactment, any person being a proponent who commences, proceeds with, carries out, executes or conducts, or causes to commence, proceed with, carry out, execute and conduct, an undertaking: (a) without an EIA license; or (b) in breach of any condition of the license, shall commit an offense...." See also The Seychelles, Section 15 (1) of the *Environmental Protection Act*(1994).

41 The last Malian enactment on EIA, the *Mali Decree* of July 5, 1999, Art. 4 provides a list of 32 activities, projects and programs which are subject to EIA. Also, Article 5 of this *Decree* states that non-prescribed projects are subject to "une notice d'etude d'impact sur l'environnement" (i.e.; environmental impact study note), which summarizes the likely impacts of the project on the environment and the mitigating measures to be undertaken and implemented to reduce or eliminate them.

42 Comoros, *Loi-cadre 94-018* which stipulates "un décret du conseil des ministres : ... arrête la liste des travaux et projets soumis à l'étude d'impact préalable, en raison de la nature des activités projetées" (article 14.1).

43 Malawi, 1996, *Environmental Management Act* s. 24. As of January 1998, no notice had been issued. It is interesting to note that in the Namibia EIA Policy, a list is provided in an Appendix, but the Policy states that this list "should be used to guide" the Board when it decides on whether a "policy, program or project requires an EA or not" (sect. 4). To be effective, this approach requires that the specifications should be broad enough and detailed to cover all projects likely to have significant harmful impacts on the environment.

44 Article 20 of the 1997 *Code de l'environnement of Burkina Faso* states that: "un décret pris en conseil des ministres ... établit la liste des travaux, ouvrages, aménagements et activités, ainsi que les documents de planification assujettis à l'étude ou à la notice d'impact sur l'environnement." However, the *Décret* is not yet enacted. Article 67 of the *Environmental Code* of Gabon states that "les travaux, ouvrages ou aménagements industriels, agricoles, urbains, ruraux, miniers ou autre, entrepris par les collectivités publiques, les entreprises publiques ou privées qui risquent, en raison de l'importance de leur dimension ou de leurs incidences écologiques, de porter atteinte à l'environnement, doivent donner lieu à une étude d'impact préalable...." The same provision is found in Togo, *Code de l'environnement de 1988*, article 22.

45 In Côte d'Ivoire, the *Décret 96-894* on EIA is followed by two annexes. Annexe I includes "projets soumis à étude d'impact environnemental" which is comprised of a wide range of activities including projects in the sectors of (a) agriculture, (b) forest management plans, (c) extractive industries, (d) energy and industry, (e) waste management, (f) food processing industry, (g) chemical industry, (h) manufacture, and (i) infrastructure, including cement plants, tourist resorts, arm industry. Annexe II includes "projets soumis au constat d'impact environnemental" comprising a list of activities in the same sectors as in one, but of smaller size and with less harmful effect on the environment. This Annex II includes also "urban planning and land use planning documents", and also projects modifying previous projects that were subjected to a full EIA. Congo (Rep. Pop. du), Décret 88/775 has a single Annex on "liste des travaux, ouvrages et aménagements soumis à l'étude d'impact" focusing on infrastructure activities. The Gambia *Environmental Management Act* includes a Schedule which Part A deals with "Projects to be

considered for environmental impact assessment" comprising a wide range of activities classified into fourteen (14) categories of sectors. In Madagascar, the *Décret 92-926* includes an Annex listing 22 categories of activities to be subject to EIAs, one of which is resettlement of population. See Nigeria, *Décret #86*, Schedule (section 13) on "Mandatory Study Activities", Uganda, *Third Schedule of the national Environmental Statutes*of 1995, South Africa, *Identification of Activities Which May Have a Detrimental Effect on the Environment under Section 21 (1) of the Environment Conservation Act*, 1989 Official Notice 1182, of 1997, and the Notices 1750, 1751, 1752 and their Explanatory Memorandum of November 1996. The case of South Africa is interesting because of the division of the activities subject to EIAs between those activities that will impact on more than one province, on an international treaty or agreement and neighboring country, and those activities that will have an impact on a single province.

46 Côte d'Ivoire, Annexe III to the above-mentioned décret which lists six (6) categories of areas including (a) protected areas, (b) wetlands and mangroves, (c) areas of scientific, cultural and/or tourist interest, (d) ecologically sensitive areas, (e) watershed buffer zones and (f) marine areas whether domestic or international.

47 South Africa, *Regulation 1182* of September 1997 on "identification under section 21 of activities which may have a substantial detrimental effect on the environment", section 4 ("the intensive husbandry of, or importation of, any plant or animal that has been declared a weed or an invasive alien species"), section 5 ("the release of any organism outside its natural area of distribution that is to be used for biological pest control") and section 6 ("the genetic modification of any organism with the purpose of fundamentally changing the inherent characteristics of that organism").

48 See, for example, Zambia, Second Schedule (regulation 7.2) of *the EPPC (EIA) Regulations*, 1997.

49 Nigeria, *Décret 86* of 1992 Schedule, Section 13 which listdetailed activities under headings such as agriculture, airport, drainage and irrigation, land reclamation, fisheries, forestry, housing, industry, infrastructure, ports, mining, petroleum, power generation and transmission, quarries, railways, transportation, resort and recreational development, waste treatment and disposal and water supply.

50 Zimbabwe, *EIA Policy*, (1994), section 6.2 and the annexed list of prescribed activities in Table 3.

51 See for example *Madagascar Law 99-021* of August 19, 1999 related to "la politique de gestion et de contrôle des pollutions industrielles" and the *Decret 99-954* of December, 15, 1999 related to EIA.

52 See for example, the requirement of the "permit de construire" (i.e.; building permit which require an evaluation of the land use change and impact on the neighborhood of all construction projects.

53 Gabon, *Loi 16/93*, article 68 states that EIA would analyze and identify the likely harmful impacts "directs et indirects, des projets ... sur la santé, la qualité de l'environnement, les ressources naturelles et les équilibres écologiques"; Gambia, *Environmental Management Act* of 1994, Section 23 (3) e "environmental impacts of the proposed activity including the direct, indirect, short term and long-term effects..."; Mauritius,*EPA* 1991 Section 14. c. "the direct and indirect effects that the undertaking is likely to have on the environment"; Côte d'Ivoire, *Décret 96-894* Article 12.2 is as a complete checklist of EIA-related issues as the one enacted in the Zambian Regulations (see Box) and may be considered as a good practice.

54 Zambia, EPPC (EIA) Regulations, 1997 Sect. 4. See box in the main text.

55 It is always interesting to evaluate the magnitude of these impacts not only in biophysical terms but also in monetary terms

56 Article 9..5 of the Decree of July 5, 1999 states that EIA should have the following elements: "une identification et une evaluation des impacts positifs et negatifs potentiels, directs et indirects, immediats et long terme, importants et secondaires, locaux et loigns du projet propos sur l'environnement". Only an estimation of the amount of wastes and emissions is required under article 9.7. There is no explicit reference to social and or human impacts of the proposed project.

57 See the legislation of Burkina Faso, Cameroon, Uganda and Kenya.

58 In general, all EIA-related legislation focuses on the external environment -that is the environment outside project site limits. This is part of the fact that African countries often copied EIA legislation from industrialized countries that have an enabling legal framework for occupational health protection and working environment. In African countries, where the legal framework

dealing with the working environment and health is generally weak, it is important to refocus EIA-related legislation on a comprehensive definition of the environment, which includes the internal and external environment. It is important to consider, for example, the change of air quality of noise level on the health of workers and their families.

59 Not to mention specifically the AIDS problem, which should not be ignored during the EIA process.

60 The World Health Organsiation (WHO), the Food and Agriculture Organisation (FAO) and other international institutions and business organisations have developed specific standards for water, waste, noise, etc.

61 LI 1652, date of Gazette Notification, 26 February, 1999

62 There are number of laws governing water resources, i.e.; *Rivers Ordinance*, 1903, *Volta River Development Act*, 1961, *Ghana Water and Sewage Act*, 1965, *Oil in Navigable Waters Act* 1964, but they re all too outdated to be of any useful and practical application.

63 Although a marginal improvement was provided under the *Mineral and Mining Law*, 1986, which require the licensee to protect the environment from adverse mining effects (Section 72 of the Law), the *Small-scale Gold Mining Law*, 1989 allows people to undertake what is usually defined as surface operations that are a real threat to the environment through chemical pollution, stream siltation, soil erosion and open ditches in the mining areas. See G.A Sarpong (1999), *The Ghanaian Environmental Law Regime*, paper presented at the Launching of the Network of African Environmental Lawyers, Nairobi, Kenya, 12 September 1999.

64 Authors mention that EIA is a process that relies on the expression of public communication. The main benefit of EIA may be to advance mutual communication and understanding. See W.A. Tilleman (1995), "Public participation in the EIA Process: A Comparative Study of Impact Assessment in Canada, the United States and the European Community", in 33 *Columbia J'l of Transitional Law* 337-439.

65 Some of the African countries' legislation state that a public inquiry may be requested by the relevant authority to confirm EIA Report conclusions and recommendations (for example Burkina Faso, Côte d'Ivoire, Zambia).

66 In many countries the right to participate in the decision-making process affecting the environment is grounded on the right of citizens to a clean environment, the freedom of expression and association. Do citizens of African countries enjoy a right to a clean and healthy environment under the public law of their respective countries? It is very difficult to answer such a question especially when it comes to differentiate between an enjoyable and enforceable right under the public law of a given African country and a commonly shared legal rhetoric, which does not confer any enforceable right to citizens. The right to participate in environmental decision-making process appears to be one of the most difficult to evaluate, because it is very narrowly drawn in the various legislation examined here. However, it is the practice that will help determine the real extent of such a right.

67 Where no legislation exists, it might be difficult to assess how governments define public participation. But in many projects and as part of donor agencies requirements an EIA has to be conducted and a public consultation process to be implemented. The consequences may be very important for both the proponent and the affected public. In the Chad-Cameroon pipeline, Chad and Cameroon have no legislative requirements for public participation. However, under Bank policy requirement, a full public consultation has been conducted during project design and the final environmental management plan for the project requires a continuing consultation process in order to ensure that the affected public would not be negatively impacted from project implementation and that mitigation measures would be defined with the participation of affected populations. See *Chad Export Project, Environmental Assessment*, disclosed on June 25, 1999 through World Bank InfoShop.

68 *Act No. 5* of 1969 last amended in 1992 and revised in 1998

69 Wamukoya, G.M. & Situma, F.D.P. (eds.) (2000) Environmental Management in Kenya: A Guide to the Environmental Management and Coordination Act, 1999 (Centre for Research and Education Environmental Law (CREEL), Nairobi,) at 2.

[70] Joseph Kessy & Ors *v.* Dar es Salaam City Council[70] and Festo Balegele and 794 Ors *v.* Dar es Salaam City Council.

[71] *Misc. Civil Application No. 28 of 1983* High Court of Tanzania at Arusha (Unreported), where the justice stated: "I will say at once that I have never heard it anywhere for a public authority, or even an individual, to go to court and confidently seek for permission to pollute the environment and endanger people's lives, regardless of their number. Such wonders appear to be peculiarly Tanzanian, but I regret to say that it is not given to any court to grant such a prayer. Article 14 of our Constitution provides that every person has a right to life and protection of his life by the society. It is therefore a contradiction in terms and denial of this basic right to deliberately expose anybody's life to danger or what is eminently monstrous, to enlist the assistance of the Court in this infringement."

72 One interesting example is Uganda where the local by-laws making process in the Local Government Act requires the dissemination of information procedures to follow the traditions of the locality where the by-laws or local law is made. This is also likely to apply to the EIA process. (Information provided by Nightingale Rukuba-Ngaize, LEGEN)

73 Section 2 of the *EPPC (EIA) Regulations*, 1997.

74 Section 10 (1) of the *EPPC (EIA) Regulations* 1997.

75 Section 16 (2) of the *EPPC (EIA) Regulations* 1997.

76 Section 17 (2) a and b. of the *EPPC (EIA) Regulations* 1997.

77 Section 3 (1) f of *Regulation 1182* of September 5, 1997 regarding activities identified under section 21 (1) of the *Environmental Conservation Act* of 1989.

78 Section 11 of *Regulation 1182* .

79 Cases include Seychelles and South Africa legislation. See Annex on Public Consultation/Participation.

80 Article 35.6 reads "toute personne a le droit d'être informée sur l'etat de l'environnement et de participer aux procedures préalables à la prise de decisions susceptibles d'avoir des effets préjudiciables à l'environnement."

81 *Decree 96-894* of 8 November 1996 on environmental impact assessment, article 12.

[82] See Algeria: article 12 of *Decree 87-143*, which states that a registry is made available to the public in order to fill comments on the proposed classified plants for a period of 45 days.

83 See Annex on "Public Consultation/Participation in EIA-related legislation of African Countries".

84 It may be interesting to look at the way industrialized countries, where EIA related Laws are more developed, have been dealing with the issue of people's right to participate in the EIA process and to stand before court to protect their rights to a clean and healthy environment; see H. Ward (1993),, "The Right to an Effective Remedy in European Community Law and Environmental Protection: A case Study of United Kingdom Decisions Concerning the Environmental Assessment Directive", 4(2) *JEL* 221.

85 Togo, article 30 of the 1988 *Code de l'environnement* which states that: "Le promoteur indiquera au bureau [des études d'impact] les noms du ou des consultants qu'il aura choisi pour effectuer l'étude.... [Le ministre] pourra recuser par une décision motivée les consultants qui auraient montré une incompétence préjudiciable à la bonne exécution des études d'impact"; South Africa, *Reg. 1183* of 1997 Sect. 3 which states that (1) an applicant (a) must appoint an independent consultant who must on behalf of the applicant comply with these regulations". The case of Côte d'Ivoire stresses the need to use national capacity for EIA, see *Décret 96-894*, article 9 which states that the proponent "peut recourir à un organisme ou consultant indépendant de son choix.... Mais l'utilisation partielle ou entière des capacités nationales est obligatoire. Elle devra, dans la mesure des comptences disponibles, être conforme à la répartition 2/3 experts et/ou consultants nationaux, 1/3 experts et/ou consultants non nationaux."

86 This is the case of Tunisia, Czech Republic, Mexico and Thailand. In a draft decree on EIA, circulated in February 2000 by the Government of Cameroon, a rough procedure for EIA consultant certification is being developed.

87 Zambia, *EPPC (EIA) Regulations*, 1997, Schedule 5 on Fees. See Côte d'Ivoire, *Décret 96-894*, article 17 which states that "l'examen des études d'impact par le Bureau d'étude d'impact donnera lieu au versement d'une taxe, au fonds de l'environnement" and Madagascar, article 12 of the *Décret* which provides that the proponent will be charged 0.5% of the amount of the investment, but that the fees are tax-deductible.

88 The IUCN (1994), *Draft International Covenant on Environmental and Development* prepared by the IUCN Commission on Environmental Law mentions a non-exhaustive list of factors that may

be taken into consideration in the definition of the scope and content of the EIA Report. These factors ensure that every possible type of significant impact is considered during the EIA process including long-term, cumulative and transboundary effects and is properly documented in the EIA report.

89 It is worth mentioning here that at the international level, it is widely accepted that the content of an EIA Report depends on whether the matter in question is an activity, project, policy or plan.

90 Article 37.2 "The Assessment shall include: (a) cumulative, long-term, indirect, long-distance and transboundary effects, (b) the possible alternative actions, including not conducting the proposed activity, and (c) measures to avert or minimize the potential adverse effects.

91 South Africa, *Environmental Conservation Act*, 1989, Regulations regarding activities identified under section 21. Regulations 1183 of1997.

92 Gabon, *Loi 16/93 relative à la protection de l'environnement*, article 68 and ff..

93 Uganda, *The National Environmental Statutes* 1995, section 21 (2) reads: "An environmental impact statement shall be made according to guidelines established by the Authority".

94 Some legislation lack provisions on review of the EIA Report. This is the case of Seychelles, *Environmental Protection Act, 1994* Section 15.6 stating that the "Authority shall be responsible for the monitoring of the conclusions of the EIA Study and to ensure that the necessary conditions are complied with." No mention of any review is made in the Act.

95 Togo, *Loi 88-14*, article 26.2.

96 South Africa, *Environment Conservation Act, 1989*. Regulations Regarding Activities identified under section 21 (1), of September 1997, see Reg. 4,5, 6,7.

97 Nigeria, *EIA Decree of 1992* provides for the option to call for mediation between the Authority and the proponent, or to establish a review panel, or to hear experts to make sure that the EIA documents reflect the reality by providing the needed information and drawing sound conclusions.

98 We should mention here that the review is not only a desk task, but involves monitoring the physical, social and economic variables associated with project impacts.To help weak or newly-established government environmental agencies, an outside and independent review process of the EIA report might always contribute to the successful completion of the review procedure, in particular in the case of large infrastructures and other development projects. However, even where such solution is an adopted effort to build capacity at central and local government and private sector levels, it is a critical exercise without which an effective EIA Report review and monitoring could not be appropriately done.

99 It is not required that such experts be permanent staff of the relevant authority. Rather it is a good option to have a list of national experts (in universities, consulting firms, industry, other government agencies) and of international experts (with a proven expertise of the country) that the relevant authority may use on contractual basis to review EIA reports and/or to provide any necessary input during the EIA process.

100 Environmental impact auditing involves comparing the impacts predicted in the EIA report and those that actually occur after implementation, in order to assess whether the impact prediction performs satisfactorily. Environmental auditing also focuses on institutions and programs for environmental management and the associated risks and liabilities.

101 The compatibility of national EIA procedures with those of international donor's agencies is intended to support funding investment initiatives. An EIA of such investments may be carried out at the time when funds are to be approved, in accordance with the Agency's own procedures. If the domestic EIA procedures are fully compatible with those of the funding agency, it may be possible to avoid the extra cost and delay of involving the funding agency in these EIA. Harmonisation may help ensure full compliance of the EIA procedures with those of the funding agencies.

102 Nigeria, *EIA Decree, 1992*, sections 11 and 49.

103 South Africa, *Environmental Conservation Act. Regulations Regarding Activities Identified under Section 21 (1)*, which clearly states that the " relevant authority" in charge of the environmental impact assessment process implementation, review and monitoring may be the Minister of Environmental Affairs, provincial or local authority. See sections 1, 3 (3), 5 (1), etc.

104 Zambia, *EPPC (EIA) Regulations*, 1997, section 7 (2) "To ensure that public views are taken ... the developer shall ORGANISE a public consultation process involving ... local authorities...."

105 MELISSA, (1999), *Knowledge & Expertise Resource Network* (KERN) (1998 Forum - The poverty, environment, governance trilogy, MELISSA (Managing Environment Locally In Sub-Saharan Africa).

106 The trend to finance projects through Social Development Funds is now established, and community action programs represent the next logical step.

107 We use the concept of "cascade" arrangement to describe those arrangements needed from the top of the government to grassroot level which include government, private sector, NGOs, communities and so on in order to organise and implement EIA capacity-building activities in the country.

108 In January 2000, a small group of mostly African EA professionals met at CSIR, Stellenbosh in South Africa to discuss a long simmering idea in the SADC region to build a regional resource center for capacity building in environmental assessments (SAIEA). The idea was hatched during the African EA Stakeholder meeting in Nairobi in July 1998, then brought to the forefront during the IAIA meeting in Glasgow last June 1999. From there the idea went for endorsement from IAIA (International Association for Impact Assessment) during a meeting in Montreal late last year, and then proceeded to Stellenbosch for a reality check and more detailed planning.

109R. Goodland, J.R Mercier, S. Mutemba (1995), *Environmental Assessment (EA) in Africa. A World Bank Commitment.* Proceedings of the Durban, World Bank Workshop, June 25, 1995.

110 GEOPLAN International (1999), *Environmental Assessment Capacity Development in Africa* (draft final report).

111 The Project is funded by the Dutch Government and located within UNEP. The Project has a steering Committee which includes UNDP, UNEP, FAO and the World Bank. It is headed by an eminent Kenyan lawyer, Prof. Charles Okidi.

112 To this end a special emphasis should be put on methodologies, techniques and skills adapted to the concerned country's needs and capacity. In particular, simpler EIA mechanisms that can determine the impact on the environment and can be implemented by local staff of environmental and line agencies at all levels should be developed. Local knowledge and experience should be taken into account, and effort should be made by environmental agencies and government to develop and release baseline data and databanks on environment.

113 The Government of Netherlands is assisting in this project. The Project is headed by Professor Charles Okidi, from the University of Nairobi.

114 As member of the steering committee for this project, World Bank lawyers assisted in various fora where EIA draft statutes for Kenya, Tanzania, Burkina Faso, Uganda and Lesotho have been discussed.

115 Other laws, statutes and regulations may contain specific provisions on EIA. This might be the case of construction laws, planning laws, forestry and water-related laws and statutes. They are not listed here above.

THE USE OF MARKET BASED ECONOMIC INSTRUMENTS FOR SUSTAINABLE ENVIRONMENTAL MANAGEMENT IN AFRICA: A CASE STUDY OF THE SADC REGION

PENDO MARO SUSSWEIN

INTRODUCTION

Since their independence, most African countries have been plagued by several problems ranging from wars and conflicts to environmental degradation, poverty and economic stagnation. Poverty has been identified as a major cause and consequence of environmental degradation and resource depletion in the region. Major environmental challenges include deforestation, soil degradation and desertification, declining biodiversity and marine resources, water scarcity, and deteriorating air and water quality (Dorm-Adzobu, 1997; SADC, 1994; UNEP, 1999). The majority of these problems although blamed on the poverty of the majority of the continent's population, have actually been found to be a result of different interacting factors such as the social-political, social-economic and the environment interactions. Countries of southern Africa, as elsewhere in Africa now face a formidable series of critical demographic, economic, agricultural, energy, technological and institutional transitions in order to move toward development that is economically, socially and environmentally sustainable (SADC, 1996; SADC, 1997).

Environmental management and the concept of sustainable development gained momentum in Africa after the United Nations Conference on the Environment and Development (UNCED) in 1992 (Rio Earth Summit). Commitment to build a new Southern African Development Community (SADC), and the independence of South Africa in 1994, paved the way for the SADC countries to better manage their multiple transitions and together move toward sustainable development (SADC, 1996; IUCN, 1998). In the SADC region as in the rest of Africa, unsustainable development has been and remains largely driven by economic and sectoral policies which are too narrowly conceived and focused and which neglect the negative consequences on the people and the environment. The challenge in Africa has been to accelerate economic growth and reduce poverty. However, this has led to more poverty and increased environmental degradation as many countries have pursued short-term growth policies, which shift ecological and economic costs to the next generation.

The current environmental policies in the region are mainly based on regulatory command and control instruments, but with the advent of globalisation, decentralisation and free market economics, most SADC States have now began to consider a broader range and mix of regulatory measures and economic instruments to facilitate and accelerate their transition toward equity-led sustainable development (UNEP, 1999). For example: new central bank legislation was enacted in Kenya, and the banking sector has been restructured in Tanzania, Mozambique and Zimbabwe. Privatisation of some State-owned companies has progressed in some countries and has been initiated in others. For example: the privatisation of the State-owned Lesotho Airways and the Central Agricultural bank in Lesotho and the proposed privatisation of the State-owned telephone company in that country. Trade and investment liberalisation also continued in several countries in Africa, for example in Egypt, trade restrictions have been eased and a new investment law was passed, offering tax incentives, protection of property rights, freedom of financial transfers and an end to restrictions on new investment. Similar measures were taken in Tanzania, in the early 1990s, as well as in Uganda and South Africa (UNEP, 1999). The reason for most of these economic reforms has been to promote equity-led sustainable development in the continent as well as to improve Africa's participation in world economics, so as to prevent the 'marginalisation' of Africa in the process of global integration and the formation of new international markets (Artyeetey et al, 1998). In addition to these policy measures, African countries have also sought to develop human resources, expand infrastructure and strengthen institutions (UN, 1998).

Most of the economic instruments in the SADC States are of the command-and control type and are contained in their numerous environment-oriented legislation and policies, without which their enacting would not be possible (Maro, personal communication, 1999). Are these instruments the solution to Africa's poverty and economic stagnation or will they result in further exploitation and plundering of the already degraded natural resources with dire consequences for the rural majority and the environment? This chapter will analyse the institutional, legal and policy frameworks for the use of market-based economic instruments in the use, management and conservation of water, marine, and soil resources and in the industrial sector, within the context of the SADC region. Comparisons of revenue or tax generations within each country will be avoided because this would depend on the marginal environmental damage and control cost. Examples of the different countries will be taken exclusively from the SADC member states.

INSTITUTIONAL, LEGAL AND POLICY FRAMEWORK FOR USING MARKET-BASED ECONOMIC INSTRUMENTS IN THE SADC REGION

Market Based Economic Instruments (MBIs) and Command And Control (CAC) regulations – A conceptualisation

A distinction can be made between economic and non-economic instruments. Non-economic instruments can be divided further into regulatory instruments – for example, ambient air quality standards, fishing permits, and quotas - and consultative instruments, such as education campaigns to highlight the effects of water and soil pollution. Regulatory instruments normally use the command and control mechanism to achieve the desired policy objectives; thus, they are referred to as Command-And-Control (CAC) instruments. On the other hand, consultative instruments use the power of persuasion to influence change in the production and consumption of natural resources.

The main difference between economic and non-economic instruments is that, unlike non-economic instruments, economic instruments use the market to allocate prices to environmental goods and services (Turner et al., 1993). Turner et al gives two ways in which this can be achieved. The first is the creation of markets in previously free services, for example, charging entrance fees in national parks, establishing exclusive economic zones in coastal waters, and changing property rights (Turner et. al., 1993). The second way is to "... 'modify' markets by centrally deciding the value of environmental services and ensuring that those values are incorporated into the prices of goods and services" (Turner et. al., 1993:144). Because they use market incentives, economic instruments are also known as Market Based Instruments (MBI's).

According to Da Motta et al. (1999), there is no standardised definition of an MBI, but a distinction is made between a 'weak' and 'strong' MBI. In this case, a weak MBI is simply a non-economic instrument, such as a CAC instrument, for example: product norms, standards, bans, restrictions and zoning (Zima et. al., 1999). While a strong MBI is one that uses the market, for example: charges for resource use and pollution tax measures and subsidies (Mabugu et al., 1999).

Ideally, policy instruments should be economically efficient, environmentally effective, equitable, acceptable, adaptable, generate revenue and modify behaviour (Ad Motto ET al., 1999; Turner et. al., 1993). The challenge is thus to design economic and non-economic policy instruments that meet most of the criteria given above. Box 1 gives a brief explanation of each policy criteria.

> Economic efficiency – policy instruments should have low implementation costs; should promote the use of the cheapest technology to bring about the desired resource management; reduce government expenditure on resource management (Artzen et. al., 1999)
>
> Environmental effectiveness – the policy instrument should contribute towards a reduction in resource consumption; generate revenue which will be used towards environmental management; should change user's perception of the resource use and promote environmental awareness (Artzen et. al., 1999).

Equity – costs and benefits of new policies should be sustainable and evenly distributed among the different groups in the society (Mabugu et. al., 1999).

Dependability – the environmental effectiveness of the instrument should be as reliable as possible (Turner et. al., 1993).

Adaptability/Flexibility – the policy instrument should have capacity to adapt to changing technology, prices and climatic conditions (Artzen et. al., 1999; Turner et. al. 1993).

Dynamic incentive – the policy instrument should modify behaviour and encourage a change in resource use technology in favour of sustainable use of resources and environmental management; should encourage technical innovation (Turner et. al., 1993; Da Motta et. al., 1999).

Political acceptability – the policy instrument should fit into existing resource rights, underlying philosophies and future practices if it is to be acceptable; it should promote the participation of stakeholders, be acceptable by the resource users and affected parties (Artzen et. al., 1999; Turner et. al., 1993).

Box 1. Selection Criteria for Policy Instruments

Revenue generation is one way in which a policy instrument can achieve economic efficiency and environmental sustainability simultaneously, while at the same time influencing people's behaviour and promoting environmental awareness and cleaner production methods. One way of achieving this is by using the 'Polluter Pays Principle' (PPP). Under this principle, those who generate pollution, such as producers and/or consumers, are the ones liable to pay for the damage costs, and not the society in general (Turner et. al., 1993). Making the polluter pay for environmental damage seems an equitable idea, but by increasing the production costs, consumers are forced to pay higher prices for goods and services. Critics have argued that the effect of higher consumer prices will be to punish the poor proportionally more than the rich (Turner et. al., 1993). Is this just?

Principal 16 of the Rio Declaration placed great stress on the use of economic instruments based on the PPP as a way for making consumption and production patterns more sustainable and as a means of generating financial resources needed to finance sustainable development worldwide, while internalising environmental costs, with regard to public interest and without the distortion of international trade and investment (UNEP, 1999). Below is Table 1 showing several Economic Instruments for Environmental Management and Protection

Property Rights	**Bond and deposit refund systems**	**Fiscal Instrument**	**Charge Systems**	**Financial Instrument**	**Market Creation**
Example of incentives: ownership use and development rights	Example of incentives: bonds for forest management, land reclamation and waste delivery	Examples of incentives: taxes on exports and imports, pollution, resource and land use charges	Examples of incentives: Pollution charges, charges for environment al impacts, road usage charges	Examples of incentives: Loans, grants, revolving funds, green funds, low interests	Examples of incentives: Tradable emission permits, catch quotas, water shares and land permits
Application: creating new and permanent property rights, temporary use rights or leases on an individual or communal level for resources such as land and water Reinvigorating or reinforcing existing rights	Application: long-term performance bonds for potential or uncertain hazards from infrastructure construction	Application: -Taxing agrochemica ls to reduce the use of fertilisers, insecticides and pesticides - Reducing tax and charges on farmers who use environment ally friendly-technology - Resources charges on mining industries	Application: -Non-compliance pollution charges -Taxes to encourage the reuse or recycling of materials such as cans and plastics -Charges on environment al impacts of different development projects	Application: -Loans or grants to companies employing 'green' technologies -Low interest rates for producers/co nsumers involved in green/clean technology	Application: -Government established tradable emission permits or resource use permits -Tradable permits for water abstraction rights, and water and air pollution emissions -Property rights attached to resources potentially impacted by urban development

Advantages: -A well defined tenure system that is secure will promote incentives for the efficient use, investment and sustainable management of natural resources	Advantages: Ensures that there is enough financial resources to counteract possible envt'l. effects	Advantages: Discourages the use of environmentally-damaging technology in the farming and agricultural sectors, while promoting 'clean/green' technology in the different sectors	Advantages: Discourages pollution of different envt'l. resources	Advantages: Promotes clean/green technology. Due to problems such as high interest rates, fixed official rates, and balance of payment problems, financial instruments are not yet popular in the SADC states	Advantages: Creates government revenue that can be used for the benefit of the people and the environment.

Table 1. Economic Instruments for Environmental Management and Protection .(Sources: SADC, 1997; UN, 1998; UNEP, 1999.)

THE SADC REGION – INTRODUCTION AND THE INSTITUTIONAL, LEGAL AND POLICY FRAMEWORK FOR SUSTAINABLE DEVELOPMENT

(a) Introduction to the SADC region

The Southern African Development Coordination Conference was established on April 1 1980, with the aim of accelerating economic growth and reducing dependency on South Africa (SADCC, 1981). The initial member countries were Angola, Botswana, Malawi, Lesotho, Mozambique, Namibia, Swaziland, Tanzania, Zambia and Zimbabwe[1]. The SADCC name changed to SADC – Southern African Development Community, in 1992, after the signing of the Declaration, Treaty and Protocol on Regional Economic Integration (SADC, 1996). The member states share a commitment towards regional economic integration, integrated management of shared resources and the achievement of regional equity-led sustainable development, among others.

In 1995 the total average population of the SADC region was estimated at 136 million people, growing at an annual average of 3.0 % (SARDC, 1995). This was before the inclusion of the populations of the D.R.C., Seychelles and Mauritius, who joined SADC after 1995[2]. Latest figures estimate the population of the SADC region at 195.262 million people (Regional Economic Review: Southern Africa Economic Summit 2000).

Economically, the SADC region is considered poor, despite the richness of its natural resources and biodiversity (SADC, 1997). South Africa is the 'economic giant' of the region, and one which can compete successfully in the global economy (Stock, 1995). Economic performance in the SADC region as a whole improved significantly in the 1996/ 97 period, with all the member countries registering positive real growth of between 2 per cent to over 10 per cent (SADC, 1996/97). The positive economic development registered in the region over the given period has been attributed to the sound market-oriented policies implemented by member states over a number of years, and the growth of the agricultural sector (SADC, 1996/97). Despite this, widespread poverty and environmental degradation continue to plague the region. The cause of this has been identified as ineffective and inefficient, as well as narrowly focused, economic and environmental polices (SADC, 1997).

(b) Laws, Policy and Institutions – A Framework for market based economic instruments in the SADC region

Over the past decades, many countries in Africa, including ones in the SADC region, have put in place a wide variety of new environmental laws and regulations[3]. Many environmental agencies in countries such as the D.R.C, Zimbabwe, Tanzania, Lesotho, South Africa, and Namibia have prepared new Conservation Strategies, Forestry Action Plans, and Plans of Action to Combat Desertification (UNEP, 1999; SARDC, 1994). All SADC countries currently have environmental agencies.

Despite these efforts at improving environmental protection and sustainable development, duplication and fragmentation of authority is emerging as major constrains to the success of different environmental policies in the region. In addition, institutional failure is caused by policies that are not backed up by legislation, and therefore cannot be legally enforced (Kumar et al, 1996; SADC, 1997; UNEP, 1999). Most of the laws are complex and overlap, making it difficult for affected people and stakeholders to understand. Gaps and deficiencies that exist in some of the resource management and conservation laws make them ineffective and do not provide incentives for individuals to assume more environmentally sustainable behaviour. Kumar et al. (1996) identified coercive rules, economic incentives, education, participation, privatisation and agreements, to be the most effective tools to promote environmental conservation and management. Looking at the SADC region, several authors reveal that the popular method of choice to achieve environmental conservation has been and continues to be the use of CAC instruments (Artzen et al., 1999; Kumar et al., 1996; Mabugu et al., 1999). Particularly, the use of coercive tools in the form of sanctions, orders and prohibition, which use penalties to implement them. Privatisation, participation and economic incentives are slowly gaining momentum in the region.

(c) Command and Control regulations versus Market Based Economic Instruments

As was stated earlier, in the SADC region, the favoured command and control regulations take the form of coercive law, which use sanctions, regulations and deregulations, and prohibition, among others, to force a person to take certain actions. Such laws employ tools such as penalties in their implementation. However, within the SADC region, the existing laws have proved ineffective in protecting the environment, promoting economic development and growth, and alleviating poverty in the rural majority of the population. The failure of CAC instruments has been attributed to the following reasons (SADC, 1996; SADC, 1997; Kumar, 1996): a lack of resources, both financial and human to policing prohibited actions[4]; inadequate awareness, especially among legal officers, as to what constitutes a prohibited activity and a lack of clarity as to the priority of environmental matters; and a lack of political power to prescribe stiff criminal sanctions.

Considering that most countries in the SADC region are grappling with the problems of poverty and external debts, among others, this explains the lack of political will in implementing economic instruments in the form of MBI's. Other problems associated with the limited use of MBI's include the difficulty to detect the appropriate level of economic incentives in the form of fees or subsidies and how to achieve compliance by a significant number of people especially in the rural areas, where environmental problems may not be perceived as urgent. The solution seems to be to combine the use of CAC and MBI's. CAC instruments, using tools such as sanctions and standards, are enforced through the use of fines, which if not paid are punishable by law; as such it forces people to adhere to the prescribed laws, and regulations, among others. The strength of regulatory/CAC instrument " ... is that it affords a reasonable degree of predictability of the results if properly implemented and enforced" (Maro et al., 2000:7). MBI's, use market forces to allocate a price to environmental goods and services, and to promote change to more environmentally friendly behaviour by resource users. MBI incentives such as the 'polluter pays principle' can generate revenue for governments to use in achieving environmental management and pollution abatement objectives elsewhere, MBI's are also flexible, allowing the polluters to choose a pollution abatement method, which is most cost-effective for them. Their strength lies in their ability to give opportunities to individuals and firms to compensate for environmental damage in market price terms. Economic instruments (such as MBI's), in order to be enforceable, must function together within legislative and policy frameworks that would give them legal powers.

EXAMPLES OF THE USE OF ECONOMIC INSTRUMENTS IN THE SADC REGION

(a) Water Resources

Lack of water is a primary constraint to development in many SADC countries such as South Africa, Namibia, Botswana, Lesotho and Zimbabwe. Within the region, water is mainly used for agricultural purposes. In South Africa, 69 % of all water consumption is used for irrigation and watering of livestock, while in

Namibia irrigation uses twice as much water as livestock, 44 % versus 23 % (Turpie et al, 1998).

A variety of instruments are applied within the SADC region, to water use and management to improve their efficiency and to promote sustainability. These instruments work by safeguarding water supply and by managing water demand (Turpie et al, 1998). The following types of economic instruments are used water supply and management in the SADC region: property rights – for example using water banks; markets and charge systems using mechanisms such as demand-side water pricing; marginal cost pricing; fiscal instruments - provision of subsidies on water-saving technologies; application of differential taxes on land uses in catchments and water-based ecosystems; taxes on pollution and waste management, and bonds and deposits levied upon pollution and waste management. In South Africa, and Namibia, past water allocation and pricing was done with little economic efficiency and equity (Lange et al., 1999). Water rights were biased towards riparian landowners and pollution control measures were lacking. In addition, heavily subsidised commercial irrigation was concentrated in the hands of the white minority, while the majority of the people had no access to a reliable source of safe water (Lange et al., 1999).

In Namibia and South Africa, resource recovery is achieved through the use of full cost recovery systems, such as provided under the new water law in South Africa (Lange et al., 1999), The National Water Act of August, 1998, developed through a highly consultative process (Maro et al., 1999). Other tools such as permits/licenses and water charges are widely used in southern Africa, especially in countries such as Botswana, Tanzania, South Africa and Zimbabwe, for water use controls (Maro et al., 2000). In Malawi, water pricing is used as a strategy to manage water consumption, while proving incentives for the sustainable use of water, and generating revenue which can be used for water management purposes (Maro et al., 1999). The Tanzania National Water Resources Management Policy (Draft, 1999), has made provision for the inclusion of the polluter pays principle (PPP) on effluent discharges, and the charge "... will be related to the environmental damage caused by the discharge, the cost of prevention and treatment to restore the resource to acceptable quality standard;" (Tanzania National Water Resources Management Policy, Draft, 1999:41-42). The Zimbabwe new water legislation and the National Water Resources Management Strategy also make provision for the use of economic incentives such as demand charges, to promote efficient water use.

Water quality control and monitoring, using quality standards, such as the 1984 WHO drinking water standards, and the use of legislation to control quality, is employed in the SADC region. These responsibilities fall principally within the mandate of the Departments of Water in the SADC countries, and partly within the mandate of the Health Departments/Ministries (Maro et al., 1999). Zoning, a regulatory approach, can be used in water resource management in the control of point-source water pollution. In southern Africa, zoning as a tool of water conservation is used indirectly through the declaration of parks, conservation areas and heritage sites, with joint initiatives such as the Gemsbok Park – between Botswana and South Africa, and the proposed parks covering parts of

Mozambique, South Africa and Swaziland, as well as the proposed Drankensberg initiative between Lesotho and South Africa (Maro et al., 2000).

In southern Africa, SADC countries occupy 9,600 km of coastal area, extending from Tanzania in the east coast to Angola in the west (SARDC et al., 1994). Eight of the fourteen SADC countries, namely, Tanzania, Seychelles, Mauritius, Mozambique, South Africa, Angola, Namibia and the D.R.C, have coastal areas. These not only provide access to the sea for other land-locked countries, but also contain important sources of marine biodiversity, and are popular tourist destinations. Marine pollution in the major cities such as Beira, Maputo, Dar-Es-Salaam, Durban and Cape Town is threatening the marine resources (Chenje, 1998). All the 14 SADC countries are signatories to the 1982 United Nations Convention on the Law of the Sea, which gives economic rights to all coastal states within an Exclusive Economic Zone (EEZ) stretching 200 nautical miles off their coasts (SARDC et al., 1994); and the 1985 Convention for the Protection, Management and Development of the Marine and Coastal Environment of the Eastern African region, (to which only Seychelles and Tanzania are signatories), which make provisions for the prevention of pollution of marine resources. In addition, Botswana, Mozambique, Tanzania, Zambia and Zimbabwe are signatories to the 1987 Agreement on the Action Plan for the Environmentally Sound Management of the Common Zambezi River System which also makes provisions for the management and protection of the shared Zambezi river. These international conventions and agreements make provisions for national frameworks for regulations to prevent pollution and the sustainable management and use of water resources in the region. Many of the SADC states are in the process of formulating new water laws that will take into account the cost of water abstraction, provision, use and pollution. The overall aim of the new water laws in the SADC states is to promote the preservation and sustainability of water and aquatic resources as well as to promote stakeholder participation especially in the rural areas, as exemplified by the Zambia Water Policy (1994), the Malawi Water Policy (1994), and the South Africa National Water Act (1998).

(b) Land resources

The 1994 United Nations Convention to Combat Desertification in countries experiencing serious drought and/or desertification particularly in Africa is the most comprehensive land use convention. It encompasses a strategy for the management and conservation of soil and land resources, preventing soil and land degradation, and provides for strategies to mitigate the effects of poverty, at the national and regional levels. Most countries in the SADC region are in the process of making national legal and institutional reforms that will embrace the objectives of the Convention.

Countries such as Botswana, Mozambique, South Africa, Zambia and Zimbabwe are all implementing land tenure reforms, with the main objective of providing security of tenure (IUCN, 1998). In Zimbabwe, the strategy focuses on resettlement, tenure security and strengthening the local institutions in land allocation and management. In order to provide security of tenure and to create a framework for the implementation of economic instruments, several SADC countries are making policy and legislative changes, which recognise the legal authority of traditional institutions in natural resource tenure regimes[5]. Traditional tenure systems, their resilience, adaptability and ability to co-exist with other tenure systems are now gaining positive recognition as a viable tenure system which can be incorporated with other modern tenure systems to promote sustainable use of land resources (Rihoy, 1999). In addition, traditional farming methods such as terracing on mountain areas to prevent soil erosion and crop rotation methods are undergoing review. The role of local chiefs in monitoring and enforcing land legislation and in promoting sustainable use of land resources is also under review (Report on the Preparatory Workshop on the Land Policy Review Process for Lesotho, 1999).

Decentralisation of authority and the co-management arrangement with local communities are proving popular in providing incentives for better management of land resources in the region. This is also taking place in the wildlife sector with programs such as the Communal Areas Management Program for Indigenous Resources (CAMPFIRE) in Zimbabwe. CAMPFIRE allows for local people to claim ownership of wildlife and to benefit from its use, provides a legislative framework for people living in communal lands to exercise the same right of proprietorship, and decentralises power to the local governments (Rhine, 1999). This program has proved successful in the development of profitable and sustainable wildlife use strategies, with an increase in the land areas available for wildlife and the numbers of wildlife in Zimbabwe (Rhine, 1999). In other SADC countries, similar programs exist. Programs such as LIFE in Namibia, ADMADE in Zambia and other similar initiatives in Botswana, Malawi and Tanzania, known as Community Based Natural Resource Management (CBNRM), aim to provide legal and institutional frameworks and economic incentives for communities to become co-managers of communal area resources (Rhine, 1999). The emphasis here is on the use of property rights as a CAC tool for land management, with emphasis on decentralisation of legal authority to local communities so they can earn financial profits from sustainable use of wildlife and conservation. Additionally, the government and non-governmental organisations (NGK's) provide training to the local communities on how to use wildlife resources sustainably, simultaneously increasing their earning power (Fakir, 1996).

Most land reform policies in southern Africa, however, have yet to adequately address the issue of women and access and ownership of land in the different countries. In Lesotho, for example, the Land Law (1979) of Lesotho states that all the land belongs to the nation and the king is responsible for the administration of the land on behalf of the nation (National Settlement Policy, 1989). The local chiefs allocate land for use on behalf of the king to any Mosotho man, but women are excluded and, therefore, have no legal access to land. This issue, together with the burning issue of landlessness which has been blamed on the Land Law (1979), is undergoing review in the preparation of a new Land Policy for Lesotho

(Report on the Preparatory Workshop on the Land Policy Review Process for Lesotho, 1999).

In the tourism sector, market-based incentives such as tourism levies and taxes are used in countries such as Tanzania, South Africa, Zimbabwe, Zambia, Namibia, Seychelles and Mauritius to generate income and to mitigate negative impacts of tourism.

(c) The Industrial Sector

Mining is a major export earner for several SADC countries, notably South Africa, Namibia and Botswana. In these countries mining is also the largest polluter of soil and water, despite the existence of several anti-pollution laws and standards. This is due, in part, to lack of enforcement and non-compliance. Analyses of countries such as Botswana and Namibia show success in the use of market based mechanisms of resources rent recovery through appropriate taxes, in the mining sectors (Lange, 1999). This could be attributed to the fact that taxes, such as export tax on precious minerals, are unavoidable when mining companies are applying for export permits to take their goods out of the countries for sale to international markets[6].

In Mauritius, The Industrial Expansion Act of 1993, provides manufacturing enterprises with duty exemptions, tax credits and other incentives to import pollution abatement and instil environmental protection facilities, new machinery and equipment in order to promote economic, industrial and technological development (GOM, 1993; UNEP, 1999). Such incentives promote environmental conservation and encourage industries to shift to cleaner technology and increase production. The polluter pays principle in the form of fines and the recovery of costs for clean up of mining sites is provided for in the Environmental Protection Act of Mauritius of 1991. In order to reflect better the extent to which economic development affects the quantity and quality of the natural resource base, countries such as Botswana, Namibia and South Africa have developed natural resources accounts or 'green' accounts for water, fisheries, land, livestock, forestry, wildlife, minerals and energy, and resource degradation, to enable a better assessment of their economic value and management (Lange, 1999; UNEP, 1999). Other SADC countries such as Seychelles have special environment protection funds in place financed by contributions from government, industry and other private sources. These funds can be re-invested in environmental cleanup campaigns, used in promoting environmental awareness or in economic activities which can generate income for natural resource conservation and management, as well as the upliftment of local communities, for example in the CBNRM initiatives. Adherence to precautionary environmental standards such as ISO 9000 production standards, by different local and multinational companies is gaining momentum in countries such as Mauritius, South Africa and Zimbabwe, although this is still on a voluntary basis. Mauritius is going a step further by planning to introduce industrial waste audit regulations which will encourage industries to self-regulate and adopt cleaner technologies in preparation for the adoption of ISO 14 000.

The International Organisation for Standardisation (ISO) is made up of members from around the globe. The standards set up by ISO come in different series and are not product specific, while their main aim is to ensure the sustainable use of natural resources. The ISO 9000 series of standards is the Management Quality standard. A company adhering to the ISO 9000 standards is one which has put in place environmental considerations within its management systems and is willing to adhere to environmental laws, regulations and standards such as the ISO 14 000. The ISO 14 000 series of standards are the Environmental Management series of specifications, while the ISO 14 001 series of standards are the Environmental Management System specification standards which were formulated in 1996 (Mabelebele, 2000). With globalisation and the desire of the SADC member states to engage more significantly in world trade and open their economies to foreign investors, it is necessary to have in place a system of standards such as the ISO standards, as well as laws and regulations which will safeguard their natural resources against overuse, abuse and pollution, in the quest for development.

CONCLUSION

The use of economic instruments is considered beneficial in promoting equity-led sustainable development worldwide (Turner et al., 1993). But, taking into account Africa's minor role in world economy and the fact that the majority of the population is poor, the question being asked in this chapter is, can such market based instruments promote natural resource conservation and environmental management in Africa, and in particular, within the SADC countries, while enhancing social development? Looking at examples from the Southeast Asian countries, market based economic instruments have been able to achieve major economic and social development, with minimum environmental degradation. The success of economic instruments in Southeast Asia can be attributed to the fact that, this has taken place within a framework of open economic and trade policies over a number of years starting from the 1970s, while in most African countries emphasis was placed on import-substitution policies in the 1970s and 1980s, and on protective inward looking economic policies, without developing appropriate infrastructure for the promotion of export-enhancing policies (Aryeetey et al, 1998). Despite this, and as has been shown in this chapter, the SADC countries, and indeed the rest of Africa is making genuine attempts to reform its economic policies, embracing free trade and strengthening their institutional, legal and policy capacity to effectively implement market-based economic instruments at the national and sub-regional levels. In order for these instruments to be efficient, they need institutional capacity, stakeholder participation, as well as the political will to enforce them. There is a danger that these instruments may retain an imbalance, concentrating too much on income generation, while providing little incentive for resource conservation and the minimisation of society cost. This will not only lead to resource degradation, but in the long run will be economically inefficient.

The main goals for using MBIs in the SADC region has been resource conservation, environmental protection and reduction of environmental impacts, revenue generation and poverty reduction (Rihoy, 1999; Lange et al., 1999; Maro et al., 2000). The majority of the economic instruments in the SADC states are command and control types or 'weak' market based instruments. This coupled with the lack of clear legislative structures has undermined their performance and their objectives. From this study it is clear that market based economic instruments are being used extensively in the SADC region in the management and use of land resources, in the use, allocation and management of water resources, and also in the industrial sectors. Within the industrial sectors, several countries were noted for making efforts to introduce different environmental standards and incentives for pollution abatement and the use of cleaner technology.

Stakeholder environmental awareness is weak in most countries, as information on the use of economic instruments and their benefits remains unknown to most stakeholders. In the SADC region, community based economic incentives have been shown to be both socially beneficial in poverty alleviation as well as environmentally beneficial, as long the local stakeholders are given enough legislative power and are involved in the whole process from the formulation to the implementation and administration of these economic incentives. Examples are programs such as CAMPFIRE and LIFE, which employ economic incentives in the management of land and wildlife resources, by empowering local communities to monitor and implement natural resource conservation laws, regulations and economic incentives, which will benefit the local communities financially while promoting environmental awareness, management and conservation. It can thus be said that in the SADC region, and in Africa, Market Based Economic Instruments (MBI's) together with local participation in resource conservation and management, can be used successfully to achieve several environmental management and protection objectives while enhancing social development, and raising environmental awareness in individuals and firms. However, their success depends, to a large extent, on the strengths of institutions, national environmental law, environmental and social policies and regulations and on their enforcement.

BIBLIOGRAPHY

C. Dorm-Adzobu, *Policy and Institutional Responses to Ecological Issues in Africa,* (Earth Council for Central and West Africa, 1997), paper presented at the African Environment Week seminar, IFA-Nord Ostev, 5 – 12 October.

C. Harvey, *Macroeconomics in Africa,* (Mc Millan Publishers, 1985).

E. Aryeetey, *Asia and Africa in the Global Economy: Economic Politics and External Performance in Southeast Asia and sub-Saharan Africa,* (UNU, 1998), paper prepared for the presentation at the UNU-AERC Conference on 'Asia and Africa in the Global Economy', held in Tokyo, 3 – 4 August.

E. Aryeetey et al., *Strengthening Africa's Participation in the Global Economy* (UNU, 1999).

E. Mabelebele, Information Specialist, South African Bureau of Standards (SABS), Pretoria, South Africa (personal communication, 2000).

E. V. Jaycox, *Proceedings of the 1985 SADC Annual Conference in Mbabane, Swaziland,* (Jongwe Publishing, 1985).

F. Ng et al., *Open Economies Work Better! Did Africa's Protectionist Policies Cause Its Marginalisation in World Trade?* (World Bank, 1996), Report 1636.

F. Ng, et al., *Good Governance and Trade Policy: Are they Keys to Africa's Global Integration and Growth?* (World Bank, 1999), Report No. 2038.

G. Lange et al., *Natural Resource Accounting Policy Brief,* (IUCN – ROSA, 1999), at 10 - 12.

Government Of Mauritius, Environmental *Protection Act, Act No. 34 of 1991* (Ministry of Environment and Quality of Life, 1991).

Government Of Mauritius, *Industrial Expansion Act,* (Ministry of Industry and Industrial Technology, 1993).

IMF, 'East Africa', 2: SURVEYS OF AFICAN ECONOMIES, (IMF, 1969).

IMF, 'BLS Countries', 5: SURVEYS OF AFRICAN ECONOMIES, (IMF, 173).

IUCN, *Environmental Strategies for Land Tenure and Community-based Natural Resources Management in Southern Africa,* (IUCN, 1998).

J. Arntzen et al., *Economic Instruments and Botswana's Environment,* (Department of Environmental Science, Department of Economics University of Botswana, 1999).

J. Turpie et al., 'Economic Rational for Environmental Management of Water Resources' in *Water Resources Management: Defining and Enhancing Environmental Sustainability,* (SADC-ELMS/IUCN-ROSA/SARDC-IMERSCA/World Bank, in press, 1999) 1, at 15.

L. Rihoy, *Natural Resources Tenure in Africa, Policy Brief,* (IUCN, 1999), at 5 - 6.

M. Chenje (ed.), *Reporting the Environment in Southern Africa*, (SADC/IUCN/SARDC, 1998).

M. Dilami et al., *What Macroeconomic Policies are "Sound"?* (World Bank, 1998), Report No. 1995.

M. A. Miller, *The Third World in Global Environmental Politics,* (Open University Press Buckingham, 1995).

National Settlement Policy, *Land Use and Land Tenure*, (Physical Planning Division, 1989), Working Paper No. 6.

P. A. Susswein, *Land Policy, Land Use and Cover Change in Southern Africa,* (Ph. D. research proposal, University of the Witwatersrand, 1999).

P. S. Maro, Technical Advisor, SADC-ELMS, Maseru, Lesotho, (personal communication, 1999).

P. S. Maro et al., 'Policy, Legislative and Institutional Framework' (Draft) in *Water Resources Management: Defining and Enhancing Environmental Sustainability*, (SADC-ELMS/IUCN-ROSA/SARDC-IMERSCA/World Bank, in press, 1999).

P.S. Maro et al., 'Policy, Legislative and Institutional Framework' (Final Draft) in *Water Resources Management: Defining and Enhancing Environmental Sustainability*, (SADC-ELMS/IUCN-ROSA/SARDC-IMERSCA/World Bank, in press, 2000), at 7.

R. Da Motta et al, 'Market-based Instruments for Environmental Policymaking in Latin America and the Caribbean: lessons from eleven countries', 19:2 ENVIRONMENT AND DEVELOPMENT ECONOMICS (Cambridge University Press, 1999), at 177.

R. F. Fuggle et al., *Environmental Management in South Africa*, (Juta and Co., 1994).

R. Mabugu et al., *Economic Instruments and Zimbabwe's Environment,* (Department of Economics, University of Zimbabwe, 1999).

Report on the Preparatory Workshop on the Land Policy Review Process for Lesotho, (Land Use Planning Division, Ministry of Agriculture, 1999), workshop held at the Maseru Sun Cabanas, Lesotho, July 8 – 9.

Regional Economic Review: Southern Africa Economic Summit 2000, (SADC Finance and Investment Sector Coordination Unit, 2000).

R. Stock, *Africa South of the Sahara*, (The Gilford Press, 1995).

S. Adriamananjara, *On the Size and Number of Regional Integration Arrangements: A Political Economy Model,* (World Bank, 1999), Report No. 2117.

SADCC, *Southern African Development Coordination Conference* (Cambridge, 1981).

SADC, *Regional Policy and Strategy for the Environment and Land Management Sector,* (SADC-ELMS, 1993).

SADC, *SADC Policy and Strategy for Environment and Sustainable Development,* (SADC-ELMS, 1996).

SADC, *Annual Report 1996/97,* (SADC, 1997).

SADC, *The Fundamentals of Natural Resources Policy Analysis for the SADC Region,* (SADC-ELMS, 1997).

SARDC, *State of the Environment in Southern Africa,* (SARDC, 1994).

S. Fakir, *The Economics, Policy and Natural Resource Management in Southern Africa,* (IUCN, 1996), workshop held at Pretoria, South Africa, September 11 – 13.

S. J. Dima et al., *Economic Instruments and Namibia's Environment,* (Department of Agricultural Economics and Extension, Department of Natural Resources and Conservation, Faculty of Agriculture and Natural Resources, University of Namibia, 1999).

Tanzania National Water Resources Management Policy, (Draft, 1999), at 41 – 42.

Turner et al., *Environmental Economics. An Elementary Introduction*, (John Hopkins University Press, 1993), at 144 and 159.

United Nations Development Program (UNDP), *Human Development Report 1997*, (Oxford University Press, 1997).

UNEP, *Global Environment Outlook 2000,* (Earthscan Publications, 1999), at 223 and 229.

UN General Assembly, 'International Economic Policy Coordination', in UN General Assembly, (UN, 1997), *Global Financial Integration,* an update report to the Secretary General, paragraph 11 – 13, at 3.

UN Economic and Social Council, 'Developing Economies: growth, reform and financial crisis', in *The World Economy at the beginning of 1998,* (UN, 1998), report to the Secretary General, paragraph 71 – 126, at 13.

U. Kumar et al., *Legal Context of Natural Resources Management,* (SADC – ELMS, 1996), Report series No. 38.

Annex 1. Polices, legislation and Institutions for Environmental Management in the SADC Countries

Country	Policy Objectives	Legislation	Institutions
Botswana	To develop new and better uses of natural resources. To diversify the rural economy. To emphasise the environmental education. To develop links with neighbouring countries concerning resource conservation. To establish a balance between population growth and natural resources availability. To conserve all main ecosystems, protect endangered species, control depletion of	Fourteen Acts (Laws) are in place to govern the regulation of utilisation of land, water and minerals, and deal with conservation and pollution. These are inadequate to address all environmental issues, and their enforcement is a problem.	National Conservation Strategy (NCS) approved. National Environmental Secretariat set up under Ministry of local Government, Lands and Housing for implementing NCS. Priority Programs of NCS reflected in 1992-1997 National Development Plan.

	non-renewable resources, restore degraded resources and prevent and control pollution.		
Lesotho	To assure all people living in the country the fundamental right to an environment adequate for their health and well being. To use and conserve the environment and natural resources for the benefit of present and future generations, taking into account the rate of population growth and the productivity of available resources. To maintain stable functioning relations between the living and non-living parts of the environment through preserving biological diversity and respecting the principle of optimum sustainable yield in the use of natural resources. To reclaim lost ecosystems where	Over sixty Acts (laws) covering land, land use, construction and mining, conservation of nature and protection of cultural heritage, health and water resources exist. The problem is that the legislation is fragmented into many pieces dealing with ecological matters. A comprehensive systematic approach towards environmental protection and sustainable use of natural resources is required.	A national environmental action plan (NEAP) and a National Environmental Management Policy have been developed. A Ministry of Environment, Gender and Youth Affairs has been developed to deal with all issues of natural resources, water, the environment, gender, and to educate the youth about such matters. Houses the SADC Environment and Land Management Coordination Sector (SADC-ELMS), which coordinates all environmental issues in the SADC member states.

	possible and reverse the degradation of natural resources. To establish adequate environmental protection standards and monitor changes in and publish data on, environmental quality and resource use. To require prior environmental assessments of proposed activities that may significantly affect the environment or use of a natural resource. To assure that environmental awareness is treated as an integral part of education at all levels, and that this knowledge is used in the planning and implementation of development activities. To ensure that the true and total costs of environmental use and abuse are borne by the user, i.e. the "polluter pays" principle. NB. More detailed policies are elaborated under each sector of the		

	economy.		
Malawi	To ensure national self-sufficiency in the supply of food and woodfuel. To safeguard the environment of Malawi from undue contamination and visual degradation. To ensure development does not ultimately affect natural Self-regulatory processes (e.g. regeneration of soil fertility and vegetation; the hydrological cycle and the self-cleaning of rivers and lakes) on which national economy depends. To maintain the natural diversity of plants and animals in order to conserve genetic material and thus retain options to develop unused plant and animal resources. To maintain selected key sites of outstanding aesthetic or scientific importance with the minimum of alteration. Under the	Over forty Acts (Laws) govern the utilisation regulation and conservation of natural resources. But as in other SADC countries, the legislation has developed in an ad hoc piece meal manner with limited environmental information. Government is undertaking the review of the environmental legislation.	Ministry of Research and Environmental Affairs coordinates all environmental issues and plays a catalytic role initiation formulation and development of policies and programs for environment and development. A National Environment Action Plan (NEAP) has been developed

	National Rural Development Program; To provide inputs and services to enable small holders to increase production per unit area. To conserve natural resource by encouraging good cultivation practices and soil conservation and by conserving key watersheds, especially in areas with fishery potential. To maintain forests by afforestation in reserves on customary and estate land.		
Mozambique	To ensure integration of environmental parameters into socio-economic development planning. To coordinate the various organs of government involved in environmental management and research. To develop legal instruments to make	Over ten Acts (Laws and Decrees) exist, but these do not address the question of sustainable development, nor do they provide security of land tenure. Hence, government is embarking on the preparation of national environmental legislation.	Ministry for Coordination of Environmental Affairs (MICOEA) has been established to handle all sustainable development and environment issues. A National Environment Management

Mozambique (cont.)	obligatory the analysis of the environmental impact of major economic projects. To ensure the inclusion of environmental topics in education curricula at all levels. To promote the acceptance of simple and creative non-pollution technology. To develop and implement an integrated management plan that will protect biological diversity especially in the islands and coastal territories. The 1990 national constitution contains eight articles that deal with the control and conservation of national resources and ecological balance for sustainable development.		Policy is under way.
Namibia	Article 91 (5) of the 1990 Constitution of the republic of	Over ten laws enacted during the colonial era cover protection of	Directorate of Environmental Affairs has been

Namibia (cont.)	Namibia states that the State shall ensure that ecosystems, essential ecological processes and biological diversity of Namibia are maintained and natural resources are utilised on a sustainable basis for the benefit of all Namibians both present and future; provide measures against the dumping or recycling of foreign nuclear toxic waste on Namibian territory. To encourage environmental awareness and education initiatives. To ensure independent environmental impact assessments form part of pre-feasibility study of all developments which will also be regularly monitored. To maintain species and ecosystems and prevent and reverse desertification. To ensure equitable and efficient use of water resources; food self-sufficiency and food security, and	wildlife mainly. More environmentally relevant legislation is being developed.	established in the Ministry of Environment and Tourism.

	greater energy self-sufficiency. Other policy objectives are mentioned under specific sectors of the economy.		
Swaziland	To protect and enhance the quality of the natural environment through good agricultural practices and soil conservation. To develop policies, plans and programs aimed at a more effective utilisation of forestry resources to prevent harmful consequences; to protect water catchment areas; and arrest soil erosion. To improve the management and utilisation of natural pastures and promote forage production, soil conservation and implementation of management systems. To minimise the incidence of disease through development of basic sanitation facilities and provision of safe water supplies.	Over thirty laws exist but their enforcement is hampered by lack of coordination among the different sectors of government empowered to implement them. The recently established Swaziland Environmental Authority will coordinate and harmonise activities of all bodies concerned with environmental issues.	The Ministry of Natural Resources and Energy is the main institution for environmental matters at national and international levels. The Ministry of Agriculture and Cooperatives also deals with environment especially its Land Use Planning. The Swaziland Environmental Authority within the MNRE plays central role in environment. A National Environmental Action Program and a National Environmental Management Policy have been

	To ensure that environmental impact assessments are carried out on all development programs and projects.		developed.
Tanzania	Emphasis is put on the need to alleviate poverty as a prerequisite for sustainable development. To formulate policy and coordinate activities of all bodies concerned with environmental matters. To advance the scientific knowledge of changes in the environment and develop technologies to minimise adverse effects on people's health and well being. To establish and operate a system of environmental information dissemination and formulation proposals for legislation and regulation. To undertake and promote a comprehensive	Over sixty legal instruments covering a diversity of environmental and economic activities exist. However, the environmental legal structure is fragmented and needs to be harmonised and give some of the institutions legal power to enforce the laws.	The Division of Environment under the Vice President's office has been established. National Environmental Management Council. National Land Use Planning Commission. Zanzibar Commission for Lands and Environment. Several Departments in different Ministries, and Research Institutes also deal with the environment. A National Conservation Strategy has been finalised. National Action Plans on forestry

	environmental education program. Specific policy objectives are detailed for each sector and sub-sector.		and desertification have been prepared. National Environmental Management Policy has been enacted. Environmental Information Centre has been established.
Zambia	To ensure the sustainable use of Zambia's natural resources. To maintain Zambia's biological diversity. To maintain essential ecological life support processes and systems including soil regeneration and protection, nutrient recycling, ensuring the availability of clear air and safe water. The National Conservation Strategy for Zambia revealed the following constraints; i.e. - the absence of strong cross-sectoral guidelines for land and natural resources allocation and conservation according to	Over forty Acts governing specific areas of environmental management and utilisation exist. The most important law on environmental issues is the 1990 Environmental Protection and Pollution Control Act which created the National Environmental Council and gave it mandate to co-ordinate all environmentally related activities.	Ministry of Environment and Natural Resources, which has a Division of Planning and Information, deal with environmental issues. Some Departments and Divisions within Ministries still deal with environment.

	productive capacities; - inadequate coordination between current conservation and development efforts.		
Zimbabwe	To integrate sustainable resource use with every aspect of the Nation's social and economic development, and rehabilitate those resources which are already degraded. To develop and implement action plans to ensure that the utilisation of Zimbabwe's natural resources is undertaken in an equitable, productive, and sustainable manner. To ensure that we live within the ecological capacity of the land. To undertake land reform and provide alternative development options to optimise sustainable yields from the land. To promote dynamic	There is comprehensive environmental legislation with over fifty Acts, but enforcement if a problem because of their cross-sectoral application and because most are out-dated.	The Ministry of Mining and Tourism and the Department of Natural Resources are the main environmental resource utilisation control agencies. Other Ministries have Departments and Divisions also responsible for environmental matters. Environmental NGO's are very active and presented on national report to UNCED. State of Environmental Report has been produced. A post-UNCED response has been held.

	public awareness, education and extension services. Preservation of biological diversity.		

Source: SADC, 1993; Maro, 1999.

[1] Currently, there are 14 member states, viz.: Angola, Botswana, Democratic Republic of Congo (D.R.C), Lesotho, Malawi, Mauritius, Mozambique, Namibia, South Africa, Swaziland, Seychelles, Tanzania, Zambia and Zimbabwe.

[2] The impact of such a growing population on agriculture in order to merely feed the people is staggering, especially given the climatic conditions and the problems of desertification (Jaycox, 1985; SADC, 1996).

[3] Annex 1 shows the different institutional, legal and policy frameworks in selected SADC countries.

[4] Additionally, there is a lack of trained staff to gather the relevant evidence for prosecution (Kumar et al., 1996)

[5] These are Botswana, Lesotho, Malawi, Namibia, Zambia and Zimbabwe.

[6] In Lesotho, diamond buyers pay a stiff export tax of 15 % on all diamonds exported.

EFFECTIVENESS OF THE ENVIRONMENTAL IMPACT ASSESSMENT PROCESS IN MANAGING TOURISM DEVELOPMENT IN THE SEYCHELLES

ROLPH PAYET[1]

INTRODUCTION

Seychelles is a very small archipelago located in the Western Indian Ocean. Consisting of over 115 islands of both continental and coralline origins, the Seychelles is renowned for its unique beauty as paradise on earth.[2] The Seychelles were colonised in the early 18th century and today has an indigenous population of about 81,995, estimated to grow to 101,987 by the year 2019.[3] Although agriculture dominated as early as 1771[4], and until the late 1960's, the inescapable temptation for Seychelles to become every tourist's dream has led to rapid development of tourism infrastructure over the last thirty years along the coastline of many of the Seychelles islands.

With the opening of the Seychelles international airport in 1971[5], the tourism industry began an upward and progressive trend to become the main pillar of the economy, in addition to the fisheries and services sectors. The unique and natural beauty of the Seychelles has provided tourism investors with the ideal setting for high quality resorts as well as abundant sun, sand and sea, which are relatively unpolluted.

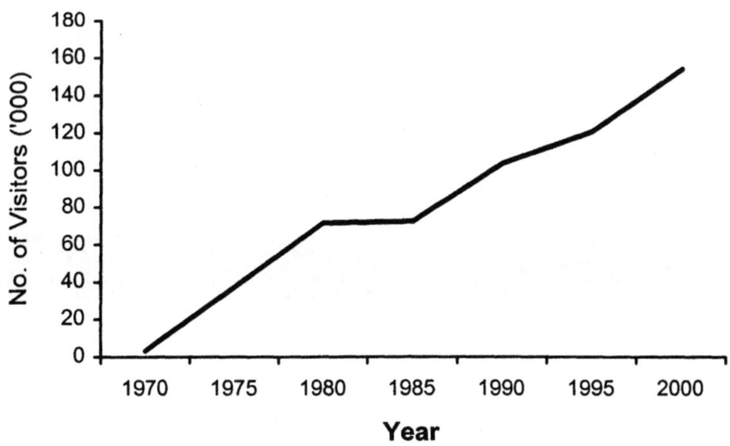

Figure 1 Visitor arrivals to the Seychelles since the opening of the International Airport (per 000's)

(Source: MTCA Data, 1999)

Europe is by far the most important market, led by France, then Germany, Italy and the United Kingdom. Figure 2 shows visitor arrivals in 1997 by region.

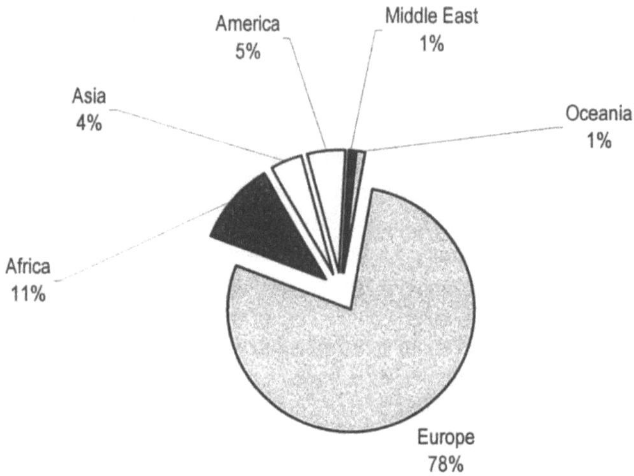

Figure 2 Visitor Arrivals in 1997 by regions of origin (MTCA Data, 1998)

The Ministry of Tourism and Civil Aviation (MTCA) recently made some projections on likely number of visitors and growth within the sector over the next seven years. The results of these projections are shown in Table 1. The assumptions used in the projection are given at the end of the table.

	Number of Visitors (1)	Growth in % (2)	Bed Nights in ('000) (3)
1998	142,000	9.2	1136
1999	154,000	8.5	1232
2000	166,000	7.8	1328
2001	178,000	7.2	1424
2002	190,000	6.7	1520
2003	202,000	6.3	1616

2004	214,000	5.9	1712
2005	226,000	5.6	1808
2006	238,000	5.3	1904
2007	250,000	5.0	2000

Table 1 Project Tourism Development (Source: MTCA, 1998)

Assumptions:

1. Visitor Arrivals: Increase by 12,000 per annum
2. Growth rates average of 6.8% (actual growth rate recorded at 7.4% for the last 10 years)
3. Bed nights – 80% of visitor nights (recorded 78.65 for last 10 years)

The tourism industry provides employment for almost 20% of the working population. In the past, Seychelles although being a relatively expensive destination, was popular among many middle-income earners from Europe. However over the last two years, a number of five star hotels have been built, and the government's policy has shifted to encourage visits from high income clientele, primarily to avoid the unnecessary strains of mass tourism. The number of rooms currently available in Seychelles totals about 1,617, implying that if Seychelles wish to experience growth in the industry as per the projections indicated in table 1, then the likelihood for more hotels to be built is very high. In 2000 alone, at least two large five star hotels are being built for opening in late 2001 on Mahe, the main island of the Seychelles.

Since most hotel developments in the Seychelles are built on the coastal plateaux and close to beaches, ribbon development dominates sections of the coastal zone on Mahe and the other granitic islands.[6] Such a rapid rate of tourism development on the coast almost completely overtook current efforts to ensure that development was integrated within coastal development plans, as well as meeting existing conditions (specified by the Planning Authority) especially those that minimise impacts on the coastal environment. Indeed, the Environmental Impact Assessment (EIA) process, put in place in 1996,[7] provided the legal framework to ensure that such rapid development in the tourism sector was integrated and implemented with minimum impacts within the ambit of the country's sustainable development policy found in the *Environment Management Plan 2000-2010*.[8]

LAND USE PLANNING IN THE PRE-EIA ERA

Jean Baptiste Philogene de Malavois, a French engineer, agronomist and geographer was perhaps one of the first colonisers in 1786 to institute orderly land tenure, allocating to each habitant at the time 112 acres of land from the coast upwards [9]. He was also instrumental in the implementation of the

'Cinquante Pas du Roi', an area of about 100m inland from high tide where no development was permitted, except for military fortifications. Unfortunately, much later in response to the Secretary of State concern over indiscriminate development, the local court in 1903 concluded that the 'Cinquante Pas du Roi' had not been respected and that there was no law in force or applicable to redress the situation.

Availability of land for development since then has become one of the major constraints of development and is subject to severe competition due to the wide range of uses, such as environment protection (43%), agriculture (22%), tourism (6.3%) and human settlements (11.7%).[10] The scarcity of coastal land for development has already led to the reclamation of over 600 ha. along the East Coast of Mahe island. Table 2 below, shows the low percentage of coastal plateaux available on the main islands.[11]

Island	Total land area (ha.)	Coastal plateau land area (ha.)	Percent Cover of total land area
Mahe	15470	744	4.8 %
Praslin	3756	193	5.1 %
La Digue	1010	165	16 %

Table 2 Coastal plateaux area on the three main islands in the Seychelles (Source: see Micock, 1998)

It is estimated that 90% to 100% of the population is concentrated on the narrow coastal strip. Mahe has a total land area of 148 square kilometres, which leads to a population density in excess of 400 persons per square kilometres. Approximately 40% of the population is located on the East Coast of Mahe from Victoria to the International Airport on a coastal belt, which is only about 7 km long by 1 km wide.

The Government of Seychelles has, despite these pressures for development and imminent encroachment on forest and sensitive habitats maintained a very tight control on the level of development as well as the types of development. Government has especially been consistent with its policy to limit the extent of housing development on the hilly slopes, preventing the construction of heavy industries. In this context, the government stood by its policy against the development of high-rise tourism infrastructures along the coast.

However, problems and ineffectiveness concerning the management of development from both planning and control perspectives still abound. Although there is over 100 years of land use legislation in Seychelles, the most modern is the *Town and Country Planning Act*, enacted in 1972. It is largely based upon the 1947 and *1962 Town and Country Planning Acts* of the United Kingdom, since at the time Seychelles was a Crown Colony of the British Empire.[12] Development is defined by the 1972 *Act* as the carrying out of building, engineering, mining, or other operations in, on, over, or under any land, and the making of any material change in the use of any buildings or land, including any subdivision of land which is or is intended to be used for any residential, commercial or industrial purposes.

The table below summarises the various legislation enacted since the 1972 *Act*, which currently provides a framework for managing development.

Legislation	**Main characteristics**
State Land and River Reserves Act, (Cap 228) 1903	Provides for the delineation of 10m vegetation buffers along named rivers.
Delineation and Classification of the Domaine Public Act (Cap 60) 1958	Management and use of public property and state lands.
Reclamation Act, (Cap 106) 1961	Permission required for any reclamation. Includes provisions for public right to object, rehabilitation of site, in case of non-compliance.
Land Survey Act, (Cap 109), 1964	Powers to conduct surveys and ensure land tenure is respected.
National Parks and Nature Conservancy Act, (Cap 141) 1969	Establishment and management of National Parks and protected areas, including marine parks.
Town and Country Planning General Development Order, S.I. 133 of 1971	Permission and list of classes of development permitted with conditions.
Town and Country Planning Act 1972, and amendments: S.I. 12 of 1974 S.I. 35 of 1975 S.I. 45, 100 of 1976 S.I. 71 of 1980 S.I. 27 of 1991	Control of development through planning applications, approval and supervision. Planning Authority mandated
Building Licenses Act 1973	
Town and Country Planning (Building) Regulations, S.I. 12 of 1975	Control and certification of building operations, including construction material quality to British Standard Code of Practice.
Decree 46 of 1978	Amendment to the Act to control sub-division of land.
Removal of Sand and Gravel Act 1982	Control and licensing of sand and gravel extraction on the coast.

Table 3 Land Use and Development Legislation enacted from 1901 to 1995 (the post-EIA era).

The 1972 *Act* and its subsidiary legislation are implemented and overseen by a statutory body: the Town and Country Planning Authority (TCPA). This is dictated in section 3.1 of the Act. In the first Schedule, the powers and responsibilities of the authority are specified. The TCPA has traditionally operated under the Ministry of Land Use and Habitat and not as an autonomous authority as originally intended. This is due to close working relationships required with the various divisions within the ministry, notably the Physical Planning and Survey Divisions. Whilst the Physical Planning Division still has the role of secretariat, processing and managing the planning applications received on a daily basis, delays in approving some applications prevail.

The 1972 *Act* is currently being reviewed to bring the legislation up to date, especially with developments in environmental law, approaches to enforcement, and to cover developments in coastal marine areas.[13] The TCPA will also be reviewed and will probably be given some degree of autonomy in practice.

HISTORICAL PERSPECTIVES IN THE DEVELOPMENT OF THE EIA REGULATIONS

The need for legislation to address emerging environmental issues and problems in Seychelles only received particular attention over the last ten years. The first *Environment Protection Act* was enacted in 1988, however this was repealed and replaced by a more comprehensive act, the *Environment Protection Act 9* of 1994 (EPA 1994).[14] The main objective of the *EPA* 1994 is to provide for the protection, improvement and preservation of the environment and for the prevention, control and abatement of environment pollution.[15] The main sections of the *EPA* 1994 are described below:

- S.1 to S.3 General and definitions
- S. 4 to S. 5 Administration of the Act and functions of bodies instituted
- S.6 to S.14 Prevention, control and abatement of Environmental Pollution - provisions for prescribing standards, subsidiary legislation, pollution areas, recycling and adoption of cleaner technologies, declare coastal zones and prepare ICZM plans, environmental research and monitoring, control, use and trade in substances, waste, or activities which may endanger the environment.
- S.15 Environment Impact Assessment - control of development and activities in a protected or ecologically sensitive area, set-up of the EIA appraisal committee, contents of the EIA, responsibilities of the developer, provision for public inspection, claims and appeals, and monitoring of development or activity for compliance with EIA recommendations.
- S.16 to S.28 Enforcement of the Act - includes provisions for serving of enforcement notice, prohibition notice, with powers given to authorised officers for entry, examining and test any equipment, record and

document, collection of samples, appointment of Government analyst and declaration of environmental emergencies.

- S.29 to S.40 Offences, penalties, procedures and other provisions - includes provisions for consultation with other bodies in implementation of the Act.

(Environment Protection Act, 1994)

Although, there was already a large number of environment related statutes, the consistency and level of fragmentation between those instruments warranted the drafting of the *EPA* 1994.[16]

Section 15 of the *EPA* 1994 provided for the enactment of the first generation of EIA regulations in Seychelles. Whilst the *EPA* 1994 provided the framework for the establishment, procedures, criteria for implementation and the administration of an EIA, the regulations were actually put into law after the *EPA* 1994 had been endorsed by the National Assembly. The *Environment Protection (Impact Assessment) Regulations*[17] (thereafter called the *EIA regulations*) consequently came into force on the 26th May 1996. It contained the following sections:

- S.1 to S.3 General and reference to section 15.1 of the EPA 1994.
- S.4 Application for authorisation - application requirements, with a specific reference to the TCPA.
- S.5 to S.6 Determination of Authority - introduces the Class I and Class II EIAs. This is linked with the TCPA (regulation 5.3).
- S.7 Submission of the EIA by the Proponent.
- S.8 Public inspection of the EIA.
- S.9 Role and function of the EIA Appraisal Committee, which shall be *ad-hoc*.
- S.10 Authorisation of a project or activity.
- S.11 Right of appeal against any decision.
- S.12 Covers other developments falling outside the scope of the TCPA, such as resource extraction, marine recreational activities, etc.
- Schedule 1 List of projects or Activities requiring environmental authorisation under regulation 3.1.
- Schedule 2 Protected and ecologically sensitive areas under regulation 3.2.
- Schedule 3 Form of Memorandum of Appeal.

(Environment Protection (Impact Assessment) Regulations, 1996)

The development of the *EIA Regulations*, as expected, was a highly contested and frequently redrafted piece of legislation. To ensure its credibility, section 15.1 of the *EPA* 1994 was added to clearly indicate that:

> "Notwithstanding any license, permit or approval granted under any other Act, any person who commences, proceeds with, carries out, executes or conducts or causes to commence, proceed with, carry out, execute or conduct any prescribed project,

activity or any project or activity within a protected or ecologically sensitive area as may be prescribed under this Act or under any other Act for the time being in force -

> a. without carrying out an Environmental Impact Assessment Study and obtaining the authorisation for the Authority; or
>
> *b.* in breach of any conditions imposed by the Authority, is guilty of an offence"

Almost every major development project in the Seychelles is subject to environmental screening and authorisation by the authority designated under the *EPA* 1994, i.e. the Ministry of Environment. Other ministries expressed several concerns at the time. These key concerns included:

(i) the powers vested in the *EPA* (1994) and subsequent EIA Regulations may compromise decisions of the Planning Authority (administrator of the *Town and Country Planning Act*, 1972);
(ii) the possible delays in processing projects;
(iii) the additional cost to development;
(iv) the technical capacity of the country to prepare and review EIA's;
(v) the availability of baseline environmental information;
(vi) the composition and independence of the EIA authorisation body; and
(vii) the relevance of EIA projects to certain types of activities.

Following many consultative meetings and draft proposals, the final text of the present regulations was agreed upon, which essentially provided a mechanism for close consultation between the key Ministries involved, i.e. the Ministry of Land Use and Habitat and Ministry of Health. A direct link and cross-reference to the *Public Health Act* 1960 and the *TCPA* 1972 was also included.

The list of projects and activities listed in Schedule 1 of the *EIA Regulations* is quite comprehensive, covering mining, agricultural production, forestry, fish processing and aquaculture, chemical industries, construction industry, transport of goods and materials, agro-industries, energy production and distribution, water, sewage and wastewater, solid waste, hotels, restaurants and tourism activities, land reclamation, housing and urban development and roads. Whilst there was much debate about the requisite threshold of each project or activity to be met for an EIA, it was agreed that through the screening process, the Authority would subject these to either an EIA Class I or II pursuant to guidelines laid down by the Authority.[18]

Schedule 2, on the other hand, covers an extensive category of legally protected and ecologically sensitive areas. These include: namely protected areas by law; historical sites; national monuments; remarkable natural landscapes; viewpoints; inter-urban buffer zones; water catchment areas; industrial risk areas; natural risk areas; steep slopes; areas prone to erosion; high elevations; and the skyline. The schedule also lists certain habitats, which are considered sensitive, and draws reference to international conventions such as *CITES*. The list also includes marshes and wetland habitats, streams and surroundings, the coastal strip

(100m from the high-water mark), beaches and intertidal zones, the seabed, small islets and all the outer islands.

The first generation of the EIA regulations were therefore very comprehensive, but involved the participation of experts and stakeholders. Contrary to the development of EIA legal frameworks elsewhere,[19] the motivation for the EIA process has been driven primarily by the need to address the issues of environmental impacts and development in a more informed and long-term manner. The present EIA regulations bring together a wider scope of issues, such as environmental, social, economic, health, cultural, resource management and risk within the entire planning process. Similarly the regulations are not only limited to the impacts of projects and activities, but look at habitat responses, biodiversity effects, and sustainability considerations.

Although many countries have opted for a separate statute for EIA (e.g. Canada and Korea),[20] using the specific regulations under the already proclaimed *EPA 1994* had its advantages. Since, the *EPA 1994* laid the basic legal principles and framework for environmental impact assessment, the development of the EIA regulations, which was part of a much longer process, was developed without revisiting the environmental principles already embedded in the *EPA 1994*.

IMPLEMENTATION OF THE ENVIRONMENT IMPACT ASSESSMENT REGULATIONS

Before the EIA regulations of 1996, a number of EIAs had already been done such as the Water Supply Options (Dams and Reservoirs) EIA[21] and the Fregate Island Resort EIA.[22] These were done within acceptable external frameworks. For example the Grand Anse EIA was modelled after *World Bank Guidelines,*[23] and the Fregate Island EIA after the South African EIA framework.[24] Both EIAs involved a scoping process, whereby views of various stakeholders were sought. Factors addressed in both EIAs covered environmental, economic and social aspects, as well as various alternatives to the project. In the case of the Grand Anse Dam development, several other options were considered on the basis of environmental impact. To reduce impact of pest introduction on Fregate Island, several scenarios for management on the island were studied. The implementation of these EIAs served as essential experience to all the institutions involved, especially in the development of practical mechanisms for implementation of the EIA regulations.

The current institutional and implementation arrangements for the EIA regulations in Seychelles are illustrated in the process flow diagram below;

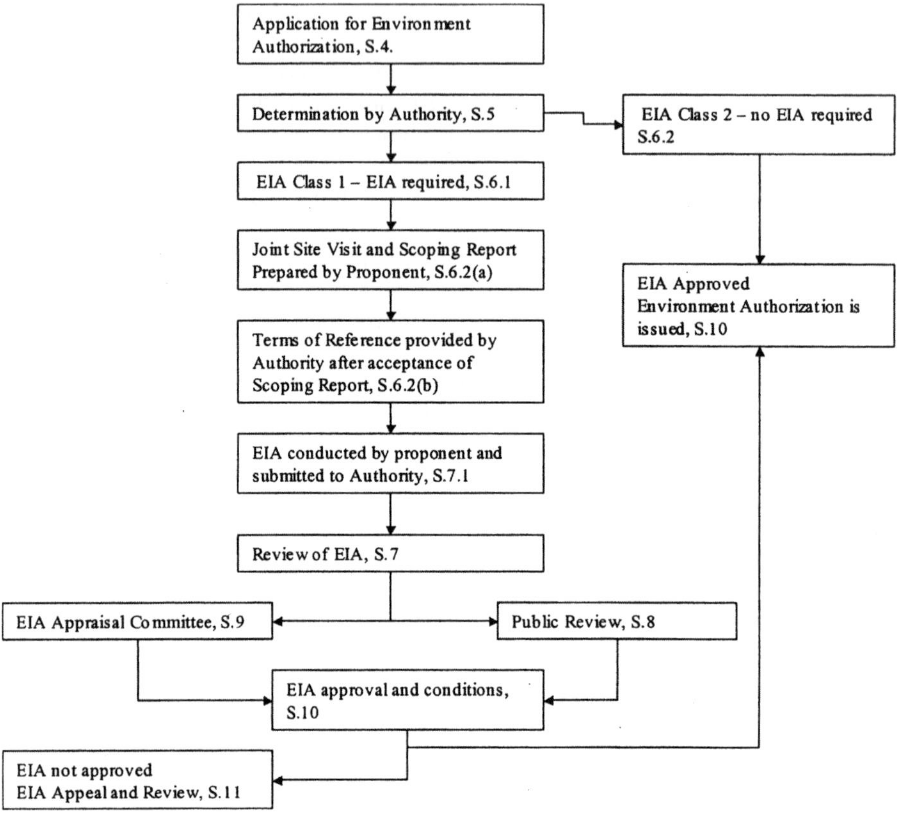

Figure 3 The EIA Implementation and Approval Process in Seychelles

A developer who wishes to undertake tourism development in Seychelles has to engage in a number of consultations before moving towards final drafting stage of the development. An initial consultation involves the presentation of the hotel tourism concept to a Project Appraisal Committee, chaired by the Ministry for Tourism and Civil Aviation, in which the Ministry of Environment and Transport is represented. This committee is responsible in giving the initial guidance on how the concept can be further developed or amended to meet national polices and guidelines. It is often after this screening process, that the Ministry of Environment decides whether the project will be subject to an EIA. Therefore a developer would be instructed to apply for environmental authorisation, and in almost all cases since the enactment of the EIA legislation, most hotel developments have proceeded with an EIA Class I. Usually to facilitate the scoping process, joint site visits with the developer; the EIA consultant; and the staff from the Environment Assessment Unit in the Division of Environment.

Scoping can be described as an open process for determining the scope of issues to be addressed in the EIA.[25] There are at this stage no clear guidelines for the scoping process but some consultants adopt the European Community

Guidance on Scoping,[26] or the *World Bank Guidelines on Public Consultation in the EA Process*.[27] Following submission of the scoping reports and verification of the issues raised by the stakeholders, the EIA Terms of reference is prepared for a set of available EIA templates for different types of development. Depending upon issues addressed in the scoping report, the TOR is modified accordingly and communicated to the proponent. According to Section 6 of the *EIA Regulations*, the Authority may undertake the EIA on behalf of the proponent or by the proponent itself. However, the former rarely occurs, except for Class II EIAs, due to lack of human resources and the potential conflict of interest. It is the responsibility of the proponent to get the study done.

Once the EIA has been completed a first draft is forwarded to the Authority for review by the EIA appraisal committee, which constitutes environment experts and representative of other sectors such as land use, the National Environment Advisory Council, tourism and social affairs, and for public review. Following review, comments and conditions attached to the proposed activity or project are communicated to the developer, who then undertakes to review the EIA accordingly. After meeting all the requirements, the developer is issued with an authorisation, and the plans are then submitted for review under the TCPA and associated building regulations. In many cases though, some projects come through a referral by the TCPA.

THE POLICY ASPECTS OF THE EIA PROCESS

Initially, Environmental Impact Assessment as a tool for assessing impacts and implementing environmentally sound measures for development projects in Seychelles was not widely accepted by policy-makers and developers. Lack of awareness was one reason. The potential for the EIA process to address the rapid pace of tourism development in a precautionary manner was not clearly understood. It was the intention that the EIA process would reinforce the existing planning regulations, as well as providing a convenient framework for investor commitment to minimise impacts on the environment. Unfortunately, many planners saw the *EIA Regulations*, as an unnecessary duplication of planning regulations and requirements, and the response 'we were doing *in promptu* EIAs before' prevailed in many circles. Some policy-makers even advocated that the EIA process would hamper decision-making procedures, and reduce investor interest.

Eventually agreement to placate that concern key legal provisions were included in the *EIA Regulations*, for example:

> " The Authority shall inform its determination under sub-regulation (1) to the proponent of the project or the activity and, where applicable, to the Minister or Town and Country Planning Authority referred to in regulation 4.1(a) within 14 days after the receipt of the application or its copy by the Authority."[28]

This provision sets clear time limits with respect to the determination of an activity or project. The EIA process has since developed and grown to become an accepted environmental management tool by the development community.

Several policy issues arose out of the implementation of this regulation in relation to tourism:

To what extent should coastal tourism development be accepted?

The extent of coastal development became increasingly identified in relation to the particular carrying capacity of the coastal area.[29] The level of tourism development and intensity of tourism was recognised as having a direct impact on coastal environmental quality. Therefore limiting coastal tourism development in certain areas became necessary, and today several beaches are not designated for any tourism development. Tourism is also becoming increasingly restricted in its sprawl and distance from the coastline. Setback lines and wetland management are increasingly becoming important criteria for tourism development. Public right of access to the beach is proscribed under law, and this is unlikely to change although there is pressure from many hotel developments that wish to have total exclusivity.

What criteria to be adopted to define the type of tourism infrastructure expected?

There have been various attempts to define the ideal architectural style that would blend into the natural environment as well as improve the aesthetic appeal of tourism development. Much has been accomplished on some islands to implement this policy. Work undertaken by Duncombe[30] illustrates in numerous situations the types of infrastructure requirements expected in the various categories of sensitive areas, for example the coastal belt, slopes and viewpoints. Although much is left to the discretion of the developer and the architects, the policy applies to a number of situations depending on the physical location of the development, the ecological and cultural importance of the area as well as the natural landscape qualities.

How will land development practices evolve to consider environmental aspects?

Practices that have evolved in developing land have been mainly destructive. Heavy machinery is utilised to change the profile of the shoreline and the hillsides imposing severe consequences on the environment. Environment Impact Assessment guidelines, published by the Ministry of Environment and Transport[31] presents a number of issues to be addressed in tourism development. Project alternatives, management and training, monitoring, implementing mitigating measures and good practices are essential requirements in the EIA process. It is expected that in time these elements will become mandatory. Furthermore, tourism infrastructure development has various upstream linkages with a number of other sectors having an impact on the environment, such as water resources, construction materials, transport, and coastal resources. Policy guidance on a number of these issues is still evolving and will likely define the nature and extent of coastal tourism in the future.

Institutional Aspects

Whilst the Ministry responsible for environment retained the portfolio responsibility for the EIA regulations, an inter-Ministerial collaboration mechanism was set-up to facilitate the co-ordination of EIA approval for development projects. For tourism projects, there is a tripartite review process involving the Ministry of Tourism and Civil Aviation, the Ministry of Land Use and Habitat and the Ministry of Environment and Transport.

The *National Land Use Plan of Seychelles (NLUP)*[32] was developed to allow land use planners to guide development within a global and integrated perspective. It provided an opportunity for long term land use planning. Although it never became a legally binding instrument due to the assumption that such plans would not be flexible and likewise constrain development options, it evolved as a useful background tool to the EIA process. It informed the developer of the types of land use anticipated within the area planned for tourism development. Due to changes in the land use policy, especially with the land bank housing projects and the need for more stakeholder involvement more detailed and focussed District Land Use Plans (DLUP) are now superseding the NLUP. However, it still remains to be seen whether such plans will be incorporated into law and become binding. The District Land Use Plan was also used in conjunction with a sensitive area map prepared by the Ministry of Environment and Transport, which was applied to development projects requiring EIAs.

The Ministry of Environment created an assessment unit to handle all administrative issues related to the EIA process. Development approval was subject to an environmental authorisation given by the Ministry of Environment through the TCPA. The capacity within the unit to review, manage, and monitor implementation of EIA requires strengthening. There is also a need for effective enforcement of the EIA recommendations. Few projects are audited during the construction phase or at completion, although there are some brief provisions in the *EIA Regulations*. An improved auditing and monitoring framework needs to be put in place to ensure compliance with the EIA recommendations.

To facilitate the application of the EIA procedure, all development projects are submitted to the respective specialist agencies and then forwarded to the Ministry of Environment for authorisation. One of the serious constraints in undertaking EIAs is the lack of adequate and reliable baseline environmental data that is area-specific. As a result, specialist studies are often necessary to determine the status of the site and address other issues identified in the scoping process. However, this is at a greater expense to the developer and may lead to delays in project implementation.

Capacity to review EIAs is limited because of unit size and availability of experts. In many instances EIAs are circulated to specialists in various departments for their views. The EIA regulation also makes provision for public review. When the EIA documents are ready, the public is informed through a notice in the national newspaper, and the EIA documents are placed at various convenient pubic places for consultation by members of the public over periods of one to two weeks. Comments from the public are evaluated and compiled. All views are then communicated to the developer, along with possible mitigating measures. By law, comments from the public are taken into consideration by the EIA Appraisal Committee and any substantial comments are issued with the recommendations.

Although there has been significant progress in the implementation of the EIA process so far, critical areas of concern such as capacity building of EIA Consultants, EIA Expert Reviewers, Government Specialists and EIA Enforcement Inspectors remain. Lack of monitoring and assessment of performance often sends the wrong message to tourism developers: EIA is probably just a certificate for construction, rather than an approach to sustainable development and responsible practices.

Limitation	**Area of Concern**	**Possible Solution**
Post-EIA changes in development plan and design.	Impacts not addressed in EIA.	EIA should account for possible alternative scenarios & establishment of environmental baseline.
Cumulative effects of later developments within the same area.	Affect environmental quality criteria set for project.	Scope of EIA to cover regional aspects, and use of GIS tools in the assessment, and monitoring.
Lack of formal land use plans and changes in land use.	Infrastructure is not located in the most sustainable manner.	Preparation of land use plans with indication of carrying capacity/ density policies.
No scenario and project alternative analysis.	Sustainable and most cost-effective alternative is not always selected.	Training in scenario building and analysis; improve environment baseline info.
Terms of Reference of EIA.	Does not address the depth of analysis and scope of EIA.	Undertake proper scoping prior to start of EIA.
EIA biased towards promoter since paid by promoter.	EIA is almost always written in favour of development.	Strengthen government reviews to enable professional review, and improve consultation process etween EIA consultant and Environment Ministry.
EIA does not go beyond mitigation	Implementation will be poor, requiring constant supervision.	Developers to prepare Project Environment Management Plan. Also build inspectorate

measures, into implementation modalities.		capacity to ensure adherence to EIA recommendations.
Project had already been through the approval system.	Compromise other EIA efforts.	Provide clear EIA guidelines to policy-makers and project managers.
Inadequate circulation and involvement of stakeholders.	Local conflicts.	An accessible information facility.

Table 4 *Some Limitations to EIA implementation*

The benefits and problems associated with EIA implementation in Seychelles have important implications for its future role in ensuring sustainable development. Efforts to progress to forge linkages between EIA and land use planning, EIA and long-term environment management plans, and strengthening the role of the EIA process in catalysing best environment practices in the tourism industry should continue.

Some Opportunities provided by the EIA Process in the Tourism Sector

The EIA process provides numerous opportunities for improving the long-term environmental sustainability of development projects and land use within coastal areas. This empowers the developer with better decision-making information that guides how and what form the development will take. The EIA process addresses issues in a preventive and proactive manner, as well as actions not anticipated by the developer and further enhances the acceptability of the project, especially by affected parties. Some of these opportunities encountered in Seychelles are discussed below.

Reduce costs

Although this is a perceived misnomer,[33] EIAs allow tourism developers to reduce costs in the short and long term. Through careful assessment and analysis of the environmental implications of a particular project, the EIA process enables subtle modifications to be made to the original design of the project, which leads to better aesthetics, water and energy conservation, reduced liability, better coastal stability and beach management, and a commitment to contribute to the overall up-keeping and maintenance of the surrounding environment and associated habitats.

Direct Environmental Benefits

The main environmental benefits arising out of the EIA process in Seychelles is the increase in species protection and habitat cover. In many cases, the extent of land cover that is developed rarely exceeds 10%, and since no further development are anticipated within the vicinity, it can be assumed that the remaining property are conserved and managed, not only for the enjoyment of the guests, but to enable past habitats to continue to thrive. By proper construction of waste treatment systems, sewage, which is otherwise disposed into the sea, can be treated leading to an increased coastal water quality. Coastal and beach habitats are also protected against sand poaching and indiscriminate trampling, as the hotel development acts as guardians for this resource. Wise land use practices are also adopted, especially in the landscaping.

Reduce risks

The EIA process can enable a tourism project to reduce future risks such as the threat of sea level rise and coastal flooding. By implementing better coastline management practices, such as protection of the dune-lands and adhering to setback lines, tourism developers can reduce insurance as well as coastal protection costs. Identification of site-specific environmental engineering issues, such as coastal plain hydraulics, provides the developer with critical information during design of the tourism infrastructure. EIA therefore provides critical information that can reduce the risks involved.

Improve Coastline/Beach Management

There are several problems associated with local sea-level rise in the Seychelles, but to date there is no conclusive evidence yet. However, what is evident is the number of beaches that are suffering from increased erosion from both natural and direct human effects.[34] The main tool for addressing issues of sea-level rise and coastal erosion has been the setback line approach discussed below.

The tourism industry places several demands on the beach front with serious environmental consequences. These can often be turned into opportunities, which not only lead to better environmental quality but also more cost-effective approaches to tourism infrastructure services, reconciling the needs of clients with that of the environment. These are summarised in the following chart.

Demands	Implications	Opportunities
I want my chalet on the beach.	Construction on the beach Beach/Dune destabilisation Removal of beach vegetation	Clients enjoy a nice nature walk to the beach (setback line). Activities on the beach do not disturb clients.
I want only my clients to use the beach.	Fencing out the beach Community conflicts with users and fisher-folk	Build partnerships with the community Allows clients to experience local culture and people.
I want a small pier for leisure yachts.	Pier perpendicular to the beach Channel through the beach	Maximise use of beachfront, offer pier facilities elsewhere or utilise public facilities Use beach-landing crafts for specific activities
I want a sandy restaurant	Construction on the beach Removal of dune vegetation	A beach restaurant does not need to be refilled with a daily load of fresh sand, and the used sand is placed back onto the coastal plain.
I do not want seaweed on the beach.	Construction of seaweed barriers at sea, interfering with movement of sand and coastal currents.	Combine daily seaweed collection with a seaweed compost and an organic garden.
I want to get rid of the beach vegetation because it blocks the sea-view.	Cutting down beach vegetation Trampling on beach vegetation	Vegetation provides beach privacy Sea-view experience can be enhanced by wildlife (such as birds).
I want to remove the beach rock to provide a nicer swimming area.	Dynamiting and excavating the rock beach erosion. Modification of coastal processes and habitats.	Be selective when buying beachfront property Match activity with natural offerings

Table 5 Addressing client demands in the tourism industry[35]

Rehabilitation of Mangroves and Wetlands

Contrary to past practices and policies,[36] mangroves can provide enhanced landscapes for tourism development. When properly managed the wetland not only improves the aesthetic quality of the environment but also fulfils a wide variety of ecological functions such as support for birdlife, improved flood control, and fish nesting habitats. However, EIAs often are not explicit in terms of sustainable management of wetlands or providing alternatives to wetland reclamation. The integration and rehabilitation of wetlands should be encouraged and developed as part of the EIA process. The maintenance of wetlands as part of the tourism development landscape can increase the value and create the wild environment sought after by many tourists.

Indigenous Landscaping Approaches

Although tourism developers are very meticulous and particular about landscaping, most often do not utilise local flora or features to enhance the natural beauty of the site. In fact, many resort to importing foreign plants and even animals, which can have impacts on the local indigenous flora and fauna. The EIA process can identify the flora needed for the landscaping activities. Therefore the EIA process can provide opportunities for the development of local landscaping businesses, which can provide the tourism developer with a wide range of indigenous and endemic flora to choose from, some of which will significantly enhance the natural setting as well as floristic importance of the establishment.

The changing nature of the EIA process

Several environment initiatives have given the EIA process the importance in terms of managing coastal resources and ensuring sustainability. A number of other actors will also continue to influence the way the EIA process will evolve in Seychelles. These actors include: international funding agencies, the private sector and the public.

The role of international agencies

One of the major incentives for the facilitation of the EIA process in Seychelles was the fact that many development banks, including the World Bank, developed policies requiring that EIAs are undertaken before funds for tourism projects are disbursed.[37] The use of these EIAs, as suggested before have been instrumental in facilitating the development of the EIA legislation in Seychelles, and with the increased emphasis on environmental economics and accounting, as well as performance auditing, it is expected that these influences will lead to changes in the way EIA are implemented and managed in Seychelles. The area of Strategic Environment Assessment is not well appreciated, but it is expected that through the implementation of climate change adaptation scenarios, the concept of integrated assessment will become more accepted.

The role of the private sector

The private sector has only been responsive to the EIA process up to what is legally required for the progress of the project, since there are costs and risks involved in conducting the EIA. Alternatively, there are many developers who have undertaken voluntary EIAs and this is increasingly being encouraged by the Ministry of Environment and Transport, because if the developer believes in the process then it is likely that the projects objectives will be successful. However, the developer faces certain risks when undertaking an EIA as part of the planning process. The table below summarises some of these risks.

Risk Area	**Cost**	**Benefit**
EIA and relevant studies	Cost of studies and consultancy fees	Better information for decision-making and risk cutting
Discovery of sensitive habitats	Project may not go ahead	Conservation of particular habitat -development priorities may change
Depth of assessment required	Cost of specialist surveys ad studies	More in depth information can be used by engineer and architect to tailor architecture to location characteristics.
Uncertainty factors in the assessment	Future claims	Consider project alternatives with less uncertainty factors.
Special land use requirements	Additional infrastructure costs	Improved long-term adaptation and stability of infrastructure
Current standards, regulations and guidelines	Additional pollution prevention costs and design costs	Maintain water quality of area
Stakeholder views	Lack of local support can reduce tourism attractiveness	Pressure to build good corporate culture with community participation
Affected parties	Relocation and Compensation costs	Build partnership and negotiate with affected parties

Interest groups with powerful lobbies	Reduce long-term return on investment	Discuss interests and address needs for mutual benefits.

Table 6 The main risks, costs and benefits associated with the EIA process

The role of stakeholders and the consultation process

Although the thought of consulting stakeholders and affected parties is not the preferred approach by developers, experience[38] shows that there are various benefits from having stakeholder consultation, especially during the early stages of the project. However, as shown in table 6, there are several costs and risks associated with the consultation process. Such costs and risks can be addressed and minimised within the EIA process by:

- Clearly Communicating the project goals and intended development;
- Providing a firm commitment to implement the recommendations the EIA
- Incorporating and addressing stakeholder interests within the development plan
- Adhering to local regulations and standards; and,
- Avoiding too much unsolicited modification to the plan.

Therefore it is important that public consultation is planned with due consideration given to attendance (who and how stakeholders are invited), timing, information to be disseminated and discussed, and choice of consensus-building techniques. Although, scoping provides the initial views of stakeholders, this process is often done by consultants involving little consultation with the developer. Close working relationships between the consultant and the developer during the consultation phase allow for effective communication of the project's objectives as well as facilitate a better evaluation of alternatives to the various project components

CONCLUSION: THE WAY FORWARD

This review of EIA implementation in Seychelles with particular reference to the tourism sector has highlighted several areas in need of strengthening and further development to reinforce the emerging need for more rationale planning, better decision-making and more stakeholder involvement at all levels of society. Whilst the current EIA requirements have not failed, there is a need to increase the quality of the EIA's as well as ensure appropriate monitoring and accountability in the implementation of the EIA recommendations.

The EIA evolution in Seychelles, although it has been an entirely Seychellois driven process, is influenced by external organisations, providing opportunities for integration of the EIA process within current land use plans and coastal development activities. In the future, the EIA process needs to address current environmental issues, as coastal erosion, habitat loss and water quality deterioration in a more integrated manner. Through improved EIA performance, the tourism industry will provide greater environmental quality and comparative advantage.

The potential for the EIA to be ignored or given lip-service by the project management team and contractors during project implementation can lead to costly liabilities at a later stage. Integrating EIA within the project management process will not only allow the environmental concerns to be adequately addressed, but provide a forum whereby the environmental performance of the works can be objectively reviewed. An Environment Management Plan is usually the document specifying the environmental requirements to be implemented at each stage of project implementation, as well as addressing roles and responsibilities for environmental monitoring and auditing.

Integrated Coastal Zone Management is a management process by which rational decisions are made concerning the conservation and sustainable use of coastal and ocean resources.[39] The EIA process should be an integrated part of the ICZM process since the majority of tourism development occurs on the coast, and it is only through an integrated approach that this development can fit within the requirement to share and sustainably manage common resources. Since EIA's are limited in scope to projects and there are currently no legal framework or in-country experience for strategic environmental assessments, local ICZM structures can be utilised to allow for greater participation from stakeholders, provide information on key linkages between uses, and serve as a basis for resource management.

The evolution of the EIA process in Seychelles, in particular with respect to tourism development, is fundamental to the government's policy for sustainable development of the tourism industry in Seychelles. As a result of the EIA process, many coastal areas have been developed in a sustainable manner. The ability to undertake consistent audits and performance review of tourism projects that are implemented, as well as include more regional factors into the EIA process, are key factors. The need for these concerns, and the emerging forces shaping the future of EIA, to be addressed within the current legal framework presents thus the challenge for the future of EIA in Seychelles.

[1] Rolph Payet is presently the Director General of the Policy, Planning and Services Division in the Ministry of Environment, Government of Seychelles. He graduated from the University of East Anglia with a BSc (Hons) in Biochemistry in 1992, attended several post-graduate courses in environmental management, and an MBA from the University of Surrey. He was closely involved in the development of the Environment Protection Act (1994) and the Environment Assessment Regulations (1996), and reviewed the very first EIA's in Seychelles. Over the last five years, he has been consultant to a number of tourism EIA projects. He was also member of the Seychelles Town and Country Planning Authority.

[2] For general reading on the Seychelles, see C. Pavard (1990), *Seychelles D'ile en Ile* (Richer/Hoa-Qui) and P. Vine (1989), *Seychelles* (Immel Publishing, UK).

[3] MISD, (1995) *Seychelles Population Projections 1996-2019* (Government of Seychelles Publication, Seychelles), at 8.

[4] J. D. Sauer (1967), *Plants and Man on the Seychelles Coast: A Study in Historical Biogeography*, (University of Wisconsin Press: Madison), at 17.

[5] Vine (1989) at 74.

[6] N. Grange and F. Odendaal (1999) *Guidelines for the Environmental Assessment of Coastal Tourism* (SEACAM publication, Mozambique), at 10.

[7] S.I. 39 of 1996, *Environment Protection (Impact Assessment) Regulations, Environment Protection Act, Act 9* of 1994.

[8] *Environment Management Plan of Seychelles, 2000-2010* (Government of Seychelles Publication, 2000 final draft), at 21-22.

[9] Sauer (1967) at 20.

[10] B. Micock (1998), *Urban Planning in Seychelles: A GIS Approach.* MSc Dissertation (University of Durham: UK), at 24-25.

[11] Table from R. Payet (2000), *Environment Cost-Benefit Analysis of the East Coast Reclamation Project Phase III* (Government of Seychelles Publication), at 7.

[12] Micock (1998) at 26.

[13] *Draft Memorandum to Revise the Town and Country Planning Act*(2000), Ministry of Land Use and Habitat at 4.

[14] *Environment Protection Act (*Act 9 of 1994) Government of Seychelles Official Gazette.

[15] Section 3.

[16] F. Ally (1996), *Compilation of Environmental Laws and Regulations* (COI Publication, Seychelles), at 1-4.

[17] *Environment Protection (Impact Assessment) Regulations* (SI 39 of 1996) Government of Seychelles Official Gazette.

[18] *EIA Regulations* (1996), sub-regulation 5.2.

[19] See P. Wathern (ed.), (1988) *Environment Impact Assessment: Theory and Practice* (Routledge: London), at 162, on how EIA regulations came about in the US.

[20] B. Sadler, *International Study on the Effectiveness of Environmental Assessment* (Web Ed.,<http://www.environment.gov.au/portfolio/epg/eianet/eastudy/final/chapter4.html>), at 12, Section 4.2.1.

[21] Anon (1992), *EIA of Selected Options of the Proposed Water Supply Program* (Government of Seychelles Publication).

[22] MER Burns (1994), *Environmental Impact Assessment, Fregate Island, Seychelles*, Vols 1 & 2 (South Africa: CSIR Publication).

[23] World Bank (1989), *Operational Directive (OD) 4.00*, including annexes A & B (World Bank: Washington, D.C.; 1989).

[24] Republic of South Africa (1992), Department of Environment Affairs, *Integrated Environment Management Procedure*.

[25] Wathern (1988) at 33.

[26] European Union (DG XI) (1996)*Environment Impact Assessment: Guidance on Scoping* (Brussels: European Commission Publication).

[27] World Bank (1999), "Public Consultation in the EA Process: A Strategic Approach", in *Environment Assessment Sourcebook Update, No. 26.* (World Bank Environment Department: Washington, D.C.) at 6.

[28] *Environment Protection (Impact Assessment) Regulations* (SI 39 of 1996) Government of Seychelles Official Gazette.

[29] B. Micock, Land Use Planning (1998),*Seychelles Tourism Master Plan,* (Ministry of Tourism and Civil Aviation) at 10.

[30] M. Duncombe (1996), *Atlas des Zones D'Environnement Sensible de Seychelles*, 2 Vols.(Government of Seychelles) at 19-23.

[31] J.R. Mercier et. al. Tourism Development (1995),*Environmental Impact Assessment Guideline 11* (Ministry of Environment) at 14 to 15.

[32] *Plan d'Amenagement du Territoire*, PAT, 1989.

[33] World Bank (1999) at 14.

[34] V. Cazes-Duvat (1999), *Les Littoraux des Iles Seychelles* (L'Harmattan) at 158.

[35] Information based upon various EIA's conducted by the author.

[36] C.J. Piggot (1968), *A Soil Survey of Seychelles*, (Directorate of Overseas Surveys) at 10.

[37] World Bank (1999).

[38] World Bank (1999) at 11.

[39] B. Cicin-Sain et al. (1998) *Integrated Coastal and Ocean Management: Concept and Practice.* (Island Press, USA) at 1.

LIST OF ABBREVIATIONS

AEETC	Asia – Europe Environmental Technology Centre
AETFAT	Association for the Taxonomic Study of the Flora of Tropical Africa
Afr. J. Ecol.	African Journal of Ecology
AFSTA	African Seed Trade Association
AIJ	Activities Jointly Implemented
AGOA	African Growth and Opportunity Act
Am. U. J. Int'l. L & Pol'y	American University Journal of International Law and Policy
AMCEN	United Nations African Ministerial Conference on the Environment
AOSIS	Alliance of Small Island States
ASEM	Asia-Europe Meeting
BIDS	Bad Ivory Database System
Biol. Conserv.	Biological Conservation
CAC	Command and Control
CAFA	Concession Area Fisheries Association
Cal. W. Int'l L. J.	California Western International Law Journal
CAMPFIRE	Communal Areas Management Programme for Indigenous Resources
CAMRE	Council of Arab Ministers Responsible for the Environment
CASS	Centre for Applied Social Sciences, Zimbabwe University
CBNRM	Community Based Natural Resource Management
CCD	Convention to Combat Desertification
CDM	Clean Development Mechanism
CECAF	Fishery Committee for the Eastern Central Atlantic
CEPMLP	Centre for Energy, Petroleum and Mineral Law and Policy
CER	Certified Emissions Reduction
CILSS	Committee on Drought Control in the Sahel
CITES	Convention on International Trade in Endangered Species
CLC	Convention on Civil Liability
Colo. J. Int'l Env'tl L & Pol'y	Colorado Journal of International Environmental Law and Policy

Colum. J. Env'tl L.	Columbia Journal of Environmental Law
DANIDA	Danish Agency for Development Assistance
DESCON	Consultative Group for Desertification Control
DFA	District Fisheries Association
Dick J. Int'l L.	Dickinson Journal of International Law
DLUP	District Land Use Plan
E. Afr. Agric. For. J.	East African Agricultural & Forestry Journal
EAP	Environmental Action Plan
ECCAS	Economic Community of Central African States
ECOSOC	Economic and Social Council
ECOWAS	Economic Community of West African States
EEZ	Exclusive Economic Zone
EFZ	Exclusive Fishing Zone
EIA	Environmental Impact Assessment
Envtl' L. Rep.	Environmental Law Reports
EPA	Environmental Protection Act
EPC	Environmental Protection Council
ETIS	Elephant Trade Information System
EUADEC	East Usambara Agricultural Development & Environmental Conservation Project
FAO	Food and Agriculture Organisation
FEPA	Federal Environment Protection Agency (Nigeria)
For. Abstr.	Forestry Abstracts
For. Ecol. & Manage	Forest Ecology & Management
Fragm. Flor. Geobot. Suppl.	Fragmenta Floristica Geobotanica Supplementum
GEF	Global Environment Facility
Geo. Int'l. Env'tl. L. Rev.	Georgetown International Environmental Law Review
GHG	Greenhouse Gases
GLOBE	Global Legislators Organisation for a Balanced Environment
GRAIN	Genetic Resources Action International
Harv. Int'l L. J.	Harvard International Law Journal
Harv. Int'l. L. J.	Harvard International Law Journal
Harvard L. Rev.	Harvard Law Review
IBRD	International Bank for Reconstruction and Development
ICCAT	International Convention for the Conservation of Atlantic Tuna
ICJ	International Court of Justice
ICSU	International Council for Science
ICZM	Integrated Coastal Zone Management

IEE	Initial Environmental Evaluation
IFAW	International Fund for Animal Welfare
IFC	International Finance Corporation
IFI	International Financial Institutions
IGAD	Intergovernmental Authority on Development
IMO	International Maritime Organisation
INCD	International Negotiating Committee (on Desertification)
IPCC	Intergovernmental Panel on Climate Change
ISO	International Organisation for Standardisation
ISP	Inter – American Strategy for the Promotion of Public Participation in Decision Making for Sustainable Development
IUCN	International Union for the Conservation of Nature
J. Nat. Resources & Envt. L.	Journal of Natural Resources and Environmental Law
J.L. & Pol'y	Journal of Law and Policy
JEL	Journal of Environmental Law
JFCM	Joint Fisheries Committee of Management
LFN	Laws Federation of Nigeria
LKFRI	Lake Kariba Fisheries Research Institute
MARPOL	International Convention for the Prevention of Pollution from Ships
MBIs	Market Based Economic Instruments
*MEA*s	Multilateral Environmental Agreements
MIKE	Monitoring Illegal Killing of Elephants
Mitt. Inst. Allg. Bot. Hamburg	Mitteilungen aus dem Institut fur Allegmein Botanik Hamburg
Monogr. Syst. Bot. Missouri Bot.Gard.	Monographs in Systematic Botany from the. Missouri Botanical Garden.
MTCA	Ministry of Tourism & Civil Aviation (Seychelles)
NARESCON	National Resources Conservation Council
NDES	Niger Delta Environmental Survey
NEAP	National Environmental Action Plan
NEPA	National Environmental Policy Act
NEPAD	New Partnership for African Development
NLUP	National Land Use Plan
NMA	Nigerian Maritime Authority
NNPC	Nigerian National Petroleum Company
OAPI	Organisation Africaine de la Propriete Intellectuelle
OAS	Organisation for American States
OAU	Organisation for African Unity
ODA	Overseas Development Assistance

OECD	Organisation for Economic Co-operation and Development
OMPADEC	Oil Mineral Producing Areas Development Commission
OPEC	Organisation of the Petroleum Exporting Countries
PACD	Plan of Action to Combat Desertification (1977)
PIC	Prior Informed Consent
PPP	Polluter Pays Principle
PRTR	Pollutant Release and Transfer Register
REC	Regional Environmental Centre for Central and Eastern Africa
Rev. Euro. Com. & Int'l L.	Review of European Community and International Law
SABS	South Africa Bureau of Standards
SADC	Southern African Development Community
SAFA	Sub Area Fisheries Association
SANP	South African National Parks
SARDC	South African Research and Documentation Centre
TANU	Tanganyika African National Union
TCPA	Town and Country Planning Authority (Seychelles)
TFAP	Tanzanian Forestry Action Plan
Trans. Am. Fish Soc.	Transactions of the American Fisheries Society Publication
Trans. Inst. Br. Geogr. N. S.	Transactions, Institute of British Geographers
TRIPS	Trade Related aspects of Intellectual Property Rights
UILR	University of IFE (Nigeria Law Reports)
UMBCP	Ulugum Mountains Biodiversity Conservation Project
UN / ECA	United Nations Economic Commission for Africa
UN/ECE	United Nations Economic Commission for Europe
UNCED	United Nations Conference on Environment and Development
UNCHE	United Nations Conference on the Human Environment
UNCLOS	United Nations Convention on the Law of the Sea
UNCOD	United Nations Conference on Desertification
UNDP	United Nations Development Programme
UNEP	United Nations Environment Programme

UNESCO	United Nations Educational, Scientific and Cultural Organisation
UNFCC	United Nations Framework Convention on Climate Change
UNGA	United Nations General Assembly
UNGASS	United Nations General Assembly Special Session
UPOV	Union for Protection of New Varieties of Plants
VMC	Village Management Centre
VMS	Vessel Monitoring Systems
WHO	World Health Organisation
WMO	World Meteorological Organisation
WRI	World Resources Institute
WTO	World Trade Organisation
ZLKFA	Zimbabwe Lake Kariba Fisheries Association
ZMC	Zonal Management Centre
ZZSFP	Zambia Zimbabwe SADC Fisheries Project

LIST OF CONVENTIONS / TREATIES:

1948
Universal Declaration of Human Rights, adopted by the U.N. General Assembly on 10 Dec. 1948,

U.N.G.A. Res. 217A (III), U.N. Doc. A/810

1966
International Covenant on Civil and Political Rights, 6 ILM. 368

1968
African Convention on the Conservation of Nature and Natural Resources, (Algiers), 1001 U.N.T.S. 3

1971
Convention on Wetlands of International Importance Especially as Waterfowl Habitat, (Ramsar) *(Ramsar Convention)* 22 ILM 698

1972
Convention on the Prevention of Marine Pollution by Dumping of Wastes and other Matter, (London), 1046 U.N.T.S. 120

Stockholm Declaration on the Human Environment, 11 ILM 1416

1973
Convention on International Trade in Endangered Species of Wild Fauna and Flora, (Washington), *(CITES)* 993 U.N.T.S. 243

1980
Convention on the Conservation of Migratory Species of Wild Animals (Bonn), 19 ILM 15

1981
West and Central African Marine Environment Convention, 20 ILM 729

1982
United Nations Convention on the Law of the Sea (Montego Bay), (UNCLOS), Dec. 10, 1982, UN Doc. A/CONF. 62/122, 1295 U.N.T.S. 211, 21 ILM 1261

African Charter on Human Rights and Peoples Rights (Banjul) 21 ILM 52
World Charter for Nature, UNGA Res. 37/7, 37 UNGAOR Suppl. (No 51) at 17, UN Doc A/37/51

1985
Convention for the Protection of the Ozone Layer, (Vienna), 26 ILM 1529

Convention for the Protection, Management and Development of the Marine and Coastal Environmental of the Eastern African Region, (Nairobi) IELMT 985:46

1987
Protocol to the Convention for the Protection of the Ozone Layer *(Montreal Protocol on Substances that Deplete the Ozone Layer),* 26 ILM 154

1989
Convention on the Control of Transboundary Movements of Hazardous Wastes and Their Disposal (Basel), 28 ILM 657

1991
Convention on Environmental Impact Assessment in a Transboundary Context, (Espoo), 30 ILM 802

Convention on the Ban of the Import into African and the Control of Transboundary Movement and Management of Hazardous Wastes within Africa (Bamako) 30 ILM 775

1992
United Nations Convention on Biological Diversity, (Rio de Janeiro), 31 ILM 818

United Framework Convention on Climate Change, (New York), 31 ILM 849

Rio Declaration on Environment and Development, (Rio de Janeiro), 31 ILM 874

1994
Convention to Combat Desertification 33 ILM 1016

Lusaka Agreement on Cooperative Enforcement Operations Directive at Illegal Trade in Wild Fauna and Fauna, (Lusaka), 1 JIWLP 155

1995
UN Agreement Relation to the Conservation and Management of Straddling Fish Stocks and Migratory Fish Stocks, 34 ILM 1542
Agreement on Conservation of African-Eurasion Migratory Water Birds, 6 YBIEL 907

1997
Convention on the Non-Navigational Uses of International Watercourses, 37 ILM 719

1998
Protocol to the Framework Convention on Climate Change, *(The Kyoto Protocol)*, 37 ILM 22

1999
Convention on Prior Informed Consent Procedure for Certain Hazardous Chemicals and Pesticides in International Trade, 38 ILM 517

Convention on Access to Information, Public participation in Decision-making and Access to Justice in Environmental Matters (Aarhus), 38 ILM 517

2000
Protocol to the Convention on Biological Diversity (Catagena), *(Cartagena Protocol on Biosafety)*, 39 ILM 1027

2001
Convention on Persistent Organic Pollutants, 40 ILM 532

2002
World Summit on Sustainable Development Plan of Implementation, (Johannesburg)
http://www.johannesburgsummit.org/html/documents/summit_docs/2309_planfinal.pdf

ENVIRONMENT & POLICY

1. Dutch Committee for Long-Term Environmental Policy: *The Environment: Towards a Sustainable Future*. 1994 ISBN 0-7923-2655-5; Pb 0-7923-2656-3
2. O. Kuik, P. Peters and N. Schrijver (eds.): *Joint Implementation to Curb Climate Change*. Legal and Economic Aspects. 1994 ISBN 0-7923-2825-6
3. C.J. Jepma (ed.): *The Feasibility of Joint Implementation*. 1995 ISBN 0-7923-3426-4
4. F.J. Dietz, H.R.J. Vollebergh and J.L. de Vries (eds.): *Environment, Incentives and the Common Market*. 1995 ISBN 0-7923-3602-X
5. J.F.Th. Schoute, P.A. Finke, F.R. Veeneklaas and H.P. Wolfert (eds.): *Scenario Studies for the Rural Environment*. 1995 ISBN 0-7923-3748-4
6. R.E. Munn, J.W.M. la Rivière and N. van Lookeren Campagne: *Policy Making in an Era of Global Environmental Change*. 1996 ISBN 0-7923-3872-3
7. F. Oosterhuis, F. Rubik and G. Scholl: *Product Policy in Europe: New Environmental Perspectives*. 1996 ISBN 0-7923-4078-7
8. J. Gupta: *The Climate Change Convention and Developing Countries: From Conflict to Consensus?* 1997 ISBN 0-7923-4577-0
9. M. Rolén, H. Sjöberg and U. Svedin (eds.): *International Governance on Environmental Issues*. 1997 ISBN 0-7923-4701-3
10. M.A. Ridley: *Lowering the Cost of Emission Reduction: Joint Implementation in the Framework Convention on Climate Change*. 1998 ISBN 0-7923-4914-8
11. G.J.I. Schrama (ed.): *Drinking Water Supply and Agricultural Pollution*. Preventive Action by the Water Supply Sector in the European Union and the United States. 1998 ISBN 0-7923-5104-5
12. P. Glasbergen: *Co-operative Environmental Governance: Public-Private Agreements as a Policy Strategy*. 1998 ISBN 0-7923-5148-7; Pb 0-7923-5149-5
13. P. Vellinga, F. Berkhout and J. Gupta (eds.): *Managing a Material World*. Perspectives in Industrial Ecology. 1998 ISBN 0-7923-5153-3; Pb 0-7923-5206-8
14. F.H.J.M. Coenen, D. Huitema and L.J. O'Toole, Jr. (eds.): *Participation and the Quality of Environmental Decision Making*. 1998 ISBN 0-7923-5264-5
15. D.M. Pugh and J.V. Tarazona (eds.): *Regulation for Chemical Safety in Europe: Analysis, Comment and Criticism*. 1998 ISBN 0-7923-5269-6
16. W. Østreng (ed.): *National Security and International Environmental Cooperation in the Arctic – the Case of the Northern Sea Route*. 1999 ISBN 0-7923-5528-8
17. S.V. Meijerink: *Conflict and Cooperation on the Scheldt River Basin*. A Case Study of Decision Making on International Scheldt Issues between 1967 and 1997. 1999 ISBN 0-7923-5650-0
18. M.A. Mohamed Salih: *Environmental Politics and Liberation in Contemporary Africa*. 1999 ISBN 0-7923-5650-0
19. C.J. Jepma and W. van der Gaast (eds.): *On the Compatibility of Flexible Instruments*. 1999 ISBN 0-7923-5728-0
20. M. Andersson: *Change and Continuity in Poland's Environmental Policy*. 1999 ISBN 0-7923-6051-6

ENVIRONMENT & POLICY

21. W. Kägi: *Economics of Climate Change: The Contribution of Forestry Projects.* 2000 ISBN 0-7923-6103-2
22. E. van der Voet, J.B. Guinée and H.A.U. de Haes (eds.): *Heavy Metals: A Problem Solved?* Methods and Models to Evaluate Policy Strategies for Heavy Metals. 2000 ISBN 0-7923-6192-X
23. G. Hønneland: *Coercive and Discursive Compliance Mechanisms in the Management of Natural Resourses.* A Case Study from the Barents Sea Fisheries. 2000 ISBN 0-7923-6243-8
24. J. van Tatenhove, B. Arts and P. Leroy (eds.): *Political Modernisation and the Environments.* The Renewal of Environmental Policy Arrangements. 2000 ISBN 0-7923-6312-4
25. G.K. Rosendal: *The Convention on Biological Diversity and Developing Countries.* 2000 ISBN 0-7923-6375-2
26. G.H. Vonkeman (ed.): *Sustainable Development of European Cities and Regions.* 2000 ISBN 0-7923-6423-6
27. J. Gupta and M. Grubb (eds.): *Climate Change and European Leadership.* A Sustainable Role for Europe? 2000 ISBN 0-7923-6466-X
28. D. Vidas (ed.): *Implementing the Environmental Protection Regime for the Antarctic.* 2000 ISBN 0-7923-6609-3; Pb 0-7923-6610-7
29. K. Eder and M. Kousis (eds.): *Environmental Politics in Southern Europe: Actors, Institutions and Discourses in a Europeanizing Society.* 2000 ISBN 0-7923-6753-7
30. R. Schwarze: *Law and Economics of International Climate Change Policy.* 2001 ISBN 0-7923-6800-2
31. M.J. Scoullos, G.H. Vonkeman, I. Thornton, and Z. Makuch: *Mercury - Cadmium-Lead: Handbook for Sustainable Heavy Metals Policy and Regulation.* 2001 ISBN 1-4020-0224-6
32. G. Sundqvist: *The Bedrock of Opinion.* Science, Technology and Society in the Siting of High-Level Nuclear Waste. 2002 ISBN 1-4020-0477-X
33. P.P.J. Driessen and P. Glasbergen (eds.): *Greening Society.* The Paradigm Shift in Dutch Environmental Politics. 2002 ISBN 1-4020-0652-7
34. D. Huitema: *Hazardous Decisions.* Hazardous Waste Siting in the UK, The Netherlands and Canada. Institutions and Discourses. 2002 ISBN 1-4020-0969-0
35. D. A. Fuchs: *An Institutional Basis for Environmental Stewardship: The Structure and Quality of Property Rights.* 2003 ISBN 1-4020-1002-8
36. B. Chaytor and K.R. Gray (eds.): *International Environmental Law and Policy in Africa.* 2003 ISBN 1-4020-1287-X

For further information about the series and how to order, please visit our Website http://www.wkap.nl/series.htm/ENPO

KLUWER ACADEMIC PUBLISHERS – DORDRECHT / BOSTON / LONDON

MIX
Papier aus verantwortungsvollen Quellen
Paper from responsible sources
FSC® C105338

If you have any concerns about our products,
you can contact us on
ProductSafety@springernature.com

In case Publisher is established outside the EU,
the EU authorized representative is:
Springer Nature Customer Service Center GmbH
Europaplatz 3, 69115 Heidelberg, Germany

Printed by Libri Plureos GmbH
in Hamburg, Germany